MINE WASTE MANAGEMENT

A resource for mining industry professionals, regulators and consulting engineers

EDITORS:
IAN P. G. HUTCHISON
RICHARD D. ELLISON

SPONSORED BY:
CALIFORNIA MINING ASSOCIATION

LEWIS PUBLISHERS
Boca Raton Ann Arbor London Tokyo

Library of Congress Cataloging-in-Publication Data

Mine waste management / California Mining Association
p. cm.
Includes bibliographical references.
1. Mineral industries--Waste disposal--Environmental aspects--California.
2. Mineral industries--Waste disposal--Environmental aspects. 3. Water quality management--California. 4. Water quality management. I. California Mining Association.
TD899.M47M57 1992
622--dc20 91-33444
ISBN 0-87371-746-5

LEWIS PUBLISHERS, INC.
121 South Main Street, Chelsea, Michigan 48118

PRINTED IN THE UNITED STATES OF AMERICA

PREFACE

Mine Waste Management is a valuable resource for mining industry professionals, regulators, and consulting engineers. This book deals with many of the important water quality and design issues at mine waste management units. Topics include an evaluation of the performance of waste containment at modern mining operations, the philosophy of waste containment, mine waste characterization, disposal facility liner and closure designs, ground water monitoring, heap leach operations, and an analysis of the cost impacts of mine waste disposal. The material presented includes technical discussions and information, as well as recommendations on how the technical issues can be accommodated in mine waste regulations.

The California Mining Association (CMA) retained several consulting firms experienced in mine waste management to prepare *Mine Waste Management*. Each of these firms provided their services at a considerable discount to CMA. The authors represent a remarkable collection of the foremost mine waste management experts in North America. In addition, each chapter has been peer reviewed by several internationally known experts in their respective technical fields.

We gratefully acknowledge the tremendous effort and support provided by the member companies of the California Mining Association, the consulting firms involved, and the peer reviewers.

We sincerely trust that this book will prove to be of value to all who are concerned with protecting the environment and supporting the mining industry, an important part of our economic base. With the diligent application of scientifically sound principles, these two objectives are well within our reach.

Ian P. G. Hutchison
Editor

Richard D. Ellison
Editor

ACKNOWLEDGEMENTS

Lead authors for the various chapters were:

Peter H. Dohms, P.G.
Richard D. Ellison, Ph.D., P.E.
Michael E. Henderson, P.E.
Ian P. G. Hutchison, Ph.D, P.E.
Don Poulter, P.E.
Andy Robertson, Ph.D., P.E.
Adrian C. Smith, Ph.D.
Mark E. Smith, P.E.
Dirk van Zyl, Ph.D, P.E.

The peer reviewers were:

Jack A. Caldwell
John A. Cherry, Ph.D.
Jean-Pierre Giroud, Ph.D.
David B. McWhorter, Ph.D.
Stuart D. Miller, Ph.D.
Herbert C. Osborne

The text editors were:

Jeani Buchanan
Ellen M. Cross
Kimberly Powers
Carolyn Trindle-Smith

A complete listing of all authors and reviewers, as well as selected résumés, is provided in Appendix A.

A committee of the following members of the California Mining Association provided information:

- James E. Good, Esq., Chairman
  Gresham, Varner, Savage, Nolan, and Tilden
  San Bernardino, California,
- John J. Fitz-Gerald, Esq.
  Gold Fields Mining Corporation
  Golden, Colorado
- Raymond E. Krauss
  Homestake Mining Company
  Lower Lake, California
- David S. Cahn, Ph.D.
  California Portland Cement Company
  Glendora, California

- John Paulsen
  U.S. Borax and Chemical Corporation
  Los Angeles, California
- Robert Sega
  Molycorp, Inc.
  Mountain Pass, California
- Richard E. Hammond, Esq. and Thomas M. Donnelly, Esq.
  Heller, Ehrman, White and McAuliffe
  San Francisco, California
- Larry R. Todd
  Gold Fields Operating Company - Mesquite
  Brawley, California

Mine Waste Management was sponsored by:

- The California Mining Association
  1121 L Street, Suite 909
  Sacramento, California 95814

With the participation of the following firms:

- ASCI, Inc.
  2555 Edgemont Boulevard
  North Vancouver, British Columbia
  Canada V7R 2M9
- Condor Earth Technologies, Inc.
  21663 Brian Lane, Suite A
  Sonora, California 95370
- D.P. Engineering, Inc.
  2 West Dry Creek Circle, Suite 250
  Littleton, Colorado 80120
- EIC, Corporation
  1355 South Colorado Boulevard, Suite 316
  Denver, Colorado 80222
- Environmental Solutions, Inc.
  21 Technology Drive
  Irvine, California 92718
- Steffen Robertson and Kirsten (U.S.), Inc.
  3232 South Vance Street, Suite 210
  Lakewood, Colorado 80227
- Vector Engineering, Inc.
  12438 Loma Rica Drive, Suite C
  Grass Valley, California 95945
- Welsh Engineering, Inc.
  5400 Equity Avenue, Suite B
  Reno, Nevada 89502

TABLE OF CONTENTS

		PAGE NO.
	PREFACE	iii
1	INTRODUCTION	1
	Background	3
	Summary	4
2	PERFORMANCE REVIEW OF MODERN MINE WASTE MANAGEMENT UNITS	11
	Introduction	15
	Current Mine Waste Management Practices in California	19
	Performance History	23
	Conclusions	50
	Regulatory Implications	61
	Bibliography	61
3	MINE WASTE MANAGEMENT REQUIREMENTS	63
	Introduction	67
	Overview of Systems Approach for Determining Mine Waste Management Requirements	70
	Protecting Beneficial Uses of Waters	71
	Identifying Potential Receiving Waters and Their Beneficial Uses	72
	Assessing Physical and Chemical Properties of Mine Waste	72
	Site-Specific Factors	97
	Waste Management Unit Containment	99
	The Systems Approach to Designing Waste Containment	101
	Regulatory Implications	114
	Bibliography	119
4	PREDICTION OF ACID GENERATION POTENTIAL	123
	Introduction	127
	Acid Rock Drainage	130
	Acid Generation Prediction	163
	Acid Generation and Leachate Migration Control Measures	184
	In-Waste Monitoring	188
	Regulatory Implications	192
	Bibliography	195

		PAGE NO.
5	ATTENUATION OF CHEMICAL CONSTITUENTS	201
	Introduction	205
	Mine Wastes and Chemistry	209
	Factors and Mechanisms Affecting Contaminant Migration	215
	Relationship Between Chemicals and Mechanisms	237
	Appropriate Levels of Verification	258
	Regulatory Implications	265
	Bibliography	268
6	CLIMATIC CONSIDERATIONS	275
	Introduction	277
	Mine Waste Pile Water Balance	279
	Infiltration and Recharge Case Histories	298
	Analytical Approach to Percolation Evaluation	308
	Prediction of Infiltration	317
	Regulatory Implications	322
	Bibliography	326
7	LINER SYSTEM DESIGN	327
	Introduction	329
	Types of Liner Systems	334
	Types of Individual Liner System Components	339
	Liner System Performance	355
	Regulatory Implications	378
	Bibliography	385
8	CLOSURE REQUIREMENTS	391
	Introduction	393
	Closure Objectives	395
	Description of Mine Waste Management Units	396
	Designing for Closure	404
	Closure Technologies and Design Elements	411
	Regulations for Closure	442
	Bibliography	448
9	HEAP LEACHING	453
	Introduction	457
	Heap Leach Practice	459
	Design Approach	488
	Heap Leach Guidelines	492
	Bibliography	497

PAGE NO.

10 GROUND WATER MONITORING 499

Introduction 503
Factors to be Considered When Designing a Monitoring System 504
Elements of a Monitoring System 524
Technical Approach to the Design of a Ground Water Monitoring System 555
Regulatory Implications 563
Bibliography 568

11 ECONOMIC IMPACT EVALUATION 571

Introduction 573
Regulatory Analysis 575
Cost Analyses 583
Conclusions and Recommendations 587
Bibliography 589

APPENDIX A: EDITORS, AUTHORS, AND PEER REVIEWERS 591

APPENDIX B: CALIFORNIA STATE AGENCIES PRIMARILY RESPONSIBLE FOR REGULATING MINE WASTE DISPOSAL 601

APPENDIX C: TAILINGS IMPOUNDMENT, CONTAINMENT ASSESSMENT CONCEPTUAL MODEL, PARAMETERS AND EQUATIONS 605

APPENDIX D: HELP MODEL RUN RESULTS 619

APPENDIX E: PROTOCOL FOR DETERMINING NATURALLY LEACHABLE RESIDUALS 623

APPENDIX F: SONORA MINING CORPORATION, JAMESTOWN MINE, DESCRIPTION AND SUMMARY OF PERFORMANCE OF THE TAILINGS MANAGEMENT FACILITY 627

APPENDIX G: WASTE MANAGEMENT COST CALCULATIONS 637

MINE WASTE MANAGEMENT

CONTENTS

CHAPTER 1
INTRODUCTION

PAGE NO.

1.1 Background		3
1.2 Summary		4
1.2.1	Chapter 2: Performance Review of Modern Mine Waste Management Units	4
1.2.2	Chapter 3: Mine Waste Management Requirements	5
1.2.3	Chapter 4: Prediction of Acid Generation Potential	6
1.2.4	Chapter 5: Attenuation of Chemical Constituents	6
1.2.5	Chapter 6: Climatic Considerations	7
1.2.6	Chapter 7: Liner System Design	8
1.2.7	Chapter 8: Closure Requirements	8
1.2.8	Chapter 9: Heap Leaching	9
1.2.9	Chapter 10: Ground Water Monitoring	9
1.2.10	Chapter 11: Economic Impact Evaluation	10

CHAPTER 1

INTRODUCTION

1.1 BACKGROUND

Natural resource development and associated environmental issues have combined to become a major force in public opinion and policy making, as reflected in the extensive array of environmental laws and regulations promulgated in recent years. Both the federal government and a number of states are in the process of adopting more comprehensive mine waste disposal regulations. Mining companies, in turn, are placing more emphasis on improving mine waste characterization and disposal technology in order to meet increasingly demanding environmental standards in a cost effective manner.

The California Mining Association (CMA) has always encouraged a proactive approach to dealing with mining environmental issues. It has played a lead role in working constructively with state regulators and in providing technical and legal input to the drafting of laws, regulations, and policies governing mine waste disposal. In sponsoring *Mine Waste Management,* the prime objective of CMA is to provide documentation on state-of-the-art practices in several key areas that are important to environmentally sound mine waste management. These areas were identified during ongoing CMA interactions with California State legislative and regulatory staff. Each chapter in this book addresses one of these areas and concludes with recommendations regarding how the technical issues can be accommodated in regulatory activities governing mine waste management.

Mine Waste Management also is intended to provide stimulation to further development and improvement of the technology of mine waste disposal. It should be a useful tool for a variety of persons involved in mining, including industry professionals, federal, state, and local regulators, and members of the consulting engineering profession. Mine waste facility designers, mine superintendents and managers, and consulting engineers will be able to utilize the technical information presented. In addition, persons involved in training seminars will be able to use this book as a reference document. Often regulators are given the responsibility of writing, implementing, and enforcing regulations without having a comprehensive source of information. Such persons should find this text a valuable resource.

Mine Waste Management draws on experiences and regulatory background of the State of California. However, since the technical issues discussed are important to mine waste management in general, and because other states and the federal government have or are developing their own mine waste regulations, the material contained herein should be of considerable benefit to mining industry professionals, regulators, and the consulting engineering profession throughout the United States, as well as in other countries.

1.2 SUMMARY

Mine Waste Management addresses a wide range of subjects concerning the management of mine wastes. It deals with waste management units such as tailings impoundments, waste piles, wastewater ponds, and heap leach and residue piles. Included are water protection considerations involved in the planning, construction, operation, closure, and abandonment of mine waste management units, with detailed lists of references. While this report focuses on the safe disposal of non-hazardous mine waste, i.e., waste that would be regulated at the federal level under Resource Conservation and Recovery Act (RCRA), Subtitle D, "State or Regional Solid Waste Plans," many of the technical aspects discussed are also applicable to hazardous waste disposal under the RCRA Subtitle C provisions.

Topics range from an evaluation of the performance of modern mine waste management units to a proposed methodology for determining containment requirements to meet water quality protection objectives. Included is a discussion of the amount of waste containment provided by both arid climate conditions and geochemical attenuation in natural geologic formations. Waste characterization, with special emphasis on identifying acid generation potential, is another important topic that is addressed. There is a discussion on the subject of liner design, which includes illustrative graphics of different liner types and data on their containment capabilities. In addition there are chapters on heap leaching practices, mine waste closure requirements, and a rational basis for the design of ground water monitoring systems. The final chapter addresses waste management cost sensitivity analyses.

1.2.1 CHAPTER 2: PERFORMANCE REVIEW OF MODERN MINE WASTE MANAGEMENT UNITS

A review of the performance of actual mine waste management units permitted during the first five years (1984 - 1989) of the 1984 mine waste disposal regulations[1] in California indicates that the waste containment provided has been satisfactory, and that the operation of these units has not resulted in impairment to the beneficial uses of the waters of the State. Incidents that have

---

(1) California Code of Regulations, Title 23, Division 3, *Chapter 15*.

occurred involved isolated spillages of fluids from ponds, piping, and pumping systems, and surface water turbidity at placer operations. In some instances, seepage through primary liners entered leachate collection systems installed between two liners and was contained by the secondary liner.

In contrast to the above, there are instances of unacceptable leakage of solutions from ponds at mines constructed in the 1970s and early 1980s. These incidents, however, are due to design and construction standards that were lesser than those currently required. There are also environmental impacts associated with acid rock drainage problems at mines put into production from the late 1800s through the 1950s and 1960s. These problems are due to disposal practices which are outdated and would not be permitted today.

Based on the above review, mine waste regulations would benefit by including the following provisions for dealing with some of the important incidents and other environmental concerns that are not addressed specifically in the current regulations:

- Realistic quantitative liner system performance standards, such as acceptable leakage rates through primary liner systems.
- Testing requirements to determine whether acid rock drainage will occur.
- Statistically-based design flood criteria to determine pond freeboard requirements.

1.2.2 CHAPTER 3: MINE WASTE MANAGEMENT REQUIREMENTS

In terms of California's Porter-Cologne Water Quality Control Act, the objective of regulating mine waste is to provide for the protection of beneficial uses of the waters of the State. Cost-effective waste disposal which meets this objective is achieved best by viewing a disposal facility as a total system. Such a system includes the climatic conditions, physical and chemical waste characteristics, methods used for placement of the waste, engineered containment, natural containment provided by underlying geologic formations, and the abundance, quality, proximity, and beneficial uses of surface and ground water. Within the context of this total system, an adequate amount of engineered containment should be provided to ensure protection of the beneficial uses of the waters of the State.

Application of such a comprehensive systems approach should be provided for in the regulation of mine waste disposal to enable the design of a disposal facility to be based on a containment analysis which considers all of the site-specific aspects referred to above. Where regulatory agencies believe that prescriptive containment requirements are necessary to simplify administration of the regulations, the systems approach should be included as an alternative method of determining the containment requirements which can be invoked by either the regulator or the operator.

1.2.3 CHAPTER 4: PREDICTION OF ACID GENERATION POTENTIAL

Acid rock drainage (ARD) potentially threatens the beneficial use of water when wastes containing reactive sulfides are disposed of in a wet climate. Mine waste characterization, therefore, should include testing for its potential to generate acid conditions. If this potential is found to be present, special waste management procedures need to be implemented.

Acid conditions are formed when there is a sufficient excess of reactive sulfides over neutralizing carbonaceous material in a waste, plus an adequate supply of oxygen and water to promote the chemical and biological reactions. ARD will occur in those instances when the climate is sufficiently wet to cause precipitation infiltration or surface water runoff to flow through the waste material. The resultant leachate or seepage that emerges from the waste will have the characteristic ARD properties of low pH, elevated total dissolved solids (TDS), and elevated sulfate and metal concentrations.

Appropriate testing protocols should incorporate static acid-base accounting tests using an appropriate safety margin between the acid generation and neutralization potential of the waste material. Where the acid generation and neutralization potentials are of similar magnitude, more accurate kinetic tests, such as the humidity cell, may be needed to reproduce the time-varying aspects of chemical reactions that cause ARD.

1.2.4 CHAPTER 5: ATTENUATION OF CHEMICAL CONSTITUENTS

Under certain conditions, the natural geologic formations underlying a waste disposal facility can attenuate the dissolved constituents in seepage emanating from waste material, thereby preventing these constituents from reaching surface or ground water. Chemical, physical, and biological mechanisms contribute to this process. Predictability of attenuation varies and depends on the constituents in question and site-specific conditions, such as climate, properties of the waste, physical and chemical characteristics of the engineered containment features, and the natural geologic formations underlying the waste management unit.

When appropriate, attenuation may be taken into account when assessing engineered containment requirements on a site-specific basis. In some cases, attenuation may be sufficient to provide primary containment. In others, it may be considered as part of the redundancy required to achieve an acceptable level of containment security.

A key aspect to consider when utilizing attenuation as a waste containment feature is verification, which can range from literature documentation, to laboratory testing, to extensive field demonstrations. The amount of effort devoted to verification generally will depend on the degree of risk posed by the waste and the extent to which attenuation will be relied upon to provide containment.

The attenuation process is sufficiently well understood to be incorporated into mine waste regulations. Its application must be judged independently for each site and will depend on the degree to which verification can be achieved. A benefit of allowing for attenuation in waste containment regulations is that it encourages the selection of waste disposal sites that have a high degree of natural geologic containment. These sites are likely to provide more secure containment in the long term than sites which rely largely on engineered containment.

1.2.5 CHAPTER 6: CLIMATIC CONSIDERATIONS

In the arid areas of California, very little or no rainfall percolates through dry waste piles, which considerably minimizes the potential for any seepage impact on ground water quality. Based on literature survey and water balance calculations using U.S. Environmental Protection Agency (EPA) infiltration models, it appears that leachate formation becomes negligible when mean annual rainfall is less than 5 to 10 inches. In some cases, therefore, arid climatic conditions can provide all or a portion of the necessary waste containment function.

Recharge, for example, is typically 10 percent of the total rainfall. Thus, in a climate where the average rainfall is 10 inches, recharge will average 1 inch/year, equivalent to approximately 10^{-7} cm/sec. In this case, the arid climatic conditions provide the equivalent containment of a single 10^{-7} cm/sec clay liner placed under a waste liquid.

Regulations should incorporate provisions for the use of arid climatic conditions as a waste containment feature. In some instances, more detailed analysis and testing may be necessary to determine the amount of containment provided by the aridity. Procedures for accomplishing this include detailed water balance analyses, detailed measurement of field soil moisture, and laboratory and field tests.

1.2.6 CHAPTER 7: LINER SYSTEM DESIGN

A range of liner types and combinations of liners can be used to contain the liquids associated with mine waste. The basic components of a liner system include an overdrain or protective layer, clay or synthetic barrier layers, intermediate leachate collection and removal drains, and underdrains or bedding layers. Liner system performance, measured as the degree to which potential seepage is minimized, depends not only on the intrinsic properties of the liner system, but also on the hydraulic loading conditions, quality of construction, and characteristics of the underlying natural geologic material.

Selection of the most appropriate liner system is best achieved using a systems approach similar to the one presented in Chapter 3. Factors to be considered are climatic conditions, waste properties, water resource values, hydrogeologic characteristics of the geologic formations underlying the site, and the proposed waste management practices. By combining these factors in a containment assessment, the liner system that provides the appropriate protection of the beneficial uses of water can be selected.

The above approach to selecting a liner system should be incorporated into mine waste regulations. Where the regulations include provisions for selecting liner systems based on a prescriptive approach, they should allow as an alternative the application of a site-specific systems approach, based on waste and site characteristics and a waste containment analysis.

1.2.7 CHAPTER 8: CLOSURE REQUIREMENTS

Effective closure of mine waste piles and impoundments can be achieved by several methods. These include as is abandonment, direct revegetation of the surface, and the placement of a variety of cover types over the waste, ranging from single soil layers to more complex cover systems incorporating clay or synthetic liners and drainage layers. Another important element of closure involves providing surface drainage controls and erosion protection measures. The design of a suitable cover involves selecting the method, or combination of methods, which provide the required degree of waste containment and long-term durability, taking into consideration the original waste disposal facility design and site-specific waste, climate, and subsurface conditions. Because they are derived from naturally occurring materials, mine wastes frequently can provide all or part of the cover material requirements.

Regulation of closure should follow the same procedure recommended for the design of liner systems, i.e., either use of a prescriptive approach or, alternately, allowance for a site-specific design based on waste and site characteristics.

1.2.8 CHAPTER 9: HEAP LEACHING

Heap leaching is a process which involves stacking the prepared ore on lined pads and allowing an extraction solution to trickle through the ore pile in order to dissolve the economic minerals from the host rock. Following completion of the metal extraction phase, the heaps usually are rinsed with water or chemical solutions to reduce the pollution potential of the residue.

During the operation of a heap leach pile, which occurs over a relatively short period of time (up to a few years), the material has not yet assumed the characteristics of a rinsed waste, so application of long-term, permanent waste disposal regulatory standards may not be suitable. Because this type of operation represents potential threats to water quality, it may be advisable to develop separate guidelines or regulations to ensure sound design and operating practices. Appropriate regulations should require site and operational information, heap leach design details, an operating plan, liner system design information, the results of a containment risk assessment, a monitoring plan, and contingency and closure plans. After the extraction and rinsing processes are complete and the spent ore becomes a waste, compliance with the requirements of the relevant waste disposal regulations becomes necessary.

1.2.9 CHAPTER 10: GROUND WATER MONITORING

A sound ground water monitoring program generally includes a combination of in-waste, vadose zone (when practical), and ground water monitoring and sampling. The design of such a program is dependent on the particular waste characteristics and the site-specific climatic and geologic conditions. Technical considerations, such as type of monitoring instruments and location and frequency of sampling, should be based on several technical criteria. These include the need for monitoring of background conditions not likely to be affected by a contaminant release, the potential routes of seepage, the requirement for early detection, and the provision of detection zones sufficiently large to intercept potential seepage pathways. In many instances, ground water monitoring alone may meet all the above criteria. Under special circumstances, in-waste and vadose zone monitoring may be effective and should be incorporated.

A comprehensive ground water monitoring program should incorporate detection monitoring to determine whether a release has occurred, provisions for evaluation monitoring to verify and quantify a release, and provisions for corrective action monitoring to demonstrate the effectiveness of any remedial program installed to correct a release.

1.2.10 CHAPTER 11: ECONOMIC IMPACT EVALUATION

Mine waste management costs represent a significant portion of the total cost of a mining operation. Examples from projects associated with the gold mining industry in California indicate that waste rock and tailings disposal costs for waste management units that contain the amount of engineered containment required by the 1984 *Chapter 15* regulations, are on the order of 1 and 2 percent, respectively, of the total gross revenue of the project. Incorporating additional containment requirements would increase these costs to between 6 and 14 percent, and 3 and 6 percent for waste rock and tailings disposal facilities, respectively. Similar cost sensitivities were evident for waste disposal facilities associated with industrial minerals projects.

Because the mining industry deals with large volumes of waste, total costs associated with waste management are extremely sensitive to the amount of engineered containment. It is important, therefore, that regulations incorporate sufficient flexibility to allow waste containment to be based on waste characteristics and site-specific conditions, rather than solely on prescriptive measures which, under certain circumstances, may include redundant and impractical requirements.

CONTENTS

CHAPTER 2
PERFORMANCE REVIEW OF MODERN MINE WASTE MANAGEMENT UNITS

PAGE NO.

LIST OF TABLES/LIST OF FIGURES 13

2.1 Introduction 15
- 2.1.1 Objective 15
- 2.1.2 Summary of Findings 15
- 2.1.3 Integration With Other Chapters 15
- 2.1.4 Study Participants 16
- 2.1.5 Background to Mine Waste Regulation in California 16
- 2.1.6 Approach and Sources of Information 16
- 2.1.7 Chapter Organization 18

2.2 Current Mine Waste Management Practices in California 19
- 2.2.1 Overburden and Waste Rock 20
- 2.2.2 Tailings 21
- 2.2.3 Heap Leach Residue Piles 23
- 2.2.4 Liquid Waste Disposal 23

2.3 Performance History 23
- 2.3.1 General 23
- 2.3.2 Region 1 - North Coast Region 24
  - 2.3.2.1 Mining in the Region 24
  - 2.3.2.2 Incidents at Waste Management Units 28
  - 2.3.2.3 Analysis of the Incidents 28
- 2.3.3 Region 2 - San Francisco Bay Region 29
  - 2.3.3.1 Mining in the Region 29
  - 2.3.3.2 Incidents at Waste Management Units 30
- 2.3.4 Region 3 - Central Coast Region 30
  - 2.3.4.1 Mining in the Region 30
  - 2.3.4.2 Incidents at Waste Management Units 30
  - 2.3.4.3 Analysis of the Incidents 33

CONTENTS

CHAPTER 2
(Continued)

PAGE NO.

2.3.5 Region 4 - Los Angeles Region 33
2.3.5.1 Mining in the Region 33
2.3.5.2 Incidents at Waste Management Units 35
2.3.6 Region 5 - Central Valley Region 35
2.3.6.1 Mining in the Region 35
2.3.6.2 Incidents at Waste Management Units 37
2.3.6.3 Analysis of the Incidents 37
2.3.7 Region 6 - Lahontan Region 38
2.3.7.1 Mining in the Region 38
2.3.7.2 Incidents at Waste Management Units 39
2.3.7.3 Analysis of the Incidents 39
2.3.8 Region 7 - Colorado River Basin 43
2.3.8.1 Mining in the Region 43
2.3.8.2 Incidents at Waste Management Units 45
2.3.8.3 Analysis of the Incidents 46
2.3.9 Region 8 - Santa Ana Region 46
2.3.9.1 Mining in the Region 47
2.3.9.2 Incidents at Waste Management Units 47
2.3.10 Region 9 - San Diego Region 47
2.3.10.1 Mines in the Region 47
2.3.10.2 Incidents at Waste Management Units 50
2.3.11 Summary of the Performance of Waste Management Units 50
2.4 Conclusions 50
2.5 Regulatory Implications 61
2.6 Bibliography 61

CONTENTS

CHAPTER 2
(Continued)

LIST OF TABLES

TABLE NO.	TITLE	PAGE NO.
2.1	Summary of WDRs in Region 1	27
2.2	Summary of WDRs in Region 2	31
2.3	Summary of WDRs in Region 3	32
2.4	Summary of WDRs in Region 4	34
2.5	Summary of WDRs in Region 5	36
2.6	Summary of WDRs in Region 6	40
2.7	EPA Guidelines for Leakage Through Primary Liners for Hazardous Waste Landfills	42
2.8	Summary of WDRs in Region 7	44
2.9	Summary of WDRs in Region 8	48
2.10	Summary of WDRs in Region 9	49
2.11	Pre-Article 7 Waste Management Units With Enforcement Actions	53
2.12	Waste Management Units Permitted Under Current Article 7 With Enforcement Actions	55
2.13	Partial List of Incidents at Waste Management Units (No Enforcement Orders Issued)	56
2.14	Recommended Regulatory Responses to the Incidents Raised in Table 2.13	59

LIST OF FIGURES

FIGURE NO.	TITLE	PAGE NO.
2.1	California Regional Water Quality Control Boards	17
2.2	Mean Annual Precipitation in the State of California	25
2.3	Mean Annual Evaporation in the State of California	26
2.4	Performance Summary	51
2.5	Summary of Enforcement Actions	52

CHAPTER 2

PERFORMANCE REVIEW OF MODERN MINE WASTE MANAGEMENT UNITS

2.1 INTRODUCTION

2.1.1 OBJECTIVE

The objective of this chapter is to provide insight into the water quality control effectiveness of mine waste management units constructed under the 1984 California *Chapter 15* regulations governing mine waste disposal.(1) From a more general perspective, the information presented in this chapter also provides an indication of probable performance of mine waste management units constructed under a formalized mine waste regulatory framework, and by mining companies aware of the need for a scientifically based approach to mine waste management.

2.1.2 SUMMARY OF FINDINGS

With one exception, none of the environmental incidents experienced at mines permitted under the current *Chapter 15* regulations, during the period 1985 - 1989, has resulted in any impairment to the beneficial uses of the waters of the State. In the one case, a minor amount of impairment occurred for a short period of time in a surface water reservoir. However, no environmental damage resulted, and the elevated constituent concentrations quickly dissipated. The major water quality impacts attributable to mining in California have been caused by historic operations which were not permitted under the current regulations. These findings confirm the conclusions drawn in the Mining Waste Study conducted by the University of California at Berkeley (U.C. Berkeley, 1988).

The data used to draw the above conclusions were based on the 113 mines permitted under *Chapter 15* since it became effective in December 1984. Although not specifically addressed, these findings are corroborated by experience at mines which have employed technology similar to that required by *Chapter 15* in California and in other states for several years prior to 1984.

2.1.3 INTEGRATION WITH OTHER CHAPTERS

This chapter provides useful background to the material presented in Chapters 3 and 7. The information on the performance of existing waste containment systems is important to the

(1) California Code of Regulations, Title 23, Division 3, *Chapter 15*.

Chapter 3 discussions dealing with a scientifically-based methodology for selecting engineered waste containment to protect the beneficial uses of the waters of the State. The information is also useful background to the detailed liner system design methodology developed in Chapter 7.

2.1.4 STUDY PARTICIPANTS

This study was conducted by the following individuals:

- Mark E. Smith, P.E., Vector Engineering
- Marc E. Orman, Vector Engineering
- Donald T. Bishop, Ph.D., P.G., Condor Earth Technologies
- Peter H. Dohms, P.G., Condor Earth Technologies
- Barry A. Hillman, Ph.D., Condor Earth Technologies

2.1.5 BACKGROUND TO MINE WASTE REGULATION IN CALIFORNIA

Water quality impacts of mine waste management in California are regulated primarily under *Chapter 15*, "Discharges of Waste to Land" (December, 1984) by nine Regional Water Quality Control Boards (Regional Boards) which fall under the jurisdiction of the California State Water Resources Control Board (State Board). A major portion of the regulations which affect mine waste management are included in Article 7, "Mining Waste Management," of *Chapter 15*. Figure 2.1 depicts the areas covered by the respective Regional Boards and Appendix B lists current addresses and phone numbers.

Chapter 15 requires that mine waste be categorized according to its potential to affect water quality. Based on this classification, prescriptive liner and closure requirements are provided for different types of mine waste management units. *Chapter 15* provides for exemptions from the prescriptive requirements if it can be demonstrated that the waste management unit site is not in hydraulic continuity with any usable ground water body. Engineered alternatives to the prescriptive requirements also are permitted if it can be demonstrated that the prescriptive requirements are not feasible and cost substantially more than engineered alternatives providing an equivalent level of containment. These requirements and exemptions are discussed further in Chapter 3.

2.1.6 APPROACH AND SOURCES OF INFORMATION

The basic approach used in developing this chapter was to employ as many different data sources as possible to determine how, why, and where water quality incidents occurred at operating mines.

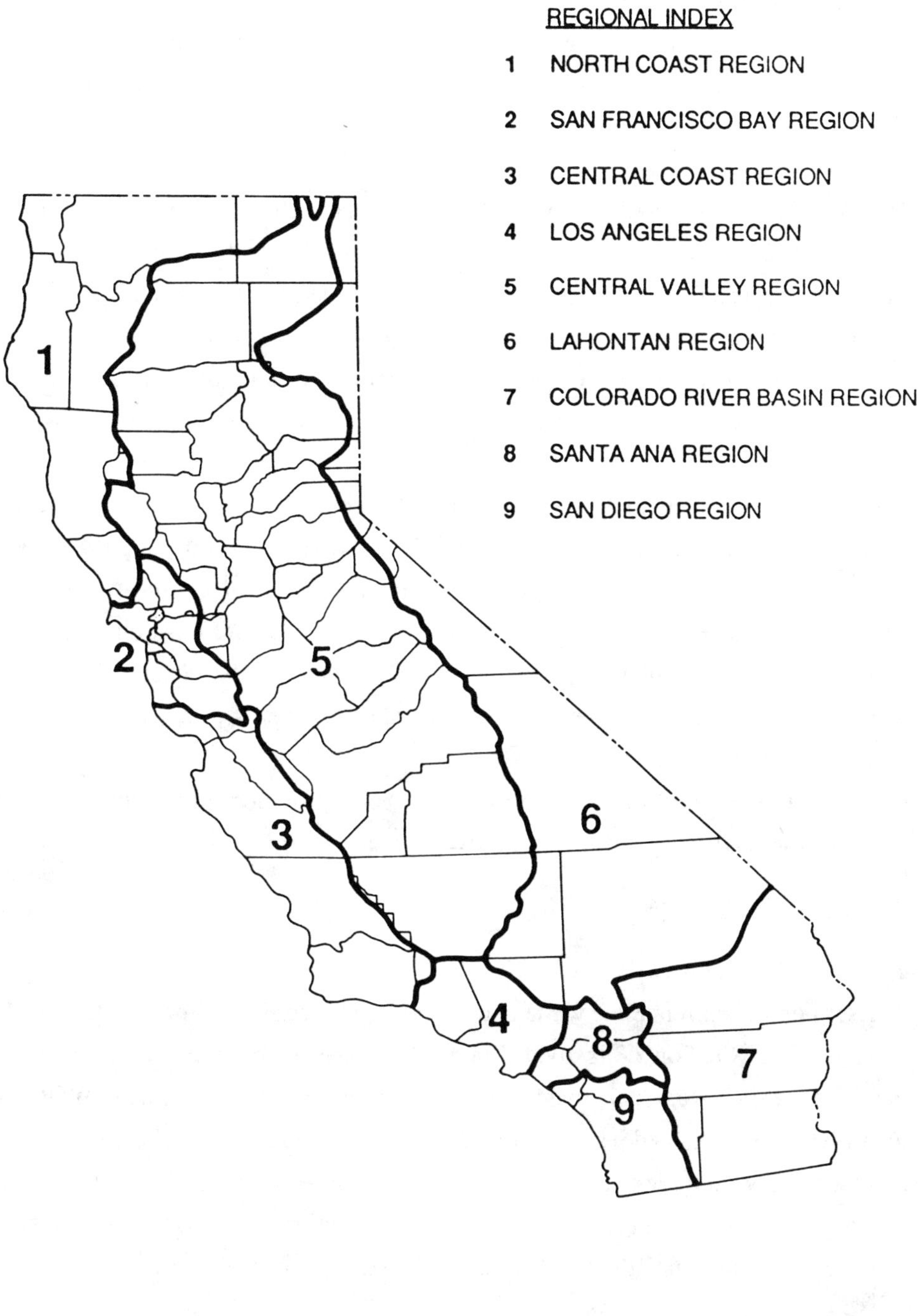

FIGURE 2.1
CALIFORNIA REGIONAL WATER QUALITY CONTROL BOARDS

For the purpose of this discussion, "incidents" are defined as events that occurred at mine waste management units, involved unanticipated releases of constituents derived from mine waste, and had the potential to affect either ground or surface water quality. The more serious incidents have led to formal enforcement actions taken by the Regional Boards. The incidents and formal actions are analyzed to determine additional design and regulatory requirements that would prevent future occurrences.

The U.C. Berkeley, 1988, study contains information on historic mines with environmental problems and evaluates the effectiveness of past waste disposal and mining practices. A telephone survey of a representative cross-section of current mine operators in the State included most of the metals mines, and a sampling of other types of mines. Telephone interviews or office visits with the Regional Boards also were conducted and included discussions with staff responsible for mine wastes, as well as review of their files.

All operating mines with Waste Discharge Requirements (WDRs)(2) were identified from the computer data base of the State and Regional Boards (State of California, 1989), and the associated information was retrieved. This database contained most of the information required for this chapter, which appeared to be accurate when compared to the information acquired from individual mine operators. The listings of the active mines contained in this chapter also were taken from this data base, which appeared to be more complete than lists kept by the individual Regional Boards.

In addition, authors of other chapters of this book provided information from project-specific investigations and designs concluded during the past several years.

2.1.7 CHAPTER ORGANIZATION

A description of current mine waste management practices in California is provided as background in Section 2.2. Section 2.3 is divided into subsections which summarize the performance history obtained for each Regional Board. For each region, the mines that have WDRs were listed and identified as being issued permits either before or after adoption of *Chapter 15*. A summary table is presented for each region, with statistics on the number of mines that have received enforcement orders, as well as incidents that did not result in any enforcement action. Information pertaining to some closed mines, including most of the old historic mines with environmental problems, also is

(2) A WDR is the terminology used for a mine waste discharge permit in California.

reviewed. Sections 2.4 and 2.5 conclude with a summary discussion of enforcement orders statewide, a listing and analysis of identified compliance incidents, and a discussion of their regulatory implications. The data sources reviewed are listed in Section 2.6, Bibliography.

2.2 CURRENT MINE WASTE MANAGEMENT PRACTICES IN CALIFORNIA

Most mining in the State of California is by open-pit methods. The wastes generated by California mines include the following (California Mining Association, 1988):

- Overburden. This is the soil and rock that is stripped from the top of the ore body. These materials are usually chemically inert and, with the exception of surface runoff turbidity, usually do not present any risk to water quality.
- Waste Rock. This material includes nonmineralized and low-grade mineralized rock removed either from above or adjacent to the ore. This material is usually placed in piles close to the mine. The nature and extent of the mineralization, climatic conditions, and buffering capacity of the foundation soil beneath the waste rock pile, determine the potential of the material to impact water quality.
- Tailings. These are the uniform, finely ground rock particles from which most of the commercial ore has been extracted in the beneficiation and extraction plant. As with waste rock, the potential for tailings to impact water quality depends on the chemistry of the material and the specific conditions at the tailings disposal site.
- Heap Leach Residue. This includes run-of-mine, crushed, and agglomerated ore piles that have been leached by allowing solutions to percolate through the material. Before closure, the ore is rinsed and, if necessary, treated to meet specified residual constituent concentration standards. The potential for water quality impacts from this material depends on site-specific conditions, as well as on chemical properties of the ore, reagent types, and the extent of rinsing and/or treatment.
- Wastewater. This includes liquid wastes that may be generated at a floatation, acid leach, or solvent extraction facility, from the operation of mechanical equipment, such as boilers, spent leach solutions remaining at the end of the active life of a heap leach facility, runoff from acid generating ore storage areas, and washdown water from recovery plant or maintenance buildings. These wastes can contain elevated concentrations of constituents that represent a threat to water quality, if the liquids are not adequately contained.

  Wastewaters are also generated in sand and gravel washing operations. However, these only contain elevated suspended solids and usually can be discharged after temporary detention in settling ponds.

Metal mining operations typically generate overburden, waste rock, and tailings or heap leach residue. In many industrial mineral operations, only overburden and waste rock are generated.

The types of mine waste management units that are employed to contain the above wastes include the following:

- Waste Piles. These units involve stacking the waste in valley fills, side hill dumps, or open piles. Leachate containment may be required sometimes, and can be provided by either the natural geologic formations or engineered systems. Where runoff discharges from the pile, sediment control facilities may also be needed. Piles are suitable for disposal of overburden, waste rock, tailings that have been dried in the recovery plant, and heap leach residue.
- Tailings Impoundments. These are impoundments that contain a slurried tailings material. They are formed by either an embarkment or a dike. Containment of the liquids associated with the tailings is achieved by either the natural geologic formations or engineered systems.
- Wastewater Ponds. These are similar to tailings impoundments in that they are designed to contain liquids. They are formed by either an embankment or dike, or by excavating a depression in the ground. Liquid containment is achieved the same as a tailings impoundment.

The following sections describe some of the current mine waste management practices in California.

2.2.1 OVERBURDEN AND WASTE ROCK

Overburden and waste rock are comprised of granular, broken rock and soil, with particles ranging in size from fine sand or clay to boulders several feet in diameter. The material is hauled by truck or conveyor from the active mining area and disposed of in waste piles. Piles are located either at the heads or along the sides of valleys to minimize interference with the natural surface water drainage. In most cases, there is no need for engineered containment.

However, when mineralized waste rock requires disposal in wetter climates, it may be necessary to provide engineered containment where disposal sites with sufficiently tight or impermeable geologic formations cannot be found. Such containment may be provided by internal drains placed within the waste pile and routed to collection basins. From there, fluids either can be recycled, evaporated, or treated and discharged. In some cases, low permeability material may have to be placed at the base of the waste.

Special waste management practices are required when the waste rock has an acid generating potential due to the presence of sulfides. These practices include lining the facility and selectively depositing the more acid generating fraction of the material within cells constructed of nonacid generating material, alternately co-mingling the acid with the nonacid generating material. The unit also may have to be covered with a low permeability layer and a vegetated surface to minimize the inflow of water and oxygen. Leachate collection and reuse or treatment also may be required. Other management options are discussed further in Chapter 4.

The slopes of a waste pile, and particularly the front face, are designed to ensure stability and to facilitate reclamation. The correct overall slope is achieved by interspersing horizontal benches between angle of repose end-dumped slopes. In high seismic zones, or in areas where reclamation may be difficult, the slopes may have to be flattened to achieve the necessary stability.

Sediment control is achieved by designing the waste pile to minimize surface erosion. Where necessary, sediment ponds are constructed to collect storm runoff that falls directly on the pile. These ponds are designed to provide sufficient retention for suspended solids to settle before the water is discharged to the natural drainage system. Ditches placed around the perimeter of the pile prevent runoff upgradient of the site from flowing onto the waste pile.

2.2.2 TAILINGS

Tailings can be discharged as a slurry or can be dried to the consistency of a moist sand prior to deposition. These two forms are referred to as "wet tailings" and "dewatered tailings" (or "dry tailings"). Wet tailings are in liquid or slurry form. Dry tailings are produced by applying mechanical pressure or a vacuum to remove the water.

The most commonly used disposal method is to deposit wet tailings in surface impoundments. These impoundments typically range in size from 5 to 500 acres, and are formed by using available valleys or depressions and by constructing embankments or dams to complete the containment provided by these natural features. In most cases, embankments are constructed from native soils found near the project site. If required, water tightness can be achieved by providing an internal clay core or other types of low permeability barriers which extend from the top of the embankment downward, and are keyed into the subsurface geologic material or connected to an impoundment liner system. In some cases, where the waste liquids are of suitable quality, the coarser fractions of the tailings themselves can be used to construct the embankments. In other cases, "leaky" embankments may be provided to promote seepage from the tailings to enhance drainage and drying. Seepage then is collected at the downstream toe of the embankment and returned to the recovery plant for reuse.

If judged to pose a potential risk to beneficial uses of receiving water, seepage from the impoundment area can be contained by several methods. First, the underlying geologic formations may suffice, if studies indicate they are of sufficiently low-permeability that they contain little or no usable ground water. A second method involves the use of one or more liners and is further discussed in Chapter 7. A third method is to use the inherent low permeability characteristics of the tailings themselves to minimize seepage from the base of the impoundment. Once the impoundment is constructed, wet tailings are slurried by pipeline and discharged in thin layers at several points around the perimeter. This allows the material to dry naturally. The dried material itself has a very low permeability, thereby limiting the downward migration of liquids. In drier climates, such practices result in a tailings impoundment containing very little moisture, further minimizing the potential for any seepage losses. Usually, modern tailings impoundments incorporate a combination of several of these containment methods.

Excess solution that may collect in the lower elevations of the tailings impoundment is returned to the process facility system for reuse. This minimizes hydrostatic pressure within the impoundment and promotes drying of the tailings solids. It also conserves water, an important consideration in drier climates. In wetter climates, where accumulation of water in the impoundment is to be minimized, a system of ditches can be provided around the perimeter of the facility in order to divert surface runoff away from the waste material.

"Dewatered" or "dry" tailings are disposed of in engineered piles on prepared ground. Dewatered tailings are transported from the process facility to the prepared site by truck or conveyor belt. Ponds sometimes are constructed around the perimeter of the site to collect runoff and seepage from the pile, thus preventing it from escaping to the environment.

Co-disposal of tailings and waste rock also can be considered. In instances where the chemical characteristics of the two materials are compatible, and where there is an economic advantage in doing so, the waste rock can be co-disposed with the tailings to provide bermed containment, to accelerate drainage of the tailings liquids, and to provide improved support for any engineered containment system that may be required at closure. There may be instances where there are geochemical advantages to co-disposal, such as when there are acid tailings with an alkali waste rock. Co-disposal of these two materials would reduce or eliminate the concentration of dissolved metals in the tailings liquid.

2.2.3 HEAP LEACH RESIDUE PILES

Heap leaching involves placement of the ore on lined pads or within lined natural valley floors. The extraction solutions are allowed to percolate through these piles, dissolving out the economic minerals. The "pregnant" solutions are collected on the liner and conveyed by drainage layers and pipes, and by surface ditch and piping systems, to lined ponds and then through a plant which extracts the mineral of interest. The liquids are recycled to the heap.

On completion of leaching, the residue may be rinsed to reduce constituent concentrations in the interstitial liquids to acceptable levels. The rinsed residue then is disposed of in place, or, where reusable pads have been constructed, moved to a prepared waste management unit.

Heap leach residue piles thus are disposed of in either lined or unlined waste management units depending on the type of leaching operation and on the chemistry of the interstitial liquids. Final closure of these piles may involve flattening of the slopes, placement of a soil and possibly clay cover material, revegetation and provision for erosion protection, and surface drainage measures. A more detailed discussion of heap leach operations is presented in Chapter 9.

2.2.4 LIQUID WASTE DISPOSAL

There are occasions when mining operations either must dispose of or temporarily store waste liquids derived either from the recovery plant or from waste pile leachate. Depending on the nature and concentration of constituents in the liquid, this waste is located in either unlined, single, or double-lined ponds. Adequate freeboard is provided to prevent over-topping and spillage during extreme precipitation events.

2.3 PERFORMANCE HISTORY

2.3.1 GENERAL

In order to evaluate the performance of mining operations with respect to water quality issues, four main categories of information were assembled: (1) mine type and product, (2) climate, (3) known or suspected water quality incidents, and (4) the nature of the incidents. Information from all types of mining operations that were issued WDRs was reviewed. Special emphasis was placed upon metals mines, not only because of the current high level of activity in that sector, but also because these operations pose more of a potential threat to water quality than other types of mines, due to the nature of the minerals processed and the higher incidence of chemical reagent usage.

Precipitation and, to a lesser degree, evaporation are important factors affecting waste containment design and the performance of waste management units. Generally, the wetter the climate, the more potential there is for liquid to accumulate and escape from waste management units. For this reason, the introductory paragraphs in each of the following sections briefly describe the climate, mean annual precipitation, and evaporation for each region. Values for the State of California are presented in Figures 2.2 and 2.3. A more detailed discussion of the role of climatological factors is presented in Chapter 6.

2.3.2 REGION 1 - NORTH COAST REGION

The North Coast Region is divided into the Klamath Mountain and Northern Coastal subregions. The climate in these two subregions ranges from alpine to coastal. Annual precipitation is approximately 50 to 60 inches, while average annual evaporation is 30 to 40 inches.

The Klamath Mountain subregion extends from the western Coast Range northward to the Oregon border and has an average elevation of 5,000 to 7,000 feet. The Klamath River drains most of this area and has many large tributaries, including the Trinity and the Scott Rivers, which flow year-round.

The Northern Coastal subregion contains the Coast Range, which stretches from just north of San Francisco Bay to the Oregon border. This range rises abruptly from sea level to approximately 6,000 feet and forms a major north-south trending barrier. The major drainages of this part of the region are the Eel and Russian Rivers.

2.3.2.1 Mining in the Region

Extensive lode and placer gold mining took place in the Klamath Mountain subregion during the 19th century gold rush in California. Placer gold miners diverted waters from the Klamath, Trinity, and Scott Rivers and other drainages in order to mine the stream gravels. Currently, the only permitted gold mines in the region are placer operations. The records show a total of 45 permitted mining operations, with 40 listed as active. Thirty-eight of these active mines are sand and gravel operations. A summary of the number of different mine types, along with statistics regarding enforcement orders, is presented in Table 2.1.

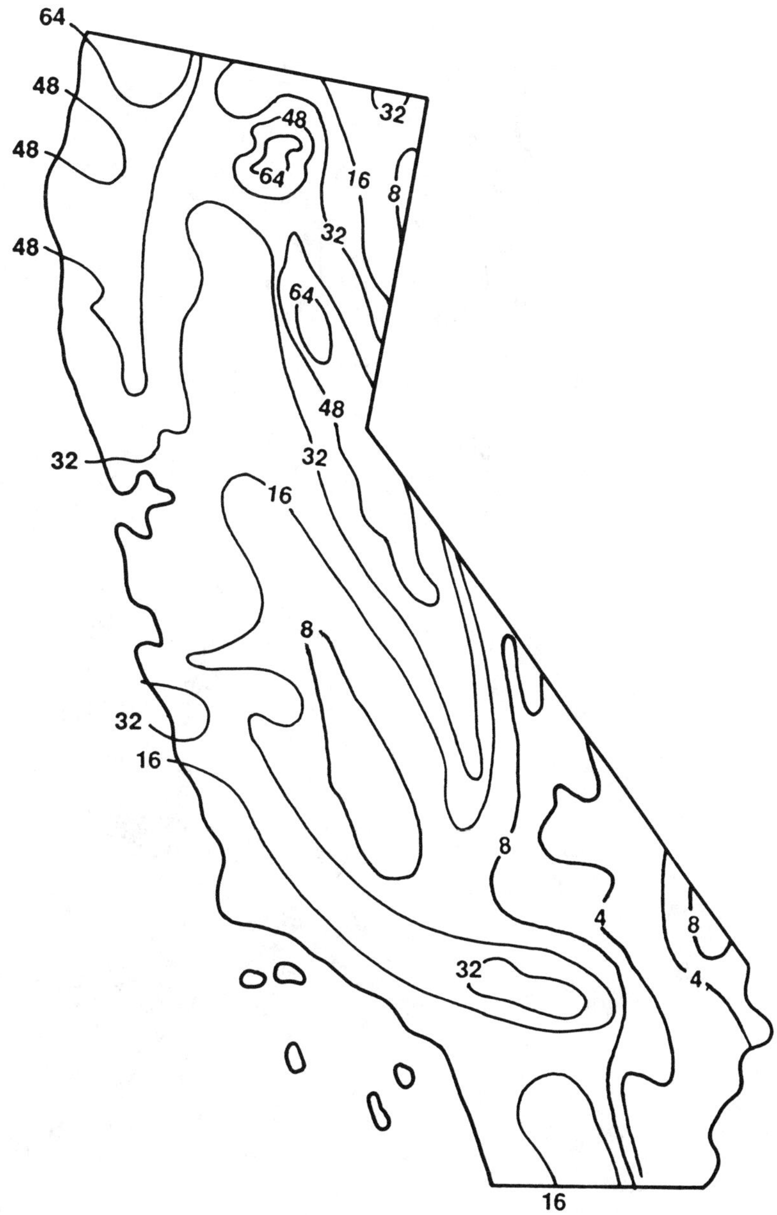

FIGURE 2.2

MEAN ANNUAL PRECIPITATION IN THE STATE OF CALIFORNIA (INCHES)

REFERENCE: *HYDROLOGY FOR ENGINEERS*, LINSLEY, ET AL. 1982.

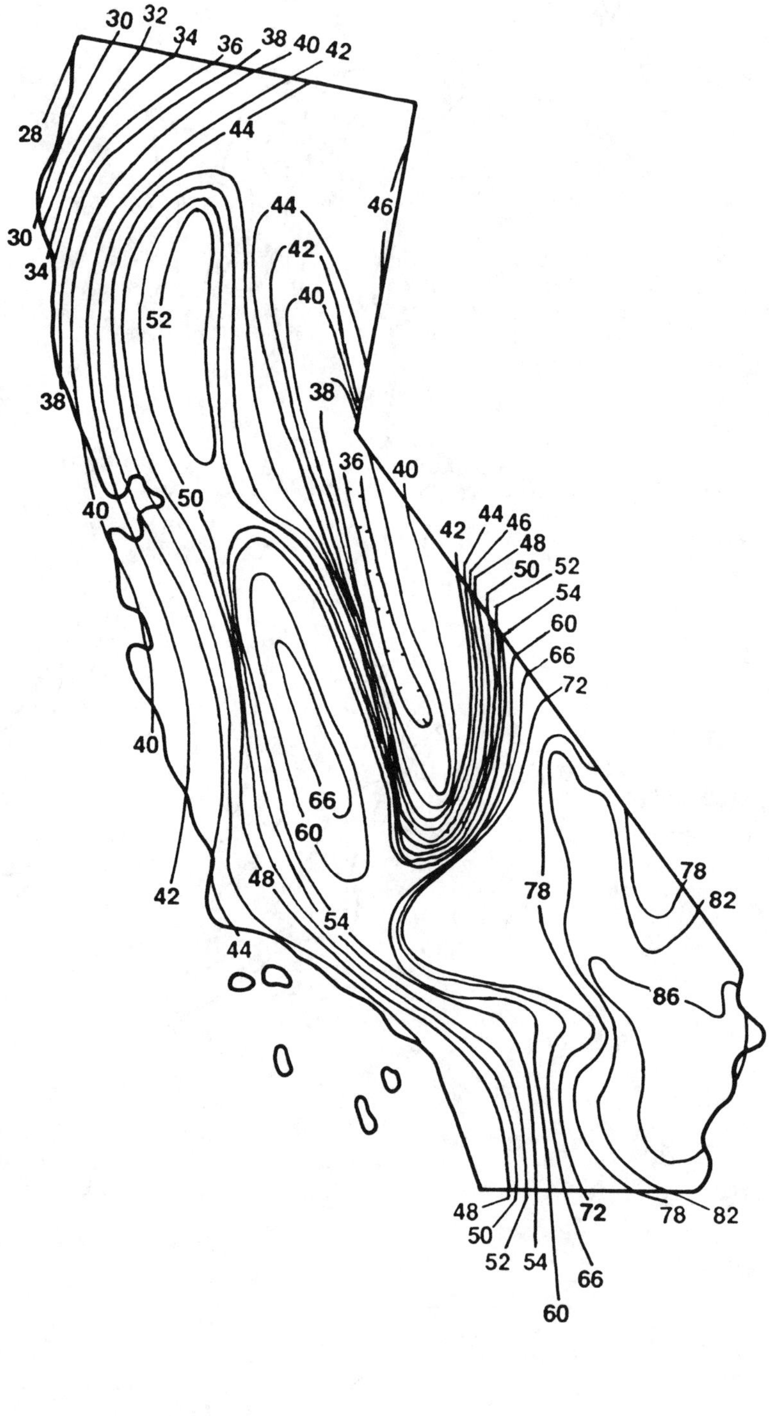

REFERENCE: *HYDROLOGY FOR ENGINEERS*, LINSLEY, ET AL. 1982.

FIGURE 2.3

MEAN ANNUAL EVAPORATION IN THE STATE OF CALIFORNIA (INCHES)

TABLE 2.1

SUMMARY OF WDRs[1] IN REGION 1

MINE TYPE	TOTAL MINES	WDRs ISSUED PRE-1984		WDRs ISSUED POST-1984	
		TOTAL MINES	MINES WITH ENFORCEMENT ORDERS	TOTAL MINES	MINES WITH ENFORCEMENT ORDERS
Sand and Gravel	38	33	2	5	0
Metallic/Conventional	3	3	3	0	0
Metallic/Placer	4	4	0	0	0
Other	0	0	0	0	0
TOTALS	45	40	5	5	0

[1] WDR = Waste Discharge Requirements.

2.3.2.2 Incidents at Waste Management Units

Water quality incidents have been associated only with mines permitted before the adoption of *Chapter 15*. The most extensively documented case is the enforcement action taken against the Noranda Grey Eagle Mine. Between 1981 and 1986, Noranda mined about 900,000 tons of gold and silver ore from an open pit. A tailings dam was constructed in a steep valley, using waste rock material for the shell and local borrow for the clay core. Ore processing consisted of agitated carbon-in-pulp cyanide leaching. The tailings then were deposited in the impoundment.

Since it was commissioned, the Grey Eagle tailings dam had problems with the seepage of tailings fluids through fractures beneath the embankment. When the seepage first appeared, some discharged into the receiving creeks. As soon as it was discovered, a pumpback system was installed, and the seepage was controlled by returning it to the tailings impoundment. The seepage problem was compounded by preexisting acid rock drainage (ARD) discharging from historic underground mines into the tailings embankment underdrains, and mixing with the tailings seepage. The mine ceased operations in 1986, and the tailings impoundment was capped and reclaimed. A treatment plant was constructed at the toe of the dam for treating the seepage before discharging it to the natural drainage system.

Recently, the cyanide levels in the seepage have dropped substantially, probably as a result of the layered cap placed over the tailings impoundment. This cap has significantly reduced or eliminated the amount of rainfall and runoff entering the impoundment. However, high metal levels in the seepage persist and are likely to continue, due to drainage from the historic underground mines.

Two other listed metals mines in this region have had incidents. These were the Humbug and Middle Earth placer gold mines, which are now inactive. Both were cited for turbidity problems. Two sand and gravel operations also have had water quality problems caused by turbidity, but continue to operate. Enforcement orders issued by the Regional Board are summarized in Table 2.1.

2.3.2.3 Analysis of the Incidents

The turbidity experienced with the placer operations was due to ineffective surface drainage and sediment control facilities. These reflect design, operational, and oversight problems rather than inadequate control technology or regulations.

The problems with the Grey Eagle Mine are twofold: (1) the acidic seepage discharging from the historic underground workings was never remediated and continues to this day, and (2) the tailings impoundment was designed as an unlined facility, even though the underlying geologic formations have fractured rock hydraulic conductivity of the order of 10^{-5} cm/sec.[(3)] A grout curtain and a clay-filled cutoff trench were constructed to provide seepage control under the embankment. However, the standard of design for construction of these facilities apparently was not adequate to prevent seepage. The design concept is unlikely to have been acceptable under the current *Chapter 15*.

The water quality impact from the Grey Eagle problem associated with seepage of tailings liquid has been negligible. The only concern was during the initial period in which cyanide solution was being discharged. Although no measurements were taken, subsequent calculation indicated that cyanide concentration levels in the creek would have been below any levels of concern. No fish kills or other environmental damage was observed.

2.3.3 REGION 2 - SAN FRANCISCO BAY REGION

The San Francisco Bay Region occupies the central portion of the Coast Range province and includes the entire bay area. Annual precipitation for this small region is approximately 30 inches. Except in local areas, precipitation is usually exceeded by an average annual evaporation of about 40 inches. In general, the climate in this area is coastal.

2.3.3.1 Mining in the Region

While there are presently no active metals mines in Region 2, there have been active operations as recently as 1950 (Clark, 1980); the San Francisco beach area produced fine-grained gold from 1938 until 1950. Mining and extraction was accomplished mostly in wash plants. The Calistoga silver-gold district produced silver and gold, with some copper and lead, from 1872 to 1893. The New Almaden mercury district, also located in this region, is one of the earliest mercury mining districts in the State.

Presently, the only type of recorded mining in Region 2 consists of 15 permitted dredging operations that are required to keep the bay's shipping lanes open. Six of these operations are

---

(3) The *Chapter 15* regulations require containment provided by, or equivalent to, a hydraulic conductivity of 10^{-7} cm/sec.

listed as inactive or historic. The active dredging operations remove sediments and screen them for resale as construction aggregate. A summary of the number of mines with WDRs in Region 2 is presented in Table 2.2.

2.3.3.2 Incidents at Waste Management Units

None of the sand, gravel, and oyster shell dredging operations in Region 2 appears to have had any water quality incidents, and no enforcement actions have been taken against them by the Regional Board.

2.3.4 REGION 3 - CENTRAL COAST REGION

The Central Coast Region occupies the California coast from north of Monterey Bay south to the Ventura Basin. The region is located predominantly in the coast ranges along the central coast and in the transverse ranges to the south. The climate is coastal-dominated throughout most of the region, becoming more arid to the south. Except for the vicinity of the Soda Lake basin, the quality of ground and surface water is suitable for both municipal and domestic use. Except for local areas, evaporation generally exceeds precipitation by an average of approximately 10 inches annually.

2.3.4.1 Mining in the Region

Sand and gravel aggregate are the most important commodities mined in the Central Coast Region. Other commodities include minor amounts of gypsum, limestone, specialty sands, and major quantities of diatomite. There are no active metal mines in the area, but there are four inactive metal mines with WDRs. There is one operating asbestos mine in the region, located in San Benito County. A summary of the number and types of mines with WDRs in Region 3 is presented in Table 2.3.

2.3.4.2 Incidents at Waste Management Units

The most significant problems in the region are associated with inactive and abandoned mine sites. Asbestos was mined by open pit at several mines in the New Idria serpentinites. Surface runoff from this region is found to contain high counts of asbestos fibers, particularly after major storms. Also of concern is ARD from abandoned mercury mines.

TABLE 2.2

SUMMARY OF WDRs[(1)] IN REGION 2

MINE TYPE	TOTAL MINES	WDRs ISSUED PRE-1984		WDRs ISSUED POST-1984	
		TOTAL MINES	MINES WITH ENFORCEMENT ORDERS	TOTAL MINES	MINES WITH ENFORCEMENT ORDERS
Sand and Gravel	0	0	0	0	0
Metallic/Conventional	0	0	0	0	0
Metallic/Placer	0	0	0	0	0
Other	15	8	0	7	0
TOTALS	15	8	0	7	0

[(1)] WDR = Waste Discharge Requirements.

TABLE 2.3

SUMMARY OF WDRs[1] IN REGION 3

MINE TYPE	TOTAL MINES	WDRs ISSUED PRE-1984		WDRs ISSUED POST-1984	
		TOTAL MINES	MINES WITH ENFORCEMENT ORDERS	TOTAL MINES	MINES WITH ENFORCEMENT ORDERS
Sand and Gravel	19	15	0	4	0
Metallic/Conventional	4	4	2	0	0
Metallic/Placer	0	0	0	0	0
Other	11	8	1	3	0
TOTALS	34	27	3	7	0

[1] WDR = Waste Discharge Requirements.

2.3.4.3 Analysis of the Incidents

Review of the asbestos data from the New Idria area indicates that the high counts occur primarily where surface runoff flows over serpentinite formations in areas that have never been mined. This indicates that the high levels of asbestos are mainly attributable to natural causes and not to past or current mining activities.

ARD containing dissolved solids and iron currently is discharging from the workings of several inactive and abandoned mercury mines. Cleanup and abatement orders have been issued to facilitate mitigation of this problem, which has resulted from out-of-date waste management practices that would not be permitted under *Chapter 15*.

2.3.5 REGION 4 - LOS ANGELES REGION

The Los Angeles Region includes the majority of Los Angeles and Ventura Counties. The region is within the Transverse Range Province. It incorporates east-west trending mountain ranges, with peaks of over 4,800 feet, interspersed with broad, alluvial valleys. The average annual precipitation in this region is approximately 20 inches and is exceeded by an average annual evaporation of about 50 inches. The climate ranges from coastal along the Pacific Ocean, to Mediterranean in the mountains, and desert in the low-lying valleys.

2.3.5.1 Mining in the Region

At one time there were many operations producing gold in this area. Presently, however, there is only one active metals mine, a tungsten operation that obtained its WDRs in 1979. Most of the historic gold districts are in the Frazier and San Gabriel Mountain areas. Some of the more famous districts include Piru, Frazier Mountain, Acton, Azwa, Mount Baldy, Mount Gleason, Newnoch, and Saugus. These districts produced gold from placer and lode mines in the 1800s and as recently as 1952 (Clark, 1980).

Sand and gravel operations currently constitute most of the active mining in this region, with a total of 26 listed mines. A summary of the number and types of mines with WDRs in Region 4 is presented in Table 2.4.

TABLE 2.4

SUMMARY OF WDRs[1] IN REGION 4

MINE TYPE	TOTAL MINES	WDRs ISSUED PRE-1984		WDRs ISSUED POST-1984	
		TOTAL MINES	MINES WITH ENFORCEMENT ORDERS	TOTAL MINES	MINES WITH ENFORCEMENT ORDERS
Sand and Gravel	26	22	0	4	0
Metallic/Conventional	1	1	0	0	0
Metallic/Placer	0	0	0	0	0
Other	0	0	0	0	0
TOTALS	27	23	0	4	0

[1] WDR = Waste Discharge Requirements.

2.3.5.2 Incidents at Waste Management Units

There are no identified incidents for any mining operation in Region 4, nor have any enforcement actions been taken.

2.3.6 REGION 5 - CENTRAL VALLEY REGION

The Central Valley Region encompasses the Great Valley of California and includes the drainage basins of the Sacramento and San Joaquin Rivers. This region includes most of the Sierra Nevada Mountains, with climatic conditions ranging from alpine to Mediterranean. Average annual evaporation exceeds precipitation in the Great Valley, although the reverse is true in most of the higher elevations of the Sierra Nevadas. Ground and surface water resources are used within the region for irrigation, municipal and industrial purposes, recreation, and wildlife habitat.

2.3.6.1 Mining in the Region

The Central Valley Region is one of the most active mining regions in the State, with a total of 198 listed mines. Mining is dominated by the production of sand, gravel, and crushed rock. Stone and limestone mining occur in the foothills of the Sierra Nevadas. Clays, specialty sands, and other industrial commodities are also important. The region has a long history of metal mining and includes the well-known Mother Lode gold mining district. Although currently inactive, the West and East Shasta districts and the western copper belt of the Sierra foothills have had important copper and zinc mines in the past.

The major producing metal mines in the region are all open pit gold operations. The largest producers are the McLaughlin Mine in Napa and Lake Counties and several mines in the Mother Lode trend, including the Carson Hill Mine in Calaveras County and the Jamestown Mine in Tuolumne County. A summary of the number and type of WDRs issued in Region 5 is presented in Table 2.5.

It is interesting to note that most of the metal mines that have been issued WDRs under *Chapter 15* have utilized the "Engineered Alternative" provisions of Article 1, or the exemption provision of Article 7, for the containment system design, and have not utilized the prescriptive standards. These engineered alternatives include the use of synthetic secondary liners instead of clay, the use of natural geologic formations instead of liners, and the construction of a primary liner consisting

TABLE 2.5

SUMMARY OF WDRs[1] IN REGION 5

MINE TYPE	TOTAL MINES	WDRs ISSUED PRE-1984		WDRs ISSUED POST-1984	
		TOTAL MINES	MINES WITH ENFORCEMENT ORDERS	TOTAL MINES	MINES WITH ENFORCEMENT ORDERS
Sand and Gravel	97	79	0	18	0
Metallic/Conventional	59	43	4	16	1
Metallic/Placer	36	27	0	9	1
Other	6	4	1	2	0
TOTALS	198	153	5	45	2

[1] WDR = Waste Discharge Requirements.

of consolidated tailings, using a managed deposition technique in a double lined tailings impoundment. This demonstrates that, in many of the new mines, it has been effective to tailor the waste containment designs to the site-specific conditions.

2.3.6.2 Incidents at Waste Management Units

Most of the problems related to mining in the Central Valley Region have been associated almost entirely with inactive or abandoned mine sites that are discharging ARD. These are discussed in the Berkeley Mining Waste Study (U.C. Berkeley, 1988).

Enforcement actions for the currently permitted mines are summarized in Table 2.5. One of these actions was levied against a cement manufacturer due to high pH/turbid runoff flowing into Stillwater Creek. All of the mines listed in the table with enforcement actions against them, except for Carson Hill and the Poison Oak Flat Mill Site, were closed or abandoned prior to adoption of the current *Chapter 15* in 1984.

The Gold Stripe and McLaughlin mines both have had water quality incidents. None resulted in any enforcement action.

2.3.6.3 Analysis of the Incidents

On May 16, 1989, a spill of pregnant (gold-bearing) cyanide solution occurred at the Carson Hill Mine in Calaveras County (permitted under the *Chapter 15* provisions). A coupling in the piping of Leach Unit 3 failed, and approximately 92,000 gallons of sodium cyanide leach solution spilled into a stream course draining to the New Melones Reservoir, located about 1,500 feet away. The solution was estimated to contain about 80 pounds of cyanide at the time of discharge.

No mortality of aquatic organisms occurred (Licari, 1989), and the contaminant levels measured during an extensive lake-water sampling program were very low. From the time sampling began, which was five hours after the spill was first noticed, the free cyanide concentrations at 10 feet, 20 feet, 30 feet, and 40 feet from the shore of the reservoir were all below the detection limit of 0.2 mg/L. Subsequent to the discharge, the coupling was replaced with an elbow and riser of new design, and additional measures were instituted to prevent further spills. Carson Hill Mine was issued an Administrative Civil Liabilities enforcement action and fined. The case is now closed.

The other site subject to an enforcement action was the Poison Oak Flat Mill Site, also known as the Minnie-Ha-Ha Placer Mine, constructed under *Chapter 15*. The operators of this site constructed a siltation pond that was not correctly engineered and failed in December 1985. The operators were cited for one unauthorized discharge that flowed into a nearby stream. This site has since been abandoned.

Incidents at the McLaughlin mine involved a spill of dilute ARD from a holding pond during the extreme precipitation events that occurred in February 1986, and the appearance of some ARD in the leachate emanating from a waste dump. The ARD leachate is being controlled by collection and pumping back to the process plant, when necessary.

Goldstripe is a heap leach operation situated in a high rainfall area. During an extreme storm event, the pregnant pond overtopped and discharged dilute cyanide solution into the nearby drainage course. The concentration of cyanide in the discharged water was highly diluted by precipitation, as well as by water in the drainage course, which was flowing near its peak discharge at the time of overtopping. These combined effects resulted in nondetectable levels of cyanide in the water course. The Regional Board elected not to issue an enforcement order.

2.3.7 REGION 6 - LAHONTAN REGION

The Lahontan Region occupies most of the State east of the Sierra Nevadas. The physiography of the region is largely basin and range, and includes the Modoc Plateau in the north and the Mojave Desert in the south. Except for the high alpine areas, the climate is arid to semi-arid, and evaporation greatly exceeds precipitation. Perennial streams are generally restricted to the alpine areas, and only the largest rivers cross the arid regions. In the basin and range, and desert areas to the south, most streams are intermittent, and most discharge to enclosed basins. Many small drainages flow only during flash flooding. In the alpine regions, ground water mostly is contained in fractured rock or restricted to small, alluvial valleys. Ground water may be abundant in alluvial fans flanking major mountain ranges, but is typically found at great depth. Ground water in the valleys and desert playas is typically abundant, but of highly variable quality. Total dissolved solids can range up to tens of thousands parts per million.

2.3.7.1 Mining in the Region

Sixty-six mines are listed for the region, including rare earth minerals, soluble salts, numerous metal mines, sand and gravel operations, limestone, clay, talc, and decorative stone quarries, and mines for other industrial minerals and commodities. Many of the old, precious metal districts are being subjected to renewed exploration activity and mining. The southern part of the Lahontan Region has a long history of tungsten, iron, rare earth minerals, and soluble salt production. The number and types of mines in Region 6 with WDRs are presented in Table 2.6.

As is the case in the Central Valley Region, most of the gold heap leach operations that have been issued WDRs under *Chapter 15* have utilized the "Engineered Alternative" provisions. The nature of these alternatives generally has been to utilize secondary synthetic liners rather than the prescriptive clay liners. There are several reasons for this. In order to meet permeability guidelines, a clay liner must be placed and compacted to stringent specifications. It has been found that clay cannot be economically placed or compacted properly on the steep slopes (greater than 3:1, horizontal to vertical) associated with heap leach ponds. Also, there is frequently a shortage of suitable clay at mine sites. In these cases, the only way to meet permeability guidelines is to use synthetic liners.

2.3.7.2 Incidents at Waste Management Units

All enforcement actions against mines in this region have been against those permitted prior to adoption of *Chapter 15*. Comments from the regional staff range from noting minor incidents, such as late reports on the results of compliance monitoring, to leaks and spills at process and waste management units.

Historically, some wet tailings and liquid wastes have been disposed of in unlined and single-lined surface impoundments, respectively, which do not meet the provisions of *Chapter 15*. Continuing surveillance of some of these facilities indicates that the surface impoundments have not contained the wastes effectively and that some migration of waste and leachate into the vadose zone and the waters of the State has occurred.

2.3.7.3 Analysis of the Incidents

An example of a pre-*Chapter 15* operation is illustrated by conditions at Mountain Pass. There, tailings and process wastewaters were disposed of in unlined and single-lined impoundments, respectively. These impoundments were found to be overlying shallow alluvium and moderately pervious rock strata. Under a Cease and Desist Order, Molycorp was required to close certain impoundments and construct new ones in compliance with the provisions of *Chapter 15*.

TABLE 2.6

SUMMARY OF WDRs[1] IN REGION 6

MINE TYPE	TOTAL MINES	WDRs ISSUED PRE-1984		WDRs ISSUED POST-1984	
		TOTAL MINES	MINES WITH ENFORCEMENT ORDERS	TOTAL MINES	MINES WITH ENFORCEMENT ORDERS
Sand and Gravel	19	16	2	3	0
Metallic/Conventional	30	19	4	11	0
Metallic/Placer	0	0	0	0	0
Other	17	14	2	3	0
TOTALS	66	49	8	17	0

(1) WDR = Waste Discharge Requirements.

By mid-1988, there were at least seven heap leach operations in the South Lahontan Region; all were issued WDRs under the current *Chapter 15*. Although heap leaching is a metal recovery process, and the ore is not a waste during leaching, the Regional Board treated each of the seven leach operations as waste disposal and required them to install double liners with leachate collection and recovery systems (LCRSs). In order to control leakage through the upper liner, the Regional Board established a maximum allowable concentration of cyanide solution in the LCRS drainage sumps of 0.2 mg/L measured weak acid dissociable (WAD) cyanide. Lahontan is the only Regional Board to set such stringent criteria.

The seven heap leach operations have had cyanide appear in the LCRS on at least one occasion. The leaks have been attributed by Regional Board staff to faults in design and construction, and damage to the liners caused by careless use of equipment, falling pieces of ore, and material sloughing at the margins of the ore heaps. All of these leaks were reported and repaired in a timely fashion. No known water quality damage has occurred due to any of these leaks, and there have been no documented cases of solution escaping the secondary containment, or outer liner. However, these incidents have resulted in a considerable amount of effort being expended by both the mining companies and the Regional Board overview staff in discussions and negotiations on acceptable leakage rates, and the extent of follow-up investigations, and repair work. These incidents also have exposed some of the mining companies to significant potential financial risks, because the Regional Board had the authority, under the WDR, to shut down operations until the repairs are made.

In all cases, the leakage rate through the primary or upper liner was well below the guidelines that the U.S. Environmental Protection Agency (EPA) has established for satisfactory performance of liner systems at hazardous waste landfills. These guidelines are discussed in the *Background Document on Proposed Liner and Leak Detection Rule* (Giroud et al., 1987), and typical responses for different leakage amounts are presented, as shown in Table 2.7. Typical leakage rates from liners in Region 6 were only a few gallons per acre per day (gpad).

Other operating mines in Region 6 that had enforcement actions taken against them were a boron mine facility and two sand and gravel operations. All predated the current *Chapter 15*. The boron mine had problems with leaks from unlined process water surface impoundments in the early 1970s. The leaks first were observed when seepage emerged from the toes of some of the containment dikes. These impoundments were abandoned at that time, and the facility has continued to comply with Regional Board recommendations for investigation and cleanup. An

TABLE 2.7

EPA GUIDELINES FOR LEAKAGE THROUGH PRIMARY LINERS FOR HAZARDOUS WASTE LANDFILLS

LEAKAGE RATE (gpad)[(1)]	REQUIRED RESPONSE
<20	Pump collected leachate.
20	Notify Regional Administrator.
20 to 250	Increase pumping and monitoring.
250 to 2,500	Implement changes in operating practices to reduce leakage to lower leakage band.
>2,500	Repair leak or close unit (or part of unit).

(1) Gallons per acre per day.

Reference: Giroud et al., 1987.

enforcement action was taken against the boron mine facility in 1985 as a result of the appearance of process water in one of the many shallow ground water monitoring wells that surround one of the clay-lined surface impoundments. The mining company determined that the incident was due to overtopping of the impoundment and subsequent infiltration of spilled liquids rather than liner failure. The impoundments nearest this well have since been abandoned. In compliance with Lahontan Regional Board recommendations, the well was pumped in an attempt to recover the spilled solutions. It is important to note that, at this site, ground water is of poor quality and at a depth of over 120 feet. It has been demonstrated that the leaks which occurred during the early 1970s have affected only this local poor quality ground water, and that there has been minimal lateral migration. None of the facilities that have had problems would have met the requirements of the current *Chapter 15*.

One of the sand and gravel operations in this region had an enforcement action against it due to turbid runoff, while another was cited due to a leaking underground petroleum storage tank.

2.3.8 REGION 7 - COLORADO RIVER BASIN

The Colorado River Basin Region occupies all of the Colorado River basin in the southeastern part of California where the climate is arid. The major surface drainages in the region are the San Gorgonio River, which drains to the Salton Sea, and the Colorado River, to which short tributaries drain from the eastern part of the State. Ephemeral streams discharge from smaller basins into closed basins. Surface water is imported into the Imperial Valley from the Colorado River for irrigation purposes.

In the mountain areas, ground water is encountered in the highly fractured bedrock, but no significant aquifers have been found. Boreholes in excess of 500 feet deep often are dry. More significant quantities of ground water are encountered in the alluvial fans flanking the mountains and in the major valleys. The quality of ground water throughout the region is highly variable and mostly of poor quality, due to high concentrations of total dissolved solids (TDS).

2.3.8.1 Mining in the Region

Thirty-one mines are listed for this region, including sand and gravel aggregate operations and producers of other industrial commodities and minerals. Gold mines in the region include the Mesquite and Picacho Peak mines in eastern Imperial County, plus numerous, smaller operations. The number of mines with WDRs is summarized in Table 2.8.

TABLE 2.8

SUMMARY OF WDRs[(1)] IN REGION 7

MINE TYPE	TOTAL MINES	WDRs ISSUED PRE-1984		WDRs ISSUED POST-1984	
		TOTAL MINES	MINES WITH ENFORCEMENT ORDERS	TOTAL MINES	MINES WITH ENFORCEMENT ORDERS
Sand and Gravel	4	3	0	1	0
Metallic/Conventional	26	12	1	14	2
Metallic/Placer	0	0	0	0	0
Other	1	0	0	1	0
TOTALS	31	15	1	16	2

[(1)] WDR = Waste Discharge Requirements.

2.3.8.2 Incidents at Waste Management Units

Records of incidents in the region range from late reports and failure to file reports on the results of mandated monitoring programs, to reports of leaks in liners and failure of process equipment. Some of these problems have led to surficial soil contamination, which was subsequently cleaned up.

At the Mesquite Mine, gold is mined and extracted by heap leaching. The pads are constructed with single synthetic liners, while the solution ponds are double lined with LCRSs. Sumps and recirculation systems are provided to return seepage collected in the pond LCRS to the process stream. Ground water in the vicinity of the mine and processing operations is deep and of limited value as a resource because a number of drinking water standards are exceeded naturally, and well extraction capacity is small. Ground water and vadose zone monitoring systems are in place. In June 1989, a leak through the upper liner was indicated by increased amounts of liquid collecting in the LCRS.

Another important gold producer in eastern Imperial County has been in operation since 1979. The producer operates heap leach facilities at three sites and each has a current WDR. Each of the sites was issued WDRs before the current *Chapter 15* was adopted; the pads have single liners and no LCRSs. Sites 1 and 2 have 20-mil liners, and Site 3 has a 40-mil, high density polyethylene (HDPE) liner. The liners are placed on a compacted soil base and covered with a perforated-pipe drainage system, plus a sand layer that provides protection and drainage. The WDR for Site 1 was issued in 1982 and updated in 1987 to comply with monitoring requirements.

Twelve ground water monitor wells were installed and are sampled on a monthly schedule. Cyanide has been found in samples from five of these wells in multiple sampling rounds over periods of more than three years.

The two mines that were issued WDRs since the adoption of *Chapter 15* and have had enforcement actions levied against them are the Rattlesnake Gold Mine and the Camp Rock Mine.

Historically, wet tailings and waste may have been deposited in facilities with less containment than required under *Chapter 15*. Many of these facilities were exempted from the provisions of *Chapter 15*, and their operations continued. However, ongoing monitoring indicates that some surface impoundments have not contained their wastes effectively and that seepage into the vadose zone and the waters of the State has occurred. The one enforcement action involved Blair Brothers

Ranch Mining operation, which was issued its first WDRs in 1979 and had six enforcement orders against it in the early 1980s. Current Colorado River Basin staff had no details of the enforcement actions against this operation.

2.3.8.3 Analysis of the Incidents

The incident at the Mesquite Gold Mine did not result in any water quality damage. Pond fluids did not escape the secondary liner. The leak was traced to a 5-inch split in the primary liner which was quickly repaired.

An investigation of the cyanide pollution experienced at Site 1 discussed above revealed that a plume emanating from the Site 1 facilities was probably due to leakage from ponds with single liners. The ground water underlying Site 1 is from 5 to 28 feet below the surface and contained in a small, natural, perched system. Because it represents a small, isolated body of poor quality ground water, this perched aquifer is not considered to represent a usable water supply. The Site 1 ponds have since been retrofitted with double liners with LCRS. No official enforcement actions have been taken against this facility, but the operator is voluntarily pumping and treating the ground water. These cleanup procedures have been approved by the Regional Board.

The Rattlesnake Gold Mine was cited due to the construction of a pregnant solution pond that was never issued a WDR. The pond has since been emptied. The Camp Rock Mine had enforcement action taken against it due to the mishandling of mercury which had been used in the plant for gold recovery. This situation has been rectified, and the enforcement action against it has been rescinded.

Experiences with the wet tailings and ponds incidents indicate that single-lined liquid containment units have not performed adequately. These types of incidents are not likely to occur in the future, however, as *Chapter 15* does not permit single lined ponds for the disposal of liquids that have the potential for impacting the beneficial use of waters of the State.

2.3.9 REGION 8 - SANTA ANA REGION

The Santa Ana Region is the smallest of the nine Regional Boards. It includes the majority of Orange County, the northwest corner of Riverside County, and the southwestern corner of San Bernardino County. The region lies in the north part of the Peninsular Range Province, which includes northwest-southeast oriented mountain ranges. The climate in this region is coastal along the Pacific Ocean, Mediterranean in the mountains, and desert in the valleys. Precipitation and evaporation are similar to Region 4, with evaporation exceeding precipitation.

2.3.9.1 Mining in the Region

There are currently no operating metal mines in Region 8. At one time there were some major gold producing mines in this area, the most recent being in the Pinacate District in the late 1930s (Clark, 1980). Other historic gold districts include the Trabuco and Menifee.

The State's database lists 11 sand and gravel operations, one of which no longer operates (Table 2.9). WDRs for these facilities were issued prior to 1986. Since then, the Regional Board has decided not to regulate sand and gravel operations.

2.3.9.2 Incidents at Waste Management Units

None of the sand and gravel operations in the Santa Ana Region has had any known environmental incidents, and no enforcement actions have been taken against them by the Regional Board.

2.3.10 REGION 9 - SAN DIEGO REGION

The San Diego Region, like the Santa Ana Region, is located in the Peninsular Range Province. It encompasses most of San Diego County, the southern edge of Orange County, and the southwest corner of Riverside County. The region has a wide range in elevation, from over 10,800 feet at San Jacinto Peak down to sea level. It has climatic conditions similar to Region 8, with variations from coastal to alpine. Evaporation usually exceeds rainfall in this region by about 35 inches in an average year.

2.3.10.1 Mines in the Region

The only active listed mines with WDRs are sand and gravel or rock quarry operations. Of the 55 listed mines, 18 are historic and no longer operate, 3 are pending, and 34 are listed as operating. A summary of the number of listed mines with WDRs is provided in Table 2.10. There are no longer any operating metals mines with WDRs, and none has operated commercially since the 1930s. Some of the historic gold mining districts in this region include the Mesa Grande, Escondido, Julian, Banner, Cuyamaca, Laguna Mountains, and Dulzura (Clark, 1980).

TABLE 2.9

SUMMARY OF WDRs[1] IN REGION 8

MINE TYPE	TOTAL MINES	WDRs ISSUED PRE-1984		WDRs ISSUED POST-1984	
		TOTAL MINES	MINES WITH ENFORCEMENT ORDERS	TOTAL MINES	MINES WITH ENFORCEMENT ORDERS
Sand and Gravel	11	8	0	3	0
Metallic/Conventional	0	0	0	0	0
Metallic/Placer	0	0	0	0	0
Other	0	0	0	0	0
TOTALS	11	8	0	3	0

[1] WDR = Waste Discharge Requirements.

TABLE 2.10

SUMMARY OF WDRs[1] IN REGION 9

MINE TYPE	TOTAL MINES	WDRs ISSUED PRE-1984		WDRs ISSUED POST-1984	
		TOTAL MINES	MINES WITH ENFORCEMENT ORDERS	TOTAL MINES	MINES WITH ENFORCEMENT ORDERS
Sand and Gravel	52	43	3	9	0
Metallic/Conventional	0	0	0	0	0
Metallic/Placer	0	0	0	0	0
Other	3	3	0	0	0
TOTALS	55	46	3	9	0

[1] WDR = Waste Discharge Requirements.

2.3.10.2 Incidents at Waste Management Units

The enforcement actions levied against the sand and gravel operations were reportedly initiated due to turbidity problems in receiving waters.

2.3.11 SUMMARY OF THE PERFORMANCE OF WASTE MANAGEMENT UNITS

As discussed in the previous subsections, there are 482 mining operations with WDRs listed on the State's data base. Of these, 266 are sand and gravel producers, 123 are conventional metals mines, 40 are placer mines, and 53 comprise all other mines. From the December 1984 adoption of *Chapter 15* through 1989, 113 operations have been issued WDRs, including 47 sand and gravel operations, 41 conventional metal mines, 9 placer mines, and 16 other types of mines. Figures 2.4 and 2.5 provide a graphical summary of these and other statistics.

Recorded enforcement actions that have been issued against mines permitted prior to the current *Chapter 15* are summarized in Table 2.11. There have been 42 enforcement orders against a total of 25 mines. As shown in Table 2.12, four mines that received their waste discharge requirements after the December 1984 adoption of *Chapter 15* have been issued enforcement actions. None of these four facilities is currently operating. In each of the four cases, the cause of the enforcement action was due to unauthorized construction, a short-term spillage incident, or improper handling of reagents. These are administrative and operational types of problems and are not related to the fundamentals of waste containment design and construction. The four incidents are summarized in Table 2.12.

The more important water quality incidents that did not lead to enforcement actions, but which provide valuable insight into improving how future mine waste management units are designed and constructed, are summarized in Table 2.13.

2.4 CONCLUSIONS

Any major water quality impacts caused by mining operations appear to have been due to older, historic operations that utilized out-dated disposal technologies. This conclusion is consistent with those drawn by the Berkeley Mining Waste Study (U.C. Berkeley, 1988). There have not been any significant environmental or water quality impacts caused by mines that were permitted since the December 1984 adoption of *Chapter 15*.

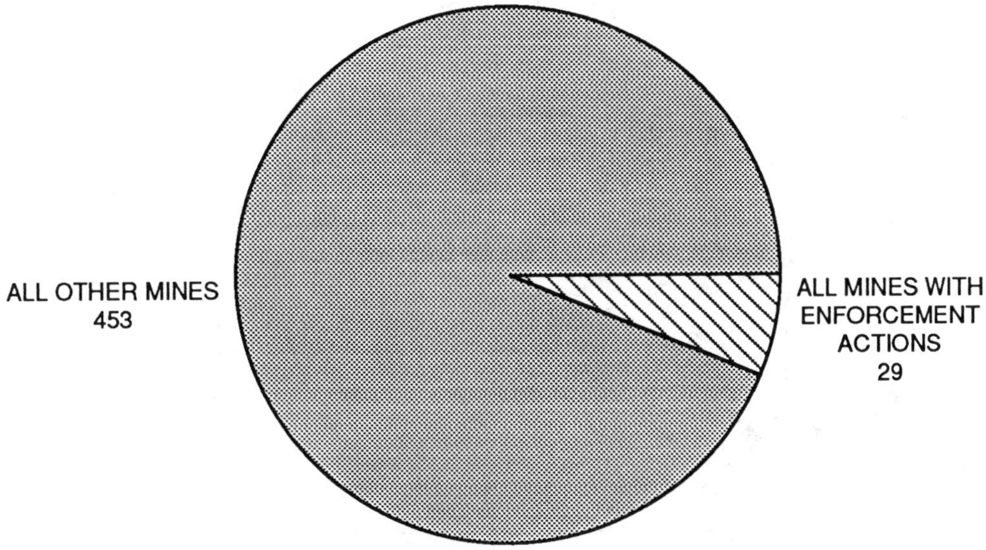

ALL MINE WASTE MANAGEMENT UNITS

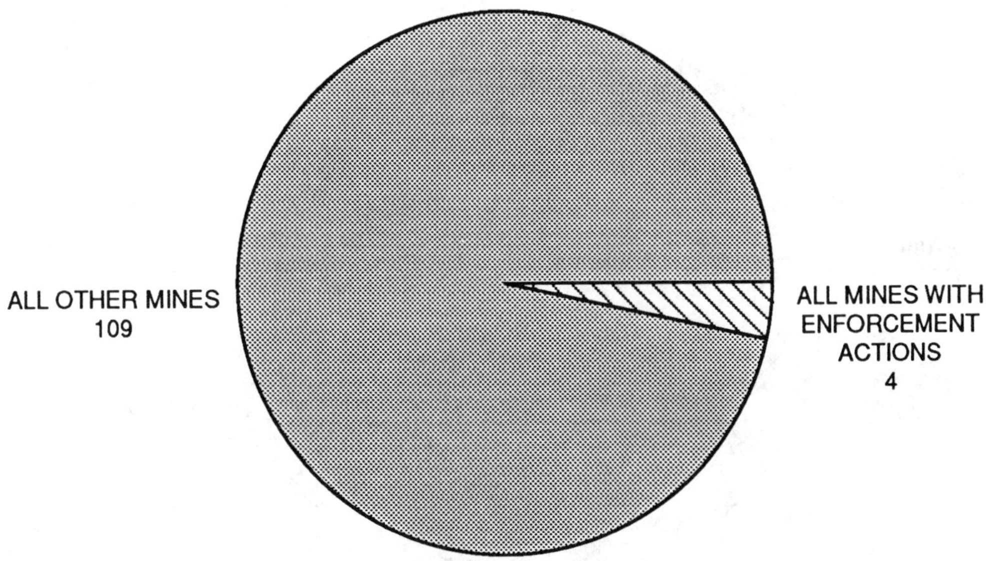

MINE WASTE MANAGEMENT UNITS PERMITTED SINCE DECEMBER 1984

FIGURE 2.4

PERFORMANCE SUMMARY

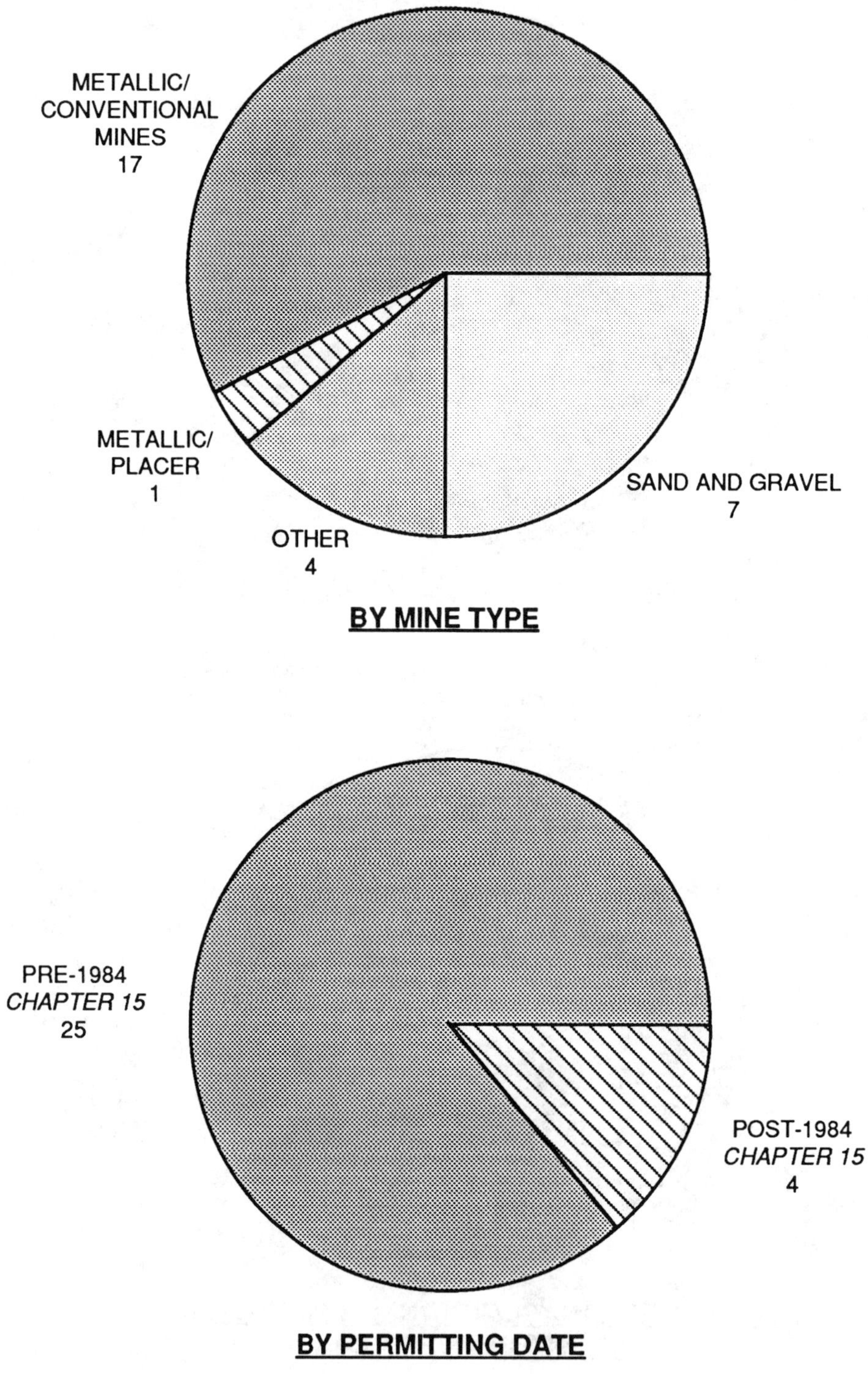

FIGURE 2.5

SUMMARY OF ENFORCEMENT ACTIONS

TABLE 2.11

PRE-ARTICLE 7 WASTE MANAGEMENT UNITS WITH ENFORCEMENT ACTIONS[1]

MINE	FACILITY STATUS	STANDARD INDUSTRIAL CLASSIFICATION	DATE WDR ADOPTED	ORDER STATUS	DATE ENFORCEMENT ORDER ADOPTED	FACILITY ID NUMBER	ORDER ACTION	NATURE OF PROBLEM
Humbug Mine-Heglar & Black	H H	1041 1041	1972 1/86	H H	4/24/74 9/24/86	1A72025OSIS 1A72025OSIS	CDO CDO	Stream Turbidity
Robinson, M.-Mid. Earth Mining	H H	1041 1041	1/86	H H	5/23/74 10/23/86	1A75061OSIS 1A75061OSIS	CDO CDO	Stream Turbidity
Noranda-Grey Eagle Mine	A	1041	1/84 1986 1/87	A	7/24/86	1A84142OSIS	ACL	Acid Rock Drainage (ARD)
Parnum Paving, Inc.	A	1442	1/80	A	3/11/85	1B80100OMEN	CAO	Stream Turbidity
Syar Indus. Inc., Doyle Plant	A	1442	1983 1987	H	5/28/87	1B83070OSON	ACL	Stream Turbidity
Buena Vista Mine	A	1092	1971 1988	A	6/10/88	3402008001	CDO	ARD
Carson Drift	A	1092	1988	A	6/10/88	3402008002	CDO	ARD
Lompoc Facility	A	1499	1975 1986	A	4/7/86	3422014001	CAO	Rescinded, Stream Turbidity
Walker Mine Tailings	A A A A	1041 1041 1041 1041	1958 1986	A A A A	2/23/87 7/13/63	5A320704003 5A320704003	ACL CDO	ARD
Cement Manufacture & Quarry	A A	1423 1423	1974 1984 1988	A A	6/24/83 5/9/86	5A452008001 5A452008001	CDO ACL	Stream Turbidity, High pH
Iron Mountain	A A	1021 1021	1/78	A A	7/24/81 7/19/88	5A459001001 5A459001001	CDO CAO	ARD
Balaklala, Keystone Mines	A A	1021 1021	1/86	A A	7/1/86 4/24/87	5A459003001 5A459003001	CDO CAO	ARD
Mammoth Mine	A A	1021 1021	1/86	A A	7/1/86 4/24/87	5A459006001 5A459006001	CDO CAO	ARD
Leviathan Mine	H H	1499 1499	1962 1988	H H	3/20/64 1/30/69	6A020006000 6A020006000	CDO CDO	ARD

(1) See Table 2.12 for legend to abbreviations used in this table.

TABLE 2.11

PRE-ARTICLE 7 WASTE MANAGEMENT UNITS WITH ENFORCEMENT ACTIONS[(1)] (Continued)

MINE	FACILITY STATUS	STANDARD INDUSTRIAL CLASSIFICATION	DATE WDR ADOPTED	ORDER STATUS	DATE ENFORCEMENT ORDER ADOPTED	FACILITY ID NUMBER	ORDER ACTION	NATURE OF PROBLEM
Doyle Plant	A	1442	1/73 1982	A	3/6/86	6A180075201	CAO	Stream Turbidity
Teichert Aggregates	A	1442	1972 1/83	A	12/6/85	6A294520011	CAO	Leaking Underground Storage Tank
Keynote Mine	H H	1041 1041	1983 1986	H H	6/20/85 3/13/86	6B148020004 6B148020004	CAO CAO	Abandoned Operation
Boron Mine Facility	A	1474	1976 1978 1980 1984 1989	A	12/31/85	6B152004001	CAO	Ground Water Contamination
Mountain Pass Mine Operations	A A A	1000 1000 1000	1981	A H H	10/3/85 8/14/86 2/9/89	6B362098001 6B362098001 6B362098001	TSO CDO CDO	Ground Water Contamination
Ivanpah Mill	H	1000	1978 1980 1/83 1/89	A	4/14/89	6B362100001	CAO	Ground Water Contamination
Telegraph Mine	H	1041	1982 1/89	A	4/14/89	6B368020006	CAO	Abandoned Operation
Blair Brothers Ranch Mining Op	H H H H H H	1000 1000 1000 1000 1000 1000	1979 1985	H H H H H H	12/23/81 11/17/82 5/3/83 5/3/83 5/18/83 5/22/85	7A362006011 7A362006011 7A362006011 7A362006011 7A362006011 7A362006011	CAO RAG CAO CAO RAG CAO	Unknown
Otay Sand & Concrete Batch Plt	A	1442	1975	H	4/10/87	9 000000158	ACL	Monitoring Report not Submitted
Marron Brothers, Inc.	A A	1442 1442	1984	H H	3/27/85 9/16/85	9 000000233 9 000000233	CAO ACL	Constructed without a WDR
Sweetwater River Dredging	A	1442	1984	H	4/9/87	9 000000291	ACL	Monitoring Report not Submitted

(1) See Table 2.12 for legend to abbreviations used in this table.

TABLE 2.12

WASTE MANAGEMENT UNITS PERMITTED UNDER CURRENT ARTICLE 7 WITH ENFORCEMENT ACTIONS

MINE	FACILITY STATUS	STANDARD INDUSTRIAL CLASSIFICATION	DATE WDR ADOPTED	ORDER STATUS	DATE ENFORCEMENT ORDER ADOPTED	FACILITY ID NUMBER	ORDER ACTION	NATURE OF PROBLEM
Poison Oak Flat Mill Site	A	1041	1985	A	1/17/86	5A042029001	CAO	Siltation and Stream Turbidity, Unauthorized Construction
Carson Hill Mine	A	1041	1986 2/88	A	6/20/89	5B052011001	ACL	Surface Spill of Process Reagents
Rattlesnake Gold Mine	A	1000	1986	A	10/23/87	7A362064002	CAO	Unauthorized Construction
Camp Rock Mine	A	1000	1988	A	10/14/87	7A362187001	CAO	Improper Handling of Mercury

LEGEND

FACILITY STATUS

- A Active
- H Historic
- P Pending
- I Inactive
- C Regulated by either City or County

ORDER STATUS

- A Active Order
- H Historic Order

ORDER ACTION

- CAO Clean-up and Abatement
- CDO Cease and Desist
- TSO Response Time Schedule
- RAG Referred to Attorney General
- RDA Referred to District Attorney
- ACL Administrative Civil Liabilities

STANDARD INDUSTRIAL CLASSIFICATION

- 1000 Some Type of Metal Ore Facility
- 1021 Copper Ores
- 1041 Gold Ores
- 1061 Ferroalloy Ores
- 1081 Metal Mining Services
- 1092 Mercury Ores
- 1094 Uranium-Radium-Vanadium Ores
- 1411 Dimension Stone
- 1422 Crushed and Broken Limestone
- 1423 Crushed and Broken Granite
- 1442 Construction Sand and Gravel
- 1446 Industrial Sand
- 1450 Clay, Ceramic and Refractory Minerals
- 1453 Fire Clay
- 1459 Clay and Related Minerals
- 1474 Potash, Soda, and Borate Minerals
- 1499 Miscellaneous Nonmetallic Minerals

TABLE 2.13

PARTIAL LIST OF INCIDENTS AT WASTE MANAGEMENT UNITS (NO ENFORCEMENT ORDERS ISSUED)

REGION	MINE TYPE/ NAME	PROBLEM	COMMENTS
5	Gold/Conventional - Gold Stripe	Heap leach pregnant pond over-topped.	Pond was designed for the 100-year, 24-hour event, which is not a worst-case storm duration.
5	Gold/Conventional - McLaughlin	Some diluted Acid Rock Drainage (ARD) drainage collected in holding pond released to receiving water during extreme rainfall in February 1986. No environmental damage resulted.	Extreme rainfall events over a period of two to three weeks exceeded the 100-year, 24-hour, flood design capacity.
		Gradual trend of increasing sulphate and calcium concentration in leachate collected at waste rock dump underflow drains.	Original design addressed this possibility and provided for collection and pumpback of leachate. Simultaneous appearance of both sulphate and calcium indicates some form of ARD and internal neutralization occurring.
6	Gold/Conventional - Snow Caps	Leaking leach pad primary liner (Stroud, 1988).	In Region 6, the allowable leak rate of a primary liner is zero. This leak was very minor, and the problem was corrected.
6	Gold/Conventional - Yellow Aster Descarga Zone	Leaking leach pad primary liner (Stroud, 1988).	In Region 6, the allowable leak rate of a primary liner is zero. This leak was very minor, and the problem was corrected.
6	Gold/Conventional - La Monte Mine	Leaking leach pad primary liner (Stroud, 1988).	In Region 6, the allowable leak rate of a primary liner is zero. This leak was very minor, and the problem was corrected.
6	Gold/Conventional - Standard Hill	Leaking leach pad primary liner (Stroud, 1988).	In Region 6, the allowable leak rate of a primary liner is zero. This leak was very minor, and the problem was corrected.

TABLE 2.13

PARTIAL LIST OF INCIDENTS AT WASTE MANAGEMENT UNITS (NO ENFORCEMENT ORDERS ISSUED)

REGION	MINE TYPE/ NAME	PROBLEM	COMMENTS
6	Gold/Conventional - Middle Buttes	Leaking leach pad primary liner (Stroud, 1988).	In Region 6, the allowable leak rate of a primary liner is zero. This leak was very minor, and the problem was corrected.
6	Gold/Conventional - Randsburg	Leaking leach pad primary liner (Stroud, 1988).	In Region 6, the allowable leak rate of a primary liner is zero. This leak was very minor, and the problem was corrected.
6	Gold/Conventional - Shaharold	Leaking leach pad primary liner (Stroud, 1988).	In Region 6, the allowable leak rate of a primary liner is zero. This leak was very minor, and the problem was corrected.
7	Gold/Conventional - Mesquite	Leakage of 5,000 gallons of cyanide solution occurred from a solution pump seal and contaminated soil in the upper 12 feet of the vadose zone.	Most of the contaminated soil was excavated and placed with the ore on the heap leach pad. Soil located under a tank that could not be excavated was monitored and eventually remediated through natural in situ cyanide degradation.
7	Gold/Conventional - Picacho Mine	Leaking single liner in pregnant solution pond.	Pond was retrofitted with double liner and LCRS. Incident resulted in minimal impact to regional ground water.

A significant number of the mine waste management units have been designed under the *Chapter 15* alternative engineering provisions, which exempt a management unit from prescriptive requirements. This has been necessary in order to construct cost-effective units with the material available at the mine site, and advantageous, as it facilitated taking advantage of the unique properties of the mine waste itself, i.e., low permeability tailings. In one instance, while producing a more cost-effective design, the exemption provisions actually encouraged the construction of a waste management unit on a geologically more suitable site. The selected site should provide more secure long-term containment than the alternative, which would have been to construct the management unit using the prescriptive liner requirements on a less suitable site. It was more economic to construct without liners on the more remote site with low permeability geologic formations than to select a lined unit on a higher permeability site close to the recovery plant.

Although no major incidents have occurred, some of the mines permitted under *Chapter 15* have had certain difficulties in complying with their WDRs. The nature of these include identifying what constitutes an acceptable leakage rate through the upper liner of a double-lined system, potential ARD problems, spillage from ponds during extreme storm events, and turbidity in receiving surface water.

A potential concern with the analyses of the effectiveness of the current mine waste disposal practices outlined in this chapter is that the period over which performance data has been gleaned is relatively short, approximately 5 years. It may be argued that this is not a long enough period for major containment failures to manifest themselves, as seepage loss from facilities can occur at very slow rates. However, the authors believe it extremely unlikely that there are hidden "latent defects" in containment that have not yet materialized. Most of the modern mine waste disposal facilities have comprehensive ground water monitoring systems, and many also have vadose zone monitoring systems, all of which increase the likelihood of detecting seepage early. Furthermore, with respect to adequate containment, the general conclusions of this chapter are borne out by experiences with new mines in many other states where similar waste disposal technology has been used over the past 5 to 10 years.

2.5 REGULATORY IMPLICATIONS

The *Chapter 15* waste containment criteria established for mine waste management units and the associated provisions for alternative containment engineering appear to be working satisfactorily. However, the California Mining Association has suggested changes which it contends would avoid

TABLE 2.14

RECOMMENDED REGULATORY RESPONSES TO THE INCIDENTS RAISED IN TABLE 2.13

REGION	MINE TYPE/NAME	RECOMMENDED REGULATORY RESPONSE
5	Gold/Conventional - Gold Stripe	The total precipitation during an extended storm event exceeded the 100-year, 24-hour design event. If ponds are to be designed for a 100-year return period, then the regulatory wording should not indicate the storm duration, but specify that the worst-case duration be evaluated for the facility under consideration.
5	Gold/Conventional - McLaughlin	The total precipitation during that two to three week period far exceeded the 100-year, 24-hour design event. If ponds are to be designed for the 100-year return period, then the regulatory wording used should not indicate the duration of the 100-year event, but specify that the worst-case duration be evaluated for the facility under consideration. The regulations would benefit by: (1) providing guidelines for the identification and characterization of acid rock drainage (ARD) in wastes, and (2) calling for special waste management requirements to deal with ARD.
6	Gold/Conventional - Snow Caps	
6	Gold/Conventional - Yellow Aster Descarga Zone	Leakage was discovered and mitigated quickly. The leakage rate was well below the guidelines used by the U.S. EPA for Resource Conservation and Recovery Act (RCRA) facilities. However, Region 6 has the policy that, since the regulations do not specify an allowable leakage rate through a primary liner, the only acceptable rate is zero. The regulations would benefit by the adoption of criteria for determining leakage rate response criteria on a site-specific basis. Such criteria would then become a part of the adopted Waste Discharge Requirements (WDRs).
6	Gold/Conventional - La Monte Mine	
6	Gold/Conventional - Standard Hill	
6	Gold/Conventional - Middle Buttes	

TABLE 2.14

RECOMMENDED REGULATORY RESPONSES TO THE INCIDENCES RAISED IN TABLE 2.13 (Continued)

Page 2 of 2

REGION	MINE TYPE/NAME	RECOMMENDED REGULATORY RESPONSE
6 6	Gold/Conventional - Randsburg Gold/Conventional - Shaharold	Leakage was discovered and mitigated quickly. The leakage rate was well below the guidelines used by the U.S. EPA for RCRA facilities. However, Region 6 has the policy that, since the regulations do not specify an allowable leakage rate through a primary liner, the only acceptable rate is zero. The regulations would benefit by the adoption of criteria for determining leakage rate response criteria on a site-specific basis. Such criteria would then become a part of the adopted WDRs.
7	Gold/Conventional - Mesquite	Soil contaminated by a leaky process pipe was required to be removed. This type of event was caused by an accident and was not related to any regulatory inadequacy.
7	Gold/Conventional - Picacho Peak	Design and construction of leach pads were completed before current Article 7. Had the pads been constructed under provisions of current Article 7, a different design would have been developed.

interpretations of the regulations that lead to requirements that are excessive, inflexible, and not cost-effective. Nevertheless, as a significant number of waste management units have been constructed under either the alternative engineering or exemption provisions of *Chapter 15*, and as there appears to be considerable merit in continuing to allow this, it would be beneficial to incorporate a more formalized approach to the selection and design of alternative containment measures. There is currently very little specific guidance provided by the regulations. The "systems approach" to containment design that is developed and discussed in Chapter 3 provides a sound basis for providing the necessary guidance.

The results of the above analyses also indicate that there may be merit in explicitly dealing with the potential problems manifested by the incidents summarized in Table 2.13. The detailed responses to dealing with these types of problems are provided in Table 2.14. A summary of the main regulatory requirements is as follows:

- There is a need to add primary (or upper) liner performance standards to *Chapter 15*. These should be either a prescribed unit leakage rate or, preferably, a methodology to allow site-specific leakage criteria to be developed.
- There is a need for an ARD testing protocol to be included in *Chapter 15*, as well as the requirement to develop special waste management procedures to mitigate its occurrence.
- There is a need to adopt more scientifically sound, risk-based flood storage criteria. These criteria should not only address single event flooding, but also cumulative water storage caused by prolonged or successive events, as well as operational water balance considerations.

The problem of turbid runoff at mine sites is best dealt with under the National Pollution Discharge Elimination System (NPDES) program and not directly in any mine waste regulations.

2.6 BIBLIOGRAPHY

California Mining Association (CMA). *Mine Waste in California - A Case for Regulation as Non-hazardous* by I. Hutchison and R. Hammond. January 1988.

California Water Resources Control Board. *Computerized Data Base of Mines with Waste Discharge Requirements (WDRs).* State Water Discharge System. August 7, 1989.

Clark, W. B. *Gold Districts of California.* California Division of Mines and Geology, Bulletin 193. 1980.

Giroud, P. P., R. Bonaparte, R. B. Wallace, C. Ah-Line, and J. Prillaman. *Background Document on Proposed Liner and Leak Detection Rule.* United States Environmental Protection Agency. EPA/530-SW-87-015. May 1987.

Licari, B. J. *Assessment of Civil Liability, Leach Unit 3 Discharge, Carson Hill Gold Mine, Calaveras County, California*. Letter to William H. Crooks, Executive Officer, California Regional Water Quality Control Board, Central Valley Region. July 18, 1989.

Linsley, R. K., Jr., Kohler, M. A., Paulhus, J. L. H. *Hydrology for Engineers*. New York: McGraw-Hill. 1982.

Stroud, D. E. *Status of Liners and LCRSs at Cyanide Heap Leach Operations*. Internal Memo to O. R. Butterfield, Executive Officer, California Regional Water Quality Control Board. July 1, 1988.

University of California at Berkeley. Mining Waste Study Team. *Mining Waste Study, Final Report*. California Water Resources Control Board. 1988.

CONTENTS

CHAPTER 3
MINE WASTE MANAGEMENT REQUIREMENTS

PAGE NO.

LIST OF TABLES/LIST OF FIGURES 65

3.1 Introduction 67
- 3.1.1 Purpose and Scope of This Chapter 67
- 3.1.2 Integration With Other Chapters 68
- 3.1.3 Study Participants 69
- 3.1.4 Sources of Information 69
- 3.1.5 Chapter Organization 70

3.2 Overview of the Systems Approach for Determining Mine Waste Management Requirements 70
- 3.2.1 Identify Potential Receiving Waters and Their Beneficial Uses 70
- 3.2.2 Assess Physical and Chemical Properties of the Mine Waste 70
- 3.2.3 Identify and Evaluate Site-Specific Factors 71
- 3.2.4 Balancing Waste Containment With Risk to Water Quality 71

3.3 Protecting Beneficial Uses of Water 71

3.4 Identifying Potential Receiving Waters and Their Beneficial Uses 72

3.5 Assessing Physical and Chemical Properties of Mine Waste 72
- 3.5.1 Introduction 72
- 3.5.2 Characteristics of Mine Wastes Common in California 77
  - 3.5.2.1 Introduction 77
  - 3.5.2.2 Sediments 77
  - 3.5.2.3 Soluble Natural Minerals 79
  - 3.5.2.4 Acid Rock Drainage 81
  - 3.5.2.5 Asbestos 82
  - 3.5.2.6 Reagents 83
  - 3.5.2.7 Nutrients 86
  - 3.5.2.8 Typical Gold Mine Waste Solution Characteristics 86
- 3.5.3 Physical Properties of Mine Waste 87

CONTENTS

CHAPTER 3
(Continued)

PAGE NO.

3.5.4 Chemical Properties of Mine Waste 89
3.5.4.1 Waste Classification 90
3.5.4.2 Waste Characterization 96
3.5.5 Other Mine Waste-Related Issues 97
3.6 Site-Specific Factors 97
3.6.1 Climate 98
3.6.2 Underlying Geologic Formations 98
3.6.3 Distance to Water Resources 98
3.7 Waste Management Unit Containment 99
3.7.1 Natural Containment 99
3.7.2 Engineered Waste Containment 99
3.7.3 Operational Aspects 100
3.8 The Systems Approach to Designing Waste Containment 101
3.8.1 Examples of the Systems Approach 101
3.8.1.1 Wastewater Pond Containment Assessments 106
3.8.1.2 Tailings Impoundment Containment Assessments 109
3.8.2 Alternative Approaches 112
3.9 Regulatory Implications 114
3.9.1 General 114
3.9.2 Current California Regulations 115
3.9.3 Recommended Regulatory Approach 118
3.10 Bibliography 119

CONTENTS

CHAPTER 3
(Continued)

LIST OF TABLES

TABLE NO.	TITLE	PAGE NO.
3.1	Typical Water Quality Standards and Goals	74
3.2	Species of Soluble Natural Minerals That Occur in California	80
3.3	Typical Reagents Used for Beneficiation of Ore at a Wide Range of Mining Operations	84
3.4	Chemical Composition of Waste Liquid Solutions Encountered at Gold Mines Throughout the World	88
3.5	California Hazardous Waste Classification Criteria	91
3.6	Federal Bevill-Excluded Wastes	94
3.7	Federal Hazardous Waste Classification Criteria	95
3.8	Containment Assessment Examples: Wastewater and Tailings Properties	102
3.9	Example 1: Results of Wastewater Pond Containment Assessments	107
3.10	Example 2: Results of Tailings Impoundment Containment Assessments	111
3.11	Current Prescriptive Waste Containment Requirements: State of California	116

LIST OF FIGURES

FIGURE NO.	TITLE	PAGE NO.
3.1	Wastewater Pond Concepts Used in Containment Assessments	103
3.2	Site Conditions in Vicinity of Tailings Impoundment	104
3.3	Impoundment Concepts Used in Containment Assessments	105

CHAPTER 3

MINE WASTE MANAGEMENT REQUIREMENTS

3.1 INTRODUCTION

3.1.1 PURPOSE AND SCOPE OF THIS CHAPTER

It is evident from the discussion in Chapter 2 that the application of sound waste management technology ensures the secure disposal of mine wastes, which otherwise have the potential to adversely affect the quality of ground and surface waters. This is in contrast to the waste containment problems associated with outdated disposal practices used decades ago at several historic mine sites (U.C. Berkeley, 1988), and illustrates the importance of encouraging the ongoing application and development of state-of-the-art mine waste management practices. The purpose of this chapter is to describe a systematic or "systems" approach that provides both a valuable guide for determining containment requirements at a mine waste management unit and a scientific basis for continuing the development of waste disposal technology.

The systems approach incorporates the unique aspects of mine waste material, primarily its large volume, generally low toxicity, predictability, homogeneity, in place management, and potential use as construction material for waste containment facilities. It also recognizes the natural site-specific conditions that affect the potential for mine waste to impact water quality. Because the volume of mine waste typically ranges from tens- to hundreds-of-millions of tons, practical considerations require that the waste management unit be located as close as possible to the source of the waste, i.e., the mine or the mineral recovery plant. Further, the characteristics of mine wastes can be reasonably well predicted using laboratory or pilot scale testing of exploration samples and can be confirmed during operations. Having this knowledge simplifies the design of the waste management unit.

Successful operation of a waste management unit is facilitated by several factors, including: (1) the unit usually receives only one waste type, (2) trained mine personnel and equipment are continually available to operate the facility, and (3) the method by which the waste is deposited can, in some instances, be used to provide containment. For example, as discussed in Chapter 2, tailings can be deposited to form a low-permeability layer and also to promote the removal of free liquids by natural evaporation.

Natural site-specific conditions that are important in waste disposal include climate, underlying geologic formations and water resources. The amount of precipitation largely determines the amount of liquid with the potential to escape from the waste. The characteristics of the underlying geologic formations determine the degree to which the liquid portion of the waste can be contained naturally. The proximity and quality of ground or surface water resources determine how susceptible they are to potential releases of liquids from the waste management unit.

The systems approach involves, first, establishing a containment performance goal for each waste management unit. A containment assessment then is utilized for determining the amount of waste containment provided by the waste itself, the disposal methods used, the site-specific conditions, and any additional engineered containment necessary. This assessment is used to compare and evaluate different waste containment alternatives and the selection of an alternative that meets the containment performance goal.

3.1.2 INTEGRATION WITH OTHER CHAPTERS

This chapter provides a systematic overview of mine waste disposal principles and approaches. As such, it serves as a useful reference for the rest of the book. Each of the chapters that follow deals with specific key components of the systems approach. The contents of these chapters, a description of how they relate to the systems approach, and the associated containment assessment are as follows:

- Chapter 4: Prediction of Acid Generation Potential - This chapter explains how acid rock drainage (ARD) can be identified and quantified. This information is important to determine the chemical properties of the waste for use in containment assessments outlined in Chapter 3.
- Chapter 5: Attenuation of Chemical Constituents - Attenuation is the mechanism whereby the concentrations of constituents in waste liquids seeping through soil or rock are reduced by chemical, physical, and biological processes. This chapter discusses approaches to evaluating the degree of attenuation or natural containment that a site may provide. This is a necessary part of understanding and evaluating the waste containment of site-specific conditions.
- Chapter 6: Climatic Considerations - This chapter describes the conditions under which waste leachate may form and how to quantify leachate flow. As with Chapter 5, the impact of site-specific conditions on waste containment is assessed.
- Chapter 7: Liner System Design - This chapter contains a detailed description of liner performance characteristics and outlines an approach to the design of liner systems. This information helps determine the amount of engineered containment necessary to meet a specific performance goal.
- Chapter 8: Closure Requirements - This chapter applies an approach, similar to the one developed in Chapter 7, to the design for closure of a waste management unit.

- Chapter 9: Heap Leaching - This chapter provides a discussion on how heap leach operations function and how they should be designed in accordance with the systems approach developed in Chapter 3.
- Chapter 10: Ground Water Monitoring - This chapter outlines the basis for establishing an effective ground water monitoring system that can be used to evaluate the performance of a waste management unit.
- Chapter 11: Economic Impact Evaluation - This chapter analyzes the potential cost impacts of different waste containment scenarios and highlights the need for the systems approach if cost-effective waste management unit designs are to be achieved.

3.1.3 STUDY PARTICIPANTS:

- The authors of this chapter include:
  - Donald T. Bishop, Ph.D., P.G., Condor Earth Technologies, Inc.
  - Peter H. Dohms, P.G., Condor Earth Technologies, Inc.
  - Kenneth J. Floom, P.E., Environmental Solutions, Inc.
  - Barry A. Hillman, Ph.D., Condor Earth Technologies, Inc.
  - Ian P. G. Hutchison, Ph.D., P.E., Environmental Solutions, Inc.
  - Gene B. Nunnelley, P.E., Condor Earth Technologies, Inc.
  - Nigel Skirmer, P.E., Steffen Robertson and Kirsten (US), Inc.
- Special input also was provided by:
  - Thomas M. Donnelly, Esq., Heller, Ehrman, White and McAuliffe
  - James E. Good, Esq., Gresham, Varner, Savage, Nolan and Tilden

3.1.4 SOURCES OF INFORMATION

Diverse sources of information were employed for the analyses presented. They include:

- Interviews with:
  - Mine operators.
  - California Regional Water Quality Control Board representatives.
  - Other consultants.
- Documentation, including:
  - Technical literature.
  - California Regional Water Quality Control Board files.
  - Private reports of mine operators.
  - File information from consultants.
- The professional experience of the primary authors.

A detailed bibliography of references used for this chapter is contained in Section 3.10.

3.1.5 CHAPTER ORGANIZATION

This chapter is divided into 10 sections, including the Introduction (Section 3.1) and an overview of the systems approach (Section 3.2). Section 3.3 outlines a waste containment performance goal based on California's Porter-Cologne Water Quality Control Act, California Water Code Sections 13000 et seq. (Porter-Cologne). Section 3.4 outlines an approach for identifying the potential receiving waters and their beneficial uses. Section 3.5 discusses the assessments necessary to determine the physical and chemical properties of typical mine wastes. Section 3.6 describes the site-specific factors that should be considered in determining the appropriate mine waste management requirements, and Section 3.7 summarizes the natural and engineered containment features that can be incorporated into a waste management unit. Section 3.8 explains the application of the systems approach to determining waste containment requirements and provides two examples of a waste containment assessment. It also briefly compares the systems approach to the alternative "prescriptive" approach. Finally, Section 3.9 discusses regulatory implications and recommends inclusion of the systems approach. A bibliography is presented in Section 3.10.

3.2 OVERVIEW OF THE SYSTEMS APPROACH FOR DETERMINING MINE WASTE MANAGEMENT REQUIREMENTS

In California, the mandate of Porter-Cologne is both protective and practical. Its purpose is to protect the beneficial uses of potential receiving waters. As discussed more fully below, the goal of mine waste management should be the same. This section enumerates in sequence the separate considerations necessary for determining mine waste containment requirements.

3.2.1 IDENTIFY POTENTIAL RECEIVING WATERS AND THEIR BENEFICIAL USES

When developing a strategy for protecting a natural resource from the potential effects of a development activity, the appropriate initial exercise is to define the natural resource that needs protection. Accordingly, both the receiving waters and the beneficial uses that could be affected by mine waste management should be identified as an important first step.

3.2.2 ASSESS PHYSICAL AND CHEMICAL PROPERTIES OF THE MINE WASTE

The next step is to assess the physical and chemical properties of the mine waste. This is central to defining the potential of the waste to adversely affect surface or ground water. There is usually considerable variation in the properties of different mine wastes, particularly in the migration potential and toxicity of liquids associated with the wastes.

3.2.3 IDENTIFY AND EVALUATE SITE-SPECIFIC FACTORS

A further step is to identify and evaluate site-specific factors that either may mitigate or exacerbate the potential water quality impacts of the waste management unit. These factors include local climatic conditions, the geochemistry of the soil and rock that underlie the waste management unit, the permeability of the underlying geologic formations, the geologic structure and seismicity of the area, potential contaminant migration pathways, the distance between the waste management unit and receiving waters, and the presence of any flood plains.

3.2.4 BALANCING WASTE CONTAINMENT WITH RISK TO WATER QUALITY

The evaluations described above provide the information necessary to assess the nature and extent of the potential threat to beneficial uses of water which may be posed by a particular mine waste. A waste management unit can be designed to appropriately balance the containment provisions with the water quality risk posed by the waste.

Site-specific conditions, such as beneficial uses of water, properties of the mine wastes, and other environmental factors vary considerably from site to site throughout the United States. The systems approach nevertheless provides a consistent standard of design and regulation under these circumstances, as it is capable of incorporating a variety of conditions into the waste containment design. Furthermore, the systems approach has the advantage of requiring that the actual risk to water quality be identified and quantified. The process of quantifying the risk results in a maximum understanding of site conditions, thereby assuring an adequate design of the waste management unit.

3.3 PROTECTING BENEFICIAL USES OF WATER

Porter-Cologne directs the Regional Water Quality Control Boards (Regional Boards) to establish "water quality objectives" that "will ensure the reasonable protection of beneficial uses" of the waters of the State (Section 13241), and to prescribe waste discharge requirements for waste management units based on such beneficial uses and objectives (Section 13263). In establishing water quality objectives, the Regional Boards are directed to consider the following factors:

- Past, present, and probable future beneficial uses of water.
- Environmental characteristics of the hydrogeologic unit under consideration, including the quality of available water.
- Water quality conditions that reasonably could be achieved through the coordinated control of the factors that affect water quality in the area.
- Economic considerations.

Therefore, the focus of mine waste management should be on protecting beneficial uses of the potential receiving waters. As stated above, this focus is both a protective and practical standard, and should serve as the performance goal for mine waste management.

3.4 IDENTIFYING POTENTIAL RECEIVING WATERS AND THEIR BENEFICIAL USES

A first step in determining the appropriate waste containment required at a mining operation is to identify the receiving waters, i.e., surface water or ground water, that need to be protected from adverse releases from the waste management unit. Next, the beneficial uses of the receiving waters should be identified by considering the following: (1) any established water quality objectives, (2) the quality of the water, (3) the quantity of the water, (4) whether it is possible to collect, store, and transport the water to a location where the beneficial use could occur, and (5) the existing and/or feasible uses of the water.

Based on these considerations, the following information is required to determine the beneficial uses of a surface water resource:

- Location of the resource and potential users.
- Average and extreme flow rates for both wet and dry cycles.
- Water quality variations with time and at different flow rates. Important quality parameters include temperature, dissolved oxygen, suspended solids, total dissolved solids (TDS), major anions and cations, and metals. Where existing polluted conditions are expected, information on biological and chemical oxygen demand, as well as a range of organic constituents, also may be required.
- River or creek bed geometry. This aspect affects the suitability of the water body for in-stream use, such as supporting a fish population.

To designate the beneficial uses of a ground water resource, the following information is necessary:

- Location of the resource, including depth to the water table and the potential users.
- Volume of water in the aquifer and amount of recharge to the aquifer.
- Storage coefficient and hydraulic conductivity of the granular material or fractured rock in which the ground water is located.
- Water quality (as for surface water).

Typical federal water quality standards and goals for a range of designated beneficial uses are outlined in Table 3.1 (Marshack, 1989a). These can be used to provide some guidance as to beneficial uses for specific water bodies.

As is evident from the data presented in Table 3.1, different water quality standards and goals apply to different beneficial uses. For instance, if a particular surface stream has a beneficial use of "propagation of fish and wildlife," then the numeric standards for water-borne copper will be more stringent than would be necessary for protection of municipal or industrial water supplies. Similarly, some beneficial uses, such as aesthetic enjoyment, would require the establishment of additional, possibly subjective, standards.

Typical beneficial uses include, but are not necessarily limited to, the following:

SURFACE WATER	GROUND WATER
Municipal Supply	Municipal Supply
Industrial Supply	Industrial Supply
Agricultural Supply	Agricultural Supply
Domestic Use	Domestic Use
Navigation	
Propagation of Fish and Wildlife	
Hydroelectric Generation	
Visual and Aesthetic Enjoyment	
Ground Water Recharge	
Recreation and Other Public Uses	

3.5 ASSESSING PHYSICAL AND CHEMICAL PROPERTIES OF MINE WASTE

3.5.1 INTRODUCTION

As a second step in determining mine waste containment requirements, it is necessary to assess the physical and chemical properties of the waste, as these properties may impact water quality if the waste is not managed properly. Such properties also may affect the choice of materials used for

TABLE 3.1
TYPICAL WATER QUALITY STANDARDS AND GOALS
(µg/L unless otherwise indicated)

Page 1 of 3

INORGANIC CONSTITUENT	STATE OR EPA DRINKING WATER STANDARDS MAXIMUM CONTAMINANT LEVELS (MCLs)			HEALTH ADVISORIES OR SUGGESTED NO-ADVERSE-RESPONSE LEVELS (SNARLs) (cancer risk not considered)	
	State levels unless indicated		EPA MCL Goal		
	Primary MCL	Secondary (1) MCL		EPA	NAS (2)
Aluminum	1,000				5,000 (7 day)
Ammonium sulfamate				1,500 (3)	
Antimony					
Arsenic	50		100 (3)		
Asbestos			7,100,000 fibers/L (3)(4)		
Barium	1,000			1,500	4,700
Beryllium			1,500 (3)		
Boron					
Bromide					2,300
Cadmium	10		5 (3)	5	5
Chloramine					166 / 581 (child/adult)
Chlorate					7 / 24 (child/adult)
Chloride		250 mg/L (5)			
Chlorine					
Chlorine dioxide					60 / 210 (child/adult)
Chlorite					7 / 24 (child/adult)
Chromium (III)					
Chromium (VI)	50 (6)		120 (3)(6)	120 (6)	
Cobalt					
Copper		1,000	1,300 (3)		
Cyanide	200 (7)			154	
Fluoride	4,000	2,000	4,000		
Iodide					1,190
Iron		300			
Lead	50 (8)/5 (3)		zero (3)	20 (3)	
Manganese		50			
Mercury	2		3 (3)	1.1	
Molybdenum					
Nickel				150	
Nitrate	45 mg/L (9)		10 mg/L (10)(3)	10 mg/L (10)	
Nitrite			1 mg/L (10)(3)	1 mg/L (10)	
pH					
Selenium	10		45 (3)		
Silver	50				
Specific conductivity (EC)		900 µmhos/cm (11)			
Strontium					8,400 (7 day)
Sulfate		250 mg/L (5)			
Thallium					
Total dissolved solids (TDS)		500 mg/L (12)			
Uranium	20 pCi/L				35
Vanadium					
Zinc		5,000			

(1) Other secondary drinking water standards include: color = 15 color units; foaming agents = .5 mg/L; odor = 3 (threshold odor number).
(2) National Academy of Sciences.
(3) Proposed.
(4) Limited to fibers longer than 10 µm.
(5) Recommended level: upper limit = 500 mg/L; short term limit = 600 mg/L.
(6) Measured as total chromium.
(7) U.S. Environmental Protection Agency, *Quality Criteria for Water, 1986* (May, 1986) plus updates. The form of cyanide is not stated, however, the standard has frequently been applied as a free cyanide standard. Further, the EPA report text implies in places that the standard should apply to free cyanide.
(8) EPA value.
(9) As nitrate (NO_3).
(10) As nitrogen.
(11) Recommended level: upper limit = 1,600 µmhos/cm; short term limit = 2,200 µmhos/cm.
(12) Recommended level: upper limit = 1,000 mg/L; short term limit = 1,500 mg/L.
(13) U.S. Environmental Protection Agency, *Water Quality Criteria, 1972.*

Reference: *A Compilation of Water Quality Goals.* Marshack, 1989.

TABLE 3.1
TYPICAL WATER QUALITY STANDARDS AND GOALS
(Continued)

INORGANIC CONSTITUENT	EPA - NATIONAL AMBIENT WATER QUALITY CRITERIA BASED ON:		AGRICULTURAL WATER QUALITY GOALS [13]	EPA - CRITERIA TO PROTECT FRESHWATER AQUATIC LIFE	
	Public Health Effects	Taste and Odor or Welfare		4-Day Average	1-Hour Average
Aluminum			5	87	750
Ammonium sulfamate					
Antimony	146				
Arsenic			0.1	190	360
Asbestos					
Barium	1,000				
Beryllium			0.1		
Boron			.7/.75		
Bromide					
Cadmium	10		.01	55	1.4
Chloramine					
Chlorate					
Chloride	250 mg/L		106 mg/L	230 mg/L	860 mg/L
Chlorine				11	19
Chlorine dioxide					
Chlorite					
Chromium (III)	170,000			98	820
Chromium (VI)	50		.1	11	16
Cobalt			.05		
Copper		1,000	.2	5.4	7.5
Cyanide	200			5.2	22
Fluoride			1		
Iodide					
Iron			5		
Lead	50		5	.99	25
Manganese		300	.2		
Mercury	0.144	50		.012	2.4
Molybdenum			.01		
Nickel	13.4		.2	73	653
Nitrate					
Nitrite	10 mg/L				
pH		5-9 pH units			
Selenium	10[10]				
Silver	50		.02	5	20
Specific conductivity (EC)			200 μmhos/cm		
Strontium					
Sulfate		250 mg/L			
Thallium	13				
Total dissolved solids (TDS)			450 mg/L		
Uranium					
Vanadium			.1		
Zinc		5,000	2	49	54

(1) Other secondary drinking water standards include: color = 15 color units; foaming agents = .5 mg/L; odor = 3 (threshold odor number).
(2) National Academy of Sciences.
(3) Proposed.
(4) Limited to fibers longer than 10 μm.
(5) Recommended level: upper limit = 500 mg/L; short term limit = 600 mg/L.
(6) Measured as total chromium.
(7) U.S. Environmental Protection Agency, *Quality Criteria for Water, 1986* (May, 1986) plus updates. The form of cyanide is not stated, however, the standard has frequently been applied as a free cyanide standard. Further, the EPA report text implies in places that the standard should apply to free cyanide.
(8) EPA value.
(9) As nitrate (NO_3).
(10) As nitrogen.
(11) Recommended level: upper limit = 1,600 μmhos/cm; short term limit = 2,200 μmhos/cm.
(12) Recommended level: upper limit = 1,000 mg/L; short term limit = 1,500 mg/L.
(13) U.S. Environmental Protection Agency, *Water Quality Criteria, 1972*.

Reference: *A Compilation of Water Quality Goals*. Marshack, 1989.

TABLE 3.1
TYPICAL WATER QUALITY STANDARDS AND GOALS
(Continued)

INORGANIC CONSTITUENT	PROP. 65 NO SIGNIFICANT RISK LEVEL (H&WA)	1-IN-A-MILLION INCREMENTAL CANCER RISK ESTIMATES	
		EPA - National Ambient Water Quality Criteria	EPA Health or Water Quality Advisories
Aluminum			
Ammonium sulfamate			
Antimony			
Arsenic	1.0	0.0022	
Asbestos		30,000,000 fibers/L	7,100,000 fibers/L(3)(4)
Barium			
Beryllium			
Boron			
Bromide			
Cadmium			
Chloramine			
Chlorate			
Chloride			
Chlorine			
Chlorine dioxide			
Chlorite			
Chromium (III)			
Chromium (VI)			
Cobalt			
Copper			
Cyanide			
Fluoride			
Iodide			
Iron			
Lead	0.05		
Manganese			
Mercury			
Molybdenum			
Nickel			
Nitrate			
Nitrite			
pH			
Selenium			
Silver			
Specific conductivity (EC)			
Strontium			
Sulfate			
Thallium			
Total dissolved solids (TDS)			
Uranium			
Vanadium			
Zinc			

(1) Other secondary drinking water standards include: color = 15 color units; foaming agents = .5 mg/L; odor = 3 (threshold odor number).
(2) National Academy of Sciences.
(3) Proposed.
(4) Limited to fibers longer than 10 μm.
(5) Recommended level: upper limit = 500 mg/L; short term limit = 600 mg/L.
(6) Measured as total chromium.
(7) U.S. Environmental Protection Agency, *Quality Criteria for Water, 1986* (May, 1986) plus updates. The form of cyanide is not stated, however, the standard has frequently been applied as a free cyanide standard. Further, the EPA report text implies in places that the standard should apply to free cyanide.
(8) EPA value.
(9) As nitrate (NO_3).
(10) As nitrogen.
(11) Recommended level: upper limit = 1,600 μmhos/cm; short term limit = 2,200 μmhos/cm.
(12) Recommended level: upper limit = 1,000 mg/L; short term limit = 1,500 mg/L.
(13) U.S. Environmental Protection Agency, *Water Quality Criteria, 1972*.

Reference: *A Compilation of Water Quality Goals*. Marshack, 1989.

constructing a waste management unit, and the determination of whether the chemical constituents of potential seepage would be attenuated by the geologic formations underlying the waste management unit.

Mine waste consists predominantly of materials that occur naturally in rock or soil. It usually is generated as a result of the physical extraction and beneficiation activities used to recover a target mineral. These activities include drilling, blasting, excavating, crushing, grinding, washing, screening, gravity separation, flotation, and leaching or solutioning. In some instances, reagents are added to assist in the recovery of the target mineral by flotation or leaching. Processing activities may follow beneficiation and generate additional wastes.

The most prevalent constituents associated with mine waste in California include sediments, naturally soluble materials, acids, heavy metals, asbestos, beneficiation or process reagents and their breakdown products, and nutrients such as nitrates that may be introduced by blasting.

3.5.2 CHARACTERISTICS OF MINE WASTES COMMON IN CALIFORNIA

3.5.2.1 Introduction

This section describes the most important physical and chemical characteristics of mine wastes encountered in California, but which might be found also in other states and mining environments. It deals with issues such as sediment transport from disturbed or fill areas associated with mine waste piles or earthen containment embankments, and soluble minerals that occur naturally and potentially could impact water quality when mined. It also describes acid rock drainage (ARD) which can occur where reactive sulfide-containing materials are mined in humid climates; asbestos that may be suspended in surface waters at mines in asbestos-bearing geologic formations; reagents which significantly can influence the chemistry of waste; and materials such as nitrates, which may be introduced by blasting. The typical chemistry of waste liquids derived from the gold extraction are discussed, because gold mining is an important part of the mineral industry in California, and because gold beneficiation wastes usually attract regulatory and public attention due to the use of cyanide as a reagent.

3.5.2.2 Sediments

Sediments are comprised of the fine-grained, easily eroded fractions of soils and certain rock formations that become suspended in surface waters. The size fractions generally include clays, silts and, occasionally, fine sands. Sediment discharges are usually associated with stormwater runoff. Sediments are by no means unique to the mining industry and can impact receiving waters at any location where barren, unconsolidated earth is present, e.g., where grading and excavation is taking place.

Sediments are also the products of natural erosion processes. A certain quantity of sediment is part of the natural bed-load of streams, and thus is necessary for the life and vitality of the stream. Poor erosion control practices, however, can result in high sediment loads that may overwhelm natural drainage systems, especially during periods of high runoff.

The most likely areas of mine sites from which sediments may be eroded by surface water runoff are: (1) excavation sites which drain to the surrounding terrain, (2) topsoil and overburden stockpiles, (3) waste rock storage and disposal areas, (4) plant sites, (5) impoundment embankments, and (6) other areas where earthen or rock materials have been disturbed or stockpiled.

The impacts to water quality posed by sediments can be measured using the following parameters:

- Total Suspended Solids (a measure of the total sediment suspended in water; usually measured in mg/L)
- Total Settleable Solids (a measure of the fraction of the suspended matter that readily settles out; mg/L)
- Turbidity (usually measured in nephelometric turbidity units [NTUs])

Sediment readily can be controlled at mining sites. In arid, unvegetated areas, runoff occurs infrequently. When it does occur, the mine waste piles usually do not cause a noticeable increase in the naturally high sediment loads. In addition, the receiving water resource usually is not sensitive to sediment load. Therefore, at mines in these areas, either very minimal or no sediment control is required.

At mining operations in wetter areas, where storm runoff is more likely, the mine operators typically must take steps to minimize runoff and erosion from mining facilities and any mine waste management units. Examples of these steps include: (1) providing diversion ditches around the upgradient perimeter of waste management units to minimize run-on to the surface of the units, (2) implementing surface and ditch erosion protection measures, (3) constructing sediment dams and holding ponds downgradient of disturbed areas or waste rock management units to allow sediments to settle out before the runoff waters are discharged, and (4) providing infiltration basins to prevent direct discharge of sediment laden runoff to surface water bodies. The control and diversion of stormwater run-on (i.e., entry onto the property of storm discharges from upgradient elevations) is becoming a more important design consideration in view of stormwater discharge requirements promulgated by the U.S. Environmental Protection Agency (EPA) in its recently published stormwater runoff regulations (EPA, 1990).

Sediment usually does not pose a threat to beneficial uses of ground water, because the unsaturated geological formations overlying ground water act as effective filters.

3.5.2.3 Soluble Natural Minerals

Soluble natural minerals include borate, chloride, sulfate, hydrous carbonate, and nitrate minerals. Table 3.2 lists some of these minerals that are known to occur in California. They usually are associated with dry lake beds in the desert areas of the State.

Several of the soluble natural minerals, such as borax, trona, and epsomite, represent economic products in California, and a number of active and formerly active mines produce or have produced them. Such operations also generate mine wastes which contain other, non-economic but water-soluble minerals and, potentially, trace amounts of heavy metals. The potential threat to beneficial uses of water presented by soluble minerals is an increase in the concentration of TDS and, potentially, increases in one or more of the major anions or cations, because soluble minerals are highly ionic. These constituents can reach both surface and ground water, either as constituents in runoff from a waste management unit or as constituents in liquids that seep from a waste management unit.

Some mining operations that generate soluble natural minerals are located in dry lake environments with associated lacustrine brines, where the underlying ground water is frequently too saline for any beneficial use other than the production of soluble minerals, unless pretreated. Therefore, an accepted disposal practice has been to return the mined material to its source, the lake beds.

In some cases, waste rock and ore residue from hard-rock(1) mining may contain soluble minerals that could increase dissolved constituent levels in liquids associated with the wastes. The principal mechanism by which mining activities may contribute to an increase in TDS is by exposing greater

---

(1) This term applies to mineral mines in the harder geologic formations. It excludes coal and placer mining.

TABLE 3.2

SPECIES OF SOLUBLE NATURAL MINERALS THAT OCCUR IN CALIFORNIA

MINERAL GROUP	MINERAL NAME	COMPOSITION	OCCURRENCE
Borates	Borax	$Na_2B_4O_7 \cdot 10H_2O$	Dry lake beds in desert environs.
Carbonates	Trona	$3Na_2O \cdot 4CO_2$	
(Hydrous)	Nahcolite	$NaHCO_3$	
Chlorides	Halite	$NaCl$	
	Sylvite	KCl	
Nitrates	Soda Niter	$NaNO_3$	
	Niter (Saltpeter)	KNO_3	
Sulfates	Chalcanthite	$CuSO_4 \cdot 5H_2O$	Active oxidizing copper deposits.
	Epsomite	$MgSO_4 \cdot 7H_2O$	Dry lake beds in desert environs.
	Glauberite	$Na_2SO_4 \cdot CaSO_4$	Dry lake beds in desert environs.

surface areas of mined rock and earth to infiltration and runoff waters. This mechanism is particularly prevalent for mine wastes that have been finely ground or contain high percentages of chemically reactive minerals.

All water naturally contains a certain amount of TDS, and substantial increases in the level can impair its beneficial uses. There are, for example, secondary drinking water standards for TDS and sulfate (Table 3.1), as water with a high TDS content can be found unsuitable for drinking, agricultural, or industrial uses. High levels of TDS also can impair the propagation of fish and wildlife.

3.5.2.4 Acid Rock Drainage

ARD can pose a threat to water quality where waste materials contain relatively high levels of reactive sulfides and where disposal occurs in a humid climate. ARD has been demonstrated to be a problem at many of the older, abandoned mines located in the wetter parts of northern California (U.C. Berkeley, 1988). Chapter 4 of this book describes in detail the ARD process, outlines methods for predicting its occurrence, and summarizes waste management alternatives for controlling ARD. For the sake of completeness, a brief summary discussion is presented here.

ARD is formed when reactive sulfides oxidize in the presence of water. Biological processes accelerate the rate of ARD formation once the pH drops below 3.5. A number of factors are required for ARD to be generated. First, the mineral deposit must contain a sufficient quantity of reactive sulfide. Typically, such deposits are tabular in nature, often comprised of multiple lenses of sulfide mineralization. Host rocks are typically silicic volcanics. Economic products recovered from such deposits include copper, zinc, silver, gold, lead, and sulfur. Sulfide minerals include, amongst others, pyrite (FeS_2), marcasite (FeS_2), pyrrhotite (ferrous sulfide of variable iron content with the formula $Fe_{1-x}S$), and arsenopyrite (FeAsS).

The second factor that contributes to the formation of ARD is the presence of air. Any mine or mineral recovery waste disposed of on the surface is exposed to air. When the ground water table in surface and underground mines is lowered by dewatering activities, mineralized material that was previously located below the water table, and hence deprived of oxygen, can be exposed to air.

Thirdly, ARD requires the presence of water to initiate the chemical and biologic reactions that are responsible for its formation. Water percolation through the waste provides the means by which ARD products are transported out of the waste to the environment.

The fourth factor is that ARD formation requires the presence of bacteria to catalyze the acid formation reaction.

Generation of ARD can be accelerated by mining and mineral recovery activities which increase the surface area of exposed reactive sulfides. In mines, ore and waste rock are broken by blasting and excavating. Mineral recovery residues are frequently crushed and, in some cases, finely ground.

ARD typically appears as a low pH seepage, leachate, or runoff. It can emanate from underground mine workings and open pits that contain residual sulfide mineralization, or from waste piles and tailings impoundments that contain sulfide-bearing material and pose a potential threat to beneficial uses of surface or ground water. The potential visual impacts include red and brown water color, similarly colored stains on stream bottoms and banks, and accumulations of "yellowboy," a colloidal yellowish brown precipitate. Potential chemical impacts include decreased pH, increased TDS, increases in anionic concentrations with sulfates dominating, and increased concentrations of metals. A list of parameters of concern and representative ranges of concentrations from hard-rock mines is presented in Chapter 4, Table 4.1.

3.5.2.5 Asbestos

Surface water runoff in asbestos mining areas may contain suspended asbestos particles. However, because asbestos is a mineral that occurs naturally at many locations in California, its presence is not confined to mining locations. It normally is found in greater or lesser quantities in all deposits of serpentine, the state rock of California. Serpentine belts underlie vast areas of the Sierra Nevada Foothills, the Coast Ranges, and the Klamath Mountains. Due to the physical and chemical properties of serpentine, these deposits typically are poorly vegetated and, therefore, susceptible to erosion.

Natural concentrations of asbestos fiber in California surface waters are extremely variable, measuring up to millions of fibers per liter (Calaveras County Planning Department, 1988). However, monitoring data from both upstream and downstream of the largest asbestos mine in California indicated that downstream concentrations were lower than upstream levels. At this mine, the sedimentation pond used to control offsite discharge of silt and sediment also acted to reduce asbestos fiber concentrations in natural storm runoff by several orders of magnitude.

No significant medical evidence has been developed to date to suggest that ingestion of water-borne asbestos poses any degree of health risk. Furthermore, California's Proposition 65[(2)] Scientific Advisory Panel has determined that exposure to asbestos by ingestion in drinking water does not pose a significant human cancer risk.[(3)]

3.5.2.6 Reagents

Reagents are chemicals used to facilitate recovery of the economic mineral product. Table 3.3 lists the reagents typically used at mines throughout California. The following discussion reviews some of the more common methods and discusses the reagents that are used. As with the other dissolved constituents, reagents have the potential to impact both surface and ground water resources.

Mineral recovery involves several different methods, depending on the target economic minerals and their physical and chemical properties. These methods range from simple physical techniques, such as the washing and screening of sand and gravel, to more complex chemical leaching or solvent extraction. Chemical reagents range from acidic, as in the strong mineral acids used in tungsten ore extraction, to alkaline, as in the caustic cyanide solutions used in the leaching of gold ores.

Flotation Reagents

Flotation, more specifically "froth flotation," is used to some extent in California. The ores are first crushed and ground to effect a physical separation between the economic minerals and their host rock, known as "gangue." The ground ore is mixed with water and flotation reagents and discharged through flotation basins or cells. The minerals are separated from the gangue when they attach themselves to the froth that floats on the solution. Both the froth and the minerals are recovered for further processing. The gangue is disposed of as a waste called "flotation tails."

---

(2) California Safe Drinking Water and Toxic Enforcement Act of 1986.

(3) Title 22, California Code of Regulations, Section 12707.

TABLE 3.3

TYPICAL REAGENTS USED FOR BENEFICIATION OF ORE AT A WIDE RANGE OF MINING OPERATIONS

REAGENT	USE
Water	Washing, wetting, and slurry transport
Calcium and sodium hydroxide	pH adjustment
Hydrochloric, sulfuric, and nitric acids	pH adjustment, acid metal/mineral leaching
Sodium cyanide	Precious metals leaching
Flotation agents[1] • Sodium carbonate • Potassium amyl xanthate • Polyanionic cellulose • Aryl dithiophosphoric acid • Cationic polymer • Anionic polyacrylamide • Pine oil • Monothiophosphate • Methyl isobutyl carbinol	Flotation of minerals

(1) Flotation agents usually are used only in the relatively few metal mines in California.

By design, flotation reagents generally remain with the mineral concentrate. However, some of the reagents and their reaction products remain with the gangue following mineral recovery and ultimately may find their way into a mine waste management unit.

Cyanide

The principal use of cyanide in the minerals industry of California is as a lixiviant in the recovery of gold. Cyanide is one of the few reagents that can dissolve gold, the others being aqua regia and thiourea. In various chemical forms, cyanide is an acutely toxic substance to humans, and aquatic and other wildlife. However, cyanide has been used in the U.S. mining industry for nearly 100 years with no known proven accidental human fatality (DuPont Corporation, 1991). Recently, however, mortality of migratory birds, has become an important issue at cyanide solution ponds (Smith and Mudder, 1991).

In mine waste disposal environments, cyanide can occur as "free" cyanide, its most chemically active and toxic form, as various metal-cyanide complexes, cyanates, and as thiocyanates (Hoye, 1987). The metal-cyanide complexes vary from strongly-bonded (iron and cobalt, for example) to weakly-bonded (zinc and cadmium) compounds. The strongly-bonded complexes can remain stable for long periods of time (tens of years) and will dissociate slowly, if at all. The weakly-bonded complexes will dissociate more easily, forming other complex ions and free cyanide.

The toxicity of the metal-cyanide complexes varies inversely with the strength of the bond, and the highest potential toxicities are posed by the most weakly-bonded complexes. For example, trout toxicity as measured by the 96-hour mean acute LC_{50}[4] for free cyanide (CN^-), is 0.045 mg/L. The 96-hour mean acute LC_{50} for cyanates (CNO^-) for different fish species ranges from about 13 to 45 mg/L.

Thiocyanate, a sulfur-bearing cyanide species, is a degradation or reaction product of cyanide found in certain gold mine leach residues. Thiocyanate (SCN^-) is much less toxic to fish and humans than the above forms of cyanide. The LC_{50} values for different fish species vary from 140 to 250 mg/L. The values cited above were taken from data presented by Smith and Mudder (1991), which contains a more detailed discussion on cyanide toxicity.

---

(4) Lethal concentration for 50 percent mortality.

Other Reagents

In the early days of California's gold recovery industry (pre-1920), much use was made of mercury due to its ability to amalgamate[5] with gold. Mercury is relatively immobile in soil environments (Leeper, 1978), so it does not tend to migrate out of waste management units. However, it does have the ability to enter and concentrate in the food chain. Today, the use of mercury in mining is believed to be restricted to recreational gold dredging and panning. Mercury is not used in any current, large scale commercial gold mining operations known to the authors. Acids and bases (highly alkaline compounds) are used in a variety of mineral recovery activities, typically to adjust the pH of the processing solutions. Generally, strong mineral acids, such as sulfuric acid, or the strong bases, such as soda ash and lime, are used.

Organic solvents are occasionally used in the more advanced stages of mineral recovery, typically in small batch operations applied to a mineral concentrate. In California, only one mine is known by the authors to use these reagents, and at that location they are contained and recycled as part of the process. The authors are not aware of any organic solvents appearing as a pollutant in the receiving waters at any mine site in California.

3.5.2.7 Nutrients

Nutrients include nitrates and phosphates. In elevated concentrations, they can promote the growth of oxygen-consuming algae in receiving waters, with consequent adverse impacts on beneficial uses of water. Historically, mining operations have not been associated with the discharge of nutrients, with the limited exception of residual nitrates from blasting or where nitric acid is used to process ore. Also, cyanide, used in agriculture as a nitrate fertilizer early in this century, presumably could degrade to nitrate in some locations. The authors know of no instance where nitrate derived from nitric acid or cyanide has caused a water quality problem at a mine site in California.

3.5.2.8 Typical Gold Mine Waste Solution Characteristics

Gold is an important component of the mining industry in California. It also involves a more complex waste chemistry than most other mining activities in the State. Some typical characteristics of waste solutions associated with gold mining, therefore, are discussed in this chapter. Typical dissolved constituents contained in liquids associated with gold mining

---

(5) Amalgamation is the blending of metals.

throughout the world are summarized in Table 3.4 (Mudder, 1989). The data presented were obtained from barren(6) solutions from heap leach operations, as well as from tailings fluids. As can be seen, typical characteristics include the potential for dissolved metals, which originate in the ore and become soluble due to natural or induced oxidation and/or the presence of reagents, various forms of cyanide and cyanate, as well as dissolved salts, particularly sulfates.

3.5.3 PHYSICAL PROPERTIES OF MINE WASTE

Physical properties of mine waste are important in determining: (1) how the material can be physically stacked, (2) its resistance to wind and water erosion, (3) the amount of free liquid it could produce, and (4) to what extent it can be used for construction purposes, particularly for waste containment facilities. Key physical attributes and the associated properties that must be known for waste containment design or operations are as follows:

- Stackability:
  - Shear strength
  - Durability
  - Hydraulic conductivity (which determines the ability of the material to drain and is a function of the grain size distribution)
- Resistance to water and wind erosion:
  - Grain size distribution
  - Cohesion
- Precipitation Infiltration Capacity:
  - Hydraulic conductivity
  - Slope of the surface of the waste pile
  - Ability to support growth of vegetation
- Amount of free-draining liquid:
  - Initial moisture content
  - Specific retention, a function of the grain size distribution (Chapter 6)
- Suitability for embankment construction:
  - Durability
  - Shear strength
  - Hydraulic conductivity
  - Chemical compatibility with fluids to be contained(7)
- Suitability for liner construction:
  - Hydraulic conductivity
  - Chemical compatibility with fluids to be contained(7)

---

(6) "Barren" applies to the extraction solution after the gold has been removed and before additional reagents have been added and the solution reapplied to the heap as leach solution (See Chapter 9 for further discussion).

(7) This "chemical" property is included with the list of physical properties for the sake of completeness.

TABLE 3.4

CHEMICAL COMPOSITION OF WASTE LIQUID SOLUTIONS ENCOUNTERED AT GOLD MINES THROUGHOUT THE WORLD

PARAMETER	RANGE OF CONCENTRATIONS (mg/L)
Arsenic	<0.02 to 10.0
Cadmium	<0.005 to 0.02
Chromium	<0.02 to 0.10
Copper	<0.10 to 400.0
Iron	<0.50 to 40.0
Lead	<0.01 to 0.10
Manganese	<0.10 to 20.0
Mercury	<0.0001 to 0.05
Nickel	<0.02 to 10.0
Selenium	<0.02 to 6.0
Silver	<0.005 to 2.0
Zinc	<0.05 to 100.0
Total Cyanide	<0.50 to 1,000.0
WAD(1) Cyanide	<0.50 to 650.0
Free Cyanide	<0.01 to 200.0
Ammonia-N	<0.10 to 50.0
Thiocyanate	<1.0 to 2,000.0
pH (in pH units)	2.0 to 11.5
Hardness (as $CaCO_3$)	<200 to 1,500
Sulfate	<5 to 20,000
Temperature (in °C)	<0 to 35
Flow (in gpm)	0 to 6,000

(1) Weak Acid Dissociable (cyanide). This method detects free cyanide plus the weakly-complexed cyanide forms.

Reference: Mudder (1989)

Physical properties of the mine waste are relatively easy to determine during the planning phase of a mining operation. Ore, waste rock, and overburden samples are usually available from the exploration program's core drilling work, underground development and/or surface sampling. Waste characteristics may be determined from bench or pilot scale tests conducted on ore samples, from metallurgical extraction testing, or they may be "synthesized" in the laboratory.

Standard laboratory soils tests can be conducted to determine parameters such as particle size distribution, moisture content, strength, durability, and hydraulic conductivity of the material. Calculations can be performed to determine, for example, the angle of repose for end-dumped waste rock or ore residues, and the slope angle required for stable, permanent slopes including earthquake loading. Similar calculations can be performed to evaluate tailings strength, densities and moisture conditions at various stages during and after deposition. In some instances, laboratory model testing of tailings deposition can be conducted to determine the rate at which free liquids would drain, tailings "beach" angles,(8) and in-place densities. This information is necessary to determine the most effective tailings deposition methods and sequences, to maximize the storage capacity of tailings impoundments, and to minimize the amount of liquid retained. Further discussion of physical properties of mine wastes is presented in Chapter 8.

3.5.4 CHEMICAL PROPERTIES OF MINE WASTE

Chemical properties of the mine waste need to be determined in order to define the characteristics of the waste at the disposal site and thus assess its potential to impact beneficial uses of water. In California, mine wastes are classified on the basis of their intrinsic chemical properties.

Classification testing generally involves simple, reproducible tests that reflect the intrinsic properties of the material and provide data for determining the appropriate regulatory category for the waste type being evaluated. The testing protocols are not intended to provide realistic, site-specific information on those characteristics that determine the waste's actual potential for impacting water quality. It is necessary, therefore, to consider waste characterization separately from waste classification testing, as each requires a different protocol. The two sections that follow discuss those aspects in more detail.

(8) Beach angle is the slope of the surface of tailings when they are deposited as a slurry, generally around the perimeter of the tailings impoundment. As the angle determines the extent to which tailings can be mounded, it is important in determining the maximum storage capacity of the impoundment.

3.5.4.1 Waste Classification

California Classification System

As discussed in Chapter 2, the regulations governing mine waste disposal in California (CCR, Title 23, Division 3, *Chapter 15*) require that wastes be classified as either Group A, Group B, or Group C mining wastes. The definition of each waste group follows:

- Group A - Mine wastes that must be managed as hazardous waste pursuant to Section 66300[9] (CCR, Title 22), provided the Regional Board finds that such mine wastes pose a significant threat to water quality.
- Group B - (1) Mine wastes that consist of or contain hazardous wastes, that qualify for a variance under Section 66310[9] (CCR, Title 22), provided that the Regional Board finds that such mine wastes pose a low risk to water quality; and (2) mine wastes that consist of or contain nonhazardous soluble pollutants at concentrations which exceed water quality objectives[10] for, or could cause degradation of, waters of the State.
- Group C - Mine wastes from which any discharge would be in compliance with the applicable water quality control plan, including water quality objectives, other than turbidity.

Determination of whether a mine waste is "hazardous" is accomplished by using the criteria adopted by the California Department of Health Services (DHS) for purposes of hazardous waste regulation in Section 66300 (Title 22) and summarized in Table 3.5. As can be seen in the table, a waste is judged to be hazardous if it fails any one of the criteria listed under the five different waste characteristic categories, i.e., toxicity, persistent and bioaccumulative toxic substance, ignitability, corrosivity, or reactivity. For mine wastes, it is most often the "persistent and bioaccumulative toxic substances" category, which includes criteria for the total and extractable metal concentrations determined by the Waste Extraction Test (WET), that determines whether or not the waste is hazardous. For the purpose of evaluating mine waste, a list of inorganic elements (principally metals) is provided in Section 66699 (Title 22). These parameters, which also are listed in Table 3.5, provide both a maximum allowable "total" content of the constituents, i.e., total threshold limit concentration (TTLC), and a maximum allowable "soluble" concentration of those same constituents, i.e., the soluble threshold limit concentration (STLC). Evaluation of the total constituent levels is by digestion, using EPA prescribed methods. The soluble levels are determined by using a prescribed 10 to 1 buffered 0.2M sodium citric acid extraction fluid at pH 5.0 (WET).

---

(9) See Article 2, "Scope and Procedures," of *Chapter 30*, "Minimum Standards for Management of Hazardous and Extremely Hazardous Wastes," in the California Code of Regulations, Title 22, Division 4.

(10) Water quality objectives reflect the potential beneficial uses of the waters of the State and are documented in published *Water Quality Control Plans* for various water basins within California.

TABLE 3.5
CALIFORNIA HAZARDOUS WASTE CLASSIFICATION CRITERIA

<table>
<tr><th>WASTE CHARACTERISTIC CATEGORY</th><th>WASTE CLASSIFICATION TESTING CRITERIA</th></tr>
<tr><td>Toxicity[1]</td><td>• Acute Oral $LD_{50} < 5{,}000$ mg/kg.
• Acute Dermal $LD_{50} < 4{,}300$ mg/kg.
• Acute Inhalation $LC_{50} < 10{,}000$ parts per million.[2]
• Acute Aquatic 96-hour $LC_{50} < 500$ mg/L.[3]
• Poses a hazard to human health because of its carcinogenicity, acute toxicity, chronic toxicity, bioaccumulation or persistence in the environment.
• Is a federally listed hazardous waste in 40 CFR 261.</td></tr>
<tr><td>Persistent and Bioaccumulative Toxic Substances</td><td>Wastes with concentrations that exceed the following standards when subjected to the Waste Extraction Test:
<table>
<tr><th>Constituent</th><th>Total Threshold Limit Concentration (TTLC) (mg/kg)</th><th>Soluble Threshold Limit Concentration (STLC) (mg/L)</th></tr>
<tr><td>Antimony</td><td>500</td><td>15</td></tr>
<tr><td>Arsenic</td><td>500</td><td>5</td></tr>
<tr><td>Barium</td><td>10,000</td><td>100</td></tr>
<tr><td>Beryllium</td><td>75</td><td>0.75</td></tr>
<tr><td>Cadmium</td><td>100</td><td>1</td></tr>
<tr><td>Chromium</td><td>2,500</td><td>560</td></tr>
<tr><td>Cobalt</td><td>8,000</td><td>80</td></tr>
<tr><td>Copper</td><td>2,500</td><td>25</td></tr>
<tr><td>Fluoride</td><td>18,000</td><td>180</td></tr>
<tr><td>Lead</td><td>1,000</td><td>5</td></tr>
<tr><td>Mercury</td><td>20</td><td>0.2</td></tr>
<tr><td>Molybdenum</td><td>3,500</td><td>350</td></tr>
<tr><td>Nickel</td><td>2,000</td><td>20</td></tr>
<tr><td>Selenium</td><td>100</td><td>1</td></tr>
<tr><td>Silver</td><td>500</td><td>5</td></tr>
<tr><td>Thallium</td><td>700</td><td>7</td></tr>
<tr><td>Vanadium</td><td>2,400</td><td>24</td></tr>
<tr><td>Zinc</td><td>5,000</td><td>250</td></tr>
</table></td></tr>
<tr><td>Ignitability</td><td>• Liquids with a flashpoint of less than 60°C (140°F).
• A nonliquid capable of causing fire through friction, absorption of moisture or spontaneous chemical changes and, when ignited, burns so vigorously and persistently that it creates a hazard.
• Is a flammable compressed gas as defined in 49 CFR 173.300(b).
• Oxidizes as defined in 49 CFR 173.151.</td></tr>
<tr><td>Corrosivity</td><td>• $pH \geq 12.5$
• $pH \leq 2$</td></tr>
<tr><td>Reactivity</td><td>• Normally unstable and readily undergoes violent change.
• Reacts violently with water.
• Forms a potential explosive mixture with water.
• Cyanide or sulfide bearing which, when exposed to pH conditions between 2 and 12.5, can generate toxic gases in quantities sufficient to endanger human health.
• Readily capable of explosive decomposition.</td></tr>
</table>

(1) LD_{50} = Lethal dose for 50 percent mortality.
LC_{50} = Lethal concentration for 50 percent mortality.

(2) Some exemptions apply.

(3) Evaluations for these criteria require bioassays using fathead minnows, rainbow trout, or golden shiners.

Reference: California Code of Regulations, Title 22, Division 4, *Chapter 30*, Article 11.

Mine wastes can range from alkaline to acidic. Alkalinity is due to natural carbonates or alkaline silicates in the host rock, the addition of caustics or lime during mineral recovery, or both. Acidity is usually due to an inorganic acid, such as sulfuric acid. In either case, the organic citric acid extraction method employed by the WET probably will not provide a representative analysis of the mobility of the metals in the waste. The DHS and its consultant (Borer, 1988) have examined this issue and have concluded that testing procedures should allow for reproducibility of results, be reliable, and be environmentally representative. To ensure maximum reproducibility, Borer recommends continued use of citric acid, in spite of it not being representative, and discourages the use of mineral acid because of the poor reproducibility of the results obtained with these acids. However, Borer also recommends that a liquid to solid ratio of 16 to 1 be used when conducting the tests, rather than the prescribed 10 to 1, to accommodate the specific nature of mine waste. Borer finally recommends that, if the waste material is not acid-generating, the WET can be conducted at a pH of 9.

The Title 22 regulations (Section 66310) allow for a variance from the prescribed test procedures listed in Table 3.5, if the applicant can show that the proposed test method or analytical method is equal or superior to the method prescribed when applied to the specific waste, with respect to accuracy, precision, sensitivity and stringency.

Variations to the WET procedures have been allowed by the Regional Boards for purposes of Porter-Cologne regulation. Where acid-base accounting has indicated an adequate net neutralization capacity, the Regional Boards have allowed the citric acid extracting fluid to be replaced by deionized water. In the authors' judgement, this methodology more accurately evaluates non-acid generating mine waste than the citric acid extraction. With the gradual acceptance of the Borer (1988) recommendations, it is expected that the 16 to 1 liquid to solid and pH 9 variations also will be utilized.

The hazardous classification of mine waste sometimes also is controlled by: (1) the toxicity criteria, which require, among other tests, a bioassay and/or (2) the corrosivity criteria, due to either a very high (equal to or greater than 12.5) or a very low (equal to or less than 2) pH.

Only a small percentage of mine waste in California has been classified as "hazardous waste" (U.C. Berkeley, 1988) and assigned to Group A by the Regional Boards. Most mine waste falls into either Group B or Group C. For purposes of distinguishing between Group B and C wastes, it is necessary to determine the potential for the waste to leach dissolved constituents at levels that could pose a threat to the beneficial uses of the waters of the State. This is achieved best through waste characterization, as discussed in Section 3.5.4.2. The State regulations do not contain any applicable testing protocols.

Federal Classification System

The impact of any federal waste classification system on mine waste management is linked to the federal waste disposal laws and regulations. It is necessary, therefore, to review these before proceeding with the waste classification discussion.

Currently, federal solid waste legislation, i.e., Resource Conservation and Recovery Act (RCRA, Subtitles C and D), and regulations do not deal with mine wastes. *Mining Extraction and Beneficiation Waste*, and 18 specific *Mineral Processing Wastes*, are excluded by the Bevill Amendment (SWDA, 1980) from hazardous waste regulation under RCRA, Subtitle C and the EPA is currently formulating regulations under Subtitle D of RCRA for these wastes. The EPA definitions and lists of Bevill-Excluded wastes are provided in Table 3.6.

While not directly applicable to mine waste regulation, the federal procedures for identifying hazardous wastes may be referenced in future RCRA Subtitle D regulations or in certain state regulations, and are therefore discussed in this chapter. These procedures are summarized in Table 3.7. Although the State of California (Table 3.5) and federal requirements are similar in most respects they do, however, have the following differences. The State's requirements include additional statutes such as having to meet TTLC's and to pass a bioassay. The federal toxicity test also employs different acids and liquid to solid ratios. The EPA currently utilizes the Toxicity Characteristic Leaching Procedure (TCLP) (EPA Method 1311) for the classification of waste and soils. It employs an acetic acid leachate at a pH of 4.5 and a liquid to solid ratio of 20 to 1.

In 1990, the EPA issued a draft modified batch test procedure, designated EPA 1312 (EPA, 1990), ultimately intended to replace EPA 1311. The draft EPA 1312 procedure employs three different leach solutions, depending on waste or soil characteristics, all at a 20 to 1 liquid to solid ratio.

Unless the material to be classified contains cyanide, the leach pH value is either pH 4.2 or pH 5, the leach solution being mineral acid (sulfuric plus nitric acid) and adjusted deionized water. Materials bearing cyanide are leached at pH 9. The federal waste classification criteria currently apply only to those mine wastes that are not excluded from RCRA Subtitle C by the Bevill Amendment. However, these classification criteria might be used in future federal mine waste regulations as a guideline to defining waste management requirements. It is prudent, therefore, for

TABLE 3.6

FEDERAL BEVILL-EXCLUDED WASTES (AS OF AUGUST 1991)

WASTE TYPE EXCLUDED FROM SUBTITLE C	DEFINITION OF WASTE TYPE
ALL MINING EXTRACTION AND BENEFICIATION WASTE(1)	Wastes that involve "crushing; grinding; washing; dissolution, crystallization; filtration; sorting; sizing; drying; sintering; pelletizing; briquetting; calcining to remove water and/or carbon dioxide; roasting, autoclaving, and/or chlorination in preparation for leaching (except where the roasting [and/or autoclaving and/or chlorination]/ leaching sequence produces a final or intermediate product that does not undergo further beneficiation or processing); gravity concentration; magnetic separation; electrostatic separation; flotation; ion exchange; solvent extraction; electrowinning; precipitation; amalgamation; and heap, dump, vat, tank, and in situ leaching."
SELECTED MINERAL PROCESSING WASTE(2)	The EPA has indicated that only the following 18 mineral processing wastes fall under the Bevill Exclusion. - Red and brown muds from bauxite refining. - Treated residue from roasting/leaching of chrome ore. - Gasifier ash from coal gasification. - Process wastewater from coal gasification. - Slag from primary copper processing. - Calcium sulfate wastewater treatment plant sludge from primary copper processing. - Slag tailings from primary copper processing. - Slag from elemental phosphorus production. - Iron blast furnace air pollution control dust/sludge. - Iron blast furnace slag. - Basic oxygen furnace and open hearth furnace air pollution control dust/sludge from carbon steel production. - Basic oxygen furnace and open hearth furnace slag from carbon steel production. - Fluorogypsum from hydrofluoric acid production. - Process wastewater from hydrofluoric acid production. - Slag from primary lead processing. - Process wastewater from primary magnesium processing by the anhydrous process. - Chloride process waste solids from titanium tetrachloride production. - Slag from primary zinc processing.

(1) CFR 261.4[b][7], January 23, 1990.
(2) Federal Register, Volume 56, No. 114, 1991.

TABLE 3.7

FEDERAL HAZARDOUS WASTE CLASSIFICATION CRITERIA[1]

WASTE CHARACTERISTIC CATEGORY	WASTE CLASSIFICATION TESTING CRITERIA
Ignitability	• Liquids with a flashpoint of less than 60°C (140°F). • Non-liquids capable of causing fire through friction or absorption of moisture, occurs as spontaneous combustion, and burns vigorously enough to create a hazard. • Ignitable compressed gas. • Oxidizes as defined in 49 CFR 173.151.
Corrosivity	• Is a liquid and corrodes steel at a rate greater than 6.35 mm per year. • pH $\geq$ 12.5 • pH $\leq$ 2
Reactivity	• Normally unstable and readily undergoes violent change. • Reacts violently with water. • Forms a potential explosive mixture with water. • Cyanide or sulfide bearing which, when exposed to pH conditions between 2 and 12.5, can generate toxic gases in quantities sufficient to endanger human health. • Readily capable of explosive decomposition at standard temperature and pressure or when subjected to a strong initiating source.
Toxicity	Wastes with concentrations which exceed the following standards[2] when subjected to the Toxicity Characteristic Leaching Procedure (TCLP)[3]

Constituent[4]	Maximum Concentration (mg/L)
Arsenic	5.0
Barium	100.0
Cadmium	1.0
Chromium	5.0
Lead	5.0
Mercury	0.2
Selenium	1.0
Silver	5.0

(1) Wastes are also judged to be hazardous if they are a listed waste in 40 CFR 261.30. Mining related wastes that appear on these lists include:
- Acid from primary copper and zinc production.
- Several smelter wastes.

(2) The list of constituents also includes a range of organic compounds. As these usually are not associated with mining activities, they are not included in the table.

(3) The extraction test procedure requires a buffered acetic acid (pH 5.0) at a liquid to solid ratio of 20 to 1. EPA intends to replace this test with a modified batch test procedure designated EPA 1312.

Reference: 40 CFR 261

owners of proposed mine waste management units to classify their wastes in terms of the federal hazardous waste criteria. The authors recommend that, for those wastes that are tested as hazardous, there should be sufficient attention to containment requirements to ensure that no future water quality problems arise which may require costly retrofits under future federal mining or environmental regulations.

3.5.4.2 Waste Characterization

Waste characterization involves determining the actual potential of the waste to impact the beneficial uses of water, and the design of the waste management unit.

Mine waste can be characterized for water quality impact evaluations by either batch extraction or column leach tests, using a chemical environment that is similar to the field conditions of actual waste disposal. Therefore, while batch tests are simpler to conduct, column tests can be made to be more representative of actual site conditions. Marshack (1986b) has published quantitative guidelines that utilize the WET batch test with deionized water as a substitute for the acetate acid to determine which wastes pose a threat to beneficial uses of water and which, therefore, are likely to be classified as a Group B waste. Other batch tests can be used, including the American Standards for Testing and Methods (ASTM) D3987-85 procedure,[11] which utilizes a 20 to 1 liquid to solid ratio. An alternative, more conservative test method, that is more representative of lower volume leachate condition, is the 5 to 1 liquid to solid soil-water extraction test procedure proposed by the U.S. Department of Agriculture (1954).

Ideally, the fluids used should closely resemble the pH of the waste itself. For neutral conditions, deionized water or a simulated rainwater using hydrochloric acid at pH 5 to 5.5 is indicated. A recommended test procedure developed and used successfully to characterize spent heap leach piles by several of the authors of this book, is discussed in Chapter 9.

One of the major problems with batch tests is determining the appropriate ratio of liquids to solids so that the tests yield representative concentrations of waste constituents. To be effective, these tests involve agitating the waste and liquid mixture for extended periods of time. It is therefore difficult to correlate test concentrations to those that would occur in the field. Where it is critical to determine representative concentration levels, second tier testing using column leaching should be considered.

---

(11) Standard test method for shake extraction of solid waste with water.

Column leach tests generally involve packing 2- to 6-inch diameter Plexiglas columns with 2 to 6 feet of waste material. Simulated rainwater, waste liquids, or both are allowed to percolate through the waste and are collected at the bottom for analysis. By insuring that the physical test conditions closely resemble the field conditions, waste constituent concentrations that are more representative than those obtained from batch tests can be obtained. However, even with column tests, caution must be exercised when interpreting the results, as it is not always possible to accurately reproduce field conditions in the laboratory, particularly the drying and wetting sequences induced by natural rainfall patterns.

Acid-generating wastes fall into a special category. Testing protocols for these wastes are discussed in Chapter 4.

3.5.5 OTHER MINE WASTE-RELATED ISSUES

Other issues at mine waste units that potentially may impact water quality include stream diversions and mass wasting. Stream diversions can result in changes to the quantity and quality of streamflow at a particular downgradient location. Mass wasting, which is defined as the gradual downslope migration of solids, or large scale erosion, can increase the amount of sediment present in stream beds or suspended in the streamflow.

The potential for the above issues to arise, and their impact on the environment, usually can be readily predicted. If required, mitigating the impact of stream diversions can be achieved by constructing aquatic habitat enhancement features, for example, artificial ponds, bank irregularities or channel bottom roughness to compensate for the loss of flow. In some instances, it also may be necessary to create new aquatic habitat at other locations to compensate for any loss of habitat caused by the waste management unit. Excessive mass wasting can be avoided by selecting the appropriate slopes for waste piles and by providing the necessary erosion protection. In some cases, catch basins can be provided to prevent the waste material from reaching receiving water bodies.

3.6 SITE-SPECIFIC FACTORS

An appropriate assessment of the potential threat to beneficial uses of water posed by mine waste necessarily involves consideration of site-specific factors. This is the third step in the systems approach. These factors include climate, geology, and distance to water resources, as discussed below.

3.6.1 CLIMATE

Rainfall that does not evaporate or run off, but rather infiltrates and percolates through the mine waste, can be a major transporter of waste constituents. Generally, in the wetter climates, percolation and subsequent leachate formation are important and must be accommodated in the design of waste management units by incorporating leachate collection and containment features. In arid climates, particularly where the mean annual precipitation is less than 10 inches and evaporation rates are high, leachate formation might not pose a problem or require any special waste containment features. Chapter 6 provides a more detailed discussion of the influence of climate on leachate formation in mine waste.

3.6.2 UNDERLYING GEOLOGIC FORMATIONS

The geologic materials overlying the ground water resource and underlying the mine waste management unit can provide natural containment for some or all of the waste constituents. The containment can be hydraulic or chemical in nature, or both. Hydraulic containment is provided by geologic formations that have a low hydraulic conductivity, i.e. typically less than or equal to 10^{-6} cm/sec,[(12)] which limits the amount of seepage that can occur. Chemical containment occurs when the concentration of the constituents in the seepage are reduced by chemical, physical, or biological processes that occur within the geologic formations. This chemical containment is typically referred to as attenuation, and is discussed in detail in Chapter 5.

3.6.3 DISTANCE TO WATER RESOURCES

Distance from the mine waste management unit to the receiving surface or ground water resource, is important in determining the potential risk to water quality. For example, surface waters located a great distance from the unit are unlikely to be impacted, as spillage is not likely to reach these waters. This is usually the case in the arid regions of Southern California where surface water bodies are infrequent. On the other hand, particularly in the wetter portions of Northern California, there are instances where waste management units have been located close to surface water bodies and consequently pose a greater potential risk. In such cases, it may be necessary to incorporate additional surface spill containment features to reduce this risk.

[(12)] This is equivalent to 1 ft/yr and is typical of relatively unfractured rock or clays in natural geologic formations.

The depth to ground water is another important parameter. The greater this depth, the longer it takes, and the less likely it becomes, for seepage to reach ground water. In the arid parts of California, where depths to ground water are usually several hundred feet, small quantities of seepage would not reach the ground water. However, where the ground water level is shallow and close to the bottom of a waste management unit, e.g., less than 10 feet, seepage could enter the ground water fairly quickly, and waste containment in excess of that required in an arid environment would be necessary to provide the same level of water quality protection.

3.7 WASTE MANAGEMENT UNIT CONTAINMENT

This section briefly summarizes the different forms of containment that can be provided at a mine waste management unit.

3.7.1 NATURAL CONTAINMENT

Natural containment may be sufficient where the waste has little or no potential to generate leachate with elevated constituent concentrations, where the geologic formations alone provide adequate containment, where receiving water constituent concentrations are higher than those in the leachate (ARD), or a combination of these factors. Examples include sites in arid climates, where an operation generates waste rock of low mineral content or dried tailings containing very small quantities of contaminated fluids, and where receiving waters are located at great distances from the mining operation. Other examples include the disposal of wastes on saline lake beds in desert regions that have naturally poor quality ground water, with TDS values of up to several hundred thousand mg/L.

Natural containment is adequate for Group C type wastes provided that, in wetter climates, provision is made for controlling sediment in surface runoff from the waste. In the case of Group B type wastes, natural containment may be sufficient under conditions similar to those described above.

3.7.2 ENGINEERED WASTE CONTAINMENT

In addition to natural containment, engineered containment features, such as liner systems (low permeability layers), seepage collection and sedimentation or liquid holding basins, may be provided. Liner systems are used to reduce or eliminate seepage losses, and are discussed in more detail in Chapter 7. They can include single or multiple layers constructed from synthetic or natural materials such as clay. A seepage collection layer(s) "sandwiched" between the waste and liners also may be included. In some instances, the waste itself can be used to provide a liner. Holding ponds or basins are used to contain liquids, particularly during storm periods, thereby preventing spillage and potential surface water pollution. Where the liquids held in these ponds may pose a threat to waters, the ponds also could be lined.

3.7.3 OPERATIONAL ASPECTS

In some instances, mine wastes are pretreated or deposited in a waste management unit in a manner that reduces the potential threat to beneficial uses of water. In these cases, the waste pretreatment or depositional practices should be viewed as constituting part of the engineered containment.

Physical pretreatment or managed deposition can be used to reduce the residual water content of the waste. For example, drying and filtering of tailings can be accomplished in the recovery plant. In arid climates, wet tailings can be deposited in the waste management unit in thin layers, each of which is allowed to dry and solidify prior to deposition of a subsequent layer. Such measures significantly can reduce or even eliminate the amount of free liquid present and hence the seepage threat to beneficial uses of water.

Chemical pretreatment of certain wastes also can be considered. For example, the destruction of cyanide in gold tailings by the addition of hypochlorite, sulfur dioxide, or chloride has been used at several mines in the U.S. In some cases, it may be appropriate to consider a cyanide recovery process such as Acidification Volatilization Reneutralization (AVR). These various forms of chemical treatment are discussed further by Mudder (1989).

In addition, there are measures that can be taken to prevent surface spillage of waste liquids from holding ponds or pumping and pipeline systems. Holding ponds should be designed to safely contain the waste liquids, excess precipitation, and runoff that could accumulate during precipitation events. In cases where several ponds are involved, and where some of the waste liquids are recycled, it usually is necessary to perform integrated, project-wide, water balance evaluations to ensure that the ponds and interconnecting pumps and pipelines are adequately sized. Furthermore, maintenance programs and operational procedures should be implemented to minimize leakage from waste conveyance facilities such as pumps and pipelines.

3.8 THE SYSTEMS APPROACH TO DESIGNING WASTE CONTAINMENT

As discussed in previous sections, the systems approach combines knowledge of site-specific factors such as climate, geology, and water resources with the known physical and chemical properties of the waste material to determine the amount of engineered containment that is required to protect the beneficial uses of water. A convenient way to combine this information is to conduct a quantitative waste containment assessment. In this assessment, the potential quantity of seepage or spillage from a waste management unit is determined. Conducted for waste disposal at a specific natural site, the assessment facilitates the evaluation of the need for engineered containment. By repeating the assessment for alternative types of containment, the most suitable engineering measures can be selected.

3.8.1 EXAMPLES OF THE SYSTEMS APPROACH

This section describes two sets of simplified containment assessments. The first set involves a pond system designed to dispose of 2.6 billion gallons of wastewater over a 10-year period. Containment assessments have been completed for ponds located at different geographic locations to illustrate how the required amount of engineered containment can be varied, depending on site conditions, while still providing adequate protection of the beneficial uses of water. Details of the ponds and the wastewater characteristics considered in the analysis are provided in Table 3.8 and Figure 3.1.

The second set involves alternative designs of an 8-million ton tailings disposal facility at a specific site. This example is intended to illustrate how the amount of engineered containment can be derived and how it can be varied for different methods of tailings deposition. To allow evaluation of the extent to which natural formations can provide containment, two types of geologic formations are considered for each site. The example also compares the performance of geomembrane and clay liners. Details are presented in Table 3.8 and in Figures 3.2 and 3.3.

The containment assessments presented in this chapter are based in part on actual case histories. Central to these assessments is the calculation of the potential seepage loss rates for the various containment scenarios. For the liner systems, these calculations are based on the information presented in Chapter 7. Simplified calculation procedures have been used for the unsaturated and saturated seepage rates and seepage travel times in natural geologic formations. For the actual design of a waste management unit, it may be necessary to perform more comprehensive and detailed seepage calculations and to verify these using laboratory or field testing.

TABLE 3.8

CONTAINMENT ASSESSMENT EXAMPLES: WASTEWATER AND TAILINGS PROPERTIES

EXAMPLE 1: WASTEWATER CHARACTERISTICS	
Delivery Rate	500 gpm
Period of Delivery	10 years
Total Quantity	2.6 billion gallons
Quality - pH - TDS - Metal A	 3 5,000 mg/L 0.5 mg/L
EXAMPLE 2: TAILINGS CHARACTERISTICS	
Delivery Rate (slurry)	950 gpm
Delivery Rate (dry)	2,300 tons/day
Slurry Composition	40% solids
Period of Delivery	10 years
Total Mass (dry)	8 million tons
Interstitial Water Quality - pH - TDS - Metal B - Reagent A	 10 3,000 mg/L 2 mg/L 50 mg/L

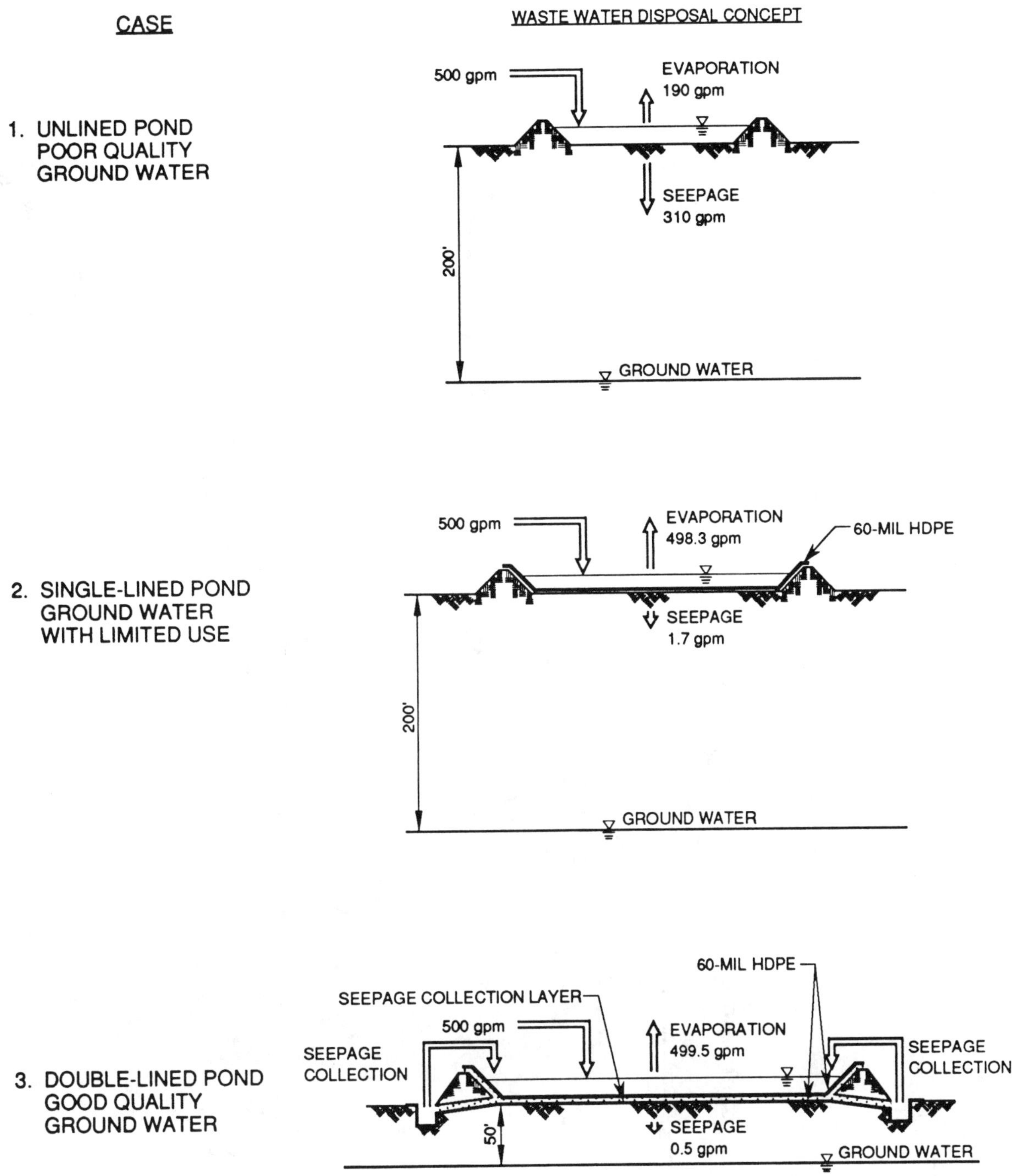

HDPE = HIGH DENSITY POLYETHYLENE

FIGURE 3.1

WASTEWATER POND CONCEPTS USED IN CONTAINMENT ASSESSMENTS

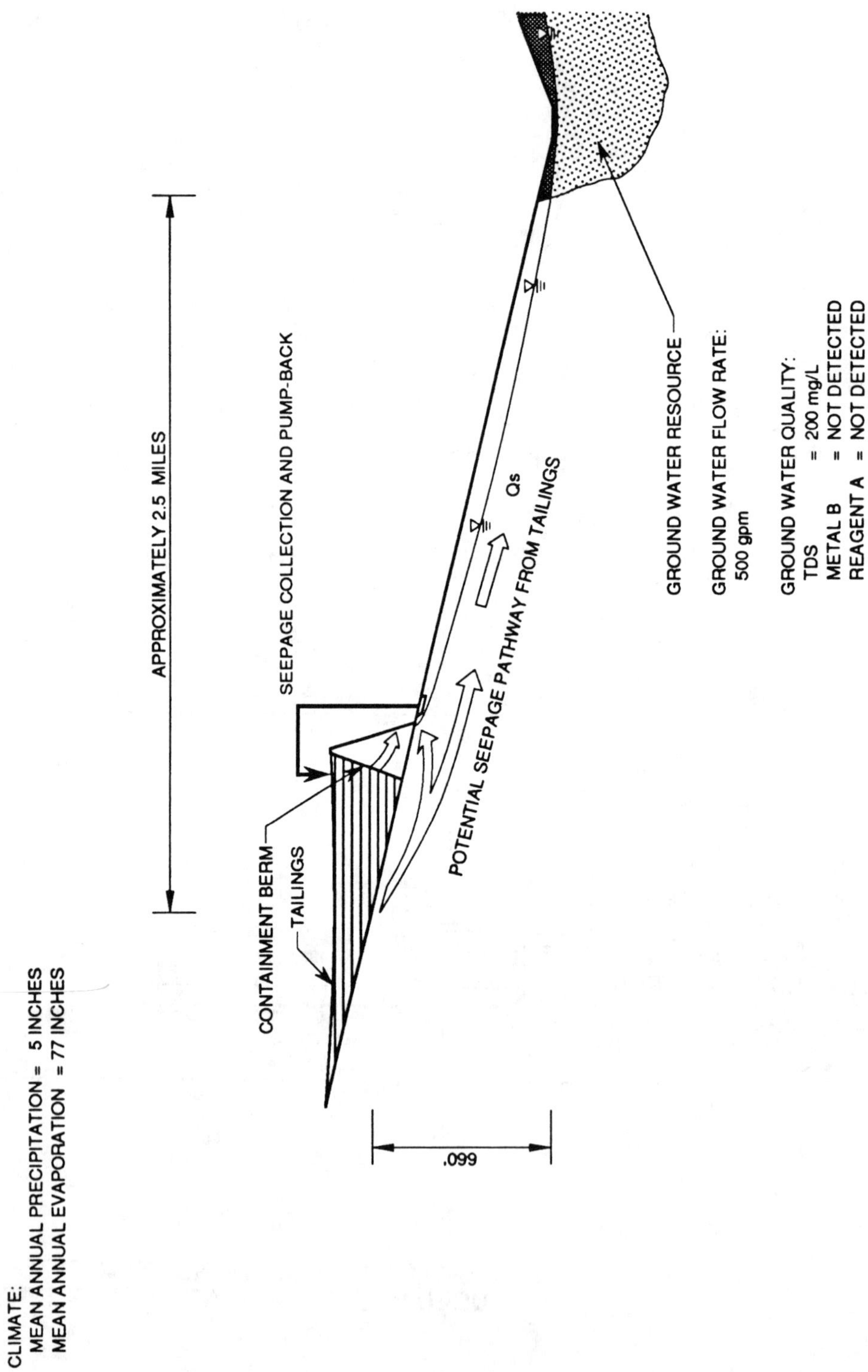

FIGURE 3.2

SITE CONDITIONS IN VICINITY OF TAILINGS IMPOUNDMENT

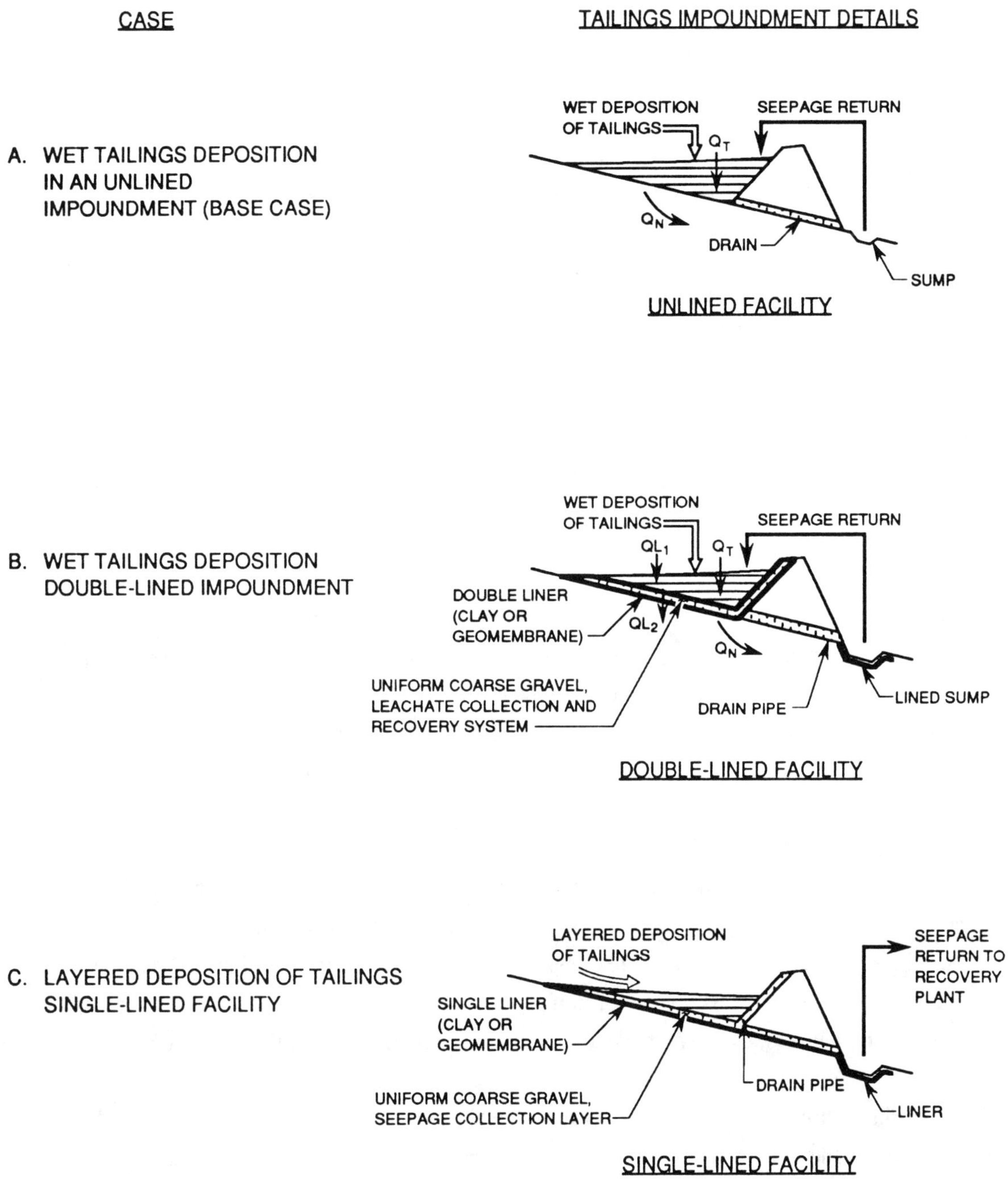

FIGURE 3.3

IMPOUNDMENT CONCEPTS USED IN CONTAINMENT ASSESSMENTS

3.8.1.1 Wastewater Pond Containment Assessments

In all three cases, the waste involves a liquid that is discharged to the pond at 500 gallons per minute (gpm). Its key chemical characteristics include TDS at a concentration of 5,000 mg/L, and Metal A, which is at a dissolved concentration of 0.5 mg/L. The maximum contaminant level (MCL) values for TDS and Metal A are 500 mg/L (secondary standard) and 0.05 mg/L (primary standard), respectively. By California standards, the waste would be classified as Group B. These three cases are discussed in detail below, while Table 3.9 summarizes the results of the containment assessments.

Case 1 - Unlined Pond

The first case involves disposing of the wastewater on a desert lake bed. Ground water at this location is 200 feet below the surface and is highly saline (i.e., 50,000 mg/L). Other pertinent site characteristics are summarized in Table 3.9 and are shown in Figure 3.1.

The containment assessment indicates that the wastewater can be disposed of in an unlined pond, approximately 50 acres in extent. Ultimate disposal occurs as seepage (310 gpm) and evaporation (190 gpm). The only containment feature that needs to be provided is a perimeter berm to prevent the wastewater from spreading over a larger area. The berm also serves to temporarily store water during wetter periods when the evaporation rate may not be high enough to immediately remove liquid that does not seep into the lake bed.

Laboratory testing on several discrete soil samples from the upper few feet of the onsite silty clays indicated that Metal A is highly attenuated and would not migrate downward by more than a few inches. These laboratory test results are considered representative of conditions under the entire pond area, as the lake deposits consist of uniform horizontal layers. The physical and chemical properties of these layers, therefore, are expected to be similar throughout the pond area.

Closure of this pond after ten years may require the removal or cover of the upper few inches of soil if it contains Metal A or other constituent concentrations that are judged to pose a risk. Lining the impoundment may not alleviate this problem, as removal of residual solids that precipitate on the liner would still be required. A lined waste management unit also would require a considerably larger pond area in order to increase evaporation losses to compensate for the reduced seepage losses. It is evident from this discussion that the impact on soil chemistry over the life of the project must be considered when designing an unlined disposal pond.

TABLE 3.9
EXAMPLE 1: RESULTS OF WASTEWATER POND CONTAINMENT ASSESSMENTS

	CASE 1	CASE 2	CASE 3
SITE-SPECIFIC CONDITIONS			
Climatic Conditions			
• Mean Annual Precipitation (inches)	5	5	30
• Mean Annual Evaporation (inches)	77	77	60
• Required Area of Pond (acres)	50	130	320
Unsaturated Zone Conditions			
• Hydraulic Conductivity (cm/sec)	10^{-5}	10^{-5}	10^{-4}
• Porosity	0.3	0.3	0.3
• Natural Moisture Content			
- Percent by weight of solids	15	10	10
- Percent by total volume	24	16	16
• Interstitial Water Quality			
- TDS (mg/L)	20,000	3,000	200
- Metal A (mg/L)	ND(1)	ND	ND
Ground Water Conditions			
• Hydraulic Conductivity (cm/sec)	10^{-5}	10^{-5}	10^{-4}
• Flow in Upper 50 Feet and Across the Width of the Pond (gpm)	0.3	4.7	64
• Water Quality			
- TDS (mg/L)	50,000	1,000	200
- Metal A (mg/L)	ND	ND	ND
- Use Designation	None	Limited to Industrial and Stock Watering	Municipal (potable)
CONTAINMENT ASSESSMENT			
Seepage Rates			
• Through Upper Liner (gpm)	310	1.2 (1.7)(2)	31 (71,000)(3)
• Through Lower Liner (gpm)	NA(4)	NA	0.46 (0.66)(5)
• Time of Seepage to Reach Ground Water Table (years)(6)	1.2	300 (200)(7)	10 (5)(7)
Volume of Seepage to Unsaturated Zone Above Ground Water Table Over 10 Years (million gallons)	1,630	6.3 (8.7)	2 (3)
Volume of Natural Interstitial Water (million gallons)	940	1,630	1,000
Ground Water Quality Impact On			
• TDS	Beneficial Impact	None	None to Negligible
• Metal A	No impact expected due to attenuation	None	None

(1) Not Detected.

(2) Calculated using equation derived by Giroud et al., 1989 and included in Section 7.5.2. An average 5-foot depth of liquid in the pond and one 0.08-inch diameter hole per acre was assumed. (The value in the brackets apply to one 0.445-inch diameter hole per acre.)

(3) Interpolated from Table 7.6. An average 5-foot depth of liquid in the pond and one 0.08-inch diameter hole per acre was assumed. (The value in the brackets applies to one 0.445-inch diameter hole per acre.)

(4) Not Applicable.

(5) Calculated using equation derived by Giroud et al., 1989 and included in Section 7.5.2. A 0.1-foot liquid level in the seepage collection layer was assumed.

(6) Using the methods outlined in Chapter 7.

(7) Interpolated from Table 7.10.

Case 2 - Single-Lined Pond

The second case illustrates how the wastewater would be disposed of in the same desert climate, but in an area that overlies a better quality ground water that is suitable for industrial and stock watering. In this case, an unlined facility would not be acceptable, as it would result in a significant increase in the TDS of the ground water and would preclude it from being used for those purposes. A single-lined pond, therefore, was selected.

In performing the containment assessment for the lined facility, it is assumed that a single, high density polyethylene (HDPE) geomembrane liner is provided. In recognition of the fact that most geomembrane liners will have some imperfection when constructed, it is assumed that there is, on the average, one 0.08-inch diameter hole per acre. As outlined in Chapter 7 (Section 7.4.2), this is considered a reasonable scenario for determining potential leakage rates from a well-constructed geomembrane system. A worst-case scenario is also evaluated, assuming the geomembrane liner has one 0.445-inch (1 cm) diameter hole per acre.

The containment assessment (Table 3.9) indicates that it would take several hundred years for any seepage to reach ground water through flaws in the liner. This represents more than 10 times the operating life of the pond. Even if the seepage does reach the ground water, it is likely to be substantially diluted by the natural interstitial water contained within the unsaturated zone. The estimated total seepage volume over the operating life of the pond is 6 million gallons (9 million gallons for the worst-case scenario), which represents less than 1 percent of the total volume of interstitial water underlying the pond. For these reasons, a single geomembrane liner is considered to provide an adequate amount of protection.

Case 3 - Double-Lined Pond

The third example assumes that the wastewater is disposed of in a wetter climate than in the first two cases and that the pond overlies better quality ground water at a shallower depth. Calculations for a single geomembrane liner in this case indicated unacceptably high seepage rates, with an arrival time at the ground water table of approximately one year. Unmitigated, these conditions would have increased the TDS concentration in the ground water from 200 mg/L to over 3,000 mg/L and resulted in a significant impact to the beneficial uses of the water, which is assumed to be municipal supply. A double HDPE, geomembrane liner, therefore, was selected.

As shown in Table 3.9, the double geomembrane liner provides adequate ground water quality protection. Seepage to the unsaturated zone is reduced to less than 1 gpm, and the calculated seepage arrival time to the ground water is 5 to 10 years. However, the amount of dilution provided by the interstitial water in the unsaturated zone, i.e., 1,000 million gallons versus the 2 to 3 million gallons of seepage, and by the regional ground water flow under the pond (64 gpm), is expected to reduce the potential impacts from salinity and Metal A to insignificant levels. This assessment is conservative, as Metal A is expected to attenuate in the upper few inches of the soil underlying the liner. Furthermore, in calculating seepage losses from the lower liner, it was assumed that the entire surface of the liner is covered by a 0.1-foot depth of liquid due to seepage from the upper liner. At an actual waste management unit, it is anticipated that these upper liner leakages would occur only over a fraction of the surface area of the lower liner and that the corresponding lower liner seepage rates, therefore, would be a fraction of those shown in Table 3.9.

3.8.1.2 Tailings Impoundment Containment Assessments

As shown in Figure 3.2, the impoundment used in these examples is located approximately 2.5 miles upgradient of a potential ground water resource that contains water of suitable quality and quantity for municipal use. The small ground water quantities located in the geologic formation between the tailings impoundment and the ground water aquifer are not considered a resource. The tailings liquid contains elevated TDS, metal, and reagent concentrations that have the potential to adversely impact water quality (see Table 3.8).

Three different tailings impoundment cases have been analyzed. These are illustrated in Figure 3.3 and include wet (or slurry) disposal in a 100-acre unlined facility (Case A), wet disposal in a double-lined impoundment (Case B), and managed tailings deposition in a single-lined impoundment (Case C).

Managed tailings deposition as envisaged in this example includes: (1) a drainage layer placed under the tailings to reduce the hydraulic head on the underlying liner, and (2) the deposition of the tailings in thin layers sequentially in different subareas, which are allowed to dry and so reduce the moisture content of the waste.

In order to evaluate the impact of site geologic conditions, two alternative types of formations were used in each of the cases. The first assumes that the site is underlain by fractured bedrock formations with a low hydraulic conductivity and porosity, i.e., 10^{-6} cm/sec and 0.1 percent, respectively. The second assumes weathered bedrock or granular formations with a moderate hydraulic conductivity and porosity, i.e., 10^{-4} cm/sec and 20 percent, respectively. As mentioned previously, both geomembrane and clay liners are considered.

For purposes of demonstrating how much higher seepage migration rates can be in fractured rock versus a granular material, Case A is also evaluated for a 10^{-6} cm/sec clay formation underlying the impoundment and surrounding area.

The results of the containment assessment evaluations are summarized in Table 3.10 and are discussed below. The analytical methods used in calculating the seepage rates are summarized in Appendix C. Further assumptions made in calculating seepage rates during the operating period of the impoundment are described in the footnotes to Table 3.10. The analytical methods used are approximate. For the actual design of facilities, it may be necessary to utilize more elaborate forms of analysis.

Case A - Wet Tailings in an Unlined Impoundment

The unlined impoundment on 10^{-6} cm/sec bedrock has an insignificant amount of seepage which should not result in any impact on the receiving ground water resource (Table 3.10). The seepage collection sump located at the downstream toe of the containment dam would collect a portion of this seepage, thereby further reducing any seepage loss from the impoundment. In addition, laboratory evaluations indicate that both Metal B and Reagent A attenuate to some degree as seepage passes through the fractures in the bedrock. Therefore, additional containment features or redundancies should not be necessary.

The adequacy of the containment system is predicated on the condition that there are no highly fractured zones that connect the site directly with the ground water resource. Detailed geologic site investigations, which included geologic mapping as well as borehole and trench logging, indicated that no such features were present and that the joint and fracture systems were of a scale small enough not to provide interconnected zones of high conductivity over more than a maximum of a few hundred feet. These investigations included installation and hydraulic conductivity testing of boreholes, joint system interpretation and mapping, and several ground water pump tests.

For the formation with the 10^{-4} cm/sec hydraulic conductivity, higher seepage rates would occur, i.e., 17 gpm (Table 3.10). As the seepage travel time to the ground water resource is estimated at more than ten times the project life, it is likely that there would be a minimal impact on the ground

TABLE 3.10

EXAMPLE 2: RESULTS OF TAILINGS IMPOUNDMENT CONTAINMENT ASSESSMENTS

CASE	SUBSURFACE CONDITIONS			LINER TYPE	MAXIMUM SEEPAGE RATES(1) FROM IMPOUNDMENT TO GROUND WATER RESOURCE DURING OPERATION (gpm)	SEEPAGE CONTROLLED BY	TRAVEL TIME(1) TO GROUND WATER BODY (years)	COMMENTS
	Hydraulic Conductivity (cm/sec)	Formation Type	Effective Porosity (%)					
A	10^{-6}	Rock	0.1	None	0.17	Geologic Formation	260	Seepage rates are controlled by geologic formations and are acceptable. No liner system is required.
	10^{-6}	Soil	20	None	0.17	Geologic Formation	53,000	
	10^{-4}	Soil	20	None	17	Geologic Formation	530	Seepage rates are high, however, travel times are long. A lined facility may therefore only be required for these site conditions, if protection of ground water quality closer to the waste management unit is necessary.
B	10^{-6}	Rock	0.1	Double Clay(2)	0.15	Geologic Formation	293	Seepage rates controlled by natural geologic formation. Clay liner system provides only a marginal benefit.
				Double Geomembrane Liner	0.01(3)	Liner System	293	Geomembrane liner system reduces seepage to a greater extent than the clay liner system.
	10^{-4}	Soil	20	Double Clay(2)	6.4	Liner System	570	The clay liner system reduces the seepage rate to less than 1/2 the unlined condition. The seepage travel time is increased slightly.
				Double Geomembrane Liner	0.3(3)	Liner System	570	The geomembrane liner system reduces seepage to a greater extent than the clay liner system.
C	10^{-6}	Rock	0.1	Single Clay(2)	0.15	Geologic Formation	293	Seepage rates are controlled by the geologic formation. Managed deposition and the liner are of limited benefit.
				Single Geomembrane Liner	0.01(3)	Liner System	293	The geomembrane liner reduces seepage significantly more than the clay liner.
	10^{-4}	Soil	20	Single Clay(2)	6.4	Liner System	570	Managed deposition and a single liner provides equivalent protection as wet disposal on a double liner system.
				Single Geomembrane Liner	0.3(3)	Liner System	570	

(1) Assumes hydraulic head within tailings remains at the level experienced during operation throughout the duration of the stated travel times.
(2) Clay hydraulic conductivity of 10^{-7} cm/sec is assumed and a thickness of 3 feet.
(3) Based on one .08-inch hole per acre, interpolated from Table 7.7 assuming a 0.2-foot liquid level in the seepage collection layer.

water resource, particularly with attenuation of the important constituents. However, because of the likely impact on the quality of ground water in the area immediately downgradient from the tailings impoundment, this case may represent an unacceptably high risk.

There is a 200 fold increase in the seepage migration rate between the 10^{-6} cm/sec clay formation. This is due to the much lower effective porosity in the bedrock. In this example, this difference in seepage travel time is not an important issue. In other situations, however, it could well be an important design consideration.

Cases B and C

Case B involves wet tailings disposal in a double-lined impoundment, while managed tailings deposition in an unlined and single-lined facility is included in Case C. Important observations that can be drawn from the results of these containment assessments, also outlined in Table 3.10, are as follows:

- The natural geologic containment provided by the formations with a 10^{-6} cm/sec hydraulic conductivity provide adequate containment. The adequacy of the containment provided by the formations with a 10^{-4} cm/sec hydraulic conductivity depends on the extent to which local ground water immediately downgradient from the impoundment requires protection.
- Geomembrane liner systems perform better than the equivalent clay liner systems, provided the geomembrane systems are well constructed and the number and size of any perforations are within accepted limits.
- Managed tailings deposition, incorporating sequential layering and an underdrain, and operated effectively, coupled with a single liner system can provide the same level of containment as a double liner system.

3.8.2 ALTERNATIVE APPROACHES

This section compares the "systems approach" to designing waste management units, as outlined above, to the alternative "prescriptive approach."

The "systems approach" requires that the potential impact to beneficial uses of water at a particular mining operation be determined before the appropriate engineered containment is selected. The containment selected may range from relying on the site's climate and natural geologic formations, to requiring a highly-engineered containment system. Beneficial uses of water can be protected by maximizing utilization of natural site conditions, providing engineered containment systems when necessary, and applying scientific waste management technology.

The advantages of the systems approach are:

- It provides the means to attain a sufficient level of protection for the beneficial uses of water.
- It favors the selection of waste disposal sites with natural waste containment features, as they are likely to be more economic to develop. In the long term, these sites are also likely to provide more secure waste containment.
- It requires the designers of the waste management unit to fully evaluate the waste characteristics and site conditions. There is thus less likelihood of design errors or omissions occurring.
- It provides financial incentives for companies to develop in "low risk" environments, such as in arid climate areas.
- As it requires a scientific approach, the systems approach stimulates the ongoing development of waste disposal technology.

This approach necessarily requires site-specific information in order to select and design the appropriate waste containment system.

This disadvantages of the systems approach are:

- It may increase the cost of designing the waste management unit.
- It will require more regulatory review effort.

The goal of the alternative "prescriptive approach" is to provide standards that describe the type of containment that should be provided for a given set of conditions, such as the results of the classification. In order to limit the given conditions to a manageable quantity, the amount of containment required frequently is based solely on the intrinsic physical and chemical properties of the waste, with limited attention to site-specific factors, such as local geology, climate, distance to water resources, and waste management practices. Furthermore, this approach usually does not require evaluation of beneficial uses of potential receiving waters.

As discussed in Section 3.9 below, the current California mine waste regulations employ a prescriptive approach with specific exceptions. This can result in situations where the same standards are applied to a waste management unit located over shallow, high-quality ground water that is used for municipal supply, as are applied to a waste management unit where ground water is located several hundred feet below the surface and is of such low quality that it cannot be utilized. The prescriptive approach also results in the same containment standards for waste management units located in a humid climate, where the waste may be close to saturated for periods of time and contain substantial quantities of free liquid, as for those in arid areas where there may be little free liquid in the waste.

The advantages of the prescriptive approach are:

- Simple, easy to administer.
- In most cases, provides at least sufficient containment.

The disadvantages of the prescriptive approach are:

- It does little to promote selection of naturally occurring "lower risk" sites for waste management units.
- The designer and the regulatory reviewer of a waste management unit frequently do not assess or understand the actual risks posed to beneficial uses of water.
- It does not necessarily promote innovation and improvement to current waste disposal technology; in some instances, it may promote the application of out-of-date technology.

From a philosophical perspective, the prescriptive approach can be viewed as a simplistic systems approach. It can be made more comprehensive by adding, for example, containment categories and requirements that deal with site-specific issues, such as site geology, remoteness of water resources, and climate. One major difference remains, however. The influence of waste type, waste management practices, and site conditions on the prescribed containment requirements is prejudged, usually conservatively, and may not reflect accurately actual site conditions and containment requirements.

Furthermore, when the more comprehensive level of detail discussed above is added to the prescriptive approach, describing the exponentially increasing number of containment categories in a regulatory document becomes cumbersome and unmanageable. The generalized containment assessment approach embodied in the systems approach simplifies this considerably.

3.9 REGULATORY IMPLICATIONS

3.9.1 GENERAL

This section examines the current status of the California mine waste regulations, outlines some federal regulatory considerations, and concludes with recommendations for the inclusion of the systems approach, incorporating a performance goal, in mine waste regulations.

3.9.2 CURRENT CALIFORNIA REGULATIONS

With the passage of Assembly Bill 1413 (Tanner), regulation of mine waste (wastes from the extraction, beneficiation, and processing of ores and minerals) in California, to the extent that such wastes also are exempted from RCRA Subtitle C regulation by the Bevill Amendment, now is conducted primarily under the jurisdiction of the State Water Resources Control Board (State Board) and the nine Regional Boards, pursuant to the Porter-Cologne Water Quality Control Act. Such wastes no longer are regulated as hazardous by the DHS. Assembly Bill 1413 provides, however, that the Regional Boards may apply DHS Title 22, CCR criteria in evaluating a mine waste's potential impact on waters of the State.(13)

Mine waste disposal in California is regulated by Articles 1, 5, 7, and 8 of Title 23, Division 3, *Chapter 15* of the California Code of Regulations (*Chapter 15*). Article 1 includes general requirements for all waste discharges under *Chapter 15*. Article 5 outlines the water quality monitoring requirements, and Article 7 contains the regulations relating to classification of mine wastes and the waste containment requirements for discharges of mine waste to land. Article 8 deals with closure and post-closure maintenance.

The major thrust of the current *Chapter 15,* Article 7 is the provision of detailed prescriptive containment standards that outline requirements for each of the three mine waste groups and each mine waste facility type, i.e., mine (water) impoundment, tailings impoundment, or waste pile (overburden, waste rock, or heap leach residue). A summary of the prescriptive containment standards for the Group A, B, and C wastes and for the different types of mine waste management units is set forth in Table 3.11.

As can be noted from Table 3.11, the prescriptive containment requirements are determined by the intrinsic properties of the waste. However, there is a provision to allow for site-specific geologic conditions. Outer liners may be replaced by natural formations if the latter have the same degree of impermeability as the prescriptive liner requirements. It is of further interest to note that the requirements favor clay liners. With the continuous advancements being made in the manufacture and construction of geomembrane liners, and the superior containment performance of these liners (see Table 3.9 and Chapter 7), this may be biasing waste containment engineering toward older, and potentially less effective, technology.

---

(13) Incorporated in part (l) of California Water Code Section 13260 (Porter-Cologne).

TABLE 3.11

CURRENT PRESCRIPTIVE WASTE CONTAINMENT REQUIREMENTS: STATE OF CALIFORNIA[1]

WASTE GROUP	WASTE MANAGEMENT UNIT	GEOLOGIC SETTING	LINER(S) HYDRAULIC CONDUCTIVITY (cm/sec)	LEACHATE COLLECTION AND REMOVAL SYSTEM
A	Waste Pile	Natural geologic material[2] or single clay liner[3]	$\leq 1 \times 10^{-7}$	Required
	Surface Impoundment or Tailings Pond	Not applicable	Double liner system, both liners $\leq 1 \times 10^{-7}$ Outer liner: clay Inner liner: clay or geomembrane	Required[4]
B	Waste Pile	Natural geologic material[2] or single clay liner[3]	$\leq 1 \times 10^{-6}$	Required
	Surface Impoundment or Tailings Pond	Not applicable	Double liner system, both liners $\leq 1 \times 10^{-6}$ Outer liner: clay or natural material $\leq 1 \times 10^{-6}$ Inner liner: clay or geomembrane, or single replaceable clay liner[5]	Required[4]
C	Waste Pile, Surface Impoundment, or Tailings Pond	Not applicable	Not applicable	Not applicable

(1) California Code of Regulations, Title 23, Division 3, *Chapter 15*, Article 7.
(2) Geologic material must be of sufficient thickness to prevent vertical movement of fluid, including waste and leachate, from the waste management unit to the waters of the State so long as such wastes pose a threat to water quality.
(3) Geomembrane liner may be used for short-term containment (see Subsection 2572 [f][1] of this article).
(4) Liner and leachate collection and removal system for tailings pond must be able to withstand the ultimate weight of wastes.
(5) Single clay liner ($\leq 1 \times 1^{-6}$ cm/sec) for surface impoundment, to be removed before the last 25 percent (minimum 1 foot thickness) of liner is penetrated by fluid, including waste and leachate.

There are also provisions for engineered alternatives to, and exemptions from, the prescriptive standards. *Chapter 15,* Article 1, Section 2510, allows for an engineered alternative to the prescriptive standards, as follows:

- *"Unless otherwise specified, alternatives to construction or prescriptive standards contained in this chapter may be considered. Alternatives shall only be approved where the discharger demonstrates that:*
  - *The construction or prescriptive standard is not feasible as provided in Subsection (c) of this section, and*
  - *There is a specific engineered alternative that (A) is consistent with the performance goal addressed by the particular construction or prescriptive standard, and (B) affords equivalent protection against water quality impairment.*
- *To establish that compliance with prescriptive standards in this chapter is not feasible for the purposes of Subsection (b) of this section, the discharger shall demonstrate that compliance with a prescriptive standard:*
  - *Is unreasonable and unnecessarily burdensome and will cost substantially more than alternatives which meet the criteria in Subsection (b) of this section; or*
  - *Is impractical and will not promote attainment of applicable performance standards.*
- *Regional Boards shall consider all relevant technical and economic factors including, but not limited to, present and projected costs of compliance, potential costs for remedial action in the event that waste or leachate is released to the environment, and the extent of ground water resources which could be affected."*

Provisions for exemption from certain prescriptive requirements are included in Article 7, Section 2570(b). This section can exempt a mine waste pile from requirements for liners and leachate collection and removal systems if it clearly can be demonstrated to the Regional Board that leachate will not form in or escape from the waste management unit. However, extensive monitoring procedures may be required by the Regional Board in lieu of certain containment features. Contingency plans also are required and would have to be implemented if monitoring indicates that the disposal procedures are inadequate.

Furthermore, Section 2570(c) of Article 7 allows Regional Boards to exempt a Group A or B mine waste management unit from certain provisions of Article 7, such as the liner and leachate collection and removal requirements, if a comprehensive hydrogeologic investigation demonstrates that:

- *There are only very minor amounts of ground water underlying the area; or*
- *The discharge is in compliance with the applicable water quality control plan; and*

- *Either natural conditions or containment structures will prevent lateral hydraulic interconnection with natural geologic materials containing ground water suitable for agricultural, domestic, or municipal beneficial uses. There is no detectable vertical hydraulic interconnection between the natural geologic materials underlying the unit and natural geologic materials containing such ground water.*

While the above provision of the regulations might allow enough flexibility for regulatory personnel to use a systems approach in the design of containment systems, there are troublesome limitations.

Before an engineered alternative can be considered, however, a prescriptive standard must be found to be unfeasible because of cost and practical considerations. Thus, selecting an engineered alternative when the cost of the prescriptive requirement is similar, may not be allowed by a strict interpretation of the regulations. Furthermore, restricting selection of alternative containment measures to only those cases where the prescriptive requirements are impractical, can inhibit the selection of an engineered alternative.

3.9.3 RECOMMENDED REGULATORY APPROACH

In keeping with the goal of the protection of the beneficial uses of the potential receiving waters, a regulatory framework specifically should allow for application of the systems approach presented in this chapter.

Whether the systems approach is applied by regulatory personnel under an exemption process, or is specifically set forth in regulations, it should contain the following elements:

- A statement of a performance goal, i.e., the protection of the beneficial uses of the potential receiving waters.
- A list of the factors that should be considered in the site-specific analysis, such as:
  - Climatic conditions.
  - Chemical and physical properties of the waste and wastewaters.
  - Chemical and physical properties of the site's geologic formations and the materials proposed for liner system construction.
  - The proposed methods of waste handling and placement.
  - The nature, extent and beneficial uses of ground and surface water bodies underlying and adjacent to the proposed waste management unit.

- A containment assessment to evaluate potential seepage losses at critical stages in the life of the waste management unit and a determination of the amount and type of engineered containment necessary to meet the performance goal.
- Evaluation of the uncertainty associated with the parameters used in the containment assessment and a demonstration that adequate containment is provided for realistic worst-case combinations of these values.

Prescriptive requirements still could be used either when resources are not available to obtain sufficient site-specific information for the systems approach, or when there is a large degree of uncertainty associated with predicting waste characteristics and site conditions.

In California, the systems approach should be incorporated directly into the current *Chapter 15* mine waste regulations, and would be the next logical step in the evolution of the regulations. Alternatively, specific provisions for its use as part of the alternative engineering provisions would give more structure to the use of these provisions. In either case, inclusion of the systems approach at a minimum, would assure that current mine waste management unit design practices achieve a uniform standard of waste containment.

3.10 BIBLIOGRAPHY

American Society for Testing and Materials. *Annual Book of ASTM Standards, Vol. 4.08, Soil and Rock.* Philadelphia: ASTM. 1989.

Borer, Londa. Department of Chemistry, California State University, Sacramento. *Report to the California Department of Health Services on the California Assessment Manual, Waste Extraction Test.* 1988.

Calaveras Asbestos Ltd. *Monitoring Data for Waterborne Asbestos in the Stanislaus River.* Report to the California Regional Water Quality Control Board, Central Valley Region. 1987.

Calaveras Asbestos Ltd. Calaveras County Planning Department. *Final Environmental Impact Report for the Proposed Asbestos Containing Waste Landfill.* 1988.

California Mining Association. *Mine Waste in California - A Case for Regulation as Non-hazardous.* I. Hutchison and R. Hammond. January 1988.

Chatwin, Dr. Terrence and John J. Trepanowski. *Utilization of Soils to Mitigate Cyanide Releases.* Proceedings of the Third Western Regional Conference on Precious Metals, Coal and Environment. Rapid City, South Dakota. Society of Mining Engineers. September 23 - 26, 1987.

Dupont Corporation. *Personal Communication.* 1991.

Federal Register, Volume 55. No. 15, January 23, 1990. *Mine Waste Exclusion, Final Rules.* 40 CFR Parts 260, 261, 262. 2322-2354.

Giroud, J.P. and Bonaparte, R. *Leakage through Liners Constructed with Geomembranes, Part I: Geomembrane Liners. Part II: Composite Liners.* Geotextiles and Geomembranes, 8(1). 1989.

Hoye, Robert. *Final Report, Gold/Silver Heap Leaching and Management Practices That Minimize the Potential for Cyanide Release.* United States Environmental Protection Agency. EPA 600/2-88-002. 1987.

Ingles, J. and J.S. Scott. *State-of-the-Art of Processes for the Treatment of Gold Mill Effluents.* Preliminary Report of the Mining, Mineral and Metallurgical Processes Division, Industrial Programs Branch, Environmental Protection Programs Directorate, Environment Canada. July 1985.

Kleinman, Robert L.P. "Acid Mine Drainage." Engineering and Mining Journal 190: (July 1989): 16I-16N.

Leeper, G.W. *Managing the Heavy Metals on the Land.* New York: Marcel Dekker, Inc., 1978.

Marshack, Jon B. *A Compilation of Water Quality Goals.* Staff Report of the California Regional Water Quality Control Board, Central Valley Region. May 1989a.

Marshack, Jon B. *Designated Level Methodology for Waste Classification and Cleanup Level Determination.* Staff Report of the California Regional Water Quality Control Board, Central Valley Region. June 1989b.

Mudder, Terry I., Editor. *The Chemistry, Analysis, Toxicity and Treatment of Cyanide Wastewater.* AIME, SME short course presented in Las Vegas. February 1989.

Porter-Cologne Water Quality Control Act. California Water Code, Sections 13000 et seq. January 1981.

Resource Conservation and Recovery Act of 1976 (PL94-580, October 31, 1976; 1/2 U.S.C. 6901 Et Seq. Subtitle C and Subtitle D).

Smith, A.C., and Mudder, T. I., *The Chemistry and Treatment of Cyanidation Waste*, Mining Journal Books Ltd., London, England (In press, 1991).

Soderberg and Busch. *Design Guide for Metal and Nonmetal Tailings Disposal.* United States Bureau of Mines IC 8755. 1977.

Solid Waste Disposal Act (SWDA) Amendments, Section 3001. October 21, 1980.

Sonora Mining Company. Letter report from Mr. Dave Skolasinski, Director of Environmental Affairs. 4 pages. September 7, 1989.

United States Department of Agriculture. *Diagnosis and Improvement of Saline and Alkali Soils.* L.A. Richards, Editor. Agricultural Handbook No. 60. February 1954.

United States Bureau of Reclamation. *Linings for Irrigation Canals.* 1963.

United States Environmental Protection Agency. *Water Quality Criteria.* EPA R3-73-033. 1972.

United States Environmental Protection Agency. *Water-Related Environmental Fate of 129 Priority Pollutants, Volume I.* EPA-440/4-79-029a. 1979.

United States Environmental Protection Agency. *Quality Criteria for Water.* 1986.

United States Environmental Protection Agency. 40 CFR Parts 123-124, *National Pollutant Discharge Elimination System Permit Application Regulations for Storm Water Discharges.* Final Rule, November 16, 1990.

United States Environmental Protection Agency. *Method 1312 - Synthetic Precipitation Leaching Procedure.* Revision 0. November 1990b.

University of California at Berkeley Mine Waste Study Team. *Mine Waste Study, Final Report.* California State Water Resources Control Board. 1988.

Vick, Steven G. *Planning, Design, and Analysis of Tailings Dams.* New York: Wiley - Interscience. 1983.

CONTENTS

CHAPTER 4
PREDICTION OF ACID GENERATION POTENTIAL

PAGE NO.

LIST OF TABLES/LIST OF FIGURES 126

4.1 Introduction 127
4.1.1 Acid Rock Drainage 127
4.1.2 Purpose 128
4.1.3 Integration With Other Chapters 128
4.1.4 Study Participants 128
4.1.5 Sources of Information 129
4.1.6 Chapter Organization 129
4.2 Acid Rock Drainage 130
4.2.1 Introduction 130
4.2.2 Sources of Acid Rock Drainage 130
4.2.2.1 Waste Rock Management Units and Ore Stockpiles 132
4.2.2.2 Tailings Waste Management Units 132
4.2.2.3 Underground Mine Workings 132
4.2.2.4 Open Mine Pits 135
4.2.2.5 Spent Heap Leach Piles 135
4.2.2.6 Timing of ARD Formation 135
4.2.2.7 California Case Histories 139
4.2.3 Acid Forming Processes 139
4.2.3.1 Pyrite Oxidation 142
4.2.3.2 Geochemical Considerations 145
4.2.3.3 Oxidation of Other Base-Metal Sulfides 149
4.2.3.4 Oxidation of Other Forms of Sulfur 150
4.2.4 Acid Consuming Processes 150
4.2.4.1 Reactions With Carbonates 151
4.2.4.2 Reactions With Aluminosilicates and Other Silicates 154
4.2.4.3 Other Reactions 156

CONTENTS

CHAPTER 4
(Continued)

PAGE NO.

4.2.5 Reaction Kinetics 158
4.2.5.1 Acid Generation Reaction Rates 159
4.2.5.2 Acid Consuming Reaction Rates 160
4.2.6 Migration of Acid Rock Drainage 161
4.2.6.1 Introduction 161
4.2.6.2 Physical Controls 161
4.2.6.3 Chemical Controls 162
4.2.6.4 Biological Controls 163
4.2.7 Impact on the Environment 163
4.3 Acid Generation Prediction 163
4.3.1 Introduction 164
4.3.2 General Approach to Prediction 164
4.3.2.1 Prediction Methods 164
4.3.2.2 Objectives and Approach 165
4.3.3 Materials Sampling 166
4.3.3.1 Geology and Mineralization 166
4.3.3.2 Objectives of Sampling 166
4.3.3.3 Application of the Sampling Program 167
4.3.3.4 Sources and Handling of Samples 167
4.3.4 Geochemical Static Tests 169
4.3.4.1 Introduction 169
4.3.4.2 Static Test Techniques 169
4.3.4.3 Comparison of Static Prediction Techniques 172
4.3.4.4 Use of Static Tests 175
4.3.5 Geochemical Kinetic Tests 177
4.3.5.1 Introduction 177
4.3.5.2 Test Methods 177
4.3.5.3 Comparison of Several Kinetic Prediction Techniques 180
4.3.5.4 Conclusions 184

CONTENTS

CHAPTER 4
(Continued)

PAGE NO.

4.4 Acid Generation and Leachate Migration Control Measures 184
4.4.1 Approach to Control 184
4.4.2 Available Control Options 186
4.4.3 Testing for Design of Control Measures 186
4.5 In-Waste Monitoring 188
4.5.1 Objectives 188
4.5.2 Concepts 188
4.5.3 Monitoring Methods 189
4.5.3.1 Temperature 189
4.5.3.2 Gases 190
4.5.3.3 Soluble Products 191
4.6 Regulatory Implications 192
4.6.1 Motivation 192
4.6.2 Recommendations 192
4.6.2.1 General 192
4.6.2.2 Suggested Regulatory Approach 192
4.7 Bibliography 195

CONTENTS

CHAPTER 4
(Continued)

LIST OF TABLES

TABLE NO.	TITLE	PAGE NO.
4.1	Examples of ARD Quality at Several Mines	131
4.2	ARD Formation Lag Time and Duration for Selected Mines in British Columbia	138
4.3	List of Inactive California Mines, Some With Evidence of ARD	140
4.4	Summary of Common Sulfide Minerals and Their Oxidation Products	141
4.5	Multi-Staged Process of Pyritic Sulfide Oxidation	143
4.6	Potential Sources of Samples for Acid Generation Prediction	168
4.7	Summary of Results of Static Prediction Procedures	173
4.8	Summary Evaluation of Static Prediction Procedures	174
4.9	Comparison of Several Kinetic Prediction Techniques	181
4.10	Summary of Results of Kinetic Prediction Procedures	183
4.11	Comparison of Kinetic Prediction Procedures	185
4.12	Available ARD Control Options	187

LIST OF FIGURES

FIGURE NO.	TITLE	PAGE NO.
4.1	Conceptual Illustration of the ARD Process in a Waste Rock Management Unit	133
4.2	Conceptual Illustration of an ARD Process in a Tailings Waste Management Unit	134
4.3	Conceptual Illustration of the ARD Process in an Abandoned Underground Mine	136
4.4	Conceptual Illustration of the ARD Process in an Open Pit Mine	137

CHAPTER 4

PREDICTION OF ACID GENERATION POTENTIAL

4.1 INTRODUCTION

4.1.1 ACID ROCK DRAINAGE

Acid rock drainage (ARD)(1) is a term used to describe leachate, seepage, or drainage that has been affected by the natural oxidation of sulfide minerals contained in rock which is exposed to air and water. The principal ingredients for ARD formation are reactive sulfide minerals, oxygen, and water. The oxidation reactions responsible for the formation of ARD are often accelerated by biological activity. These reactions yield low pH (acidic) water that has the potential to mobilize heavy metals that may be contained in the geologic materials that are contacted. ARD can cause a detrimental impact on the quality of ground or surface water to which it discharges.

ARD likely has been associated with mines since mining began. For example, Norway, which has been a major producer of copper and sulfide concentrates for over 300 years, has a large number of mines with ARD problems. The coal mining industry in the eastern United States has been an identified source of ARD for decades. Also, ARD problems have been encountered at hard rock mines in western Canada, particularly British Columbia, at some of the historic mines in the northwestern United States, including Northern California, and in mountain states such as Colorado.

ARD is not necessarily confined to mining activities, but can occur naturally wherever sulfide-bearing rock is exposed to air and water. For example, there is documentation of naturally occurring springs that are acidic. These usually are located in the vicinity of outcrops of sulfide-bearing rock.

It is important to recognize that not all operations that expose sulfide-bearing rock will result in ARD. For example, acid drainage will not occur if the sulfide minerals are nonreactive, the rock contains sufficient alkaline material to neutralize any acid generated, or the climate is arid and there is insufficient rainfall infiltration to generate leachate.

---

(1) In many publications, ARD is also referred to as acid mine drainage (AMD).

There are other cases where ARD formation occurs, but then comes into contact with sufficient alkaline material to raise the pH as a result of neutralizing reactions. This "neutralized" ARD still may contain elevated total dissolved solids (TDS) and metal concentrations. However, the metal concentration levels generally will be lower than for the original ARD.

4.1.2 PURPOSE

This chapter discusses potential sources of ARD and the processes by which it is formed. The chapter describes test procedures that can be used to determine the ARD potential of mine wastes. The objective of these procedures is to determine the extent of acid generation in order to assist in characterizing the waste and to determine the degree of containment or management required when planning or designing waste management units. The types of containment or waste management measures that can be utilized to control ARD are also briefly summarized. Finally, procedures are recommended for identifying and regulating mine waste which has ARD potential.

The emphasis of this chapter is on characterizing ARD conditions. Therefore, the evaluation of the types of containment or management measures that can be implemented to reduce or eliminate ARD is not covered in a great amount of detail. For more information on these aspects of ARD, the reader is referred to the references listed in the bibliography, Section 4.7.

4.1.3 INTEGRATION WITH OTHER CHAPTERS

The information contained in this chapter can be used for determining the threat to water quality posed by ARD. It can be used in the "systems" approach to designing a mine waste management unit, outlined in Chapter 3. Characterization of ARD leachate quality also is required when determining the degree to which natural site conditions can mitigate the impacts of seepage by attenuation, as discussed in Chapter 5. Finally, information on ARD generation is required when designing for closure of a waste management unit, as outlined in Chapter 8.

4.1.4 STUDY PARTICIPANTS

This study was conducted by the following individuals:

- Adrian Smith, Ph.D., ASCI
- Andy Robertson, Ph.D., P.E., Steffen Robertson and Kirsten (B.C.), Inc.
- John Barton-Bridges, P.E., Steffen Robertson and Kirsten (U.S.), Inc.
- Ian P. G. Hutchison, Ph.D., P.E., Environmental Solutions, Inc.

4.1.5 SOURCES OF INFORMATION

The information presented in this chapter has been gathered from published technical papers, reports produced by government agencies, universities and consultants, mining associations, and data obtained from operating mines. A detailed bibliography is included in Section 4.7.

Government and Regulatory Agencies

- United States Environmental Protection Agency (EPA)
- United States Bureau of Mines (USBM)
- Environment Canada, Environmental Protection Services (EPS)
- Energy, Mines and Resources Canada

Universities and Research Organizations

- University of South Carolina
- University of West Virginia
- University of California, Berkeley
- British Columbia Research, Inc.

Mining Associations

- American Mining Congress (AMC)
- West Virginia Surface Mine Drainage Task Force
- West Virginia Mining and Reclamation Association
- Province of British Columbia Acid Mine Drainage Task Force (B.C. AMD)

4.1.6 CHAPTER ORGANIZATION

Section 4.2 of this chapter deals with different aspects of the ARD generation process and includes descriptions of acid generation, acid consumption, reaction kinetics, leachate migration processes, potential impacts on the environment, and case histories. Prediction of acid generation is described in Section 4.3, which includes a description of a sampling methodology and suitable geochemical static and kinetic laboratory tests. Section 4.4 contains a summary of acid generation and leachate migration control measures, while Section 4.5 discusses the monitoring of acid generation and migration. Possible approaches for incorporating requirements for the testing of ARD in mine waste management regulations are outlined in Section 4.6, and a bibliography is included in Section 4.7.

4.2 ACID ROCK DRAINAGE

4.2.1 INTRODUCTION

This section on ARD is divided into a series of subsections which deal with the important aspects of the ARD process. Subsection 4.2.2 provides a general description of the types of mines and mine wastes that represent potential sources of ARD. Subsection 4.2.3 presents details of the acid generating process, including chemical and biochemical oxidation of sulfides, anaerobic reactions, and oxidation of other forms of (nonsulfide) sulfur. The acid consuming processes are described in Subsection 4.2.4, and include reactions with carbonate, silicates, and other substances that can lead to a reduction or elimination of the acid formed. Subsection 4.2.5 discusses reaction kinetics, which are important because the relative rates at which the acid forming and acid consuming reactions occur play a major role in the net amount of acid that ultimately is generated. The final subsections, 4.2.6 through 4.2.7, deal with the ARD migration process, and the potential impacts of ARD on the environment.

4.2.2 SOURCES OF ACID ROCK DRAINAGE

ARD may occur naturally, or as a result of human intervention, where sulfidic rock has been exposed as a result of construction, mining, or other activities. The potential sources of ARD from mining operations include:

- Waste rock management units
- Ore stockpiles
- Tailings waste management units
- Underground mine workings
- Open mine pits
- Spent heap leach piles

ARD typically occurs when all of the following conditions are met:

- The waste contains enough sulfide to react chemically and biologically and form acid leachate at a rate faster than it can be neutralized by any alkaline compounds contained in the waste.
- The physical properties of the waste are such that they allow both water and oxygen to permeate in sufficient quantities to support the chemical and biological reactions.
- The climate is humid enough to cause rainfall infiltration and/or the waste is located such that surface water passes through it and transports the acidic drainage into the environment.

Typical constituent concentrations associated with ARD are provided in Table 4.1.

TABLE 4.1

EXAMPLES OF ARD QUALITY AT SEVERAL MINES[1]

PARAMETER[2]	SEEPAGE FROM AN ABANDONED URANIUM MINE TAILINGS POND IN ONTARIO, CANADA	WASTE ROCK DUMP SEEPAGE FROM AN ACTIVE SILVER MINE IN BRITISH COLUMBIA, CANADA	MINE WATER FROM UNDERGROUND COPPER MINE IN BRITISH COLUMBIA, CANADA
pH	2.0	2.8	3.5
Sulfate	7,440.0	7,650.0	1,500.0
Acidity (as $CaCO_3$)	14,600.0	43,000.0	--
Iron	3,200.0	1,190.0	10.6
Manganese	5.6	78.3	6.4
Copper	3.6	89.8	16.5
Aluminum	588.0	359.0	--
Lead	0.67	2.0	0.1
Cadmium	0.05	0.5	0.143
Zinc	11.4	53.2	28.5
Arsenic	0.74	25.0	0.05
Nickel	3.5	8.0	0.06

(1) Reference: Paine, 1987.
(2) Units are mg/L except pH.

4.2.2.1 Waste Rock Management Units and Ore Stockpiles

Waste rock management units, ore stockpiles, and spent ore piles which contain reactive sulfides can be significant sources of ARD if water is available to transport the products of acid generation to the environment. This can occur in humid climates where the piles are exposed to precipitation, runoff, and possibly even ground water flow. The chemical and physical properties of the waste rock will affect the quality of the ARD that is formed, as well as changes in quality over time.

A conceptual illustration of the ARD process in a typical waste rock management unit is illustrated in Figure 4.1. The figure shows typical water flow pathways through the waste rock due to rainfall infiltration, surface runoff and ground water inflow. A typical air flow pattern is also shown. The microscopic view of the individual rock particles illustrates how water flow pathways can pass through zones of acid generation and neutralization.

4.2.2.2 Tailings Waste Management Units

The potential for acid generation in sulfide-rich tailings has long been recognized. However, this potential may not always be evident during the operating period of a mine. This is due to: (1) the fact that tailings often are deposited under water or in a saturated condition, thus limiting the supply of oxygen, (2) the low permeability of tailings deposits which may limit the passage of water and air through the waste, and/or (3) alkaline reagents which may be used in the extraction process that initially may neutralize any ARD that is formed. Therefore, the critical period frequently occurs after closure, when the tailings may be drained, allowing oxygen entry. ARD can develop once the sulfide materials have overcome the neutralization capacity of the tailings. A typical ARD formation process in a tailings waste management unit is shown in Figure 4.2. Water and oxygen infiltrating from the surface result in sulfide oxidation that gradually penetrates into the tailings material. The resultant ARD passes through tailings material below this oxidizing zone and may be neutralized by the residual process reagents. Significant ARD seepage from the tailings unit will occur when the ARD process overwhelms the neutralization process. This can occur many years after mine closure.

4.2.2.3 Underground Mine Workings

Underground mines likely to result in ARD are those where substantial amounts of sulfide minerals are left in the ground during mining and are located above a water table which has been lowered by the mining activities. The mines in which these conditions are most prevalent are those located in

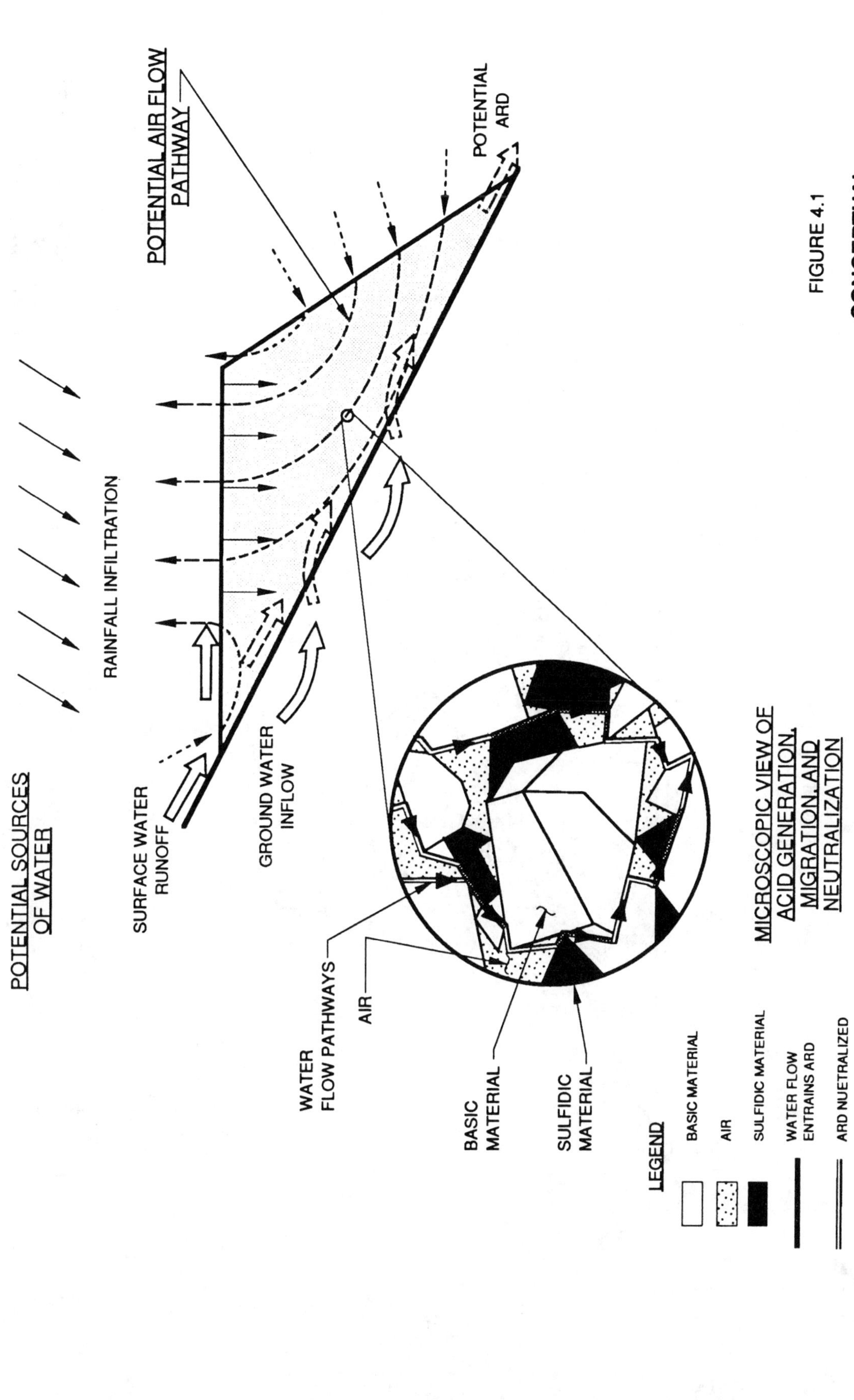

FIGURE 4.1

CONCEPTUAL ILLUSTRATION OF THE ARD PROCESS IN A WASTE ROCK MANAGEMENT UNIT

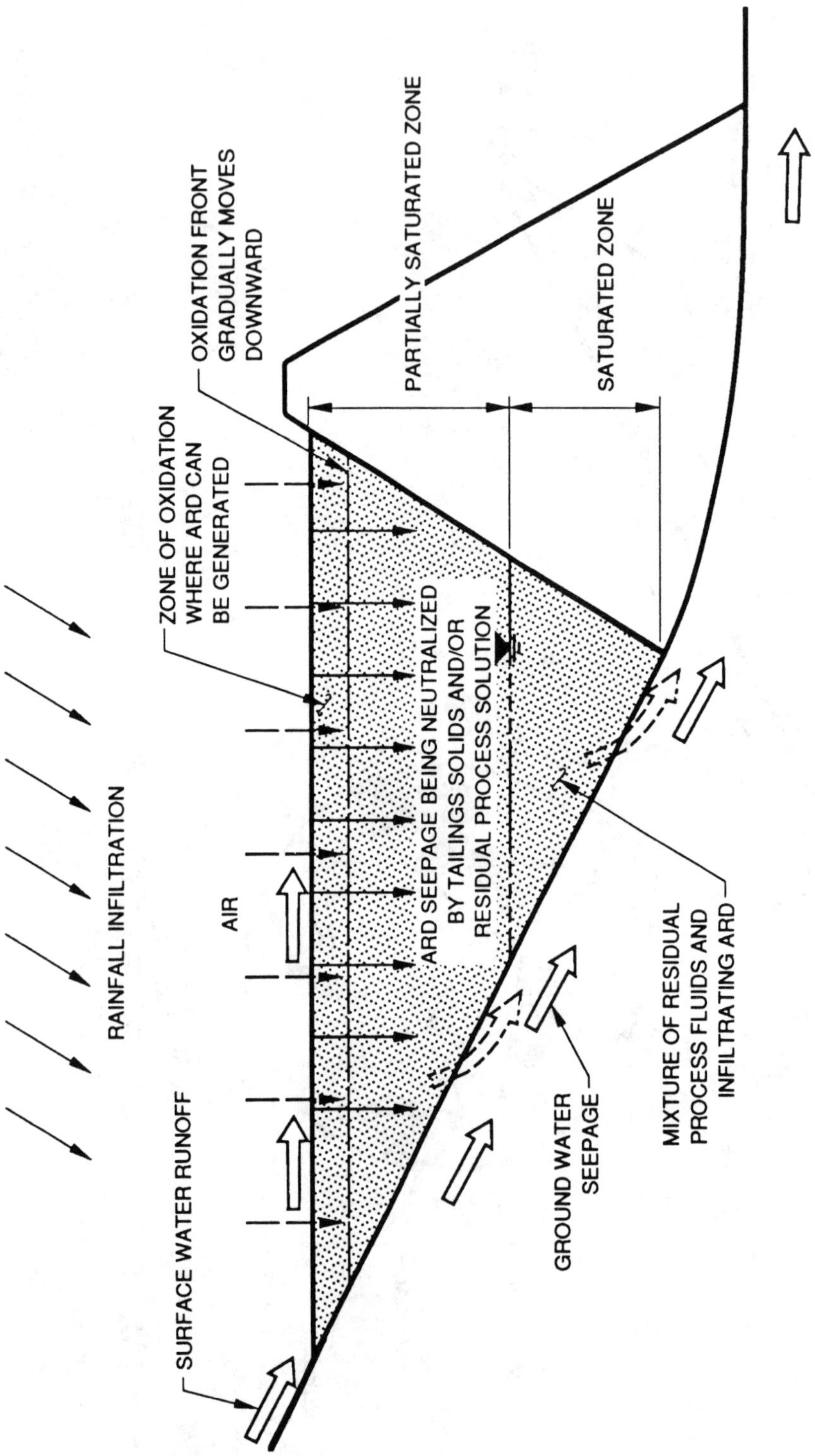

FIGURE 4.2

CONCEPTUAL ILLUSTRATION OF AN ARD PROCESS IN A TAILINGS WASTE MANAGEMENT UNIT

mountainous terrain, above a valley floor. Underground workings usually result in a ground water table that has been lowered significantly and permanently. They also can expose sulfides to oxygen and infiltrating water. Furthermore, mining contributes to the fracturing of the rock, exposing additional surface area to oxidation and promoting infiltration of rainfall.

A conceptual illustration of the ARD formation process in an underground mine is shown in Figure 4.3.

4.2.2.4 Open Mine Pits

The large areas of exposed rock inherent in open pits also can result in ARD if lowering of the ground water level has occurred and if sulfide rock is exposed to air and infiltration. Long-term slope deterioration can result in continual exposure of fresh rock to the natural elements and, hence, ARD generation. Figure 4.4 shows some typical conditions under which ARD can form in a pit.

4.2.2.5 Spent Heap Leach Piles

Heap leach operations include cyanide-leach for gold recovery, acid-leach for base-metal recovery, and bioleach processes. These operations result in piles of spent ore after operations have ceased. The spent heap leach piles can be a source of ARD, particularly those associated with low pH leach solutions. Cyanide leach operations are less likely to present ARD conditions, as ores containing sulfide are not usually amenable to cyanide leaching. Furthermore, if sulfides are leached, the process is conducted with a high pH solution, which introduces a significant neutralizing factor.

4.2.2.6 Timing of ARD Formation

The initial release of ARD can occur anywhere from a few months to many decades after the initial disturbance or deposition of the sulfide-containing material. The time lag is a function of the amount of water that passes through the material, the rate at which the sulfide oxidizes, the amount of neutralizing material available, and the neutralization reaction rates. Table 4.2 describes the timing associated with some of the mines in British Columbia and lists the known durations of ARD.

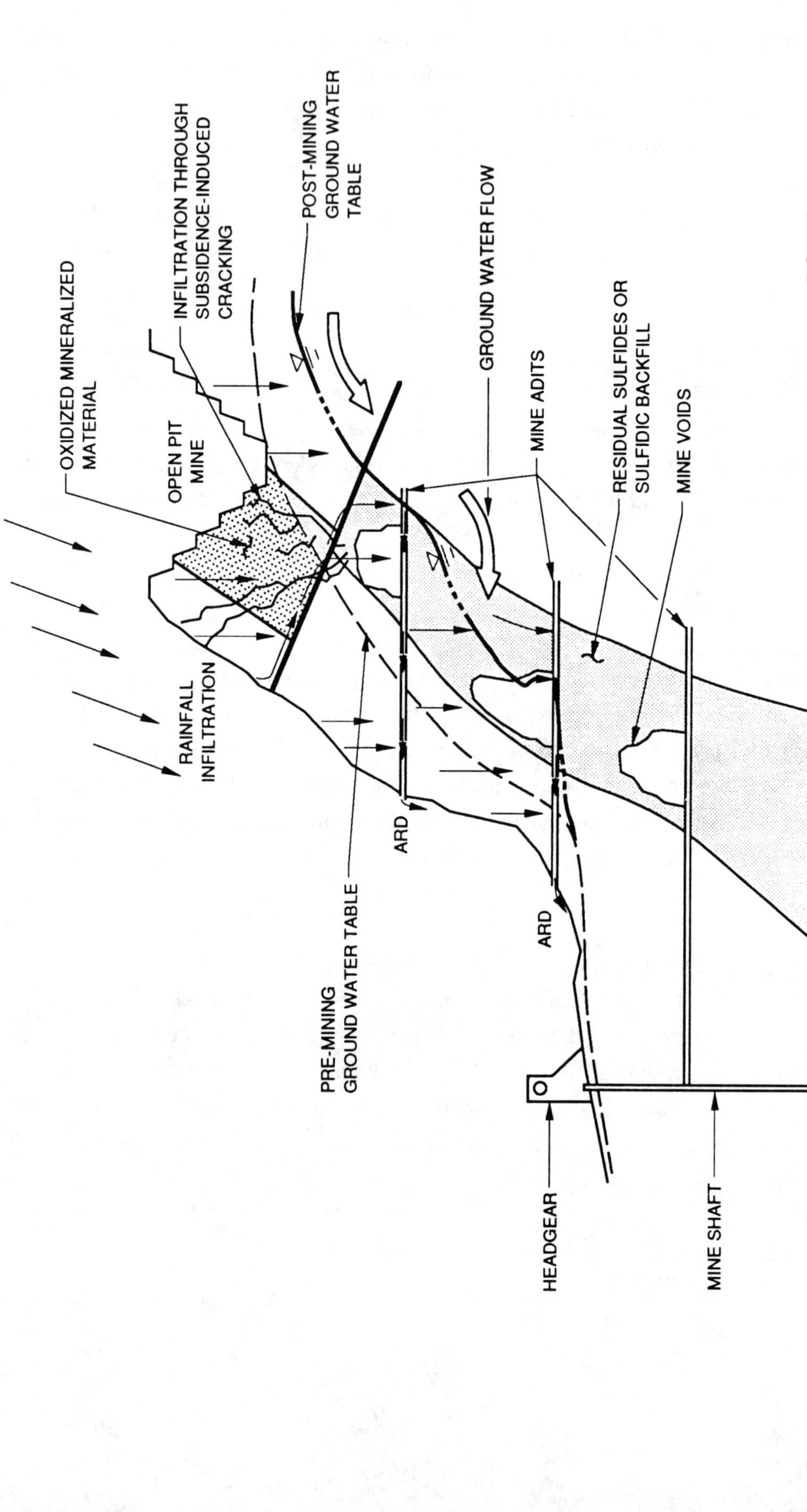

FIGURE 4.3

CONCEPTUAL ILLUSTRATION OF THE ARD PROCESS IN AN ABANDONED UNDERGROUND MINE

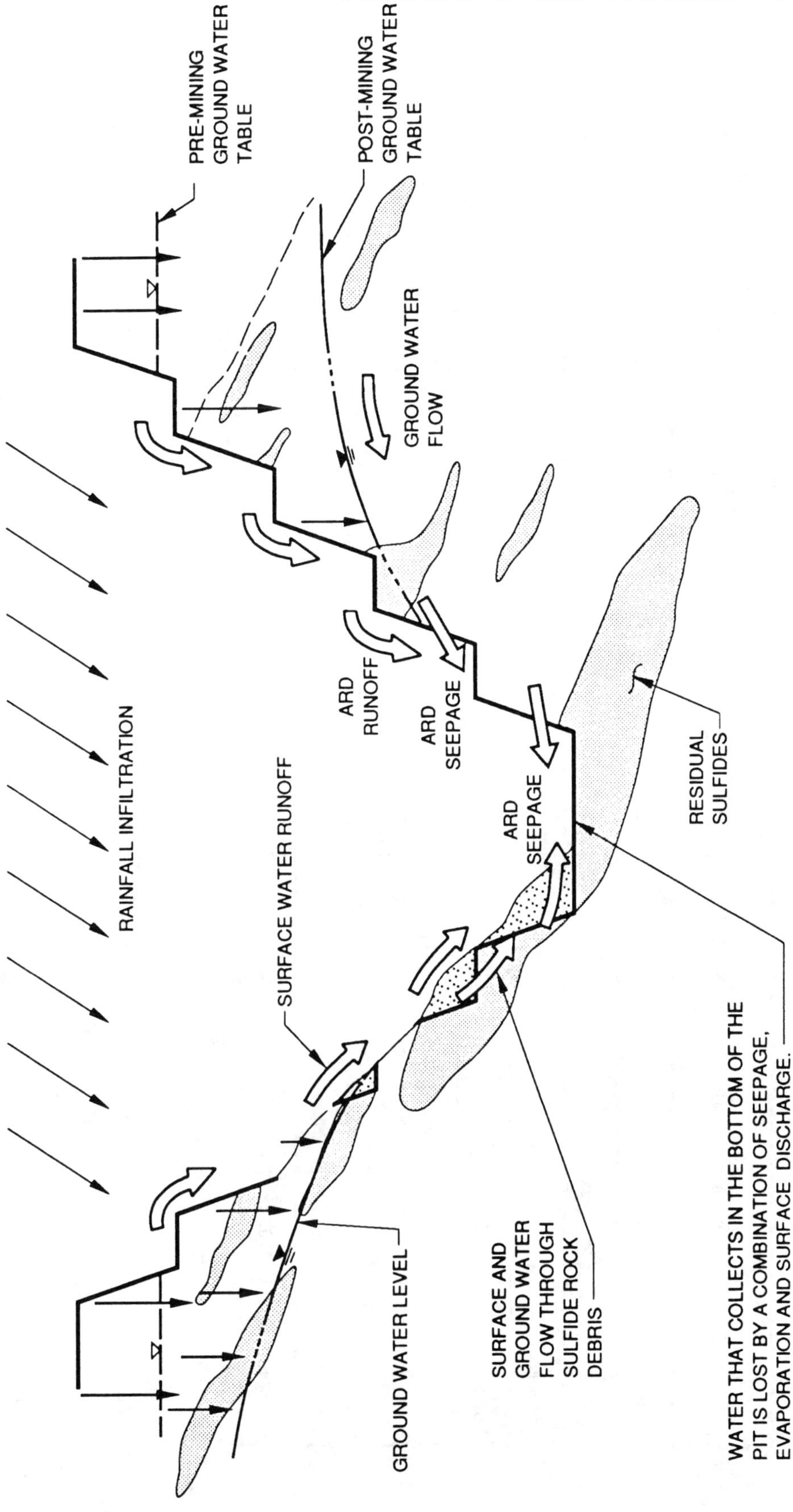

FIGURE 4.4

CONCEPTUAL ILLUSTRATION OF THE ARD PROCESS IN AN OPEN PIT MINE

TABLE 4.2

ARD FORMATION LAG TIME AND DURATION FOR SELECTED MINES IN BRITISH COLUMBIA[(1)]

MINE	TYPE OF ARD	MINE START-UP DATE	MINE ABANDONMENT OR CLOSURE DATE	DATE OF ARD APPEARANCE	TIME OF ARD APPEARANCE (years)	DURATION OF ARD (years)
Sullivan	Minewater	1909	Operating	Before 1950	Less than 41	Greater than 39
Anyox	Minewater	1914	1936	Before 1972	Less than 36	Greater than 17
Brittania	Minewater	1905	1974	Before 1928	Less than 23	Greater than 61
Duthie	Tailings seepage	1952	1954	1980	Less than 26	Greater than 61
Westmin	Waste rock leachate	1966	Operating	Before 1979	Less than 13	Greater than 9
Equity	Waste rock leachate	1980	Operating	1981	1	8
Gibraltar	Waste rock leachate	1971	Operating	1982	11	7
Mt. Washington	Waste rock leachate	1964	1966	Before 1972	Less than 6	Greater than 17

(1) Reference: Errington and Ferguson, 1987.

4.2.2.7 California Case Histories

Table 4.3 provides a list of historic mines in California where there is evidence of adverse impacts to ground or surface water quality from ARD. This information was obtained from a statewide study of environmental issues associated with mining, conducted by the University of California at Berkeley, Mine Wasting Study Team (U.C. Berkeley, 1988). This evaluation indicated that 8 of the 21 listed mines had water quality problems due to ARD. These problems at the historic mines resulted from outdated mine waste management practices and so are not indicative of conditions at modern mines. These problems do, however, reinforce the need for a sound understanding of ARD so that it can be predicted, included in mine planning, and appropriately managed.

4.2.3 ACID FORMING PROCESSES

The acid generation process in mine waste commonly is stated to be "pyrite plus air and water yields acid." However, this is a broad simplification of a relatively complex chemical and biological process. Despite a large amount of research effort, as illustrated by Gleeson and Russell (1976), not all of the relationships between the various controls on sulfide oxidation are as yet fully understood.

Pyrite and marcasite are abundant sulfides in most sulfide-bearing mineral deposits. The oxidation of these sulfides is the major contributor in acid generation from mine wastes. However, as summarized by Smith and Barton-Bridges (1991), studies over the past 5 to 10 years have indicated that other processes involving reactions of other sulfides and sulfur forms, such as thiosalts, and anaerobic reactions also play significant roles in acid generation. Table 4.4 contains a summary of the common sulfide minerals and their oxidation products.

The aspects that need to be addressed in order to cover the broad framework of acid generating processes include the following:

- Oxidation of iron sulfide (pyrite and marcasite).
- Oxidation of other base-metal sulfides.
- "Anaerobic" acid generating chemical reactions.
- Oxidation of other sulfur forms (e.g., sulfur, thiosulfates and polythionates).

These are discussed in more detail in the following sections.

TABLE 4.3

LIST OF INACTIVE CALIFORNIA MINES, SOME WITH EVIDENCE OF ARD [(1)]

INACTIVE MINES REVIEWED IN THE STUDY	ENVIRONMENTAL ISSUES IDENTIFIED	ENVIRONMENTAL ISSUES DUE TO ARD
Atlas	●	
Coalinga	●	
Iron Mountain	●	●
Balaklala	●	●
Mammoth	●	●
Golinsky	●	●
Sutro	●	●
Leviathan	●	●
Walker	●	●
Sulphur Bank	●	
Wright Mountain		
Engel		
Grey Eagle	●	●
Buena Vista		
Klau		
Bully Hill/Rising Star		
Afterthought		
Greenhorn		
New Idria		
Green Mountain		
Miracle		

(1) Reference: U. C. Berkeley, 1988.

TABLE 4.4

SUMMARY OF COMMON SULFIDE MINERALS AND THEIR OXIDATION PRODUCTS[1]

MINERAL	COMPOSITION	AQUEOUS END PRODUCTS OF COMPLETE OXIDATION[2]	POSSIBLE SECONDARY MINERALS FORMED AT NEUTRAL PH AFTER COMPLETE OXIDATION AND NEUTRALIZATION[3]
Pyrite	FeS_2	Fe^{3+}, SO_4^{2-}, H^+	Ferric hydroxides and sulfates; gypsum.
Marcasite	FeS_2	Fe^{3+}, SO_4^{2-}, H^+	Ferric hydroxides and sulfates; gypsum.
Pyrrhotite	$Fe_{1-x}S$[4]	Fe^{3+}, SO_4^{2-}, H^+	Ferric hydroxides and sulfates; gypsum.
Smythite, Greigite	Fe_3S_4	Fe^{3+}, SO_4^{2-}, H^+	Ferric hydroxides and sulfates; gypsum.
Mackinawite	FeS	Fe^{3+}, SO_4^{2-}, H^+	Ferric hydroxides and sulfates; gypsum.
Amorphous	FeS	Fe^{3+}, SO_4^{2-}, H^+	Ferric hydroxides and sulfates; gypsum.
Chalcopyrite	$CuFeS_2$	Cu^{2+}, Fe^{3+} SO_4^{2-}, H^+	Ferric hydroxides and sulfates; copper hydroxides and carbonates; gypsum.
Chalcocite	Cu_2S	Cu^{2+}, SO_4^{2-}, H^+	Copper hydroxides and carbonates; gypsum.
Bornite	Cu_5FeS_4	Cu^{2+}, Fe^{3+} SO_4^{2-}, H^+	Ferric hydroxides and sulfates; copper hydroxides and carbonates; gypsum.
Arsenopyrite	FeAsS	Fe^{3+}, AsO_4^{3-}, SO_4^{2-}, H^+	Ferric hydroxides and sulfates; ferric and calcium arsenates; gypsum.
Realgar	AsS	AsO_4^{3-}, SO_4^{2-}, H^+	Ferric and calcium arsenates; gypsum.
Orpiment	As_2S_3	AsO_4^{3-}, SO_4^{2-}, H^+	Ferric and calcium arsenates; gypsum.
Tetrahedrite and Tennenite	$Cu_{12}(Sb,As)_4S_{13}$	Cu^{2+}, SbO^{3-}, AsO_4^{3-}, SO_4^{2-}, H^+	Copper hydroxides and carbonates; calcium and ferric arsenates; antimony materials; gypsum.
Molybdenite	MoS_2	MoO_4^{2-}, SO_4^{2-}, H^+	Ferric hydroxides; sulfates; molybdates, molybdenum oxides; gypsum.
Sphalerite	ZnS	Zn^{2+}, SO_4^{2-}, H^+	Zinc hydroxides and carbonates; gypsum.
Galena	PbS	Pb^{2+}, SO_4^{2-}, H^+	Lead hydroxides, carbonates, and sulfates; gypsum.
Cinnabar	HgS	Hg^{2+}, SO_4^{2-}, H^+	Mercuric hydroxide; gypsum.
Cobaltite	CoAsS	Co^{2+}, AsO_4^{3-}, SO_4^{2-}, H^+	Cobalt hydroxides and carbonates; ferric and calcium arsenates; gypsum.
Niccolite	NiAs	Ni^{2+}, AsO_4^{3-}, So_4^{2-}, H^+	Nickel hydroxides and carbonates; ferric, nickel and calcium arsenates; gypsum.
Pentlandite	$(Fe,Ni)_9S_8$	Fe^{3+}, Ni^{2+}, SO_4^{2-}, H^+	Ferric and nickel hydroxides; gypsum.

(1) Reference: B.C. AMD Task Force, 1989.

(2) Intermediate species such as ferrous iron (Fe^{2+}) and $S_2O_3^{2-}$ may be important.

(3) Depending on overall water chemistry, other minerals may form with, or instead of, the minerals listed here.

(4) Where x = 0 to 0.2.

4.2.3.1 Pyrite Oxidation

Introduction

Pyrite oxidation can occur directly from the reaction with air and water, or indirectly from the reaction with ferric iron. At mining sites, direct oxidation processes and the associated biochemical reactions that occur are considered to be the major contributors to ARD. However, depending on site-specific conditions, the indirect oxidation process also can influence the production of ARD. The direct and indirect oxidation and biochemical processes are discussed below. In the context of this discussion, pyrite is assumed to be hydrogeochemically and geochemically synonymous with all other iron disulfides (e.g., marcasite).

Direct Oxidation

The direct air and water oxidation of pyrite is described by the first three interdependent reactions included in Table 4.5, which have become the "classic" explanation of the process.

Indirect Oxidation

There is an additional reaction (Williams et al, 1982) where trivalent ferric iron in solution will itself oxidize pyrite, resulting in acid generation, but without oxygen involvement. The equation for this process is listed as Reaction 4 in Table 4.5.

The reaction products include divalent ferrous iron in solution, acidity, and sulfate. It is noted that, while trivalent iron in solution is limited, except in low pH conditions (less than pH 4), a little ferric iron is a powerful pyrite oxidant. Comparison of the air/water and ferric iron oxidation of pyrite shows that the ratio of pyrite to acidity (H+ ions) is substantially different in the two reactions. Air/water oxidation produces four moles of H+ (two moles of sulfuric acid) for each mole of pyrite (Reaction 1), while ferric iron oxidation produces 16 moles of H+ (eight moles of sulfuric acid - Reaction 4). Therefore, if ferric iron oxidation is at all possible, it can produce substantially more "acidity" when compared to air/water oxidation.

Previous explanations of the ability of ferric iron to oxidize pyrite have concentrated on the interdependent system of pyrite-divalent iron-trivalent iron (Kleinmann et al., 1981). As illustrated in Table 4.5, Reaction 4 is limited by the rate at which Reaction 3, ferrous iron oxidation to ferric iron, can produce quantities of ferric iron in solution. This "closed-loop" approach assumes there is no external source of ferric iron, and that ferric iron can be generated only from ferrous iron in the interconnected reaction (Singer and Stumm, 1970). However, it is evident from visual

TABLE 4.5

MULTI-STAGED PROCESS OF PYRITIC SULFIDE OXIDATION[1]

REACTIONS

1. $2FeS_2 + 7\,O_2 + 2H_2O = 2Fe^{2+} + 4SO_4^{2-} + 4H^+$

2. $4Fe^{2+} + 10\,H_2O + O_2 = 4Fe(OH)_3 + 8H^+$

3. $2Fe^{2+} + O_2 + 2H^+ = 2Fe^{3+} + H_2O$

4. $FeS_2 + 14\,Fe^{3+} + 8\,H_2O = 15\,Fe^{2+} + 2\,SO_4^{2-} + 16\,H^+$

STAGED PYRITIC SULFIDE OXIDATION

STAGE 1

Reaction 1 proceeds both abiotically and by direct bacterial oxidation.
Reaction 2 proceeds abiotically and slows as pH falls.

Chemical Conditions: pH above approximately 4.5, high sulfate and low iron concentrations, little or no acidity.

STAGE 2

Reaction 1 proceeds both abiotically and by direct bacterial oxidation.
Reaction 2 proceeds at a rate determined primarily by the activity of T. ferro-oxidans.

Chemical Conditions: Approximate pH range of 2.5 to 4.5, high sulfate levels, acidity, total iron increasing, and a low Fe^{3+}/Fe^{2+} ratio.

STAGE 3

Reaction 3 proceeds at a rate totally determined by activity of T. ferro-oxidans.
Reaction 4 proceeds at a rate primarily determined by rate of Reaction 3.

Chemical Conditions: pH below approximately 2.5, high sulfate levels, acidity, total iron and high Fe^{3+}/Fe^{2+} ratio.

(1) Reference: Kleinmann et al., 1981.

observation, chemical analysis of waste rock and tailings, and analysis of water samples from mining areas, that there can be many other sources of iron. The above limitations, therefore, frequently may not apply at mining sites.

A further aspect that has not received much attention is the effect of suspended, colloidal, ferric iron in solution. It is common for ferric iron, formed at values above approximately pH 3, to remain in suspension as sub 0.45 micron, colloidal particles which do not readily nucleate and precipitate out. Such particles can be in intimate contact with pyrite and promote oxidation.

Work by Sugio et al. (1985), reviewed by Silver (1987), showed that it was possible to produce acid by ferric chloride oxidation of sulfur, namely:

$$S + 6\,FeCl_3 + 4\,H_2O = H_2SO_4 + 6\,FeCl_2 + 6\,HCl$$

Elemental sulfur and its analogs are intermediate products of many of the reactions or partial reactions in the oxidation of sulfides, while chlorides are ubiquitous in mine wastes. Spent ore from heap leach operations can contain elevated chloride levels due to solution recycling, as will tailings where the decant water is reclaimed from the tailings and recycled to the mill. The use of hypochlorite to destroy cyanide in gold recovery also results in increased chloride concentrations. Ferric chloride solutions are very acidic, however, and are likely to be of relevance only in low pH systems. However, the acid production process is anaerobic and can occur without the presence of oxygen in the system.

The implications of the above discussion are as follows. When interpreting acid generation data it is important to consider acidity, loss of alkalinity, pH, and sulfate, as these are the results of sulfide oxidation. However, it also must be recognized that the behavior of iron may modify the pyrite oxidation process. This may result in a differing manifestation of the data and a need for reinterpretation.

The discussion so far has considered pyrite oxidation only in terms of its basic acid-producing reactions. These reactions may be influenced, modified, or inhibited by numerous factors, primarily biochemical or geochemical, as discussed below.

Biochemical Process

Earlier studies (Walsh and Mitchell, 1975) indicated that pyrite oxidation proceeded as a chemical process until the system pH dropped to about 3.5 to 4. At this level, the process becomes biotically catalyzed by the bacterium Thiobacillus ferro-oxidans (T. ferro-oxidans). It later was demonstrated (Kleinmann et al., 1981) that this bacterium was also effective in the first stage of the reaction (see Table 4.5) at pH values greater than 3.5 to 4, and that it catalyzed (or accelerated) the reaction. Knapp (1987) showed that at lower pH values of around 3 to 3.5 the rate of the catalyzed reaction is much faster than the abiotic rate. At higher pH values its effect was still present, but the rate of increase was less than at the higher pH values.

The bacteria T. ferro-oxidans appears to be widespread in natural environments. It therefore can be expected to play a role wherever acid generation occurs. However, as with other bacteria, it is likely to be adversely affected by low temperatures which limit both its efficiency and effect.

4.2.3.2 Geochemical Considerations

This section discusses how factors such as pH, oxygen, alkalinity, pyrite abundance and surface area of pyrite grains, temperature, microenvironments, pyrite structural defects, and trace elements affect acid generation from pyrite. In general, these discussions also would apply to the other, less investigated, forms of pyrite.

Much of this discussion is based on the work of Williams et al. (1982) and subsequent developments and modifications by other investigators. A fundamental premise developed by these investigations is that the initial rate of acid production at a pH greater than about 4.5 appears to determine whether significant levels of acid ultimately will be produced. A high initial rate indicates a high ultimate rate of production. While this relationship may be intuitive, it cannot be considered self evident, as the individual factors that influence acid generation vary with time. As the development of the ultimate acid generation rate can involve a significant number of years, this relationship is extremely useful in obtaining an early understanding of the ultimate acid generation potential of a waste.

pH Value

There is no clear understanding of whether pH actually can control or prevent pyrite oxidation, whether it merely acts as a factor which affects the rate of reaction, or whether it is only an indicator of buffering effects.

Some of the views on pH expressed by various investigators are as follows:

- Knapp (1987): The chemical oxidation rate is proportional to pH value. As the pH decreases from 7 to 3.5, the rate of pyrite oxidation effectively decreases by 50 percent. Below pH 3, the rate of decrease slows considerably, so at pH 2 it is some 40 percent of the rate at pH 7.
- Williams et al. (1982): Work by Kleinmann et al. (1981) indicated, for example, that if the pH is kept in the range of 6 to 8, the dissolution of pyrite is very slow, producing negligible amounts of acid, despite the presence of bacteria. Furthermore, the imposition of a uniform, high pH condition appears to override any microenvironments which could be acid generating.

Oxygen

As discussed above, all reactions in the "classic" sulfide oxidation sequence (Reaction 1) require oxygen from the air. Typically, the actual amount of oxygen required is low, less than 10 percent of the oxygen content of the air. Furthermore, where water represents a source of oxygen, dissolved oxygen down to 0.5 mg/L, 5 percent of its nominal saturation in surface waters, can be sufficient for the oxidation process.

Alkalinity

The general effects of alkalinity on solution pH and, by extension, on pyrite oxidation, have been documented by Williams et al. (1982). The following three cases describe the effects in a simple calcite system, a more complex but realistic soil and ground water system, and an unbuffered system without any alkalinity:

- Using a completely unbuffered (zero alkalinity), oxygen-saturated water to oxidize pyrite, the resultant pH will be about 3.2, given no external source of oxygen.
- In the presence of, and at equilibrium with, calcite, the oxygen in oxygen-saturated water would oxidize pyrite, but the final pH, once all the oxygen was used, would be about 7.4. The alkalinity would not only consume the acid produced, but would leave residual alkalinity capable of further acid consumption.
- In soil and ground water, the carbon dioxide concentrations are commonly 10 to 100 times higher than in surface waters (Buckmann and Brady, 1960). Hence, both the final pH and the residual alkalinity would be higher than in the second example above. The pH, for example, might approach 8.2 to 8.4, as discussed in Section 4.2.4.1 of this chapter.

The net effect of the alkalinity is to keep the system pH closer to neutral for a longer time than might be expected in an acid generating system. Whether or not this is of benefit in reducing the rate of pyrite oxidation depends on the relative merits of either the Knapp (1987) thesis, which favors higher reaction rates at near-neutral pH, or the Williams et al. (1982) thesis, which indicates that reaction rates are lower at near-neutral pHs. Field observations generally support the Williams et al. thesis in systems containing neutralizing potential. The pH values tend to remain neutral for some length of time; the pH drops when the acid generating process starts to dominate. It would seem that if the neutral pH conditions did promote rapid pyrite oxidation, then all the alkalinity, including the residual alkalinity, would be rapidly consumed and the pH would drop much earlier.

Pyrite Abundance and Surface Area of Pyrite Grains

Based on work by Caruccio and Geidel (1981), there is an interrelationship between pyrite content, the surface area of the pyrite grains, and the acid generating capability of mine waste. While much of this work was done for coal mine waste in the eastern United States, the general conclusions also would apply to hard rock minerals:

- The mass rate of acid generation is directly related[(2)] to the mass of pyrite, all other aspects being equal.
- The mass rate of acid generation is directly related to the surface area of pyrite, all other aspects being equal.

Some of the above factors can work against each other. For example, certain wastes with a low total sulfide content of finely disseminated pyrite, which results in a large exposed surface area, may generate acid at a higher rate than those with a higher content of massive pyrite.

Therefore, it is important to consider the following two additional factors:

- The form of sulfide: Disseminated sulfides generally will be stronger acid generators than massive sulfides. Furthermore, the mineral form of the sulfide will influence its reactivity.
- The extent of sulfide encapsulation: It is common to see sulfides encapsulated in a silicate matrix in mine wastes, particularly in unweathered waste rock. The result is that the oxidation reactants cannot reach the sulfide grain. Therefore, despite elevated sulfide levels, the sulfide may be nonreactive due to lack of physical accessibility.

---

(2) The mass rate reflects the amount of acid that is generated over a specified period of time. This is different from the chemical reaction rate which is not related to the same parameters.

The corollary to the second factor is that, due to destruction of the encapsulating layers, weathered or partially weathered pyrite-bearing wastes may be more efficient acid generators than unweathered wastes, even though their sulfide levels may be lower (Smith, 1987).

Temperature

A change in temperature changes the rate of chemical or biochemical reactions. In general, reaction rates double for a 10° C rise in temperature, albeit over a limited range of temperature in some cases. This has been shown to be true for pyrite oxidation (Smith and Shumate, 1971). The implication is that, in order to achieve consistency, temperature control would appear to be necessary in diagnostic testing for acid generation.

Microenvironments

The geochemical systems associated with mine wastes, particularly waste rock material, can be extremely heterogeneous and include a wide variety of microenvironments. In these cases, it will be the geochemical activities that occur within certain microenvironments that control the acid generating potential of the waste and not the "average" properties of the waste. Some of the more important microenvironments include: (1) the formation of local acidic "hot spots" in waste rock piles, despite the alkaline nature of "average" system, (2) the inhibition of bacterial action where local acidity due to acid generation is too high (Kleinmann et al., 1981), and (3) formation of gypsum coatings on calcite grains or jarosite on pyrite grains, despite the average chemistry of seepage from the wastes being below saturation or not in equilibrium with the solid phase which is formed.

The formation of gypsum and jarosite coatings is discussed further in Section 4.2.4.3.

Pyrite Structural Defects and Trace Element Contents

Evidence for the effects of trace elements incorporated in the pyrite crystal lattice is inconclusive and seemingly contradictory. For example, Richardson (1979) associates elevated trace element incorporation in pyrite with a decrease in the oxidation rate of pyrite. However, Hawley (1977) implies that incorporation of transition elements in pyrite may distort its crystal structure (perfection) and that such distortion leads to physical stresses which may make the pyrite more susceptible to chemical attack, i.e., oxidation. The latter explanation would appear to be more reasonable.

4.2.3.3 Oxidation of Other Base-Metal Sulfides

The biochemical oxidation of base-metal sulfides (and other metal-bearing compounds) has been reviewed extensively by Silver (1987). In general, it is suggested that most base-metal sulfides are oxidizable, either directly by bacteria in the presence of air and water, or indirectly by ferric iron as either Fe^{3+} or ferric sulfate. For certain sulfides, such as galena (PbS) and sphalerite (ZnS), the rate of direct oxidation is sufficiently slow that ferric iron oxidation is likely to be the only process which causes oxidation.

As mentioned previously, Table 4.4 provides a list of the different minerals, as well as the associated end products, of complete oxidation and the possible secondary minerals that can be formed at a neutral pH. The oxidation of some typical copper-iron sulfides is discussed in more detail below.

The direct oxidation process by bacteria is thought to be a series of reactions (Silver, 1987), as illustrated by the oxidation of chalcopyrite, a copper-iron sulfide. A summary equation for the five-stage process is:

$$60\ CuFeS_2 + 25\ O_2 + 90\ H_2O = 60\ CuSO_4 + 20\ H[Fe(SO_4)_2.2\ Fe(OH)_3] + 20\ H_2SO_4$$

The net result of oxidation by bacteria is the production of the metal sulfate (copper sulfate), acid (sulfuric), and either a jarositic type of compound or ferrous sulfate and/or ferric hydroxide, depending on redox potential or pH of the overall system.

Indirect oxidation by ferric iron, as ferric sulfate, of a copper sulfide mineral such as bornite would be:

$$Cu_5FeS_4 + 6\ Fe_2(SO_4)_3 = 5\ CuSO_4 + 13\ FeSO_4 + S$$

Both the ferrous sulfate and the elevated sulfur produced can be the subject of further oxidation. The ferrous sulfate will oxidize to the ferric state and result in ferric hydroxide precipitation under the appropriate pH conditions (above 3.5 to 4), with the associated generation of acid. As discussed below, sulfur also can be oxidized to yield acid.

4.2.3.4 Oxidation of Other Forms of Sulfur

There is an increasing awareness that secondary sulfur-bearing species can be of importance in acid generation from mine wastes, not only as intermediate products of reactions, but as sources of acid. Hawley (1977) noted that, under certain conditions, sulfides can be partially oxidized to sulfites and thiosulfates, even at neutral or alkaline pH values. These are then subsequently oxidized to polythionate and eventually sulfuric acid, in the series:

$$S_2O_3 \rightarrow S_nO_6 \rightarrow S_3O_6 \rightarrow H_2SO_4$$

Where "n" is a number from 4 to 8.

The chemistry of the oxidation reactions in this series is:

$$4\,S^{2-} + 3\,O_2 = 2\,S_2O_3^{2-}$$

$$S_2O_3^{2-} + H_2O + 2O_2 = 2\,SO_4^{2-} + 2\,H^+$$

$$2S_nO_6^{2+} + 2(n-1)\,H_2O + (3n-5)\,O_2 = 2n\,SO_4^{2-} + 4(n-1)\,H^+$$

For the last equation, if "n" were 4, the equation would be:

$$2\,S_4O_6^{2+} + 6\,H_2O + 7\,O_2 = 8\,SO_4^{2-} + 12\,H^+$$

As can be seen by the production of 12 moles of H+ (6 moles of sulfuric acid) from 2 moles of a polythionate, these reactions have the potential to be significant acid generating processes.

A second potential source of acidity is the resolution of jarosites, which are secondary potassium-iron sulfate minerals. The chemistry of these minerals is discussed in Section 4.2.4.3.

4.2.4 ACID CONSUMING PROCESSES

The presence of either sulfide or reactive sulfide in mine waste does not necessarily indicate that acid will be generated, even if there is sufficient infiltration to cause a discharge from the material. Acid generation can be inhibited or its products modified by reaction with other components in the waste or with water infiltrating the waste. This section addresses the processes that inhibit or prevent the formation of acidic conditions or delay the onset of acid conditions.

There are three groups of reactions which are important in terms of their effects on acid generation and its products. These are:

- Reactions with carbonates.
- Reactions with aluminosilicates.
- Reactions with, or formation of, other compounds.

The sections that follow describe the nature and characteristics of the materials involved in these reactions, the reaction chemistry, and the resulting implications to the interpretation of data from acid generation potential testing, which is discussed later in this chapter.

4.2.4.1 <u>Reactions with Carbonates</u>

<u>Nature and Characteristics of Carbonates</u>

Carbonates are some of the most widely distributed minerals that occur in most rock types. They commonly are located in, or associated with, ore deposits. Kraus et al. (1959) list three major mineralized groups of carbonates - the calcite, aragonite and malachite groups - all of which include a range of carbonate minerals. The most ubiquitous of the minerals are calcite and dolomite which are from the calcite group. There is a solid solution series between calcite and dolomite, with increasing substitution of magnesium for calcium occurring.(3)

Carbonates, particularly calcium carbonate, tend to be sparingly soluble in water. Typical solubility products in pure water would be:

MINERAL	MINERALOGICAL GROUP	FORMULA	SOLUBILITY
• Dolomite	Calcite	$MgCO_3.3H_2O$	2.0×10^{-5}
• Calcite	Calcite	$CaCO_3$	4.5×10^{-9}
• Strontianite	Aragonite	$SrCO_3$	2.5×10^{-10}
• Siderite	Calcite	$FeCO_3$	3.0×10^{-11}
• Rhodochrosite	Calcite	$MnCO_3$	6.0×10^{-18}

(3) The pure magnesium end member would be magnesite ($MgCO_3$), but this is far less common in nature than is dolomite.

Reaction Chemistry of Carbonates

The "carbonate" can be available to consume acid, either as analog in solution or as the solid-phase mineral itself.

Consider the simple solution of calcite to give calcium and carbonate ions:

$$CaCO_3 = Ca^{2+} + CO_3^{2-}$$

Then this will be modified by the hydrolysis of the carbonate ion to yield bicarbonate (Hawley, 1977):

$$CaCO_3 + H_2O = Ca^{2+} + OH^- + HCO_3^-$$

However, if carbon dioxide is present and dissolved as carbonic acid, the process would be:

$$CaCO_3 + H_2CO_3 = Ca^{2+} + 2\ HCO_3^-$$

Given equilibrium with the atmospheric carbon dioxide concentration, the net result of this reaction is a solution with a pH of about 8.4 and a predominance of bicarbonate ions. Hence, the solution itself will have the capacity to consume hydrogen ions.

The acid consuming reactions of solid-phase carbonates can be expressed as:

$$2\ H^+ + CaCO_3 = Ca^{2+} + CO_2 + H_2O \text{ (for calcite) or}$$

$$2H^+ + [CaMg\ (CO_3)_2] = Ca^{2+} + Mg^{2+} + CO_2 + H_2O \text{ (for dolomite)}$$

where there is solution contact with the atmosphere and the pH is generally below 7.

Under saturated conditions, the loss of the gas phase as carbon dioxide is prevented, and bicarbonate ions are formed in solution, i.e.:

$$H^+ + CaCO_3 = Ca^{2+} + HCO_3^- \text{ (for calcite), or}$$

$$2H^+ + [CaMg\ (CO_3)_2] = Ca^{2+} + Mg^{2+} + 2\ HCO_3^- \text{ (for dolomite)}$$

While calcite and dolomite have similar reactions, siderite (iron carbonate) displays different behavior and does not produce a similar acid consuming response. Initially, iron is taken into solution as divalent, ferrous iron:

$$2H^+ + FeCO_3 = Fe^{2+} + CO_2 + H_2O$$

However, in relatively oxidizing and/or nonacidic pH conditions, there is oxidation of divalent to trivalent ferric iron as:

$$4\ Fe^{2+} + 4\ H^{+} + O_2 = 4\ Fe^{3+} + 2\ H_2O$$

The latter subsequently precipitates as ferric hydroxide, upon which further hydrogen ions are released:

$$Fe^{3+} + 3\ H_2O = Fe(OH)_3 + 3\ H^{+}$$

Combining the three equations for siderite behavior, the net result is:

$$4\ FeCO_3 + O_2 + 6\ H_2O = 4\ Fe(OH)_3 + 4\ CO_2$$

Therefore, it is apparent that hydrogen addition and removal cancel each other, the result being ferric hydroxide precipitation.

Williams et al. (1982) describe the combined reaction of acid generation from pyrite by air/water oxidation and reaction with calcite using the following equation:

$$4\ FeS_2 + 8\ CaCO_3 + 15\ O_2 + 6\ H_2O = 4\ Fe(OH)_3 + 8\ SO_4^{2-} + 8\ Ca^{2+} + 8\ CO_2$$

Effectively, it requires two moles of calcite to neutralize the acid generated by a single mole of pyrite, equivalent to 200 gm of calcite for 120 gm of pyrite on a mass basis.

The products of the above reaction have an important influence on the effectiveness of acid consuming materials and, hence, on the interpretation of the results of acid generation potential predictive tests. The formation of a relatively insoluble iron (ferric) hydroxide, and potentially of gypsum (calcium sulfate), by precipitation from solution, has the potential to coat or "blind" surfaces of reacting materials, particularly the carbonates. This coating has the potential to stop the acid consuming reactions, or at least limit their effectiveness, as the reactions would have to continue by diffusion through the precipitate coating.

In the case of sulfides, if there is any acid generation at all, it will occur at the surface of the sulfide particles. This, in turn, will result in the solution of any ferric hydroxide to yield ferric ions at the low pH conditions, or result in increased calcium solubility in cases where gypsum formation is a potential. Hence, a coating is not likely to be formed.

However, the microchemical environment adjacent to the carbonate will be different in the case of calcite, for example, than sulfides. In the case of the carbonate, a "high" pH condition will occur (pH 8 or greater) with a corresponding "high" availability of calcium ions in solution. The high pH will enhance ferric hydroxide precipitation on or around the carbonate itself, producing a coating. In addition, when the concentration of sulfate from sulfide oxidation in the vicinity of the carbonate is high enough to exceed the local calcium sulfate solution product, a gypsum precipitate will be formed.

Importance of Carbonates in Acid Generation Potential Testing

The potential for coating acid consuming carbonate materials, has more important implications than the coating of acid generating materials, when interpreting the results of acid prediction tests. A simple assessment of the total neutralizing capacity can result in a serious underestimation of the acid generating potential. While the total acid consuming potential may ultimately become available, its rate of reaction would be limited as it would be controlled by diffusion through, or the actual physical or chemical loss of, the coating.

4.2.4.2 Reactions With Aluminosilicates and Other Silicates

Nature and Characteristics of Silicates/Aluminosilicates

Many silicate minerals and aluminosilicates (aluminum-bearing silicates) have characteristics which make them or their degradation products acid consuming. These characteristics are:

- In contact with water they tend to produce an alkaline pH.
- In contact with acids, they tend to degrade, consume hydrogen ions, and produce clay minerals.
- Some silicates, notably aluminosilicates such as mica, clay minerals, etc., are capable of removing hydrogen ions by ion exchange reactions.

These processes are less effective acid consuming reactions than the carbonate systems described in the previous section. However, if acid generation is slow, the effect is significant due to the sheer mass of silicate minerals in any mine waste or natural environment (Williams et al., 1982).

Data provided by Hawley (1977) show ranges of moderate to high pH values from paste pH tests on silicates. pH ranges for various silicates include:

- Pyroxenes 8 to 11 (depending on the magnesium content)
- Micas 7 to 9
- Olivine 10 to 11 (depending on the magnesium content)
- Feldspars 8 to 10
- Amphiboles 10 to 11

By way of comparison, clay minerals have paste pH values in the range of 5 to 7, and aluminum hydroxides range from pH 6 to 7.

The above test results indicate that silicate decomposition in natural systems will result in higher pH values. The results, however, cannot be directly transferred to natural systems, as the test methods involved pulverizing the silicate samples that were used.

Reaction Chemistry of Silicates/Aluminosilicates

The general acid consuming reactions of aluminosilicates are described by the equations for calcic and potassic feldspars (Lapakko, 1987; Williams et al., 1982) as:

$$2\,K\,AlSi_3O_8 + 2\,H^+ + H_2O = Al_2Si_2\,O_5\,(OH)_4 + 4\,SiO_2 + 2\,K^+ \text{ and}$$
$$CaAl_2Si_2O_8 + 2\,H^+ + H_2O = Al_2Si_2O_5(OH)_4 + Ca^{2+}$$

In both cases, the cation (calcium, potassium) is freed into solution, and an alumino-rich clay mineral is produced. The rate of these processes has been shown by Busenberg and Clemency (1976) to be relatively slow and to proceed in the following two stages:

- An initial ion exchange reaction on the mineral surface. The hydrogen ions in solution exchange for an alkali ion (K, Na, etc.) from the solid phase.
- Dissolution and physical breakdown of the original structure to yield a clay mineral.

Lapakko (1987) has demonstrated that decomposition of plagioclase feldspar, olivine, and pyroxene from the Duluth Complex in Minnesota, was rapid enough to consume acid generated by iron sulfide oxidation. This occurred in several cases where the initial sulfur content of the waste was less than 0.8 percent, which is equivalent to a pyrite content just over 1.5 percent.

A further study at a site in Halifax, Nova Scotia (Albright, 1987), revealed that the waste shale materials had a significantly greater acid consuming capacity than the carbonates in those shales. This acid consumption was apparent only at low pH values in the range 2.5 to 2.8. Albright attributed this effect to the production of "additional alkalinity" due to breakdown of the silicate minerals.

Implications of Silicates/Aluminosilicates to Acid Generation Potential Testing

The ability of silicate minerals to consume acid, in addition to consumption by carbonate minerals, has obvious implications for interpretation of acid prediction tests. For slow acid producing natural systems, silicate decomposition and its acid consumption can act as an effective control. However, short term[(4)] acid generating potential tests may not incorporate this acid consuming ability.

4.2.4.3 Other Reactions

Two other processes which have been described as acid consumption processes are the removal of sulfate as gypsum and the removal of sulfate and ferrous iron in jarosites (generally: hydrated potassium/sodium, iron sulfates). Both processes are the subject of differing interpretations as to their actual reactions and relative effectiveness in hydrogen ion consumption.

Gypsum

Gypsum formation and its chemistry were described under the "coating" discussion in Section 4.2.4.1. The net effect of the sulfate removal associated with the formation of gypsum from a solution containing other species will be a corresponding increase in pH value. It should be noted, however, that gypsum precipitation is limited to local areas adjacent to calcite or dolomite particles where there is the necessary solubility control. Williams et al. (1982) suggest the local sulfate values in the presence of calcite would be limited to as low as 400 mg/L, according to the equation:

$$CaCO_3 + SO_4^{2+} + 2H^+ + H_2O = CaSO_4\ 2\ H_2O + CO_2$$

(4) As explained in Section 4.3 of this chapter, the short-term tests are also referred to as "static" tests.

In free solution within the waste, sulfate values can be much higher (900 mg/L) without any gypsum precipitation, based on data by Tanji and Doneen (1966). Increasing levels of other ions such as magnesium in solution, will increase sulfate solubility further, due to the common ion effect.

Jarosite

Jarosites are a group of secondary iron minerals which appear commonly as precipitates and coatings associated with acidic seeps or drainages. The most prevalent jarosite is a potassium type, formed from the hydration of ferric sulfate, with available potassium in the system at very low pH values (Silver, 1987):

$$3\ Fe_2(SO_4)_3 + 2\ KOH + 10\ H_2O = 2\ K[Fe(SO_4)_2\ 2Fe(OH)_3] + 5\ H_2SO_4$$

The potassium (or other monovalent species) needed for jarosite formation comes either from the dissolved ions in solution or from the decomposition minerals, such as feldspar. In the case of dissolved ions, there will be a decrease in both ferric iron and monovalent ions in solution, with a relative increase in sulfate concentration; this implies a decrease in solution pH (addition of hydrogen ions) as the above reaction indicates. Therefore, this reaction appears to be an acid generating process and not an acid consumption process. Williams et al. (1982), however, imply that the precipitation of iron and sulfate as jarosites may result in an increase in pH in a system, although such an increase is likely to be small, given the low concentrations of potassium normally available. With the low pH condition and the limited pH range (1.7 to 2.0) under which jarosites are said to be stable, it is apparent that, in most natural systems, jarosites will be metastable and will degrade, going back into solution. Redissolved jarosite solutions will be highly acidic and have a low pH in the absence of any buffering.

In conclusion, jarosite formation should not be considered an acid consuming process. At best, it could act as a very temporary storage for acidity (Williams et al., 1982), to be released in subsequent solution.

The formation of jarosites does have some potential implications when interpreting acid prediction tests. If jarosites form on sulfide particles, they will tend to inhibit oxidation processes, as they are stable in the pH range at which sulfide oxidation is very active. Hence, due to formation and subsequent resolution of jarosites, high potassium solutions may produce some anomalous results during longer term tests.(5)

4.2.5 REACTION KINETICS

The net acid generating potential of a waste is the difference between the amount of acid generated and the amount of acid consumed. There is usually an interest in the net rate of acid generation at a particular time or the net amount of acid generated over a period of time. Both of these are determined by the difference in the rate of acid generation and consumption, either at a particular time or integrated over a time period. Therefore, it is evident that the total potential for acid generation or acid consumption of a waste has only an indirect influence on the net rate of acid generation.

An appreciation of the relative rates of reaction also has a significant bearing on the interpretation of results of tests for acid generation potential. For example, the short-term static tests, which are conducted to determine the acid generating or acid consuming potential, usually will provide only an indirect assessment of the net acid generating potential. A longer term kinetic test, which allows reactions to occur, will generally provide a more direct assessment. However, even a kinetic test may not predict the net acid generation potential accurately if, for example, there are long-term effects which are not evident during the test period. For example, the test may only simulate the first part of an acid generation process, where acid consumption is dominant due to alkalinity in the system. In the longer term, however, this alkalinity could be consumed by additional acid or be leached from the waste material by natural processes. The ongoing sulfide oxidation would be able to generate an acidic system which manifests itself after an extended period of time and may not be simulated in the test.

Based on the above discussion, it is clear that some understanding of reaction rates, as well as the total potential, is necessary for a more informed interpretation of the results of acid generation tests. The reactions described in Sections 4.2.3 and 4.2.4 are therefore important to determining whether or not a waste will have a net acid generation potential. In addition, it is necessary to understand the reaction rates of these processes. The following sections describe the current state of knowledge on the rates of the acid generating and acid consuming reactions.

---

(5) As explained in Section 4.3 of this chapter, the long-term tests are also referred to as "kinetic" tests.

4.2.5.1 Acid Generation Reaction Rates

It is convenient to discuss the reaction rates in terms of the two stages of pyrite oxidation discussed in Section 4.2.3.1:

- The reactions involving dissolution of pyrite to form ferrous iron and sulfate; and
- The reactions involving ferrous iron oxidation to ferric iron or to ferric hydroxide.

Pyrite to Ferrous Iron Reactions

As discussed in Section 4.2.3.1, varying rates of oxidation have been observed during the initial chemical oxidation of pyrite to ferrous iron. In the authors' general experience, rates have typically varied from as low as 0.1 mg/gm/day (ferrous iron/gm pyrite) through values about two orders of magnitude greater. Factors affecting this oxidation rate include the nature of the sulfide and its form, structure, and trace element incorporation into the sulfide lattice, as discussed in Section 4.2.3.1.

Trivalent ferric iron availability, either as Fe^{3+} or the products of its hydrolysis and secondary reactions ($Fe[OH]_3$, $Fe_2[SO_4]_3$), is suggested to be the rate-controlling factor in pyrite oxidation. For example, Kleinmann et al. (1981) observed that oxidation of pyrite by ferric iron can be rapid if the ferric/ferrous ratio is 0.01 or greater.

Catalyzation by T. ferro-oxidans increases the reaction rate to a factor of 5 to 20 times the abiotic chemical rate (B.C. AMD Task Force, 1989; and Kleinmann et al., 1981). The biological oxidation rate itself is influenced by pH, being most rapid at pH 3 to 3.5 and decreasing with both increasing and decreasing pH value. It is many orders of magnitude greater at pH 3 than at pH 6. Also, both the biotic and abiotic rates are altered by changes in temperature, as described in Section 4.2.3.1.

Ferrous Iron Oxidation

Within the concept of a "closed-loop" system, Singer and Stumm (1970) have suggested that ferrous iron oxidation is relatively slow and is the controlling process of the acid generation rate from pyrite oxidation. Clearly, with an external source of even a small amount of ferric iron, the oxidation rate would be considerably increased, as it would no longer be controlled by ferrous iron oxidation.

In an abiotic system, pH has a marked influence on ferrous iron oxidation (Williams et al., 1982). For example:

- At pH 7, ferrous iron is oxidized to ferric iron in a matter of minutes.
- At pH 4.5, the process takes 300 days.
- At pH 3.5, the process takes about 1,000 days (3 years).

Therefore, the ferrous iron oxidation process is negligible at low pH values. However, even at these low values, the rate of biological oxidation of ferrous iron is rapid and can be up to 1 million times greater than abiotic processes (Singer and Stumm, 1970; Kleinmann et al., 1981). Furthermore, the rate of ferric iron formation from pyrite oxidation also can be rapid and become the rate control on acid formation from pyrite.

4.2.5.2 Acid Consuming Reaction Rates

Acid consuming reaction rates are discussed below within the same three groupings used in Section 4.2.4, i.e., reactions with carbonates, reactions with aluminosilicates and silicates, and reactions with or formation of other compounds.

Reactions With Carbonates

The rate of reaction between acids and carbonates is normally considered to be rapid, such as with fresh calcite and sulfuric acid (Williams et al., 1982). However, in most mine wastes, the acid consuming material is usually limestone ($CaCO_3$). Therefore, the reaction rate is much slower and may take several days or more to reach equilibrium (Hawley, 1977).

The susceptibility of carbonates, particularly calcite and dolomite, to coating or blinding by gypsum and iron salts will result in a substantial decrease in reaction rates, as the process will become controlled by diffusion of carbonate through the coating.

Reactions With Aluminosilicates and Silicates

Dissolution of aluminosilicates was discussed in Section 4.2.4.2 as a two-stage process of ion exchange followed by actual dissolution of the material. It has been demonstrated that the initial ion exchange reaction is relatively rapid, as would be expected from classical chemical explanations of ion exchange processes (Busenberg and Clemency, 1976). However, the rate of dissolution of the silicate structure itself is slow, based on the work of these authors and on general observations of silicate behavior in waste rock stockpiles.

Reactions With, or Formation of, Other Compounds

There appears to be little information on reaction rates for the formation of gypsum or iron hydroxide coatings on calcite, other than field observations of the gradual buildup of such a deposit on fresh limestone surfaces. In limestone acid consumption processes, the formation time for a coating may be on the order of weeks to months. Jarosite formation appears to be rapid in the pH range 1.8 to 2, but very slow outside this range (Silver, 1987).

4.2.6 MIGRATION OF ACID ROCK DRAINAGE

4.2.6.1 Introduction

The low pH ARD, generated as pore water within mine waste, has the potential to dissolve heavy metals and other soluble constituents that may be contained in the waste. The impact of ARD on receiving ground or surface water, however, can only be realized if the pore water migrates from the site of generation to the receiving water bodies.

A number of physical, chemical, and biological characteristics associated with mine waste units control metal solubility and the migration of ARD. The mobilization of metals is principally controlled by chemical factors and, to a lesser extent, by physical and biological factors. The processes that occur along the ARD migration route are controlled by both physical and chemical factors. The following sections describe these controls and how they influence the timing and the quality of the discharge of ARD from a waste.

4.2.6.2 Physical Controls

There are certain physical characteristics that influence metal solubility, such as waste particle size and shape, ambient temperature, and pressure of the pore space gases. These properties are secondary to the chemical reactions that cause metals or other constituents to dissolve in ARD.

The influence of these physical characteristics is more important to the rate of ARD migration. The important characteristics include the amount of rainfall infiltration, waste permeability, availability of pore water, pore water pressure, and the mechanisms by which ARD is transported. These mechanisms can include discrete channel flow through the more porous areas of the waste, porous media flow, and/or molecular diffusion. These characteristics control the rate of ARD movement, the amount of dilution, and the degree of mixing that occurs as the ARD moves from the source to the receiving water environment.

The physical properties in the waste material are likely to be different from those in the subsurface strata which separates the waste from receiving water bodies. ARD migration may, therefore, occur at significantly different rates. There is also the potential for the geologic formations through which the ARD migrates to reduce the concentration of dissolved constituents by attenuation. This is discussed in detail in Chapter 5. Where there is potential for ARD to migrate via surface flow pathways directly to surface water bodies, the migration rates may be extremely rapid, with less opportunity for chemical changes to occur.

4.2.6.3 Chemical Controls

The solubility of metals is generally controlled by the pH of ARD. Other chemical controls include redox potential,[(6)] adsorption characteristics of the solids, the type of metal being mobilized, and the chemical composition of the ARD. The water quality of the drainage entering the receiving environment is a function of these chemical controls, as they occur at the source of acid generation and along the route of migration. As ARD moves away from the sulfide source through the waste material, more acid generating material may be encountered, causing a further reduction in pH. Conversely, the drainage may come into contact with alkaline material, in which case partial or complete neutralization may take place.

With decreasing pH, the dissolved metal load generally increases. However, a combination of chemical conditions could occur where metals are mobilized at neutral or even alkaline conditions. During neutralization of low pH drainage, precipitation of many of the soluble metals may occur, with the resultant drainage containing the residual metals and products of the buffering reactions. This has been observed with copper and massive sulfide ores in western Canada where elevated zinc loadings occur in neutral drainages (Steffen Robertson and Kirsten, 1986). While dissolved copper largely precipitates out as ARD is neutralized, the zinc remains at a relatively high concentration until the pH is raised to values above 9.5.

---

(6) Reduction-oxidation potential.

4.2.6.4 Biological Controls

Biological activity along the route of migration may influence the quality of ARD. Metal leaching occurs where iron oxidizing bacteria are present with iron and metal sulfides. Biological species also can attenuate the mobility of metals by adsorption or, possibly, precipitation. Biological activity may be a factor if the drainage passes through wetland conditions or areas of biological activity on the mine site before entering the receiving environment.

4.2.7 IMPACT ON THE ENVIRONMENT

The natural environment, such as the geologic formations and ground and surface water bodies, usually has a certain capacity to neutralize ARD as it migrates from its source. For example, ARD seeps from a waste will pass through soils or rock which may have a neutralizing capacity. ARD that also mixes with surface and ground water can be neutralized and diluted. As alkaline materials close to the source of the ARD are consumed, the ARD migrates further from its source, resulting in an increasing zone of impact. At any time, there is a point some distance from the waste facility at which the effect of ARD is abated by the neutralization and dilution capabilities of the environment. In some areas, natural streams can be acidic due to the weathering and oxidation of natural sulfide outcrops. In such case, the impact of ARD could be less pronounced than on neutral or basic streams. The environmental impact of ARD is therefore a function of time and is dependent on the quality of the ARD, the natural dilution and neutralizing capacities of the environment surrounding the waste, and the quantity and quality of potentially affected water bodies.

For each mine waste deposit which is prone to ARD, there is a certain level of ARD release that can be sustained by the environment without significant damage. Some environments have an extremely high load of alkaline materials in the soils, surface water, and ground water. This is the case in many of the old, deeply weathered landscapes with well developed old soils. The dilution capacity of the environment depends on the magnitude of flow of the receiving waters in relation to the polluted drainage.

4.3 ACID GENERATION PREDICTION

4.3.1 INTRODUCTION

The ability to predict acid generation from mine wastes allows the appropriate control measures to be designed and implemented to either prevent or inhibit the formation of ARD, to contain it, or to achieve a combination of inhibition and containment. The development and application of techniques for the prediction of acid formation has been one of the major areas of work in mine waste geochemistry for the past 10 years.

4.3.2 GENERAL APPROACH TO PREDICTION

4.3.2.1 Prediction Methods

Prediction techniques related to ARD were initially developed through the efforts of the coal mining industry, particularly in the eastern United States. Many of these techniques have, with modifications, been applied to hard rock metal mining. Methods to predict ARD have been divided into five groups (Ferguson and Erickson, 1986), as follows:

- Geographical comparisons (Method 1) - This method relies on extrapolating waste geochemistry from one mine to others in the same geographic area.
- Paleoenvironmental and geological models (Method 2) - The rationale behind this method assumes that areas with similar paleoenvironments and geologic formations have similar potential for acid generation.
- Geochemical static tests (Method 3) - These are short-term batch laboratory tests that compare the acid generation and the neutralization capacity of a material.
- Geochemical kinetic tests (Method 4) - These are longer term laboratory tests which produce a leachate for analysis and which reflect the rate of acid generation of a material.
- Mathematical models (Method 5) - Mathematical models can be used to extrapolate the rates of acid generation and neutralization beyond the time frame of the laboratory tests described above.

While Methods 1 and 2 are useful for obtaining a preliminary indication of acid generation potential in the coal mining industry, they are not adequate for reliable prediction in hard rock mines.

Static and kinetic tests, Methods 3 and 4, form the most reliable basis for characterizing waste materials and are commonly used to predict whether or not a particular waste will generate acid. These types of tests can also be used in defining the level of control measures required to prevent, control, or contain acid generation. A number of different static and kinetic test methods are currently used, and are discussed in the following sections.

Mathematical models, Method 5, range from simple to complex. The simple models include best fit curves which can be extended beyond the time range of the test results. The more complex models theoretically simulate the acid generation and consumption processes and are calibrated using laboratory or field test data. More information on modeling is contained in the *Draft Acid Rock Drainage Technical Guide* prepared by the British Columbia Acid Mine Drainage Task Force (B.C. AMD Task Force, 1989).

4.3.2.2 Objectives and Approach

To be useful, the acid generation prediction procedure should meet the following two objectives:

- Predict whether a particular waste will generate acid at some time and, if so, the rate at which it is likely to occur.
- Predict the quantity and quality of the drainage leaving the waste, under both controlled and uncontrolled site conditions.

The first objective is generally achieved by characterization of waste materials utilizing simple laboratory geochemical testing and/or simulation of actual waste disposal conditions. The latter is accomplished by means of more elaborate laboratory and, possibly, small-scale field tests. The laboratory tests that can be used to characterize materials in terms of their acid generation potential are described later in this chapter.

The second objective is somewhat more difficult to achieve with a high degree of confidence, particularly over the long term due to the many factors that influence net acid generation. The general approach is to extrapolate the results of short-term laboratory and field tests. This can be accomplished using mathematical computations or models.

The steps required to predict the acid generating potential of the waste can involve considerable expense. They include not only the testing described above, but also analyzing the mineralogy of the ore body and obtaining representative samples for testing. It is therefore advisable to proceed with a carefully planned and phased program. Phasing involves starting with a smaller amount of information representing the worst-case conditions to establish whether ARD is likely to be a problem or not. If needed, more detailed testing would then be conducted to quantify an indicated potential and/or to increase the level of confidence in the preliminary prediction.

A general phased approach involves the following steps:

- Define the geologic units likely to be encountered during mining, and define their mineralization.
- Design and implement a materials sampling program.
- Conduct laboratory and field testing, as necessary.

- Evaluate the potential for ARD formation.
- Conduct further sampling and testing, if required.
- Characterize the waste materials in terms of acid generation potential.

The following sections describe the major elements involved in the above approach, i.e., the materials sampling and the geochemical static and kinetic testing procedures.

4.3.3 MATERIALS SAMPLING

4.3.3.1 Geology and Mineralization

The prediction of acid generation potential begins with an understanding of the geology and mineralization of the ore body and surrounding waste rock scheduled for removal. It is necessary to characterize the ore after recovery of the economic minerals, low grade ore that could be mined and subsequently disposed of waste rock, and residual mineralized rock that remains behind in either an open pit or underground mine.

The first step in characterizing these materials is to identify the lithologic units that are likely to be encountered during mining. The next step is to describe the basic geochemical properties and the range and type of mineralization that are present in each of the lithologic units, as these usually have different characteristics in terms of acid generation and neutralization potential. It is important to recognize that, depending on the type of mineralization, the acid generation potential of waste rock may be related to the lithology, to the distance from the ore body, or both.

It is necessary to identify the types of sulfide minerals present and the form in which the sulfide mineralization occurs (e.g., mineral particle sizes, whether disseminated or in vein), as these factors influence the reactivity of the minerals. If information is available from previous exploration, nearby mines, or from similar paleoenvironmental and geological areas, comparisons should be made and used as initial indicators of potential acid generation.

4.3.3.2 Objectives of Sampling

Initially, the objectives of sampling are to obtain materials that represent the worst-case, or the maximum, acid generating potential. As the evaluation program evolves, the objective of subsequent sampling programs would be to obtain representative samples from a range of geochemical groups within each of the lithologic units to be affected by mine development. This will provide information so that the percentage of waste likely to generate acid can be determined, the extent of acid generation can be quantified, and the remainder of the material can be evaluated in terms of its suitability for mitigating ARD. When designing the sampling and analysis programs, considerations should include the cost of sampling and testing, the risk of incorrectly predicting the waste characteristics, and the potential cost of remediating unanticipated ARD problems. The determination of sample size should also include factors such as the nature and distribution of sulfide materials, waste type, and particle size.

Although the collection of representative samples from lithological units is simple in concept, implementation can be more complicated and may require several iterations. Laboratory tests on samples from a defined lithologic unit may indicate significant variability in acid generation or neutralization potential. This variability may indicate that samples from more or different geochemical groups are necessary. These groups should then be redefined, the sampling plan revised, and further samples collected to confirm homogeneity within the groups.

The design of the sampling program should be related to the proposed mine plan, recognizing that mine plans change during mining due, for example, to assay data obtained during mining that differs from exploration results, or changes in the cut-off grade as a result of economic circumstances. Therefore, an ongoing sampling and testing program also may be necessary during the operational phase of the mine.

4.3.3.3 Application of the Sampling Program

The project geologist and metallurgist should guide the initial identification of lithologic units and geochemical groups, as well as the selection of representative samples. Logging of available drill core will usually provide sufficient information for this task. Depending on the size and complexity of the proposed mining operation, the number of samples required may be in the tens for the initial program and could range as high as one or two hundred for the more detailed program.

4.3.3.4 Sources and Handling of Samples

The potential sources of samples for acid generation prediction testing include pit walls, rock surrounding underground workings, waste rock and overburden, ore, tailings, and spent ore piles. These are summarized in Table 4.6. The main source of samples at proposed mines is the drill core from exploratory drilling. Samples also can be obtained from spent ore residues generated during laboratory or pilot scale metallurgical testing. Where metallurgical extraction involves

TABLE 4.6

POTENTIAL SOURCES OF SAMPLES FOR ACID GENERATION PREDICTION[1]

MINE COMPONENT FOR WHICH ARD TESTING MAY BE REQUIRED	SOURCES OF SAMPLES FOR ACID GENERATION PREDICTION
Pit Walls	Drill Core and Cuttings Underground Exploration Passages Trenches Pit Walls (Existing Mines Only)
Rock Surrounding Underground Workings	Drill Core Underground Exploration Adits Mining Faces and Walls (Existing Mines Only) Excavated Waste Rock (Existing Mines Only)
Waste Rock and Overburden Piles	Drill Core Underground Exploration Adits Waste Rock Piles (Existing Mines Only)
Ore Stockpiles	Drill Core Underground Exploration Adits Ore Stockpile (Existing Mines Only)
Tailings From Milling Operations	Residue From Bench Scale Metallurgical or Pilot Plant Tests Tailings Impoundment (Existing Mines Only)
Spent Ore From Heap Leach Operations	Residue From Laboratory Column or Pilot Scale Heap Leach Metallurgical Tests Heap Leach Residues (Existing Mines Only)

(1) Adapted from B. C. AMD Task Force, 1989.

substantial reagent addition that modifies the geochemical properties of the waste, it is important to obtain residue samples while the process is operating at equilibrium. For existing mines, samples can be obtained onsite from each mine component or waste type.

The method of sampling, length of time after sampling, and conditions under which the samples are stored can affect their physical and geochemical characteristics. Furthermore, if fluid with different properties than the ground water is used during drilling, the properties of the samples can be altered. These potential issues should all be considered and, if any are judged to affect the outcome of the acid generation testing, appropriate measures should be taken to compensate for these effects. These measures include sample preservation, minimizing delays in testing, and use of different sampling methods and sample sources.

The mass of each sample required for most static and kinetic tests is approximately 1 pound. However, the mass of the sample required to assure it is representative of a specific geochemical group may be significantly larger. This would be the case where small-scale heterogeneities such as veins, are present, as they can control acid generation. Under these conditions, the larger sample can assure that a representative quantity of the vein material is included. If pilot scale tests are contemplated, larger samples of waste will be required.

4.3.4 GEOCHEMICAL STATIC TESTS

4.3.4.1 Introduction

The static test is usually the first step in the analysis of acid generation potential. It estimates the balance between the acid generating and acid neutralizing capacity of a sample. There are many tests available, including the so-called "acid-base accounting" techniques. Each test uses separate methods to determine the capacity of acid generation and neutralization.

The advantages of the static tests are that they are quick and economic to perform and can be reliably used as an initial screening process to separate materials which clearly are not acid generating from those requiring more thorough, kinetic testing before final conclusions can be drawn. The shortcomings of the static tests are that their assumptions on the instant availability of neutralizing potential, reactivity of sulfur, and conversion of all sulfur (either total or pyritic) to acid can lead to an inaccurate or incorrect estimation of net acid generation potential in cases where the balance between acid generation and acid neutralization is close. Some of the available static tests are discussed in the following sections.

4.3.4.2 Static Test Techniques

B.C. Research Initial Test

This test provides a quantitative measure of the net acid producing potential by comparing the maximum acid generating potential, based on the assumption of complete sulfur oxidation, to the capacity of the sample to consume acid.

The acid potential (AP) is calculated from a total sulfur analysis and is expressed as an equivalent sulfuric acid calcium carbonate weight percentage, e.g., kg of H_2SO_4 or $CaCO_3$ per ton of waste. Total sulfur analyses are conducted in a Leco furnace which roasts the sample to liberate and measure all sulfur-bearing forms, including organic sulfur, sulfides, and sulfates.

The neutralization potential (NP) is determined by sulfuric acid titration of a slurry of finely ground waste material from its natural pH to a pH of 3.5. The NP is expressed in the same units as the AP. The acid generation potential of a sample is assessed by subtracting these two values.

Sobek Test

The Basic Sobek Test (Sobek et al., 1978) is similar to the above procedure, but incorporates some refinements to the total sulfur analysis, potentially increasing the value of AP estimated. These refinements include the addition of excess hydrochloric acid and the heating of the sample to assure a complete reaction. Any potentially unconsumed acid is determined and added to the AP by titration with sodium hydroxide to pH 7.0.

The Modified Sobek Test only uses the iron sulfide content to determine the AP of the material. This test usually results in lower and more realistic AP values. It is conducted by first leaching the nonsulfide iron from the sample with a dilute hydrochloric acid. This is followed by the oxidation of the sulfides with a hot strong oxidizing acid. The equivalent amount of sulfur is then calculated from the amount of liberated iron and the known iron-sulfur relationship.

Alkaline Production Method

In this test the NP of a waste is evaluated by determining its alkaline production potential. This is accomplished by measuring the amount of hydrochloric acid (20 mL of 0.1M) consumed by 500 mg of the finely ground sample (less than 23 μm) in two hours at room temperature. Acid consumption is determined by back titration of the excess acid to an end point pH of 5.0. AP is determined by the total sulfur content.

The originators of this technique (Caruccio et al., 1981) calibrated it against the results of longer term kinetic tests and then used it to provide a rapid indication of acid generation potential. The alkaline production potential values obtained for a number of samples from related geologic units were plotted against total sulfur contents. By superimposing the results of a kinetic test on the Humidity Cell Test,[(7)] which allows weathering of the sample being tested to occur, weathering charts were generated which indicated the changes in the alkalinity production potential versus sulfur ratios that could occur with time and, hence, the longer term acid generation potential. These charts were then used on other waste samples to determine the potential for acid production from the Alkaline Production Method, without the need for a more time consuming kinetic weathering test.

Hydrogen Peroxide Test

The purpose of the Hydrogen Peroxide Test is to determine the amount of pyrite in a sample. This is done by comparing the rate of chemical oxidation by hydrogen peroxide to a standard curve, prepared by observing the oxidation rates of material with known pyrite content. The rate of chemical oxidation is determined by measuring the rate at which the pH of the reagent in contact with the sample changes.

Net Acid Production Test

This test was developed by Coastech Research, Inc. (1989). It is the only static test method discussed in this chapter that directly evaluates the net acid generating potential. The other test methods described above evaluate the AP and NP separately.

This method uses hydrogen peroxide to oxidize the sulfides contained in the waste. All or part of the acid generated by this oxidation is then consumed by the alkaline constituents in the sample. The net amount of acid remaining after the reaction is determined by titration to pH 7.0, and is usually expressed as kg $CaCO_3$ equivalent per ton of waste.

---

(7) This test method is described in Section 4.3.5.

4.3.4.3 Comparison of Static Prediction Techniques

The Canada Center for Mineral and Energy Technology (CANMET) commissioned a study of the different tests. This study was part of the Mine Environment Neutral Drainage (MEND) program and was conducted by Coastech Research, Inc. in Vancouver, British Columbia (CANMET, 1989).

Tests were conducted on eight tailings samples and four waste rock samples obtained from Canadian mines. All of the waste rock samples had generated acid in the field, and five of the eight tailings samples were from sites where acid generation had been observed. The field assessments of ARD referred to in the study were based on operators' observations.

A summary of the results of static test procedures is given in Table 4.7. The table summarizes the net NP of the eight tailings and four waste rock samples as determined from the test methods discussed above, with the exception of the Hydrogen Peroxide Test which was not included in the study. A negative net generating potential indicates that the AP exceeds the NP.

With the exception of the Net Acid Production Test, all the tests provide net NP values that are of similar magnitude. The Net Acid Production Test indicates a significantly lower net AP (higher net NP) for those wastes which are acid generating. A summary evaluation of the test methods provided in Table 4.8 includes an evaluation of the simplicity of the testing procedure, the time required for running the tests, typical costs, and the ease of interpreting the test results.

For the tailings sample, when comparing the test values in Table 4.7 to the field observations of ARD conditions, it is evident that the tests provided a conservative indication of the net acid generating potential, i.e., they tended to indicate one when none was in fact observed. The results of the tests conducted on the waste rock revealed that the tests may not always provide a conservative result. The Modified Sobek Test indicated a positive net NP, while the other tests and the field observations indicated a net AP for the same material. These conclusions must be tempered by the fact that the field observations had not covered a very long period of time, and the potential for acid generating conditions to develop in the future in certain wastes cannot be ruled out.

Coastech Research, Inc. concluded that Static Prediction Methods have merit in providing a simple, rapid, and low cost "first pass" indication of wastes with acid generating potential. The Net Acid Production Test was rated the best. Use of sulfide sulfur analyses to calculate AP was

TABLE 4.7

SUMMARY OF RESULTS OF STATIC PREDICTION PROCEDURES[1]

SAMPLE IDENTIFICATION NUMBERS	NET NEUTRALIZATION POTENTIAL[2] - KEY COLOR EQUIVALENT/TON					
	BC RESEARCH INITIAL TEST	SOBEK	MODIFIED[3] SOBEK	ALKALINE PRODUCTION TEST[4]	NET ACID PRODUCTION	FIELD OBSERVATION OF ARD
TAILINGS:						
CUT	-737.1	-773.5	-750.2	-775.8	-82.3	Yes, after 15 years
ELT	-94.4	-119.0	-103.1	-94.4	-20.6	Yes
EST	-76.3	-68.3[5]	-59.0	-79.2	-7.2	No
HST	-1217.5	-1275.5	-1217.0	-1261.4	-69.2	Yes
KLT	- 44.7	-96.7	38.5	-84.0	0.0	No
NBT	5.0	-6.4	6.1	- 43.3	0.0	No
WAT	-396.8	- 409.2	- 400.9	- 409.6	- 41.2	Yes
WMT	-678.4	-701.6	-677.3	-698.6	- 48.4	Yes, after 1 year
WASTE ROCK:						
CUWR	-226.5	-175.6	-160.7	-246.0	-9.6	Yes
ESWR	-573.7	-596.6	-616.4	-595.0	-84.9	Yes, after 6 months
HSWR	-157.9	-178.7	-171.6	-177.1	-7.3	Yes
INWR	-7.2	-8.5	7.3[6]	-9.1	-6.8	Yes

(1) Reference: CANMET, 1989.
(2) Positive values designate an acid consuming material; negative values indicate the potential for acid generation.
(3) Modified procedure: 24-hour digestion at ambient temperature and sulfide sulfur analysis.
(4) Net neutralization potential values were calculated as for the standard Acid Base Account test, using titration data and total sulfur assays.
(5) Where the values are enclosed in boxes, the test method provides an incorrect, but conservative, indication of the potential for acid generation.
(6) Where the boxes are shaded, the test method provides an incorrect and unconservative indication of the potential for acid generation.

TABLE 4.8

SUMMARY EVALUATION OF STATIC PREDICTION PROCEDURES[1]

PREDICTION PROCEDURE	SIMPLICITY OF TEST[2]	TIME REQUIRED[3]	SPECIAL EQUIPMENT REQUIREMENT[4]	APPROXIMATE COST PER SAMPLE[5]	EASE OF INTERPRETATION	CORRELATION OF TEST RESULTS WITH FIELD OBSERVATIONS (Table 4.7)
STATIC TESTS						
B.C. Research Initial	Yes	2 days *	None	\$75 - \$200	Easy	Two conservative[6] errors in evaluating tailings.
Sobek	Yes	4 hours*	None	\$50 - \$130	Easy	Three conservative errors in evaluating tailings.
Sobek Modified	Yes	1 day*	None	\$50 - \$130	Easy	One conservative error in evaluating tailings and one unconservative[7] error in evaluating waste rock.
Alkaline Production	Yes	4 hours*	None	\$50 - \$130	Moderate	Three conservative errors in evaluating tailings.
Net Acid Production	Yes	4 hours	None	\$40 - \$80	Easy	One conservative error in evaluating tailings.
OTHER TESTS						
Hydrogen Peroxide Test	Yes	4 hours*	None	\$80 - \$120	Moderate	Comparison not performed. However, the test method is known to provide a poor estimate of pyrite content.

(1) Reference: CANMET, 1989.
(2) This rating assumes that the test is carried out by a trained laboratory technician or technologist.
(3) For tests marked with *, additional time may have to be allowed for sulfur assay turnaround.
(4) This assumes that the tests are carried out in a chemical laboratory equipped to undertake environmental testing.
(5) Costs will depend on the laboratory used, the number of samples tested at the same time, and on the number and cost of the sulfur species assays required. Costs are estimates as of March 1989.
(6) "Conservative" means the test predicted the potential for the waste to be acid generating when none was actually observed.
(7) "Unconservative" means the test does not predict acid generation which is known to have occurred.

recommended over the use of total sulfur in order to avoid errors in assessment due to the presence of sulfates such as gypsum. The Hydrogen Peroxide Test for pyrite estimation was not recommended.

4.3.4.4 Use of Static Tests

Static tests can be used to provide a preliminary indication of the net acid generating potential of a waste by allowing the material to be categorized as one of the following:

- Nonacid generating waste.
- Possibly acid generating waste. The net AP or NP predicted by the test is less than the reliability of the test. Further testing is required to suitably categorize this waste.
- Acid generating waste. The waste would be acid generating, provided the site-specific conditions are favorable for the process to occur. Further testing may be required to confirm the acid generation potential and to determine the timing and quality of the ARD.

The most important aspects in establishing this categorization system are numerical criteria, which are required to determine the category into which a waste sample falls. The "art" in establishing these criteria is to make sure that the boundaries between the categories are larger than the errors associated with the static test used. However, they should not be so wide that a large proportion of the waste falls into the least definitive category, i.e., "possibly acid generating," which requires further testing.

Two approaches which incorporate the above categorization system have been developed and are outlined below. Both of these involve phased, sequenced testing. Initially, simple static tests are conducted. Subsequent, more detailed phases of static or kinetic testing are required if the initial test results do not allow the waste to be put into a specific category, or if the waste is acid generating and more quantitative information on reaction rates and leachate quality is required.

The first approach is outlined in the *Draft Acid Rock Drainage Technical Guide* (B.C. AMD Task Force, 1989). It involves conducting static tests to determine the NP of a waste and to express it in terms of kg $CaCO_3$ equivalent per ton of waste. The criteria recommended for the three categories are as follows:

Net Neutralization Potential (kg $CaCO_3$/ton)	Category
Greater than 20	Nonacid generating waste.
Between 20 and -20	Possibly acid generating waste.
Less than -20	Acid generating waste.

These numeric criteria are based on general experience in conducting tests and comparing the results to known acid conditions. The B.C. AMD Task Force (1989) does not recommend any specific test but does indicate that use of the sulfide species analysis is more reliable in determining the acid generating capacity than is the total sulfur analysis. This is based on the work conducted by Coastech (1989).

A second approach to acid generation prediction has been proposed by Smith and Barton-Bridges (1991). It includes the use of static and kinetic tests in three stages, as follows:

- Stage 1: Determine the AP using total sulfur (LECO furnace method) and the acid NP. Convert both to common units and calculate the ratio of NP to AP. If the NP/AP ratio is greater than 3, the waste should be nonacid generating. If the ratio is less than 3, proceed with Stage 2 testing.
- Stage 2: Determine the forms of sulfur in the waste rock, in particular the sulfide sulfur. Recalculate the AP on the basis of sulfide sulfur values and the NP/AP ratio. If the NP/AP ratio does not exceed 3, the waste may be acid generating, and Stage 3 testing should be conducted.
- Stage 3: Perform a kinetic test using humidity cell or column/lysimeter (see Section 4.3.5). Determine concentrations of all leachable species of interest on leach samples, plus those of potential products of acid generation.

The Stage 1 testing is a very economic test to perform and allows wastes that are very obviously not acid generating to be screened out easily and conservatively. Where it is known that the balance between acid generation and neutralization is close, and that Stage 2 testing will be required, it may be appropriate to bypass Stage 1 testing.

The major difference between the two approaches described above is the use of an absolute value of net NP for the first, and a ratio between the total NP and AP for the second. While both have their merits, it is believed that the second approach (Smith and Barton-Bridges, 1991) has an advantage, as it provides a constant factor of safety, irrespective of the amount of acid generation. This allows a higher degree of confidence in prediction of nonacid generation from those samples that meet the selection criteria. The first approach results in a variable factor of safety. Which decreases as the proportions of reactive sulfides in waste increase.

Use of the Smith and Barton-Bridges (1991) criteria may result in more samples being subject to kinetic testing than with the first approach. This is a small disadvantage when compared to the increased confidence and consistency in the factor of safety in the results.

Both of the approaches described above require kinetic tests for the final phases of testing. These tests are described in the following section.

4.3.5 GEOCHEMICAL KINETIC TESTS

4.3.5.1 Introduction

Kinetic tests involve aging or "weathering" of samples under laboratory controlled or onsite conditions to simulate the time-varying chemical changes that occur in waste. The tests are conducted to confirm the potential to generate net acidity, determine the rates of sulfide oxidation and neutralization, determine typical constituent concentrations that occur in the acid drainage, and evaluate the effectiveness of proposed ARD control and treatment methods. The results of these tests would indicate, for example, if the rate of acid generation is significant or negligible, over what period it may be severe and, hence, over what time period(s) controls would be required.

In order to provide meaningful information, kinetic tests need to be conducted over longer periods of time than some of the other tests, up to and possibly exceeding several months. Laboratory short-term tests generally only provide qualitative information on drainage water quality. As indicated previously, kinetic testing can also be conducted for a period of months or even years on pilot scale waste piles or impoundments at the mine site.

It is usually necessary to obtain additional information on the waste material characteristics, such as specific surface areas, mineralogy, and the total content of the metals of interest. These characteristics can affect to the interpretation of the results from kinetic tests. They provide insight into the chemical process and the metal balance between leachate and waste, thereby facilitating extrapolation of the test results to much longer term field conditions.

4.3.5.2 Test Methods

Kinetic tests can be conducted within containers designed to allow: (1) addition of chemicals to control reactions or reaction rate, (2) inoculation by bacteria, and (3) control over environmental parameters such as atmospheric gases, temperature, and humidity. There are several different standardized protocols that have been established for conducting this type of test. These are discussed in the following sections.

B.C. Research Confirmation

This test is a kinetic, biological oxidation test designed to determine if sulfide oxidizing bacteria can produce more acid from oxidation of pyrite in the sample than can be consumed by an equal quantity of the sample (Duncan and Bruynesteyn, 1979). The test involves inoculating a pre-acidified (pH 2.0 to 2.4) pulp containing the test sample with an active culture of bacteria such as T. ferro-oxidans. The pulp pH is monitored, and the test is terminated when oxidation ceases, as indicated by the attainment of a stable pH. A weight equivalent to the original sample is then added in two increments after 24 and 48 hours, and the pH is measured 24 hours after each addition. If the pH is above 3.5 on either occasion, the sample is classified as a nonacid producer. If the final pH remains below 3.5, potential acid generation is indicated. Typically, the test requires 3 to 4 weeks to complete.

Modified Confirmation

A procedure similar to the tests carried out at B.C. Research is used routinely at Coastech (Lawrence and Sadeghnobari, 1986).

This procedure differs from the B.C. Research test in that the weight of the sample used is generally smaller and is based on the sulfide sulfur content. Also, and more significant is that, once the additional sample has been added and the reaction is complete, sodium hydroxide is added if the pH is still below 3.5. The amount of sodium hydroxide is sufficient to neutralize the acid added in the initial stage of the test. This removes the bias toward an acid generating classification that is provided by the initial acidification procedure. The classification of the sample is then based on the final pH. This modified procedure is particularly relevant to those samples where the initial amount of added acid is substantial and can significantly affect the final predictive result (Lawrence and Sadeghnobari, 1986).

Shake Flask

The Shake Flask Test (Halbert et al., 1983) is intended to evaluate leachate quality as a sample of waste is weathered. This test involves leaching waste samples in water in Erlenmeyer flasks placed on a reciprocating shaker. Usually a series of identically prepared flasks is set up for each sample, and one flask is removed for leachate analysis on a weekly or alternate schedule. The type of analyses conducted are similar to those in the humidity cell test, discussed below. The effect of temperature, initial pH, and the presence of bacteria can all be evaluated by establishing different series of flasks with varying starting pHs, bacterial populations, or test temperatures.

Soxhlet Extraction

This test uses Soxhlet apparatus to simulate geochemical weathering (Singleton and Lavkulich, 1978; Sullivan and Sobek, 1982). Two procedures are generally used: one, referred to as the "standard test" for the purposes of this book, uses the calibrated Singleton and Lavkulich Soxhlet, and both acetic acid and distilled water as leachate. The other, referred to as the "modified test," uses distilled water leachate as described by Sobek (1982). In the standard method, the test sample is leached with hot acetic acid or water at 70° C. The modified method uses a different Soxhlet arrangement so that the temperature of the leachate is reduced to 25° C.

Humidity Cell

The Humidity Cell Test models the processes of geochemical weathering of a waste material (Sobek et al., 1978). A special apparatus is used to provide simple control over air, temperature, and moisture, while allowing for the removal of oxidation products which are collected and monitored.

The test procedure involves subjecting a bed of the waste material in a humidity cell to alternating cycles of dry air (3 days), moist air (3 days), and leaching (1 day). The leachate generated is collected and analyzed for a number of parameters, including pH, redox (mV), acidity, alkalinity, sulfate, conductivity, and dissolved metals, utilizing the inductively coupled plasma (ICP) techniques. The test is generally run for 10 weeks, although a longer time period might be required to induce sufficient weathering in some samples.

Column Test

This test involves stacking the waste material in a cylindrical column and allowing water to trickle through at a predetermined pattern to simulate, for example, wet and dry weather cycles. There are no established protocols for Column Tests, and their design is usually customized to suit site conditions and to address specific waste management considerations.

Pilot Scale

This is the most elaborate form of kinetic test. It involves placing a relatively large sample of the waste, i.e., tons to thousands of tons, over liners and leachate collection systems. The waste pile is appropriately instrumented and sampled at frequent intervals to track both the weathering process and the generation and quality of any acid leachate. As with Column Tests, there are no established protocols for Pilot Scale Tests.

Carbon Pressure

This is a specialized test which is used on a limited basis to determine what proportion of the carbonate in a pyrite mine waste will be available to neutralize acidity. Its purpose is to determine the rates of reaction of carbonates, including calcite, dolomite, and siderite. The rates of dissolution of the carbonates in hydrochloric acid are determined by measuring the pressure of the CO_2 generated. Evangelou et al. (1985) claim the test is able to differentiate and quantify the three carbonate species. However, evaluation of this test procedure by Coastech indicated that the results were inconclusive.

4.3.5.3 Comparison of Several Kinetic Prediction Techniques

The procedures and results of different kinetic tests were reviewed by Ferguson (1985) and Coastech Research, Inc. (1989). These are discussed in the following paragraphs.

A summary of the kinetic test comparison undertaken by Ferguson (1985) is presented in Table 4.9. Ferguson concluded that the humidity cell technique was successful in assessing the potential for acid production in samples which exhibited a balance between AP and NP according to static acid-base accounting. It was also found that decreasing sample particle size increased the pyrite oxidation rate exponentially. Acidity production was significantly greater for the smaller size particles and varied linearly with the specific surface area of the pyrite particles.

The study by Coastech Research, Inc. (1989) compared the results of four kinetic tests conducted on the same tailings and waste rock samples discussed under the static test comparisons in Section 4.3.4.3.

The test procedures used by Coastech generally followed the procedures outlined in Section 4.3.5.2. Leachate in the Humidity Cell Tests was developed by pouring distilled water on top of the sample bed, allowing it to stand for one hour, and then decanting it. The Shake Flask Tests were conducted at room temperature using 250 gm samples in 500 mL distilled water with no

TABLE 4.9

COMPARISON OF SEVERAL KINETIC PREDICTION TECHNIQUES[1]

KINETIC TEST	USE	ADVANTAGES	DISADVANTAGES
B.C. Research Confirmation	Frequently used for base metal and gold mines in Canada.	• Relatively simple to use. • Relatively low cost. • Allows assessment of potential for biological leaching.	• Bacteria may not be acclimated to sample. • May be difficult to interpret if pH does not strongly rise or fall. • Does not simulate initial ARD production step. • May take several weeks for sample pH to stabilize.
Shake Flasks	Infrequently used for base metal mines in Canada.	• Simulates acid production and consumption reactions. • Assesses effect of several controlling conditions on acid production.	• May be difficult to interpret results. • Moderate cost. • Involves large amount of data to compile and interpret. • Requires long time to complete testing. • Not practical for large number of samples. • Limited in considering different waste particle sizes.
Soxhlet Extraction	Infrequently used in coal fields of Appalachia, United States.	• Relatively simple to use. • Results are available in relatively short time period. • Allows assessment of interaction between acid production and consumption reactions.	• Still in the development stages. • Results are difficult to interpret. • Relationship of results to actual site-specific conditions is not clear.
Humidity Cell	Relatively common in coal fields of Appalachia, United States.	• Simulates acid production and consumption reactions kinetically. • Can approximately simulate wet and dry weather cycles.	• May be difficult to interpret results. • Moderate cost. • Involves large amount of data to compile and interpret. • Requires long time to complete testing. • Not practical for large number of samples. • Limited in considering different waste particle sizes.
Columns	Gradually being used more extensively on gold mine projects in the western United States.	• Simulates acid production and consumption reactions. • Can simulate effect of different rock types within the same column. • Can simulate wet and dry weather cycles. • Can consider different waste particle sizes within column size constraints.	• May be difficult to interpret results. • Relatively high costs. • Involves large amount of data to compile and interpret. • Requires long test period. • Not practical for large number of samples. • May have problems with uneven leachate application and channelization.
Pilot Scale Waste Piles	Infrequently used.	• Most realistic simulation of waste and site-specific climatic conditions. • Can model relevant natural conditions in a single test.	• High cost. • Involves large amount of data to compile and interpret. • Very long test period. • Not practical for more than a very small number of individual tests.

(1) Reference: Fergusen, 1985.

bacteria added. The waste rock samples were preprocessed by grinding to a minus-200 mesh size. No preprocessing of the tailings samples was conducted. The Soxhlet Extraction Tests were run using the modified procedure and distilled water.

A summary of the results of the Coastech Research Inc. (1989) study of kinetic test procedures for predicting acid generation is given in Table 4.10. Field observations are also included in the table. The principal observations made by Coastech for each method are as follows:

- Modified Confirmation Test: With the exception of tailings samples ELT and EST (Table 4.10), the method produced results that were consistent with the field observations of ARD. The initial acidification of samples makes some results unrealistic. The modified procedure which compensates for the initial acid addition is an improvement, although it still does not provide the opportunity to investigate the potential for the development of ARD in the higher pH ranges. For some samples, the reactions taking place in the pH range from 7 to 2 might be important in determining the formation of ARD.
- Humidity Cell Test: The humidity cell correctly predicted field results on all eight tailings samples and three of the four waste rock samples. The one incorrectly classified waste rock sample was shown not to generate ARD in the lab, although it did in the field. The test was judged to be well suited to the evaluation of waste materials deposited on land, where they are alternately subjected to sequences of infiltrating water, drying, and flushing.
- Shake Flask Test: The Shake Flask Test correctly predicted the behavior of all eight tailings samples, but was incorrect for three out of four waste rock samples which were shown not to generate ARD in the lab, but did in the field. The good prediction for tailings could be expected from this test, which has a constant leaching action, as it simulates the behavior of tailings, which are often saturated and have low oxygen permeability. For the same reason, the test might not be well suited to waste rock, which would be more likely to be subjected to alternate wetting, drying, and flushing cycles as simulated by humidity cells.
- Modified Soxhlet Extraction Tests: Although this test procedure is seen to be less realistic than some other weathering tests, such as humidity cells, the modified method using water was successful in predicting the field behavior of all eight tailing samples and three of the four waste rock samples.

  The standard method using acetic acid was also tested and predicted all samples to be sources of ARD. It is believed that the results of this procedure are unrealistic.

  The modified method has the disadvantage of being more complex than the Humidity Cell or Shake Flask tests and requires more sophisticated equipment. However, the test is of shorter duration and may be capable of simulating a long weathering sequence in a relatively short time.

TABLE 4.10

SUMMARY OF RESULTS OF KINETIC PREDICTION PROCEDURES[1]

SAMPLE	KINETIC TEST				FIELD OBSERVATION OF ARD
	MODIFIED CONFIRMATION	HUMIDITY CELL	SHAKE FLASKS	MODIFIED SOXHLET EXTRACTION	
TAILING:					
CUT	Yes, strong	Yes, moderate	Yes, at end of test	Yes, strong	Yes, after 15 years
ELT	No, very marginal[2]	Minor at end of test	Very low activity	Yes	Yes
EST	Yes, strong[3]	No	No	No	No
HST	Yes, strong	Yes, strong	Yes, strong	Yes, strong	Yes
KLT	No	No	No	No	No
NBT	No	No	No	No	No
WAT	Yes, strong	Yes	Yes, moderate	Yes, strong	Yes
WMT	Yes, strong	Strong at end of test	Yes, moderate	Yes	Yes, after 1 year
WASTE ROCK:					
CUWR	Yes, strong	No	No	No	Yes
ESWR	Yes, strong	Very low production	Low production	Yes, strong	Yes, after 6 months
HSWR	Yes, strong	Very minor at end of test	Insignificant activity	Yes, low production	Yes
INWR	Yes, marginal	Very minor at end of test	Insignificant activity	Yes, low production	Yes

(1) Reference: Coastech Research, Inc, 1989.
(2) Where the boxes are shaded, the test method provides an incorrect and unconservative indication of the potential for acid generation.
(3) Where the values are enclosed in boxes, the test method provides an incorrect, but conservative indication of the potential for acid generation.

A summary comparison of the various kinetic test procedures for prediction of acid generation, as evaluated in the Coastech Research, Inc. study, is presented in Table 4.11. This table provides a comparison on the basis of simplicity of the test, time required, special equipment needs, cost, ease of interpretation of the data collected, and correlation of the results with field data.

4.3.5.4 Conclusions

The most suitable test methods for determining the ARD potential of waste rock or tailings disposed of on land are the Humidity Cell or the Modified Soxhlet Extraction methods using distilled water. The advantage of the Humidity Cell Test is that it can simulate the wetting and drying cycles to which a waste is naturally exposed. The advantage of the Modified Soxhlet Extraction Test is that it provides a more rapid assessment of ARD conditions. The Modified Confirmation test is not generally recommended. Even though it provided a reliable indication of ARD formation, it does not address the weathering that can occur in a pH range of 7 to 2, and it does not provide information on leachate quality.

Shake Flask Tests are not likely to be as reliable or useful as Humidity Cell or Soxhlet Extraction tests for wastes stored on land. They are, however, likely to be more reliable for evaluating ARD generation where underwater storage of tailings and waste rock is contemplated.

4.4 ACID GENERATION AND LEACHATE MIGRATION CONTROL MEASURES

4.4.1 APPROACH TO CONTROL

The economic and environmentally sound development of mines in sulfide-bearing rock is dependent on reliable prediction of the ARD potential of the ore and mine wastes, as well as the application of suitable control measures where ARD is expected to occur. Although the control of ARD is not the focus of this chapter, this section provides an overview of control technologies to facilitate the discussion of ARD testing procedures for control alternatives that appears in Section 4.4.3, and also to provide the reader with a more complete understanding of how ARD can be dealt with in the mining industry.

The purpose of ARD control measures is to both prevent acid drainage at new facilities, and to abate acid drainage where it occurs at existing facilities. In this chapter, the term "prevention" is used to describe the control measures selected for proposed waste management units with the potential to produce ARD. "Abatement" is used to describe controls implemented at an existing facility which is generating ARD.

TABLE 4.11

COMPARISON OF KINETIC PREDICTION PROCEDURES [(1)]

PREDICTION PROCEDURE	SIMPLICITY OF TEST [(2)]	TIME REQUIRED	SPECIAL EQUIPMENT [(3)]	COST PER SAMPLE ($)[(4)]	EASE OF INTERPRETATION	CORRECT CORRELATION WITH FIELD DATA	
						Tailings	Waste Rock
Modified Confirmation	Moderate	3 to 4 weeks	Moderate	200 - 400	Moderate	6 out of 8	4 out of 4
Humidity Cell	Moderate	More than 10 weeks	Moderate	500 - 1,000	Moderate	8 out of 8	3 out of 4
Shake Flasks	Moderate	More than 10 weeks	Moderate	500 - 1,000	Moderate	8 out of 8	1 out of 4
Modified Soxhlet Extraction	Moderate	More than 3 weeks	Yes	250 - 500	Moderate	8 out of 8	3 out of 4

(1) Reference: Coastech Research, Inc., 1989.
(2) In the Simplicity of Test category, rating assumes test carried out by a trained laboratory technician or technologist.
(3) In the Special Equipment category, rating assumes test carried out in a normally equipped chemical laboratory.
(4) In the Cost Per Sample category, costs will vary with the laboratory used, will depend on number of samples tested at the same time, and on the number and cost of the sulfur species assays required. For kinetic tests, costs are based on a weekly sampling regime. Actual costs will depend on the number and frequency of leachate analyses performed, including ICP multi-element analyses. Costs are estimated as of March 1989.

The technical options available for the prevention of acid generation and leachate migration are considerably greater than those available for abatement of existing conditions. Implementation of abatement measures is limited to those that can be retrofitted and that conform to any physical constraints posed by the existing waste management unit site. Therefore, abatement measure costs are generally significantly higher than equivalent prevention measures.

An important criterion to be identified before designing control measures is the length of time that the measure must remain effective. Certain measures or techniques are effective short-term controls but may not be suitable over the long term. Conversely, the most suitable long-term control method may not be the most cost-effective or practical short-term measure.

There are three generally accepted approaches to the prevention or abatement of acid generation and leachate migration. These are:

- Control of the acid generation process.
- Control of migration of the leachate.
- Collection and treatment of acid drainage.

The above approaches are listed in order of preference. It is best to control acid generation and prevent the formation of ARD. Where this is not possible, leachate migration control is the next best approach and is intended to prevent leachate from being formed or if formed, from escaping the waste management unit. If neither of these control measures is effective or possible, the final alternative approach is to collect and treat the ARD. Frequently, a combination of measures may provide the most secure control.

4.4.2 AVAILABLE CONTROL OPTIONS

Control options for each of the three categories are listed in Table 4.12 (Barton-Bridges and Robertson, 1989). These measures include both short- and long-term controls. Each option may experience a different level of success, depending on the project and site-specific conditions.

4.4.3 TESTING FOR DESIGN OF CONTROL MEASURES

An essential step in the design of appropriate measures to control ARD is to conduct tests to evaluate the effectiveness of the control options being considered. Kinetic tests conducted in the laboratory or in the field are the most appropriate available techniques for this purpose. The tests

TABLE 4.12

AVAILABLE ARD CONTROL OPTIONS[1]

CONTROL OF ACID GENERATION

- Pre-treatment of tailings or waste rock to remove or exclude sulfide minerals.
- Covers and seals to exclude water.
- Covers and seals to exclude oxygen (including a water cover).
- Waste segregation and blending to control pH.
- Base additives to control pH.
- Bactericides to control bacterial oxidation of sulfide minerals.
- Avoid exposing reactive materials to conditions which lead to acid generation, e.g., modify mine plan or avoid mining the deposit or portions of the deposit.

CONTROL OF ARD MIGRATION

- Covers and seals to exclude infiltration of precipitation.
- Controlled placement of waste to minimize infiltration.
- Interception and diversion of surface water.
- Interception and diversion of ground water.

COLLECTION AND TREATMENT OF ARD

- Surface water and ground water collection and treatment systems, utilizing:
  - Active treatment systems, e.g., treatment by a physical-chemical plant.
  - Passive treatment systems, e.g., treatment by wetlands.

(1) Modified from Barton-Bridges and Robertson (1989).

should be designed to simulate the site-specific climatic conditions, the method of waste deposition, and the physical and chemical impacts of the control measure being evaluated. For example, if blending of potentially acid generating waste with acid consuming waste (or other imported material) is a control option, the tests could be conducted with appropriately blended samples. Such tests could be conducted with different ratios of blending to assist in predicting the required quantity of acid consuming materials.

It is important to recognize that, in most cases, the control measure will have to be in effect for periods greater than the test duration. For this reason, it is important to determine the relative rates of acid generation and neutralization, plus total potential acidity and neutralization capacity in order to extrapolate the test results into the future. While these tests are conducted under the proposed conditions to inhibit acid generation, parallel tests under ideal conditions for acid generation should be conducted to serve as a control.

4.5 IN-WASTE MONITORING

4.5.1 OBJECTIVES

Monitoring forms an important and integral part of any mine waste management system. This section deals with the in-waste monitoring that can be implemented to provide site-specific and real-time information on acid generating conditions, and the performance of measures aimed at controlling or eliminating the generation of ARD. A detailed discussion of other forms of in-waste monitoring, and of vadose zone and ground water monitoring, is provided in Chapter 10.

The objectives of in-waste monitoring of ARD are as follows:

- To provide confirmation that the ARD control measures are operating as intended.
- To provide an early warning of any problems that may arise so that appropriate modifications to the proposed control measures can be made.
- To provide information to refine the plans or designs for control measures to be implemented, if necessary.

4.5.2 CONCEPTS

In-waste monitoring is intended to identify the presence or changes of any of the products of the acid generation process. These products can be identified by one or more of the following parameters:

- Temperature - The oxidation of sulfides involves exothermic, heat producing reactions. Hence, indications of temperature change can be symptomatic of acid generation.

- Gases - The acid generation process is oxygen-consumptive. The depletion of oxygen within the gas phase (or as dissolved oxygen in the aqueous phase) can, therefore, be indicative of acid generation. Furthermore, the carbonate acid neutralization process produces carbon dioxide in some cases. Therefore, enrichment of carbon dioxide in the gaseous phase of a waste can be indicative of acid generation.
- Soluble products - Acid generation can be identified by soluble products such as acidity, sulfate, iron, and a range of secondary trace metals associated with the mineralization.

4.5.3 MONITORING METHODS

4.5.3.1 Temperature

The effects of heat generation can be assessed by a number of different methods, including:

- Direct observation.
- Remote detection and measurement.
- In situ temperature sensing.

These methods are discussed in the following sections.

Direct Observation

The heat produced by acid generation in waste rock piles can be observed in some cases and is manifested by the following processes:

- "Hot spots" on the waste surface that are warm to the touch.
- The appearance of "steam" from sections of the dump, particularly during or after precipitation events.
- The presence of bare spots on the dump surface during and following snowfall.

Such simple observational methods are useful in identifying acid generation in a waste facility and its location within the area of the facility. However, they are not reliable or sufficiently sensitive to be relied upon as the only monitoring method.

Remote Detection and Measurement

A more sophisticated approach to identification of these heat producing processes is to use remote imaging techniques such as thermal infrared (thermography). The contrast produced by differences in the heat-radiating properties of surface and shallow subsurface materials may be indicative of zones of acid generation. Such methods have been used in aerial surveys of three waste rock piles and open pits in British Columbia to identify acid generating zones (Ferguson and Erickson, 1987). It is possible that thermal infrared scanning can be conducted from the ground. To date, this method does not appear to have been tried at mine waste piles.

For remote imaging techniques to be effective, there needs to be sufficient thermal contrast between the acid generating and nonacid generating materials. However, there is not always sufficient contrast between these materials. This method is, therefore, probably best suited to identifying zones of high acid generation and is not appropriate for detecting the onset of acid generating conditions.

In Situ Temperature Sensing

Thermistors (temperature sensitive electrical probes) can be used to detect temperatures and temperature gradients when placed in series, or "strings." They are applicable to both tailings and waste rock piles and can be used to assess changes during either the operating phase of a waste management unit or after it is closed.

Currently there are several relatively comprehensive monitoring programs in British Columbia aimed at measuring thermal gradients and changes in temperature caused by acid generating processes which are being remediated. For example, temperatures up to 45° C were recorded by thermistor strings buried in a very active acid generating waste rock pile in British Columbia (Patterson, 1986). In another example, a significant drop in temperature following reclamation of a waste rock pile was recorded by Harries and Ritchie (1985), using thermistor probes installed in monitoring wells.

4.5.3.2 Gases

Gas sampling, use of sampling tubes, lysimeters, etc., can be performed in the partially-saturated zones in both waste rock piles and tailings impoundments. In addition, if there are dissolved gas concentrations, they can be measured in the aqueous phase from tailings, pore water, or from below the water table surface. Careful purging and sampling protocols are needed to avoid pollution of gas samples due to the ingress of oxygen to the sample during collection and handling.

Some of the potential problems with free gas sampling can be avoided by collecting dissolved gas samples, where possible. Immediate preservation and procedures to minimize aeration are necessary to avoid changes in the dissolved gas content of the samples. This is particularly important for samples collected at depth in a waste facility, where the equilibrium between dissolved gas and fluid may be different from that at the surface where the sample is handled.

Interpretation of the analysis of gas samples requires some basic understanding of the gas mixtures and concentration profiles in the waste. Some typical oxygen partial pressure profiles have been summarized by Ferguson and Erickson (1987) and are described below.

When oxygen to the interior of a waste pile is supplied by diffusion which is induced by concentration differential from the atmosphere, the profile within the pile shows decreasing oxygen with increasing depth below the surface. Under these conditions, active oxidation zones associated with acid generation can be detected, as they result in sharp partial pressure gradients. When advection, which is induced by thermal or pressure differential, supplies oxygen to the mine waste, high partial pressures are observed at intermediate depths. Partial pressure gradients are, therefore, less easily correlated with active oxidation. However, the availability of oxygen in advection zones indicates that, if present, pyritic material will likely undergo oxidation.

4.5.3.3 Soluble Products

Samples of soluble ARD products in the waste and the underlying geologic formation can be collected and analyzed. In the unsaturated zones, which generally occur within waste rock piles and some tailings piles, samples of fluids adjacent to waste particles can be obtained using specialized equipment (described in Chapter 10) or by direct sampling of solid material followed by a laboratory extraction of the fluids attached to the particles. In saturated zones, such as those within "wet" tailings impoundments, samples can be obtained using piezometers or wells. Once collected, these samples are analyzed for the soluble products of the acid generation process. When sampling in the unsaturated zone, particularly in waste rock, it must be recognized that the waste is heterogeneous and that there will be preferential flow paths acting as hydraulic conduits for these soluble products. The chemical characteristics of the liquids in these conditions may be significantly different than liquids located in low flow or "dead" areas within the waste. Therefore, the interpretation of the onset, occurrence, and magnitude of acid generation should be based on a sufficiently large number of samples collected from these different areas within the unsaturated zone. In tailings, due to their more uniform nature and "relative" hydraulic homogeneity, a lesser number of samples usually would be required.

4.6 REGULATORY IMPLICATIONS

4.6.1 MOTIVATION

ARD is a recognized potential environmental problem associated with certain mine wastes, particularly those located in wetter climates where significant volumes of leachate can be produced. Sound waste management requires sufficient waste sampling and characterization testing to identify whether ARD will occur, and appropriate waste management alternatives to alleviate or deal with ARD, if necessary.

The current mine waste regulations in California and many other states do not explicitly require identification of potential ARD conditions in either waste or remnant mineralized rock in mines. To promote a consistent approach to identifying ARD problems, there would be considerable merit in providing general guidelines for testing waste for the potential presence of ARD, as well as for the control or management of ARD in mine waste regulations.

4.6.2 RECOMMENDATIONS

4.6.2.1 General

Mine waste regulations should incorporate the following major components that influence the ability to identify and manage ARD:

- Waste Sampling - Representative sampling of the waste is required for reliable predictive testing for the potential presence of ARD.
- Waste Testing - General guidelines for waste testing should be provided to ensure that consistent approaches are adopted by mining companies. The most suitable approach to testing involves a phased program which starts with the simpler static tests and, for those materials which have ARD potential, requires subsequent kinetic testing. Specific test methods can be prescribed, but operators should be allowed the flexibility to proceed with alternative methods if indicated by site-specific conditions or other constraints.
- Waste Characterization and Management - Site-specific waste containment practices should be determined for wastes or specific portions of a waste that have been demonstrated to generate ARD and are to be disposed of at locations and climates where leachate is formed.
- Mine Operator Requirements - A special waste management plan should be prepared for those waste management units that are designed and operated to contain materials capable of generating ARD.

4.6.2.2 Suggested Regulatory Approach

This section outlines a conceptual approach for dealing with the ARD issue in mine waste regulations. This approach is based on the four major components discussed above.

Waste Sampling

Mine wastes should be characterized for acid generating potential using samples which represent the range of geochemical extremes likely to be encountered within the various lithologic units that comprise the waste. In selecting these samples, the following factors should be considered:

- Lithology of host rock.
- Extent of sulfide mineralization.
- Type or reactivity of sulfide mineralization.
- Extent of the material to be mined.
- Schedule for mining and wasting of the different types of material and the extent to which blending will occur.
- Physical and chemical processing which affects the acid generating potential and leachate quality.

Waste Testing

The following sequence of waste testing should be conducted on the set of samples in order to determine the presence and extent, if any, of net acid generating potential. Where commingling of different types of waste will occur, samples of the blended material should be tested. Only samples which represent more than 1 percent of the total waste material require testing.(8)

- Stage 1 Testing - Determine the total sulfur by the Leco furnace(9) method and the acid NP using acid titration. Express results in terms of percent Calcium Carbonate Equivalent.

  Samples which have NP greater than three times the AP can be considered to be nonacid generating. For the other samples, proceed to Stage 2 Testing.

- Stage 2 Testing - Determine the sulfide sulfur content of the sample. Express the results in terms of percent Calcium Carbonate Equivalent.

  Samples that have a neutralization capacity (determined in Stage 1) greater than three times the sulfide sulfur content can be considered to be nonacid generating. For the other samples, proceed to Stage 3 Testing.

- Stage 3 Testing - Perform humidity cell testing, or the equivalent, and collect weekly leachate samples for analyses over a period of not less than one month. Use the results of the leachate analyses as an indicator of waste leachate characteristics. Alternative testing methods can be applied where it is judged that they would provide a more realistic determination of the net acid generating potential.

---

(8) Very small, isolated quantities of material with ARD potential will not affect the chemical characteristics of either the waste as a whole or major parts of it. It will therefore not be necessary to characterize these.

(9) See Section 4.3.4.

In the above testing sequence, Stage 1 testing can be omitted for a particular sample, provided that Stage 2 or 3 testing is conducted. Stage 1 and 2 testing can be omitted, provided that Stage 3 testing is conducted.

Waste Characterization and Management

In cases where a fraction of the waste material is judged to be acid generating, special waste management practices should be utilized to isolate or neutralize the impact of these waste portions. These special measures will depend on site-specific conditions, and could include treatment by chemical, physical, or biological means, commingling with nonacid generating waste materials, and/or providing containment or encapsulation within the main bulk of the waste. Waste stream separation, with the potential acid generating material being disposed of in its own facility, can also be considered. The fraction of waste material that can be accommodated by special measures, or by co-disposal with the remaining waste material, will depend on the geochemical properties of the various waste fractions.

In cases where a major portion of the waste material is judged to be potentially acid generating, the entire waste mass should be disposed of as if it were acid generating. The geochemical mitigating effects of the nonacid generating material can be taken into account when evaluating the total net acid generating potential of the waste. The required degree of treatment in containment required to assure the protection of the beneficial uses of the waters of the State should be determined based on site-specific conditions.

Mine Operator Requirements

Where there is an indication that even a fraction of the waste to be disposed of in any single waste management unit is acid generating, the operator should prepare a special waste management plan that specifically incorporates measures for dealing with ARD. The following information should be provided as part of the plan:

- The results of test data and analyses which define both the short- and long-term quantity and quality of ARD likely to be produced from the geochemically similar acid generating fractions of the waste.
- Geochemical information on the nonacid generating waste also should be provided to indicate its potential to neutralize ARD, or to prevent or impede fluid or gas migration.
- Finally, the plan should incorporate a detailed description of how the waste will be managed, including the proposed ARD control and/or containment measures, the anticipated production schedules for the different waste categories, and monitoring, interim and final closure, and contingency plans.

4.7 BIBLIOGRAPHY

Albright, R. *Prediction of Acid Drainage in Meguma Slates*. Proceedings Acid Mine Drainage Workshop, Environment Canada, Transport Canada. 1987. 146-62.

Barton-Bridges, J.P. and A. MacG. Robertson. *Design and Reclamation of Mine Waste Facilities to Control Acid Mine Drainage*. Proceedings Joint Conference of CLRA and ASSMR, Calgary. August 1989.

Brady, K.B.C., J.R. Shaulis and V. W. Skema. *A Study of Mine Drainage Quality and Prediction Using Overburden Analysis and Paleoenvironmental Reconstructions, Fayette County, Pennsylvania*. Proceedings Conference on Mine Drainage and Surface Mine Reclamation, Vol. I, United States Bureau of Mines IC 9183. Pittsburgh. April 19-21, 1988. 33-43.

British Columbia Acid Mine Drainage Task Force. *Draft Acid Rock Drainage Technical Guide, Volume 1*. Prepared by Steffen Robertson and Kirsten (B.C.), Inc. August 1989.

Bruynesteyn, A. and R.P. Hackl. "Evaluation of Acid Production Potential of Mine Waste Material." Minerals and the Environment. (1984). 5-8.

Buckman, H.O. and N.C. Brady. *The Nature and Properties of Soils*. 6th Edition. New York: Macmillan, 1960. 567.

Busenberg, E. and C. Clemency. "The Dissolution Kinetics of Feldspars at 25° C and 1 Atmosphere CO_2 Partial Pressure." Geochim. Cosmochim, Acta 40 (1976): 41-49.

Canect Environmental Control Technologies. *Field Sampling Manual for Reactive Tailings*. Prepared for MEND Program, CANMET. 1989 (in preparation).

CANMET (Canada Centre for Mineral and Energy Technology; Energy, Mines and Resources Canada). *Investigation of Prediction Techniques for Acid Mine Drainage*. MEND Project 1.16.1a. DSS File No. 30SW.23440-7-9187. March 1989.

Caruccio, F.T. and G. Geidel. *Estimating the Minimum Acid Load That Can Be Expected From a Coal Strip Mine*. Symposium on Surface Mining Hydrology, Sedimentation and Reclamation. Lexington, N.Y. 1981.

Caruccio, F.T. and G. Geidel. *Reclamation Strategies as Applied at the DLM Properties*. Proceedings of the Fourth West Virginia Surface Mine Drainage Task Force Symposium. Morgantown, W.V. 1986.

Caruccio, F.T. and G. Geidel. *The Effect of Plastic Liner of Acid Loads/DLM Site*, W.V. Proceedings of the Fourth West Virginia Surface Mine Drainage Task Force Symposium, Clarksburg, W.V. 1983.

Caruccio, F.T., G. Geidel, and A. Pelletier. *The Assessment of a Stratum's Capability to Produce Acid Drainage*. Proceedings Symposium on Surface Mining Hydrology, Sedimentology and Reclamation. University of South Carolina. 1983. 437-443.

Caruccio, F.T., G. Geidel, and M. Pelletier. *Occurrence and Prediction of Acid Drainages*. Journal of the Energy Division American Society of Civil Engineers, 107, No. EY1. May 1981.

City Resources (Canada) Inc. *Volume V: Environmental Research and Special Studies*. Cinola Gold Project Stage II Report. 1988.

Coastech Research, Inc. *Investigation of Prediction Techniques for Acid Mine Drainage*. Study Conducted for CANMET, Energy, Mines, and Resources Canada. DSS File No. 30SZ.23440-7-9178. March 1989.

diPretoro, R.S. and H.W. Rauch. *Use of Acid-Case Accounts in Premining Prediction of Acid Drainage Potential: A New Approach From Northern West Virginia*. Proceedings Conference on Mine Drainage and Surface Mine Reclamation, Vol. I, USBM IC 9183. Pittsburgh. April 19-21, 1988.

Duncan, D.W. and A. Bruynesteyn. *Determination of Acid Production Potential of Waste Materials*. Met. Soc. AIME, paper A-79-29, 1979. 10.

Erickson, P.M. and R.S. Hedin. *Evaluation of Overburden Analytical Methods as Means to Predict Post-Mining Coal Mine Drainage Quality*. Proceedings Conference on Mine Drainage and Surface Mine Reclamation, Vol. I, USBM IC 9183. Pittsburgh. April 19-21, 1988.

Erickson, P.M., R.W. Hammack, and R.L.P. Kleinmann. *Prediction of Acid Drainage Potential in Advance of Mining Control of ARD*. United States Bureau of Mines IC 9027. Pittsburgh Research Center. 1985.

Errington, J.C. and K.D. Ferguson. *Acid Mine Drainage in British Columbia: Today and Tomorrow*. Proceedings of the Acid Mine Drainage Seminar/Workshop. Nova Scotia: Environment Canada Catalogue En 40-11-7. March 23-26, 1987.

Evangelou, V.P., K. Roberts, and G.W. Szekeres. *The Use of an Automated Apparatus for Determining Coal Spoil Carbonate Types, Content and Reactivity*. Symposium on Surface Mining, Hydrology, Sedimentology and Reclamation, Lexington, KY. December 9-13, 1985.

Ferguson K.D., and P.M. Erickson. *Pre-Mine Prediction of Acid Mine Drainage*. Salomans W. Forstner U. (eds) Environmental Impact and Management of Mine Tailings and Dredged Materials. Springer Verlag, Heidelburg. 1986.

Ferguson, K. and P. Erickson. *Will It Generate Acid? An Overview of Methods to Predict Acid Mine Drainage*. Proceedings ARD Seminar/Workshop, Environment Canada. Halifax, Nova Scotia, March 1987.

Ferguson, K.D. and J.C. Errington. *Planning for Acid Mine Drainage*. Canadian Institute of Mining and Metallurgy, District 6, 11th Annual Meeting. Vancouver. October 30-31, 1987.

Ferguson, K.D. *Static and Kinetic Methods to Predict Acid Mine Drainage*. International Symposium on Biohydrometallurgy. Vancouver. August 22-24, 1985.

Finkelman, R.B., and D.E. Giffin. *Hydrogen Peroxide Oxidation: An Improved Method for Rapidly Assessing Acid-Generating Potential of Sediments and Sedimentary Rocks*. Recreation and Revegetation Research 5. 1986. 521-34.

Freeman, J., J.W. Sturn, and R.M. Smith. *Acid/Base Accounting - U.S. Coal Mine Spoils and Preparation Plant Refuse*. Proceedings ARD Seminar/Workshop, Environment Canada. Halifax, Nova Scotia. March 1987.

Gleeson, V. and H.R. Russell. *Mine Drainage Bibliography 1910-1976*. Coal and the Environment Abutment Series, Bituminous Coal Research Institute. 1976.

Halbert, B.E., J.M. Schaver, P.A. Knapp, and D. M. Gorber. *Determination of Acid Generation Rates in Pyritic Mine Tailings*. 56th Annual Conference of WPCF. Atlanta. October 2-7, 1983.

Hammack, R.W. *Evolved Gas Analysis - A Quick Method for Identifying Toxic Overburden*. 7th Annual Symposium of Surface Mine Drainage. Morgantown, W.V. 1986.

Harries, J.R. and A.I.M. Ritchie. "Pore Gas Composition in Waste Rock Dumps Undergoing Pyritic Oxidation." Soil Science 140. 1985. 143-152.

Hawley, J.R. *The Problem of Acid Mine Drainage in Ontario*. The Ontario Water Resources Commission. 1977.

Hedin, R.S., and P.M. Erickson. *Relationships Between the Initial Geochemistry and Leachate Chemistry of Weathering Overburden Samples*. Proceedings Conference on Mine Drainage and Surface Mine Reclamation, Vol. I, United States Bureau of Mines IC 9183. Pittsburgh. April 19-21, 1988.

Kleinmann, R., P. Crerar, and R. Pacelli. "*Biogeochemistry of Acid Mine Drainage and a Method to Control Acid Formation*." Mining Engineering. March 1981.

Knapp, R. *Bio/Geochemistry of Acid Generation in Sulphides Tailings and Waste Rock*. Proceedings of the ARD Seminar/Workshop. Environment Canada, Halifax, Nova Scotia, 1987.

Kolpack, R.L. and S.A. Bell. "Gasometric Determination of Carbon in Sediments by Hydroxide Absorption." Journal of Sedimentary Petrology 38, (1968), No. 2.

Kraus, E.H., W.F. Hunt, and L.S. Ramsdell. *Mineralogy*. New York: McGraw-Hill. 1959.

Lapakko, L. *Prediction of ARD From Duluth Complex Mine Waste in North Eastern Minnesota*. Acid Mine Drainage Workshop. DSS Cat. No. En. 40-11-7 11987E. 1987. 187-221.

Lawrence, R.W. and A. Sadeghnobari. *In-house Development of a Modified Biological Confirmation Test*. Coastech Research. 1986.

Lawrence, R.W., S. Jaffe, and L. M. Broughton. *In-house Development of the Net Acid Production Test Method*. Coastech Research. 1988.

Lutwick, G.D. *Personal Communication* . Nova Scotia Research Foundation. 1988.

Miller, S.D. and G.S. Murray. *Application of Acid-Base Analysis to Wastes From Base Metal and Precious Metal Mines*. Proceedings Conference on Mine Drainage and Surface Mine Reclamation, Vol. I, United States Bureau of Mines IC 9183. Pittsburgh. April 19-21, 1988.

Morin, K.A. *Groundwater Contamination From Precious-Metal, Base-Metal, Uranium, Phosphate, and Potash (KCl) Mining Operations*. C.L. Lin (ed.): 165-174. International Groundwater Symposium, International Association of Hydrogeologists. Halifax, Nova Scotia. May 1-5, 1988.

Paine, P.J. *An Historic and Geographic Overview of Acid Mine Drainage*. Proceedings of the ARD Seminar/Workshop. Halifax. March 1987.

Patterson, R.J. *Environmental and Reclamation Measures at Equity Silver Mines Ltd.* Presented at XCIM District Six Meeting. Victoria, British Columbia. October 4, 1986.

Richardson, D.L. *A Study of the Occurrence of Pyrite in Coal and its Relationship to Liberation in Coal Preparation and Acid Mine Drainage Formation.* M.S. Thesis. Pennsylvania State University. 1979. 126.

Silver, M. *Aquatic Plants and Bog Covers to Prevent Acid Generation - Base Metal Tailings.* Proceedings of the ARD Seminar/Workshop. Halifax. March 1987.

Singer, P.C. and W. Stumm. "Acidic Mine Drainage: The Rate-Determining Step." Science 167, (1970): 1121-1123.

Singleton, G.A. and L.M. Lavkulich. "Adaption of the Soxhlet Extractor for Pedologic Studies." Science Society of American Journal. 42 (1978): 984-6.

Skousen, J.G., J.C. Sencindiver, and R.M. Smith. *A Review of Procedures for Surface Mining and Reclamation in Areas With Acid-Production Materials.* West Virginia Surface Mine Drainage Task Force, West Virginia Energy and Water Research Center, West Virginia and Reclamation Assoc. Morgantown, W.V. 1987.

Smith, A. *Some Implications of Characteristics, Acid Generation and Feasibility Test Data to Waste Rock and Spent Ore Disposal.* Proceedings AIME/SME Symposium, Las Vegas, NV. February 1989.

Smith, A.C.S. "The Disposal of Acid-Bearing and Acid Generating Sludges in the Fertilizer Manufacturing Industry." Canadian Journal of Civil Engineering. (1987): 1-6.

Smith, Adrian and J.B. Barton-Bridges. *Some Considerations in the Prediction and Control of Acid Mine Drainage Impact on Groundwater From Mining in North America.* Proceedings EPPIC Conference, Johannesburg, South Africa. May 1991.

Smith, E.E., and K.S. Shumate. *Sulfide to Sulfate Reaction Mechanism. A Study of the Sulfide to Sulfate Reaction Mechanism as It Relates to the Formation of Acid Mine Waters.* Report by Ohio State University Research Foundation to the United States Federal Water Pollution Control Administration, PB-213 876. 1971.

Smith, E.E., K. Svanks, and K.S. Shumate. *Sulfide-to-Sulfate Reaction Studies.* Second Symposium on Coal Mine Drainage Research, Mellon Institute. Pittsburgh, PA. 1968.

Sobek, A.A., W.A. Schuller, J.R. Freeman, and R.M. Smith. *Field and Laboratory Methods Applicable to Overburden and Minesoils.* United States Environmental Protection Agency EPA 600/Z-78-054. 1978.

Steffen Robertson and Kirsten. *Acid Mine Drainage in British Columbia.* Analysis of Results of Questionnaire from Acid Mine Drainage Task Force. Report 66001/1. Prepared for the Province of British Columbia Acid Mine Drainage Task Force. 1988.

Steffen Robertson and Kirsten. *Studies Related to Evaluation of Alternative Abandonment Measures for Faro Mine Tailings.* Curragh Resources, November Report 60601. 1986.

Steffen Robertson and Kirsten. *The Technology of Uranium Tailings Covers.* Research report prepared for national program. CANMET, EMR, Ottawa. 1986a.

Stumm, W. and J.J. Morgan. *Aquatic Chemistry. An Introduction Emphasizing Chemical Equilibria in Natural Waters.* New York: John Wiley. 1981.

Sugio, T., et al. "Role of Ferric Iron-Reducing System in Sulphur Oxidation of Thiobacillus Ferro-Oxidans." Journal of Bacteriology 44. (1985). 1401-1416.

Sullivan, P.J. and A.A. Sobek. "Laboratory Weathering Studies of Coal Refuse." Minerals and the Environment 4 (1982): 9-16, No. 1.

Tanji, K.K. and L.D. Doneen. "Predictions of the Solubility of Gypsum in Aqueous Salt Solution." Water Resources Research. 2. (1966). 543-548.

United States Environmental Protection Agency. American Public Health Association. *Standard Methods for the Examination of Water and Wastewater*. 15th Edition. Washington, DC.: APHA. 1981.

United States Environmental Protection Agency. *Inactive and Abandoned Underground Mines - Water Pollution Prevention and Control*. EPA-440/9-75-007. 1975.

United States Environmental Protection Agency. *Paleoenvironment of Coal and Its Relation to Drainage Quality*. EPA-600/7-77-067. 1977.

University of California at Berkeley Mining Waste Study Team. *Mining Waste Study, Final Report*. California State Water Resources Control Board. 1988.

Walsh, F.M. and R. Mitchell. "Mine Drainage Pollution Reduction by Inhibition of Iron Bacteria." Water Research, IX (1975): 525-528.

Watzlof, G.R., and K.J. Ladwig. *Electromagnetic Conductivity Surveys to Identify Acid Sources in U.S. Coal Mines*. Proceedings of the Acid Mine Drainage Symposium, Halifax, Nova Scotia. March 1987.

White, A.F. and H.C. Classen. *Dissolution of Silicate Rocks - Application to Solute Modelling*. In Chemical Modelling in Aqueous Systems. Washington, D.C.: American Chemical Society. 1979. 447-473.

Williams, E.G., et al. *Factors Controlling the Generation of Acid Mine Drainage*. Report to the United States Bureau of Mines, Research Grant No. G5105086. 1982. 256.

CONTENTS

CHAPTER 5
ATTENUATION OF CHEMICAL CONSTITUENTS

PAGE NO.

LIST OF TABLES/LIST OF FIGURES 204

5.1 Introduction 205
5.1.1 Background 205
5.1.2 Objectives 207
5.1.3 Integration With Other Chapters 207
5.1.4 Study Participants 208
5.1.5 Sources of Information 208
5.1.6 Chapter Organization 209
5.2 Mine Wastes and Chemistry 209
5.2.1 General 209
5.2.2 Mine Wastes 210
5.2.3 Chemical Characteristics of Mine Waste Types 211
5.2.4 Geochemical Changes in Mine Wastes 213
5.2.5 Predictability 215
5.3 Factors and Mechanisms Affecting Contaminant Migration 215
5.3.1 General 215
5.3.2 Mechanisms of Attenuation 217
5.3.2.1 Physical Mechanisms 221
5.3.2.2 Physiochemical Mechanisms 221
5.3.2.3 Chemical Mechanisms 221
5.3.2.4 Biological Mechanisms 222
5.3.3 Physical Mechanisms 222
5.3.3.1 Filtration 222
5.3.3.2 Dispersion 223
5.3.3.3 Dilution 226
5.3.3.4 Volatilization 226
5.3.4 Physiochemical Mechanisms 226
5.3.4.1 Adsorption 227
5.3.4.2 Fixation 230

CONTENTS

CHAPTER 5
(Continued)

PAGE NO.

5.3.5 Chemical Mechanisms 230
5.3.5.1 Direct Precipitation 232
5.3.5.2 Adsorption-Coprecipitation-Precipitation 232
5.3.5.3 Precipitation Preceded by Hydrolysis 232
5.3.5.4 Precipitation Preceded by Complexation 233
5.3.5.5 Precipitation Preceded by Reduction 233
5.3.5.6 Toxicity Reduction 234
5.3.6 Biological Mechanisms 235
5.3.6.1 Biodegradation 235
5.3.6.2 Cellular Uptake 236
5.4 Relationship Between Chemicals and Mechanisms 237
5.4.1 General 237
5.4.2 Predictability by Chemical Type 237
5.4.2.1 Asbestos 237
5.4.2.2 Metal Cations 239
5.4.2.3 Major Anions 244
5.4.2.4 Cyanides 245
5.4.3 Nature and Importance of Environmental and Operating Conditions 249
5.4.3.1 Vadose Zone Conditions 249
5.4.3.2 Other Site Characteristics 254
5.4.3.3 Waste Management Unit Practices 254
5.4.4 Data Requirements 255
5.4.4.1 Information About the Unsaturated Zone 255
5.4.4.2 Information About the Waste 257
5.5 Appropriate Levels of Verification 258
5.5.1 General 258
5.5.2 Predictions Based on Basic Technology, Prior Experience, and Research 259
5.5.3 Laboratory Testing 260
5.5.4 Modeling 264
5.5.5 Field Demonstrations 264

CONTENTS

CHAPTER 5
(Continued)

PAGE NO.

5.6 Regulatory Implications 265

5.6.1 The Appropriate Role of Attenuation in Mine Waste Regulations 265

5.6.2 Recommended Regulatory Approach 265

5.7 Bibliography 268

CONTENTS

CHAPTER 5
(Continued)

LIST OF TABLES

TABLE NO.	TITLE	PAGE NO.
5.1	Potentially Mobile Chemical Species in Mine Waste Pore Waters	212
5.2	Illustrative Pore Water Quality Stages	214
5.3	Soil Characteristics Which Affect the Behavior of Chemicals Species	219
5.4	Attenuation Mechanisms of Chemical Species	220
5.5	General Ranges for Cation Exchange Capacities of Soil Clays, Soil Organic Matter, and Several Soil Types	228
5.6	Common Ionic Forms of Chemicals Found in Mining Operations	238
5.7	Examples of the Nature and Importance of Vadose Zone Properties Relative to Attenuation	250

LIST OF FIGURES

FIGURE NO.	TITLE	PAGE NO.
5.1	An Illustration of Attenuation	218
5.2	Mechanisms Causing Dispersion	224
5.3	Variable Fluid Velocities in Porous Media Flow	225
5.4	Illustrations of Chemical Attenuation Mechanisms	231
5.5	Example of Attenuation at Excavation Beneath Tailings Facility	243
5.6	Cyanide Attenuation Tests (Sodium Cyanide Solution)	248
5.7	Example Sequential Batch Test Procedure	261
5.8	Example Unsaturated Column Percolation Test Procedure	263

CHAPTER 5

ATTENUATION OF CHEMICAL CONSTITUENTS

5.1 INTRODUCTION

5.1.1 BACKGROUND

Attenuation is defined as the reduction in quantity or toxicity of a chemical constituent in liquids that may be released from a mine waste management unit. Attenuation of the constituent could occur because of physical, chemical, and/or biological mechanisms along the liquid migration pathway.

During the past several years, a great deal of discussion has been devoted to the questions of:

- Should attenuation of dissolved chemicals in the vadose (unsaturated) zone, separating waste material from ground water (saturated zone), be considered in determining the types of engineered containment systems required for specific mine waste management units?
- If the attenuation potential is included in the design of the unit, what procedures are appropriate for verifying its applicability?
- If an engineered containment system fails, what is the potential for attenuation processes to reduce concentrations of dissolved chemicals passing through the vadose zone?
- Are designs of waste management units, which rely on natural attenuation rather than engineered containment, more protective of the environment over the long term?

The discussion in this chapter deals primarily with attenuation mechanisms that occur in the waste, liner materials, and/or vadose zone, because the design goal will usually be to eliminate or control the level of constituents that could reach ground water. Additional attenuation, especially dilution, can also occur in the ground water. Attenuation in ground water may be an appropriate consideration for sites where there are substantial travel distances between points of potential release and points of potential use.

Even though attenuation can occur in certain natural systems for most constituents when conditions are appropriate, it has not been readily allowed as a containment feature in mine waste regulations. There are probably four factors which have contributed to this. First, because attenuation processes are fairly complex, concern has been expressed that such processes cannot reliably be predicted. Second, concern has been expressed that if attenuation could be shown to occur for a given site condition, it may not be possible to verify sufficient site uniformity to be sure that such attenuation would occur throughout the site. Third, in many early studies, there has been a lack of quantification of the effects of attenuation in specific systems. This has led to incorrect applications of attenuation as a waste containment feature. Finally, because dealing with attenuation mechanisms often requires special expertise (e.g., in both geochemistry and hydrogeology), regulating bodies may not have the in-house capabilities to adequately evaluate data presented by an applicant to allow consideration of attenuation in the permitting process.

Indeed, there are many circumstances where attenuation should not be relied upon and others where extensive investigations would be required to quantify the level of water resource protection provided by attenuation. However, these factors should not "a priori" preclude consideration of attenuation, because:

- The degree to which attenuation and site uniformity predictions can be made vary greatly. In some cases the potential benefit from attenuation will be speculative, while in others the potential will be evident. There also will be a variety of conditions between these two extremes.
- The degree to which vadose zone attenuation may be considered in an engineered containment system design can vary greatly. It can range from:
  - The primary mechanism, which entirely eliminates the need for engineered containment; to
  - A redundant factor where vadose zone attenuation provides an additional level of containment that is not required unless problems occur with the basic containment system.
- A large amount of research and actual project experience data on attenuation already exist and have been used by the regulatory community. This existing data base will continue to grow as special studies are undertaken for both mine waste and other waste sites. Our understanding of attenuation and our ability to utilize it for waste containment will, therefore, continue to improve.
- There may be cases where natural attenuation is preferable to engineered containment of semimobile constituents over long periods of time in a system with a significant demonstrated attenuation capacity.

Therefore, the regulations should allow for opportunities to consider, evaluate and, if appropriate, use attenuation as a waste containment feature. The determination of its suitability must be tailored to specific waste types, their constituents, available data, site-specific conditions, and design considerations, as well as the consequences of containment failure.

5.1.2 OBJECTIVES

The primary objectives of this chapter are:

- To summarize available attenuation data and to determine the various degrees to which attenuation may or may not be quantified and predicted for various mine wastes, constituents, and site conditions.
- To examine the degree to which attenuation can be included within containment design for a variety of mine wastes and site conditions.
- To suggest regulatory wording which will allow attenuation to be used as a factor in designing waste management units, when appropriate.

This chapter is also intended to provide a general understanding of geochemical conditions and reactions within the vadose zone and ground water that may influence attenuation processes. For more detailed information on chemical behavior and reactions, the reader should consult the following references:

- Dragun, 1988
- Krauskopf, 1979

5.1.3 INTEGRATION WITH OTHER CHAPTERS

Considerations of potential chemical attenuation during migration within, or from, a waste management unit has direct relationship to the following chapters:

- <u>Chapter 3 - Mine Waste Management Requirements</u>: Attenuation potential is considered in determining the need for, and type of, engineered containment and in evaluating risks associated with possible liner leaks.
- <u>Chapter 4 - Prediction of Acid Generation Potential</u>: The potential for acid generation may be important, because in some cases the acidic conditions will be the initial cause for metals migration. Attenuation in the waste and natural geologic formations in turn, contributes to a reduction in these metal concentration levels.
- <u>Chapter 6 - Climatic Considerations</u>: Climatic conditions have a direct relationship to the potential for adequate attenuation. In arid areas, there may be only small amounts or very slow migration of water into the vadose zone. In this case, there is the potential for significant attenuation. In high rainfall areas, the extent and rate of fluid migration may reduce the potential for attenuation to be effective.
- <u>Chapter 7 - Liner System Design</u>: When appropriate, consideration of attenuation potential can have a direct effect upon the liner system selected for a specific waste unit. In some cases, attenuation potential may be sufficient to completely eliminate the need to use a liner for containment. In others, a less conservative liner may be appropriate if attenuation would protect against degradation of water resources in the event of some leakage from the liner system.
- <u>Chapter 8 - Closure Requirements</u>: Potential attenuation capacity as a mechanism for long-term protection of water resources should be

considered when selecting the type of closure to be used at a site. The degree to which attenuation retards chemical migration can affect the degree of conservatism required for a waste management unit closure design.

- Chapter 10 - Ground Water Monitoring: If attenuation is used as a major factor in determining the design of containment requirements, monitoring systems may require procedures to verify anticipated behavior. This determination will depend upon the potential threat posed by the waste unit and the degree to which attenuation is relied upon as a primary or redundant containment.

5.1.4 STUDY PARTICIPANTS

- Primary Authors
  - Richard Ellison, Ph.D., P.E., Environmental Solutions, Inc.
  - John Ellis, MICE, Chartered Engineer, Environmental Solutions, Inc.
  - Jeff Hess, R.G., Environmental Solutions, Inc.
  - Lisé Rendina, Environmental Solutions, Inc.
- Support Authors
  - Donald T. Bishop, Ph.D., P.E., Condor Earth Technologies, Inc.
  - Terry Chatwin, Ph.D., Resource Recovery Conservation Consultants, Inc.
  - Peter H. Dohms, P.G., Condor Earth Technologies, Inc.
  - Roman Pyrih, Ph.D., GEOCHEM, Inc.
  - Jim Rouse, GEOCHEM, Inc.

5.1.5 SOURCES OF INFORMATION

There are many technical papers and books that discuss attenuation mechanisms in soil. Those used for this chapter are believed to provide a cross-section of viewpoints and useful technical data. Most of the data sources reviewed are listed in Section 5.7. Some of the most useful include:

- Dragun (1988) for the overall concepts.
- EPA (1980), Griffin et al. (1977), DHS (1986), Battelle Laboratories (1976), and Marshack (1989) to indicate regulatory viewpoints.
- Environmental Solutions, Inc. (1985), IT Corporation (1985), Perkens and Dohms (1989), Geochem, Inc. (1989), Cherry et al. (1986), Contra Costa Times (1989), and Thorsen and Stensby (1982) to illustrate published and unpublished case histories.
- Hornick (1976) and Leeper (1978) for experience from land application of wastes.

5.1.6 CHAPTER ORGANIZATION

The remainder of this chapter is divided into the following discussions, intended to provide the framework for analyzing the role of attenuation in mine waste management regulation:

- Section 5.2 describes types of mine wastes, chemical characteristics, and potential for chemical changes over time.
- Section 5.3 provides a general introduction to the various types of chemical, physical, and biological mechanisms that can result in constituent attenuation, both in the liquid phase and by contact between liquid and solid phases in the vadose zone.
- Section 5.4 describes which mechanisms potentially can have an attenuating effect on the dissolved species of interest and their quantification, including brief descriptions of case histories where attenuation or the lack of attenuation has been demonstrated.
- Section 5.5 discusses the various levels of verification (literature review, laboratory testing, and/or field demonstration testing) which may be applicable for attenuation mechanisms.
- Section 5.6 summarizes the data needed to address the concerns of both the regulatory staff and proponents of attenuation and provides a suggested approach for inclusion in *Chapter 15*, Article 7.[(1)]

5.2 MINE WASTES AND CHEMISTRY

5.2.1 GENERAL

Chemical constituents of potential concern to water resources can be derived from the extraction of minerals because either: (1) the constituent naturally associated with the ore becomes available for transport to the environment due to the mining or processing activities, or (2) the chemical is added to the ore as a part of the process to extract the mineral of interest. It is therefore important to identify the mineral types of the waste and their constituent chemical species in order to assess their likely mobility in the environment. Chemicals that are added in the processing of ores include flotation agents, acids, alkalis, and lixiviants such as cyanide. Not all mineral extraction and processing operations use chemicals to recover their economic products. For example, construction minerals such as sand and gravel are processed by simple physical means. Chapter 3 contains a more detailed discussion of mineral types encountered in California and the reagents that are commonly used in mineral processing.

Furthermore, mine waste deposits are often dynamic geochemical systems; their chemical characteristics and pore water quality will change over time. Pore water is the water contained within the interstitial spaces of the waste material and has the potential to migrate to the

(1) California Code of Regulations (CCR) Title 23, Division 3.

environment. Basic geochemical principals, combined with laboratory and field testing where necessary, can be used to understand these processes and to predict the quantity and quality of liquids likely to migrate from the waste.

5.2.2 MINE WASTES

Mine wastes can be divided into a number of groups and subgroups which have generally similar characteristics. These are discussed in Chapter 2, Section 2.2 and include:

- Overburden
- Waste Rock
- Tailings
- Heap Leach Residue
- Wastewater

The residual natural materials do not usually have the potential to generate soluble constituents of concern and are therefore not discussed further in this chapter. In the rare incidents where the rock from a quarry, for example, may contain mineralization which could be of concern, it should be appropriately evaluated and managed as a waste rock material.

The principal factor in determining the nature, extent, and dissolved constituent mass loading from a particular waste rock material will be whether it will generate acid. This consideration is also important when evaluating the geochemical behavior of tailings, but becomes the prime consideration in the case of waste rock, as the potential chemical load derived from acid generating materials far exceeds the normally insignificant or low levels mobilized from nonacid generating waste rock.

The tailings group has several important geochemical subgroups with significantly different chemical and geochemical characteristics relevant to determining the attenuation behavior of the soluble constituents. These are:

- Flotation tailings: These are residuals from flotation processes. These additives physically interact with the comminuted ore to cause separation of key minerals from the remainder of the ore, "gangue," but do not chemically decompose the minerals. The tailings produced by this process can be relatively benign and usually have a low sulfide content. The residual soluble chemicals will be largely determined by the inherent leachability of the natural material.

- Concentrate tailings: These are tailings produced by the chemical extraction of the economic mineral from concentrate produced by an initial flotation process. The remaining minerals in the concentrate will have been subject to reaction with the chemical reagents used in the extraction process and, due to the nature of the concentration process, will carry elevated levels of residual metals.
- Undifferentiated tailings: This type of tailing is produced when either the entire ore feed is subject to the mineral extraction process or when the concentrate and flotation tails are recombined at the end of processing for discharge to the tailings impoundment. The chemical characteristics of these tailings will be influenced by the same mechanisms discussed under the concentrate tailings. Metals and other species concentrations in the liquid portions of the waste would, however, be lower than in the concentrate tailings due to the lower concentration of minerals.

5.2.3 CHEMICAL CHARACTERISTICS OF MINE WASTE TYPES

This section describes the behavior of inorganic chemical species derived from mine wastes. This information is provided as background to subsequent discussions on the natural attenuation of these species. As discussed in Chapter 3, the organic chemical content of mine wastes is not of environmental concern and so is not addressed in this chapter.

Table 5.1 summarizes the more likely mobile chemical species in the pore water which can be derived from the various mine waste types described in Section 5.2.2 and which could be subject to natural attenuation if migration from the waste occurs. There is obviously overlap in the chemical species associated with the various types of mine wastes. An important similarity is the fact that they all are derived from naturally occurring material. However, there are also some important differences, such as those related to the presence of reagents, and the mobility and the relative amounts of the various chemical species.

Table 5.1 is designed to highlight the more significant species that may be found in mine wastes. Some of the more important aspects included in the table are:

- For tailings, concentration and mass loads of the chemical species are likely to increase from flotation tailings to concentrate tailings.
- Where cyanide is present in the waste, metal species may be complexed as metalo-cyanide, which modifies normal mobility as long as the complex remains stable. This applies to both the tailings and the spent heap leach ore.
- Amphoteric species, i.e., those which may be mobile at both low and high pH conditions (although in different chemical forms), may be present in all mine wastes. Consideration of the attenuative abilities in any given geochemical environment requires recognition of this dual mobility.

TABLE 5.1

POTENTIALLY MOBILE CHEMICAL SPECIES IN MINE WASTE PORE WATERS

CHEMICAL GROUPS	MINE WASTE TYPE					
	Mine Tailings			Waste Rock		Spent Heap Leach Ore
	Flotation	Concentrate	Undifferentiated	Acid-Generating	Nonacid-Generating	
Cations and Metal Cations	Calcium, ammonia, transition metals,(1) lead, mercury, and barium.	Calcium, transition metals, lead, mercury, and barium.	Calcium, transition metals, lead, mercury, and barium.	Calcium, ammonia, transition metals, lead, mercury, and barium.	Calcium, ammonia	Calcium, ammonia, transition metals and metals that form cyanide complexes, e.g., mercury.
Anions	Nitrate, sulfate	Nitrate, chloride, sulfate	Nitrate, chloride, sulfate	Nitrate, sulfate	Nitrate, sulfate	Nitrate, sulfate
Amphoteric Species	Arsenic, antimony, chromium, molybdenum, selenium	Arsenic, antimony, chromium, molybdenum, selenium	Arsenic, antimony, chromium, molybdenum, selenium	Arsenic, antimony, chromium, selenium	Arsenic, antimony, chromium, molybdenum, selenium	Arsenic, antimony, molybdenum, selenium
Cyanide Complexes (where cyanide is used as a process reagent)	--	• Free cyanide (CN and HCN). • Weak metal cyanide complexes (e.g., zinc, copper, nickel cyanides). • Strongly complexed cyanides (e.g., iron, cobalt complexes).		--	--	--

(1) Transition metals include: Chromium, cobalt, copper, nickel, zinc, iron, manganese

A working knowledge of the chemistry of the waste and the potential interactions between the different chemical constituents present is an important prerequisite in the understanding, prediction, and quantification of attenuation. Chemical constituents must be viewed, however, as part of a dynamic geochemical system in which the chemical conditions, and hence constituent concentrations, can change over time. This is discussed in Section 5.2.4 below.

5.2.4 GEOCHEMICAL CHANGES IN MINE WASTES

Mine wastes represent a geochemical system that is in some degree of disequilibrium with the surrounding environment. With time, therefore, the geochemical system can change. The extent and rate at which this change over time occurs, depends on the initial waste geochemistry and mineralogy. These changes documented in the literature, are well understood, and can be satisfactorily predicted (Cherry et al., 1980; Dave and Lim, 1985).

Depending on the type of mine waste, the geochemical changes can be divided into a number of stages, the characteristics of which can be predicted (Clifton and Smith, 1988). In the case of mine tailings, these stages are:

- A process controlled stage, where the pore water chemistry is almost entirely controlled by the milling process and the chemical added or formed by reaction in the process.
- If the tailings are acid generating, the second stage will be the acid generation process where the acidic pore water dissolves species from the tailings solid phase and from minerals formed as secondary reaction products in the tailings mass.
- If acid generation does not occur, the second stage involves a re-solution phase, in which the secondary products and the more soluble components of the initial tailings source material are removed.
- Finally, there is a longer term leaching phase by infiltration of natural precipitation as the more resistant minerals degrade. The release rate of chemical species in this phase is relatively slow, and the mass of the chemical species removed is small.

Table 5.2 provides water chemistry from an illustrative example derived from actual data for different stages in the pore water chemistry in tailings (Clifton and Smith, 1988).

The chemical changes that occur in waste rock are different. There is no initial process controlled stage. Chemical species mobility occurs with either the onset of acid generation or a solution process, which in the case of relatively inert silicate rocks is similar to the long-term leaching condition.

TABLE 5.2

ILLUSTRATIVE PORE WATER QUALITY STAGES[1]

PARAMETER (mg/L)	PROCESS CONTROLLED STAGE	ALTERNATIVE 1		ALTERNATIVE 2	LONG-TERM[4] INFILTRATION STATUS
		Sulphide Oxidation Stage[2]	Re-Solution Stage[3]	Re-Solution Stage[3] Under Nonacid Generating Conditions	
pH (units)	10	2	3.5	7.5	5
TDS	4,500	30,000	2,500	2,500	100
Ca	600	500	400	400	20
SO_4	3,000	20,000	1,800	1,200	10
NH_3	300	<10	<10	<10	<1
Na	100	100	20	20	10
Cl	200	200	20	20	10
Fe	<1	10,000	20	5	<1
Mg	100	200	500	<5	10
As	1	1	<1	<1	<0.1
Al	<5	20	20	<5	5
Ni	<.01	1	<0.1	<0.1-1	<0.1
Si	<5	20	20	<5	5

(1) Clifton and Smith, 1988.
(2) If pyrite or iron-bearing sulfide oxidizes.
(3) May be accompanied by alteration of jarosite, in iron-bearing tailings.
(4) Applicable to both Alternatives 1 and 2.

Geochemical changes on heap leach residues are similar to those that occur in tailings. The initial leaching involves a stage which is process controlled. After that, rinsing represents the re-solution stage, which is followed by the infiltration stage.

5.2.5 PREDICTABILITY

The chemical systems discussed above are generally predictable using knowledge of chemical reactions and the results of laboratory and metallurgical testing. The basic chemical reactions are used to hypothesize the likely reaction products, results, and concentrations. The hypotheses are then tested by a tiered series of laboratory tests, pilot studies and, where appropriate, field trials. For example, batch attenuation tests may follow postulation of attenuative mechanisms for a soil material, which themselves precede column tests designed to assess reaction kinetics once the principal attenuative processes have been confirmed. In many cases, the reactions are sufficiently well understood and predictable that field trials are not necessary.

5.3 FACTORS AND MECHANISMS AFFECTING CONTAMINANT MIGRATION

5.3.1 GENERAL

Attenuation of chemical constituents in soil and rock is a well recognized phenomenon. The concept has been and is currently being used by federal and state regulators for: (1) establishing general standards, (2) determining cleanup criteria for polluted sites on a case-by-case basis, (3) providing variances for the disposal of wastes into landfills, (4) regulating the disposal of treated waste water sludge to land, and (5) regulating ground water recharge projects. California regulatory agencies have considered attenuation as a factor for evaluating potential environmental impacts of chemicals for a number of years. For example, the Soluble Threshold Limit Concentrations (STLC) used by the Department of Health Services (DHS) to establish hazardous criteria for soils (see Chapter 3) was based upon earlier work performed by Battelle Laboratories (1976) and the EPA (1980) and was directed toward determining appropriate relationships between acceptable soil pollution levels and ground water quality drinking standards. Those prior studies recognized that attenuation is dependent on site-specific characteristics, and concluded that the amount of attenuation (reduction in pollutant concentrations) expected under a wide variety of conditions could vary from about 4 to 1,000.

The attenuation value of 100 was chosen to develop the STLC for inorganic chemicals as a conservative representation of the average amount of attenuation to be anticipated for constituent concentrations as leachate migrates toward ground water. In practice the DHS provides nonhazardous designations for constituents above the STLC when shown that the waste has physical and chemical characteristics that promote attenuation in site-specific conditions. For example, attenuation was a major factor in determining that wastes containing concentrations of lead in excess of the STLC level are not hazardous at the Molycorp Mountain Pass Mine because of the mitigating chemical and physical characteristics of the site conditions (IT Corporation, 1985; DHS, 1986; DHS, 1989). These characteristics include: (1) the high overall net neutralization potential of the tailings, indicating that acid leaching conditions would not occur under normal environmental conditions, and (2) the very high adsorption capacity of lead which renders the metal highly immobile in the tailings and bedrock environments at the site.

As an additional example, the DHS' California Site Mitigation Decision Tree method for determining cleanup standards applicable to toxic waste sites (DHS, 1986), allows certain types of attenuation to be considered in determining the potential threat to ground water posed by a chemical constituent in soil. In all cases, the site-specific attenuation model must be based on detailed site investigation data to characterize vadose zone and ground water conditions. Supporting data concerning general knowledge about attenuation of constituents under certain types of conditions (e.g., lead attenuation in soils with normal pH conditions) are considered in the evaluation. When possible, measurements illustrating the degree to which attenuation has occurred have been used to demonstrate field applicability of the theories.

As outlined in Chapter 3, the staff of the Central Valley Regional Water Quality Control Board (Marshack, 1989) prepared a report titled *Designated Level Methodology for Waste Classification and Cleanup Level Determination* to provide guidance for determinations of the classification of mine and other wastes. Two of the major elements of this report are: (1) both waste- and site-specific factors are important considerations for evaluating the threat of "nonhazardous" wastes, and (2) evaluation of attenuation is a key requirement for determining the concentrations of constituents that have the potential to degrade beneficial uses of water resources. The report provides the following equation for determining, on a site-specific basis, the particular waste characteristics that should not exceed water quality goals:

$$\begin{array}{c}\text{Soluble Concentration of}\\ \text{Constituent in Waste (mg/kg of waste)}\\ \text{that could degrade water quality}\end{array} = \begin{array}{c}\text{Water}\\ \text{Quality}\\ \text{Goal (mg/L)}\end{array} \times \begin{array}{c}\text{Environmental}\\ \text{Attenuation}\\ \text{Factor}\end{array}$$

The purpose of the above discussion is to indicate that attenuation is a valid consideration that can be factored into site-specific evaluations, as illustrated in Figure 5.1. The major questions in this chapter, therefore, are directed toward determining the mechanism of attenuation, the conditions under which attenuation may occur, the estimation/quantification of attenuation, and the degree to which its beneficial results can be predicted.

5.3.2 MECHANISMS OF ATTENUATION

This section identifies attenuation mechanisms for the chemical species frequently found at mines in California. Subsequently, Section 5.4 discusses: (1) the relationships between various chemicals and attenuation mechanisms, and (2) the degree to which each may be predictable on a case-by-case investigation basis.

Table 5.3 summarizes the types of mechanisms that have the greatest attenuation potential and indicates the soil properties that may affect these mechanisms. This table is divided into the four major categories of:

- Physical Mechanisms
- Physiochemical Mechanisms
- Chemical Mechanisms
- Biological Mechanisms

Attenuating mechanisms can occur in the waste material itself, the containment system (e.g., liners, embankments), the vadose zone, and surface and ground waters. The potential for biological activity is usually greatest in an oxygen-rich environment, such as waste piles and the upper portion of the vadose zone. Physical and chemical mechanisms can occur throughout the waste material or in certain types of soil or rock.

Table 5.4 provides a generalized summary of the most significant attenuation mechanism for the chemical species present in mine wastes. The table categorizes the species based on the predictability for attenuation to occur. It also indicates which types of attenuation mechanisms occur and whether they are of primary and secondary importance. As discussed in Section 5.2.4, it is important to remember that mine wastes are physiochemically dynamic and that the physiochemistry of their pore water changes with time in a predictable manner. This information is very useful in predicting attenuation potential for the different chemicals of concern.

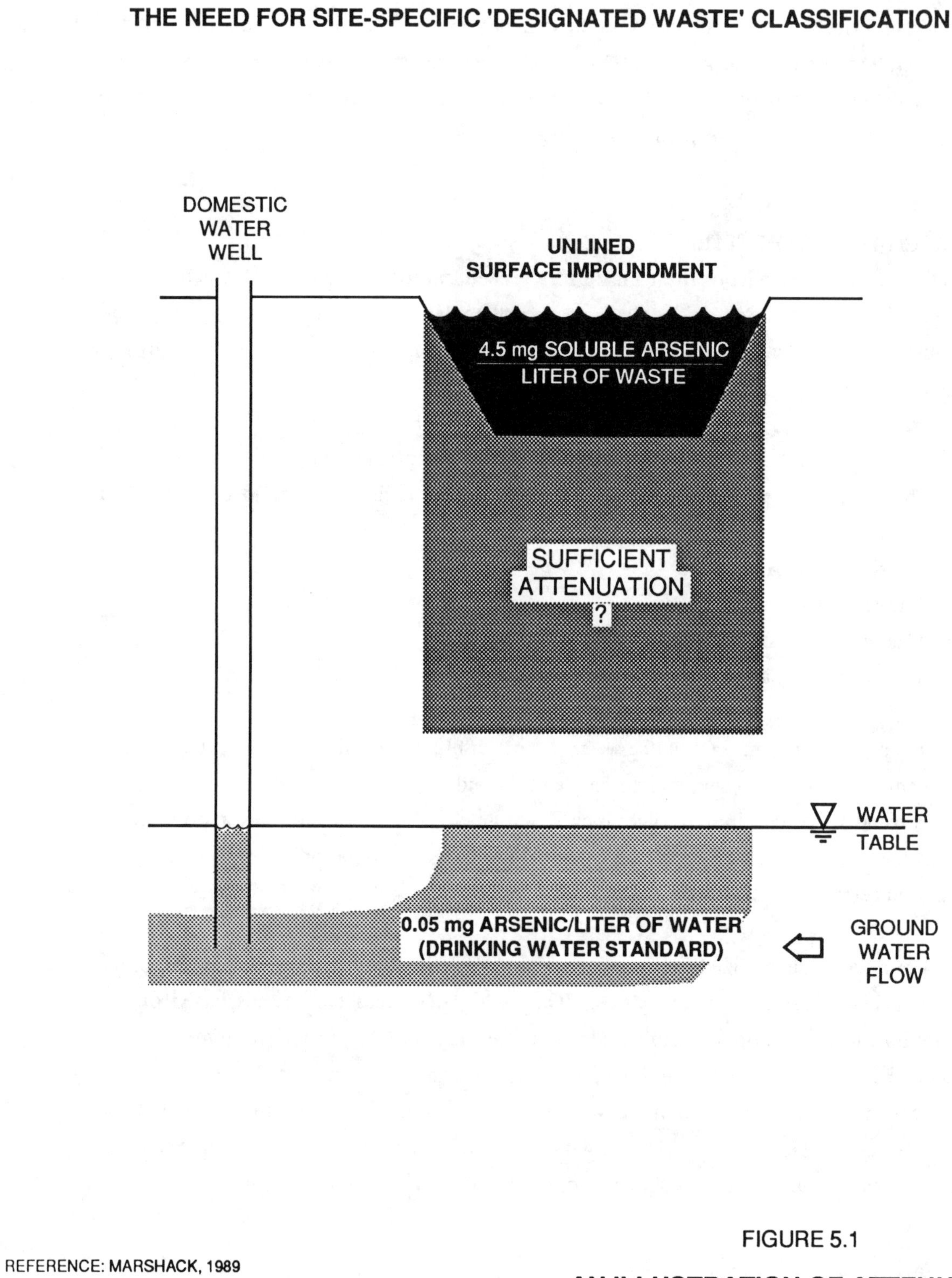

FIGURE 5.1

AN ILLUSTRATION OF ATTENUATION

TABLE 5.3

SOIL CHARACTERISTICS WHICH AFFECT THE BEHAVIOR OF CHEMICAL SPECIES

TYPE	MECHANISM	SOIL PROPERTY
Physical	Filtration	Porous Media
	Dispersion	Porous Media
	Dilution	Soil Water
	Volatilization	Soil Gas Atmospheric Connection
Physiochemical	Adsorption	Clay Content Type of Clay Organic Content Soil Water pH, Eh
	Fixation	Selected Soil Minerals Soil Water pH, Redox Potential Excess Cations/Anions
Chemical	Precipitation	Selected Soil Minerals Soluble Compounds Soil Water pH, Redox Potential Excess Cations/Anions
	Hydrolysis	Soil Water Soil Water pH, Redox Potential
	Complexation	Soluble Compounds Organic Content Soil Water pH, Redox Potential Excess Cations/Anions
	Oxidation	Soil Water pH, Redox Potential
	Reduction	Soil Water pH, Redox Potential
Biological	Biodegradation	Organic Content Nutrients Oxygen
	Cellular Uptake	Flora Organic Content

TABLE 5.4
ATTENUATION MECHANISMS OF CHEMICAL SPECIES

	CHEMICAL SPECIES	ATTENUATION MECHANISM												
		PHYSICAL				PHYSIO-CHEMICAL		CHEMICAL					BIOLOGICAL	
		Filtration	Dispersion	Dilution	Volatilization	Adsorption	Fixation	Precipitation	Hydrolysis	Complexation	Oxidation	Reduction	Bacterial Reaction	Cellular Uptake
VERY PREDICTABLY ATTENUATED	Asbestos	●	○	○ (1)					○					
	Chromium III	●	○	○				●	●		○	○		○
	Lead		○			●	○	○	○	○				
	Mercury	●	○		○	●		○			○	○		○
	Beryllium	●				○		●	●					○
	Cyanide, Free		○	○	●	●		●	●	○	○		●	○
STRONGLY ATTENUATED	Arsenic		○	○		●	○	●	○		○	○		
	Barium		○	○		●	○	●						
	Copper		○	○		●	●		○	○	○	○		
	Nickel		○	○		○	○	○	○					○
	Zinc		○	○		●	○	●	○					○
	Cadmium		○	○		○			○	○				○
	Cobalt		○	○		●	○			○				
ATTENUATION UNCERTAIN	Antimony		○	○		○	○	○	○		○			
	Cyanide, Strongly Complexed		○	○		○		●	○					○
	Cyanide, Weakly Complexed		○	○	○	○		○	●	○	○		●	○
	Sulfate		○	○				●						
	Nitrate		○	○						○		○	●	●
	Chromium VI (2)		○	○					○	○		●		○
	Boron		○	○					○					

LEGEND

● Primary Attenuation Mechanism

○ Secondary Mechanism

(1) With sufficient water flow, Mg will gradually leach out, thus changing the nature of the asbestos with loss of its fibrous structure.

(2) Due to the high solubility of Cr^{+6} it is not easily attenuated unless reduced to Cr^{+3}.

5.3.2.1 Physical Mechanisms

As summarized in Table 5.3, physical mechanisms include filtration, dispersion, dilution, and volatilization. In general, physical mechanisms have a lower attenuation potential, although they can be important in support of physiochemical and chemical mechanisms. For example, dispersion can increase the effectiveness of adsorption because "spreading" of the migrating zone exposes the pollutants to a greater number of soil particles. In addition, dispersion can increase the potential for biological reactions which rely upon exposure to nutrients, bacteria and/or oxygen. Furthermore, filtration is a factor in chemical precipitation because the soil pores can stop or retard migration of the precipitates, although most chemical precipitation occurs on the surfaces of existing soil minerals.

5.3.2.2 Physiochemical Mechanisms

Physiochemical mechanisms are dependent on both physical (intermolecular forces) and chemical (Redox and pH of soil water) conditions of the environment and can include adsorption and fixation. Adsorption is the accumulation of an element at the surface of soil particles with a corresponding decrease in the concentration of the dissolved element in the pore water. It is primarily a function of the ionic charge associated with soil particles and the type of ion in solution. As it is the result of relatively weak van der Waals attractive forces, it represents an equilibrium condition which is reversible.

In this chapter, fixation is defined to include both chemisorption and solid state diffusion. Chemisorption is the formation of a covalent bond between an adsorbed element and a mineral surface which renders the element immobile. Unlike physical adsorption, the chemical bonding is usually irreversible. Solid state diffusion involves the irreversible penetration of an element into the pore spaces of a mineral's structure. The degree of reversibility of these processes decreases from adsorption to chemisorption to solid state diffusion.

5.3.2.3 Chemical Mechanisms

Chemical mechanisms are dependent on the chemical interaction of an element or mineral with the soil or pore water. These mechanisms can include: (1) precipitation of compounds, or (2) the reduction in toxicity of a constituent by changing its valence state, by hydrolysis through oxidation or reduction reactions, or by changing its molecular form through complexation. Precipitation actually "fixes" the constituent in the soil matrix, preventing it from migrating. It is an equilibrium process and therefore the reaction may be reversed if the physiochemical conditions (i.e., Redox Potential and pH) of the pore waters change.

Toxicity reduction occurs when a change in valence state causes a chemical species to be less toxic. This change in valence can also affect the mobility of the species. In the case of hexavalent chromium, for example, reduction to trivalent chromium results in a decrease in both migration potential and toxicity.

5.3.2.4 Biological Mechanisms

Biological mechanisms can include: (1) biodegradation of a chemical into the basic oxidation products (carbon dioxide, water, and nitrogen), (2) bacterial consumption of the chemical as a nutrient by microorganisms, or (3) cellular uptake, for example by plants, if migrating water is in the root zone.

5.3.3 PHYSICAL MECHANISMS

The following sections describe the characteristics of the potential physical mechanisms discussed above. Sections 5.3.4 through 5.3.6 describe the physiochemical, chemical, and biological mechanisms.

5.3.3.1 Filtration

Filtration is the simple removal of particulates flowing through a porous soil. The particulate matter can be from:

- Insoluble minerals such as asbestos present as suspended fibers.
- Chemical precipitates.

Physical filtration occurs when solid matter is removed from a suspension because the particulate matter cannot flow through smaller pore spaces.

5.3.3.2 Dispersion

Dispersion is the tendency of the solution containing a chemical to spread out and reduce its net concentration as it is transported through the vadose zone and in ground water. Figure 5.2 illustrates different mechanisms of dispersion. These include mechanical mixing of the solution as it is transported through the soil column and ground water by the following three mechanisms:

- Mechanical spreading of a solution as it migrates through either a saturated or unsaturated granular media occurs because of the way the particles are packed and their various geometric shape. This causes the width of the migrating solution front to increase as the solution spreads out and is diluted by other pore water. The spreading is caused primarily by variable fluid velocities in the individual pores. These variable velocities are caused by fluid interactions with the different particle surfaces and varying pore sizes. They are also caused by variable flow directions due to the meandering pathways provided by the interfingered pores (Freeze and Cherry, 1979) (shown in Figures 5.2 and 5.3).
- The effect of this mechanical dispersion can be increased for thick soil strata which contain intermittent layers of higher and lower soil permeabilities. At each zone of higher permeability soil, the migrating solution has a tendency to flow laterally until the head buildup on the subsequent low permeability soil is sufficient to create an adequate downward flow. Mechanical spreading can also occur in a saturated or unsaturated fractured rock media where the multiple fractures conduct solution. If there are a large number of interconnected fractures, the solution will spread out when migrating by finding its way into an increasing number of fractures. Dispersion can have an important benefit of exposing the chemical to a larger volume of soil and pore water, with the resulting potential for increased attenuation by adsorption or precipitation.
- Capillary suction in the unsaturated zone can also cause the solution to spread laterally as it migrates downward due to gravity. The amount of capillary suction increases with decreasing grain size of the soil and the degree to which the moisture content is below saturation. (Chapter 10 contains more details on the theory of fluid migration in the unsaturated zone.) Capillary suction can also cause spreading of migrating solution in unsaturated fractured rock. The degree to which it occurs will depend on the amount and distribution of fractures with opening sizes that allow capillary forces to develop.
- The amount of spread of a chemical species can also be increased by a complementary phenomenon, referred to as diffusion. Unlike the previous two mechanisms, which rely on the dispersion of the solution, diffusion involves the migration of only the chemical species within the solution from an area of higher concentration to one of lower concentration. Diffusion usually has a smaller effect than mechanical spreading or capillary action, and is negligible where solution migration is rapid.

Chapter 7 examines the fate of seepage through a hole in a containment liner and illustrates the effect of flow spreading due to either chemical or physical factors. An example is discussed

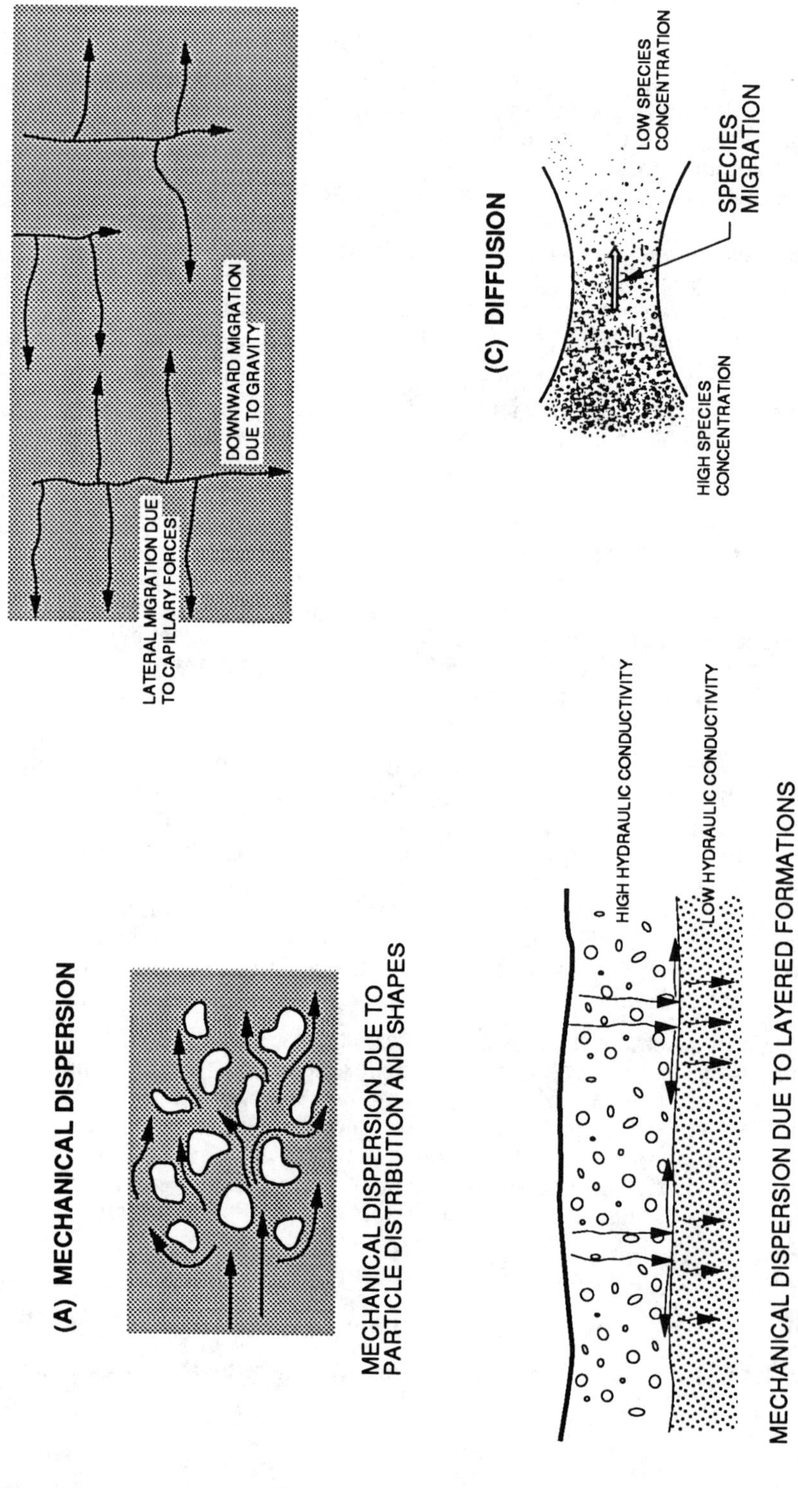

FIGURE 5.2
MECHANISMS CAUSING DISPERSION

(A) VELOCITY VARIATION WITHIN INDIVIDUAL MIGRATION PATHWAYS

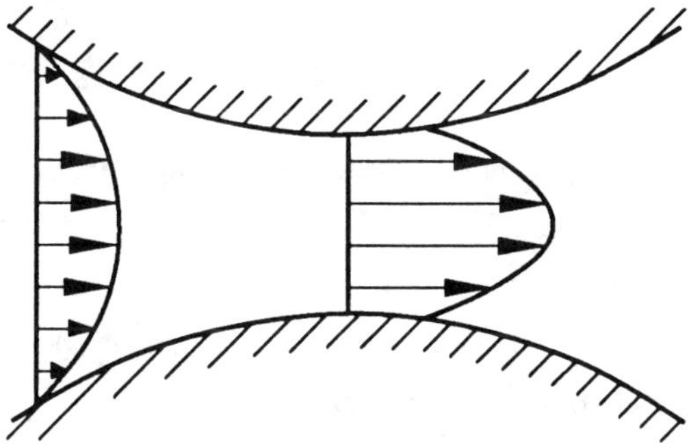

(B) VARIATIONS IN VELOCITY BETWEEN DIFFERENT MIGRATION PATHWAYS

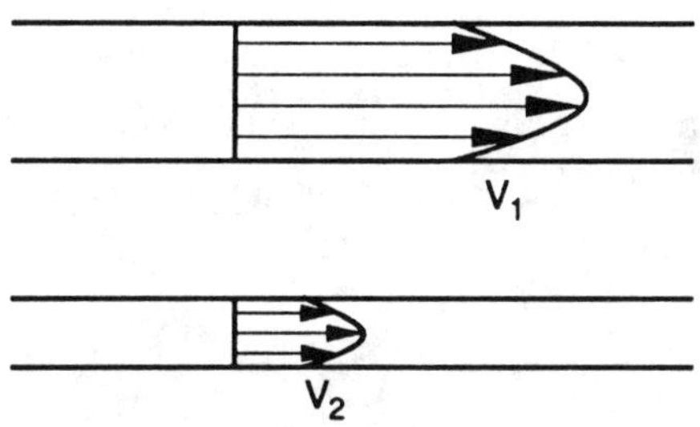

FIGURE 5.3

VARIABLE FLUID VELOCITIES IN POROUS MEDIA FLOW

REFERENCE: ADAPTED FROM FREEZE AND CHERRY (1989)

which illustrates how the volume of exposed soil is increased several thousands of times for a travel depth increase from 10 to 200 feet. The added available time for attenuation to occur under these conditions could be years.

5.3.3.3 Dilution

Dilution in ground water is technically a physical attenuation mechanism since it results in reduced concentration levels of dissolved chemical species. However, this mechanism is not discussed in detail because this chapter is primarily concerned with attenuation mechanisms that occur in the unsaturated zone, before the constituents reach ground water. Dilution is included in Table 5.4 to illustrate its general effect on the chemicals of interest.

5.3.3.4 Volatilization

Volatilization is potentially important for free cyanide removal, due to the equilibrium between the cyanide ion (CN^-) and undissociated hydrocyanic acid (HCN) which is highly volatile. This equilibrium is forced in the direction of HCN of pH values below about 8. The volatilization mechanism also has the potential to attenuate weak cyanide complexes that break down to HCN during migration through the unsaturated zone. The weak complex will be in equilibrium with free cyanide in solution. If the free cyanide is removed by volatilization, the complex will dissociate further. The rate of HCN volatilization will depend primarily on: (1) the ability of the soil to buffer the pH to an appropriate level, and (2) the rate at which previously volatilized HCN is replaced with air, the rate at which HCN diffuses as a gas through the pore spaces, or the rate at which HCN is redissolved in pore water containing lower concentrations of HCN in solution. For waste management units that cover large areas, the rate at which pore air is replaced may represent the limiting factor for the amount of attenuation that may occur by volatilization.

5.3.4 PHYSIOCHEMICAL MECHANISMS

As discussed above and shown in Table 5.4, physiochemical mechanisms can be divided into two categories: adsorption and fixation. These mechanisms are influenced by both intermolecular forces and chemical conditions such as pH and the redox potential. Theoretically, physiochemical and chemical mechanisms are distinct and can be described as separate phenomena. However, in field studies it is difficult to separate the significance of each mechanism. In literature, when data is not available to assess the extent and magnitude of each, the attenuation of a chemical species to the solid phase has been referred to as "sorption."

The adsorption-coprecipitation-precipitation reaction involving cadmium ions and calcite is an example of the complexity of the soil water system. This reaction involves the adsorption of cadmium ions on calcite precipitates. As the concentration of the cadmium ions increases on the surface of the precipitates, cadmium begins to coprecipitate with the forming calcite. If the concentration of cadmium reaches saturation, then cadmium carbonate will begin to precipitate. Some of the factors controlling the direction of the reaction include pH, redox potential, physiochemical properties of the forming precipitate surfaces, and concentrations of the ions in solution.

5.3.4.1 Adsorption

Adsorption is the accumulation of elements or compounds on the surface of soil particles, with a corresponding reduction in concentration in the soil pore water. Adsorption by ion exchange is primarily a function of the ionic charge associated with the soil particles and the ion in solution. In addition, adsorption-desorption reactions are influenced by: (1) solution concentration of the ion, (2) concentration of competing ions in the solution, (3) pH of the soil water, and (4) formation and destruction of organic chelates and inorganic complexes. Depending on mineral type, each soil particle will have a net negative or positive electrical charge. Most soils possess a net negative charge which can attract positively charged metal cations (e.g., lead, mercury) in seepage. However, there are some, such as iron, manganese, and aluminum hydrous oxides, that may have a net positive charge. This mechanism also is referred to as cationic adsorption.

The total amount of cations adsorbed by negative charges on a unit mass of soil is known as the cation exchange capacity (Dragun, 1988) and is dependent on a number of factors, including the amount and type of clay and the organic content in the soil. Table 5.5 illustrates the general range of cation exchange capacities for clays, organic matter, and several soil types.

The major anions in seepage, which possess a negative charge, can also undergo adsorption in soils that have particles with a positive change. This phenomenon, referred to as anionic adsorption, is less predominant and usually not as predictable as cation adsorption.

The cationic exchange capacity (CEC) is a measure of the quantity of cations that are exchangeable per unit weight of soil, expressed as milliequivalents (meq) per 100 grams of soil. One meq per 100 grams is the adsorption of 1 mg of hydrogen ion, or the equivalent weight of any other ion,

TABLE 5.5

GENERAL RANGES FOR CATION EXCHANGE CAPACITIES OF SOIL CLAYS, SOIL ORGANIC MATTER, AND SEVERAL SOIL TYPES

SOIL TYPE	CEC[1] (milliequivalents/100 grams)
SOIL TYPE	
• Sand	2 - 7
• Sandy Loam	2 - 18
• Loam	8 - 22
• Silt Loam	9 - 27
• Clay Loam	4 - 32
• Clay	5 - 150
CLAY MINERALS	
• Chlorite	10 - 40
• Illite	10 - 40
• Kaolinite	3 - 15
• Montmorillonite	80 - 150
• Oxides and Oxyhydroxides	2 - 6
• Saponite	80 - 120
• Vermiculite	100 - 150
SOIL ORGANIC MATTER	>200

[1] Cation exchange capacity.

Reference: Dragun, 1988

per 100 grams of soil. A common value in soils is on the order of 10 meq per 100 grams of soil. Some high clay content soils have been reported at over 50 meq per 100 grams, but these are unstable and exist only in dry climates. (Dragun, 1988).

Several procedures are available for measuring the CEC of a soil in the laboratory. It is important to specify the conditions of the CEC analysis since the interpretation of the results will depend upon these conditions. For example, the CEC of a soil at near-neutral pH is not necessarily indicative of the CEC of that soil at pH 4 or pH 9.

Anion exchange capacity (AEC) is a measure of the quantity of anions that can be exchanged with a specified weight of soil. Unfortunately, the anion exchange of a soil is not routinely measured. Furthermore, during the analysis for anion exchange capacity, decomposition of the clay minerals in the soil is always a possibility, and results in unreliable evaluations. AEC is sometimes estimated by choosing a factored value based on the CEC measurement. An average ratio of cation- to anion-exchange capacity is reported by Grim (1968) for kaolinite (0.5), for illite (2.3), and for montmorillonite (6.7) clays.

The following example illustrates the significance of the CEC of a particular soil. One acre of land one foot deep represents 43,560 cubic feet, or about 2,000 tons of soil. If the soil has a CEC of 10 meq per 100 grams, this is the same as 0.010 ton equivalent per 100 tons of soil. Thus, for 2,000 tons of soil, there would be 0.20 ton equivalents adsorbed. For the case of Cu^{2+}, this would amount to $0.20 \times \frac{63.5}{2} = 6.35$ tons of copper per acre-foot of soil, $\frac{63.5}{2}$ being the equivalent weight of copper. If the copper is adsorbed as the hydrated complex $CuOH^+$, twice as much copper could be adsorbed.

If there were an infiltration rate of 2 inches per year of leachate containing copper at 25 mg/L, and the maximum permissible concentration limit for copper in drinking water is 1 mg/L (Secondary U.S. EPA Maximum Concentration Level [MCL]), then 24 mg/L of copper has to be adsorbed by the soil to comply with the established MCL. An infiltration rate of 2 inches per year over an acre amounts to:

$$\frac{2}{12}\left(\frac{\text{ft}}{\text{year}}\right) \times 43{,}560\left(\frac{\text{ft}^2}{\text{acre}}\right) \times 7.48\left(\frac{\text{gals}}{\text{ft}^3}\right) \times 3.78\left(\frac{\text{L}}{\text{gal}}\right) = 205{,}300\ \text{L/year} \cdot \text{acre}$$

At a concentration of 24 mg/L, this computes to:

$$205{,}300 \left(\frac{L}{\text{year} \cdot \text{acre}}\right) \times 24 \left(\frac{mg}{L}\right) \times \frac{1}{1000} \left(\frac{g}{mg}\right) \times \frac{1}{453.6} \left(\frac{lb}{g}\right)$$

$$\approx 11 \frac{lb}{\text{year} \cdot \text{acre}} \text{ of copper to be adsorbed}$$

Since the adsorption capacity of the soil is 6.35 tons of copper for an acre of soil one foot deep, the time required to reach the adsorption capacity of one foot of soil would be:

$$6.35 \text{ (tons of copper)} \times 2{,}000 \left(\frac{lb}{ton}\right) \times \frac{1}{11.86} \left(\frac{\text{year} \cdot \text{acre}}{lb}\right) \approx 1{,}070 \text{ years}$$

5.3.4.2 Fixation

As previously indicated, in this text fixation refers to chemisorption and solid state diffusion, both of which immobilize an element within the structure of a mineral or at the mineral surface. Chemisorption is the formation of a covalent bond between an adsorbed element and a mineral surface which results in element immobilization. These bonds are much stronger than the van der Waals type bond characteristic of adsorption, and thus chemisorption is considered as a stronger irreversible attenuating mechanism. Since it occurs at specific sites, the adsorbed molecule is not free to migrate. Solid state diffusion refers to the penetration of an element into pore spaces of a mineral's structure and is also considered an irreversible attenuating mechanism.

5.3.5 CHEMICAL MECHANISMS

Precipitation is an important mechanism, as it immobilizes many chemicals by formation of an insoluble solid (precipitate) comprised of ions which were previously dissolved in water. The precipitates may be retained within fine-grained soils or rocks. Precipitation is an equilibrium reaction which is controlled by several factors, such as concentration of the ions, soil water pH and redox potential, and solubility constants. If one or more of these conditions changes, the precipitates may be released back into the soil water (dissolution). An equilibrium is established when the rate of precipitation equals the rate of dissolution. Precipitation may result from several potential reactions, as illustrated in Figure 5.4. The different types of reactions that cause precipitation are discussed in the following sections.

(A) PRECIPITATION

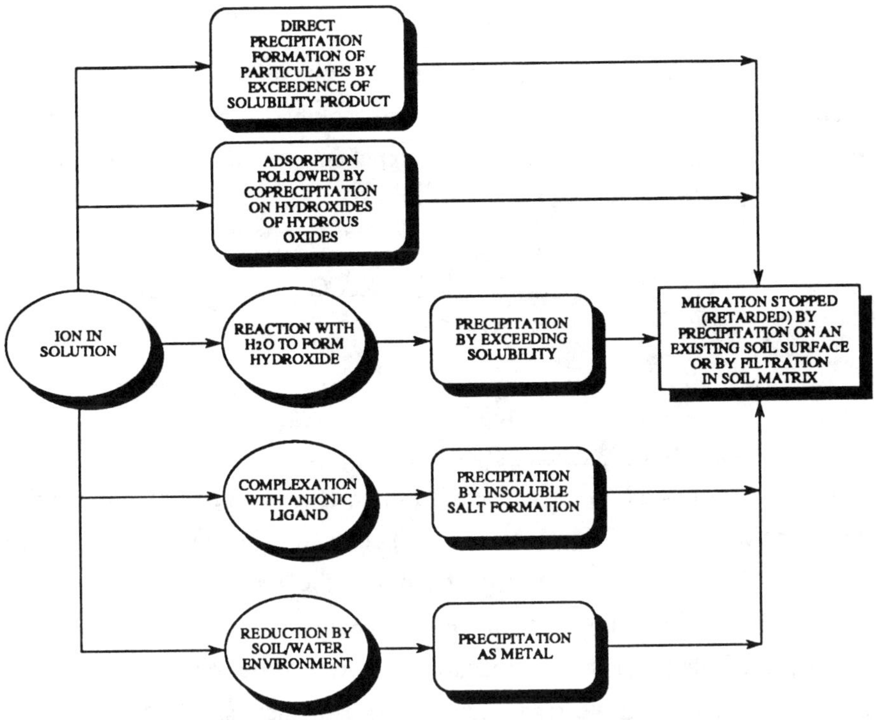

(B) TOXICITY REDUCTION

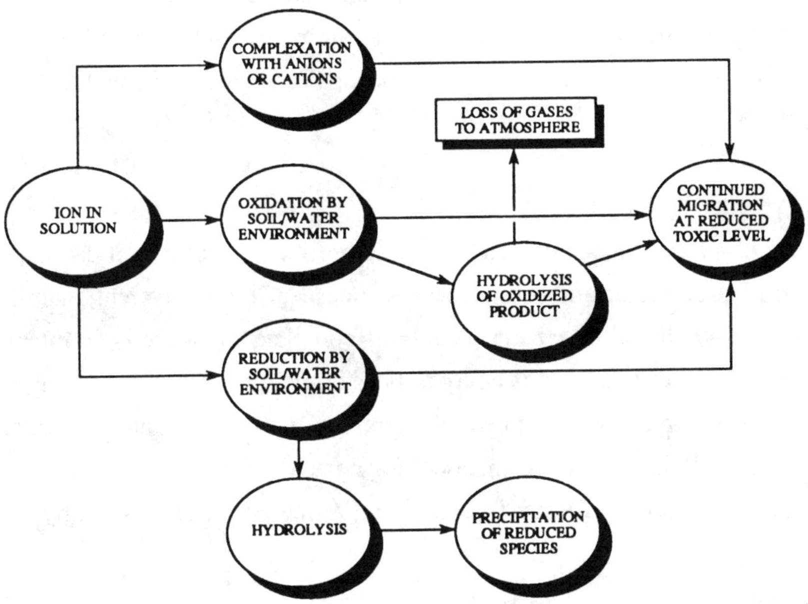

FIGURE 5.4

ILLUSTRATIONS OF CHEMICAL ATTENUATION MECHANISMS

5.3.5.1 Direct Precipitation

Direct precipitation occurs when the constituent forms an insoluble (solid) compound simply by exceeding its solubility limits. The particulate can remain insoluble indefinitely unless conditions in the soil water (redox potential, pH, ion concentrations) change to dissolve the compound. An example of direct precipitation is:

$$Ca^{2+} + CO_3^{2-} \underset{\text{dissolution}}{\overset{\text{precipitation}}{\rightleftharpoons}} CaCO_3$$

5.3.5.2 Adsorption-Coprecipitation-Precipitation

As previously discussed, this process involves the adsorption of ions on growing precipitates. As the concentration of the absorbed ion increases, the ion begins to coprecipitate with the growing mineral crystal. If the solution becomes super saturated with the ion, it will begin to precipitate. This reaction is controlled by both physiochemical and chemical conditions of the soil water environment. Another example of this type of reaction involves hydrous oxides of aluminum, iron and manganese, all of which constitute significant "sinks" for metals. These hydroxides and oxides readily adsorb and coprecipitate cations and anions.

Practical application of the attenuation ability of hydrous metal oxides has been made in water and waste water treatment, particularly by the use of iron and aluminum salts for the removal of phosphorus compounds, organic contaminants, and trace metals (Forstner and Wittman, 1979).

5.3.5.3 Precipitation Preceded by Hydrolysis

Water can act as an acid or a base. This is known as an amphiprotic solvent, i.e., it can accept or reject protons. For salts of strong acids and strong bases, such as NaCl, if equivalent quantities of acid and base are used to make the salt, the neutralization is complete and there is no interaction between the solvent and the salt; that is, there is no hydrolysis of the salt. If the equivalent weight of a strong acid and a weak base react, then the neutralization is not completed, and the water will act as a base. Thus, if the base is weak, it forms a conjugate acid BH+ which has appreciable strength. This will react with water to form a free base and a free acid. This is the hydrolysis reaction.

$$\underset{\text{Acid}}{BH^+} + \underset{\text{Base}}{H_2O} \rightleftharpoons \underset{\text{Acid}}{H_3O^+} + \underset{\text{Base}}{B}$$

For metal salts in solution in water, it is probable that the conjugate acid is the hydrated ion of the metal, e.g., Fe $(H_2O)_6^{+++}$. The hydrolysis can then be represented by:

$$M(H_2O)_n^+ + H_2O \rightleftharpoons H_3O^+ + M(H_2O)_{n-1}\ OH$$

Many metal hydroxides are insoluble and thus will precipitate as soon as they are formed by hydrolysis.

5.3.5.4 <u>Precipitation Preceded by Complexation</u>

Precipitation can occur when the chemical of interest in solution forms a complex with an anionic ligand. For example, many metal ions complex with phosphate, sulfate, carbonate, chloride ions, free cyanide, hydroxyl ions, or other inorganic solutes. If the complex remains soluble, it will continue to migrate with water. Or, the complex may form an insoluble particulate with other metal cations in solution. Whether or not an insoluble complex is formed depends on the concentration of the fulvic acid[(2)] and cations present in the soil. Metal ions can also complex with organic molecules and form an organo-metallic complex. This reaction is termed chelation when the complex takes the form of a soluble organic compound with a metal at its center.

5.3.5.5 <u>Precipitation Preceded by Reduction</u>

The final precipitation pathway illustrated in Figure 5.4 occurs when the chemical of interest gains an electron in a reducing environment, typical of conditions that occur in deep soil and rock. A number of metals and metal complexes will readily reduce in an anoxic environment and may then stay in solution or precipitate as a solid.

Table 5.4 shows that the most important metal for attenuation by reduction is hexavalent chromium, a very toxic ion. Hexavalent chromium reduces to trivalent chromium, which is far less toxic, and at pH conditions above about 4, forms an oxide, carbonate, or sulfide precipitate. The reducing behavior of this metal was demonstrated at a site in Wisconsin (Thorsen and Stensby, 1982) where chrome plating solution, containing hexavalent chromium, had been discharged onto the soil. Frequent discharge of the plating solution resulted in pollution of soils and shallow ground water with hexavalent chromium. The natural capacity of the soil to reduce the

---

(2) High molecular weight organic acid.

hexavalent chromium to trivalent chromium was incorporated in the final remedial action plan for the site, in which excavation of the polluted soils was not required, and a simple ground water recovery system was installed.

5.3.5.6 Toxicity Reduction

Figure 5.4 shows three attenuation mechanisms which can reduce the toxicity of a chemical, even if the chemical remains in solution in its modified form and continues to migrate through the soil or rock to surface or ground water. These mechanisms are complexation, oxidation, and reduction and are discussed below.

Complexation

Complexation involves the formation of a coordination complex by the union of a metal ion with nonmetallic ions. The resultant complex may not exhibit the toxicity of either of the complexed chemicals. For example, the strongly bound cobalt cyanide complex is relatively inert and non-toxic, whereas both free cyanide and cobalt are toxic at high concentrations.

Oxidation

Oxidation involves the loss of electrons by an element, resulting in a change from a metallic state to a soluble ion, or a change in valence of an ion from a lower to a higher valence state. Changing the oxidation state of an ion may in turn cause it to hydrolyze and precipitate without any change in pH. Table 5.4 shows that oxidation is a possible attenuation mechanism for cyanides and is accomplished by reducing its toxicity by the formation of cyanate. Cyanate can hydrolyze to carbon dioxide and ammonia.

Reduction

Reduction, which can decrease the toxicity of hexavalent chromium, was previously discussed in relation to precipitation (Section 5.3.5.4). The resulting chromium reduction product may remain in solution (may not precipitate) only if conditions are extremely acidic, beyond the normal conditions for soils.

5.3.6 BIOLOGICAL MECHANISMS

5.3.6.1 Biodegradation

Biodegradation is a process where biological organisms modify or degrade chemicals into basic oxidation products such as carbon dioxide, water, and nitrogen. These reactions are catalyzed by enzymes produced by the bacteria. Biodegradation reactions, for example, can have a profound effect on the concentrations of cyanides and nitrate in some soils, due to indigenous microorganisms. Soils, especially with organic materials, can contain a large and diverse population of microorganisms, including bacteria, antinomycetes, molds, and algae (Dragun, 1988). Bacteria, the most abundant microorganisms in the soil, also possess the ability to rapidly reproduce.

Biodegradation (denitrification) of nitrate results in nitrite and nitrogen gas in the presence of organic matter and denitrifying bacteria in an anoxic environment. Most of these bacteria use nitrate as an electron acceptor in the oxidation of organic material. Factors influencing denitrification include the presence of oxygen, type of substrate, temperature, pH, and the presence of trace minerals (Rouse, 1988). Anoxic or anaerobic biological decomposition of nitrate is slow and requires a source of organic matter. Therefore, once nitrate reaches the ground water or deeply penetrates the vadose zone, it is very stable. Denitrification processes currently are used in sewage works, and ground water is being treated by ion exchange to remove nitrate.

Under aerobic conditions, cyanide can be catalytically oxidized to cyanate, by the enzymes excreted by bacteria. The cyanate is then catalytically hydrolyzed to carbon dioxide and water.

$$HCN + \frac{1}{2}\,O_2 \xrightarrow{\text{Enzymes}} HCNO$$

$$HCNO + H_2O \xrightarrow{\text{Enzymes}} CO_2 + NH_3$$

Ammonia (NH_3) is then oxidized to nitrate:

$$NH_4^+ + 2O_2 \rightarrow NO_3^- + H_2O + 2H^+$$

This latter reaction is catalyzed by the chemolithotrophic bacteria, nitrifying organisms such as Nitrosomonas and Nitrobacter.

Under anaerobic conditions, biological decomposition of cyanides is very slow. There are reports that the bottom of stagnant ponds do represent environments where cyanide is reduced to nitrogen and carbon dioxide. However, the bacteria are not tolerant to high levels of cyanide and will only work below a concentration of about 2 mg/L (Fuller, 1984).

Various reactions are also promoted or take place due to bacterial action. These reactions are activated by enzymes which catalyze the breakdown of substances in the water or soil so that they can be utilized in the building of bacterial cells and the production of energy. As an example, bacteria can break down nitrate by an anaerobic process such as:

$$5S + 2H_2O + 6NO_3^- \rightarrow 5SO_4^{2-} + 3N_2 + 4H^+$$

Bacteria can also decompose cyanide to carbon dioxide and ammonia. The ammonia can be converted to nitrate.

Other, more complex, bacterial reactions are possible. They would involve the synthesis of substances containing metals, such as methyl mercury, dimethyl arsenate, methylated lead, and possibly methylated antimony.

5.3.6.2 Cellular Uptake

Cellular uptake is the process by which the concentration of metal cations and major anions can be attenuated in near-surface soils by absorption into certain types of plants, in the same manner as soil nutrients and fertilizers. The uptake of chemicals in the soil by plants occurs primarily at their roots. Once in the plant root system, the chemical can then be transported throughout the plant. This process is discussed in considerable detail by Brady (1974).

Cellular uptake has the potential to be effective only for a small portion of mine waste units where leachate flow could be shallow enough so that the chemicals are actually exposed to plant root systems. However, the concept has sufficient merit that various plants and man-made wetlands are being considered as possible tools in the control and cleanup of elevated concentrations of metals in surface waters and soils. For example, a species of wild mustard grass imported from Pakistan is being tested by the U. S. Department of Agriculture in Fresno for possible use in the clean up of selenium from soils in central California. The wild mustard grass appears to remove up to 60 percent of the selenate, the most accessible form of selenium, from the soils tested (Contra Costa Times, 1989). Also, man-made wetlands are being used in a pilot study to treat metals from mine drainage near Idaho Springs, Colorado (EPA, 1988).

5.4 RELATIONSHIP BETWEEN CHEMICALS AND MECHANISMS

5.4.1 GENERAL

This section discusses: (1) attenuation predictability as a function of each chemical species of interest, and (2) site and environmental factors that affect attenuation potential and its predictability.

Table 5.4 indicates the potential importance of the various physical, physiochemical, chemical, and biological processes for the chemicals of interest, and the table divides the elements into three groups based on their predictability for attenuation to occur (i.e., very predictably attenuated, strongly attenuated and attenuation uncertain). The table generally shows that the highest level of predictability is associated with filtration of asbestos fibers and precipitation or adsorption of metal cations. The predictability potential for attenuation of major anions and complexed cyanide is relatively lower. These differences often have a direct effect on the level of investigation required to determine the potential for attenuation in the selection of containment systems for mine waste management units. Table 5.6 aids the discussion by showing the most common form each chemical takes when contacted with water and the types of ions found for various environmental conditions.

The attenuation predictability for each chemical is discussed in Section 5.4.2. Section 5.4.3 discusses the effects of environmental and waste management unit operational conditions on attenuation performance and prediction. Section 5.4.4 summarizes the types of data required in order to initiate an attenuation evaluation.

5.4.2 PREDICTABILITY BY CHEMICAL TYPE

Discussion of behavior of chemicals in the vadose zone is simplified by grouping the chemicals into the following categories:

- Asbestos
- Metal cations
- Major anions
- Cyanides

5.4.2.1 Asbestos

Asbestos refers to a number of naturally occurring minerals which have a fibrous structure and are composed of hydrated metal silicates of iron, magnesium, sodium, and calcium. Asbestos can occur naturally in process liquids due to its presence in the ore being processed, especially from

TABLE 5.6
COMMON IONIC FORMS OF CHEMICALS FOUND IN MINING OPERATIONS

ION		ENVIRONMENT	IONIC CHARGE		
			NEUTRAL[(2)]	POSITIVE	NEGATIVE
METAL CATIONS	Beryllium	All[(1)]	$Be(OH)_2(s)$	Be_2OH^{+3} $Be_3(OH)^{+3}$	
	Chromium III	All[(1)]	$Cr_2O_3(s)$ $Cr(OH)_3(s)$		
	Barium	All[(1)]		Ba^{+2}	
	Cadmium	All[(1)]	$CdCO_3(s)$ CdS	Cd^{+2} $Cd(OH)^+$	
	Cobalt	Oxidizing	$Co(OH)_2(ss)$ $Co_3O_4(s)$ $Co_2O_3(s)$	Co^{+2} Co_2OH^{+3}	$CoCl_4^{-2}$ $Co(CN)_5^{-3}$
		Reducing	Co(s) CoS(ss)		$CuCl_2^{-2}$ $CuCl_5^{-3}$
	Copper	Oxidizing	$Cu_2O(s)$ $Cu(OH)_2(s)$ $CuCO_3(s)$	Cu^{+2} $Cu(OH)^+$ $Cu_2(OH)_2^{+2}$	
		Reducing	Cu(s) CuS(s)		
	Lead	Oxidizing	$PbO_2(s)$ $PbSO_4(ss)$ PbO(s) $PbCO_3(ss)$	Pb^{+2} $PbOH^+$	
		Reducing	Pb(s) PbS(s)		
	Mercury	Oxidizing	HgO(ss) $HgCL_2(aq)$ $Hg_2CL_2(ss)$	Hg^{+2} $HgOH^+$	
		Reducing	Hg(1) HgS(s)		
	Nickel	Oxidizing	$Ni(OH)_2(ss)$ $Ni_3O_4(s)$	Ni^{+2} $NiOH^+$	
		Reducing	Ni(s) NiS(s)		
	Zinc	Oxidizing	$Zn(OH)^2(s)$ $ZnCO_3$	Zn^{+2} $Zn_2(OH)^+$	
		Reducing	ZnS(s)		
MAJOR ANIONS	Antimony	Oxidizing	$Sb_2O_3(s)$		SbO_3^{-3}
		Neutral	$Sb_2O_3(s)$		
		Reducing	SbO(s)		
	Arsenic	Oxidizing	H_3AsO_4 (aq)		H_2AsO_4 $HAsO_4^{-2}$
		Neutral	As_2O_3 (aq) H_3AsO_3		$HAsO_4^{-2}$
		Reducing	AsO(s)		$H_3AsO_3^{-1}$ $H_2AsO_3^{-2}$ $Ha_sO_3^{-2}$ AsO_3^{2-}
	Boron	All[(1)]			$B_4O_7^{-2}$
	Chromium VI	Oxidizing			$HCrO_4^{-1}$ CrO_4^{-2}
	Chloride	All[(1)]			Cl^-
	Nitrate	All[(1)]			NO_3^-
	Sulfate	Neutral and Oxidizing			SO_4^{-2}
CYANIDES	Cyanide, Free	All[(1)]			CN^{-1}
	Cyanide, Weakly Complexed	All[(1)]			$Zm(CN)_4^{-2}$ $Cd(CN)_3^{-1}$
	Cyanide, Strongly Complexed	All[(1)]			$Fe(CN)_6^{-3}$ $Fe(CN)_6^{-4}$

(1) All includes oxidizing, neutral, or reducing environments.
(2) l = liquid phase
s = solid phase
ss = suspended solid
aq = aqueous phase

serpentinite bodies. Asbestos is also mined and can be beneficiated by wet or dry processing which produces waste liquid streams containing asbestos fiber. The primary mechanism for attenuation of asbestos is the filtration of the fibers in the pores of the soil. This is a highly predictable and effective mechanism for retaining asbestos in the soils.

5.4.2.2 Metal Cations

Beryllium and Chromium III - Under nonacidic conditions, beryllium and chromium III both react with water to form an insoluble hydroxide precipitation. This reaction is highly predictable. The insoluble precipitates are filtered out of migrating waters in fine-grained soils. The filtration process is also readily predictable, if chemical and site conditions are well established.

Barium - Barium is strongly adsorbed by clay minerals, replacing calcium and magnesium in their exchangeable sites (Dragun, 1988). For this condition, some consideration about the ultimate fate and impact of the calcium and magnesium is required in the evaluation of potential water resource impacts. Barium will also form carbonate and sulfate precipitates, which are so stable that they are considered to be common minerals in arid areas and can be economically mined in parts of California.

Cadmium and Zinc - Cadmium and zinc are adjacent in the periodic table and have similar chemical properties. Their main mode of attenuation is by adsorption in nonacidic conditions. Cadmium and zinc also form hydroxide or carbonate precipitates as an important attenuation mechanism.

Cobalt and Nickel - Cobalt and nickel have similar properties and usually occur together in nature. They both are adsorbed on the surface of soil minerals as the primary attenuation mechanism. Nickel is probably the most mobile of the heavy metals and is often used as a precursor or an indicator of pollution fronts. Cobalt and nickel can form insoluble hydroxide participates, but at pH conditions which usually are above those found in natural soils.

Copper, Lead, and Mercury - These metals are all very strongly retained in the soil under nonacidic pH conditions. Copper usually precipitates as a hydroxide, oxide, or carbonate. Lead also precipitates as a hydroxide, oxide, or carbonate. Both of these metals are also cations and would become strongly adsorbed on the surface of clay minerals if precipitation did not occur. Adsorption is the primary mechanism which retains mercury in soil, although this metal will also precipitate as an oxide under soil conditions with neutral to basic pH. Mercury also is strongly adsorbed on the surfaces of soils.

Case Histories - Several examples of metal attenuation in soils are discussed below. The first involves soil samples collected and analyzed from a scrap yard as part of the closure of a major steel mill in Southern California (Environmental Solutions, Inc., 1985). The scrap yard had been in use for more than 60 years, completely open to rainfall and infiltration. The concentrations indicated below represent total metal concentrations. Subsequent analyses indicated that approximately 75 percent of the total lead concentration and 90 percent of the total zinc concentration were in a soluble form. Conversely, more than 90 percent of the copper was found to be insoluble. Solubility data for the other metals is not available (Environmental Solutions, Inc., 1985).

	SAMPLE DEPTH		
	SURFACE (mg/kg)	1 FOOT (mg/kg)	2 FEET (mg/kg)
Copper	12,900	282	28
Lead	2,130	36	<2
Mercury	2.7	1.3	0.17
Cadmium	59	6.9	0.62
Nickel	1,520	81	36
Zinc	4,550	1,850	114

The concentrations of copper and lead detected in soils at a depth of 2 feet were on the order of 1,000 times less than concentrations found in the surface soils. Concentrations of cadmium, nickel, and zinc were on the order of 100 times less, and mercury was on the order of 10 times less. These data indicate the effectiveness of soluble metal attenuation mechanisms in the top 2 feet of soil at this particular site.

Another example of soil attenuation involves soil samples collected from a metal plating facility located in California. These samples were analyzed for total and soluble heavy metal content as part of hydrogeological investigations at the site (Geochem, Inc., 1989). Some of the analytical results are presented below and show the effects of heavy metal attenuation in the top 4 feet of soil at the site. The concentrations of copper, lead, and zinc found in the soil samples were dependent upon the pH and total iron levels in the soil. The data indicate that the mechanisms responsible for attenuation were neutralization of acidity, precipitation or coprecipitation of the heavy metals, and adsorption of the heavy metals onto iron-hydrous oxide material.

HEAVY METAL CONCENTRATION IN SOIL
AS A FUNCTION OF DEPTH AND SOIL pH

Sample Depth (feet)	Soil pH	Total Concentrations (mg/kg)				Soluble Concentrations (Net) (mg/L)			
		Fe	Cu	Pb	Zn	Fe	Cu	Pb	Zn
0.2	2.8	63,000	48	130	440	440	<2.5	<0.5	<25
0.9	2.9	9,400	24	16	2,600	168	<2.5	1.1	220
1.1	5.7	26,000	68	74	14,000	34	2.7	2.9	560
1.4	6.0	23,000	27	12	10,400	12.8	<2.5	<0.5	400
3.0	7.0	23,600	34	10.5	28	7.7	<2.5	<0.5	<25
4.0	7.4	22,000	34	10.5	51	3	<2.5	<0.5	<25

In addition, several studies have shown that lead has a very high adsorption capacity in the tailings, soil liners, and bedrock environment at the Molycorp Mountain Pass Mine in California. Dissolved lead at the site, which has the potential to migrate to ground water, is attenuated substantially (retardation factor is estimated to range from 5,000 to 10,000) and thus behaves as though it were highly immobile. Also, calculations show that the adsorption capacity of the bedrock is orders of magnitude greater than the amount of possible metal migration. The situation, therefore, does not represent a significant threat to ground water because of the mitigating effect of the natural environmental conditions.

The hydrology of the tailings impoundment and underlying rocks at the Molycorp mine showed that flow is predominantly vertical in the tailings and lateral in the bedrock. Although large quantities of lead were found in the solid matrix of the tailings and the bedrock immediately beneath the tailings facility, the concentration of dissolved lead and copper in the migrating pore water was very low. High distribution coefficients and retardation factors for lead and copper result in the fixation of these metals in the bedrock and in prevention of their migration downgradient. For example, even though water flows at an average velocity of about 20 feet per year, the migration of lead and copper was estimated to be only about 0.004 to 0.002 feet per year. Also, although the lead concentration in the pore water in the bottom of the tailings pile averages 0.54 mg/L, the concentration found in downstream monitoring wells was below the detection limit of 0.08 mg/L (IT Corporation, 1985; DHS, March 1986; DHS, July 1989).

Adrian Smith (1989) has provided another example of attenuation of metals in a mining environment involving a metal-bearing seepage from a low pH, base metal tailings facility in the southwestern United States. Seepage from the facility had reached ground water, as evidenced by increased sulfate levels which were characteristic of the tailings pore water, although the sulfate levels were significantly attenuated by available calcium carbonate in the vadose zone. Travel time to the ground water was estimated to be approximately 3 years. The facility had been in operation for 15 years when the study was performed.

The pH values and metal concentrations in both the initial and attenuated seepage are shown below, and are compared to the background (upgradient) and receiving (downgradient) ground water quality.

Parameter (mg/L)	Background Ground Water Quality	Seepage Quality	Attenuated Seepage Quality	Receiving Ground Water Quality
Iron	<1.05 - 0.25	750	0.1	<0.05 - 0.13
Manganese	<0.01 - 0.17	600	0.1	<0.01 - 0.08
Copper	<0.01 - 0.01	200	0.01	<0.01 - 0.02
Zinc	0.01 - 0.42	400	0.02	<0.01 - 0.03
Arsenic	<0.01 - 0.01	0.2	0.005	<0.01 - 0.02
Chromium	<0.01 - 0.01	5	0.01	<0.01 - 0.01
Lead	<0.05	1.2	<0.05	<0.05
Cobalt	No Data	2	0.02	No Data
Sulfate	50 - 150	11,000 - 30,000	2,000 - 3,000	50 - 250
pH	6.9 - 8.4	<1 - 2	6.8 - 7.5	7.1 - 7.7

These data indicate:

- High metal levels and a low pH value in the seepage.
- Very substantial attenuation in the vadose zone materials.
- Despite the arrival of seepage in the ground water (based on both chemical identification and calculated seepage velocities), there is no impact on ground water quality by metals.

Figure 5.5 has been sketched from a photograph of the profile observed immediately below the tailings within the vadose zone material when a portion of the tailings impoundment was excavated. The photograph illustrates distinct zones of chemical interaction between the seepage and the vadose zone materials (a calcareous sand with no soil development) and is concurrent with

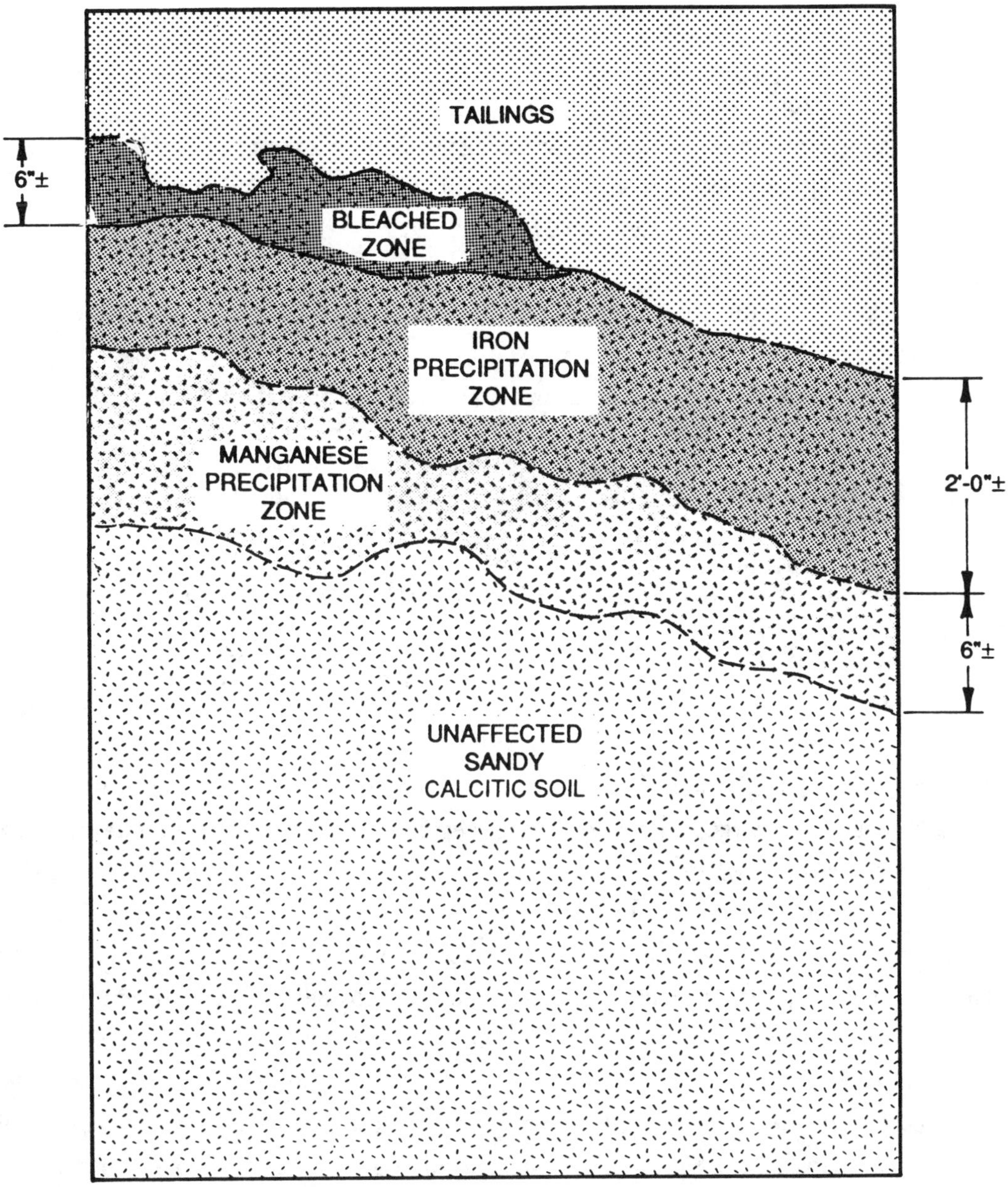

REFERENCE: ADRIAN SMITH, 1989.

FIGURE 5.5

EXAMPLE OF ATTENUATION AT EXCAVATION BENEATH TAILINGS FACILITY

accepted geochemical theory on metal behavior in soils. The total depth of the profile was about 3 feet and indicated the following zones in order of occurrence:

- A leached/bleached zone, white in color, due to acid solution of the vadose zone material (± 6 inches).
- An iron precipitation zone, brown-orange in color (± 2 feet).
- A manganese precipitation zone, essentially black in color (± 6 inches).
- A rapid return to the yellow sandy calcitic materials.

The field and laboratory data demonstrate an excellent cross-correlation of the attenuative processes. In particular they indicate that:

- Despite having a large, long-time source of seepage, attenuation will occur.
- Attenuation can be predicted; the laboratory testing being capable of predicting what was observed in the field profile and shown by the monitoring data.
- The waste tailings were effectively contained in terms of impact on a ground water system, and there was no impact on the water supply wells from the system.

5.4.2.3 Major Anions

Antimony and Arsenic - These elements commonly exist in the form of arsenates and antimonates and are negatively charged. Because the net charge of most soils is predominantly negative, antimony and arsenic will not be as readily adsorbed as the metal cations. However, since soil particles do contain some positively charged surfaces which may be dominant, under some conditions, antimony and arsenic can also be adsorbed over some distance of travel through soil. The rate of adsorption of arsenic will be increased substantially if there are hydrous iron, aluminum, and manganese oxides in the soil (Hornick, 1976).

In the presence of excess iron or magnesium, and with low pH and high redox potential, arsenic may form insoluble arsenate precipitates which could result in attenuation by soil filtration. These types of conditions can exist at tailings facilities. As an example, the uncontrolled discharge of arsenic-bearing tailings for over 100 years in the Whitewood Creek drainage basin in South Dakota has been recently investigated by Cherry et al. (1986). These investigations have shown that large amounts of arsenic are present as solid-phase arsenic adsorbed on inorganic solids, primarily hydrous aluminum and iron oxides, which are present in the alluvium and tailings in the drainage basin. Concentrations of arsenic in the vadose zone typically are in the range of 500 to 5,000 mg/kg.

Geochemical conditions in the alluvium and tailings limit the concentration of soluble arsenic in an adjacent creek to levels typically less than 0.04 mg/L (below the EPA primary drinking water standard for arsenic of 0.05 mg/L) even though there are local areas within the tailings and alluvium where soluble arsenic concentrations are appreciably higher.

Boron - Boron is adsorbed in soils, especially by hydrous iron and aluminum oxides under neutral to basic pH conditions (Hornick, 1976). The soluble form of boron in water is boric acid, H_3BO_3, which is weakly acidic. Boron is soluble in water and, as a result of capillary action, may become concentrated at the soil surface in arid climates where the water table is shallow.

Chromium VI - Chromium VI is present in water as negatively charged chromate (CrO_4^{2-}) or hydrochromate ($HCrO_4^-$). Both forms are soluble but can be adsorbed on positively charged sites and/or on hydrous metal oxides in soils. The main mechanism for the immobilization of chromium VI is by reduction to the trivalent state which renders the chromium insoluble. The presence of ferrous iron, sulfides, or organic matter can promote a reduction reaction.

Chlorides, Nitrates, and Sulfates - These are common anions found in water and are similar in that each can be adsorbed to some extent in slightly acidic soils. However, their solubility characteristics vary significantly. Chlorides and nitrates are generally soluble anions, whereas sulfates can form insoluble salts with barium and sparingly soluble salts with calcium. Nitrate and sulfate can be biologically reduced, while chloride is already at its simplest state. Chloride, nitrate, and sulfate can be absorbed by plant roots and are essential to plant life.

5.4.2.4 Cyanides

Because of the rapid increase in cyanide use for gold processing during the past decade, a substantial level of effort has been devoted toward determining its fate in the environment. Emphasis was originally placed on measurement of free cyanide, which is cyanide's most toxic form. More recently, added attention has been directed toward Weak Acid Dissociable (WAD) cyanide. WAD cyanide includes both free cyanide and weak complexes (e.g., copper, zinc) which can be easily dissociated. Stronger complexed iron cyanides, which are significantly less toxic than free cyanide, are included in a total cyanide measurement. Therefore total cyanide is not a good indicator of potential toxicity. Although gold and cobalt form very strong complexes, they are not measured in common laboratory analyses techniques, including those for total cyanide.

Most investigators and regulators accept the concept that free cyanide will not be a persistent ion in nature because it will eventually complex or dissociate to nontoxic forms. However, many investigators are reluctant to rely upon natural degradation because estimates of time and migration distances cannot always be accurate. The period for detoxification of WAD cyanide is more difficult to determine than for free cyanide.

During the past several years, a large effort toward improving knowledge about the attenuation/degradation of cyanide has been conducted by Resource Recovery and Conservation Consultants (Chatwin and Trepanowski, 1987). Additional important work has been accomplished by Rouse and Pyrih (1985), and Environmental Solutions, Inc. (1988).

The results of these investigations have confirmed that, under certain circumstances, free cyanide will attenuate within the vadose zone. Generic predictive techniques have not been established; quantitative estimates of the potential for free cyanide attenuation must be determined on a detailed, site-specific basis.

These studies have demonstrated that, under appropriate conditions, free cyanide can be attenuated within a short time and travel distance by the following mechanisms:

- Volatilization
- Adsorption
- Direct or Complexation Precipitation
- Oxidation to Cyanate
- Biodegradation and Cellular Uptake

It also can be attenuated by several less common reactions, such as thiocyanate formation or the slower hydrolysis reaction involving HCN which produces ammonia and formic acid.

Volatilization and biodegradation are generally considered to have high potential as attenuating factors for free cyanide in the vadose zone. Volatilization will occur if the soil acts to buffer the high pH process solutions to below about 8. As discussed previously, the limiting factor concerning time will be the rate at which air migrates through the soil pores to remove the vaporized HCN gas. Biodegradation rates will depend on the amount of oxygen, nutrients, and bacteria available in the soil. In some desert areas, the relatively inert alluvial soils may not be conducive to biodegradation. In areas with significant organic material in the soil, attenuation may occur over depths of only a few feet.

The attenuation of free cyanide in the presence of organic matter has been demonstrated in an extensive series of column tests conducted by Environmental Solutions, Inc. (1988). A series of 6- to 12-foot columns were packed with clean silty sands mixed with specified percentages of organic matter, iron minerals, or salts. A solution of 400 mg/L sodium cyanide was applied at a rate of about 0.0005 gpm/ft.^{2} for a period of approximately 100 days (approximately 33 pore volumes). This rate of application was selected to represent a leakage rate from a small hole (the size of a pinhole) in a liner system. The concentration of free cyanide was monitored in the effluent. The results of the tests indicated that: (1) the free cyanide content of effluent from columns not containing added organic matter initially rose to as much as 75 percent of the source solution and then declined to steady-state conditions varying from 0 to about 60 percent of the source solution, and (2) every column containing 5 percent of organic matter showed complete attenuation of the free cyanide for the entire test duration. Figure 5.6 illustrates the results of one of these tests.

The behavior or WAD cyanide is similar to free cyanide. The weak complexes readily dissociate with changing conditions to form free cyanide, which can then be attenuated, as discussed above. WAD cyanide may also form an insoluble salt by reaction with other metals in the soil. These complexes may be more resistant to dissociation to free cyanide and would be retained in the soil as a precipitate.

The table that follows presents WAD cyanide attenuation results from one long-term column test on a sample of clay-liner material to be constructed from a Nevada soil (Geochem Inc., 1989). The particular soil possessed chemical properties which effectively attenuate WAD cyanide, namely the presence of:

- Acidic soil pH
- Smectite and illite clays
- Traces of organic carbon
- Traces of iron hydrous-oxides

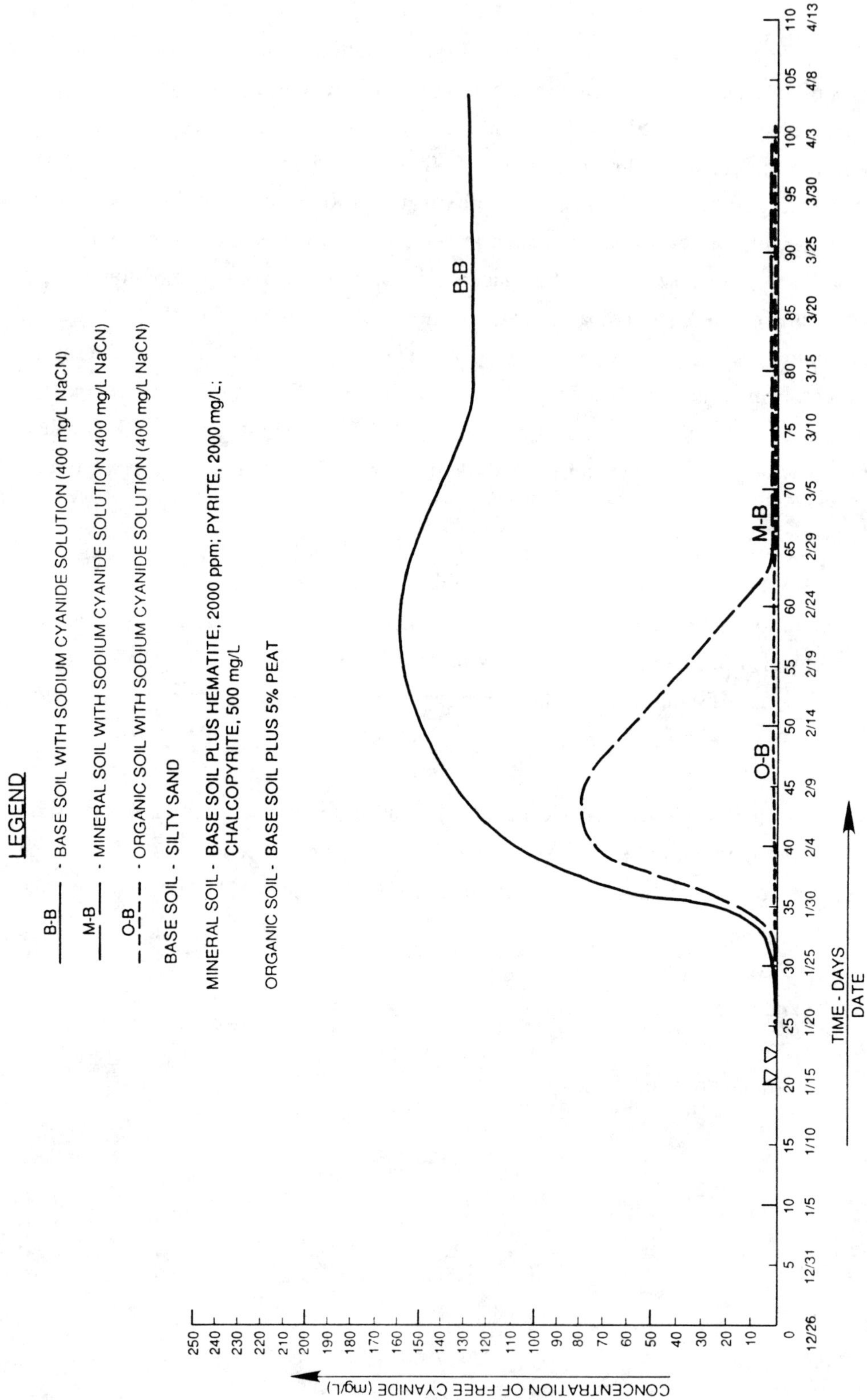

FIGURE 5.6

CYANIDE ATTENUATION TESTS (SODIUM CYANIDE SOLUTION)

REFERENCE: ENVIRONMENTAL SOLUTIONS, INC. 1988

COLUMN EFFLUENT ANALYSES - NEVADA SITE

Solution Feed Expressed as a Fraction of the Pore Volume	pH	WAD Cyanide Concentrations (mg/L)
Column Feed Solution	9.8	510.00
0.0 - 0.5	7.6	0.56
0.5 - 0.8	7.4	0.40
0.8 - 1.1	7.8	0.57
1.1 - 1.3	7.8	0.66
1.3 - 1.7	7.9	0.26
1.7 - 2.1	8.0	0.14

The iron cyanide complex is very stable in the soil system and mechanisms applicable to free and weakly complexed cyanide, such as volatilization or hydrolysis, will not apply. However, it can be adsorbed on soil minerals or form insoluble salts such as ferric ferro cyanide or zinc ferro cyanides and precipitate out.

5.4.3 NATURE AND IMPORTANCE OF ENVIRONMENTAL AND OPERATING CONDITIONS

The potential for attenuation is dependent upon the physical and chemical conditions that exist at each site, including the vadose zone. Table 5.7 illustrates some of the ways that vadose zone conditions can affect attenuation for the various chemicals, as discussed in Sections 5.4.3.1 and 5.4.3.2. Additional factors which can be important for evaluating the applicability of attenuation on a site-specific basis also are addressed; climatic conditions, vadose zone travel times, and attenuation capacity in Section 5.4.3.2, and waste unit management practices in Section 5.4.3.3.

5.4.3.1 Vadose Zone Conditions

Table 5.7 indicates vadose zone characteristics which have the potential to enhance attenuation. The larger the circles, the greater the potential for attenuation to occur. In addition, the table illustrates the degree of verification required for specific characteristics of the vadose zone. Those characteristics requiring relatively more verification to demonstrate the effectiveness of attenuation are indicated by increasingly darker shades. Observations from the data provided for each of the main headings in the table are discussed in the following paragraphs.

TABLE 5.7

EXAMPLES OF THE NATURE AND IMPORTANCE OF VADOSE ZONE PROPERTIES RELATIVE TO ATTENUATION

(Symbols in the table are written as Potential to Enhance Attenuation / Relative Degree of Verification Required, following the key below: H = High, M = Moderate, L = Low.)

	INORGANIC CHEMICAL ELEMENTS AND MINERALS	SOIL/ROCK PHYSICAL PROPERTIES: ROCK: Highly Fractured	ROCK: Micro fractured	SOIL: Organic Content	SOIL: Clay	SOIL: Silt	SOIL: Sand	SOIL: Gravel	SOIL: "Fractured"(1)	SOIL/ROCK/ PORE WATER CHEMISTRY: pH: pH <6	pH: 6 < pH < 8	pH: pH > 8	Eh: Oxidizing	Eh: Reducing	Cation Exchange Capacity: High	Cation Exchange Capacity: Low	Hydrous Metal Oxides	Cation Availa-bility: High	Cation Availa-bility: Low	Anion Availa-bility: High	Anion Availa-bility: Low	Organic Content: High	Organic Content: Low	Nutrient Content: High	Nutrient Content: Low	PORE WATER VOLUME: Below Field Capacity	PORE WATER VOLUME: At Field Capacity
	Asbestos	L/H	M/M		H/L	M/M	M/M	L/H	M/H																	H/L	
METAL CATIONS	Beryllium	L/H	M/M		H/L	M/M	L/H	L/H	M/H	M/H	M/M	H/L			H/L		H/L									H/L	L/L
METAL CATIONS	Chromium III	L/H	M/M		H/L	M/M	L/H	L/H	M/H	M/H	M/M	H/L	L/L	H/L	H/L		H/L					L/H				H/L	L/L
METAL CATIONS	Barium	L/H	M/M	M/H	H/L	M/M	L/H	L/H	M/H	L/H	M/M	H/L			H/M	L/H	H/L	L/H	H/M	H/M	L/H					H/L	L/L
METAL CATIONS	Cadmium	L/H	M/M	M/H	H/L	M/M	L/H	L/H	M/H	L/H	M/M	H/L			H/M	L/H	H/L	L/H	H/M	H/M	L/H					H/L	L/L
METAL CATIONS	Cobalt	L/H	M/M	M/H	H/L	M/M	L/H	L/H	M/H	L/H	M/M	H/L			H/M	L/H	H/L	L/H	H/M	H/M	L/H					H/L	L/L
METAL CATIONS	Copper	L/H	M/M	M/H	H/L	M/M	L/H	L/H	M/H	L/H	M/M	H/L	L/L	H/L	H/M	L/H	H/L	L/H	H/M	H/M	L/H					H/L	L/L
METAL CATIONS	Lead	L/H	M/M	M/H	H/L	M/M	L/H	L/H	M/H	L/H	M/M	H/L			H/M	L/H	H/L	L/H	H/M	H/M	L/H					H/L	L/L
METAL CATIONS	Mercury	L/H	M/M	H/H	H/L	M/M	L/H	L/H	M/H	L/H	M/M	H/L			H/M	L/H	H/L	L/H	H/M	H/M	L/H					H/L	L/L
METAL CATIONS	Nickel	L/H	M/M	M/H	H/L	M/M	L/H	L/H	M/H	L/H	M/M	H/L			H/M	L/H	H/L	L/H	H/M	H/M	L/H					H/L	L/L
METAL CATIONS	Zinc	L/H	M/M	H/H	H/L	M/M	L/H	L/H	M/H	L/H	M/M	H/L			H/M	L/H	H/L	L/H	H/M	H/M	L/H					H/L	L/L
MAJOR ANIONS	Antimony	L/H	M/M	M/H	H/L	M/M	L/H	L/H	M/H	H/L	L/H	L/H	L/H	L/H	M/H	L/H	H/L	M/M	L/H	L/H	M/H					H/L	L/L
MAJOR ANIONS	Arsenic	L/H	M/M	H/H	H/L	M/M	L/H	L/H	M/H	H/L	L/H	L/H	L/H	L/H	M/H	L/H	H/L	M/M	L/H	L/H	M/H	L/L				H/L	L/L
MAJOR ANIONS	Boron	L/H	M/M		H/L	M/M	L/H	L/H	M/H	M/H	M/H	H/L	M/H	L/H	M/H	L/H	H/L	M/M		L/H	M/H					H/L	L/L
MAJOR ANIONS	Chromium VI	L/H	M/M	M/H	L/L	M/M	L/H	L/H	M/H	M/M	L/H	L/H		H/L	M/H	L/H	L/H	M/M		L/H	M/H	H/M	M/H			H/L	L/L
MAJOR ANIONS	Chloride	L/H	M/M		H/L	M/M	L/H	L/H	M/H	M/M	L/H	L/H			M/H	L/H	H/L	M/M		L/H	M/H					H/L	L/L
MAJOR ANIONS	Nitrate	L/H	M/M	M/M	H/L	M/M	L/H	L/H	M/H	M/M	L/H	L/H		H/M	M/H	L/H	H/L	M/H		L/H	M/H	M/M	M/H	M/M	M/H	H/L	L/L
MAJOR ANIONS	Sulfate	L/H	M/M	M/H	H/L	M/M	L/H	L/H	M/H	M/M	L/H	L/H		H/M	M/H	L/H	H/L	M/H		L/H	M/H	M/H				H/L	L/L
CYANIDES	Cyanide, Free	L/H	M/M	M/M	H/L	M/M	L/H	L/H(2)	M/H	H/L	M/L	L/L	H/M	L/H	H/H	L/H	H/L	M/H	L/H	L/H	M/H	H/M	L/H	H/M	M/H	H/L	L/H
CYANIDES	Cyanide, Weakly Complexed	L/H	M/M	M/M	H/L	M/M	L/H	L/H(2)	M/H	H/L	M/L	L/L	M/M	L/H	H/H	L/H	H/L	M/H	L/H	L/H	M/H	H/M	L/H	H/M	M/H	H/L	L/H
CYANIDES	Cyanide, Strongly Complexed	L/H	M/M	M/H	H/L	M/M	L/H	L/H	M/H						L/H	L/H	H/L	M/H	L/H	L/H	M/H	M/M		M/M		H/L	L/L

IMPORTANCE OF PROPERTY TO ATTENUATION OF CHEMICAL

Potential to Enhance Attenuation \ Relative Degree of Verification Required	Low	Moderate	High
High	H/L	H/M	H/H
Moderate	M/L	M/M	M/H
Low	L/L	L/M	L/H

(1) "Fractured" soil may relate to soil conditions such as very deep dessication cracks or active fault areas which could disrupt normal sedimentary patterns. Fractured limestone could be an exception to this general pattern, since limestone tends to neutralize acidic solutions promoting precipitation and adsorption mechanisms.

(2) Gravel conditions could potentially assist attenuation of free and WAD cyanide by volatilization, if the pH is below about 9.5.

Soil/Rock Physical Properties - One of the most important issues when considering attenuation as a containment feature is to adequately characterize the site and identity any preferential flow paths that may exist and which could eliminate or reduce the exposure time and conditions required for attenuation to occur. Typical site conditions under which preferential flow paths could occur are illustrated in Table 5.7, under the headings of Highly Fractured Rock, Sand, Gravel, and Fractured Soil. The table shows that, when these soil/rock physical properties exist, the potential for most attenuation mechanisms is reduced, and greater verification effort would be required to demonstrate the applicability of attenuation as a containment feature. Exceptions for these conditions can exist however. For example, free or WAD cyanide could volatilize while "trickling" through unsaturated sand or gravel if pH conditions were below about 9.5; or the rock or permeable soil could buffer pH conditions to eliminate acidic conditions and then promote precipitation and adsorption mechanisms.

Sandy soil may be less effective for attenuation than fine grained silt or clay, depending upon: (1) the amount of fine-grained particles in the sand, (2) the chemical characteristics of the sand particles, and (3) existing pore water fluids.

Low permeability rock, which has only small microfractures, often will have greater potential for attenuation than openly fractured rock. The site investigations in such cases will require sufficient information to verify that large fractured zones do not exist. If they cannot be ruled out, reliance on attenuation for waste containment usually will not be appropriate.

The Grey Eagle mine in Northern California illustrates how soil/rock physical properties can determine the effectiveness of attenuation. In 1981, the tailings impoundment for that facility was constructed to contain the cyanide-bearing tailings. The impoundment was constructed over a series of fractures in bedrock, running perpendicular to the dam, which were grouted to cut off the potential for underseepage. However, because of an apparent failure in the grout curtain, a permeable conduit was created for leachate to flow from the tailings impoundment beneath the dam and enter into a downgradient leachate collection system. Attenuation did not occur, and would not be expected, in such a highly permeable situation. For further discussion of the Grey Eagle mine, see Chapter 2.

The term fractured soil is included primarily as a warning that even at sites underlain by soils, conditions might exist which include preferential flow pathways to ground water. Most mine sites, however, are not likely to have such conditions. Instead, many mine waste management unit sites are located on fairly thick alluvium or colluvium which was deposited in a series of layers over a long period. Any given layer may have localized, shallow discontinuities, but there is a very low probability that they are continuous for significant vertical or lateral distances. Instead, these types of soil systems can actually enhance attenuation potential because the layered conditions increase dispersion and, therefore, expose migrating liquids to a much larger volume of soil in which the various attenuating mechanisms can occur.

Residual soils, formed from weathering of rock in place, do not have the layered structure that is characteristic of transported soils. The grain size characteristics of residual soils reflect the type of rock being weathered, e.g., clay from claystone, sand from sandstone. Fracturing can be more prevalent in residual soils if the rock was highly fractured and the soil is not thick. However, weaker zones in the soil can frequently close as the more plastic soil (in comparison to rock) deforms with time and weathering. In general, verification of the potential for attenuation to occur uniformly in residual soils will require more field verification than for deep, layered soils.

In general, organic matter in the soil will increase the potential for attenuation of several of the metals, especially by suction mechanisms. Soil organic matter possesses functional groups that adsorb H^+ and OH^- to form charged surfaces, and thus it behaves in a manner similar to metal hydroxide surfaces:

$$R-OH + OH^- \rightleftharpoons R-O^- + H_2O$$

$$R-OH + H^+ \rightleftharpoons R-OH^+$$

$$R-NH_2 + H^+ \rightleftharpoons R-NH_3^+$$

The organic surface charge will be pH dependent. If acidic, it will be positive and attract anions. If neutral-to-basic, it will attract metallic cations. Soils containing a large amount of organic matter may tend to be acidic, due to the formation of organic acids by biodegradation. This may reduce the attenuation of metals. Therefore, since soils containing an abundance of organic matter have properties that both enhance and impede attenuation, they generally will require thorough investigation.

Table 5.7 indicates that clay is often the best type of soil to promote attenuation, although silty soil (especially with some clay fraction) may also function well. This is because of the highly reactive characteristics of the surfaces of clay minerals, as discussed in Section 5.3.

Soil/Rock/Pore Water Chemistry - Many of the relationships between vadose zone chemistry and potential attenuation mechanisms are shown in Table 5.7. The following key observations from the data are worthy of special note:

- Attenuation potential for the metal cations is reduced by low pH conditions. This is because low pH causes (or results in) an abundance of positive hydrogen ions on soil surfaces, tending to decrease the adsorption of positive metal ions.
- Predictability of metal cation attenuation is relatively high and can be a useful design consideration for many site conditions. The factors that could adversely affect this mode would be associated with materials which have unusually low cation exchange capacity or low pH conditions.
- In general, there are fewer unsaturated zone chemical characteristics which promote attenuation of the major anions than the metals. Higher pH conditions tend to restrict attenuation of major anions (in general, pH dependence of metal cations and anions are inverse).
- Neutral to slightly acidic pore water can be beneficial for the adsorption of free cyanide or complexed cyanide. At pH of less than about 9.5, free and WAD cyanide is hydrolyzed to HCN, which is volatile. Also, a high organic content is beneficial both to adsorb cyanide and to provide microbiological activity which will rapidly decompose cyanide, especially in an oxidizing environment. Nutrients such as phosphate, potassium, sulfur, and micronutrients also serve to encourage microbiological activity.

Pore Water Volume - The potential for attenuation often will be enhanced if the pore water volume is below field capacity, because this condition will slow the migration rate and will promote dispersion in the vadose zone. The added time and volume of exposure provide additional attenuation capacity. Larger natural water content can be beneficial, however, if the pore water volume itself promotes attenuation by dilution or beneficial chemical changes to the waste unit seepage liquids.

Ratio of Waste Unit Width to Ground Water Depth - For most conditions, the hydrogeochemistry and thickness of the unsaturated zone will be the important characteristics, independent of the geographic size of the waste unit. However, size may potentially have some effect on attenuation capacity or rate if constituent reduction is dependent upon air flow through the soil or rock pores (e.g., for volatilization). This potential factor is noted to indicate that the verification effort for large waste units may have to consider waste unit size in some instances.

5.4.3.2 Other Site Characteristics

The following paragraphs introduce additional site considerations potentially affecting the amount of seepage infiltration which may have to be attenuated.

Climatic Factors - Climatic factors, discussed in detail in Chapter 6, can have an important effect on the behavior of chemicals in the soil. Rainfall is particularly important for waste piles and "dry" portions of tailings because it is the only potential driving force causing constituents to migrate into the soil. In arid and low rainfall areas, the low infiltration rates will proportionally increase the factor of safety provided by the attenuation potential.

Calculations summarized in Chapter 6 show that in areas of very high evaporation and very low rainfall, essentially no water even reaches the base of a waste pile or into the vadose zone. In those cases, the attenuation potential is very high and may be the only factor necessary to assure that beneficial uses of ground water are not degraded.

Travel Time to Surface or Ground Water - It is evident that attenuation potential will increase with increased seepage travel time because:

- Time related mechanisms will have longer periods to react.
- More spreading (dispersion) usually will occur with increased travel time.

The lower limit of attenuation potential will be site-specific. The extent of verification of attenuation will increase with shorter time periods, especially when the estimated seepage time is less than the operating life of the waste management unit.

Attenuation Capacity - The reliability of attenuation becomes greater as the depth to ground water increases. This effect could be quantified by calculating a factor of safety equal to the overall estimated attenuation capacity of the unsaturated zone, divided by the quantity of the constituent of concern in the seepage. The desired degree of safety can be evaluated on a case-by-case basis, depending on the reliability of the mechanism of attenuation and the degree to which site conditions can be verified.

5.4.3.3 Waste Management Unit Practices

There are a number of direct and indirect activities which can be incorporated into a waste management unit design and operating practices. These activities can improve the potential for attenuation to occur to the extent desired to protect beneficial uses of the waters of the State. Examples of direct activities include:

- Providing lining or containment systems which have good attenuation properties for the chemical species of interest. For instance, liners or constructed foundation materials with a basic pH may be very effective for retarding metal cation migration, even if a small amount of liquid seepage would occur over time. Also, the inclusion of foundation or liner soil with some natural organic material can increase the potential for free cyanide attenuation.
- Combining wastes with differing chemical or physical properties in the same waste management unit. For instance, if waste with the potential to generate acid is combined with other wastes with a buffering potential, the potential for the formation of acid rock drainage (ARD) can be reduced or eliminated; and the potential for attenuation of metals within the waste unit can be enhanced. This type of combination could actually include concurrent disposal of both material types, or the placement of one material type above the other.
- Locating certain waste units within specific areas that have conditions amenable for attenuation to occur, which can be verified with confidence.

Examples of indirect activities include:

- Controlling of the amount of liquid contacting the waste to minimize infiltration with constituents requiring attenuation. Dry disposal of tailings represents this type of operating control.
- Optimizing the metallurgical processing to minimize the levels of required reagents in liquids. This can occur by increased preparation of the ore for processing or by using processing circuits which can extract minerals with lower reagent concentrations.

5.4.4 DATA REQUIREMENTS

In evaluating a mine site for its potential to affect the behavior of the chemical of concern, it is necessary to collect the following types of data: (1) physical and chemical properties of the aqueous phase in the source, (2) the chemicals which may leach out of the source, (3) the characteristics of the unsaturated zone beneath the site, including the geochemical, mineralogical, and physical properties of the foundation material, and (4) the redox potential and pH conditions that are established during contact of the seepage with the foundation material.

5.4.4.1 Information About the Unsaturated Zone

To accurately predict the potential for attenuation, it is essential to understand both physical and chemical characteristics of the unsaturated zone. The degree of knowledge must be adequate to confirm that conditions are laterally consistent over portions of the waste management unit which have the potential for seepage, to assure that the attenuating mechanism will not be bypassed. All or portions of the following information should be obtained, depending on the attenuation mechanism being studied:

- Stratigraphic conditions, including the general degree of strata continuity.
- Grain size and plasticity index properties of key soil strata, to be used to estimate permeability and porosity.
- Mineralogy, pH, redox potential, and percentage of organic materials in key strata, especially for materials that will exist immediately below the waste unit.
- Cationic and anionic exchange capacities of key strata.
- Degree and orientation of microfracturing in consolidated formations.
- Depth to ground water.
- Existence of perched zones.
- Existing field water content, including variations with depth.

Usually the first, and often the most important, site investigation activity should be development of regional and site geologic and hydrogeologic models using existing data. Often this can be done from the literature, records of wells and borings in the area, aerial photo interpretations, and site mapping. Mining sites usually provide a great deal of additional data from exploration and condemnation drill holes used to locate and evaluate the mineral reserve.

This initial characterization step usually will provide the information needed to determine the degree to which waste unit site conditions will be uniform and predictable or random and unpredictable. For example, if the site is unfaulted and overlies thick soil deposited in layers over geologic time, large, continuous vertical discontinuities are not likely to exist. Therefore, significant vertical pathways to ground water (unless very shallow) are unlikely. The investigation in this case would concentrate on: (1) understanding the layering to estimate the manner in which downward migration could occur, and (2) assuring that there are not shallow lateral pathways (e.g., gravel channels) which could cause undesirable lateral migration.

On the other hand, if hard bedrock is at or near the ground surface, and the area is highly faulted and folded, there will be high potential for the existence of fractured zones which could represent preferential seepage pathways. Under such conditions, key evaluations will involve attempts to locate fractured zones and, if located, determining their permeability and mineral characteristics.

The initial site characterization data should be used to design the most appropriate field and laboratory investigation program to gather data that will indicate the degree of containment required, considering the waste management facility and site conditions. The field and laboratory investigations may include:

- Mapping surface conditions to assess: (1) drainage patterns, including potential pathways for leachate migration, (2) soil/rock type, (3) vegetation type, and (4) surface gradient.
- Excavating test pits or trenches to assess: (1) the lateral continuity of relatively shallow lithologic strata in the vadose zone, (2) the presence of soil/rock structures such as fractures and/or faults, and (3) soil/rock types and properties. For example, many mines in California are located in the desert where thick layers of caliche (indurated calcareous layers) commonly form, providing a natural barrier to downward migration of liquids. Site investigations performed to evaluate containment requirements for a gold mining operation have included excavating a series of trenches to determine the lateral continuity of these layers.
- Conducting seismic investigations to assess lateral continuity of lithologic strata and the presence of soil/rock structures such as fractures and/or faults.
- Drilling soil borings to assess the vertical profile of subsurface conditions, including soil/rock properties, and depth to ground water.
- Installing ground water monitoring wells to assess the hydrologic conditions at the site, such as: (1) depth to ground water, (2) soil permeability, (3) ground water flow direction and velocity, and (4) ground water gradient.
- Conducting percolation tests and ground water pump tests to assess the hydraulic conductivity and the extent of higher conductivity for fracture zones.
- Characterizing the mineralogical and geochemical conditions of the geologic materials to determine the extent of attenuation that can be expected from individual strata.

5.4.4.2 <u>Information About the Waste</u>

Once the information about the unsaturated zone is determined, the characteristics of the potential leachate from the spent ore or other source should be determined. These characteristics will include:

- Chemical analysis of the pore water in the mine waste or other source.
- Physical characteristics of the pore water, such as density, viscosity, pH, and the vapor pressure over the pore water of any volatile constituents.
- The concentration of important constituents under various environmental conditions, including the undeveloped site, as well as operating and closure conditions.
- Column adsorption testing, using the source solution and typical unsaturated zone soil to determine the behavior of the chemical(s) in the site soils. These tests should be done at infiltration rates representative of site conditions.

- Laboratory tests to determine the interactions that will occur between chemical constituents in seepage solutions and material in the unsaturated zone. Such tests should include sequential batch-contact and/or column percolation procedures using source solutions and typical vadose zone soils from the site. These tests should be performed using varying conditions, such as pH, ionic strength, and concentrations of dissolved constituents.

5.5 APPROPRIATE LEVELS OF VERIFICATION

5.5.1 GENERAL

The discussion in Sections 5.3 and 5.4 show that, although attenuation mechanisms are known for each of the chemicals of interest, the degree to which each can be predicted varies greatly. The variability occurs because some mechanisms are relatively simple and occur under normal vadose zone conditions, while others either require abnormal vadose zone conditions or a combination of conditions that occur infrequently or may be difficult to verify.

Because there are likely to be many sites at which attenuation will provide reliable total or partial containment of certain chemical constituents, waste management regulations should include some procedure to allow it to be considered and utilized if appropriate. Because of the large differences in the predictability in attenuation for different sites, tests used to determine the applicability of attenuation should not be the same for each chemical onsite. Also, the degree of verification should vary depending upon the level of potential threat posed by the waste management unit. Alternative verification procedures can vary within the following types of activities:

- Verification through experience provided by theory, case histories, and previous studies (university research, reports for other projects, etc.). Examples of this type of verification could be for relatively straightforward attenuation of metal cations such as lead, copper, or zinc within several feet of nonacidic clayey or silty soil. This type of "literature" verification would be supplemented by substantial field and laboratory testing to verify the existence of appropriate conditions for the desired attenuation mechanisms.
- Verification through laboratory testing, for those cases where the principal attenuation mechanism is assured, but data for quantifying rates and capacities are needed for safety assurance. For example, special laboratory testing of the attenuation of arsenic may be required because this metal is more sensitive to geochemical properties and conditions than many of the other metals. Another example is laboratory tests used to quantify the capacity of a clay liner system to attenuate cyanide.

- Verification through detailed field demonstration tests may be required where in situ subsurface conditions can have a major impact on attenuation effectiveness, or if field conditions cannot be adequately simulated in the laboratory.
- Long-term monitoring and comparison to conceptual and mathematical models of contaminant behavior.

In addition to determining the reliability of an attenuation mechanism, the waste facility designer must also consider the cost and schedule implications of undertaking a special attenuation investigation. A high-cost field test, with only limited potential for providing reliable verification data, may not be justified in many cases. In these instances, it may be more appropriate not to include attenuation as a waste containment feature.

The following sections discuss the types of activities which could be carried out for each of the types of verification programs.

5.5.2 PREDICTIONS BASED ON BASIC TECHNOLOGY, PRIOR EXPERIENCE, AND RESEARCH

Some chemicals have well known behaviors that are dependent on a few common properties. Many of the heavy metals behave quite similarly, dependent primarily on the pH conditions of the soil. For instance, soluble lead introduced into a soil of normal to basic pH will react with the soil water to form lead hydroxides or lead carbonates, depending on the pH. As an example:

$$Pb^{+2} + 2H_2O = Pb(OH)_2 + 2H^+$$

This reaction is controlled by the pH, the presence of soil water, and the presence of lead. Given an unsaturated zone where there is a sufficient soil column of neutral to basic pH, lead would react and precipitate. Verification of the predicted behavior of lead should require only a basic knowledge of the unsaturated zone at the specific site, calculations to assure adequate attenuation capacity or direct correlation to a prior project or research experience, and verification of sufficient acid neutralization capacity to maintain near-neutral pH conditions.

Other chemicals have well known behaviors but may require more site-specific information to predict the effectiveness of their attenuation. For example, free cyanide is easily volatilized under certain pH conditions. Under different redox conditions, it can be oxidized or hydrolyzed to NH_3 by bacterial action, or react with copper and zinc to form insoluble salts. Detailed site information on the soil water pH and chemistry, along with the organic content of the soil, is important to evaluate the potential effectiveness of these competing reactions. Also, reliance on these mechanisms, at least for the present time, usually will require some site-specific attenuation testing.

5.5.3 LABORATORY TESTING

The initial level of site-specific testing usually will occur in the laboratory to verify the predicted behavior of the specific chemical for given site conditions. This level of verification may be needed for several reasons, including: (1) the possibility of competing reactions which would adversely affect the predicted behavior, (2) the need to determine the capacity of the soils to react where there may be a finite amount of reactant or a finite adsorption capacity, (3) the need to determine if the zone of attenuation has a limited thickness which would require that the readily predictable behavior must occur within a specified time, (4) the need to determine if chemicals displaced by ion exchange could have any secondary, undesirable impacts, or (5) the need to assess attenuation mechanisms in a chemical system that changes with time.

For example, chemicals such as barium, which replace calcium in the structure of clay minerals, may have a finite capacity in a given soil, depending on the amount and type of clay in the soil column. Laboratory tests may be required to determine the capacity of the soils to adsorb barium and to verify the efficiency of the reaction. As another example, a clay liner with 5 percent organic material may be effective for retarding free cyanide migration. A laboratory test could be useful in determining if this physiochemical attenuation mechanism has limitations with regard to the magnitude of flow that could be released from a mine waste management unit. These tests can include sequential batch extraction tests and/or column tests, as described below. Each test must be designed to suit the chemicals and conditions being simulated.

Sequential batch extraction tests (Figure 5.7) can simulate years of potential seepage in the field through a soil column or clay liner in a few days of laboratory testing. The procedure simulates field conditions by subjecting the seepage solution to a series of fresh batches of native soil. Known amounts of soil are placed in containers, and measured volumes of solution are introduced to reach a prescribed liquid to solid ratio. The resulting slurries are gently agitated for a specified period of time, then the solutions are decanted and the solids filtered. The solutions and soils are then advanced according to a previously determined matrix similar to the illustration in the figure (Pyrih and Rouse 1989).

Column tests simulate the solution percolating through the soil, using native material from the site. Solution is percolated through the column, and effluent from the bottom of the column is

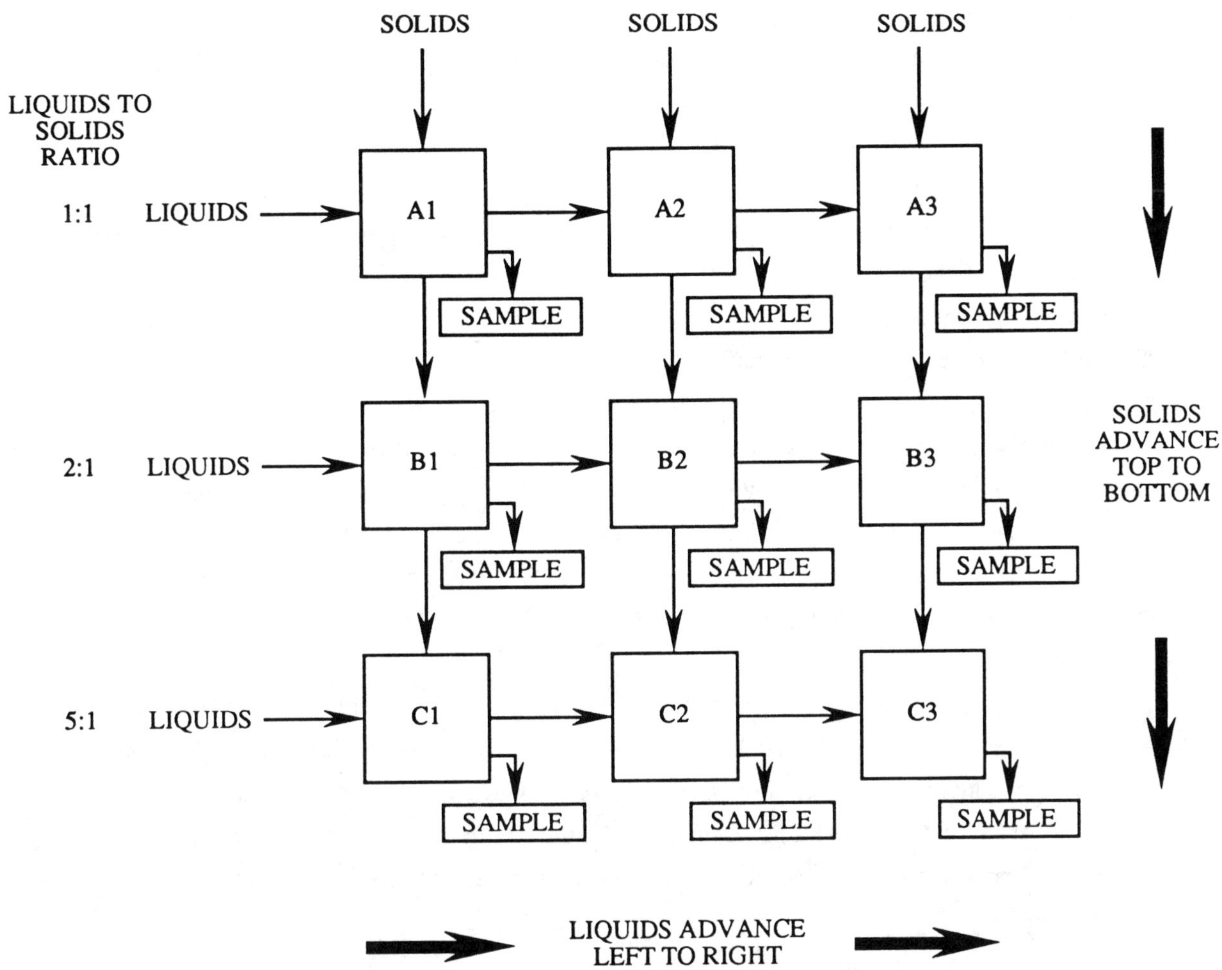

FIGURE 5.7

EXAMPLE SEQUENTIAL BATCH TEST PROCEDURE

REFERENCE: MODIFIED FROM PYRIH AND ROUSE, 1989

monitored. The solution can be monitored as it percolates through the soil by withdrawing samples from access points spaced down the side of the column. These tests must be designed to assure that scale effects at the column walls do not generate misleading results. A schematic of a typical down-flow column test is shown in Figure 5.8.

The results of the above tests can reveal both the capacity of the soil to adsorb or react with the chemical and the relative efficiency of the adsorption or reaction process in reducing the concentration of the chemical in solution. Column studies utilizing a continuous source of solution will determine the breakthrough time of the chemical in solution, the removal efficiency of the predicted process, and the estimated capacity of the soil to react. The designer of a column study must recognize, however: (1) the length of time required to demonstrate behavior, (2) the parameters which must be tested in order to simulate the appropriate range of field and environmental conditions which may be encountered, (3) the potential limitations associated with column geometry, potential edge effects, etc., and (4) that the effects of excavating, handling, and loading the soil into the column may alter its characteristics. The potential use of these tests can be seen in the following example from Pyrih and Rouse (1989).

The spent ore from a proposed mine site in Idaho contains levels of soluble arsenic above the state water quality standards. Geochemical analyses of the colluvium from the footprint of the proposed spent ore impoundment revealed properties that were favorable to the immobilization of arsenic. These properties included: (1) a generally acidic soil pH ranging between approximately 4.4 and 6.2, (2) the presence of illite as the principal clay mineral, and (3) the presence of abundant limonite staining.

To determine the capacity of the colluvium to immobilize the soluble arsenic, column tests were performed on the native material from the footprint area. Percolating a solution representing leachate from the spent ore through the colluvium resulted in a decrease in pH from approximately 8.3 to approximately 6.0. The concentration of soluble arsenic in the leachate solution was reduced over 85 percent, to less than 0.2 mg/L. These changes were seen for a throughput in excess of 8 pore volumes of solution, representing more than 100 years of unabated seepage from the spent ore.

To confirm the findings of the column tests, sequential batch tests were run. These tests resulted in a reduction of the arsenic concentration in the leachate solution by over 95 percent. Based on the positive results of these tests, provisions for natural attenuation were incorporated into the National Pollution Discharge Elimination System (NPDES) permit for the site. These provisions included

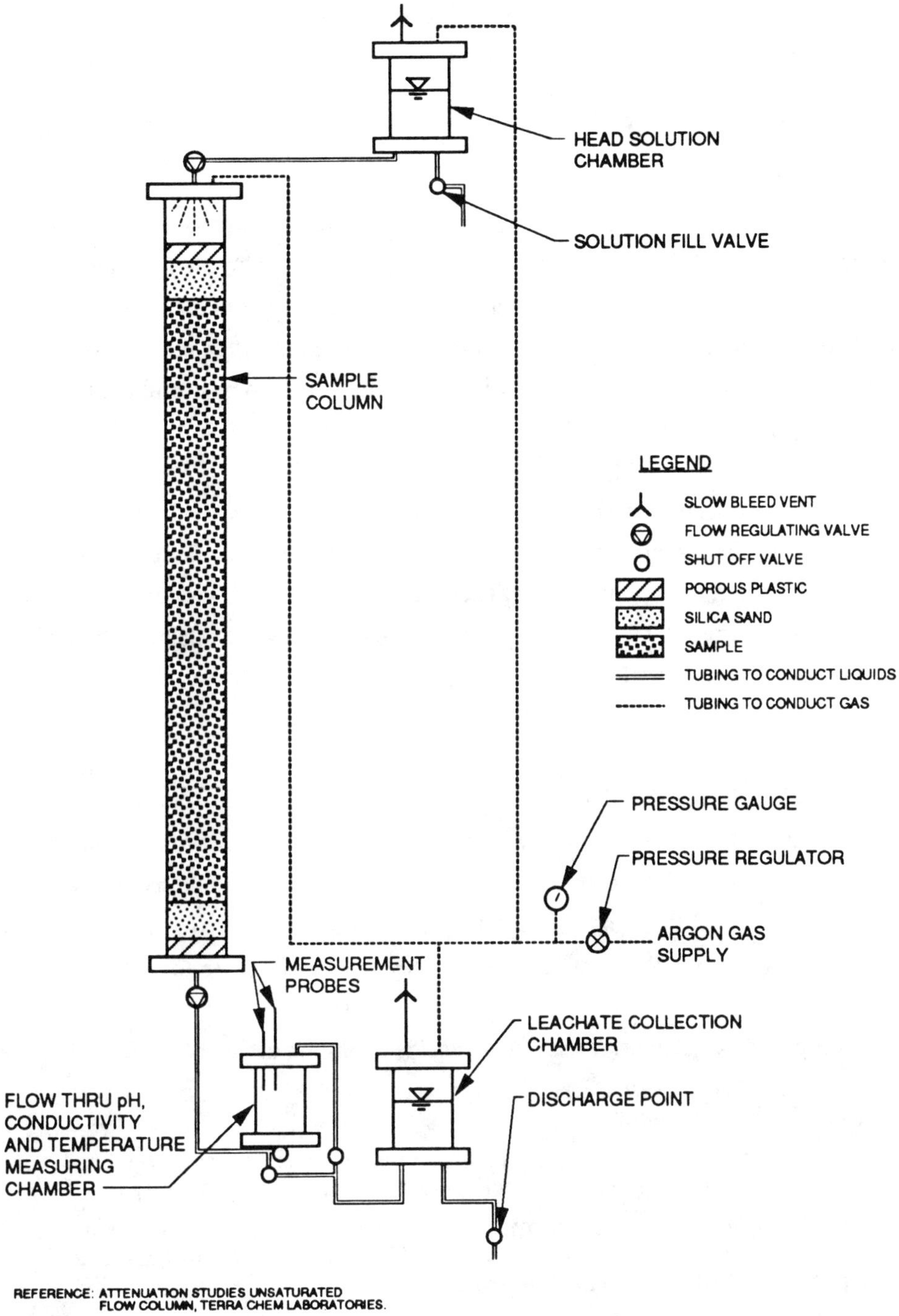

FIGURE 5.8

EXAMPLE UNSATURATED COLUMN PERCOLATION TEST PROCEDURE

construction of four clay berms across the disposal area in order to provide additional leachate/soil interaction by increasing the leachate infiltration rate. Extensive vadose zone and ground water monitoring programs were designed to assess whether attenuation proceeds as predicted.

5.5.4 MODELING

Simple models that are used, for example, to determine the total attenuation capacity of a geologic formation from limited laboratory or field data can be reliably used in the design of waste management units. However, more complex mathematical models of the time varying behavior of chemical solutions that migrate through the unsaturated zone are still in the development stage. It is likely that these models will improve with time, and they will become useful tools in evaluating the attenuation capacity of sites.

Even though there is a considerable body of literature on the development of models to predict the attenuation of solutes, there has been little interaction between model prediction and field evaluation. This may be because it is difficult to conduct a field test of a predictive behavior without causing environmental damage. The literature also shows that there is still uncertainty over the validity of particular modeling approaches (Nielson, 1990).

5.5.5 FIELD DEMONSTRATIONS

Field demonstrations will be required where: (1) the predicted behavior of the chemical in the vadose zone at the site is dependent on properties of the soil which could have considerable lateral or vertical variation, (2) there may be significant soil structures to provide pathways through the soil column, or (3) laboratory test procedures may not be adequate to demonstrate conditions which will be encountered in the field. Field demonstrations will usually have a number of potential difficulties which must be carefully considered during design of the test. These include:

- Assuring that the test is large enough to test a representative zone where verification is required.
- Assuring that the amount of chemical released to the subsurface would not have the potential to degrade beneficial uses of surface or ground water.
- Setting up an experimental configuration which represents field conditions and avoids the potential for test elements (including monitoring) to create test results that do not reflect full-scale operations.

Most often, when a field test is required, a parallel laboratory program also will be appropriate. Then, the field test can be directed specifically toward evaluating a particular site characteristic.

If the comparable results can be demonstrated in both field and laboratory tests, results of the laboratory tests can be utilized with greater confidence.

5.6 REGULATORY IMPLICATIONS

5.6.1 THE APPROPRIATE ROLE OF ATTENUATION IN MINING WASTE REGULATIONS

The discussions provided in this chapter provide substantial data indicating that attenuation of chemical species can occur in many vadose zone situations. A first step toward determining attenuation potential is to assure that site conditions are suitable or that conditions do not exist which preclude effective attenuation. The data also show that many attenuation mechanisms are readily predictable. Others cannot be relied upon without site-specific verification, which could vary from relatively simple laboratory tests to an extensive field testing program.

The extent to which attenuation may be applicable will vary depending upon whether or not its consideration would have a significant effect on containment operations or closure activities. If those factors would not be changed appreciably, the costs to verify attenuation may not be justified.

There will be situations, however, especially for low risk conditions and chemicals with high prediction potential, where attenuation could become either the major containment feature or a redundant backup to a primary, engineered containment system. In the simplest of cases, attenuation may preclude the need for a liner. In other cases, attenuation may become a consideration for designing a "leak-free" liner without excessive redundancy. In this latter case, it would be necessary that minor leakage not create an unacceptable threat, in part due to attenuation.

Attenuation, therefore, should be permitted a legitimate role in the design and regulation of mine waste units.

5.6.2 RECOMMENDED REGULATORY APPROACH

A discharger should be permitted to consider attenuation as a potential method of preventing the migration of mine waste or its constituents, either in conjunction with, or in lieu of, a liner system.

The use of attenuation in the design of a mine waste management unit should be based on one or more of the following steps:

- Item A - Documentation by the discharger that attenuation of the constituents in question will occur under the site-specific conditions. Data to be provided for this documentation must include:
    - Detailed physical and chemical analyses of leachate expected to be generated within the waste unit.
    - Detailed chemical and physical analyses of the liner or natural soil/rock material in which the predicted attenuation will occur.
    - Description of the attenuation mechanisms that will occur, and verification that the attenuating capacity (i.e., mass loading capacity) will be at least 10 times greater than the mass of each constituent to be attenuated over the life of the waste management unit.
    - Description of prior experience, including project examples where the attenuating mechanism being relied upon has been successfully demonstrated.
    - Certification of the responsible geologist or geotechnical engineer responsible for the work, and résumés of hydrogeologists, geochemists, and geotechnical engineers who have participated in the study.
    - Certification of a qualified reviewer of the data, with expertise in hydrogeology, geochemistry or geotechnical engineering, who is independent of the organization that performed the basic work.
- Item B - The same as Item A, in addition to laboratory test data which verify the magnitude and capacity of the anticipated attenuation. The laboratory tests must be conducted according to an approved plan, using samples of leachate representative of that expected from the waste unit and representative site soil or rock materials. Laboratory test work will consider both mass loading and system dynamics. Data to be provided for this documentation must be the same as that required for Item A, plus a complete description of the laboratory tests performed and analyses of the data which verify attenuation magnitude and capacity. The description must provide the technical reasons for concluding that laboratory test data are applicable for predicting conditions that will occur in the field.
- Item C - The same as Item A or B, plus field demonstration tests designed to show that the magnitude and capacity of attenuation will occur to an extent equal to containment achieved in a liner system. The field test program must be conducted according to an approved plan, in an area of up to 5,000 square feet. It will be contingent upon, but not limited to, the following:
    - A field attenuation demonstration test which shall reliably and accurately predict the spatial location, areal extent, and concentration of all leakage, and attenuation thereof, beyond well defined and surveyed boundaries of a demonstrated test waste management unit. If the proposal for the demonstration test is approved, detailed criteria by which the success or failure of the research demonstration project shall be evaluated shall be stipulated prior to discharge of waste to the test waste management unit. All liquids discharged to the test unit

shall be listed and characterized in the proposal and shall include a geochemically mobile constituent that has the ability to act as a tracer which can be easily monitored and distinguished from waste and background constituents.

- The demonstration test shall include a comprehensive hydrogeological characterization report of the proposed test site prepared by a registered geologist or geotechnical engineer having professional expertise in hydrogeology. The hydrogeological characterization report shall be based on data collected by investigations which identify and characterize all geologic materials (e.g., hydrostratigraphy, geologic structure, geochemistry, geotechnical properties, etc.), all possible migration pathways for leakage from the demonstration test waste management unit, potentiometric data, and background concentration data for onsite soils and waters, and shall quantify all recharge and discharge parameters of the site hydrologic balance. Data for site-specific attenuation mechanisms used in models resulting from the demonstration test shall be based on verifiable field evidence that demonstrates a clear understanding of subsurface geologic heterogeneities at the proposed site.
- The field demonstration test shall include proposed designs for a reliable and comprehensive water quality monitoring system for all possible migration pathways for leakage from the research waste management unit, including the unsaturated zone underlying and surrounding the unit, surface waters, and uppermost ground water. If unsaturated zone monitoring is not considered to be technically feasible or reliable at a particular proposed site, then the program shall include sampling of soil in the vadose zone. If it is determined by the regulatory authority that an existing monitoring system has provided all necessary potentiometric and background concentration data, and will provide reliable periodic representative samples for analyses suitable for accurate measurement of predicted attenuation of leakage along all migration pathways, this can be the basis for approval of the field demonstration test.
- The field demonstration test shall include contingency plans for corrective actions, with assurances of financial responsibility, to be implemented in the event that attenuation is less effective than predicted, or waste constituents are detected in surface or ground water. Contingency plans shall include provisions for removing waste from the test area to acceptable concentrations.

The regulatory authority will determine whether a literature study, laboratory investigation, and/or field demonstration test will be required in order to allow attenuation to be considered in the design of a proposed waste management units.

If it is determined that specific site characteristics can be relied upon to prevent the migration of mine waste or its constituents to ground or surface water in a predictable manner, a full-scale containment unit which uses attenuation as a design factor can be approved.

5.7 BIBLIOGRAPHY

Battelle Pacific Northwest Laboratories. *Toxicologic Criteria for Defining Hazardous Wastes*. Developed for the Minnesota Pollution Control Agency, Roseville, Minnesota. September 30, 1976.

Brady, N.C., The Nature and Properties of Soils. 8th edition. New York, MacMillan Publishing Company, Inc. 1974. 639.

California Department of Health Services. Letter to Mr. George Duker, Molycorp. March 10, 1986.

California Department of Health Services. Letter to Mr. Jim Strong, Molycorp. July 21, 1989.

California Department of Health Services. *The California Site Mitigation Decision Tree Manual*. May 1986.

California Regional Water Quality Control Board. Central Valley Region. *The Designated Level Methodology for Waste Classification and Cleanup Level Determination*. October 1986.

California State Water Resources Control Board. *Amendment of Article 5 (Water Quality Monitoring for Classified Waste Water Management Units)*. July 21, 1989.

Chatwin, T.D., and J.J. Trepanowski. *Utilization of Soils to Mitigate Cyanide Releases, Resource Recovery and Conservation Consultants*. Proceedings of the Third Western Regional Conference on Precious Metals, Coal and Environment. September 23-26, 1987.

Cherry, J.A., F.M. M.Morel, J.V.Rouse, J.L. Schnoor, and M.G. Wolman. *Hydrogeochemistry of Sulfide and Arsenic-Rich Tailings and Alluvium Along Whitewood Creek, South Dakota (Part 3 of 3)*. Mineral and Energy Resources, Colorado School of Mines, 29, (November 1986).

Clifton, A.W. and A. Smith. "Stability of Natural Materials Exposed to Uranium Mill Wastes". Proceedings of the Canadian Geotechnical Society Annual Meeting. Saskatchewan, Canada, September 1988.

Contra Costa Times. *Wild Mustard Might "Eat" Selenium*. September 4, 1989.

D'Appolonia Consulting Engineers. *Draft Report - Geochemical Considerations for Hazardous Waste Facility Design*. Las Vegas, Nevada: The Second Geotechnical Conference and Exhibit on Design and Construction. April 27, 1982.

Dave, N.K. and T.P. Lim. "Hydrogeochemical Evolution of Inactive Pyritic Tailings and Remediation of Contaminant Migration in a Surrounding Aquifer." International Symposium of Mine Waste Management. Albuquerque, New Mexico. 1985.

Dragun, J. *The Soil Chemistry of Hazardous Materials*. Silver Spring: Hazardous Materials Control Research Institute. 1988. 156.

Drever, J.I. *The Geochemistry of Natural Waters*. Englewood Cliffs: Prentice-Hall, Inc. 1988. 310-317.

Environmental Solutions, Inc. *Site Characterization and Feasibility Study. Prepared for Bethlehem Steel Corporation*. August 1985.

Environmental Solutions, Inc., *Natural Cyanide Degradation*. 1988.

Ewing, R.E. *Cyanide Removal/Destruction Technologies Currently Used in the United States*. December 2, 1986.

Forstner, U. and Wittman, G.T.W. *Metal Pollution in the Aquatic Environment*. New York: Springer-Verlag. 1979.

Freeze, R. A., and J. A. Cherry. *Groundwater*. Englewood Cliffs: Prentice-Hall, Inc. 1979. 604.

Fuller, W.H. "Cyanides in the Environment With Particular Attention to the Soil - Cyanide and the Environment." Proceedings of a Conference. Tucson, AZ; December, 1984.

Geochem, Inc. Personal Communication. 1989.

Griffin, R. A., K. Cartwright, N. F. Shimp, J. D. Steele, R. R. Ruch, W. A. White, G. M. Hughes, R. H. Gilkeson. *Attenuation of Pollutants in Municipal Landfill Leachate by Clay Minerals: Part 1 - Column Leaching and Field Verification*. Environmental Geology Notes, Illinois State Geological Survey, Number 78. November 1976.

Griffin, R.A., R.R. Frost, A.K Au, G.D. Robinson, and N.F. Shimp. *Attenuation of Pollutants in Municipal Landfill Leachate by Clay Minerals: Part 2 - Heavy Metal Adsorption*. Environmental Geology Notes, Illinois State Geological Survey, Number 79. April 1977.

Grim, R.E. *Clay Mineralogy*. New York: McGraw-Hill. 1968. 596.

Hornick, S. B. "The Interaction of Soils With Waste Constituents," in Soil Chemistry, Volume A: Basic Elements. G. H. Bolt and M. G. M. Bruggenwert, editors. New York: Elsevier Scientific Publishing Company. 1976. 4-19.

Hoye R. *Gold/Silver Heap Leaching and Management Practices That Minimize the Potential for Cyanide Releases*. United States Environmental Protection Agency. EPA 600/2-88-002. September, 1987.

IT Corporation. *Characterization of the Tailings*. Mountain Pass, California: Molycorp Mountain Pass Operations. June 1985.

Jury, W.A., A. Khan, and G. Butters. *The Difficulty in Predicting Ground Water Contamination*. University of California, Riverside. Undated paper. 16.

Jury, W.A., W.F. Spencer, and W.J. Farmer, "Behavior Assessment Model for Trace Organics in Soil: IV. Review of Experimental Evidence". Journal of Environmental Quality 13, (1984), No. 4.

Jury, W.A., W.F. Spencer, and W.J. Farmer. "Behavior Assessment Model for Trace Organics in Soil: I. Model Description". (1983), No. 12.

Jury, W.A., W.F. Spencer, and W.J. Farmer. "Behavior Assessment Model for Trace Organics in Soil: III. Application of Screening Model". Journal of Environmental Quality 13, (1984), No. 4.

Jury, W.A., W.J. Farmer, and W.F. Spencer. "Behavior Assessment Model for Trace Organics in Soil: II. Chemical Classification and Parameter Sensitivity". Journal of Environmental Quality 13, (1984), No. 4.

Krauskopf, K. B. *Introduction to Geochemistry.* New York: McGraw-Hill. 1979. 617.

Leeper, G.W. *Managing the Heavy Metals on Land.* Pollution Engineering and Technology. New York: Marcel Dekker, Inc. 1978.

Lyman, W. J., W. F. Reehl, and D.H. Rosenblatt. *Handbook of Chemical Property Estimation Methods.* New York: McGraw-Hill. 1982.

Marshack, J.B. *Designated Level Methodology for Waste Classification and Cleanup Level Determination.* California Regional Water Quality Control Board, Central Valley Region. June 1989.

Mattigod, S.V. and G. Sposito. *Chemical Modeling of Trace Metal Equilibria in Contaminated Soil Solutions Using the Computer Program GEOCHEM.* Chemical Modeling in Aqueous Systems: Washington, D.C.: American Chemical Society. Speciation, Sorption, Solubility and Kinetics. Edited by E. A. Jenne. 1979.

Nielson, D.R., M. Van Genuchten, and W.A. Jury. "Monitoring and Analyzing Water and Solute Transport in the Vadose Zone". International Symposium on Ground Water Monitoring and Management. March 23-28, 1987.

Nielson, D.R., M. Van Genuchten, and J.W. Biggar. "Water Flow and Solute Transport Processes in the Unsaturated Zone". Water Resources Research, 22, (1986): 895-1085, No. 9.

Perkins, K., and Dohms. "Using Wood Ash for Landfill Cover." *Solid Waste Power*, August 1989: 32-39.

Pyrih, R. Z., and J. V. Rouse. *Attenuation Processes: A Viable Regulatory Alternative.* Environmental Hazards Conference and Exposition, Bellevue, Washington. Preprint. February 14-16, 1989.

Rouse, J.V. *Natural and Man-Aided Attenuation of Contaminants at In Situ and Hazardous Waste Sites. In Situ Recovery of Minerals.* Edited by K.R. Coyne, and J.B. Hiskey. 1987.

Rouse, J.V. and R.Z. Pyrih. *Natural Geochemical Attenuation of Contaminants Contained in Acidic Seepage.* International Conference on New Frontiers for Hazardous Waste Management. September 1985.

Rouse, J.V. *Nitrate Sources, Mobility and Transformations in the Ground Water Environment.* Geochem, Inc. May 1988.

Smith, Adrian. Personal Communication. 1989.

Thorsen, J.W., and D. G. Stensby. *Impact of Chromium Waste Spill in Glacial Till Soils.* 1982 Hazardous Material Spills Conference, April 19-22, 1982, Milwaukee, WI, Government Institutes, Inc. 1982. 305-310.

Tuolomne County (CA) Health Department. *Ash Disposal - A Local Success Story.* November 1988.

United States Environmental Protection Agency. Office of Solid Waste. *Background Document; Resource Conservation and Recovery Act, Subtitle C - Hazardous Waste Management, Section 3001 - Identification and Listing of Hazardous Waste, Section 261.24 - EP Toxicity Characteristic.* May 2, 1980.

United States Environmental Protection Agency. *The Superfund Innovative Technology Evaluation Program: Technology Profiles.* EPA 540/5-88/003. November 1988.

United States Environmental Protection Agency. *Water-Related Environmental Fate of 129 Priority Pollutants.* Volume I EPA-440/4-79-029a. December 1979.

United States Environmental Protection Agency. *Wetlands for Attenuation.* 1988.

University of California at Berkeley. Mining Waste Study Team. *Mining Waste Study, Final Report.* California State Water Resources Control Board, 1988.

Wilson, L.C. and J.V. Rouse. *Geohydrological and Geochemical Evaluation of Existing and Potential Contaminant Transport From Dawn Mining Co. Tailings Pond.* Envirologic Systems, Inc.: Ford, Washington. 1980.

CONTENTS

CHAPTER 6
CLIMATIC CONSIDERATIONS

	PAGE NO.
LIST OF TABLES/LIST OF FIGURES	275
6.1 Introduction	277
6.1.1 Purpose	277
6.1.2 Integration With Other Chapters	277
6.1.3 Study Participants	277
6.1.4 Sources of Information	278
6.1.5 Definitions	278
6.2 Mine Waste Pile Water Balance	279
6.2.1 Introduction	279
6.2.2 Operational Water Cycle	281
6.2.3 Natural Hydrologic Cycle	281
6.2.4 Natural Water Balances	281
6.2.4.1 Waste Moisture Storage	283
6.2.4.2 Surface Runoff	286
6.2.4.3 Infiltration	286
6.2.4.4 Evaporation	291
6.2.4.5 Evapotranspiration	293
6.2.4.6 Percolation	293
6.2.5 Leachate	294
6.2.6 Waste Pile Effects On Percolation	294
6.2.6.1 Surface Slope and Grading	294
6.2.6.2 Surface Texture	296
6.2.6.3 Vegetation	296
6.2.6.4 Layering	296
6.2.6.5 Grain Size Effects	298
6.2.6.6 Moisture Content	298

CONTENTS

CHAPTER 6
(Continued)

PAGE NO.

6.3 Infiltration and Recharge Case Histories 298
6.3.1 Recharge Case Studies 298
6.3.1.1 Foster, 1987 298
6.3.1.2 Houston, 1987 300
6.3.1.3 Bredenkamp, 1988 300
6.3.1.4 Sniff, Davis, and McWhorter, 1988 300
6.3.1.5 Other References 302
6.3.2 Waste Pile Water Balance Assessments 302
6.3.2.1 ASARCO Waterloo Project 302
6.3.2.2 UMTRA Program 305
6.3.2.3 Imperial County, California 307
6.4 Analytical Approach to Percolation Evaluation 308
6.4.1 Introduction 308
6.4.2 Hydrologic Evaluation of Landfill Performance (HELP) Model 309
6.4.2.1 General Description 309
6.4.2.2 Method of Solution 309
6.4.3 Application of the HELP Model 310
6.4.3.1 Waste Pile Soil Profile 310
6.4.3.2 Input Data 310
6.4.3.3 Summary of HELP Results 313
6.4.3.4 Conclusions on Reliability 313
6.4.4 Application of the Thornthwaite Method 317
6.5 Prediction of Infiltration 317
6.5.1 Percolation in Perspective 317
6.5.2 Guidelines 318
6.5.3 Evaluation of Percolation 321
6.5.3.1 Initial Evaluation 321
6.5.3.2 Detailed Evaluations 321
6.6 Regulatory Implications 322
6.6.1 Recommended Regulatory Approach 323
6.6.1.1 Background 323
6.6.1.2 Regulatory Approach 323
6.7 Bibliography 326

CONTENTS

CHAPTER 6
(Continued)

LIST OF TABLES

TABLE NO.	TITLE	PAGE NO.
6.1	Typical "Dry" Mine Wastes	280
6.2	Typical Soil Moisture Conditions	285
6.3	Runoff Curve Numbers for Hydrologic Soil-Cover Complexes	288
6.4	Hydrologic Soil Group Classification	289
6.5	Depth of Soil in Inches Required to Accommodate Infiltration for a Specified Change in Moisture	290
6.6	Extreme Precipitation and Evaporation Values	295
6.7	Results of Runoff/Recharge Simulation Model for an Idealized Basement Aquifer in Victoria Province, Zimbabwe	301
6.8	Results from Other References	304
6.9	Department of Energy UMTRA Study Results	306
6.10	HELP Model Data Input Sources	314
6.11	Results of HELP Model Runs	315
6.12	Water Balance Data for Los Angeles, California	316

LIST OF FIGURES

FIGURE NO.	TITLE	PAGE NO.
6.1	Hydrologic Cycle For a Typical Waste Pile	282
6.2	Moisture Storage Conditions in a Granular Waste	284
6.3	Rainfall - Runoff Relationship	287
6.4	Typical Moisture Suction Versus Unsaturated Hydraulic Conductivity Curves for Granular Material	292
6.5	Layered Waste System	297
6.6	Typical Ranges of Saturated Hydraulic Conductivities	299
6.7	Distribution of Recharge Fractions	303
6.8	Soil Profile A: Nonvegetated	311
6.9	Profile B Reclamation: Vegetated Topsoil Cover	312
6.10	Time of First Appearance of Leachate (Water Balance Method)	319
6.11	Areas of Percolation (Tentative)	320

CHAPTER 6

CLIMATIC CONSIDERATIONS

6.1 INTRODUCTION

6.1.1 PURPOSE

The purpose of this chapter is to investigate the conditions under which rainfall and snowmelt result in infiltration, percolation, and leachate formation, and to develop guidelines for incorporating these processes into the mine waste disposal regulations. This is important because in mine waste, and under certain circumstances, these processes can result in conditions which pose a threat to surface and ground water quality.

This chapter provides a general overview of infiltration, percolation, and leachate formation. It incorporates a discussion of the methods that can be used to quantify infiltration and the climatic and physical site and waste conditions under which percolation and leachate formation occur. Reference is made to case histories on infiltration, ground water recharge (a natural infiltration phenomenon), and analytical procedures for calculating infiltration. An approach to infiltration prediction is outlined, and the chapter concludes with a discussion on how climatic factors and prediction of infiltration could be incorporated into the regulations.

6.1.2 INTEGRATION WITH OTHER CHAPTERS

The chemical properties of the solid waste material and the associated interstitial water determine the quality of the leachate that emerges from a waste pile. The quantity and quality of this leachate will determine the risk that the waste poses to water quality. The risk aspect is discussed in Chapter 4, Prediction of Acid Generation Potential. Chapter 5, Attenuation of Chemical Constituents, addresses the degree to which the dissolved constituents in the leachate can be contained by the natural geologic formation located below the waste disposal facility. Finally, Chapter 7, Liner System Design, discusses engineering measures that can be applied to provide sufficient containment of leachate.

6.1.3 STUDY PARTICIPANTS

This chapter was prepared by:

- Ian P. G. Hutchison, Ph.D., P.E., Environmental Solutions, Inc.
- Eric Dillé, Steffen Robertson and Kirsten (U.S.), Inc.

6.1.4 SOURCES OF INFORMATION

The general sources of information used in compiling this chapter include:

- California Department of Water Resources (CDWR) publications.
- Climatic data from the National Oceanographic and Atmospheric Agency (NOAA).
- Technical papers and climatic data from the United States Geological Survey (USGS).
- Climatic data from the United States Department of Agriculture (USDA).
- U.S. Environmental Protection Agency (EPA) documents on methods of calculating infiltration.
- Technical documents from the Department of Energy (DOE) Uranium Mill Tailings Remedial Action (UMTRA) Program.
- General literature on ground water recharge and near surface water balance.
- The "ASARCO" design documents for a dry tailings disposal facility in the Lahontan Water Quality Control Board Region, California.

Most of the data sources reviewed are listed in Section 6.7, Bibliography.

6.1.5 DEFINITIONS

For purposes of this chapter, the following definitions have been adopted:

- Infiltration - The movement of water below the surface of the waste material and into the "active layer."
- Active Layer - The near surface material from which evaporative losses of interstitial water can still occur to a measurable degree, over a reasonable period of time.
- Percolation - The vertical downward movement of water through the bottom of the active layer. By definition, this is too far below the surface to be significantly affected by evaporation.
- Leachate - The liquid, derived from percolation, that exits at the bottom of the waste and that may contain dissolved constituents derived from the waste.

Infiltration is equal to precipitation minus the sum of surface evaporation and runoff. Percolation is equal to the infiltration rate minus the surface evaporation loss rate. Under steady state conditions, the rates of percolation and leachate flow are equal. However, there are processes that can occur which can result in differences between these two rates. An example of this is percolation in a very dry waste. Initially, and possibly even for several months or years, the percolation will go toward raising the moisture content of the waste to levels at which leachate flow can ultimately occur.

Generally in arid and semi-arid environments, infiltration and percolation rates are very low, and the water moves through an unsaturated waste (i.e., the voids between solid waste particles are not completely filled with water). Under these conditions, no measurable quantities of leachate are generated. It is only in wetter climatic conditions that saturated flow occurs in a waste, developing when the infiltration rate equals or exceeds the saturated hydraulic conductivity of the material. Saturated water migration rates are usually several orders of magnitude higher than unsaturated rates.

6.2 MINE WASTE PILE WATER BALANCE

6.2.1 INTRODUCTION

A mine waste or tailings pile is generally subjected to two different types of water balance cycles. The first is the artificial or mining related cycle, which it includes water addition or subtraction as part of the mining or metallurgical extractive process. The second is the natural cycle and includes rainfall, runoff, and infiltration.

Consideration of climatic factors in designing a waste disposal facility becomes important when the waste or disposed material is at a moisture content which is low enough that continued drain-down of interstitial water does not occur and the climate is sufficiently arid so that significant infiltration does not occur. In these cases, which generally involve overburden waste rock, heap leach residue, and dried or subareally deposited tailings, the waste does not generate leachates, and the arid climatic conditions may provide adequate containment. These conditions would not apply to wet tailings slurry, usually discharged to the waste disposal facility at densities of 25 to 50 percent solids irrespective of the climatic conditions, or to the above listed wastes in a humid climate. These wet tailings are likely to have a liquid component which must be adequately contained by the site geologic conditions or by engineered systems such as liners.

Table 6.1 summarizes the waste types for which climatic considerations are an important factor and discusses the artificial and natural water balance cycle components. It also lists the typical saturated flow hydraulic conductivities.

TABLE 6.1

TYPICAL "DRY" MINE WASTES

WASTE TYPE	DESCRIPTION	MOISTURE CONDITION	OPERATIONAL WATER CYCLE COMPONENTS	NATURAL WATER CYCLE COMPONENTS
Overburden	Topsoils, clays, sands, and gravels.	Usually relatively dry. At or below specific retention.	Air drying during mining and transportation; some wetting if watered for dust control or irrigated for vegetative growth.	Precipitation, runoff, and infiltration.
Waste Rock	Blasted or otherwise excavated bedrock. Size ranges from fine-grained clay size particles to boulders several feet in diameter.	Usually relatively dry. At or below specific retention.	Air drying during mining and transportation, some wetting if watered for dust control or irrigated for vegetative growth.	
Heap Leach Residue	Initially a process material. Run-of-mine, crushed, or crushed and agglomerated ore that has been piled up and leached with a lixiviant to extract minerals. Generally particle sizes vary from a few inches down to silt size. May have cement or lime agglomerated particles and contain residue of process solutions. Frequently rinsed with water or chemical reagent to remove process solutions prior to it becoming a waste.	Usually at specific retention. Near surface material can be drier than specific retention due to evaporation.	Leaching, rinsing, and draindown cycles.	
Dry Tailings	Finely ground mined rock from which the economic minerals have been extracted. Frequently in the clay, silt to fine sand size range. Contains residuals of the process solutions.	Clay and silt sized materials usually are saturated. Sand sized and coarser material may be at a lower moisture content as it can drain rapidly.	Some air drying may occur in transporting the dried tailings to the pile. Also, if deposited in layers, the tailings can dry out naturally.	

6.2.2 OPERATIONAL WATER CYCLE

The operational water cycles are summarized in Table 6.1. These are under the control of the mine operator and are predictable. If required, a water cycle can be managed to achieve a dry waste on disposal. A dry waste has a moisture content sufficiently low that infiltration and percolation will not occur alone. Therefore, the potential for leachate formation becomes a function of the natural hydrologic cycle and the physical properties of the waste pile. On the other hand, a wet waste (where moisture content exceeds specific retention) will act as a limited source of water for infiltration and percolation, independent of climatic factors, thereby enhancing the potential for leachate formation. These are discussed in the following sections.

6.2.3 NATURAL HYDROLOGIC CYCLE

Figure 6.1 shows the hydrologic cycle associated with a typical waste pile. The primary components of this cycle are precipitation, vegetation, and climate. Precipitation occurs as either rain, sleet, or snow, and is dispersed in several ways. The runoff component runs off the surface and discharges to the surrounding environment. The remainder either evaporates from the surface of the waste or infiltrates. All, or a portion, of the infiltrate is temporarily retained in the waste near the surface and is ultimately returned to the atmosphere through evaporation and/or transpiration. The amount of infiltrate retained and returned to the atmosphere through evaporation directly from the surface of the waste or transpiration through surface vegetation, is determined by climate, characteristics of the waste, and amount and characteristics of vegetation on the waste pile. Infiltrate that is transpired through vegetation is subsequently evaporated in the atmosphere (evapotranspiration). Any portion of the infiltrate that does not reach the atmosphere through the surface percolates downward.

6.2.4 NATURAL WATER BALANCES

Thornthwaite and Mather (1957) developed a water balance method to evaluate the amount of infiltration and percolation that occur in natural soil systems. Because mine waste piles are very similar to natural soil systems, this method is applied here to describe the water balance components in a waste pile. Three critical factors that must be considered in the water balance method are soil or waste moisture storage, evapotranspiration, and surface runoff. These are discussed below.

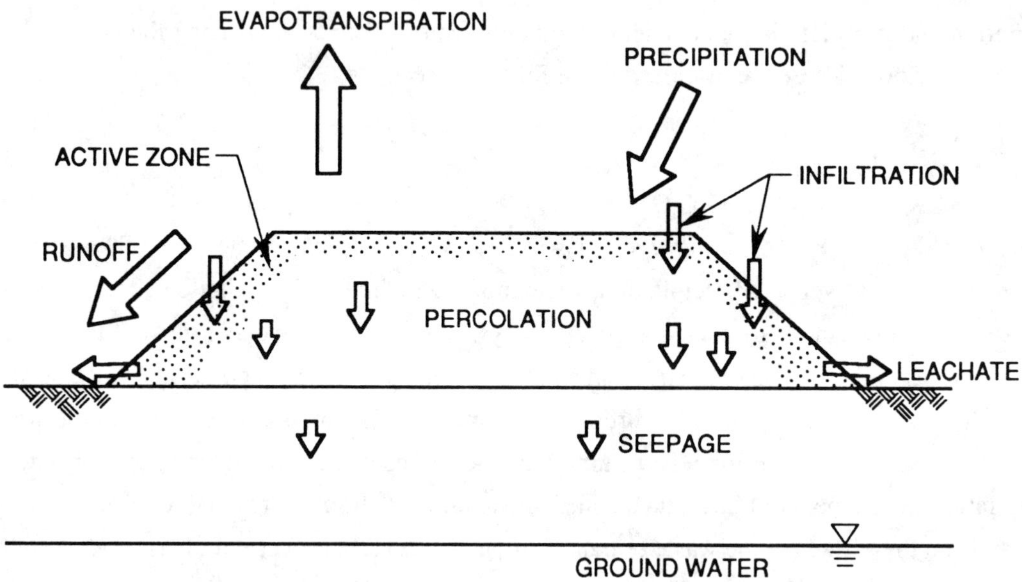

FIGURE 6.1

HYDROLOGIC CYCLE FOR A TYPICAL WASTE PILE

6.2.4.1 Waste Moisture Storage

The terminology that describes the different moisture conditions in a granular waste material are shown in Figure 6.2. Three important moisture conditions are defined as follows:

- Saturation - The moisture condition at which all pore spaces are completely filled with liquid. At this moisture content, as well as lower moisture contents down to the specific retention, downward percolation of liquids in the pores occurs predominantly due to gravitational forces.
- Specific Retention - This is a lower moisture content than saturation and corresponds to a condition at which downward migration of fluids by gravitational forces cease. The fluid is held in place by capillary forces.
- Wilting Point - This corresponds to yet a lower moisture content at which plant roots can no longer extract water from the waste material.

Some residual migration of moisture does occur at moisture contents lower than specific retention, due to differences that may occur in the soil moisture suction at various locations in the material. Moisture will generally migrate from areas of lower suction to those of higher suction. These migration rates are also controlled by the unsaturated hydraulic conductivity of the material. This type of moisture migration is generally extremely low and not of interest in the design of conventional surface waste management units such as those discussed in this book.

One way in which the near surface waste material influences the amount of percolation is through its capacity to store water in the active layer. The amount of storage mainly depends on the waste grain size and structure and the presence and type of vegetation. These, in turn, determine the characteristic specific retention, as well as depth, of the active layer.

As shown in Table 6.1, the amount of water present in the waste can vary from virtually zero to the saturated moisture content. Moisture conditions below specific retention are frequently found in dry waste rock, tailings deposited by layered deposition, the upper layers of drained heap leach piles, mechanically dried tailings, or tailings deposited in thin layers in an arid climate.

Typical moisture conditions for a wide range of soil types are described in Brady (1974) and Hanks and Ashcroft (1980), and are summarized in Table 6.2, where they are expressed in percent by weight and can be used as a first approximation to characterize a granular mine waste. They indicate, for example, that a fine-grained tailings pile (silt, clay, loam) may have a specific retention of as much as 30 percent. For waste rock, the value may be 10 percent or lower (see value for sand in Table 6.2).

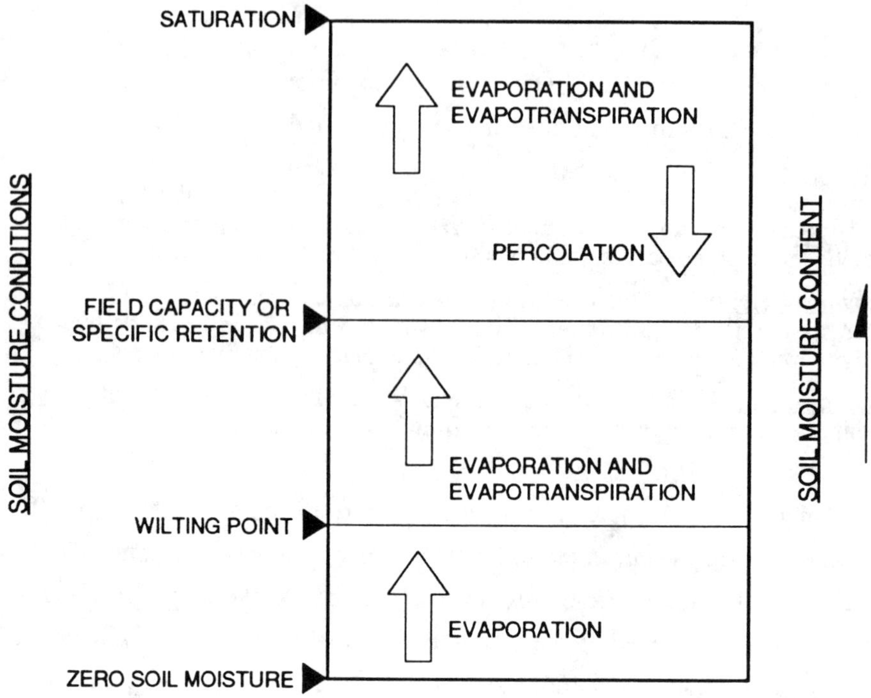

A. MOISTURE CONDITIONS AND THE ASSOCIATED SIGNIFICANT PROCESSES CAUSING MOISTURE CHANGES

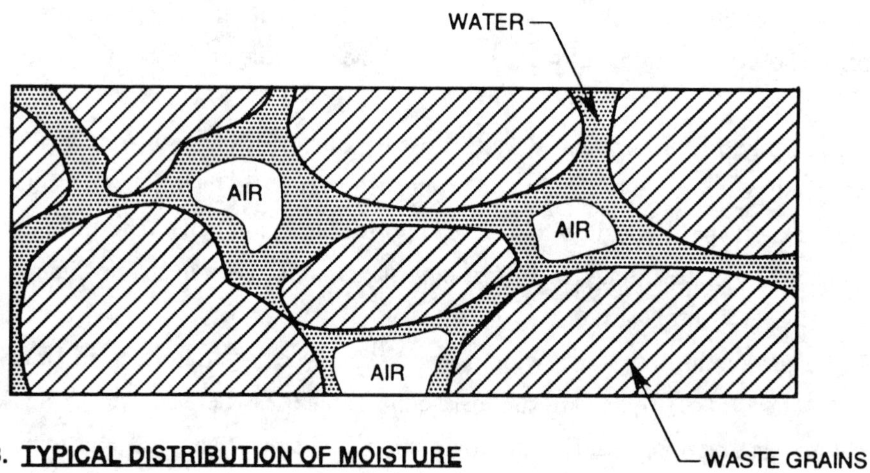

B. TYPICAL DISTRIBUTION OF MOISTURE

FIGURE 6.2

MOISTURE STORAGE CONDITIONS IN A GRANULAR WASTE

TABLE 6.2

TYPICAL SOIL MOISTURE CONDITIONS

SOIL TYPE	MOISTURE CONTENT (% BY WEIGHT)	
	SPECIFIC RETENTION	OTHER
Brady (1974)		HYGROSCOPIC [(1)]
Sandy Soil	7.9	3.3
Red Loam	19.2	10.0
Silt Loam	27.8	10.2
Blank Adobe	25.8	12.9
Hanks and Ashcroft (1980)		AIR DRIED
Sand	10.0	.5
Loam	20.0	1.0
Silty Clay Loam	30.0	3.0

(1) Hygroscopic moisture content represents very low moisture conditions established under hot, arid conditions in which a water-vapor equilibrium tends to become established between the ambient air and the surfaces of the soil particles. As a result only thin films of moisture remain adsorbed on the surfaces of the soil grains.

6.2.4.2 Surface Runoff

A fraction of the incident precipitation will run off the waste pile and be lost to overland flow before it has a chance to infiltrate. The amount of surface runoff depends upon several factors, including the intensity and duration of the storm, the antecedent near-surface waste moisture condition, the permeability and infiltration capacity of the near-surface waste material, the surface slopes, and the amount and type of vegetative cover.

Surface runoff is usually expressed as a fraction of the amount of precipitation falling on the waste. Without actual laboratory or field measurements, it is usually difficult to estimate this fraction accurately. However, approximate values can be obtained using the Soil Conservation Service (SCS) methodology (USDA, 1972) which is summarized in Figure 6.3, and Tables 6.3 and 6.4.

6.2.4.3 Infiltration

The rate of infiltration depends on grain size distribution, the surface texture and geometry of the waste pile, the moisture content of the waste material, and the rate at which rainfall occurs. Coarser wastes, with rough textures and flat slopes, tend to increase infiltration rates. Fine-grained wastes, with a smooth surface texture and well-drained slopes, tend to reduce infiltration. During intense storm periods, the rainfall rate may exceed the infiltration capacity of the waste; rain water that does not infiltrate runs off. During periods in which the rainfall rate is low, the infiltration capacity of the waste dominates. The infiltration capacity of the waste will vary over time, according to precipitation, climate, and surface vegetation. In general, the drier the waste, the greater its infiltration capacity.

Table 6.5 shows the relationship of total infiltration to waste moisture content over a period of time. It shows the soil or granular waste depth required to store a given amount of infiltration for given water moisture content changes. For example, it illustrates that 1 inch of infiltration will raise the moisture content by 10 percent over a 6.3-inch depth of waste material. This means that, for a waste material which has a specific retention which exceeds the actual in situ moisture content by more than 10 percent, the depth to which the infiltration will initially penetrate is 6.3 inches. Provided the depth of the active layer of the waste is more than 6.3 inches, the infiltrated water can be subsequently removed by evaporation or evapotranspiration.

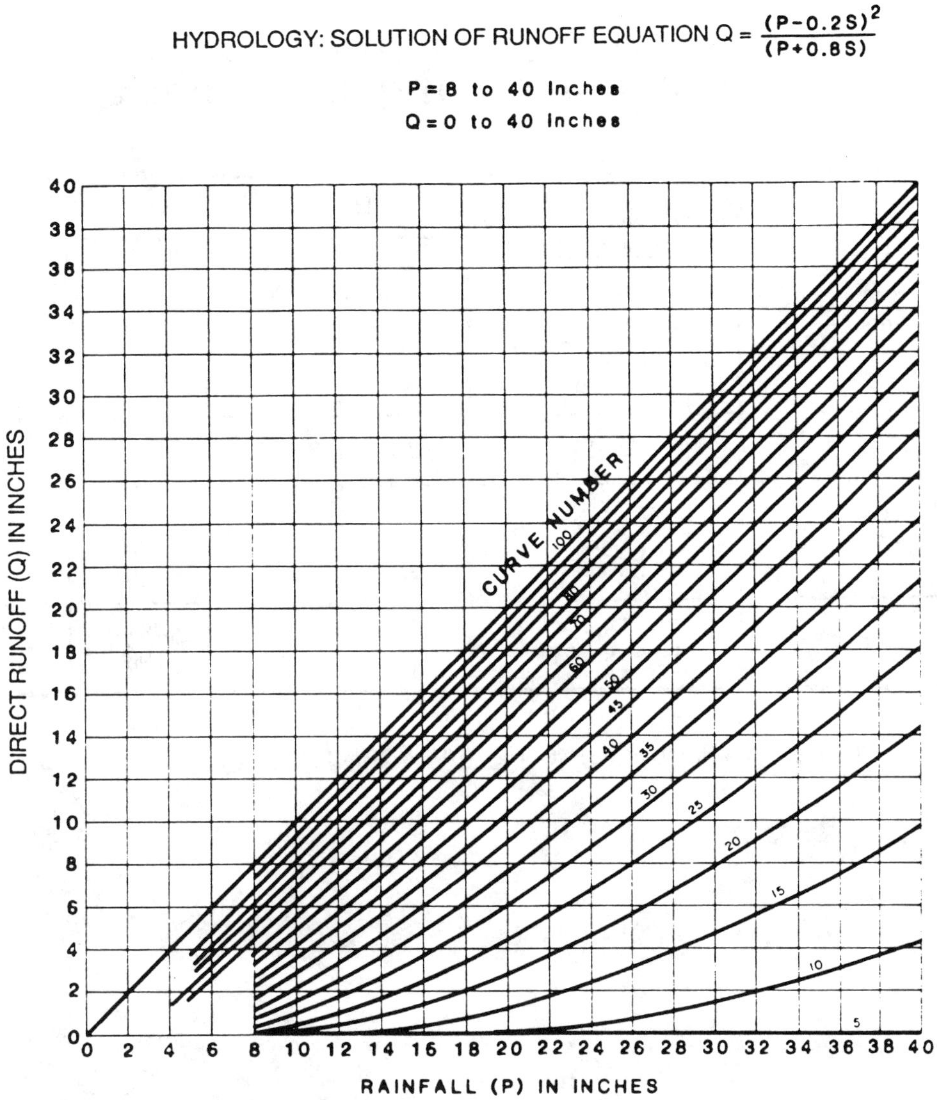

FIGURE 6.3

RAINFALL - RUNOFF RELATIONSHIP

REFERENCE: U.S. DEPARTMENT OF AGRICULTURE,(1972).

TABLE 6.3

RUNOFF CURVE NUMBERS FOR HYDROLOGIC SOIL-COVER COMPLEXES[(1)]

LAND USE	HYDROLOGIC CONDITION	HYDROLOGIC SOIL GROUP			
		A	B	C	D
Pasture/Range	Poor	68	79	86	89
	Fair	49	69	79	84
	Good	39	61	74	80
Meadow	Good	30	58	71	78
Woods	Poor	45	66	77	83
	Fair	36	60	73	79
	Good	25	55	70	77
Roads	Dirt	72	82	87	89
	Hard-Surface	74	84	90	92

(1) Reference: Table 9.1, SCS Handbook on Hydrology, (USDA, 1972).

TABLE 6.4

HYDROLOGIC SOIL GROUP CLASSIFICATION[1]

HYDROLOGY SOIL GROUP	DEFINITION
A	Low runoff potential. Soils having high infiltration rates even when thoroughly wetted. Consisting chiefly of deep, well- to excessively-drained sands or gravels.
B	Soils having slow infiltration rates when thoroughly wetted. Consisting chiefly of soils with a layer that impedes downward movement of water, or soils with fine to moderately fine to fine texture.
C	High runoff potential. Soils having very slow infiltration rates when thoroughly wetted. Consisting chiefly of clay soils with a high swelling potential, soils with a permanent high water table, soils with a claypan or clay layer at or near the surface, and shallow soils over nearly impervious material.

(1) Reference: Table 7.0, SCS Handbook on Hydrology, (USDA, 1972).

TABLE 6.5

DEPTH OF SOIL IN INCHES REQUIRED TO ACCOMMODATE INFILTRATION FOR A SPECIFIED CHANGE IN MOISTURE

SOIL MOISTURE STORAGE CHANGE (% BY WEIGHT)	INFILTRATION DEPTH (INCHES)					
	0.2	0.5	0.75	1.0	1.5	2.0
2	6.3	15.6	23.4	31.3	46.9	62.5
5	2.5	6.3	9.4	12.5	18.8	25.0
10	1.3	3.1	4.7	6.3	9.4	12.5
15	0.8	2.1	3.1	4.2	6.3	8.3
20	0.6	1.6	2.3	3.1	4.7	6.3
25	0.5	1.3	1.9	2.5	3.8	5.0
30	0.4	1.0	1.6	2.1	3.1	4.2

6.2.4.4 Evaporation

Evaporation occurs directly and entirely from the waste, or soil material, and can result in considerable losses of water (Hillel, 1982). Three conditions are necessary for the evaporation process from a given body to persist. These include:

- Continued supply of heat.
- The vapor pressure in the atmosphere over the evaporating waste material must remain lower than the vapor pressure at the surface of that body, and the vapor must be transported away.
- There must be a continual supply of water from or through the interior of the body of waste. This depends on the moisture content of the waste as well as its water conducting properties, which together determine the maximal rate at which the material can transmit water to the evaporation site.

The actual evaporation rate is determined either by external evaporation forces or by the soil's own ability to deliver water, whichever is the lesser and, hence, the limiting factor. In arid climates, the external evaporative forces (i.e., heat supply from the sun and moisture removal through wind action) generally far exceed the ability of the waste to deliver moisture. As the top layers of waste dry through evaporation, the soil moisture suction in these layers increases. This, in turn, causes soil water to be drawn upward from the layers below. Fine-grained wastes which have larger hydraulic conductivities at higher moisture suction pressures than coarser wastes will, therefore, tend to evaporate water at greater rates (see Figure 6.4).

Very little data appear to be available on the active depth from which evaporation has been observed in various soil types. Hillel (1982) indicates that evaporation from a sandy loam could occur down to a depth of 6 feet. Hillel further defines three evaporation stages:

- Initial, Constant-Rate Stage: This occurs initially in wet soil (or waste) when the material is wet enough to supply water to the site of evaporation commensurate with the evaporative demand.
- Intermediate, Falling-Rate Stage: The evaporation rate falls progressively below the potential rate as it is limited by the rate at which the gradually drying soil profile can deliver moisture to the evaporation zone.
- Residual, Slow-Rate Stage: This stage apparently comes about after the surface-zone has become so desiccated that further liquid-water conduction through it effectively ceases. Water transmission through the desiccated layer, therefore, occurs primarily by the slow process of vapor diffusivity of the dried surface zone and by adsorptive forces acting over molecular distances at the particle surfaces. This stage may persist for many days, weeks, or even months.

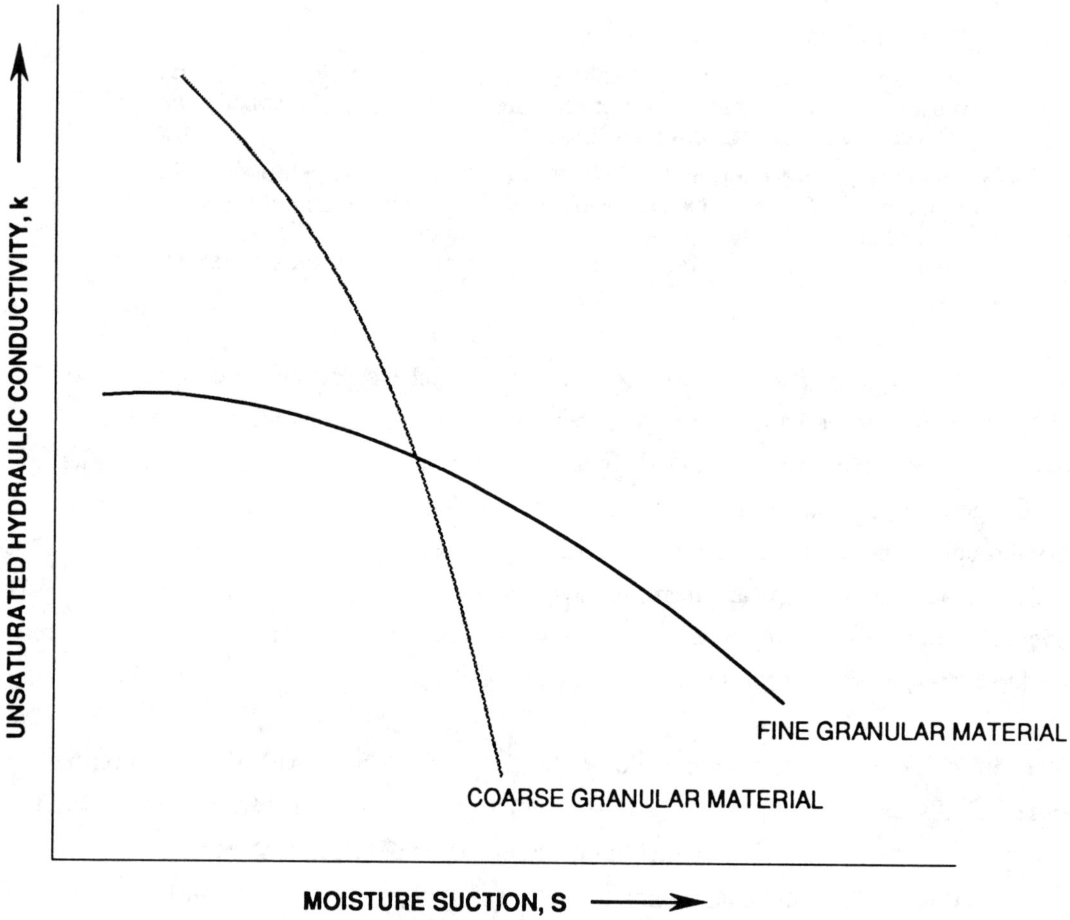

FIGURE 6.4

TYPICAL MOISTURE SUCTION VERSUS UNSATURATED HYDRAULIC CONDUCTIVITY CURVES FOR GRANULAR MATERIAL

6.2.4.5 Evapotranspiration

Evapotranspiration occurs as the result of evaporation from the soil and transpiration by the vegetative cover. Actual measurements using soil lysimeters have shown that the rate of evapotranspiration drops as soil moisture is depleted (Mustonen and McGuinness, 1968). Under fully revegetated conditions, when the soil moisture is at or near specific retention, evapotranspiration occurs at its maximum potential rate. However, as the soil moisture content approaches the wilting point, the amount of available water begins to restrict the rate of evapotranspiration, resulting in reduced actual water losses. Also, sparse or dormant vegetative covers, during winter months, for example, also result in evapotranspiration rates that are lower than the maximum.

Generally, field data show that, for humid areas, there is little difference between the potential and actual evapotranspiration rates during the wet season when sufficient water is available in the soil and there is an adequate vegetative cover. However, the actual evapotranspiration rate will decrease if the soil moisture becomes depleted.

6.2.4.6 Percolation

Percolation usually occurs when the moisture content at the bottom of the active layer exceeds the specific retention. This condition results when the precipitation occurs for a sufficiently long duration that an adequate amount of water can infiltrate, yet the time following cessation of precipitation is not sufficient for evaporation losses to prevent the moisture buildup in the active layer.

Some continued downward movement will occur so long as the moisture content and, therefore, water pressure in the active zone is greater than in the underlying material, thereby creating a downward gradient. In unsaturated conditions, however, these rates would be extremely low, as the unsaturated hydraulic conductivities at this moisture content are typically orders of magnitude lower than the saturated hydraulic conductivity values.

In wet climates, percolation conditions typically occur during rainy months. In drier areas, percolation may only occur during certain months or only during unusually wet periods. In arid regions, it may not occur at all.

Given the above considerations, analyses of water balances should include not only average climatic conditions, but extreme precipitation events, as well. Furthermore, as percolation conditions can develop over a few days, weeks, or months, precipitation events over these same periods of time should be used.

Table 6.6 compares some of the extreme wet precipitation events for various regions in California with the potential evaporation values for the same periods. As shown in Table 6.6, extreme precipitation and evaporation values vary widely, with 60-day precipitation extremes ranging from 33.5 inches for Guadalupe Reservoir to 2.65 inches for Death Valley, and 60-day evaporation rates ranging from 1.97 inches for Guadalupe Reservoir to 8.82 inches for Death Valley. In the more arid parts of the State, extreme precipitation values are relatively low compared to the potential evaporation for the same periods. The amount of moisture that can be stored in the active layers of the waste pile is greater in the arid areas than in those where precipitation is high compared to evaporation.

6.2.5 LEACHATE

The term "leachate" applies to liquids that have passed through a waste pile and exit along the sides or at the bottom of the pile. As with percolation, leachate is produced frequently or continuously during rainfall periods in wetter climates, infrequently for short durations in drier climates, and not at all in very dry climates.

The quality of the leachate is dependent on the chemical properties of the waste material and the interstitial water chemistry. The leachate will contain dissolved levels of any constituents that are solubilized, as well as any remnant soluble reagents or their reaction products that may be present.

6.2.6 WASTE PILE EFFECTS ON PERCOLATION

Climate is the primary driving force behind the infiltration, percolation, leachate formation process. The physical properties of the waste, however, exert an important secondary impact, as discussed below.

6.2.6.1 Surface Slope and Grading

The smoother the waste surface is and the steeper the surface slope, the greater the runoff and, hence, the lower the rate of infiltration. Sloped piles without any flat areas will tend to minimize the potential for infiltration and leachate formation, all other aspects being equal. Poorly drained surfaces, particularly if hollows are present, will promote infiltration.

TABLE 6.6

EXTREME PRECIPITATION AND EVAPORATION VALUES

PRECIPITATION STATION NAME (Low Evaporation Months)	60-DAY PRECIPITATION EXTREME WITH A RETURN PERIOD OF 50 YEARS (inches)	LOW 60-DAY LAKE EVAPORATION (1) (inches)	ANNUAL MEAN PRECIPITATION (inches)	ANNUAL MEAN PAN EVAPORATION (1) (inches)
Guadalupe Reservoir (Dec., Jan.)(2)	33.50	1.97	27.31	45.0
Napa State Hospital (Dec., Jan.) (2)	25.36	2.99	23.65	64.0
Riverside (Dec., Jan.)	13.85	5.51	10.36	69.0
Palm Springs (Jan., Feb.) (2)	9.84	7.64	5.81	73.0
La Quinta SDF (Dec., Jan.) (2)	4.53	4.80	2.55	90.0
Indio (Dec., Jan.)	4.29	5.24	2.37	84.0
Death Valley (Dec., Jan.)	2.65	8.82	1.8	160.0

(1) These values represent the average 60-day values for the two lowest evaporation months of the year.
(2) Where evaporation data were not available at a specific site, data from the closest measuring station were used.

6.2.6.2 Surface Texture

Waste rock from mining operations will vary in texture from finely ground tailings material which is typically 80 percent finer than the No. 200 mesh sieve (0.0075 mm) to roughly boulder size rocks (average diameter more than 12 inches). Surface porosity in the larger waste rock is greater, thereby decreasing the runoff component of precipitation and capturing a larger percentage of water, making it easier for infiltration to occur. Coarser material also has an effect on the rate of evaporation from the waste material. Pea gravel layers 6 to 12 inches thick are known to virtually prevent evaporation and, at low moisture contents, have a very low permeability that inhibits the upward migration of water (McWhorter, 1990) (see Figure 6.4). Rocks of larger size may not have the same effect. If they are large enough to provide air circulation, evaporation may still continue from underlying finer grained material.

6.2.6.3 Vegetation

A uniform layer of vegetation over the waste or soil will, in many cases, increase the rate of evaporation. In coarser waste it may extend the depth of the active layer, thus increasing the capacity for temporary storage of infiltration. Therefore, the overall effect of vegetation is to reduce the potential for percolation.

6.2.6.4 Layering

Alternate layering of low and high permeability material near the surface of the waste pile has the impact of reducing the amount of percolation that occurs. In humid climates, where the materials may be close to saturation, the low permeability layers reduce infiltration, while the higher permeability layers can be used to carry away any resulting infiltration before it has a chance to percolate into the lower lying wastes. Figure 6.5 illustrates this concept.

In arid and semi-arid regions, where the surface materials are unsaturated, layering may have a different effect. The unsaturated permeability of the fine material may be orders of magnitude greater than that for the coarser drainage material (see Figure 6.4). Water permeation with this layer is greatly retarded, and water tends to accumulate in the finer-grained layer and move downslope, as if the drainage layer were impermeable. This process has the advantage of holding more water in the fine layer, making it more available for evapotranspiration.

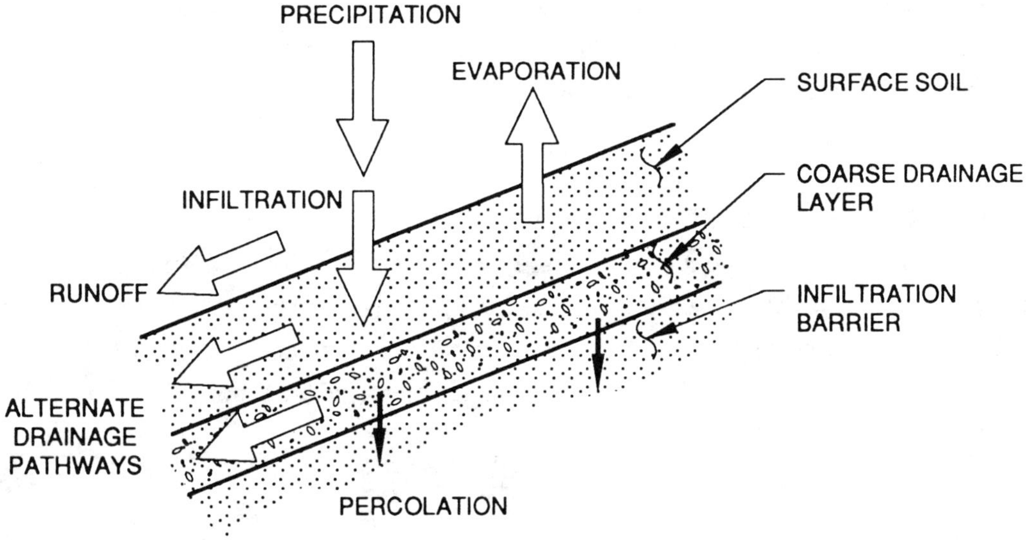

FIGURE 6.5

LAYERED WASTE SYSTEM

6.2.6.5 Grain Size Effects

Fine-grained material has a larger particle surface area than does the coarser material. The fine-grained material, therefore, has a greater capacity to store water (i.e., has a higher soil moisture retention capacity). As discussed earlier, it also has a deeper active layer. Finally, finer-grained material has a lower hydraulic conductivity than coarse-grained material. Figure 6.6 illustrates some typical examples.

6.2.6.6 Moisture Content

As discussed above and as illustrated in Figure 6.4, the unsaturated percolation rate is very strongly dependent on moisture conditions. The fine-grained materials may have a higher percolation rate than coarse materials at very low moisture contents.

6.3 INFILTRATION AND RECHARGE CASE HISTORIES

This section outlines the results of infiltration in recharge evaluations from the professional literature. Some of the data pertain to waste disposal units, while other data deal with rainfall recharge to ground water systems. Recharge is that portion of the rainfall that ultimately reaches the ground water table by percolation through the natural geologic formations above the ground water table.

6.3.1 RECHARGE CASE STUDIES

The following sections discuss case studies at various locations throughout the world at which estimates of ground water recharge were made.

6.3.1.1 Foster, 1987

This reference provides a simple categorization of natural aquifer recharge in desert regions, with a basic subdivision into two types. Diffuse recharge from rainfall is comprised of excess precipitation over soil moisture deficits and short-term vegetation needs, and infiltrates directly. Localized recharge results from infiltration through the strata of perennial or, more typically, ephemeral water courses and other features such as fractures and faults.

Foster states that localized recharge occurs more often than diffuse recharge in the desert environment, indicating that the major component of the average rate is due to local recharge.

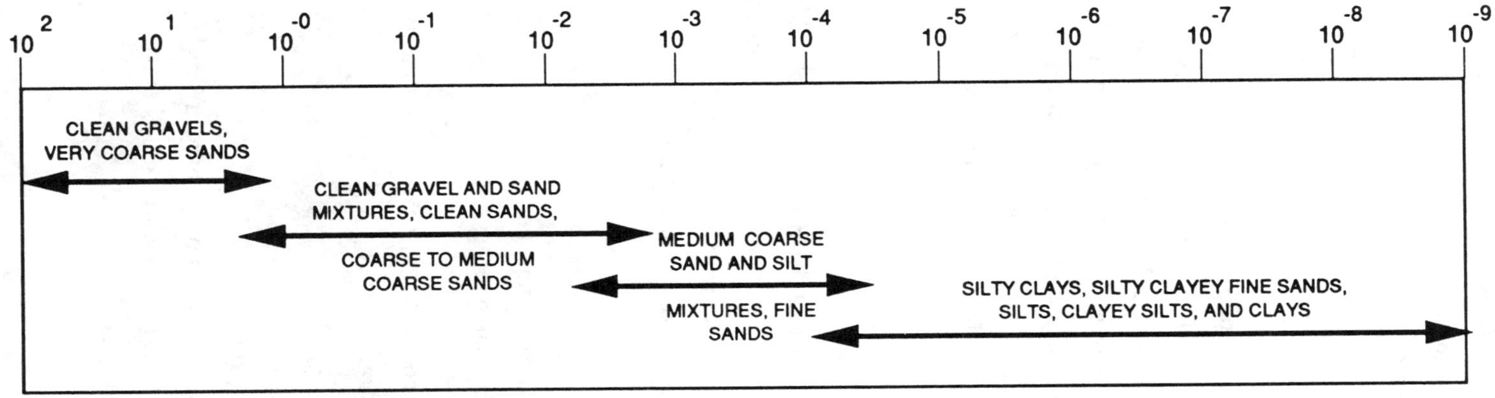

REFERENCE: MODIFIED FROM FREEZE AND CHERRY (1979).

FIGURE 6.6

TYPICAL RANGES OF SATURATED HYDRAULIC CONDUCTIVITIES (IN CM/SEC)

Therefore, extrapolation of these recharge values could lead to overestimating percolation rates for waste piles, which are normally designed and constructed specifically not to incorporate any features which could promote local recharge.

6.3.1.2 Houston, 1987

Houston presents a study in the grasslands of Zimbabwe which suggests that recharge amounts to between 2 and 5 percent of annual rainfall. The area examined is the Victoria Province which has a mean annual precipitation of 19 inches and a recharge of between 0.4 and 1.0 inches (see Table 6.7). Houston concludes that areas receiving less than 15 inches (400 mm) of precipitation annually are unlikely to produce recharge. For areas receiving more than 15 inches, the amount of recharge depends upon soil and vegetation characteristics. Houston's conclusions were based on a runoff-recharge mass balance evaluation for basin catchments larger than 2,500 square miles.

6.3.1.3 Bredenkamp, 1988

Bredenkamp studied the hydrologic balance of the Bo Molopo dolomitic region in the Western Transvaal, South Africa. Recharge values were determined from a sinkhole hydrograph and from water balance calculations. Excellent results were obtained for predicting the average annual recharge, using the average annual rainfall in different dolomitic areas and the following empirical equation:

$$RE = 0.3\ (RF - 12.3) \text{ inches}$$

Where: RE = average annual recharge in inches
RF = average annual rainfall in inches

From the above investigations, it was concluded that there was a threshold precipitation, in this case 12.3 inches, below which no recharge occurred.

6.3.1.4 Sniff, Davis, and McWhorter, 1988

This study involved the review of technical papers on recharge from 19 countries throughout the world. The objective was to show what kinds of correlations exist between precipitation and recharge, particularly in arid and semi-arid regions.

TABLE 6.7

RESULTS OF RUNOFF/RECHARGE SIMULATION MODEL FOR AN IDEALIZED BASEMENT AQUIFER IN VICTORIA PROVINCE, ZIMBABWE[(1)]

YEAR	RAINFALL (inches)	RECHARGE (inches)	RECHARGE AS % OF RAINFALL
1975	60.0	3.0	5.4
1976	28.0	0.4	1.5
1977	30.0	1.5	5.1
1978	36.0	1.0	2.9
1979	21.0	0.3	1.5
1980	25.0	0.3	1.4
1981	35.0	0.5	1.3
1982	17.0	0.3	1.6
1983	8.0	0.1	1.4

(1) Houston, 1987.

A total of 116 precipitation and recharge values were obtained for average annual precipitation conditions varying between 2 and 55 inches. A bar graph depicting the number of different recharge values as a percentage of precipitation for the data base is shown in Figure 6.7.

Some of the conclusions drawn by this study include:

- In low precipitation areas (i.e., areas that receive less than 12 inches annually), recharge rates range from less than 0 to over 50 percent of precipitation.
- A single relationship regarding annual precipitation and recharge could not be developed.

In the above study, it was not possible to separate the effect of surface conditions from soil grain size and texture. It is, therefore, not possible to evaluate why some of the sites had relatively high recharge rates, while others had low to zero recharge rates.

6.3.1.5 Other References

Other references and results that were reviewed as part of this study are listed in Table 6.8. In this evaluation, an attempt was made to describe the physical conditions of the basins for which recharge was evaluated. The results show the reduction in recharge due to vegetation and also increases in recharge caused by localized surface features, such as sinkholes.

As can be seen, for average annual precipitation values ranging from 3 inches to over 30 inches, the recharge ranged from 0 percent to over 30 percent.

6.3.2 WASTE PILE WATER BALANCE ASSESSMENTS

This section presents water balance studies conducted at actual mine waste disposal sites.

6.3.2.1 ASARCO Waterloo Project

The ASARCO Waterloo project is a proposed precious metals project near Calico in the Lahontan Water Quality Control Board Region of California (ASARCO, 1989). Mean annual precipitation at the site is approximately 4 inches (CDWR, 1980), and the mean annual A pan evaporation is 110 inches (National Weather Service, 1982).

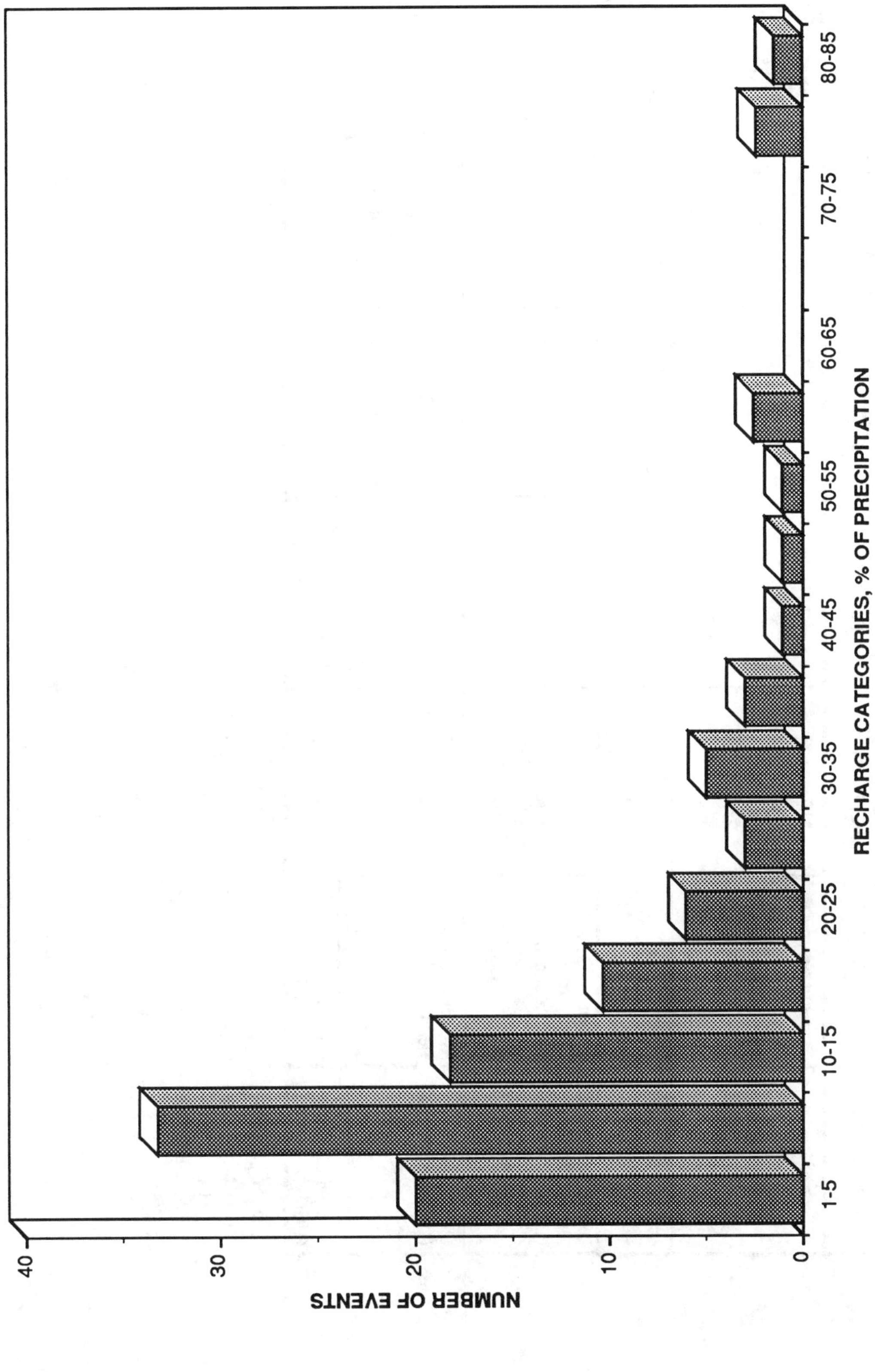

FIGURE 6.7

DISTRIBUTION OF RECHARGE FRACTIONS

REFERENCE: SNIFF, DAVIS, & McWHORTER, 1988.

TABLE 6.8

RESULTS FROM OTHER REFERENCES

REFERENCE	LOCATION	CLIMATE	TERRAIN	ANNUAL INCHES	PERCENT OF MAP	METHOD
Allison, Stone, Hughes, (1985)	Southern Australia	MAP 11.8"	Calcrete Flats/ Sinkholes	3.9	33	Geochemical Profiles in Vadose Zone (55 ft. to 95 ft.)
	Murray Basin	MAPE 71.0"	Sand Dunes/ Revegetated	0.0(1)	0	
			Sand Dunes/Cleared	0.6	5	
Dincer, Al-Mugrin, Zimmerman (1974)	Saudi Arabia	MAP 3.2" MAPE (A) 141"	Dunes	Less Than .8(2)	Less Than 25	Geochemical Profiles in Vadose Zone (Surface to 20 ft.)
Kitching, Edmunds, Shearer, Walton (1980)	Cyprus: (1) Akrotin (2) Paralimni (3) Mesaona	MAP 18" MAP 15.4" MAPE 57.1" MAP 15.4"	NA(3) NA NA	2.0 0.4 0.2	11 2 1	NA
Sharma, Hughes (1985)	Western Australia Perth	MAP 30.5" MAPE (A) 91"	Swan Coastal Plan Woodland		15 (Note: 7-1/2 through soil matrix, and 7-1/2 through preferred pathways)	Geochemical Profiles (65 ft. to 90 ft.)
Stephens, Knowlton (1986)	Socorro, New Mexico, USA	MAP 7"	Sparsely Vegetated Dry Sand Site	0.3 to 1.5	4 21	Soil Moisture Profiles (Surface to 2.44 mm)

(1) Value quoted is mm/year.

(2) Measured within the active zone, i.e., still influenced by evaporation.
MAP = Mean Annual Precipitation
MAPE = Mean Annual Potential Evaporation
A = A Pan

(3) NA = Not Available

The tailings waste disposal was designed as an unlined, zero infiltration system to be achieved by depositing dried tailings in an extremely arid climate. Drying of the tailings to a residual moisture content just below the specific retention of the material was to be accomplished in the mill by mechanical mean. The dried tailings were to be trucked to the unlined disposal site and disposed of by spreading in layers up to a few feet thick. This process was to further reduce the moisture content of the material.

Field tests were conducted for a 14-week period, from December through April, using tailings placed in a pit 3 feet deep and 5 feet square. These tests indicated that the tailings could be adequately "dried" and that they continued to dry once deposited in layers. The tests showed that approximately two-thirds of the water which was initially in the tailings material was removed by evaporation after placement in the pit. Because the test took place in winter, the climatic conditions during the test period were significantly "wetter" than average.

The above design concept and field test results were accepted by the Lahontan Regional Water Quality Control Board which issued waste discharge requirements for the site.

6.3.2.2 UMTRA Program

As part of the Department of Energy (Caldwell, 1989) UMTRA Program, field investigations were carried out to determine rates of infiltration at several reclaimed uranium tailings piles (U.S. Department of Energy, 1989 and 1991).

The uranium mill tailings covers studied consisted generally of, from top to bottom, a 1-foot layer of rip-rap, 6 to 12 inches of sand filter, and a 3- to 7-foot thick radon barrier of compacted fine-grained material. The saturated hydraulic conductivities of the sand layer were in the 10^{-3} to 10^{-2} cm/sec (1,000 to 10,000 ft/year) range, while the radon barrier were less than 10^{-7} cm/sec (.1 ft/year).

The field studies included taking core samples and analyzing them for moisture content, and laboratory testing to determine relationships between soil moisture tension and soil moisture at three facilities located near Shiprock (New Mexico), Clive (Utah), and Burrell (Pennsylvania). Climatic data for these sites are provided in Table 6.9. Instruments for continuously monitoring soil moisture and tension were installed and operated for over six months at the Shiprock facility. Some of the key findings from these studies are discussed below.

TABLE 6.9

DEPARTMENT OF ENERGY UMTRA STUDY RESULTS

FACILITY	LOCATION	MEAN ANNUAL INCHES		CONTAINMENT EFFECTIVENESS EXPRESSED IN TERMS OF HYDRAULIC CONDUCTIVITY
		PRECIPITATION	A PAN EVAPORATION	
Shiprock	New Mexico	6	68	10^{-8} cm/sec[(1)]
Clive	Utah	5	60	Not reliably evaluated but expected to be significantly less than the measured saturated hydraulic conductivity of 5×10^{-8} cm/sec[(2)]
Burrell	Pennsylvania	44	39	Not reliably evaluated, but expected to be less than the measured saturated hydraulic conductivity of 4×10^{-7} cm/sec[(2)]

(1) These values correspond to the unsaturated hydraulic conductivity. The measured saturated hydraulic conductivity is 6×10^{-7} cm/sec.

(2) Field measurements and model projections indicate that the infiltration/radon barriers are unsaturated and are expected to remain unsaturated.

Soil moisture in the 6-inch sand layer drained or evaporated within a few days after a rainstorm. Fluctuations in soil moisture tension experienced in the upper 6 to 12 inches of the radon barrier did not propagate downward. It was concluded that evaporation through rock layers is an effective mechanism for removal of moisture from typical UMTRA covers in semi-arid climates. The rate and depth of drying following a precipitation event indicated that evaporation losses are significant.

These studies concluded, more specifically with respect to the Shiprock tailings pile, that "in no case was a front of saturation observed to be propagating downward through the radon barriers," and "... moisture that has infiltrated the radon barriers is held within the top 12 inches of the barrier until evaporated."

At the Burrell site, located in a wetter climate, the radon cover also was not saturated. The DOE concluded that, even in wetter climates, covers may not, in fact, become saturated.

For design and permitting purposes, the DOE evaluated the potential long-term ground water impact from the reclaimed tailings piles. For this purpose, they estimated unsaturated flow migration rates through the radon barrier for the Shiprock and Clive facilities using the data collected. The resulting design migration rates were 10^{-9} cm/sec and 10^{-13} cm/sec, respectively, equivalent to 0.012 and .0000012 inches/year.

6.3.2.3 Imperial County, California

Sargent, Hauskins, and Beckwith (1984) conducted site characterization studies for the Mesquite mining project, located 35 miles east of Brawley in Imperial County, California. The mean annual precipitation at the site is approximately 4 inches, and lake evaporation is 96 inches.

As part of this investigation, the soil moisture conditions in the upper 50 feet of natural soils at the site were measured. These moisture contents were compared to the calculated wilting point and soil moisture retention capacities using Gupta and Larson (1979), Salter et al. (1966), and U.S. Department of Agriculture (1977).

The results of these investigations are as follows:

SOIL (feet)	MEASURED ACTUAL MOISTURE CONTENT (%)	CALCULATED WILTING POINT (%)	SPECIFIC RETENTION (%)
0 to 10	.75 - 4	NA	NA
10 to 25	1 - 7	5 - 7.5	10 - 15
25 to 50	2 - 9	NA	NA

The results indicate a large deficit between the actual moisture content and the specific retention. Under these conditions, noticeable vertical migration of infiltrating rainwater would not occur. In fact, a considerable amount of seepage from a water pond located on the surface could be stored in the vadose zone before any noticeable seepage reached the ground water table.

Similar analyses for the proposed heap leach piles resulted in the conclusion that "seepage (percolation) extending through the pile should not be anticipated."

6.4 ANALYTICAL APPROACH TO PERCOLATION EVALUATION

6.4.1 INTRODUCTION

This section describes and applies one of the analytical approaches available for predicting infiltration and percolation rates in waste piles. A water balance approach has been adopted because of its ease in application and usefulness as a design tool, and a computerized calculation tool, the Hydrologic Evaluation of Landfill Performance (HELP) model (EPA, 1983) was selected.

An alternative approach (DOE, 1991) would be to use an unsaturated ground water flow model such as UNSAT-H. While it is more complex to apply, it may provide a more accurate result in arid climates. This is primarily due to its ability to model the upward movement of water due to capillary rise. In the HELP model, all water movements below the evaporation depth is downward only resulting in a possible overestimation of the infiltration rate.

6.4.2 HYDROLOGIC EVALUATION OF LANDFILL PERFORMANCE (HELP) MODEL

6.4.2.1 General Description

The HELP program was developed to facilitate estimation of the amounts of surface runoff, subsurface drainage, and leachate that may be expected to result from a wide variety of possible waste pile designs. The program simulates the effects of hydrologic processes, including precipitation, surface storage, runoff, infiltration, percolation, evapotranspiration, soil moisture storage, and lateral drainage, using a two-dimensional approach. Waste management systems, including various combinations of vegetation, cover soils, waste cells, drainage layers, and relatively impermeable barrier soils, as well as synthetic membrane covers and liners, may be modeled. The infiltration routine relies on the Hydrology Section of the National Engineering Handbook (USDA, 1972).

The model was specifically designed to perform hazardous waste disposal evaluations as required by the Resource Conservation and Recovery Act (RCRA) and expanded upon in 40 CFR 264. Among other things, these regulations deal with design specifications for the use of liners to minimize the production of leachate and the buildup of leachate head over the liner.

6.4.2.2 Method of Solution

The HELP model was developed to estimate daily water movement on the surface and through a waste pile. Precipitation is partitioned into runoff, evapotranspiration, percolation, and subsurface lateral drainage. The model computes runoff by the Soil Conservation Service (SCS) Runoff Curve Number method (USDA, 1972) and percolation by Darcy's Law for saturated flow (Freeze and Cherry, 1979), with modifications for unsaturated conditions. Lateral drainage is computed analytically from a linearized Boussinesq equation, corrected to agree with numerical solutions of the non-linearized Boussinesq for the range of design specifications used in hazardous waste landfills (Skaggs, 1982). Evapotranspiration is determined by a modified Penman method developed by Ritchie (1972).

The HELP model is deterministic in concept insofar as the model treats all variables and their relationships as being known, although it often uses empirical relationships.

6.4.3 APPLICATION OF THE HELP MODEL

The HELP Model was used to examine the particular climatic and waste pile conditions which may or may not produce leachate from a typical waste pile. Three application sites were selected based on the particular climatic conditions they represent. The three cases are:

- Case I - A site in central northern California.
- Case II - A site near Riverside, California.
- Case III - ASARCO Waterloo Project in southwestern California near Calico.

Case I was chosen as a site representative of known leachate production from a waste rock pile. Mean annual precipitation (30 inches) and evaporation (50 inches) are relatively close in this case. Case III represents an arid climate where leachate production is not expected. Climatic conditions are such that mean evaporation exceeds precipitation by more than an order of magnitude (113 versus 4 inches). Case II represents those areas in which leachate production may be marginal. Depending on the waste pile characteristics and precipitation events during any one day, month, or year, leachate may or may not be produced. For this case, the mean annual precipitation is 13 inches, and pan evaporation of 82 inches exceeds the precipitation by just under an order of magnitude.

6.4.3.1 Waste Pile Soil Profile

Figures 6.8 and 6.9 present the waste pile profiles used in the HELP model as "typical" for the model runs. Figure 6.8 illustrates a waste pile without vegetation. Figure 6.9 shows the same waste pile with a 24-inch topsoil/vegetation cover. A third profile was used to evaluate the effect of fine-grained material. This profile was the same as the one shown in Figure 6.8 with the exception that the 100 feet of waste rock were replaced by low permeability tailings with a saturated hydraulic conductivity of 10^{-6} cm/sec.

By using these waste piles as "standards" for all three cases, the actual climatic influences from precipitation and evaporation events can be more readily evaluated. Characteristics of the waste pile such as texture, grain sizing, slope, and layering that affect leachate production are held constant.

6.4.3.2 Input Data

The HELP model provides default climatic data such as daily precipitation, mean monthly temperature, mean monthly solar radiation, leaf area indices, and ground cover coefficients. The

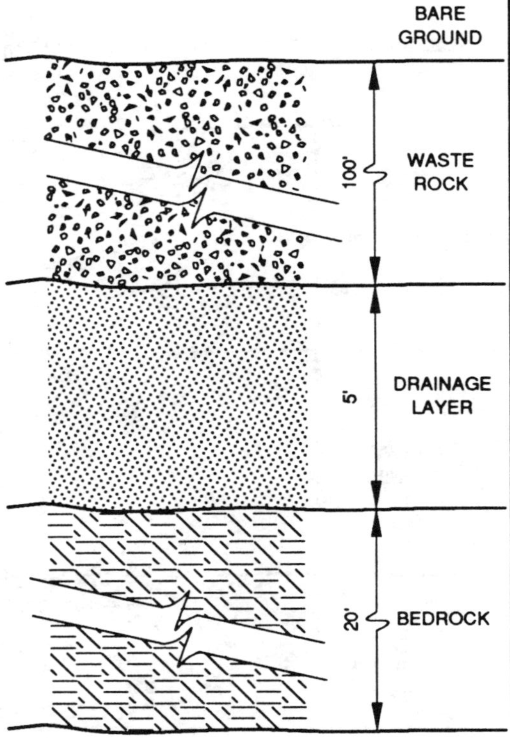

USCS SOIL[1] CLASSIFICATION	KEY PARAMETERS				
	CN	HYDRAULIC CONDUCTIVITY CM/SEC	POROSITY	FIELD CAPACITY	WILTING POINT
GRAVELLY SAND	80	1.0×10^{-3}	.35	.17	.11
SILTY SAND	—	1.0×10^{-1}	.40	.13	.08
—	—	5.0×10^{-7}	.05	—	—

[1] USCS = UNIFIED SOIL CLASSIFICATION SYSTEM

FIGURE 6.8

SOIL PROFILE A NONVEGETATED

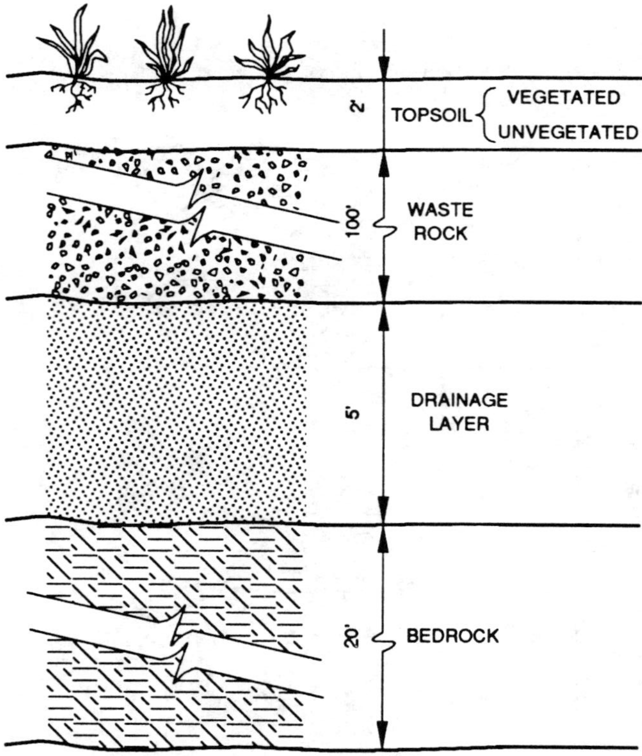

USCS SOIL[(1)] CLASSIFICATION	KEY PARAMETERS				
	CN	HYDRAULIC CONDUCTIVITY CM/SEC	POROSITY	FIELD CAPACITY	WILTING POINT
NON-PLASTIC SILT	90	2.0×10^{-4}	.54	.42	.22
NON-PLASTIC SILT	90	0.8×10^{-4}	.54	.42	.22
GRAVELLY SAND	—	1.0×10^{-3}	.35	.17	.11
SILTY SAND	—	1.0×10^{-1}	.40	.13	.08
—	—	5.0×10^{-7}	.05	—	—

[(1)] USCS - UNIFIED SOIL CLASSIFICATION SYSTEM

FIGURE 6.9

PROFILE B
RECLAMATION:
VEGETATED TOPSOIL COVER

cities in California for which default data are available include Fresno, Los Angeles, Sacramento, San Diego, and Santa Maria. The modeler can choose to use the default data for all or part of the various parameters listed.

For Case I, daily precipitation values came from the CDWR for an average year (32.3 inches precipitation) and a slightly wetter than average year (35.2 inches). Case II precipitation values came from the Riverside County Flood Control and Water Conservation District for years 1980 to 1984. The wet year was 1983. Solar radiation and temperature data came from the meteorological station at the University of California at Riverside. Precipitation data for Case III came from CDWR and the weather station at Daggett Airport. The years used for Case III were from 1981 to 1986, with 1983 being the wet year.

Table 6.10 presents the source of input data for each case. Whenever possible, data from the actual site locations were used in modeling. Mean monthly solar radiation is not measured at many of the weather stations in California. When these data were not available, default values from a nearby representative city were used. For example, Phoenix, Arizona default solar radiation data were used for the southeast desert areas of California.

6.4.3.3 Summary of HELP Results

Tables 6.11 and 6.12 present summaries of the HELP model run results. Appendix D summarizes the parameters used in these HELP procedures. With the exception of the bare waste rock scenario, Case III yielded no percolation under the conditions simulated. The bare waste rock yielded a wet year percolation rate of 1.9 inches, which is equivalent to 1.6×10^{-7} cm/sec and an annual average rate of 0.6 inches or 5×10^{-8} cm/sec. Case II produced extremely low percolation values for the vegetated waste rock case. The cm/sec equivalents for these values are approximately 3×10^{-7} cm/sec. It produced no percolation for the bare (unvegetated) tailings profile. Case I yielded low to moderate infiltration rates. As can be seen, vegetation significantly reduced the amount of percolation.

6.4.3.4 Conclusions on Reliability

Many of the parameters used in the model were based on professional judgement. While it is likely that the model needs further refinements, with additional field data on waste properties, measurements of waste moisture changes, and leachate observations, the addition of actual site data into the model will result in greater accuracy and predictive reliability.

TABLE 6.10

HELP MODEL DATA INPUT SOURCES

CASE NUMBER	YEAR(S) OF DATA	MEAN ANNUAL PRECIPITATION (inches)	MEAN MONTHLY TEMPERATURE	MEAN MONTHLY SOLAR RADIATION	LEAF AREA INDICES	WINTER COVER FACTOR	APPROXIMATE EQUIVALENT A PAN (inches)
I	1989-1990	Clear Lake, California (32)	D-Sacramento, California	D-Sacramento, California	D-Sacramento, California	D-Sacramento, California	50
II	1983-1986	UCR-Riverside, California (13)	UCR-Riverside, California	UCR-Riverside, California	UCR-Riverside, California	UCR-Riverside, California	82
III	1981-1986	Daggett Airport, California (5)	Daggett Airport, California	D-Phoenix, Arizona	D-Phoenix, Arizona	D-Phoenix, Arizona	113

D - Default Data from Named City
UCR - University of California, Riverside

TABLE 6.11

RESULTS OF HELP MODEL RUNS

CASE NUMBER	ANNUAL PRECIPITATION (inches)	ANNUAL PERCOLATION (inches)			RUN NUMBERS
		Bare Waste Rock	Vegetated Waste Rock	Bare Tailings	
I	35.2	17.8	7.6	0.0	1,2,3
	32.3	19.9	8.5	0.0	4,5,6
II	22.9 Wet	10.5	4.6	0.0	7,8,9
	13.4 Average[(1)]	6.2	2.9	0.0	7,8,9
III	9.4 Wet	1.9	0.0	0.0	10,11
	5.0 Average[(2)]	0.6	0.0	0.0	10,11

(1) Average for the period 1980 to 1984.
(2) Average for the period 1981 to 1986.

TABLE 6.12

WATER BALANCE DATA FOR LOS ANGELES, CALIFORNIA (Values in mm Unless Otherwise Specified)

PARAMETER	JAN.	FEB.	MAR.	APR.	MAY	JUNE	JULY	AUG.	SEPT.	OCT.	NOV.	DEC.	ANNUAL (mm)	TOTAL (inches)
PET	34	36	49	59	76	94	117	115	96	73	52	39	840	33
P	78	79	66	27	9	2	0	1	5	14	29	68	378	15
$C_{R/O}$	0.15	0.15	0.15	0	0	0	0	0	0	0	0	0.15		
R/O	12	12	10	0	0	0	0	0	0	0	0	10	44	2
I	66	67	56	27	9	2	0	1	5	14	29	58	334	13
I-PET	+32	+31	+7	-32	-67	-92	-117	-114	-91	-59	-23	+19	-506	-20
Sum of (NEG (I-PET)			-39	-71	-138	-320	-347	-461	-522	-611	-634			
ST (Table B)	52	83	90	70	40	19	7	3	1	1	1	20		
ST	+32	+31	+7	-20	-30	-21	-12	-4	-2	0	0	+19		
AET	34	36	49	47	39	23	12	5	7	14	29	39	334	13
PERC	0	0	0	0	0	0	0	0	0	0	0	0	0	0

PET = Potential Evapotranspiration
P = Precipitation
$C_{R/O}$ = Runoff Coefficient
R/O = Runoff
I = Infiltration, equals P-R/O
ST = Soil Moisture Storage, obtained from Table B in Fenn et al., (1974)
AET = Actual Evapotranspiration
PERC = Percolation

6.4.4 APPLICATION OF THE THORNTHWAITE METHOD

In a study conducted by the Office of Solid Waste Management Programs (Fenn et al., 1975), the Thornthwaite water balance method was used to predict leachate generation in a sanitary landfill. Three cases were selected to reflect various climatic and soil conditions. One of the case studies chosen was Los Angeles, California. Los Angeles represents a dry climate (15 inches average annual precipitation) with a fine-grained soil. The surface of the landfill is assumed to be fully vegetated with moderately deep-rooted grass.

The results, as shown in Table 6.12, indicate that no percolation was predicted.

6.5 PREDICTION OF INFILTRATION

6.5.1 PERCOLATION IN PERSPECTIVE

In terms of waste piles, the amount of percolation that occurs, coupled with the chemical properties of the waste material, is an important characteristic that should be considered when selecting or designing the level of containment for the waste. Significant amounts of percolation, of a quality that poses a threat to beneficial uses of ground or surface water, must be dealt with by providing adequate natural or engineered containment and, in some cases, leachate removal coupled with recycling or treatment and discharge. Limited amounts of percolation may require some form of containment and removal, depending on the site-specific conditions and the leachate quality. In cases where there are very minor amounts or no percolation, leachate containment and/or removal is unlikely to be required. In cases of minor quantities of percolation, the containment offered by the natural geologic formation below the wastes may be sufficient to prevent any impact to beneficial uses of the waters of the State.

Determining what the likely percolation rate is going to be for a given waste pile is important. However, there are several other issues that must be recognized and factored into the design and permitting of the waste disposal facility. These include:

- Relative Magnitude - For disposal of liquid and/or wet tailings, a containment requirement of between 10^{-6} cm/sec and 10^{-7} cm/sec is generally regarded as adequate and is incorporated in the current Article 7[(1)] mine waste regulations in California. These are equivalent to 12 inches and 1 inch, respectively, of allowable downward seepage over a year, for a typical case of saturated tailings with a unit vertical hydraulic gradient or a clay liner containing a small quantity of waste fluid. An

(1) California Code of Regulations, Title 23, Division 3, *Chapter 15*, Article 7.

inch of percolation per year through an unlined waste pile would therefore provide the same degree of containment, for example, as a liquid waste impoundment lined with 10^{-7} cm/sec material.

- As discussed in Section 6.2.4.6, there may be a small amount of residual percolation associated with the vertical moisture content profile through the waste. In semi-arid and arid and some humid climates, this migration will occur at rates determined by the unsaturated hydraulic conductivity, and will therefore usually be several orders of magnitude lower than the approximate magnitudes discussed in the preceding paragraph.
- Travel Time - As illustrated in Figure 6.10, it can take a great deal of time for water to travel through a waste pile at low percolation rates. This is an important concept.

  The active life of a waste pile can vary from a few months to several tens of years. During this active period, the pile may be subjected to slightly higher infiltration rates than will occur later, due to surfaces which are not optimally graded for drainage and which have not been covered or revegetated. In some cases, for example, the initial moisture content may be elevated due to infiltration of precipitation through hollow or flat areas on a waste rock pile. This may also generate some short-term percolation. However, when such percolation rates are coupled with relatively short operating periods, acceptably short migration distances may result, thus alleviating the need for any additional containment.

  Once the pile is regraded, covered and revegetated, the percolation rate would be reduced or eliminated. Thus, also in the long term, additional containment may not be required.

6.5.2 GUIDELINES

From the material presented in this chapter, it is apparent that precipitation is the dominant and primary driving force behind percolation. However, the physical characteristics of the waste, as well as the extent of vegetative cover, play a very important secondary role.

Figure 6.11 summarizes some of the results contained in this chapter and provides a tentative boundary line between climatic conditions that are conducive to leachate formation and those that generate zero or negligible leachate formation. The conditions that produce no leachate or negligible quantities are also those cases where percolation rates are extremely low (i.e., less than 10^{-7} cm/sec). There is obviously a relatively large "grey area" in between these two climatic conditions where, either due to technical inaccuracy or the secondary influence of waste pile physical properties, there is no clear cut distinction. This is illustrated by the shaded area.

Using 10 inches of mean annual precipitation as the threshold below which no leachate or a negligible quantity is produced, the portion of California that would not be expected to experience

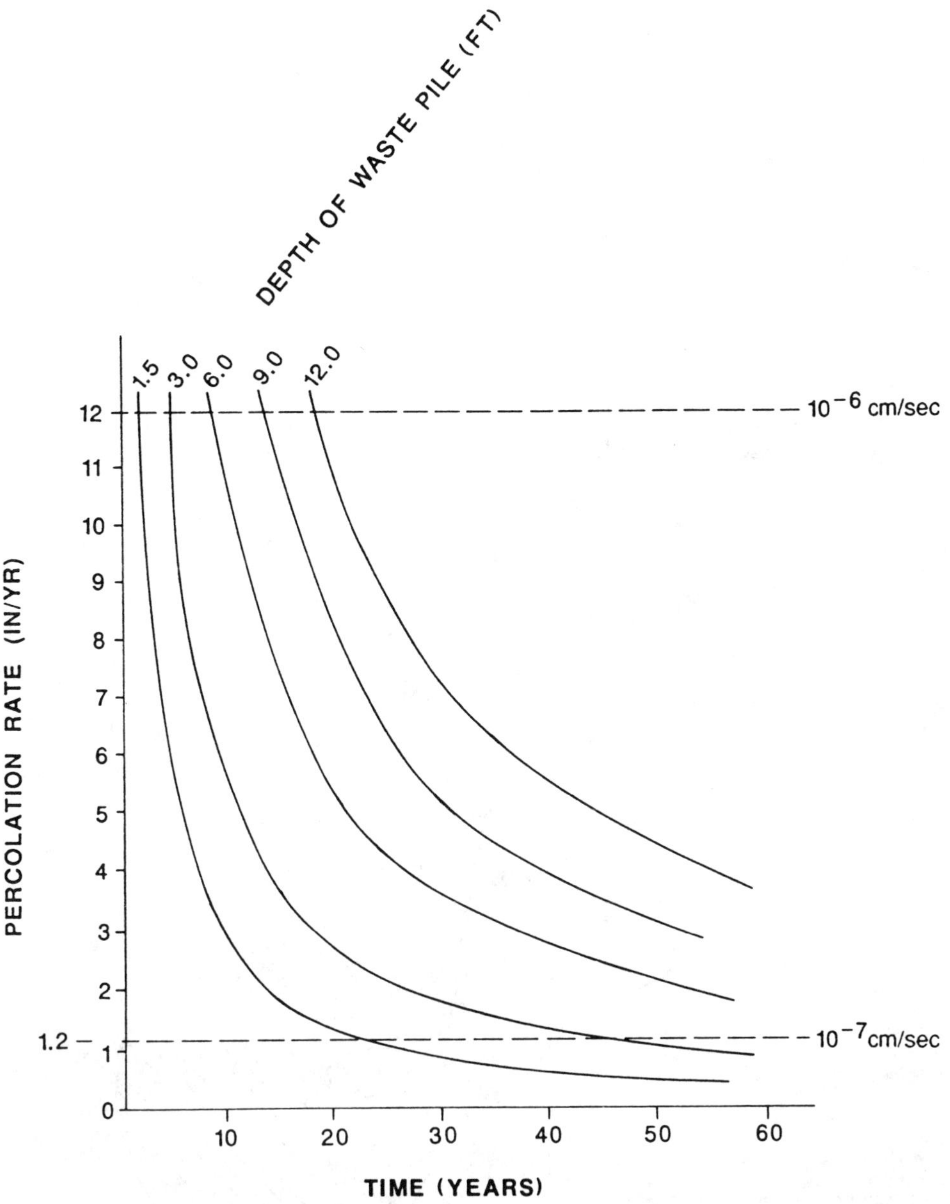

REFERENCE: FENN ET AL., 1975

FIGURE 6.10

TIME OF FIRST APPEARANCE OF LEACHATE (WATER BALANCE METHOD)

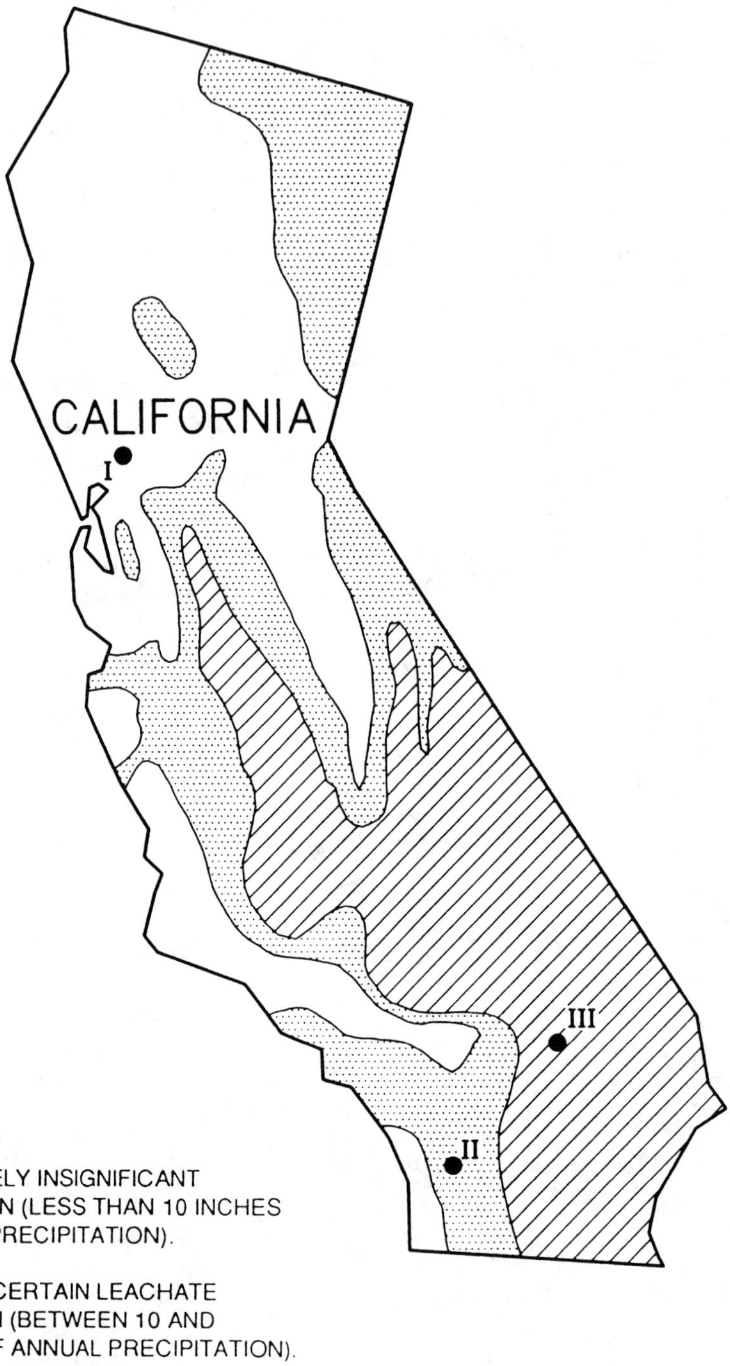

FIGURE 6.11

AREAS OF PERCOLATION (TENTATIVE)

leachate problems covers the southeastern quadrant of the State. The areas in which leachate generation may be a problem are judged to be those which have a mean annual precipitation of between 10 and 20 inches.

6.5.3 EVALUATION OF PERCOLATION

6.5.3.1 Initial Evaluation

The initial step in evaluating likely infiltration and percolation rates should be based on the climate and the initial moisture content of the waste as delivered to the waste disposal facility. Climatic data in terms of mean annual precipitation and lake or pan evaporation should be collected or developed for the site. In cases where the mean annual precipitation is less than 10 inches and the A pan evaporation rate is above 75 inches, it is very likely that natural percolation will not occur in measurable quantities. In areas where precipitation exceeds 20 inches, it can be assumed that some measurable percolation will occur.

In reviewing the moisture content of the waste material, it is important to compare the moisture content of the delivered waste to the specific retention capacity of the waste material. In some cases, laboratory tests would have to be conducted to determine what the specific retention is. Where the delivered moisture content is lower than the specific retention, it can be assumed that no percolation will be induced by the water contained in the waste. However, values at or above the specific retention could pose problems. It is important to recognize that the method of waste placement can also influence the potential for leachate formation. For example, waste that has a moisture condition a few percentage points above the specific retention of the material may not cause percolation if it is spread out in layers of less than a few feet thick and allowed to dry under the influence of wind and sun.

6.5.3.2 Detailed Evaluations

Should the initial evaluation prove to be inconclusive, then the following types of analyses and tests are available:

- Conduct water balance calculations along the lines presented in this chapter (i.e., using the HELP model or the Thornthwaite method). Use extensive sensitivity runs to fully explore the impact on the percolation prediction of realistic ranges for the model parameters.
- Conduct onsite sampling and analyses of the soil moisture profiles in the geologic formations to determine the natural recharge rates.

- Conduct onsite small-scale field tests to determine the potential for leachate formation using lysimetry.
- Conduct laboratory column leach tests to predict the quality of any leachate that may form.

Depending on the certainty of the results of the above analyses, it may be appropriate to consider incorporating an in-waste monitoring program and a contingency plan as part of the final full-scale waste disposal plans. Monitoring would involve sampling for moisture content on a routine basis to confirm that the moisture conditions in the pile are not increasing and threatening to produce leachate. Should problems be experienced, then the contingency plan should be implemented. For dry tailings disposal, typical contingency plans could include disposing of the waste material in a small lined area during short, extremely moist periods, or adding additional drying capacity in the mill. Covering and relocating to a lined area are other alternatives.

6.6 REGULATORY IMPLICATIONS

As demonstrated in this chapter, there is evidence indicating that leachate formation does not occur, or occurs at insignificant rates, in "arid" areas for "dry" types of mine waste management units. These units include heap leach residue piles, dry tailings or tailings deposited in thin layers, and overburden and waste rock piles. A typical dry waste disposed of in arid areas obviously would pose a significantly lower risk of impact to ground or surface water than if the same waste were disposed of in a facility in a wetter climate. The waste disposed of in an "arid" climate requires far less and, in some cases, no additional engineered containment to ensure protection of the beneficial uses of water.

By adopting regulations that recognize the effect of climate on leachate generation in wastes, depending on other project factors, mining companies will have an incentive to switch to drying the wastes before disposal. It is obvious that tailings waste management units, for example, which are filled with dry material and are not subjected to significant percolation, represent a much lower water quality risk than a wet facility that has to rely, for an extended period of time, on engineered liners. Furthermore, with such regulations in place, there may be an economic advantage to the development of properties in arid rather than wet climates.

6.6.1 RECOMMENDED REGULATORY APPROACH

6.6.1.1 Background

The concept of "arid climate" should be used in regulations as part of the decision-making process to determine the appropriate degree of waste containment. Where the climate is arid and the moisture in the waste being deposited is at or below specific retention, the operator should be allowed to utilize the fact that no leachate is likely to form when designing the containment system of the waste management units.

For example, in a wet climate, the requirement might be for a single liner with a leachate collection system under a waste rock pile that has the potential to generate leachate with elevated constituent concentrations. For the same waste disposed of in an arid climate, the containment requirements might be reduced to reliance on the natural geologic formations.

6.6.1.2 Regulatory Approach

Arid climatic conditions should be considered in the design of a waste management unit. In determining whether the climate is arid, the operator should demonstrate one of the following:

- The mean annual precipitation at the site is less than 8 inches and the A pan evaporation is greater than 50 inches.(2)
- Water balance calculations and/or moisture profile measurements and evaluations indicate that no significant percolation will occur. Significant percolation is defined as an average annual amount that exceeds the containment requirements for the given waste (i.e., 10^{-6} or 10^{-7} cm/sec). Calculations will use field data obtained from either lysimeter test plots or other field percolation tests, or natural recharge determinations.

Arid climatic conditions should be used in the design of a waste management unit, provided the operator can demonstrate that the waste will be deposited at a moisture content below specific retention, or that it will be deposited in a manner that will cause the waste to dry to specific retention or a lower moisture content without forming leachate. Where the prescriptive approach is being used for containment design, the waste category selected should take into account the fact that there are no free liquids present in the waste material. Similarly, if a containment assessment approach is being used, potential migration scenarios should take into account that no leachate will

---

(2) These values are based on the results contained in this chapter which indicate leachate formation is not likely to occur where mean annual precipitation is less than 10 inches. The actual value is based on the assumptions that infiltration should be limited to 10^{-7} cm/sec (1.2 inches/year), and that 15 percent of the mean annual precipitation infiltrates. The annual evaporation is set as a value that should be adequate for the removal of water that does accumulate in the upper layer of the waste.

be formed. The containment assessment also should evaluate such conditions as waste deposition methods and climatic extremes under which leachate potentially could be generated and the fate of such leachate, in order to determine the potential risk to water quality, if any.

In cases where the performance of the proposed waste management unit is uncertain, it may be necessary to require the operator to conduct monitoring of the moisture conditions in the waste, and to have a contingency plan in the event that the actual performance is judged to pose a threat to the beneficial uses of water.

6.7 BIBLIOGRAPHY

Allison, G. B., W. J. Stone, and M. W. Hughes. "Recharge in Karst and Dune elements of a Semi-Arid Landscape as indicated by Natural Isotopes and Cholide." *Journal of Hydrology 76* (1985):1-25.

ASARCO. Design documents for a dry tailings disposal facility. 1989.

Brady, N.C. *The Nature and Properties of Soils. 8th Edition.* New York: MacMillan Publishing Company, Inc. 1974.

Bredenkamp, D. B. *Quantitative Estimation of Ground-Water Recharge in Dolomite.* Pretoria, South Africa. 1988.

Caldwell, J.A. *Groundwater Protection and Restoration at Uranium Mill Tailings Disposal Sites.* Fourth Annual Hazardous Waste and Hazardous Materials Management Conference, San Francisco, CA. September 27-29, 1989.

California Code of Regulations. Title 23, *Chapter 15.* 1988.

California Department of Water Resources. *California Rainfall Summary, Monthly Total Precipitation 1849-1979.* Sacramento, CA. 1980.

Chow, V. T. *Handbook of Applied Hydrology; a Compendium of Water Resources Technology.* New York: McGraw-Hill. 1964.

Dincer, T., A. Al-Mugrin, and U. Zimmerman. "Study of the Infiltration and Recharge Through the Sand Dunes in Arid Zones with Special Reference to the Stable Isotopes and Thermonuclear Tritium." *Journal of Hydrology 23*, (1974): 79-109.

Fenn, D. G., J. Hanley, and T. V. DeGeare. *Use of Water Balance Method for Predicting Leachate Generation from Solid Waste Disposal Sites.* United States Environmental Protection Agency. EPA 530/5W-168. 1975.

Foster, S. S. D. *Quantification of Groundwater Recharge in Arid Regions: A Practical View for Resource Development and Management.* Estimation of Natural Groundwater Recharge. Amsterdam: Reidel Publishing Company. 1987.

Freeze, R. A. and J. A. Cherry. *Groundwater.* Englewood Cliffs: Prentice-Hall. 1979.

Gardner, W. H. *Water Content in 1965. Methods of Soil Analyses.* Agronomy, 9 (1965): 82-125.

Glendon, W. Gee and Daniel Hillel. *Groundwater Recharges in Arid Regions: Review and Critique of Estimation Methods - Hydrological Processes.* Volume 2, 255-266. 1988.

Gupta, S. C. and W. E. Larson. *Estimating Soil Water Retention Characteristics from Particle Size Distribution, Organic Matter Percent, and Bulk Density.* Water Resources Research, 15 (December 1979). No. 6.

Hanks, R. J. and G. L. Ashcroft. Applied Soil Physics. Berlin: Springer-Verlag. 1980.

Hillel, D. *Introduction to Soil Physics.* New York: Academic Press. 1982.

Holmes, J. W., S. A. Taylor, and S. J. Richards. *Irrigation of Agricultural Lands.* Argronomy 9 (1967): 275-298.

Houston, J. *Rainfall-Runoff-Recharge Relationships in the Basement Rocks of Zimbabwe.* Estimation of Natural Groundwater Recharge. Amsterdam: Reidel Publishing Company. 1987.

Jensen, M. E. and H. R. Haise. "Estimating Evapotranspiration from Solar Radiation". *Journal of the Irrigation and Drainage Division* (1963):15-41.

Kitching, R., W. M. Edmunds, T. R. Shearer, and N. R. G. Walton. *Assessment of Recharge to Aquifers.* Hydrological Sciences - Bulletin - des Sciences Hydrologiques 25 (1980).

McWhorter, D. B. *Personal communication.* 1990.

Merrill, S. D. and S. C. Rawlins. "Field Measurement of Soil Water Potential with Thermocouple Psychrometers." Soil Science 113 (1972): 102-109, No. 2.

Mustonen, S. E. and J. L. McGuinness. *Estimating Evapotranspiration in a Humid Region.* United States Agriculture Research Service Technical Bulletin No. 1389: 123. 1968.

Nielsen, D. R., J. W. Biggar, and K. T. Erh. "Spatial Variability of Field Measured Soil-Water Properties." Hilgardia 43 (1973): 215-260, No. 7.

Rawlins, S. L. and F. N. Dalton. *Psychrometric Measurement of Soil Water Potential Without Precise Temperature Control.* Soil Science Society of America. 1967.

Ritchie, J. T. "A Model for Predicting Evaporation from a Row Crop with Incomplete Cover." Water Resources Research 8 (1972): 1204-1213, No. 5.

Salter, P. J., G. Berry, and J. B. Williams. "The Influence of Texture on the Moisture Characteristics of Soils: III Quantitative Relationship between Particle Size, Composition, and Available-Water Capacity." *Journal of Soil Science* 17 (1966), No. 1.

Sargent, Hauskins, and Beckwith. *Supporting Technical Report for the Environmental Impact Report/Environmental Assessment for the Proposed Mesquite Project.* Prepared for Gold Fields Operating Co. March 1984.

Sharma, M. L. and M. W. Hughes. "Groundwater Recharge Estimation Using Chloride, Deuterium and Oxygen-18 Profiles in the Deep Coastal Sands of Western Australia." *Journal of Hydrology* 81 (1985): 93-109.

Siegel, Joel. *Numerical Simulation of Seepage Beneath Lined Ponds*. Dissertation. New Mexico Institute of Mining and Technology. 1980.

Skaggs, R. W. *Modification to DRAINMOD to Consider Drainage from and Seepage Through a Landfill*. Draft Report. Cincinnati, Ohio: U.S. Environmental Protection Agency. 1982.

Sniff, T. L., L. A. Davis, and D. B. McWhorter. *An Overview of Recharge Estimates*. Technical Report 7. Contract No.: RS-NMS-85-009, Subtask 1.5. February 1988.

Stephens, D. B. and R. Knowlton, Jr. "Soil Water Movement and Recharge Through Sand at a Semiarid Site in New Mexico." Water Resources Research 22 (1986): 881-89, No. 2.

Thornthwaite, C. W. and J. R. Mather. "Instructions and Tables for Computing Potential Evapotranspiration and the Water Balance." Publications in Climatology 10 (1957): 185-311, No. 3.

Thornthwaite, C. W. and J. R. Mather. "The Water Balance." Publications in Climatology 8 (1955): 104, No. 1.

United States Department of Agriculture. Soil Conservation Service. *Arizona, Central Part: Soil Survey of Maricopa County*. September 1977.

United States Department of Agriculture. Soil Conservation Service. *National Engineering Handbook, Section 4, Hydrology*. 1972.

United States Department of Energy. *Analysis of Infiltration through a Clay Radon Barrier at an UMTRA Disposal Cell*. UMTRA Project Office. Albuquerque Operation Office. January 1991.

United States Environmental Protection Agency. Office of Research and Development. *User Guide for the Hydrologic Evaluation of Landfill Performance (HELP) Model*. Cincinnati, Ohio. 1983.

United States National Weather Service. *Evaporation Atlas for the Contiguous 48 United States*. Report NWS33. Washington, D. C.: National Oceanic and Atmospheric Administration. 1982.

van Bavel, C. H. M. *Neutron Scattering Measurement of Soil Moisture: Development and Current Status*. Washington D.C.: Proceedings of the International Symposium on Humidity and Moisture. 1963.

CONTENTS

CHAPTER 7
LINER SYSTEM DESIGN

PAGE NO.

LIST OF TABLES/LIST OF FIGURES 328

7.1 Introduction 329

7.1.1 Background and Objective 329

7.1.1.1 Background 329

7.1.1.2 Objectives 331

7.1.2 Integration With Other Chapters 331

7.1.3 Study Participants 332

7.1.4 Sources of Information 333

7.1.5 Chapter Organization 333

7.2 Types of Liner Systems 334

7.2.1 Single Liner Systems 334

7.2.2 Single Liner Systems With Overlying Hydraulic Head Control 337

7.2.3 Composite Liners With and Without Overlying Head Control 337

7.2.4 Double Liners With an Intervening Leachate Collection and Removal System 338

7.2.5 Drainage System Above Double Liners With an Intervening Leachate Collection and Removal System 338

7.3 Types of Individual Liner System Components 339

7.3.1 Low Hydraulic Conductivity Liners 341

7.3.1.1 Basic Liner Properties 341

7.3.1.2 Potential Problems With Low Hydraulic Conductivity Liners 346

7.3.2 Cushioning or Load-Bearing Protection Layers 348

7.3.3 Hydraulic Head Control Layers 351

7.3.4 Leachate Collection and Removal Systems 353

7.4 Liner System Performance 355

7.4.1 General 355

7.4.2 Leakage Potential 355

7.4.3 Leakage in the Unsaturated Zone 365

7.4.4 Liner Stability Considerations 374

7.5 Regulatory Implications 378

7.5.1 General 378

7.5.2 Alternative Regulatory Approaches 380

7.5.3 Recommended Regulatory Approach 383

7.6 Bibliography 385

CONTENTS

CHAPTER 7 (Continued)

LIST OF TABLES

TABLE NO.	TITLE	PAGE NO.
7.1	Available Materials or Procedures for Liner System Components	336
7.2	Typical Hydraulic Conductivity Values	342
7.3	Typical Physical Properties for Pre-Formed Geomembranes	343
7.4	Potential Problems for Low Hydraulic Conductivity Liners	344
7.5	Estimated Leakage and Travel Times Through Clay Liners and Permeation Through Geomembrane and Composite Liners	357
7.6	Liner Leakage for Various Size Holes and Slits (in VLDPE and HDPE Liners Without a Soil Backing)	362
7.7	Estimated Leakage Through Holes in Geomembrane Liners in Gallons Per Day	363
7.8	Summary of Leakage Rates Through Clay, Geomembrane and Composite Liners for Wetted Front Travel Time Calculations	366
7.9	Estimated Time of Arrival of the Wetted Front (Years) for Leakage Through Clay Liners	368
7.10	Estimated Time of Arrival of the Wetted Front (Years) for Holes in Geomembrane	369
7.11	Summary of Material Properties	372
7.12	Comparison of Transit Time Through Hole in Liner for Different Assumed Spreading Angles	373
7.13	Waste Pile Stability Considerations in Liner System Design	376
7.14	Site-Specific Factors to Consider in Liner System Design	381

LIST OF FIGURES

FIGURE NO.	TITLE	PAGE NO.
7.1	Types of Liner Systems Potentially Applicable to Mine Wastes	335
7.2	Estimated Geosynthetic Sales in North America	340
7.3	Comparative Liner Penetration Study	349
7.4	Schematic Profile of Seepage Model	370

CHAPTER 7

LINER SYSTEM DESIGN

7.1 INTRODUCTION

7.1.1 BACKGROUND AND OBJECTIVE

7.1.1.1 Background

One of the most important regulatory and design decisions is determining the type of liner system. The liner system includes a combination of low hydraulic conductivity and leakage control materials to be provided beneath a mine waste management unit to avoid seepage losses, which could result in an unacceptable threat to beneficial uses of ground water. This is more difficult for mine wastes than for other types of waste disposal because:

- The physical and chemical properties of mine wastes vary widely. Physical properties vary from blasted rock to finely ground sand, silt, or clay tailings required for some processing procedures. The chemistry can be practically inert or may contain reagents which are considered to be potentially hazardous. As discussed in Chapter 2, most mine wastes present a relatively small threat to water quality, particularly in relation to other types of wastes.
- The sizes (volume and areal extent) of mine waste management units is often very large so that the costs of liners can impact economic feasibility of some operations.

The U.S. Congress considered the differences between mine wastes and other types of wastes when it passed the Bevill amendment to the Resource Conservation and Recovery Act (RCRA) in 1980. That amendment exempted most mine wastes from hazardous waste regulation until the United States Environmental Protection Agency (EPA) conducted a study to determine the appropriate degree of regulation for mine wastes. In 1986, the EPA issued a report recognizing that, with a few exceptions for certain processed materials, mine wastes do not present the same level of threat as other wastes and therefore should be regulated differently. The EPA currently is developing regulations for mine wastes under the nonhazardous Subtitle D program of RCRA. Chapter 3 contains further information on the federal regulation of mine waste. An additional important factor which differentiates mine waste disposal management units from other solid waste disposal units is that, except in unusual circumstances, mine and process facilities are located where the mineral resource is being extracted. Therefore, the location of the mine waste disposal facilities cannot solely be based upon a site selection study. As a result, some mines are located where the distance or depth to a valuable water resource is relatively small, while others are located in remote desert areas with no contiguous surface water resources, and deep ground water of limited quantity and/or quality. When variable conditions occur within or near a mining site, it may be possible to select the locations of some types of waste units (e.g., tailings) to take advantage of the best available subsurface conditions. The incentive for selecting this best site can be provided by regulations which appropriately recognize site-specific conditions in the determination of liner requirements.

The locations of mines also substantially affect the types of materials naturally available to construct mine waste management unit liner systems. Sometimes the site area includes ample natural materials (e.g., clay, sand, gravel, etc.) to construct liner components. In other cases, such natural materials are not available within hundreds of miles. In these cases, alternative (synthetic) liner system components become relatively attractive. These conditions, coupled with significant and continuing advances in the use and technology of synthetic materials during the past several years, have resulted in an increasing number of liner system designs based primarily on man-made "plastic" materials for each component of the liner system. As a result, it will become increasingly important that mine waste unit regulations allow the use of synthetic materials as well as natural clay and other soil materials.

A final introductory observation is that mine operators are increasingly developing mine waste management unit techniques which reduce the potential for leakage through liner systems. Management improvements include:

- Reducing the amount of time that any particular mine waste management unit area may be exposed to liquids which must be contained.
- Reducing the hydraulic head acting on the liner system to reduce the driving force which could cause leakage.
- Using attractive properties of waste materials themselves to reduce the potential for leakage to occur. This can include such applications as:
    - Using predeposited fine-grained wastes (e.g., tails) to create the low hydraulic conductivity portion of the liner system.
    - Using granular portions of waste materials to encourage free drainage to occur so that hydraulic head does not buildup on the low hydraulic conductivity portion of a liner system.
    - Using wastes with basic (high pH) characteristics to buffer those with acid-generating potential.

These types of management practices greatly reduce the inherent short- and long-term risks associated with mine wastes. Therefore, regulations should reflect and encourage good management practices by providing incentives when overall risks are reduced.

The above background discussion suggests that liner regulations for mine waste management units need to be flexible to allow the design for each facility to include those elements necessary to protect water resources based on actual characteristics of the mine wastes, site conditions, and management practices to be employed. Such factors can be considered in the design only when the mine operator is prepared to undertake sufficient investigations to quantitatively predict the degree to which water resources actually will be protected. Since such investigations may not always be warranted, or an operator may simply choose not to undertake such engineering analyses, the regulations also should specify minimum "standards," which assume both unfavorable site conditions and no special management practices.

7.1.1.2 Objectives

The primary objective of this chapter is to provide suggestions for how the liner selection and design requirements should be written into regulations to allow for either:

- Specified standards when site-specific data show a high potential threat to surface and/or ground water, or when site-specific studies are not undertaken by the applicant.
- Flexibility in the degree of liner system redundancy when the potential threat can be shown to be acceptably low due to waste material characteristics, site conditions, and/or the manner in which the mine waste management unit, will be managed.

Secondary objects are to quantitatively compare liner system component performance for various material options and the relative threat to ground water posed by comparable liner systems under a variety of site conditions. The component material evaluations compare different types of natural materials (e.g., clay and silt) and synthetic materials. The site conditions evaluated relate primarily to the relative threat to ground water due to its depth and a range of potential liner imperfections. The relative threat associated with arid versus more temperate climatic conditions, in terms of the moisture content of the waste and underlying soils, also is evaluated. These evaluations form a major part of containment assessment approach to the design of a waste management unit that is outlined in Chapter 3.

7.1.2 INTEGRATION WITH OTHER CHAPTERS

Considerations of liner system requirements relate directly to the following chapters:

- Chapter 3 - Mine Waste Management Requirements: The relationship to this chapter is important for two reasons. Chapter 3 provides a basis for determining how to evaluate the potential threat and impact to beneficial uses of water. The chapter also discusses characteristics of mine wastes that need to be evaluated.

- Chapter 5 - Attenuation of Chemical Constituents: Discussions of attenuation and liner requirements are related for two reasons. In some instances, attenuation characteristics of the unsaturated zone may entirely eliminate the need for an engineered containment system. In other cases, the potential for attenuation to successfully function as redundant backup protection for a liner will depend upon the amount of leakage which may occur from the engineered component.
- Chapter 6 - Climatic Considerations: Especially in arid areas, climate affects decisions regarding liner requirements because the rainfall and evaporation characteristics, together with waste material properties, will determine the amount and characteristics of liquids that must be contained.
- Chapter 8 - Closure Requirements: The types of closure activities for a waste management unit can have a direct relationship on the longevity requirements of individual liner components. For example, if closure will greatly reduce or eliminate infiltration to the waste management unit, the long-term exposure of liquids to the underlying containment system will be diminished, thereby reducing longevity requirements.
- Chapter 9 - Heap Leaching: Liner design and performance are major factors in the design and operation of heap leach facilities. Even if the operation itself is not regulated, performance considerations for containment are similar to those discussed in this chapter. Furthermore, if the heap leach pile is closed in place after leaching has been completed, then the pile becomes a waste and the liner system must then comply with the waste management regulatory requirements.
- Chapter 10 - Ground Water Monitoring: Monitoring systems will often be designed to specifically reflect the configuration of engineered containment used to separate a waste management unit from the unsaturated zone or water resources. Sometimes monitoring may actually be incorporated into the containment system design, for example by measuring the flow volume and quality of any fluids found in the leachate collection system between two liners.

7.1.3 STUDY PARTICIPANTS

- Primary Authors
  - Richard D. Ellison, Ph.D., P.E., Environmental Solutions, Inc.
  - Vincent Ip, P.E., Environmental Solutions, Inc.
  - Kerry Parkinson, P.E., Environmental Solutions, Inc.
  - Donald Shaw, Environmental Solutions, Inc.

7.1.4 SOURCES OF INFORMATION

Section 7.6 provides a bibliography of the sources of information used in the preparation of this chapter. The most important references are:

- The Background Document on Proposed Liner and Leak Detection Rule, prepared by GeoSyntec, Inc. Consulting Services for the EPA, 1987.
- Design, Construction, and Evaluation of Clay Liner in Hazardous Waste Facilities, EPA, 1985.
- 40 CFR Parts 257 and 258, the EPA's Solid Waste Disposal Facility Criteria; Proposed Rule.
- EPA's Criteria for Municipal Solid Waste Landfills, Design Criteria (Draft Background Document) (Subpart D), 1988.
- Designing with Geosynthetics, Robert M. Koerner, 1990.
- Unsaturated Flow Beneath Tailing Impoundments, David M. McWhorter and John D. Nelson, ASCE Geotechnical Division, November 1979.

7.1.5 CHAPTER ORGANIZATION

The remainder of this chapter is divided into the following discussions which provide information on liner system performance in a manner that will support decisions regarding appropriate regulation of liner system requirements for mine waste management units:

- Section 7.2 introduces the types of liner systems which may be employed for mine waste management units where special protection is required to avoid an unacceptable threat to the beneficial uses of waters of the State.
- Section 7.3 describes the types of materials most frequently used for each component of liner systems. Natural soil (clay to gravel) and geosynthetics are discussed for each of the following:
    - Low hydraulic conductivity liners.
    - Cushioning or load-bearing protection layers.
    - Hydraulic head control layers.
    - Leachate control and recovery systems (LCRS).
- Section 7.4 discusses liner system performance as it relates to controlling the potential threat of leakage to water resources. The discussion initially evaluates the potential for fluids to escape from the liner system. Then the rate of fluid leakage within the unsaturated zone is considered in order to provide a means of evaluating the potential adverse effect to beneficial uses of water to be adversely affected.
- Section 7.5 summarizes regulatory implications and provides a recommended approach to the regulation of liner systems.

7.2 TYPES OF LINER SYSTEMS

Design and performance requirements of a liner system are determined by the following:

- Waste Material Characteristics
  - Chemical composition
  - Grain size distribution
  - Hydraulic conductivity
  - Availability of free liquids
- Waste Management Characteristics
  - Slope of liner
  - Height and slope of waste placed on the liner
  - Waste placement method
  - Hydraulic head controls
  - Duration of operation (for portions or all of the facility)
- Site Characteristics
  - Location (including depth) to beneficial water resources
  - Unsaturated zone conditions
  - Climatic conditions

The following sections discuss general liner system performance requirements in relation to the above factors. This section initially introduces types of systems (Figure 7.1) available for consideration on a case-by-case basis. Table 7.1 summarizes the types of materials and/or procedures available for the individual components that comprise a liner system. Section 7.3 discusses materials frequently used for each component.

7.2.1 SINGLE LINER SYSTEMS

Single liner systems are designed to restrict the downward (or outward) seepage of water (leachate) at the base (perimeter) of the mine waste management unit. The potential leakage through the liner will depend on:

- The magnitude of hydraulic head above the liner.
- The thickness and effective hydraulic conductivity of the liner material. The effective hydraulic conductivity is dependent upon the properties of the liner material, plus the size and frequency of holes, cracks, or other discontinuities which may exist in the liner.
- The length of time the hydraulic head is applied to the liner.

These factors (head, liner hydraulic conductivity, and time of application) can be controlled for most mine waste management units and should be considered in designing a project-specific containment system. Materials and design considerations appropriate for low hydraulic conductivity liner components are discussed in Section 7.3.1. Cushion or load-bearing material

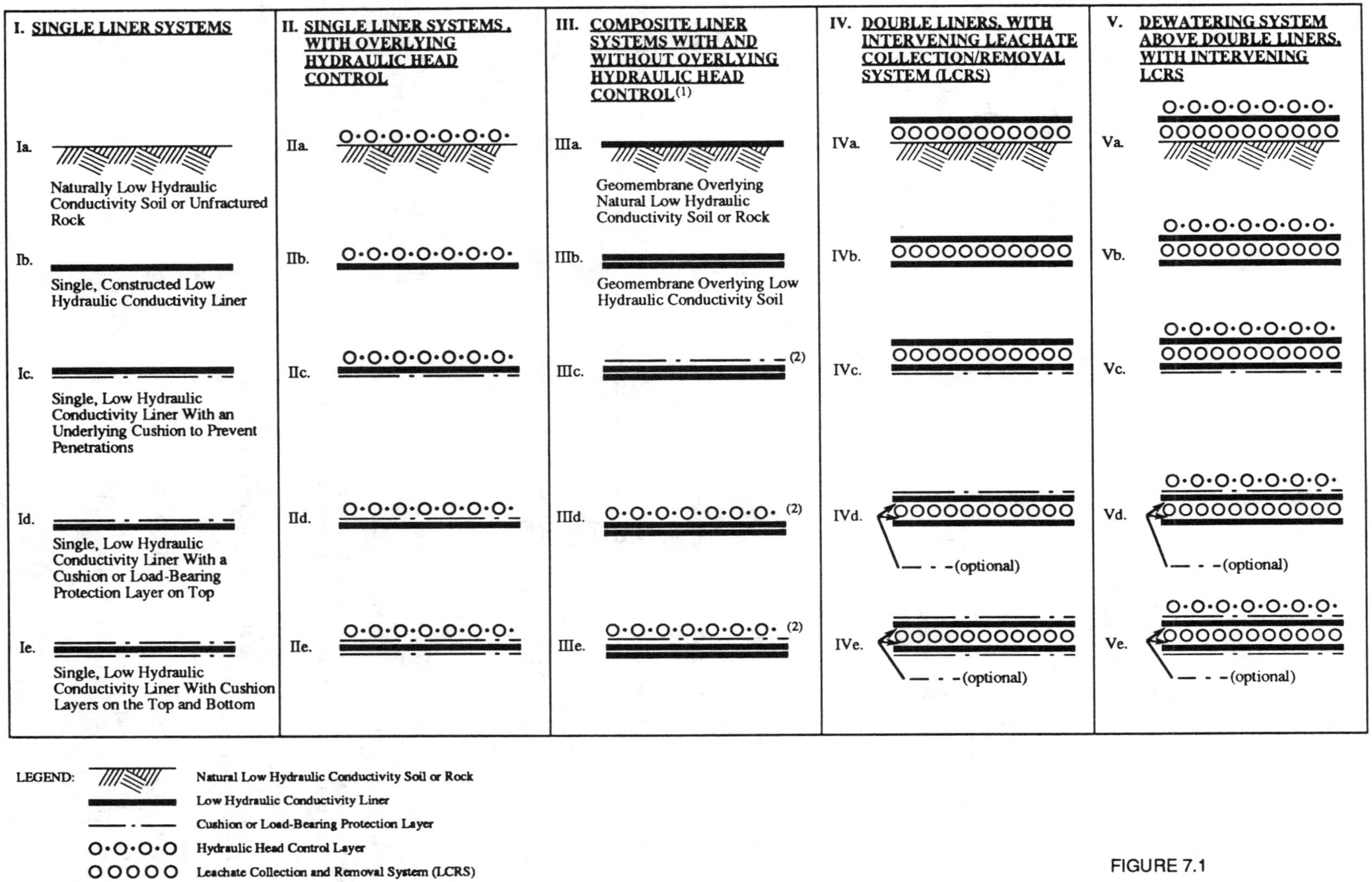

(1) The most common composite liner system involves a geomembrane overlying a low hydraulic conductivity soil.

(2) These composites also include a geomembrane overlying natural low hydraulic conductivity soil or rock.

FIGURE 7.1

TYPES OF LINER SYSTEMS POTENTIALLY APPLICABLE TO MINE WASTES

TABLE 7.1

AVAILABLE MATERIALS OR PROCEDURES FOR LINER SYSTEM COMPONENTS

A. LOW HYDRAULIC CONDUCTIVITY LINERS

A.1 Low Hydraulic Conductivity Natural Soil or Rock - Natural soils or rock may be used as a low hydraulic conductivity liner so long as it is possible to demonstrate by field investigations that the material is uniform over the entire area requiring the liner. This demonstration may be particularly difficult for rock that is extensively fractured.

A.2 Constructed Low Hydraulic Conductivity Liners - Low hydraulic conductivity liners which are constructed beneath a mine waste management unit may consist of any of the following materials:

- Compacted, low hydraulic conductivity soils (e.g., clayey-silt to clay depending upon required hydraulic conductivity).
- Soil and bentonite or cement mixtures.
- Pre-formed flexible membrane liners, made from a variety of available polymer material, generally called geomembranes; varying in thickness from about 20 to 100 mils.
- Field-applied liners, varying from about 80 mils of spray-on asphaltic materials to 6 inches of conventionally placed asphaltic materials.
- Composite liners, consisting of combinations of soil and geomembrane low hydraulic conductivity layers.

A.3 Waste Material - Settled or mechanically placed tailings often have a low hydraulic conductivity and can be used as part of the long-term liner system, provided the tailings serve one of the low hydraulic conductivity liner functions illustrated in Figure 7.1.

B. CUSHION OR LINER PROTECTION MATERIALS

B.1 Geotextile - Synthetic geotextile materials varying in weight from 4 to 20 ounces per square yard may be used above or below geomembranes to protect against penetrations from rock particles due to loads from construction activities or the weight of the waste material. The suitability of a geotextile material varies with its method (density) of fabrication.

B.2 Fine-Grained Soil for Geomembrane Protection - Soils varying from clay to sand can also be used to protect most geomembranes from equipment traffic or static loading of the waste material. Small gravel size material has also been used to protect thick geomembranes. The protective soil must be relatively free of large rock particles which could cause stress concentrations on the liner.

B.3 Coarse-Grained Material for Clay Liner Protection - Gravel protection may also be required for a compacted soil liner, if the liner may be subjected to extreme loads such as impacts from large boulders placed by the end dump method.

C. HYDRAULIC HEAD CONTROL COMPONENTS O·O·O·O

C.1 Free-Draining Gravel Layer - Several inches of free-draining gravel (including coarse sand) are usually adequate to rapidly remove small volumes of leakage. However, thicker layers (8 to 18 inches) are usually placed to facilitate construction. The waste material itself may function for this purpose if the material is granular and relatively free draining.

C.2 Perforated Pipes - Closely spaced perforated pipes can be used to control hydraulic head above a liner. The required spacing is calculated considering the maximum desired head, and the flow rate and hydraulic conductivity of the waste material between the pipes.

C.3 Geocomposite Systems - Composite systems consisting of synthetic drainage associated with geotextile filters have recently been developed for a wide range of drainage control functions. Presently these systems have limited load-carrying capacity, but are routinely being improved for wider use.

D. LEACHATE COLLECTION AND REMOVAL SYSTEMS (LCRS) OOOOO

D.1 Synthetic Geonet Materials - Geonets are net-like polymer products designed to allow high rates of transverse flow. Typical thicknesses of these materials vary from 0.16 to 0.30 inches.

D.2 Free-Draining Gravel Layer (See Item C.1)

which may be required beneath or above the low hydraulic conductivity liner to protect against penetration by projecting rock materials and loading conditions are discussed in Section 7.3.2.

7.2.2 SINGLE LINER SYSTEMS WITH OVERLYING HYDRAULIC HEAD CONTROL

This type of system is similar to the single liner except that the hydraulic head buildup is controlled by a pervious layer above the low hydraulic conductivity liner. To be effective, the pervious layer must be free-draining and direct the liquids away from the liner system. Often, the mine waste material itself is sufficiently porous so that it alone can act as the free-draining overliner material. For finer-grained wastes such as tailings, the free-draining layer must be constructed from imported materials such as those discussed in Section 7.3.3.

In design, control of the hydraulic head can be an important consideration in determining the potential for leakage to occur. Often it is necessary to establish a trade-off between head control and liner hydraulic conductivity, especially at sites where low hydraulic conductivity materials are not readily available.

Figure 7.1 illustrates that cushions or liner protection layers may also be appropriate for this type of system, if required to avoid liner penetrations from underlying conditions or from the free-draining material itself.

7.2.3 COMPOSITE LINERS WITH AND WITHOUT OVERLYING HEAD CONTROL

This type of liner system usually involves the placement of a geomembrane liner directly on top of a low hydraulic conductivity soil or clay layer. To be effective, the two layers must be in close contact so that any leakage through imperfections in the geomembrane liner must also pass through the soil or clay layer as a limited point source. Without this contact, leakage through the geomembrane can spread and hence leak through a large soil or clay area, thus increasing the total leakage through the liner system. Usually in mine waste applications, there is sufficient load placed on the upper geomembrane to assure good contact between the two layers.

As for single liner systems, liner protection and/or hydraulic head control layers can be located above the composite liner system (see Figure 7.1).

7.2.4 DOUBLE LINERS WITH AN INTERVENING LEACHATE COLLECTION AND REMOVAL SYSTEM

This type of system assures a very low hydraulic head on the bottom liner. This is provided by a free-draining LCRS medium above the bottom, low hydraulic conductivity liner, plus a low hydraulic conductivity top liner. LCRS materials are discussed in Section 7.3.4.

Acceptable quantities of leakage through the top liner and into the LCRS should avoid the potential for unacceptable leakage through the bottom liner. Leakage through the bottom liner depends upon the rate at which the LCRS material can remove any leakage which does occur and the hydraulic conductivity of the bottom liner. In many cases, it is possible to remove this leakage so that the head on the bottom liner is just a fraction of an inch in those areas exposed to top liner leakage, and negligible or zero elsewhere. In every case, it is important that the LCRS removal rate be adequate to avoid the buildup of a continuous column of water between the top and bottom impervious liners. Otherwise, the hydraulic pressure above the top liner would be transmitted directly through to the bottom liner, and the double-lined system would have essentially no benefit toward reducing the potential for leakage through the bottom liner as a result of the hydraulic head.

Another important consideration for double liner systems is the manner in which leachate is collected and removed from the LCRS. Sometimes the collection point can be free-draining by gravity. In most cases, however, a collection sump with pumping capacity is required between the two liners. Special considerations for the design of these types of sumps are discussed in Section 7.3.4.

Robertson (1989) reports about a recently developed, patented, double lined system where the primary and secondary membranes are sealed around their perimeters to form independent liner cells. Tubing connected to each of the liner cells allows a vacuum to be applied in the LCRS space. This potentially has technical advantages of: (1) being able to test a system immediately after installation, and (2) creating a suction gradient so that leakage in either the primary or secondary liner is directed into the LCRS. The vacuum can also be used to collect leakage which may occur. This type of arrangement is a technological innovation applied to geomembrane systems.

7.2.5 DRAINAGE SYSTEM ABOVE DOUBLE LINERS, WITH AN INTERVENING LEACHATE COLLECTION AND REMOVAL SYSTEM

This system provides an additional level of assurance that the liner system will not be exposed to excessive hydraulic head. The top, free-draining layer, which can be the waste itself, reduces the hydraulic pressure on the top liner and therefore reduces the potential for leakage to occur into the LCRS between the liners. The reduced leakage into the LCRS, in turn, reduces both the thickness of the film of water which may flow above portions of the bottom liner, and the resulting potential for leakage to occur into the underlying unsaturated zone.

7.3 TYPES OF INDIVIDUAL LINER SYSTEM COMPONENTS

This section discusses alternative materials available for each liner system component illustrated in Figure 7.1 and summarized in Table 7.1. Both natural soil materials (e.g., compacted clay) and various available synthetic membranes are addressed. Consideration of the synthetic materials is particularly important because:

- Many mining sites are located where low hydraulic conductivity clay soils do not exist.
- Mine waste management units often cover very large areas so that the importation of clay soils is impractical and infeasible.
- Large portions of the liner for mine waste management units may remain exposed for long periods as the waste management unit is developed, which maximizes the potential negative aspects of natural soil materials (e.g., desiccation cracking).
- The rate at which synthetic materials are being used is increasing dramatically because of: (1) improvements in material properties for specific applications, (2) improved quality control procedures, and (3) the confidence gained from longer term observation of their successful applications. Figure 7.2 illustrates the estimated growth of geosynthetic material applications in North America (Koerner, 1990).

The use of waste materials for portions of liner systems is also considered. This potential is more appropriate for mining than for most other disposal activities because of the large amounts of soil and rock materials available as a normal function of mining operations.

The discussion is divided into the following topics:

- Section 7.3.1 - Low Hydraulic Conductivity Liners
- Section 7.3.2 - Cushioning or Load-Bearing Protection Layers
- Section 7.3.3 - Hydraulic Head Control Layers
- Section 7.3.4 - Leachate Collection and Removal Systems

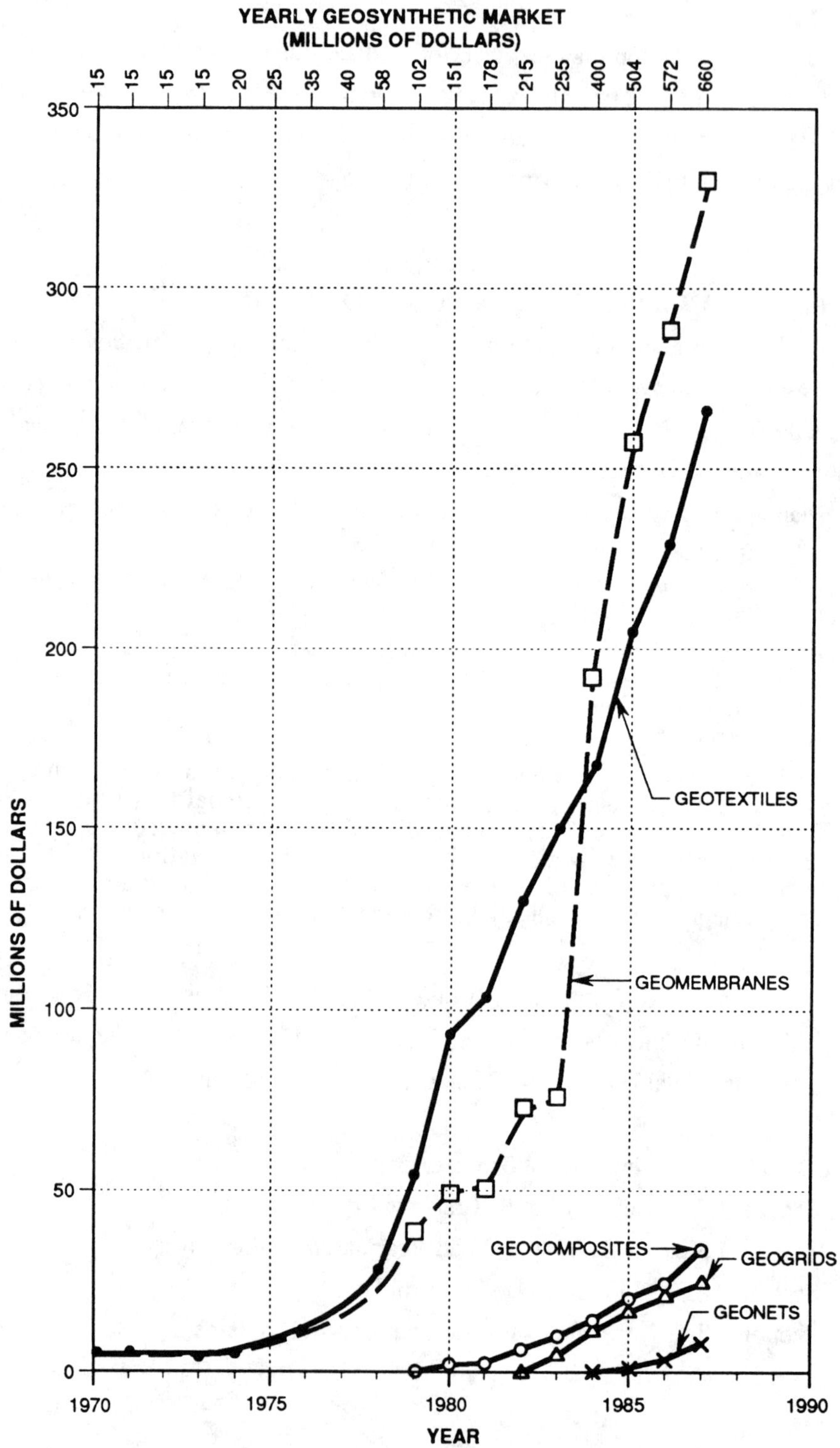

FIGURE 7.2

ESTIMATED GEOSYNTHETIC SALES IN NORTH AMERICA

REFERENCE: ADAPTED KOERNER (1990).

7.3.1 LOW HYDRAULIC CONDUCTIVITY LINERS

Tables 7.2 through 7.4 provide summary data for a variety of low hydraulic conductivity liner materials which can be applied to mine waste management units. The liner types are divided into categories of low hydraulic conductivity soils, geomembranes, field-applied liners, and composite systems. The data in each table are discussed in separate subsections below.

7.3.1.1 Basic Liner Properties

Table 7.2 summarizes typical hydraulic conductivity values for a variety of soil and synthetic materials. Table 7.3 indicates typical physical properties for commonly-used low hydraulic conductivity synthetic liner materials. Table 7.4 qualitatively indicates problems which can occur for each liner type and indicates the potential degree of difficulty that may be realized in attempting to solve each type of problem.

Hydraulic conductivity is the most important property of a liner and is discussed in detail in Section 7.4. The hydraulic conductivity values presented in Table 7.2 are typical for materials without defects from either construction or operational activities (e.g., punctures or seam openings in synthetic materials) or natural causes (e.g., desiccation cracks in soil materials). Table 7.2 includes the conventional clay and geomembrane liners as well as components such as fabric-contained bentonite panels. These parameters may be appropriately used for predicting the "expected" liner behavior under normal conditions if very good quality control is used during construction and if environmental damage is not allowed to occur in unused portions of soil liners. It is also necessary to consider the degree of risk that may be associated with defects which may occur under certain site-specific conditions. The more detailed discussion of liner performance in Section 7.4 considers the rate of leakage and liquid migration for a reasonable range of defects.

Two types of tensile strengths are typically discussed in reference to geomembrane liners:

- Tensile strength at yield.
- Tensile strength at break.

The strengths are similar in value for nonreinforced geomembranes. For reinforced geomembranes (i.e., those which have a scrim between two separate molded thin liners), the tensile strength at yield is higher than tensile strength at break. Tensile strength at yield is the load above which "the material exhibits limiting deviation from the proportionality of stress to strain" (ASTM D 638). Tensile strength at break is the load on the geomembrane at failure. For design purposes, tensile

TABLE 7.2

TYPICAL HYDRAULIC CONDUCTIVITY VALUES

MATERIALS	HYDRAULIC CONDUCTIVITY (cm/sec)
In Situ Rock	1×10^{-8} - 1×10^{-11}
In Situ Soil	1×10^{-5} - 1×10^{-8}
Compacted Clay	1×10^{-6} - 1×10^{-8}
Compacted Silt	1×10^{-5} - 1×10^{-7}
Soil Bentonite Mixture	1×10^{-7} - 1×10^{-9}
Soil Cement Mixture	1×10^{-4} - 1×10^{-7}
Pre-Formed Geomembrane	5×10^{-11} - 5×10^{-14}
Field-Applied Geomembrane	1×10^{-9} - 1×10^{-11}
Fabric-Contained Bentonite Panel	Approximately 10^{9}

TABLE 7.3

TYPICAL PHYSICAL PROPERTIES FOR PRE-FORMED GEOMEMBRANES[(1)(2)]

GEOMEMBRANE[(3)] TYPE	THICKNESS (mils)	BREAKING STRENGTH (lb/in width)	ELONGATION AT BREAK (%)	YIELD STRENGTH (lb/in width)	ELONGATION AT YIELD (%)	STRENGTH AT 100% ELONGATION (lb/in width)	BONDED SEAM STRENGTH (lb/in width)	TEAR RESISTANCE (lb)	PUNCTURE RESISTANCE (lb)	THERMAL EXPANSION COEFFICIENT (cm/cm/°c)
CPE - Reinforced[(4)]	30 - 45	120 - 200	25 - 35	(5)	(5)	Not Relevant	96 - 160	25 - 60	50	Low
CPE - Unreinforced[(4)]	20 - 36	30 - 40	250 - 300			20 - 35	25-30	7 - 10	20 - 40	Not Available
CSPE - Reinforced	30 - 45	60 - 200	20	100 - 200	13	Not Relevant	80 - 180	10 - 70	25 - 65	Low
HDPE[(6)]	40 - 80	165 - 330	700 - 750	95 - 190	13	75 - 120	65-105	30 - 55	50 - 100	$1x10^{-4}$ to $3x10^{-4}$
VLDPE and LDPE	20 - 40	75 - 140	1,000	(7)	(7)	35	Not Available	10 - 20	45- 75	$1x10^{-4}$to $3x10^{-4}$
EIA - Reinforced	30 - 40	300 - 400	Low	(5)	25	Not Relevant	240 - 320	60 - 125	300 - 350	$1.5 x 10^{-5}$
PVC	20 - 45	45 - 100	300	(7)	(7)	29 - 35	38 - 80	5 - 10	20 - 40	

(1) It is likely that different test methods were used for the tensile characteristics and puncture resistance reported in this table. Caution should therefore be exercised when using this information.

(2) Sources:
- Geotechnical Fabric Report, December, 1989
- Standard Number 54, November, 1985
- Gundel Lining Systems Inc.
- Poly-America (Poly-Flex) Reference Manual

(3) CPE = Chlorinated Polyethylene
CSPE = Chlorosufonated Polyethylene (Hypalon)
HDPE = High Density Polyethylene
VLDPE = Very Low Density Polyethylene
LDPE = Low Density Polyethylene
EIA = Ethylene Interpolymer Alloy
PVC = Polyvinyl Chloride

(4) CPE geomembranes are rarely used.
(5) Same as breaking conditions due to stiffness of reinforcement.
(6) The most frequently used HDPE thickness is 60 mils. 80 mils is also used, while 40 mils is finding less application and also presents seaming problems.
(7) No yield point is identifiable.

TABLE 7.4

POTENTIAL PROBLEMS FOR LOW HYDRAULIC CONDUCTIVITY LINERS

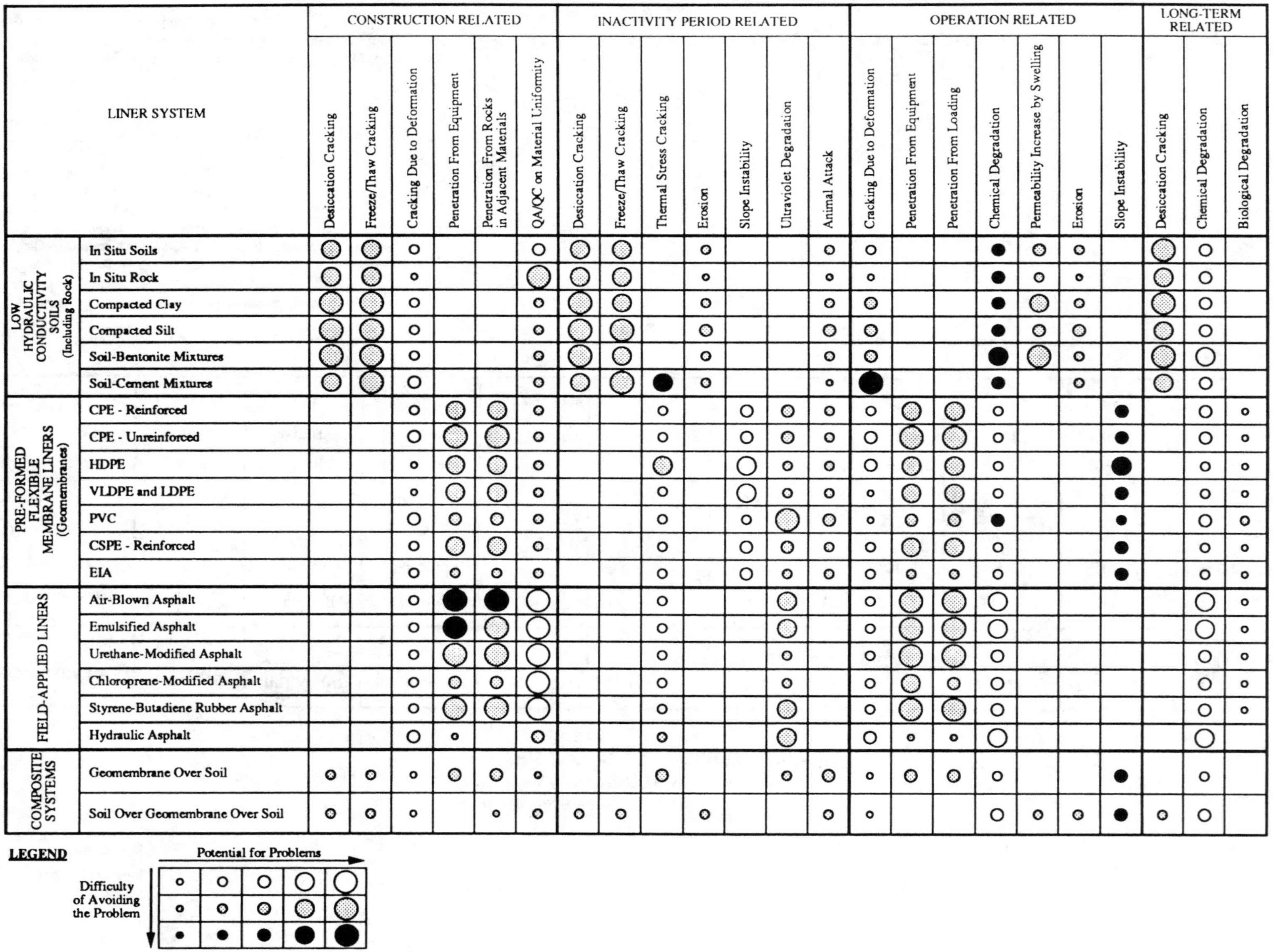

strength at yield should generally be used to assess the geomembrane thickness required. Where the tensile strength at yield is not available, the tensile strength at break with a factor of safety can be used.

The elongation potential is also a consideration at sites or for facility details with high potential for differential movement. Elongation could occur in areas where soft or saturated subsurface materials could experience significant settlement under the load of the wastes, or at locations where liners are anchored or connected to appurtenances such as pipes protruding through the liner system. Generally, unreinforced liners have a greater elongation potential than reinforced liners, which can tear if larger differential displacements occur. The authors have noted that the elongation benefits may not be as effective for impact loadings as for slow stretching with time, especially for the more "rigid" HDPE liner material. In cases of impact loading, tearing may occur before the material has had time to fully elongate.

Soil liner materials with elongation potential are difficult to quantify and are not included in Table 7.3. When large differential movements are expected, soils will be most flexible if they are compacted at a water content slightly above optimum. Even then, compacted soil will not adjust to large differential movements nearly as well as will synthetic materials.

Thermal expansion and contraction properties of the various liner materials can be especially important for portions of lined waste management units that will be exposed for significant periods. Of the geomembrane materials, thermal expansion is most important for HDPE. Because of its relative rigidity, the high thermal expansion properties of unloaded HDPE cause liner bulging to occur during warmer daytime periods, but the liner becomes taut during periods of cooler temperatures. This effect is especially evident in high desert areas where summer high temperatures exceed 100° F and winter cold periods are well below freezing. Although HDPE has been successfully used for many projects with large temperature ranges, care must be taken to assure that the material properties and installation procedures will accommodate the anticipated temperature ranges. The combined action of thermal contraction and increase in geomembrane brittleness due to cold temperature is known to have caused shattering failures in HDPE geomembranes. Expansion and contraction should therefore always be a design consideration in locations of known thermal extremes.

7.3.1.2 Potential Problems With Low Hydraulic Conductivity Liners

Table 7.4 qualitatively illustrates potential problems which may be realized with various low hydraulic conductivity liners. The size of the circle in the matrix indicates the relative potential for the identified problem to adversely affect liner performance. The darkness of shading in each circle indicates the degree to which special efforts may be required to resolve the difficulty. Each of these relationships will not be discussed in this chapter. However, several general trends are noted, primarily to: (1) illustrate the most important areas for special attention to soil and synthetic materials, and (2) indicate how problem conditions can be reduced by using composite liners when appropriate materials are available.

The most consistent potential problem with low hydraulic conductivity soils is associated cracking, from the effects of either desiccation (drying) or freeze/thaw cycles. Desiccation cracking can occur up to several feet deep with a resulting substantial increase in material hydraulic conductivity. In some materials, cracking will partially or completely close when subjected to water, but this characteristic is not sufficiently predictable to rely upon resealing to achieve the original hydraulic conductivity. Instead, desiccation cracking should be avoided by providing a cover material above the liner which will not allow moisture to escape, or by frequently wetting the liner surface. Frequent wetting usually is not attractive because of quality assurance/quality control (QA/QC) difficulty and related operating costs.

Freeze/thaw cracking is more difficult (costly) to protect against because a greater cover thickness is required. For this reason, compacted soil liners may not be appropriate when large portions of liner will be left exposed throughout freezing winter conditions.

Additional important potential problems with low hydraulic conductivity soil/rock liners are:

- QA/QC verification of liner continuity for natural, in situ soil and rock liner conditions. This is particularly important for rock formations that may be fractured due to factors such as stress-releasing tectonic folding, differential weathering, etc.
- Soil and rock hydraulic conductivity can be unaffected or greatly increased due to the chemistry of the liquid to be contained. High pH liquids or liquids which contain high levels of dissolved salts which could exchange with ions in the soil, can cause hydraulic conductivity increases in soil liners. The most severe potential impacts are associated with high concentrations of organic solvents. However, this is not generally important, as most mine waste management units are not exposed to organics.

The most important potential problems for most of the geomembrane liners are associated with the possibility for penetration of the liner as a result of construction activities or waste material loading. These types of problems are most easily mitigated by the provision of cushioning (or protection layers) below and/or above the synthetic material to reduce the effects of concentrated loading conditions, as discussed in Section 7.3.2.

Polyvinyl chloride (PVC) liners also can be subjected to accelerated deterioration if exposed for long periods to sunlight or to certain organic chemicals. The sunlight exposure problem can be mitigated by covering the PVC, often with the same material used for cushioning, provided this material is not also subject to ultraviolet degradation. The chemical effect is not severe for most mining applications.

QA/QC controls have traditionally represented a significant problem with synthetic liners, particularly at field seams. This potential problem is not shown with large circles in Table 7.4, however, because substantial improvements in both construction and QA/QC procedures during the past several years have greatly improved seam reliability. Presently, installations do not suffer from severe potential seam opening problems when reliable workmanship and adequate quality assurance are used for liner installation.

Relatively thin, field-applied liners generally have the same potential for penetration problems as the pre-formed geomembranes. An exception is hydraulically-applied asphalt which has a greater thickness, on the order of 6 to 8 inches. QA/QC procedures for field-applied liners are more difficult, however, because verification of acceptable applications requires consideration of thickness and mixture consistency at all locations. Field-applied liners are not frequently used for large mine waste management units.

Composite systems, which consist of a combination of compacted soil and a geomembrane, have less potential for either penetration or cracking problems than either material alone because the dissimilarities in the two provide redundancy. The system least likely to suffer from potential penetrations is one in which a geomembrane is sandwiched between two compacted soil layers. This type of system can only be economically feasible, however, when the low hydraulic conductivity soils are readily available and a reasonably thin geomembrane is utilized. Construction of this liner system is likely to require special compaction methods for the upper soil liner.

7.3.2 CUSHIONING OR LOAD-BEARING PROTECTION LAYERS

As discussed in Section 7.3.1.2, thin geomembrane liners are susceptible to being penetrated by sharp objects, especially under heavy loads. This penetration can occur as a result of rock particles in the subsurface material, or from sharp objects or rock particles above the liner.

Holes from base irregularities can occur during construction (e.g., when workers step on a portion of liner directly above a sharp protruding rock) or due to stresses when the overlying waste load is applied. The two most frequent procedures for protecting against such penetrations are to:

- Assure the subbase soil is smooth, either by conditioning (e.g., removing rocks and/or compacting the soil) the natural soils, or by the placement of a supporting natural or screened clayey or silty soil. If possible, it is desirable that the base soil have some degree of cohesion so, that when it is compacted to a dense and stable condition, it will not shift after the liner is placed.
- Utilize a layer of geotextile to act as a cushion between the liner and the subsurface.

Geotextiles are pervious fabrics, varying in thickness typically from 100 to 200 mils. They were originally developed in the 1950s to function as filters between layers where flow occurred from small grain size soils into seepage collection and transport materials with higher hydraulic conductivity. Geotextiles now have a much wider variety of applications in geotechnical engineering.

Cushioning above a geomembrane can also be provided by a quality geotextile liner, or by a carefully placed soil or gravel layer. Laboratory liner penetration tests conducted by Environmental Solutions, Inc. (1989) have shown that, for liner applications, a high quality 110-mil thick (12-ounce per square foot) needle-punched nonwoven geotextile will provide much greater resistance to penetrations than any of the available geomembrane liners, with the geotextile being either below or above the geomembrane. Examples from the many tests conducted are shown in Figure 7.3 and illustrate that:

- With a geotextile, the penetration resistance is similar regardless of liner type.
- The benefit is highly dependent upon the characteristic of the geotextile. Nonwoven type geotextiles with lower masses per unit area did not exhibit good penetration resistance properties and provided much less strength to the geomembrane materials than did those with higher masses per unit area.
- The added penetration resistance from using a 110-mil (12-ounce per square foot) nonwoven geotextile varies from about three to four times the penetration resistance of 20- to 40-mil PVC and two and one half to three times the penetration resistance of 40- and 60-mil HDPE.

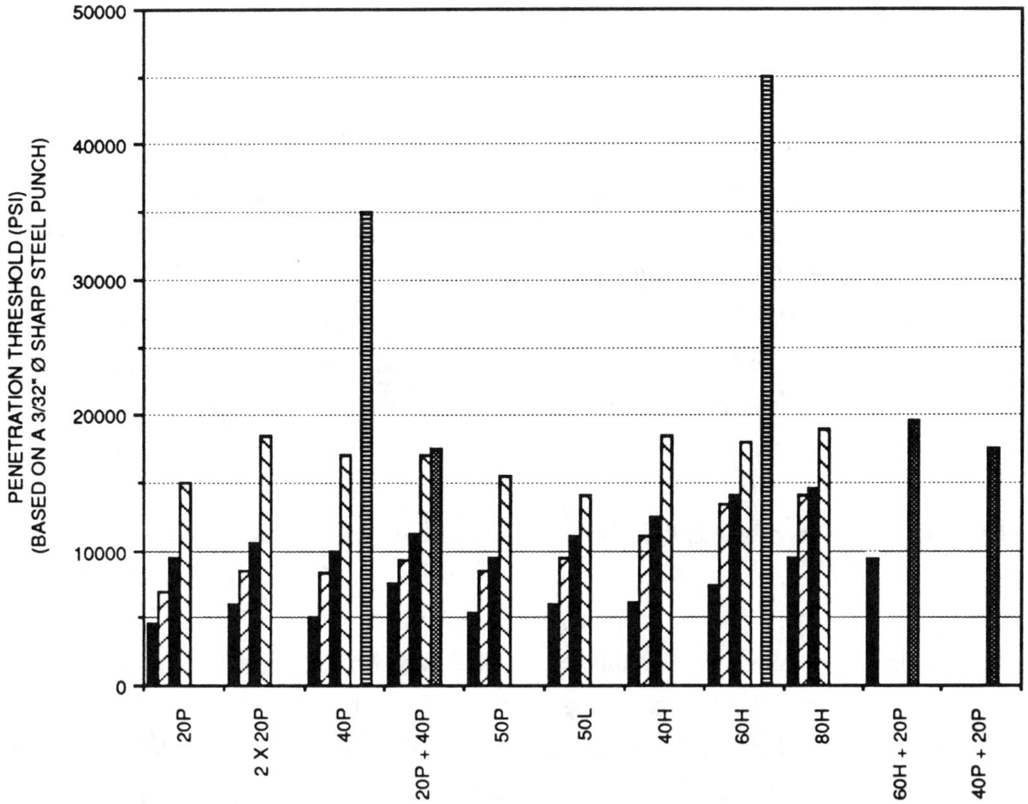

LINER TYPES

20P = 20-MIL PVC
40P = 40-MIL PVC
50P = 50-MIL PVC
50L = 50-MIL VLDPE
40H = 40-MIL HDPE
60H = 60-MIL HDPE
80H = 80-MIL HDPE

PROTECTION TYPES

- LINER ALONE
- 40-MIL LOW DENSITY WOVEN GEOTEXTILE ON TOP
- 100-MIL LOW DENSITY WOVEN GEOTEXTILE ON TOP
- 110-MIL NONWOVEN GEOTEXTILE ON TOP
- 110-MIL NONWOVEN GEOTEXTILE ON BOTTOM
- 2 LAYERS OF 110-MIL NONWOVEN GEOTEXTILE ON TOP

NOTE: THESE RESULTS ARE EXAMPLES OF MANY LABORATORY TESTS CONDUCTED USING VARIABLE TYPES OF PUNCHES AND SUBBASE STIFFNESS CHARACTERISTICS.

REFERENCE: ENVIRONMENTAL SOLUTIONS, INC. (1989).

FIGURE 7.3

COMPARATIVE LINER PENETRATION STUDY

- The resistance was approximately doubled for tests using two 110-mil geotextile layers, indicating an approximate linear relationship between penetration resistance and thickness of the cushioning geotextile.

Sergent, Hauskins and Beckwith (SHB, 1988) conducted a series of large diameter laboratory tests to evaluate the height to which heap leach piles could be safety placed over a 40-mil PVC liner (protected with nonwoven geotextile) for the Mesquite gold mine in Imperial County, California. These tests were conducted in 7- by 8-inch cells with actual subbase soil placed beneath the liner and representative samples of the crushed ore above the liner and geotextile cushion. The results showed that, for 2-inch sized crushed ore, the pile height could extend to a height of 150 feet on a 12-ounce geotextile protection layer, with no indications of even pinhole penetrations in the PVC liner.

The cushioning benefits of geotextiles for protecting geomembranes have also been demonstrated by Koerner (1990) and EPA (1988d). Koerner shows the puncturing resistance of an unreinforced membrane increases by the addition of a geotextile behind, in front of, or on both sides of a geomembrane. The EPA test work conducted by the Southwest Research Institute, also concludes that geomembrane puncture resistance performance can be increased when a continuous filament, nonwoven, needle-punched geotextile is placed between the geomembrane and the supporting subgrade. The data presented by both Koerner and the EPA compare favorably with the work performed by Environmental Solutions, Inc. (1989).

When soil is used as the cushioning layer above a geomembrane, it is mandatory that the construction process be carefully planned and monitored to assure that the moving equipment does not damage the geomembrane. Special precautions should include:

- Assurance that the equipment will always be working sufficiently far from the end of the soil protection so that accidental tearing of the liner cannot occur.
- Controls to avoid rapid turning of the equipment tires or treads, which could cause large soil displacements and resulting stresses and tearing of the geomembrane.
- Routing of trucks or scrapers in changing pathways over the cushion layer to avoid "pumping" of the protection soil or underlying support soil, which could lead to large deformations and the potential for geomembrane tearing.

Generally, cushioning layers of soil with thicknesses of 18 inches or more are used. However, cushioning layers as thin as 12 inches have been successfully used for protection from normal construction equipment that is operated carefully. At 12 inches, however, continuous observation of the operation is required to assure that procedures do not damage the liner.

The cushion material is usually specified to be rounded, with particles no larger than sand-size passing a 1/4-inch sieve. However, a number of heap leach gold mines in Nevada have successfully used crushed gravel (road base sized material) above 60- and 80-mil HDPE to act as both a cushion and free-drainage layer. Special, field demonstration tests may be justified if angular crushed gravel is to be placed directly above the liner. This can usually be accomplished in less than one day by making a small (e.g., 50 x 50 foot) test plot over which the gravel is placed in a manner similar to that which will be used for production. The test should also include a specified number of loaded truck passes to simulate construction conditions.

If relatively low hydraulic conductivity (e.g., less than 10^{-5} cm/sec) soil is available, its use as the cushioning layer above a geomembrane can provide an added advantage of retarding leakage of liquids which might otherwise reach holes in the liner as a result of fabrication and/or installation. This procedure has been successfully used for the Chimney Creek gold mine in Nevada (Environmental Solutions, Inc. 1987a). For that application, a 20-mil PVC liner is sandwiched between a bottom 6-inch layer and top 12-inch layer of compacted natural silt which exists over most of the site area. Initially, a 100- by 100-foot test area was built to evaluate the constructability of this composite system using 6-, 20-, and 40-mil membrane thicknesses. Several small holes occurred in the 6-mil material; both the 20- and 40-mil liners functioned with no damage when exposed at numerous locations. Subsequently, the composite system has been successfully installed for several hundred acres of liner.

Sometimes substantially more cushioning is required if run-of-mine rather than crushed rock is being deposited on the liner system, using an end dumping placement method. In these instances, large boulders can roll down the end of the dump face and cause large dynamic loading to the liner system. Successful applications have used between 2 and 5 feet of soil as cushioning for this purpose.

7.3.3 HYDRAULIC HEAD CONTROL LAYERS

Provisions to control hydraulic head above a low hydraulic conductivity liner are important in minimizing the potential for excessive leakage. This is accomplished by providing a free draining system above the liner, with the primary function of removing liquids without the need to establish a large hydraulic gradient. Table 7.1 provides descriptions of three basic types of hydraulic head control systems commonly used in mine waste management practice: free-draining gravel, perforated pipes, and geocomposite systems. The rate at which liquids are removed from any of these systems depends on: (1) the system's hydraulic conductivity, (2) the slope of the waste management unit base, and (3) the volume of liquid involved.

A free-draining gravel layer provides a "blanket" flow system directly above the liner. For most applications, flow through the gravel occurs as a very thin sheet (e.g., a fraction of an inch) directly above the underlying low hydraulic conductivity liner). The thickness of the gravel layer usually varies from 12 to 18 inches, depending on requirements for construction and protection of the underlying materials. If necessary, an 8-inch layer can be used very successfully both from a placement and functional perspective provided special precautions are taken during construction (Environmental Solutions, Inc., 1987b). The hydraulic conductivity for gravel varies from 1.0 to 100 cm/sec (Terzaghi and Peck, 1967), being higher for uniform size gravels than for well-graded gravels. Waste materials with similar hydraulic conductivity which are placed directly on a liner can be relied upon to maintain a low hydraulic head, without the need for a layer of gravel.

Protection of geomembranes from being damaged by a free-draining gravel layer may be required in some instances and can be accomplished by using geomembrane that is thicker than otherwise required, or by placing a cushion layer of finer grained soil or a geotextile between the gravel and geomembrane. The benefit and effectiveness of using geotextile as a cushioning layer is discussed in Section 7.3.2.

With the development of inexpensive, corrugated HDPE pipe during the past decade, perforated pipe systems have more frequently been used in mine waste management practice to control the hydraulic head above liners. A pipe system usually consists of perforated pipes placed above the liner (or cushioning layer) at relatively close spacing. Spacing of the pipes is estimated by considering the maximum desired hydraulic head, the anticipated flow rate, and the hydraulic conductivity of the waste materials between the pipes. The maximum hydraulic head will occur midway between pipes, while the hydraulic head in each pipe will be approximately at the level of the perforations.

The type of pipe used is important because it must withstand the load of the waste material, be resistant to chemical degradation, and be economical. The corrugated plastic pipe usually satisfies each of these criteria. The load-carrying capacity is achieved because the pipe easily deforms so that the waste material "arches" around the pipe without excessively closing the pipe opening.

Other types of pipes also may be appropriate, depending on site-specific design requirements. In each case, however, calculations or tests should be done to verify that the selected pipe will not collapse.

Geocomposite systems consist of a variety of combinations of pervious "geonets" and geotextiles. The geotextile is placed against the waste to act as a filter to limit solids migration, and the very pervious "geonet" acts as the medium to transport the fluid away without excessive hydraulic buildup. This type of system is being used increasingly for LCRS (Section 7.3.4), but geocomposites are not widely used directly beneath waste materials due to their limited resistance to damage by concentrated loads. However, because of their low cost and ease of installation, increasing use of geocomposite systems for all drainage applications can be expected in the future, as materials are developed to accomplish required engineering functions.

7.3.4 LEACHATE COLLECTION AND REMOVAL SYSTEMS

The function of an LCRS is similar to that explained in Section 7.3.3 for the hydraulic head control layer. However, instead of controlling the liquids above the low hydraulic conductivity liner systems, the primary function of an LCRS is to detect and collect liquids that may pass through a top, low hydraulic conductivity liner. As shown in Figure 7.1, an LCRS is placed immediately between two low hydraulic conductivity liners. Therefore, liquids that pass through the top liner can be removed with only a fraction of an inch of hydraulic head buildup on the lower liner.

An LCRS usually consists of the following components:

- The drainage layer itself.
- A sump, or depression, for collecting the liquids.
- A liquids removal system, either mechanical or by gravity.

The drainage layer of the LCRS can be constructed of either granular (sand or gravel) or synthetic materials. Sand has the advantage of providing cushioning protection for the low hydraulic conductivity liners, but has the disadvantage of not being completely free draining. Gravel will provide free drainage, but may require thicker geomembranes or protection layers to avoid penetration damage. A combination of sand and perforated pipes can represent a compromise where the capacity of the sand layer is not adequate. The use of limestone sands or gravels should be evaluated carefully to assure that precipitation of calcium will not cause clogging of the system, or that the material will not be excessively dissolved by acidic liquid.

The use of synthetic materials (e.g., geonets) as the LCRS drainage layer can be very attractive because of its high flow capacity, ease of installation, and QA/QC. Geonets are especially functional as the LCRS for ponds, which have high leakage potential but low loading conditions. Geonets may have limitations beneath high waste piles and high tailings because of limited load-carrying capacity.

In the early 1980s, thick geotextiles were sometimes used as the LCRS drainage medium between two low hydraulic conductivity layers. This material has the advantage of being inexpensive and capable of carrying large loads. Its use is often discouraged, however, because it is very compressible and, as a result, its flow-carrying capacity under load is very small, so that a larger hydraulic head is required to remove liquids.

In the near future, geocomposites, consisting of both a geotextile and geonet may be suitable for many installations to take advantage of the beneficial properties of both materials. If the LCRS is between two geomembranes, a geonet alone will have a greater flow-carrying capacity than a geonet bonded to one or two geotextiles to form a geocomposite. As shown by Williams, et al. (1984), the presence of a geotextile in contact with a geonet decreases the hydraulic conductivity significantly where geotextile has to be used between a geomembrane and a geonet in order to:

- Protect the geomembrane from damage by the geonet under load; and/or
- Increase interface shear strength (friction).

Sumps or depressions are at the lowest location of the LCRS, to allow free draining of liquids by gravity. The sumps should be sized appropriately to prevent liquids from backing up into the drainage layer. If this is accomplished, it often will not be necessary for flow in the LCRS to be zero, because the hydraulic head from small amounts of flow will be only a fraction of an inch above the underlying low hydraulic conductivity liner. As discussed in Section 7.4, this type of condition poses essentially no threat to underlying ground water uses at many sites.

Removal of leachate from a sump can be accomplished by a pump located in the sump bottom and connected to riser pipes which are either vertical or riser pipes located between the two liners. In some instances it may be possible to gravity drain a sump to a low spot located outside the waste management unit. In order to be most effective, the sump must usually be filled with a very pervious material (e.g., crushed rock) so that liquids can be removed without a large hydraulic head buildup within the sump itself. Otherwise, the small sump area may require an additional liner with LCRS to avoid the potential for excessive seepage through the sump bottom.

The pumping arrangement for an LCRS sump collection system can be used as an important monitoring element to evaluate performance of the overlying low hydraulic conductivity liner. The pump can be activated by level switches in order to frequently remove liquids entering the sump. The resulting flow should be passed through a monitoring device so that an average rate can be measured. If necessary, permitting requirements can place limits on the LCRS flow to require operating restrictions and/or repairs at certain trigger flows, determined on the basis of liner properties or a containment assessment.

7.4 LINER SYSTEM PERFORMANCE

7.4.1 GENERAL

Liner systems installed for mine waste management units must function adequately under the various physical and chemical conditions they are exposed to during construction and operation, and for the conditions which will exist at closure. Adequacy is determined in this instance by the degree to which the liner system prevents leakage which could threaten beneficial uses of water resources. The following sections discuss procedures for evaluating liner system performance:

- Section 7.4.2 - Leakage Potential
- Section 7.4.3 - Rate of Leakage to the Unsaturated Zone
- Section 7.4.4 - Liner Stability Considerations

7.4.2 LEAKAGE POTENTIAL

A key question that must be addressed in the design of any liner containment system is related to the acceptable level of risk of some liquids leaking from the waste management unit. If the waste will not create any undesirable constituents in the liquids, no containment is required, and it is acceptable for the liquids to percolate directly to the ground water. However, if liquid that contains high concentrations of toxic constituents is to be stored for long periods, it is necessary to minimize the potential for liquid to leak from the containment system. Both ends of the spectrum represent relatively simple decision-making conditions.

The vast range of conditions between these two extremes presents the more challenging technical and regulatory requirements of determining the level of redundancy appropriate to each potential risk. The rational method for evaluating the risk is to: (1) examine the potential for leakage to occur from containment, (2) determine the potential pathways of travel of fluid leakage toward beneficial water resources, and (3) evaluate the resulting potential for impacts to water use. This section evaluates the potential for leakage to occur from liner systems. The next section discusses the potential fate of that liquid in the unsaturated zone.

This section evaluates the potential for leakage to occur from liner systems. The next section discusses the potential fate of that liquid in the unsaturated zone.

The volume of leakage that may escape from a liner system will be dependent upon the following parameters:

- The height of liquid or water above the overlying liner component.
- The hydraulic conductivity of the continuous liner material.
- The size and spacing of defects (e.g., holes, slits, cracks) in the liner.
- The hydraulic conductivity, porosity, and natural water content of the soil or rock underlying the liner.

The height of water (hydraulic head) above the liner can usually be controlled by waste management unit design, facility operation, and/or facility management practices. For example, if a free-draining material (Section 7.3.3) is provided above a liner system, the hydraulic head usually can be maintained well below 1 foot and often to less than an inch (0.08 foot). Further, in double-liner systems with a free-draining LCRS (Section 7.3.4), the flow can be controlled so that the average hydraulic head above the underlying liner will be only a fraction of an inch (e.g., 0.01 foot).

Therefore, in order to begin an evaluation of potential risks from liner system leakage, it is necessary to initially compare the effects of a reasonable range of hydraulic heads (e.g., 10 feet, 1 foot, and 0.1 foot). Table 7.2 shows the hydraulic conductivity for a variety of liner materials. With no defects, compacted clay liners usually fall within the 10^{-6} to 10^{-8} cm/sec hydraulic conductivity category. Synthetic materials with no defects, and composite liners, have equivalent hydraulic conductivities which are at least several orders of magnitude lower than clay.

To compare a basic performance of various liners, Table 7.5 summarizes the leakage rates and travel times for water to pass through various clay, geomembrane, and composite liners. The leakage rates for the clay liner have been calculated both for the initial (maximum) conditions during the wetting phase when the soil moisture suction below the wetting front can be fairly high, and for the long-term steady state (minimum) conditions which assume negligible suction in and below the liner.

The following equations, based on the model developed by McWhorter and Nelson (1979) for one dimensional downward seepage, are used to estimate the leakage rate and travel times for saturated and partially saturated wetted fronts through and below a liner.

TABLE 7.5

ESTIMATED LEAKAGE AND TRAVEL TIMES THROUGH CLAY LINERS AND PERMEATION THROUGH GEOMEMBRANE AND COMPOSITE LINERS

LINER	LINER CONDITION		HYDRAULIC HEAD (ft)	ESTIMATED STEADY STATE LEAKAGE VALUE (gpad) (1)		TRAVEL TIME OF WETTED FRONT TO BASE OF LINER (years) (4)
	THICKNESS (ft)	HYDRAULIC CONDUCTIVITY (cm/sec)		MINIMUM (2)	MAXIMUM (3)	
CLAY	1	1×10^{-7}	0.1 1.0 10.0	100 180 890	230 310 1,110	1 .9 .3
CLAY	2	1×10^{-7}	0.1 1.0 10.0	90 130 510	160 200 600	3 3 1
CLAY	2	1×10^{-6}	0.1 1.0 10.0	930 1,280 3,500	1,580 1,980 6,000	.3 .3 .1
CLAY	10	1×10^{-6}	0.1 1.0 10.0	900 970 1,620	1,030 1,100 1,910	2 2 1
40-mil HDPE GEOMEMBRANE, AND A GEOMEMBRANE AND LOW HYDRAULIC CONDUCTIVITY SOIL COMPOSITE (5)			0.1 1.0 10.0	0.0015 0.1 7.0		NA

(1) Gallons per acre per day.

(2) The minimum steady state value corresponds to negligible pore water suction.

(3) The maximum steady state value corresponds to a pore water pressure of -1.4 feet. This value was derived assuming the pore water pressure (h_f) is equal to the residual water content pressure (h_d) and using the equation (McWhorter and Nelson, 1979) of:

$$h_d = \frac{(1 + 3\lambda)}{(2 + 3\lambda)} h_c$$

A value of 50 cm was used for h_c (Table 7.11).

(4) Correspond to the maximum leakage rates.

(5) Flow rates were adopted from Table 2.2-8 of Background Document on Proposed Liner and Leak Detection Rules by GeoSyntec, 1987. Values apply to permeation through the geomembrane and do not take into account leakage through any imperfections. Due to low rates of permeation influence of soil hydraulic conductivity has been assumed to be negligible.

The unit area leakage through clay liner overlain by waste with a low hydraulic conductivity is estimated by the following equation:

$$q = k_\ell \left(\frac{D_t + D_\ell - h_f}{D_\ell + D_t \frac{k_\ell}{k_t}} \right)$$

where

q = leakage rate (length/time)
k_ℓ = hydraulic conductivity of the liner (length/time)
D_ℓ = thickness of liner (length)
k_t = hydraulic conductivity of waste (length/time)
D_t = thickness of saturated waste (length)
h_f = soil pore-water pressure below liner (length)

For a clay liner which is not overlain by waste the equation is as follows:

$$q = k_\ell \left(\frac{H + D_\ell - h_f}{D_\ell} \right)$$

where

H = depth of ponded liquid or hydraulic head above the top of the liner (length)

The travel time for the saturated seepage time through a liner takes the form of:

$$T = \frac{(n - \varnothing_i)}{k_\ell} \left[D_\ell - (H - h_c) \ln \left(\frac{2 D_\ell + H - h_f}{D_\ell + H - h_f} \right) \right] \quad \text{(Saturated Wetted Front)}$$

where

n = effective porosity of the material underlying the liner (dimensionless)
$\varnothing_i$ = initial in situ volumetric moisture content below the wetting front (dimensionless)

Saturated seepage conditions will occur below the liner if q is greater than k_f, the saturated hydraulic conductivity of the subsoils. If this occurs, the liner seepage rate should be recalculated incorporating the subsoil layer and using the following equation:

$$q = k_f \left(\frac{H + D_f + D_\ell}{D_f + D_\ell \frac{k_f}{k_\ell}} \right)$$

The travel time, T, from the bottom of the liner to the edge of the seepage or wetted front below the liner can be evaluated as follows:

$$T = \frac{(n - \varnothing_i)}{k_f}\left[L + \left(D_t\frac{k_f}{k_t} + D_\ell \frac{k_f}{k_\ell} - D_t - D_\ell + h_c\right) \ln\left(\frac{L + D_t + D_\ell - h_f}{D_t + D_\ell - h_f}\right)\right]$$

(Saturated Wetted Front)

$$T = \frac{L}{q}\left[(n - \varnothing_r)\left(\frac{q}{k_f}\right)^{\frac{\lambda}{(2 + 3\lambda)}} + (\varnothing_r - \varnothing_i)\right]$$

(Partially Saturated Wetted Front)

where

$\varnothing_r$ = residual volumetric moisture content above the wetting front, approximately equal to the specific retention (dimensionless)

k_f = saturated hydraulic conductivity of the subsoils (length/time)

L = depth of the leading edge of the wetted front below the liner (length). Maximum value of L is equal to total depth of unsaturated zone, D_f.

λ = pore size distribution index (dimensionless; typically between 1 and 3)

The estimates assume that the clay is continuous, without defects such as desiccation cracks. Such defects would, of course, allow greater leakage. Based on the results presented in Table 7.5, it is reasonable to conclude that composite liners without defects, pose much less risk than compacted soil liners.

These data show that geomembrane liners without defects can perform very well for controlling leakage, especially when a low overlying hydraulic head is maintained. The main concern with geomembrane liners is associated with the potential for defects to exist due to punctures, inadequately sealed seams, or inadequate seals at penetrations, such as pipes. The major means by which such defects can be minimized are by:

- Implementing continuous QA/QC procedures during construction for both the quality of seams and the general surface of the liner.
- Providing adequate protection to the geomembrane to avoid loading conditions that could create defects during construction and operational activities.
- Providing redundancy; for example, in the form of a composite liner, or additional geomembrane liner at critical locations.

Experience by the authors of this chapter has been that the risk of significant seam defects can be greatly reduced by careful testing of every inch of the seam. Seams should initially be tested by the contractor, and any areas of leakage should be repaired. Then the final seam should be completely checked again by QA/QC personnel independent of the contractor. Field seams should also be arranged to overlap in the downslope direction. For LCRS flows with small hydraulic head, this will reduce the tendency for water to enter an incompletely sealed seam.

Holes in the liner itself are minimized by requiring: (1) installation personnel to avoid walking on the geomembrane except as absolutely necessary, and (2) all personnel working near the installation to wear flat soled shoes with minimum potential for causing damage to the geomembrane. Before placing the top cushion or protection layer over the liner, the entire surface should be inspected for any holes of visual size, which can be appropriately patched. In this way, only very small pinholes, not visible to the naked eye, are potentially remaining in the geomembrane.

As a basic design goal, it may be appropriate to assume that QA/QC procedures will be accomplished adequately to eliminate all significant defects. With a geomembrane liner, this would essentially eliminate the potential for leakage to occur. However, a worst-case risk assessment must consider that some defects at seams or holes in the geomembrane may exist.

In preparing backup documentation for proposed liner and leak detection rules for the EPA, GeoSyntec (formerly GeoServices, 1987) developed several scenarios concerning the size and spacing of holes which might be expected to occur in geomembranes. Tentative conclusions were that with reasonably good installation, adequate quality assurance, and repair of noted defects, one defect can be expected per 1,000 feet of seam. Even better results, up to one defect per 3,000 feet of seam, may be possible with the very best installation procedures and intensive quality assurance. If 20- to 30-foot wide liner panels are used, the reasonably good installation case would result in about one to two seam defects per acre. A high level of QA/QC should result in less than one seam defect per acre. For evaluation purposes, GeoSyntec used an arbitrary frequency of one seam defect per acre for estimating leakage rate potentials.

The study also considered estimates of hole sizes and determined that a hole diameter of 0.04 to 0.08 inch is appropriate for best available demonstrated installation techniques. As a conservative standard for performing leakage detection capacity calculations, the study recommended the use of a "standard" hole of 0.445 inch in diameter (0.16 inch^2) at a frequency of one per acre. Experience of the authors is that this standard hole is conservative for geomembrane installations with appropriate quality control procedures. Such holes would normally be found by seam testing and can be easily observed by visual inspection of the liner surface. For this study, it is concluded that one hole per acre, with a diameter of 0.08 inches, should be considered as having a reasonable potential to exist.[1] If considered at all, the occurrence of a 0.445-inch diameter hole per acre of liner should be evaluated as a worst-case condition. Even the 0.08-inch diameter hole (0.005 $inch^2$) should only occur at seams, because this size opening would be visible on surface inspection. The size of the hole which is not readily visible, based on the work by GeoSyntec, is considered to be about 0.012-inch in diameter.

Table 7.6 summarizes laboratory tests conducted by Environmental Solutions, Inc., where water was placed over slits and drilled holes in HDPE and Very Low Density Polyethylene (VLDPE) with no soil backing. One important observation was that no leakage occurred through the 0.016-inch holes even with hydraulic head up to 10 feet. The absence of leakage for very small holes is explained by the surface tension characteristics of water and the "elastic" rebound of liner materials which tend to "close" the hole. Both of these flow-reducing effects would be enhanced for holes caused by a soil particle, because the small hole would be further restricted by partial penetration of the soil provided there is no movement between the hole and the soil particle. Because of this behavior, very small hole conditions are not considered further for evaluating risks associated with potential defects in geomembranes. The potential leakage rates for larger, completely open holes are much more important from a risk evaluation viewpoint.

The results shown in Table 7.7 are consistent with Bernoulli's orifice flow equation. This justifies the use of Bernoulli's equation which was recommended by Giroud (1984), and Giroud and Bonaparte (1989). Table 7.7 summarizes estimated leakage rates through holes of various size in a geomembrane overlying soils: (1) with a hydraulic conductivity in the 10^{-4} to 10^{-6} cm/sec range, and (2) for hydraulic heads of 0.1 to 10 feet. These leakage rates are estimated by two methods, one using an empirical approach and the other based on the theoretical development of seepage into partially saturated soils.

The empirical approach is based on equations derived by Giroud et al. (1989) from experimental data reported by Giroud and Bonaparte (1989). These authors consider two contact conditions between the geomembrane and the underlying soil, i.e., good and poor. The condition of good contact between the geomembrane and underlying soil was considered because the liner for mine

---

(1) As of 1991, GeoSyntec recommends a hole of 1 cm^2 (0.16 $inch^2$) for design, i.e., sizing the leakage collection layers, and a hole of 0.1 cm^2 (0.016 $inch^2$) for liner system performance evaluations. While the hole size recommended for liner system performance evaluations is larger than that proposed by the authors above, the actual difference in leakage rates, as illustrated in Table 7.7, is not substantial.

TABLE 7.6

LINER LEAKAGE FOR VARIOUS SIZE HOLES AND SLITS (IN VLDPE AND HDPE LINERS WITHOUT A SOIL BACKING)[1]

HOLE SIZE	HYDRAULIC HEAD (Feet)				
	0.01	0.1	0.5	1.0	10.0
VLDPE					
1/4-inch (0.25) Slit	N/L[2]	N/L	0.14[3]	0.72	12
1/2-inch (0.5) Slit	0.14	1.1	8.6	5.8	370
1/64-inch (0.016) Hole	N/L	N/L	N/L	N/L	N/L
5/64-inch (0.078) Hole	0.14	9	22	29	100
7/16-inch (0.44) Hole	300	850	1,650	2,500	>4,000[4]
HDPE					
1/4-inch (0.25) Slit	5.8	17	48	66	380
1/2-inch (0.5) Slit	5.8	17	70	90	570
1/64-inch (0.016) Hole	N/L	N/L	N/L	N/L	N/L
5/64-inch (0.078) Hole	2.9	14	40	60	210
7/16-inch (0.44) Hole	410	1,100	2,700	3,500	>5,000[4]

(1) Based on Environmental Solutions, Inc. (1989).
(2) N/L = No Leakage.
(3) All numbers are expressed in gallons per day (gpd).
(4) Leakage exceeds laboratory experiment flow measurement capacity.

TABLE 7.7

ESTIMATED LEAKAGE THROUGH HOLES IN GEOMEMBRANE LINERS IN GALLONS PER DAY (gpd)

ASSUMED CONDITIONS		SUBSOIL HYDRAULIC CONDUCTIVITY (cm/sec)					
		1×10^{-4}		1×10^{-5}		1×10^{-6}	
HOLE DIAMETER	HYDRAULIC HEAD (feet)	Empirical(1)	Theoretical(2)	Empirical(1)	Theoretical(2)	Empirical(1)	Theoretical(2)
0.08 inch	0.1	2.1	2.7	0.4	0.8	0.1	0.5
	1.0	16.9	5.3	3.1	1.2	0.6	0.6
	10.0	134.3	63.7	24.4	7.9	4.4	1.6
0.445 inch	0.1	3.0	5.6	0.5	1.5	0.1	0.9
	1.0	23.8	10.3	4.3	2.2	0.8	1.0
	10.0	189.4	110.3	34.5	13.7	6.3	2.7

(1) Based on Giroud et al. (1989).

(2) Based on Aravin and Numerov (1965).

waste management units normally functions under loaded conditions. The equation for leakage in this case is given by the following:

$$q = 0.21a^{0.1} k_n^{0.74} H^{0.9}$$

where

q = leakage rate per hole (m^3/sec)
a = area of the hole (m^2)
k_n = hydraulic conductivity of the subsoil (m/sec)
H = the hydraulic head on the liner (m)

The theoretical approach is based on equations developed by Aravin and Numerov (1965) to estimate seepage due to capillary action which could occur from a ditch used for irrigation flow. The model was modified by the authors to include hydraulic head above the liner. The modified equation takes the form of:

$$d = \frac{4}{\pi^2} q_r \operatorname{Exp}\left[\frac{-\pi\,(H - h_f)}{q_r}\right]$$

where

d = diameter of hole (m)
h_f = soil pore water pressure below the liner (m)
H = hydraulic head on top of the liner (m)
q_r = reduced specific leakage (m)

The above equation is solved for q_r by iteration. The total estimated leakage q is related to q_r by the equation to account for the axisymmetric nature of the flow:

$$q = \frac{\pi k}{4} q_r^2$$

where

k = hydraulic conductivity of the subsoil (m/sec)

For calculating the leakage shown in Table 7.7, soil pore water pressures of -50, -100, and -300 cm were used for subsoils with permeabilities of 10^{-4}, 10^{-5}, and 10^{-6} cm/sec, respectively. Either metric or English units can be used in the above two equations so long as the units are consistent throughout the calculation.

The empirical and theoretical results shown are consistent, and agreement for all the estimates is considered reasonable for evaluating behavior of the potential leakage into the unsaturated zone, as discussed in the next section. To be conservative, the largest of the two estimates is used in each case.

7.4.3 LEAKAGE TO THE UNSATURATED ZONE

The prior section illustrates the approximate leakage estimated for clay, geomembrane, and composite liners. These are summarized in Table 7.8. In transferring the information to Table 7.8, the maximum clay liner leakage rates calculated in Table 7.5 have been assumed.

These data illustrate the following important factors for consideration in the design and construction of liners:

- Given the same operating conditions, the leakage through a clay liner will be much greater than through a properly designed and installed geomembrane, or composite liner system consisting of a geomembrane on top of a low hydraulic conductivity soil layer.
- Maintenance of a low hydraulic head on the liner will substantially reduce the potential for seepage.

From a regulatory viewpoint, the most important consideration concerning leakage through liners is the potential for the seepage to reach ground water and adversely impact potential beneficial uses. Mitigating factors which can protect ground water, even if leakage through the liner occurs, can be:

- A reduction in chemical constituents, due to attenuation, as seepage occurs through the unsaturated zone. The potential for attenuation of various chemicals common to the mining industry are discussed in Chapter 4.
- Long seepage periods through the unsaturated zone, especially with respect to the operational period of the waste management unit.

This section discusses the length of travel time which may be expected related to the rates of leakage through the liner types and unsaturated zone conditions discussed above.

The unsaturated zone properties are an important consideration for determining liner requirements, so long as the conditions are relatively uniform and predictable. Such conditions occur at many locations where:

- Thick alluvium formations have been deposited in a series of many layers. These types of formations usually have a much lower vertical than horizontal hydraulic conductivity, because the grain size characteristics of individual layers vary depending upon the type of geologic environment existing at the time of deposition. The finer-grained layers restrict vertical movement and cause horizontal spreading in the more pervious layers.
- Lake bed formations where thick silt and/or clay formations have been deposited over large, continuous areas.
- Relatively young and weakly cemented, fine-grained sedimentary bedrock, with no major fracture or folded zones.

TABLE 7.8

SUMMARY OF LEAKAGE RATES THROUGH CLAY, GEOMEMBRANE, AND COMPOSITE LINERS [(1)] (GPAD)

HYDRAULIC HEAD ABOVE LINER (ft)	CLAY LINER				COMPOSITE LINER WITH NO GEOMEMBRANE DEFECTS	GEOMEMBRANE LINER ABOVE UNSATURATED ZONE SOIL/ROCK OR COMPOSITE LINER					
						One 0.08-Inch Hole Per Acre			One 0.448-Inch Hole Per Acre		
						Unsaturated Zone Hydraulic Conductivity			Unsaturated Zone Hydraulic Conductivity		
	One-Foot Thick (10^{-7} cm/sec)	Two-Foot Thick (10^{-7} cm/sec)	Two-Foot Thick (10^{-6} cm/sec)	Ten-Foot Thick (10^{-6} cm/sec)		1×10^{-4}	1×10^{-5}	1×10^{-6}	1×10^{-4}	1×10^{-5}	1×10^{-6}
0.1	230	160	1,580	1,030	0.0015	3	0.8	0.5	6	2	0.9
1.0	310	200	1,980	1,100	0.1	17	3	0.6	24	4	1
10.0	1,110	600	6,000	1,910	7.0 [(2)]	134	24	4 [(2)]	189	35	6 [(2)]

[(1)] Based on values presented in Tables 7.5 and 7.7 (maximum values).

[(2)] The leakage for a geomembrane with a hole (4 and 6 gpad) is greater than that from the composite liner without defects (7 gpad), because the case with the holes does not include very slow leakage through the remainder of the liner.

Both lake bed and sedimentary rock will often have some degree of layering so that vertical hydraulic conductivity is lower than horizontal hydraulic conductivity, and spreading of seepage occurs with downward migration.

If the waste management unit is located directly above very pervious soil or highly fractured rock, seepage rates through the unsaturated zone could be faster than those discussed below. The information presented below can be representative of rock which has only microfractures and therefore a hydraulic conductivity and porosity which is comparable or even lower than that of fine-grained soil. In the case of highly fractured rock, the average leakage rate for a clay liner would depend upon the portions of liner above the more pervious zones. For a geomembrane liner case, the leakage rate would depend on whether or not any hole in the liner were in the same location as an open fracture. At sites where locations of variable subsurface conditions can be predicted, several liner types may be appropriate to maintain overall potential leakage below rates that could affect beneficial uses of water.

The time required for seepage to migrate downward through the unsaturated zone has been calculated using the same model developed by McWhorter and Nelson (1979) as discussed in Section 7.4.2. The results are summarized in Tables 7.9 and 7.10. However, to account for leakage through holes in the geomembranes, and lateral spreading of the wetted front, the travel time equation has been modified by the authors to:

$$T = \frac{V}{Q}\left[(n - \emptyset_r)\left(\frac{Q}{Ak_f}\right)^{\frac{\lambda}{2+3\lambda}} + (\emptyset_r - \emptyset_i)\right]$$

where:

- T = time (time)
- V = volume of a frustrum of a cone with diameter q_r at the top (length^3)
- A = cross section of a frustrum of a cone with diameter q_r at the top (length^2)
- Q = total leakage through a hole in the liner (length^3/time)
- k_f = hydraulic conductivity of the subsoil (length/time)
- n = effective porosity of the material underlying the liner (dimensionless)

A schematic profile of the seepage model is shown in Figure 7.4.

Tables 7.9 and 7.10 show estimates of travel times for ground water to depths of 200 feet for clay and holes in geomembrane, respectively, considering the leakage rates provided in Table 7.8. Typical performance of a composite liner which incorporates a geomembrane located on top of a

TABLE 7.9

ESTIMATED TIME OF ARRIVAL OF THE WETTED FRONT (YEARS) FOR LEAKAGE THROUGH CLAY LINERS (1)

CLAY LINER CONDITION	UNSATURATED ZONE HYDRAULIC CONDUCTIVITY (cm/sec)	HYDRAULIC PRESSURE (ft)	DEPTH OF UNSATURATED ZONE (ft)(2)					
			10	20	50	100	150	200
Thickness: 1 Foot; Hydraulic Conductivity: 1 x 10^{-7} cm/sec	1 x 10^{-4}	0.1	3	6	14	28	42	56
		1.0	2	4	11	21	32	43
		10.0	0.7	1	3	7	10	13
	1 x 10^{-5}	0.1	6	11	28	56	83	111
		1.0	4	8	21	42	63	84
		10.0	1	3	7	13	19	26
	1 x 10^{-6}	0.1	10	21	51	103	154	205
		1.0	8	16	39	78	117	156
		10.0	2	5	12	25	37	49
Thickness: 2 Feet; Hydraulic Conductivity: 1 x 10^{-7} cm/sec	1 x 10^{-4}	0.1	4	8	20	39	59	79
		1.0	3	6	16	32	48	64
		10.0	1	2	6	12	18	23
	1 x 10^{-5}	0.1	8	16	39	78	117	156
		1.0	6	13	32	63	95	126
		10.0	2	5	11	23	34	46
	1 x 10^{-6}	0.1	14	29	72	143	215	286
		1.0	12	23	58	117	175	233
		10.0	4	9	21	43	64	86
Thickness: 2 Feet; Hydraulic Conductivity: 1 x 10^{-6} cm/sec	1 x 10^{-4}	0.1	0.5	1	2	5	7	10
		1.0	0.4	0.8	2	4	6	8
		10.0	0.2	0.3	0.8	2	2	3
	1 x 10^{-5}	0.1	0.9	2	5	9	14	19
		1.0	0.8	2	4	8	12	15
		10.0	0.3	0.6	1	3	4	6
	1 x 10^{-6}	0.1	2	4	9	18	27	36
		1.0	1	3	7	15	22	29
		10.0	0.6	1	3	6	8	11
Thickness: 10 Feet; Hydraulic Conductivity: 1 x 10^{-6} cm/sec	1 x 10^{-4}	0.1	0.7	1	4	7	11	14
		1.0	0.7	1	3	7	10	14
		10.0	0.4	0.8	2	4	6	8
	1 x 10^{-5}	0.1	1	3	7	14	21	28
		1.0	1	3	7	13	20	26
		10.0	0.8	2	4	8	12	16
	1 x 10^{-6}	0.1	3	5	13	26	40	53
		1.0	2	5	12	25	37	50
		10.0	2	3	8	15	23	30

(1) See Tables 7.5 and 7.8 for estimated leakage rates. Estimates do not include travel time through the liner.

(2) Unshaded values indicate where the travel time for the clay liner is longer than for the geomembrane liner with one 0.08-inch hole, as shown in Table 7.10.

TABLE 7.10

**ESTIMATED TIME OF ARRIVAL OF THE WETTED FRONT (YEARS)
FOR HOLES IN GEOMEMBRANE
(20 DEGREE SPREADING ANGLE)**

HOLE DIAMETER	UNSATURATED ZONE HYDRAULIC CONDUCTIVITY (cm/sec)	HYDRAULIC PRESSURE OF LIQUID ABOVE THE HOLE (ft)	ESTIMATED LEAKAGE RATE (gpd)[1]	DEPTH OF UNSATURATED ZONE (ft)					
				10	20	50	100	150	200
0.08 inch	10^{-4}	0.1	2.7	0.2	0.8	10	70	220	510
		1.0	16.9	<0.1	0.2	2	13	40	90
		10.0	134.3	<0.1	<0.1	0.4	2	7	14
	10^{-5}	0.1	0.8	1.0	6.0	67	460	1,480	3,400
		1.0	3.1	0.4	2.0	19	130	400	930
		10.0	24.4	0.1	0.5	4	22	64	140
	10^{-6}	0.1	0.5	7.0	28.0	230	1,420	4,300	9,600
		1.0	0.6	6.0	25.0	200	1,220	3,700	8,200
		10.0	4.4	2.0	6.0	40	210	620	1,320
0.445 inch	10^{-4}	0.1	5.6	<0.1	0.5	5	36	110	260
		1.0	23.8	<0.1	0.2	1	10	30	67
		10.0	189.4	<0.1	<0.1	0.3	2	5	10
	10^{-5}	0.1	1.5	0.8	4.0	39	261	820	1,880
		1.0	4.3	0.4	2.0	15	100	300	690
		10.0	34.5	0.2	0.5	3	17	50	105
	10^{-6}	0.1	0.9	6.0	20.0	150	880	2,600	5,700
		1.0	1.0	5.0	19.0	140	800	2,350	5,100
		10.0	6.3	2.0	6.0	34	170	470	1,000

[1] Gallons per day.

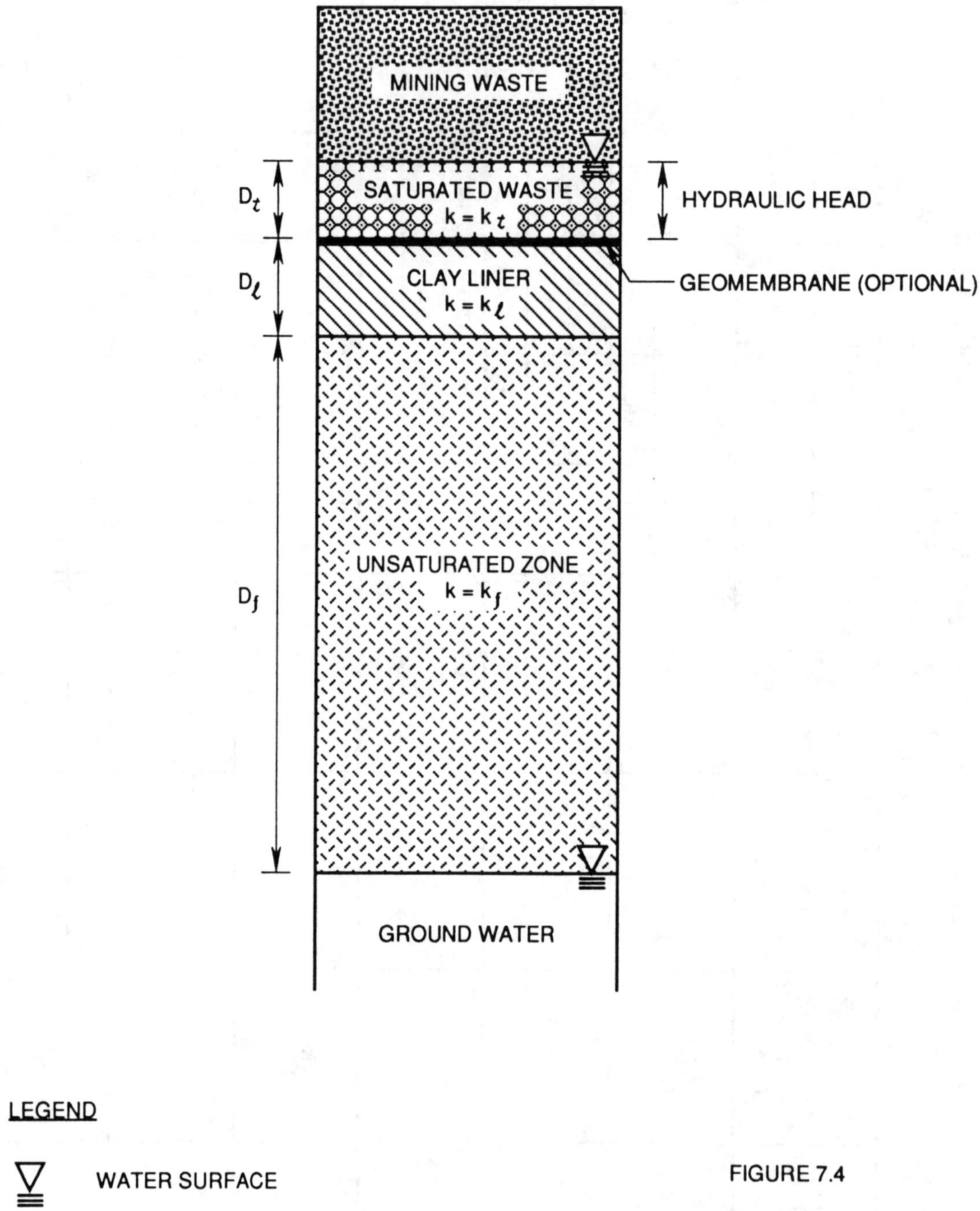

FIGURE 7.4

SCHEMATIC PROFILE OF SEEPAGE MODEL

low hydraulic conductivity soil layer also can be inferred from the information presented. The travel time estimates are based on the unsaturated zone having an initial water content (by volume) ranging from 10 to 20 percent, and a field moisture water content between 15 and 35 percent, as summarized in Table 7.11. The estimated travel times in Table 7.10, for holes in geomembrane liners, are based on an assumed seepage spreading angle of 20 degrees from the vertical. For unsaturated zones with lower hydraulic conductivities such as fine sands and silts, and layered conditions, the seepage spreading angle from an individual hole could be substantially greater than 20 degrees, when both capillary and variable hydraulic conductivity conditions cause horizontal seepage to occur. Table 7.12 illustrates that travel times are approximately doubled if the assumed spreading angle is increased from 20 to 30 degrees.

Important observations from the data presented in Tables 7.9 and 7.10 are:

- The arrival times for seepage from a clay liner is generally longer than for a geomembrane or composite liner with a hole, for site conditions involving (1) low conductivity clay liners, (2) shallower depths to ground water, and (3) higher unsaturated zone hydraulic conductivities. (See the shaded values in Table 7.9.) Geomembrane and composite liners have a longer travel time for lower unsaturated zone hydraulic conductivities and deeper ground water depths. The relative differences in travel times for clay and geomembranes increase with unsaturated zone thickness. This occurs because the leakage associated with the clay occurs at all locations, and therefore does not realize horizontal spreading such as that associated with leakage from a single, small hole in the geomembrane liner.
- When the hydraulic head on the liner is kept low, travel times for small holes in a geomembrane or composite liner are very long. Considering the 0.08-inch hole, which occasionally could be overlooked during QA/QC activities, the time required for migration to 100 feet under a head of 0.1 foot is estimated to be greater than 50 years, even for an unsaturated zone with a 10^{-4} cm/sec hydraulic conductivity. For lower hydraulic conductivity, (i.e., 1×10^{-5} cm/sec), the travel time to 100 feet is estimated to be more than 400 years.
- For geomembrane and composite liners, travel times to a depth of 200 feet are very long, even for larger holes, when a low hydraulic head of less than 1 foot is maintained. For these conditions, there would essentially be no threat to adversely affect ground water. However, travel times are significantly shortened if the hydraulic head is large, i.e., 10 feet.
- Estimated travel times to depths of 10 or 20 feet are relatively short for all the liner systems considered (e.g., less than 5 to 10 years) in comparison with the duration of many mining operations. Conservative liner designs may be required for these types of shallower ground water conditions if constituents may be released from the waste management unit and migrate through the unsaturated zone.

TABLE 7.11

SUMMARY OF MATERIAL PROPERTIES

HYDRAULIC CONDUCTIVITY (cm/sec)	EFFECTIVE POROSITY	RESIDUAL MOISTURE CONTENT	IN SITU MOISTURE CONTENT	PORE SIZE DISTRIBUTION INDEX	SOIL MOISTURE PRESSURE (cm)
1×10^{-4}	0.25	0.15	0.10	2.0	-50
1×10^{-5}	0.35	0.25	0.15	1.5	-100
1×10^{-6}	0.50	0.35	0.20	1.0	-300

TABLE 7.12

COMPARISON OF TRANSIT TIME THROUGH HOLE IN LINER FOR DIFFERENT ASSUMED SPREADING ANGLES[1]

Hole Diameter = 0.08 inch
Subsoil Hydraulic Conductivity = 1×10^{-5} cm/sec
Hydraulic Head = 1.0 feet

DEPTH (ft)	TRANSIT TIME (years)	
	$\beta = 20°$ [2]	$\beta = 30°$ [2]
10	0.4	0.7
20	2	4
50	20	40
100	130	300
150	400	970
200	930	2,220

(1) Modified McWhorter and Nelson method (McWhorter and Nelson, 1979).
(2) β = Wetting Front Spreading Angle

Another important parameter to be considered when evaluating liner system performance is the relative amount of unsaturated zone that is potentially affected by leakage. It is obvious considering the geometrical shapes of the leakage pattern from clay versus geomembrane or geocomposite liner systems that leakage from a clay liner has the potential for affecting a considerably larger proportion of the unsaturated zone. Where this is an important consideration, the above approach can be used to determine the relative magnitudes of affected unsaturated zone.

The above approach to travel time calculations while very approximate, is usually consistent with the accuracy with which the hydraulic conductivity and moisture conditions in the unsaturated zone are known. Care should therefore be taken when using these values and it is advisable to use ranges of known conditions to calculate ranges of potential travel times. Where more accuracy is required field data collection and laboratory testing is necessary to improve the accuracy of the relevant physical properties of the unsaturated zone. Where this approach is adopted it will be necessary to refine the calculation procedures as well. This can be accomplished by utilizing the equations developed by Phillip (1968), Raats (1971), Warrick (1974), and Lorentz et al. (1990). Alternatively numerical solutions to the unsaturated flow equations can be utilized. Existing computer codes that could be considered include those developed by United States Geologic Service (USGS 1987, 1990, and 1991), the United States Bureau of Mines (USBM, 1991), and Davis et al. (1983).

Overall, these above analyses confirm the importance of maintaining low hydraulic heads on any type of liner system. They also illustrate the importance of engineering designs and QA/QC procedures to minimize the potential for larger holes in geomembrane liners to occur. For deep ground water conditions, it appears reasonable that a single, properly constructed geomembrane or composite liner would provide adequate protection to beneficial uses of ground water. Conversely, travel times to relatively shallow ground water could be relatively short. Therefore, additional protection may be warranted unless it could be convincingly demonstrated that undesirable chemicals would be attenuated if a liner leak were to occur. Finally, the data in this section clearly illustrate that the unsaturated zone thickness can be a very important parameter for many sites. This parameter should be included as one of the key factors in determining site-specific liner requirements.

7.4.4 LINER STABILITY CONSIDERATIONS

The structural stability of a liner system is an important element of the design of a waste management unit. The design should prevent movement or displacement of waste, which overlies a liner system, and could cause excessive deformation and/or damage to the liner system. The types of waste stability issues that could cause liner damage are associated with waste piles and include:

- Sloughing of loose uncompacted fill from the surficial zones of the waste.
- Static slope instability causing failure surfaces within the waste along which the waste moves vertically and laterally.
- Block failure of a mass of waste moving laterally with shearing occurring predominantly within the liner system. This type of slope stability issue is an important consideration with liner systems that include geomembranes and in cases where waste is placed on sloping liner systems.
- Dynamic slope instability and permanent displacement related to earthquake and other seismic loading along failure surfaces within the waste or along the liner system.

Mine waste management units such as tailings impoundments and wastewater ponds usually receive slurries or liquids and therefore, do not pose any stability problems to liner systems provided the containment dikes are stable.

Table 7.13 summarizes examples of stability issues that may need to be considered for waste piles. It also lists typical design or waste management features that can be included to mitigate any potential for, or effects of, slope movement.

Potential slope failure modes can be evaluated using conventional engineering methods which include the following:

- A limit equilibrium slope stability analysis in which the resisting shear forces or movements (which include cohesion and friction) to the driving gravitational forces or moments which have the potential to cause downward lateral displacement of the waste are determined. The factor of safety of the slope against failure is then determined as the ratio of the sum of the resisting forces or moments to the sum of the driving forces or moments.
- A pseudo-static stability analysis in which the factor of safety or ratio of resisting forces to applied forces from dynamic (earthquake) ground motion is determined.
- Permanent displacements under dynamic loading which can be calculated using approximations similar to that described by Newmark (1965).

There are many accepted methods of calculating the factor of safety, but all generally involve dividing the area above the failure surface into vertical slices, then calculating the driving and resisting forces on these slices. The basic principles and simplifying assumptions used in

TABLE 7.13

WASTE PILE STABILITY CONSIDERATIONS IN LINER SYSTEM DESIGN

CONDITION AT PLACEMENT	STABILITY ISSUES	DESIGN OR MANAGEMENT FEATURES FOR CONTROLLING STABILITY
• Angle of repose slopes that have a very low factor of safety • Uncompacted, i.e., material has a low density and hence, low strength • Possible high moisture conditions which reduce shear strength of the material	• Sloughing • Permanent displacement along circular arc surface from static or earthquake loading • Block slippage along bottom liner system (especially important when geomembranes are used) • Tearing or shearing of liner system during movement of waste	• Buttress along the toe of the waste pile • Benched construction to flatten overall slopes • Internal drainage to prevent pore pressure buildup • Maintenance to repair drainage as it occurs

calculating slope stability are well described in geotechnical texts such as Lambe and Whitman (1969). More advanced discussions are found in Huang (1983), Janbu (1973), Spencer (1967), Morgenstern and Price (1965), and Bishop (1955).

In a stability evaluation, several failure surfaces are postulated, and the minimum calculated factor of safety is determined. The designer then determines the acceptability of the minimum calculated factor of safety. Minimum criteria for earth dams (outlined in U.S. Army Corps of Engineers [1982] and Wilson and Marsal [1979]) list minimum factors of safety of 1.3 to 1.5 for static conditions and 1.0 for seismic conditions. A slightly higher minimum value of 1.1 for seismic stability of earth dams (under certain conditions) is proposed by Seed (1979). A factor of safety of 1.0 for the more severe Maximum Credible Design Earthquake (MCE) and 1.1 for the less severe Maximum Probable Earthquake (MPE)(2). If the seismic factor of safety is less than 1.0, the analysis of permanent displacements can be conducted to determine whether the deformation that occurs during the seismic event would result in damage to the containment system's integrity which could not be readily repaired and which would result in significant and permanent water quality impact.

The cohesion and friction angles (shear strength parameters) of both the waste and liner materials are important parameters for conducting slope stability analyses. As indicated below, the parameter values reported in the literature vary widely, particularly for the geosynthetic liner component interfaces, and reflect that the values are highly dependent on site-specific conditions and materials. The approach that should be adopted therefore is either to use very conservative (i.e., low) values derived from the literature or, if these are not representative use values derived from site-specific testing. Typical ranges of values as reported in Martin, et al., Seed et al. (1988), EPA (1988e), and Williams et al. (1987) are presented below:

Internal Strength Characteristics of Waste Materials:		Residual Cohesion (lb/ft^2)	Residual Friction Angle(3)
	Waste Rock	0	35° to 40°
	Residual Crushed Ore	0	35° to 40°
	Tailings	Usually negligible	25° to 35°

(2) The MCE is the theoretical maximum earthquake event that can occur. The MPE is an earthquake of lesser magnitude that usually has a frequency of occurrence of 100 years.

(3) In some instances, particularly with roughened geomembranes, the peak friction angle can be higher. Designing with peak friction angles is however, not recommended unless field testing or site-specific field analysis indicate it is appropriate.

		Residual Cohesion (lb/ft^2)	Residual Friction Angle[3]
Internal Strength Characteristics of Liner Materials:	Clay	1,000 to 2,500	0° to 28°
Strength Characteristics of Interfaces Between Liner System Layers:	Geomembrane on Soil for the following materials:		
	- Smooth and Roughened[4] PVC	100 to 300	6° to 25°
	- Smooth HDPE	10 to 300	11° to 27°
	- Roughened HDPE	230 to 380	25° to 35°
	Geomembrane on Geomembrane for the following materials:		
	- Smooth HDPE	NA[5]	6° to 13°
	Geomembrane on Geotextile for the following materials:		
	- Smooth PVC	20 to 50	10° to 24°
	- Roughened PVC	NA	11° to 28°
	- Smooth HDPE	20 to 40	6° to 15°
	Geotextile on Soil	NA	23° to 30°

The cohesion values reported above for the liner systems involving geomembranes are small and are not usually taken into account in stability analyses.

7.5 REGULATORY IMPLICATIONS

7.5.1 GENERAL

The literature evaluations and analyses performed for this chapter, and the discussions about climate, attenuation, and closure in other chapters, have indicated that:

- There are many liner systems that can be safely used for mine waste management units. Each has certain cost, construction, or performance advantages, and different types of potential problems that must be accounted for in design, construction, and operation.
- Construction materials available at mine sites can vary greatly. Often, natural clay materials will not be available within reasonable distances for use in construction. In these cases especially, synthetic materials such as flexible membrane liners (geomembrane) should be considered as the primary containment system for sites where some type of liner will be required. The studies in this chapter show that a properly designed and constructed geomembrane or composite liner will provide substantial protection to beneficial uses of water, especially for ground water at

---

(3) In some instances, particularly with roughened geomembranes, the peak friction angle can be higher. Designing with peak friction angles is however, not recommended unless field testing or site-specific field analysis indicate it is appropriate.

(4) Due to limited extent of roughening that is usually provided, the roughness does not appear to significantly influence the strength characteristics.

(5) Not available.

depth. For a given hydraulic head and unsaturated zone characteristic, a geomembrane or composite liner will perform better than a single clay (compacted low hydraulic conductivity soil) liner. Where practical, composite liners, consisting of both geomembrane and clay, provide greater protection than either material alone.

- A liner system performance can be predicted best by its overall design and construction QA/QC procedures, and not simply by its component specifications (e.g., thickness). For example, if a liner will be subjected to construction or operational loads, geomembrane components may require protective layers below and/or above the liner, to avoid penetrations. In those cases, the properties of the protective layers and not the geomembrane thickness often will be the most important design criteria.
- Clay liner performance does not increase proportionately to its thickness, but potential leakage does decrease as the thickness is increased. The thickness of the clay liner should be designed to be sufficient to avoid leakage breakthrough which poses a threat to the beneficial uses of ground water during operation of the waste management unit. At sites with limited availability of clay, the minimum practical thickness of compacted soil is warranted. Based on experience, this is considered to be 1 foot, although with special construction, considerations 8 to 9 feet can be utilized.
- The potential threat to beneficial uses of water can vary greatly, depending on operational factors such as the duration of waste unit operation and controls employed to reduce the hydraulic head pressure exposed to the liner. Low hydraulic heads can be maintained by providing a control component above the low hydraulic conductivity layer, or a double liner with an intervening LCRS.
- The potential threat to beneficial uses also will vary depending on liner system design and the condition of the unsaturated zone below the waste management unit. Potential migration periods through the unsaturated zone can be tens or hundreds of years with a properly designed and constructed geomembrane liner and verification by state-of-the-art QA/QC procedures, especially when ground water is deep and the unsaturated zone material has relatively low hydraulic conductivity. The rate of migration and probability of seepage reaching ground water depend upon the hydraulic conductivity of the unsaturated zone and the depth to water.
- The need to provide redundancy within the liner system or to rely upon natural subsurface conditions will depend on waste and site-specific factors. At some sites, liner system construction and operating procedures may provide basic containment so that the unsaturated zone can be relied upon to provide redundancy, with an adequate factor of safety. For example:
  - Double liners may not be necessary at many sites if hydraulic heads are maintained sufficiently low.
  - Small rates of leakage into the LCRS for double liners can be acceptable so long as it is collected and removed to avoid significant head buildup on the lower liner.

These factors, combined with the range of chemistries associated with mine wastes, indicate the necessity for flexibility and consideration of site-specific conditions in designing engineered liner (containment) systems. Otherwise, uniform design specifications either are likely to be inappropriate, with insufficient containment for sites with unfavorable conditions, or too restrictive, resulting in over-regulation of sites with favorable conditions. This need for flexibility is consistent with the conclusion reached by the EPA (1988b) in the development of draft criteria for designing Municipal Solid Waste Landfill (MSWLF) containment systems. The EPA studies showed that a site-specific, risk-based approach would be most appropriate. In reaching its conclusion, the EPA specifically recognized the importance of climate and geologic site factors, including the effects of these factors on travel time from the waste management unit to ground water. Table 7.14 summarizes factors that should be considered for site-specific liner system designs.

Alternative regulatory approaches are evaluated in the next section, with respect to the ability of each approach to incorporate site-specific variables and still remain manageable within the regulatory process.

7.5.2 ALTERNATE REGULATORY APPROACHES

The following options represent a range of regulatory approaches in determining appropriate mine waste management unit liner systems:

- Specification of uniform design criteria.
- Establishment of categorical design criteria, based on a range of waste characteristics and site conditions.
- Satisfaction of performance criteria based on site-specific waste containment assessments.
- A combination of the above approaches.

Specification of uniform design criteria that are adequate and feasible for the wide range of mining and site conditions is not practical. As discussed by EPA (1988c), if uniform design criteria were to be established for the "worst" site conditions, they would provide the greatest amount of protection to the environment. However, because this approach does not consider site-specific factors, such as climate, depth to ground water, unsaturated zone hydraulic conductivity, and special management practices, it would result in unnecessary and costly over-regulation of facilities located in more favorable site conditions.

TABLE 7.14

SITE-SPECIFIC FACTORS TO CONSIDER IN LINER SYSTEM DESIGN

Potential Waste Material Toxicity

- Chemical Properties of the Waste
- Physical or Chemical Changes Resulting from Mining or Ore Extraction
- Net Acid Generation Potential
- Soluble Constituents for Anticipated Environmental Conditions
- Special Treatment or Neutralization Procedures Utilized
- Total Resulting Mass of Soluble Constituents Which Could Be Mobilized Under Site Conditions

General Water Resource Values at the Site

- Adequate Quality for Beneficial Use
- Sufficient Quantity for Beneficial Use
- Existing or Identified Beneficial Uses
- Probable Locations of Future Beneficial Uses

Leachate Availability to the Environment

- Waste Material Characteristics
  - Hydraulic Conductivity Based on Direct Measurement, Laboratory Tests, or Grain Size and Density
  - Moisture Retention Capacity
- Thickness of the Waste
- Site Climatic Conditions
- Provisions at Closure to Restrict Infiltration

Site Factors

- Topography
- Geology, Including Predictability of Uniformity and/or the Potential for Discontinuities
- Unsaturated Zone Thickness, Continuity, Hydraulic Conductivity and Natural Water Content
- Potential Migration Time for Seepage to Ground Water
- Effects of Climatic Conditions on Long Term Unsaturated Zone Migration Characteristics
- Constituent Attenuation Potential

Waste Unit Management Practices

- Facility Type
- Waste Placement Method
- Protection of Liner Systems From Environmental or Physical Damage
- Controls on the Hydraulic Head
- Risk Reduction Practices, Such as Placement of Underdrains, Sub-Aerial Depositions, Limited Time of Operations
- Non-Liner Barriers, Such as Cutoff Walls
- Installation of Special Early Warning Monitoring Systems

If appropriately structured, the categorical design criteria approach offers a number of advantages over the uniform design criteria approach. Theoretically, this approach can allow the regulations to be somewhat prescriptive, but still avoid unnecessary over-regulation by allowing the operator to design each facility according to site-specific factors. This approach could be especially useful for regulating smaller mining operations where the costs involved in performing major investigations to justify site-specific designs are not warranted. However, in order for the categorical approach to be sufficiently flexible to cover the wide range of variables associated with mining, it would have to incorporate into its regulations fairly complex systems.

As discussed in Chapter 3, California's existing regulations (Article 7) are based on a form of the categorical design criteria approach. The mine waste materials are divided into Groups A, B, and C, depending on the material's relative potential to adversely affect water uses and separate design criteria are provided for each of the groups or categories. This system has presented some problems to both the regulatory community and the mining industry. In part, these difficulties stem from limitations on the opportunity to incorporate site-specific factors for individual projects. This has resulted in the need for many facilities to be designed and permitted in accordance with the *Chapter 15*, Section 2510 and 2570 exemption provisions (Chapter 3). These exemptions allow variations from the prescriptive requirements provided these are infeasible, or there are only minor amounts of ground water underlying the site, or there is not detectable vertical hydraulic connection between the waste management unit and ground water. It is therefore appropriate that more flexibility be provided within Article 7 to also allow and provide performance criteria for designs which take into account the site-specific conditions at various mine waste management units. A proposed containment design may offer the best approach, not because a prescribed design is infeasible, but rather because the site-specific design is more appropriate.

The performance criteria approach, using site-specific waste containment assessment, is theoretically the procedure that can most effectively provide the desired level of ground water protection. For Article 7, the fundamental performance criterion would be protection of the potential beneficial uses of waters of the State (see Chapter 3). As discussed by EPA (1988b), in developing recommended design criteria for solid waste landfills, the performance criteria approach also provides allowance for innovative technologies developed in the future. The greatest disadvantages of this approach are:

- Relatively complex investigations are required on a site-specific basis to support each unit design proposal. This may not be justified for small mining operations.
- Regulators must have the level of expertise necessary to evaluate each design on a case-by-case basis.

The first disadvantage can be overcome by also providing a conservative categorical approach which can be used if desired by the applicant. The second disadvantage can be mitigated by specifying the information required by the regulatory agency for its site-specific risk evaluation in order to confirm that the "custom" designed liner system will satisfy required performance goals.

Based on this evaluation, it is concluded that providing a choice of either the categorical and performance goal approaches offers the best procedure to serve both regulators and project proponents. The categorical approach could be provided as a separate decision tree, with consideration of the major factors presented in Table 7.14. The resulting prescribed liner system for each category should be reasonable, yet somewhat conservative, for each set of site-specific conditions. Site data requirements for this approach would be those sufficient to assign the appropriate category. At the applicant's discretion, a more conservative categorical design could be used in lieu of developing detailed site information. The categorical designs would specify minimum material standards, (e.g., liner thickness), but would allow alternatives to be used if performance adequacy were demonstrated by the applicant.

7.5.3 RECOMMENDED REGULATORY APPROACH

Regulations should allow the applicant to design liner systems by either one of the following two approaches:

- Approach 1 - Categorizing the waste management unit characteristics and then designing the liner system in compliance with prescribed regulatory containment requirements based on:
  - The type of facility (e.g., waste pile, tailings impoundment containing slurried waste, or a wastewater pond).
  - Site and operational factors, including:
    - Potential waste toxicity
    - Water resource values
    - The amount of leachate available
    - Site factors
    - Proposed management practices
- Approach 2 - Conducting a site-specific waste containment assessment and demonstrate that the waste management unit, with the application of liner systems or other controls, if necessary, does not pose a threat to the beneficial uses of water.

The minimum data to be provided for the first approach should be: (1) waste chemistry data, including its acid generation potential; (2) whether the waste management unit is a pile, tailings slurry impoundment, or wastewater pond, with data describing the size of the unit and the rate of waste discharge, and (3) how the facility will be constructed and operated to assure integrity of the liner system. The facility then should be placed into a category that provides a conservative liner system designation based on the factors itemized in Approach No. 1, above. Additional data concerning water resource values, site climate, hydrogeologic information, and special operating properties may be provided to allow the facility to be placed into a more appropriate category.

If conditions vary within the waste management unit, the applicant may explain the differences so that the appropriate liner may be used at each location. Otherwise, the system for the most restrictive conditions should be assumed and applied throughout the waste management unit.

The applicant should be allowed to select alternative materials as components of the prescribed liners, so long as it can be demonstrated that such alternative materials will not compromise function.

Regulations specifically should allow an applicant to propose an alternative to the categorical specifications, so long as it can be demonstrated that the engineered alternative will be consistent with applicable performance goals for protection of beneficial uses of water.

For Approach 2, the applicant should provide at least the following data:

- Waste properties, including grain size, acid generation potential, and soluble constituents for expected environmental conditions.
- Ultimate size and configuration of the waste management unit, and procedures to be used for waste material placement.
- General geologic characteristics of the area, extending an appropriate distance in all directions from the waste management unit.
- Subsurface conditions at the waste management unit, including types of soil and rock, stratigraphic conditions, and the potential existence of discontinuities (e.g., fractured or faulted zones). Unsaturated zone property data should include natural hydraulic conductivity, natural water content, and estimated field moisture capacity.
- Depth to ground water and estimated extraction rate potential per individual well and per acre of site area.
- Existing ground water (and surface water when appropriate) quality, existing and identified beneficial uses, and potential applicable beneficial use objectives.
- The locations of all existing wells or open boreholes to ground water within 1,000 feet of the facility boundary.
- A detailed description of the proposed liner systems and operational procedures including:
    - The manner in which conditions will vary throughout the waste management unit, and variations in liner systems to be used to accommodate such variability.

  - Low hydraulic conductivity components of each type of liner system.
  - The maximum hydraulic head to be applied to the liner, including methods for maintaining a low level and, when appropriate, liquid removal.
  - The operating duration of the waste management unit.
  - The manner in which each liner system component will be constructed, including QA/QC procedures, and the methods to be used to load the waste onto each liner system.
  - The ultimate height of the waste above each liner system.
  - Special procedures to protect liner components, including tests or data to verify protection performance.
  - Verification of chemical compatibility between liner system components and: (1) the waste or subsurface materials, and (2) environmental conditions to which the system will be exposed.
- Design calculations showing the performance of the individual liner system components and demonstrating that the integrated system will operate as intended.
- Calculations to estimate the potential time for migration of any undesirable constituents to ground water, based on conservative estimates of liner system performance under specified design, construction, and operational conditions (see Chapter 3). If constituent attenuation is used in this estimate, the method for verifying attenuation potential should be provided (see Chapter 5).
- Where necessary, waste and liner stability calculations should be provided to demonstrate that waste movement which could damage the liner system's integrity would not occur.
- A description of the monitoring system to be provided to satisfy regulatory requirements, and/or to demonstrate liner system performance.
- A site-specific waste containment assessment for the proposed design, evaluating the risk posed to beneficial uses of ground or surface water that could be affected by the waste management unit.

7.6 BIBLIOGRAPHY

Aravin, V.I. and N. Numerov. *Theory of Fluid Flow in Undeformable Porous Media.* Israel Program for Scientific Translation. 1965.

Bishop, A.W. "The Use of the Slope Circle in the Stability Analysis of Slopes", Geotechnique, 5. (1955): 7-17.

Bonaparte et al. *Rates of Leakage through Landfill Liners, Geosynthetics.* Conference Proceedings. 1989.

Bouwer, Herman. *Unsaturated Flow in Groundwater Hydraulics.* ASCE Hydraulic Division. September 1964.

Brooks, Royal H. and Arthur T. Corey. *Properties of Porous Media Affecting Fluid Flow.* ASCE Irrigation and Drainage Division. June 1966.

Cary, H.C. *Geosynthetic Design Guidance for Hazardous Waste Landfill Cells and Surface Impoundments*. United States Environmental Protection Agency, EPA 600/2-87-009.

Chambers, Carlon C. *Seepage Control Using SBR/Asphalt Hot Sprayed, Elastomeric Membranes*, ASME Uranium Mine Waste Disposal. 1982.

Davis, L.A. and S.P. Neuman. *Documentation and User's Guide: UNSAT2-Variably Saturated Flow Model*. NUREG/CR-3390, 1983c. Update in progress.

Demetracopoulos, Alexander C. "Overview of Landfill Bottom Liner Hydraulics." Water Resources Bulletin. February 1988.

Environmental Solutions, Inc. *Final Design Heap Leach Facilities Permitting Report*. Chimney Creek Gold Mine, Humboldt County, Nevada. 1987a.

Environmental Solutions, Inc. *Internal File Information Regarding Laboratory Liner Penetration Tests*. 1989.

Environmental Solutions, Inc. *Tailings Disposal Facility Design Concept*. Chimney Creek Gold Mine, Humboldt County, Nevada. 1987b.

Freeze and Cherry. *Ground Water*. Englewood Cliffs: Prentice-Hall, Inc. 1979.

Geotechnical Fabric Report, December 1989.

Giroud, J.P., and Bonaparte, R. "Leakage Through Liners Constructed with Geomembranes, Part I: Geomembrane Liners." Geotextiles and Geomembranes. (1989): 27-67, No. 1.

Giroud, J.P., and Bonaparte, R. "Leakage Through Liners Constructed with Geomembranes, Part II: Composite Liners." Geotextiles and Geomembranes. (1989): 71-111, No. 2.

Giroud, J.P., *Impermeability: The Myth and a Rational Approach*, Proceedings of the International Conference on Geomembranes. Vol. 1, Denver, Colorado. June 1984. 157-162.

Giroud, J.P., Khatami, A., and Badu-Tweneboah, K., "Evaluation of the Rate of Leakage Through Composite Liners." Geotextiles and Geomembranes. (1989): 337-340, No. 4.

Goode, Daniel J. *Evaluation of Simplified Techniques for Predicting Moisture Breakthrough of Soil Liners*, National Conference on Management of Uncontrolled Hazardous Waste Sites. 1983.

Gundle Lining Systems, Inc., Product Calatogue.

Harr, M.E. *Groundwater Seepage*. McGraw-Hill Book Company. 1962.

Huang, Y. H. *Stability Analysis of Earth Slopes*. New York: Van Nostrand Reinhold Company. 1983.

Janbu, N. *Slope Stability Computations*, Embankment-Dam Engineering. Casagrande Volume. New York: John Wiley. 1973. 47-86.

Koerner, Robert M. *Designing with Geosynthetics*. New York: Prentice-Hall. 1990.

Lambe, T.W. and R.V. Whitman. *Soil Mechanics*. New York: John Wiley. 1969.

Lorentz, S.A., D.S. Dumford, and A.T. Corey. *Liquid Retention Measurements on Porous Media Using a Controlled Outflow Cell*. ASA-CSSA-SSSA 1991 Annual Meeting. November 1991.

Lorentz, S.A., T.D. Chatwin, and D.B. McWhorter. *A Screening Model to Assess Cyanide Migration in the Vadose Zone*. Presented at the Nineteenth Annual Rocky Mountain Ground Water Conference. Butte, Montana. October 1990.

Martin, J.P., R.M. Koerner, and J.E. Whitty. *Experimental Friction Evaluation of Slippage Between Geomembranes, Geotextiles and Soils*. International Conference on Geomembranes. Denver, Colorado.

McEnroe, Bruce M. "Discussion of Overview of Landfill Bottom Liner Hydraulics." Water Resources Bulletin. April 1989.

McWhorter, D.B. and J.D. Nelson. *Unsaturated Flow Beneath Tailing Impoundments*. American Society of Civil Engineers. Journal of Geotechnical Engineers Division. November 1979.

Morgenstern, N.R. and V.E. Price. "The Analysis of the Stability of General Slope Surfaces". Geotechnique, 15 (1965): 79-93.

National Standards Foundation. Standard Number 54. *Flexible Membrane Liners*. 1985.

Newmark, N.M. "Effects of Earthquakes and Embankments." Geotechnique. (January 1965), No. 2.

Oakley, Richard E. *Design and Performance of Earth-Lined Containment Systems*. ASCE Waste Disposal Practice. 1987.

Phillip, J.R. *Steady Infiltration From Buried Point Sources and Spherical Cavities*. Water Resources Research Journal. Vol. 4, No. 5. October, 1968.

Phillip, J.R. *General Theorem on Steady Infiltration From Surface Sources With Application to Point and Fine Sources*. Soil Sciences Society of America. Proceedings. Volume 35. 1971.

Poly-America "Poly-Flex Polyethylene Geomembrane References Manual," Grand Prairie, Texas.

Raats, P.A. *Steady Infiltration From Point Sources, Cavity and Basins*. Soil Sciences Society of America. Proceedings, Volume 35. 1971.

Robertson, Barrier Systems, Inc., Product Catalogue.

Seed, H.B. "Considerations in the Earthquake-Resistant Design of Earth and Rockfill Dams". Geotechnique, 29. 1979: 215-263, No. 3.

Seed, Raymond B., Mitchell, James K., Seed, J. Bolton. *Slope Stability Failure Investigation: Landfill Unit B-19, Phase 1-A, Kettleman Hills, California*. University of California, Berkeley, CA. July 1988.

Sergent, Hauskins and Beckwith (SHB). *Liner Testing Communications*. 1988.

Spencer, E.A. "Method of Analysis of the Stability of Embankments Assuming Parallel Interslice Forces". Geotechnique, 17. (1967): 11-26.

Terzaghi, K. and R.B. Peck. *Soil Mechanics in Engineering Practice*. New York: John Wiley. 1967.

United States Army Corps of Engineers. *Engineering and Design Stability for Earth and Rockfill Dams*. EM 1110-2-1902. 1982.

United States Bureau of Mines. *TWOD: Two-dimensional Finite Element Analysis of Variably Saturated Flow*. Bureau Information Circular (IC) 9296. 1991.

United States Environmental Protection Agency. *Criteria for Municipal Solid Waste Landfills, Design Criteria* (Draft background document) (Subpart D). 1988b.

United States Environmental Protection Agency. *Design, Construction, and Evaluation of Clay Liner in Hazardous Waste Facilities*. 1985a.

United States Environmental Protection Agency. *Lining of Waste Containment and Other Impoundment Facilities*. 1988c.

United States Environmental Protection Agency. *Loading Point Puncturability Analysis of Geosynthetic Liner Materials*, Southwest Research Institute. 1988d.

United States Environmental Protection Agency. *Solid Waste Disposal Facility Criteria.* 40 CFR Parts 257 and 258; proposed rule. 1988a.

United States Environmental Protection Agency. *Stability of Lined Slopes at Landfills and Surface Impoundments*. by D.M. Mitchell, M.A., McLean and T.E., Gates. August 1988e.

United States Environmental Protection Agency. *The Background Document on Proposed Liner and Leak Detection Rule*. GeoServices (Currently GeoSyntec, Inc.), Consulting Engineers.

United States Environmental Protection Agency. *"Wastes from the Extraction and Beneficiation of Metallic Ores, Phosphate Rock, Asbestos, Overburden from Uranium Mining and Oil Shale.* Report to Congress. 1985b.

United States Geological Survey. *Documentation of Computer Program VS2D to Solve the Equations of Fluid Flow in Variably Saturated Porous Media.* Water Resources Investigations Report 83-4099. 1987.

United States Geological Survey. *Simulation of Solute Transport in Variably Saturated Porous Media With Supplemental Information Modifications to the U.S. Geological Surveys Computer Program VS2D*. Water Resources Investigations Report 90-4025. 1990.

United States Geological Survey. *A Finite-Element Simulation Model for Saturated-Unsaturated, Fluid-Density-Dependent Ground Water Flow with Energy Transport a Chemically Reactive Single-Species Solute Transport*. SUTRA. June 1991.

Warrick, A.W. *Time-Dependent Linearized Infiltration: 1. Point Sources*. Soil Sciences Society of America Proceedings. Volume 38. 1974.

Williams, M.N., and M. Houlihan. *Evaluation of Friction Coefficient Between Geosynthetics and Soils*. Geosynthetics Conference. New Orleans, Louisiana, February 24-25, 1987.

Williams, N., J.P. Giroud, and P. Bonaparte. *Properties of Plastic Nets for Liquid and Gas Drainage Associated with Geomembranes*. Proceedings of the International Conference on Geomembranes. Vol. II. Denver, Colorado. June 1984. 399-404.

Wilson, S.D. and R.J. Marsal. *Current Trends in the Design and Construction of Embankment Dams*. New York: American Society of Civil Engineers. 1979.

Wong, J. *The Design of a System for Collecting Leachate from a Lined Landfill Site,* Water Resources Bulletin. April 1977.

CONTENTS

CHAPTER 8
CLOSURE REQUIREMENTS

PAGE NO.

LIST OF TABLES/LIST OF FIGURES 391

8.1 Introduction 393
- 8.1.1 Background 393
- 8.1.2 Integration With Other Chapters 394
- 8.1.3 Study Participants 394
- 8.1.4 Sources of Information 395
- 8.1.5 Chapter Organization 395

8.2 Closure Objectives 395

8.3 Description of Mine Waste Management Units 396
- 8.3.1 Introduction 396
- 8.3.2 Waste Piles 398
  - 8.3.2.1 Overburden and Waste Rock 398
  - 8.3.2.2 Dried Tailings 399
  - 8.3.2.3 Heap Leach Residue 400
- 8.3.3 Tailings Impoundments 401
- 8.3.4 Wastewater Ponds 404
- 8.3.5 Conclusions 404

8.4 Designing for Closure 404
- 8.4.1 Introduction 404
- 8.4.2 Closure Performance Goals and Approaches 405
  - 8.4.2.1 Alternatives for the Mining Industry 405
  - 8.4.2.2 Practices at Hazardous Municipal, and Uranium Tailings Waste Management Unit 406
  - 8.4.2.3 Relevance of Other Industry Practices to Mine Wastes 409
- 8.4.3 Factors To Be Considered for Closure Design 410

8.5 Closure Technologies and Design Elements 411
- 8.5.1 Introduction 411
- 8.5.2 No Action 412
- 8.5.3 Institutional Controls 412

CONTENTS

CHAPTER 8
(Continued)

PAGE NO.

8.5.4 Conditioning or Treatment 414
8.5.5 Encapsulation 415
8.5.5.1 Introduction 415
8.5.5.2 Covers 415
8.5.5.3 Surface Water Management 433
8.5.5.4 Enhancement of Waste Containment 433
8.5.6 Waste Removal for Offsite Disposal 434
8.5.7 Closure Design 434
8.5.7.1 Important Perspectives 435
8.5.7.2 Special Considerations 435
8.5.8 Case Histories 436
8.5.8.1 Case History No. 1 - Closure of an Abandoned Dry Tailings Pile 436
8.5.8.2 Case History No. 2 - Closure of an Active Waste Rock Pile 436
8.5.8.3 Case History No. 3 - Closure of an Abandoned Tailings Impoundment 440
8.5.8.4 Case History No. 4 - Closure of a Heap Leach Pile 441
8.6 Regulations for Closure 442
8.6.1 General 442
8.6.2 Current California Regulations 442
8.6.2.1 Current Regulations 443
8.6.2.2 Evaluation of Current Regulation 445
8.6.3 Alternative Regulatory Approaches 445
8.6.4 Recommended Regulatory Approach 446
8.7 Bibliography 448

CONTENTS

CHAPTER 8
(Continued)

LIST OF TABLES

TABLE NO.	TITLE	PAGE NO.
8.1	Characteristics of Mine Waste Management Units at Closure	397
8.2	Design Guidelines for Cover Elements	420
8.3	Filter Criteria	424
8.4	Typical Closure Plans	426

LIST OF FIGURES

FIGURE NO.	TITLE	PAGE NO.
8.1	Closure Technologies, Alternatives, and Design Elements	413
8.2	Typical Mine Waste Management Unit Covers	416
8.3	Examples of Gravel Envelopes for Various Base Materials	428

CHAPTER 8

CLOSURE REQUIREMENTS

8.1 INTRODUCTION

8.1.1 BACKGROUND

Closure of a waste management unit can be either permanent or temporary. Permanent closure may be due to:

- Economic factors which make it uneconomical to mine the remaining minerals.
- Depletion of mineral resources.
- Physical site constraints that preclude further mining and beneficiation.
- Environmental, regulatory or other requirements that make it uneconomical to continue to develop the resources.

Temporary closure can occur for a period of several months to several years, and may be caused by factors such as:

- Periods of high rainfall or snowfall which prevent mining and waste disposal.
- Economic circumstances which temporarily make it uneconomical to mine the target mineral.
- Labor problems requiring a cessation of operations for a period of time.
- Construction activities that are required to upgrade project components such as the process facilities and waste management units.
- Mine or process plant failures that require extensive repairs.

Permanent closure of a mine waste management unit involves the provision of durable surface containment features to protect the waters of the State in the long-term. Temporary closure may involve activities that range from ongoing maintenance of the existing facilities to the installation of several permanent closure features in order to reduce ongoing maintenance. This chapter deals with the permanent closure features.

The closure objective for a waste management unit is the long-term protection of the beneficial uses of water. Appropriate methods for closure should be selected on the basis of the waste characteristics, as well as local climatic and site conditions. The operational history of the unit and the engineered containment features, such as a liner system, also should be taken into consideration. This is the same approach that is recommended for waste containment design in Chapter 3 and liner system design in Chapter 7.

The objectives of this chapter are to describe mine waste management units with an emphasis on the characteristics that are important to closure, outline performance and design criteria for closure, and describe the range of available closure technologies and design elements. Several case histories of closure also are provided. The California regulations dealing with closure are reviewed, and recommendations are provided on how the approach and criteria described in this chapter could be included in a regulatory framework.

8.1.2 INTEGRATION WITH OTHER CHAPTERS

The methodologies outlined in this chapter are related to information presented in several other chapters. As discussed above, the containment assessment approach developed in Chapter 3 is also applicable to closure design. The acid generating potential testing outlined in Chapter 4 is used to determine the likely short- and long-term chemical constituent concentrations that must be controlled by the closure plan for a waste management unit containing acid generating waste. Chapter 5, which deals with the attenuation mechanisms, contains information that is useful for determining the ultimate fate of any residual migration of constituents from a unit, and facilitates a closure design that prevents this migration from impacting the beneficial uses of water.

The information on infiltration in Chapter 6 is also relevant to the design of cover systems required for closure to ensure that these systems adequately control long-term infiltration through the waste. Chapter 7 contains liner system performance information which can be applied to the design of cover systems, as both liners and cover systems contain the same layer types. Ground water monitoring system design, discussed in Chapter 10, is important to the establishment of post-closure monitoring programs that are required to confirm the performance of closure activities.

8.1.3 STUDY PARTICIPANTS

The following persons are the principal authors of this chapter:

- Michael E. Henderson, P.E., Welsh Engineering
- Ian P. C. Hutchison, Ph.D., P.E., Environmental Solutions, Inc.
- Don Poulter, P.E., DP Engineering
- Dirk van Zyl, Ph.D., P.E., EIC Corporation

8.1.4 SOURCES OF INFORMATION

The information sources utilized for this chapter include the California Mine Waste Regulations[(1)], case histories, discussions with various state regulators, and United States Environmental Protection Agency (EPA) guidance documents for other waste types.

8.1.5 CHAPTER ORGANIZATION

Section 8.2 addresses the objectives of closure and describes a general approach that can be adopted to develop a closure plan. Characteristics of typical mine waste management units at the time of closure are described in Section 8.3. Designing for closure is discussed in Section 8.4, which outlines performance goals, design approaches, and other factors to be considered. Section 8.5 describes appropriate closure technologies and design elements for waste piles, tailings impoundments, and ponds. It also includes a description of four case histories. Section 8.6 considers the regulation of closure. The current California regulations are reviewed, and recommendations are made for additional regulatory considerations which would be specific to mine wastes.

8.2 CLOSURE OBJECTIVES

The objective of closure is to provide for the continued protection of the beneficial uses of the waters of the State. This is consistent with the waste containment objectives discussed in Chapter 3. Closure therefore should be planned and designed to eliminate or minimize, to the extent necessary, the potential for release of waste constituents to surface and/or ground water systems.

In order to satisfy these closure objectives, the following two potential waste constituent migration pathways should be evaluated and controlled, to the extent necessary, by the closure plan: (1) wind and surface water erosion of surface materials, and (2) seepage caused by excess interstitial liquids contained within the waste, precipitation infiltration, or both. While the existing waste containment system, which may include a liner system, plays the major role in seepage control, any surface preparation of the waste or cover placement that is included in closure will provide important supplementary control by eliminating or minimizing infiltration and, hence, potential moisture accumulation in the waste.

[(1)] California Code of Regulations, Title 23, Division 3, *Chapter 15*, Article 7.

Air transport and surface erosion by water can disperse a mine waste. Air transport becomes significant for particle sizes below .01 to 0.1 millimeter (mm). Surface water flow, on the other hand, is capable of transporting particles sizes up to approximately 1/8 inch (3 mm) over substantial distances. The resulting potential water quality impacts include: (1) sedimentation of surface water bodies, and (2) increases in dissolved constituent levels either due to water leaching the surface of the waste management unit or mine waste that is dispersed outside the waste management unit. Dispersion of the waste also can result in impacts to surrounding ecological systems or human health due to adsorption, ingestion, and/or inhalation of the particles.

Seepage from a mine waste management unit has the potential to impact the quality of ground water below and/or downgradient from the facility, as well as the quality of downgradient surface waters.

The general approach to closure is to develop measures that control the wind and surface water erosion, and the seepage migration pathways to the extent necessary, to prevent impacts to the beneficial uses of water. Measures for controlling the surface pathways can include stabilizing the surface of the waste and providing an erosion resistant cover. Measures to control seepage supplement the existing containment measures and can include surface grading, placement of a cover to reduce infiltration or, on occasion, providing supplementary waste containment.

As the objective is to achieve effective closure for the long-term, the emphasis should be on "passive-care" approaches such as the provision of a durable cover. Closure features should be designed for a substantial life and to minimize maintenance. "Active-care" approaches such as collection and chemical treatment of contaminated water, should be avoided, if possible, or limited to short-term applications, and then replaced by passive-care approaches.

8.3 DESCRIPTION OF MINE WASTE MANAGEMENT UNITS

8.3.1 INTRODUCTION

Mine waste includes waste piles, tailings impoundments, and wastewater ponds. The characteristics of these wastes and of waste management units at the time of closure are outlined in Table 8.1 and described in the following paragraphs. The potential closure issues associated with each type of waste management unit also are presented in Table 8.1 and the discussion below. The information presented here supplements the description of mine waste and mine waste management units provided in Chapters 2, 3, and 5.

TABLE 8.1

CHARACTERISTICS OF MINE WASTE MANAGEMENT UNITS AT CLOSURE

WASTE UNIT AND CONTAINED WASTE	TYPICAL CHARACTERISTICS OF WASTE	TYPICAL WASTE MANAGEMENT UNIT CONFIGURATIONS	POTENTIAL WATER QUALITY ISSUES AT CLOSURE
Waste Piles • Waste rock and overburden	• No "free" water. • Gravel to boulder size material. • May include some finer grained material due to soils in source, e.g. overburden or weathering of some of the waste rock constituents.	• Three typical configurations: - Valley pile - Sidehill pile - Open pile • Usually unlined unless designed to contain acid rock drainage (ARD) in a humid climate. • Sediment control basins provided in humid climates.	• Erosion, with consequential sediment discharge to surface water. • Acid leachate generation and metal transport to surface and ground waters. • Slope failure, leading to sediment discharge and waste dispersion. • Windblown transport of fines.
• Dried tailings	• Dry to wet, depending on deposition method and climate. • Silt- to sand-sized. • May have the potential to leach waste constituents or residual reagents.	• Lined or unlined area. • Sediment detention basins incorporated in, or attached to, the disposal area. • May have lined runoff collection ponds.	• Erosion, with consequential sediment discharge to receiving waters. • Flood water accumulation, resulting in overtopping and breaching of embankments. • Potential leaching of metals and residual chemical constituents to surface or ground water.
• Heap Leach Residue	• Moist material, usually no free liquid present. • Sand- and gravel-size, with fines. • May have the potential to leach waste or residual reagents.	• For a reusable pad operation, the typical configurations would be as for waste rock and overburden. Disposal units may be lined. • For a dedicated pad operation the waste would be located in an open or valley pile on a liner system. • Lined solution ponds containing remnant beneficiation solutions. • Lined or unlined overflow ponds designed to temporarily contain diluted beneficiation solutions during extreme flood events.	• Erosion, with consequential sediment discharge to receiving waters. • Leaching of metals and residual chemical constituents to surface or ground water. • Slope failure, leading to sediment and dissolved constituent discharge to surface or ground waters. • Windblown transport of fines.
Tailings Impoundments • Tailings deposited as a slurry. • Tailings deposited as a slurry in a managed manner to promote in situ drying.	• Free water present in slurried tailings. • Little to no free water available for ponding and seepage in dry or managed tailings. • Crushed and ground to a sandy silt- or clay-size material. • May have the potential to leach waste or residual reagents.	• Types of tailings impoundments: - Downstream construction - Centerline construction - Upstream construction - Water-retaining embankment • Containing embankment constructed with tailings, natural materials, or both. • Clay, synthetic, or natural liner often present.	• Leachate formation and seepage of waste constituents to surface and ground waters. • Erosion, with consequential sediment and dissolved constituent discharge to surface and ground waters. • Slope failure, leading to sediment or dissolved constituent discharge to surface or ground waters. • Windblown transport of fines. • Embankment failure of slurried tailings, particularly during storm events.
Wastewater Ponds • Solutions	• Water containing few or no solids. • Dissolved beneficiation reagents and constituents present in the ore can occur.	• Unlined ponds for wastewater that do not pose a threat to beneficial uses of water. • Lined ponds (in many cases double lined) with leak detection systems between the liners for wastewaters that pose a potential threat to beneficial uses of water.	• Seepage of solutions into surface and ground waters. • Residue of sludges or precipitates that may impact beneficial uses of water. • Flood water accumulation, resulting in overtopping and breaching of embankments.

8.3.2 WASTE PILES

8.3.2.1 Overburden and Waste Rock

Characteristics

Overburden is derived from soils and soil-like materials. Typically, the particle size of overburden ranges from boulder size to gravel or sand size. Finer fractions such as silt and clay also can be present, sometimes in significant quantities. This material usually is chemically inert. Physical characteristics of overburden are typical of soil materials in terms of weight, moisture content, degree of consolidation, and strength characteristics.

Waste rock is material from uneconomical zones in the mine pit and usually consists of lithologic units that generate gravel- to boulder-sized material. The nominal rock size and percentage of fines is a function of the method of mining, such as blasting or ripping, haulage and dumping techniques, and the competency of the host rock. The percentage of sands and fines is usually minimal unless some of the lithologic units weather rapidly to form a clay. The material's strength is typically controlled by the gravels and boulders. The material is usually fresh or unweathered and can be acid generating (see Chapter 4).

Waste Management Units

Waste rock and overburden may be disposed of in valley fills, sidehill dumps, or open piles (Table 8.1). The type of waste management unit is a function of the volume of waste, site topography and drainage, and haul distance from the mine. Unless located in an arid climate, these waste management units will have flood runoff detention ponds for purposes of sediment control, and surface runoff diversion ditches to minimize natural runoff intruding on the pile. Even in arid climates there may be a need to install flood diversions to deal with flash flood events.

Disposal of waste rock and overburden is by end-dumping from trucks, or stacking using conveyors. The material is dumped over the face of a lift. The lift may be the full height of the dump, i.e., up to several hundred feet, or a 50- to 100-foot intermediate bench. The working face of a lift is at the angle of repose of the material. As the dumped material cascades down the face, segregation of particles occurs. The large boulders roll to the bottom and the gravels, sands, and fines remain near the top. The higher the working face of a lift, the more prominent the segregation. As a result, the bottom of disposal facilities is frequently free-draining rock, while the upper elevations and working surfaces are less permeable.

The upper surfaces of the pile are usually flat. They are gently sloped away from the dump face toward the natural terrain to enhance dump truck safety during operation, and also to prevent surface runoff during storm events from cascading over the edge of the piles.

Valley fills usually are started at the upstream end of the valley. Dumping progresses downstream, with increasing lift thickness. Sidehill piles are formed by dumping along hillsides or valley slopes so that the toe of the dump does not cross the drainage. Dumping is continued along the natural contour and out from the slope. Surfaces are graded to keep runoff away from the exposed face of the dump. An open pile may be constructed over relatively flat topography or as a ridgeline fill that straddles the crest of a ridge. One or both sides of the ridgeline fill can be extended to form a sidehill dump.

Potential Environmental Issues

Two primary potential environmental issues of closure of waste rock and overburden piles are: (1) erosion and subsequent sediment discharges into the receiving waters, and (2) acid leachate of dissolved metals to receiving waters.

Other potential water quality issues which also may have to be addressed in closure are long-term slope stability, prevention of windblown dust, and the long-term utility and stability of sediment control structures at the toe of the waste management unit.

8.3.2.2 Dried Tailings

Characteristics

The maximum particle size of tailings is coarse sand or smaller (i.e., less than 4.75 mm). Often, the percent of material finer than the No. 200 U.S. Standard sieve size (0.074 mm; i.e., silt and clay) is greater than 50 percent by weight, and may be as high as 80 to 90 percent.

"Dewatering" or "drying" of the tailings occurs in the metals recovery plant using unit processes, such as thickening, pressure filtration, vacuum filtration, and a combination of these. The residual moisture content is usually at or just below saturation, and generally varies from 15 to 25 percent[(2)], however for very fine grained material higher values can occur (Robertson, et.al.,

---

(2) For the purposes of *Mine Waste Management*, the moisture content is expressed as the weight of the water as liquid component over the weight of the solids component.

1982). The dry density of these materials will generally vary from 75 to 100 lb/ft^3[(3)] depending on how much compaction occurs when the material is deposited.

Waste Management Units

Deposition of dried tailings is accomplished by either truck dumping and spreading with a bulldozer, or by conveyor transport and stacking. This type of deposition results in a pile with no pools of free-standing solution. Direct precipitation onto the pile is allowed to drain off and collect in a separate holding pond. Depending upon the degree of dewatering, residual seepage of solution can occur as a result of gravity drainage and consolidation of the tailings. Where the tailings contain a high proportion of silt and clay sized fractions, some form of containment may be necessary to prevent the tailings from spreading.

Potential Environmental Issues

Leachate formation and potential seepage of contaminants from the impoundment area into the ground waters and downgradient surface waters are usually the most significant issues when closing a tailings pile. Also important is the long-term stability of the tailings pile particularly under storm conditions which could result in erosion damage and/or floodwaters overtopping and washing out portions of the pile or any containment embankments. Partial or complete failure of the pile or the embankments could lead to adverse sediment loads and contaminant concentration in downstream surface waters. Air transport of the dried tailings particles is also an issue at those waste management units where the tailings do not exhibit cohesion or chemical cementation properties.

8.3.2.3 Heap Leach Residue

Characteristics

Generally, spent heap leach ore can be described as a coarse gravel material. However, it can vary from a sandy gravel to a waste rock type material with small boulders. The actual grain size distribution is a function of the method of ore preparation prior to its placement on the leach pads. The moisture content of a rinsed ore is a function of the grain size distribution of the ore, the amount of cement or lime that was added during ore precipitation or the leach operation, and the

(3) Densities are expressed as a dry weight and apply to materials that have a specific gravity of 2.6.

time since leaching and rinsing was terminated. Moisture contents immediately following the cessation of rinsing can be as high as 15 to 20 percent by weight. Well-drained materials probably will have a residual moisture content of 5 to 15 percent.

Waste Management Units

The most common way of handling the residue is to use a dedicated or expandable pad system, where ore is placed on a pad, leached, and left in place. Additional ore then is stacked in subsequent lifts, or on new pad(s) that are constructed. An alternative is the reusable pad, where the residue is removed upon completion of rinsing and disposed of in a separate waste management unit. For both methods, the heaps usually are constructed on relatively flat ground. In mountainous terrain, however, they may be constructed as a valley fill. Heap leach operations are discussed in detail in Chapter 9.

In the case of a dedicated pad, the ore pile and pad used for leaching becomes the waste management unit. Its configuration is generally that of a stacked or open pile. The pad usually is lined with a geomembrane and in some cases with clay. The pad liner is either a single to a double liner with a leak detection system (see Chapters 7 and 9).

Reusable pads usually are lined with asphalt, and disposal areas for the spent ore are separate from the leach area. As for waste rock (Section 8.3.2.1), valley fills and sidehill piles are common types of disposal, with leachate collection and sediment control usually included. The area may be lined to reduce infiltration of leachate to ground water.

Potential Environmental Issues

Environmental issues that should be addressed when closing heap leach facilities include seepage of dissolved constituents from the spent ore, the disposal of remnant beneficiation solutions that may be contained in the ponds, as well as the residual pond sludges and precipitates. Erosion control and dust generation by wind action also may have to be managed.

8.3.3 TAILINGS IMPOUNDMENTS

Characteristics

The grain size distribution of deposited tailings are similar to those for dried tailings discussed under 8.3.2.2.

The moisture content and density of deposited tailings are functions of the ore extraction process, method of deposition, grain size distribution, and specific gravity, and can vary from 20 to over 60 percent. For slurried deposition moisture contents are saturated and typically vary from 40 to 60 percent and dry densities from 60 to 80 lb/ft^3. The moisture content is typically higher and the density lower than for dried tailings. At the time of closure the density of a tailings deposit generally increases with depth. The lower tailings have usually consolidated, due to the weight of the upper tailings and a long residence time, while the upper tailings are still in the process of consolidating.

Where managed,(4) e.g. layered, deposition of tailings has occurred, the densities may be higher than the slurried deposition, i.e., typically 80 to 90 lb/ft^3. As a result there is likely to be less settlement of the surface of the tailings after closure. With lower moisture contents, typically 20 percent to 40 percent, and no standing water, the surface of managed tailings becomes accessible for closure activities in a shorter period of time than slurried tailings, and usually without the need for special construction techniques.

The actual values of the deposited tailings moisture content and densities usually will be different for each impoundment. Therefore, if closure design relies on an accurate knowledge of these parameters, field sampling and testing may be required.

Waste Management Units

Tailings are typically deposited in an impoundment formed by an embankment or dike. The embankment can be constructed of fill material or, in some instances, of the tailings themselves. Construction with tailings material involves controlled spigotting(5) of the tailings and allowing the solutions to drain from the solids. In some instances, the tailings may be cycloned to separate the coarser fraction, which then is used for embankment construction.

An impoundment can be formed by constructing a single embankment across a valley or a three-sided embankment on the side of a valley or on sloping ground. On flat ground, an embankment or a "ring" dike that completely surrounds the deposited tailings materials, is required.

---

(4) Managed deposition is also referred to as sub-areal deposition.

(5) Spigotting refers to the discharge of tailings through several small pipes or spigots.

Deposition

Slurried deposition is the most common and practical means and can be accomplished by the methods discussed in the following paragraphs. Each method may result in different closure plan requirements.

- Subaqueous Deposition involves discharge under water in the impoundment. Deposition is uncontrolled, and the tailings settle as a soft bottom sediment, or are transported and dispersed over a large area. The resulting low density material is generally very soft and may require special construction techniques should a cover be required during closure.
- Managed Deposition is typically accomplished by discharging the slurry from one or more points around the perimeter of the impoundment area. As the slurry drains toward the low point in the impoundment, the solids spread out over the bottom to form beaches or deltas. The free water drains to a central location, or low point along the perimeter, where it either is recycled into the processing system, decanted to a separate containment facility for reuse or evaporation, or evaporated in place. The point or points of discharge are moved around the impoundment to allow the exposed solids to dry and solidify. Subsequent layers are then deposited, and the cycle is repeated.
- Conventional Point Discharge is similar to managed deposition, except that no effort is made to add additional discharge points or frequently move the discharge points. As a result, the tailings are usually not as dense or as dry as with managed deposition. Also, with conventional point discharge, the finer portion of the solids tends to settle out in the same area and create a soft area that is difficult to access, and subject to additional settlement/consolidation over time.
- Cycloned Deposition involves gravity separation of the coarser tailings fraction prior to deposition. This type of deposition creates a zone or pile of coarse solids relatively free of excess solutions, plus a pool of solutions and wet fine solids which may not drain readily and consolidate. Operation of a cycloning[(6)] system requires planning and a significant amount of management if the system is to function properly. It usually is done, at least in part, to provide coarse tailings for use as fill in raising the embankment.

Potential Environmental Issues

The environmental issues are similar to those that occur at a dried tailings with the exceptions that pile stability is usually a lessor concern, and seepage issues may potentially be of greater concern. Pile stability is generally not a significant issue as tailings are very seldom stacked to heights that exceed the crest of the containment embankments. Potential seepage issues may be greater due to the presence of more free liquids in the waste than for dried tailings.

---

(6) Cycloning uses centrifugal forces to separate the fine tailings fraction from the coarser fraction. The coarse fraction drains and consolidates more readily than the fines and therefore is a more useful construction material.

8.3.4 WASTEWATER PONDS

Waste Management Units

Waste solution ponds are frequently single- or double-lined and may include leak detection systems and solution return sumps and ponds. These ponds either are excavated below surface or constructed using earthen embankments.

Exceptions are ponds constructed for purposes of sediment control only. These ponds generally are unlined and designed to discharge the water to satisfy National Pollution Discharge Elimination System (NPDES) permit conditions. These sediment control facilities are usually earthen embankments designed with outlet works and emergency overflow spillways.

Potential Environmental Issues

The environmental considerations that must be addressed when closing ponds involve removing the solutions contained in the ponds, and dealing with any sludges or precipitates which may have deposited on the liner system. Clay liners also may contain constituents that have seeped into the liner material during the operation of the pond.

Closure of sediment control ponds requires dealing with the accumulated fine-grained sediment which, if not contained, could erode and impact downgradient surface water bodies, as well as maintaining the long-term stability of any embankments that are not removed as part of closure.

8.3.5 CONCLUSIONS

The information presented above illustrates that mine waste management units at closure and the associated mine waste characteristics are extremely variable, and that they are largely dependent on site-specific conditions at the various mine sites. This indicates that the potential water quality impacts, which are a function of the waste characteristics, the type of mine waste management unit, its performance, and site conditions, also are variable. These differences make it necessary to consider the site-specific conditions when planning, designing, and regulating closure.

8.4 DESIGNING FOR CLOSURE

8.4.1 INTRODUCTION

This section describes performance goals which are developed from the closure objectives outlined in Section 8.2, and discusses different approaches that can be used to develop closure design standards, and describes the factors to be considered when designing for closure. Also included is a brief review of the approaches adopted in other industries.

The timing of the planning and design for closure is also an important issue. As a minimum, conceptual level plans and designs should be completed during the initial planning and design of the mining operation to ensure that the proposed waste containment features are consistent with cost-effective closure methods.

8.4.2 CLOSURE PERFORMANCE GOALS AND APPROACHES

8.4.2.1 Alternatives for the Mining Industry

To meet the closure objectives stated in Section 8.2, the following performance goals should be established for the closure of a waste management unit:

- Provide adequate encapsulation of the waste material to avoid impacting beneficial uses of water by leachate or seepage.
- Provide adequate surface drainage and erosion protection to prevent surface water transport of waste materials beyond the management unit boundary.
- Condition the surface of the waste to prevent windblown transport of waste materials that could impact water quality.
- Design and construct the closure elements to maximize durability and minimize ongoing maintenance requirements.
- Minimize need for post-closure active-care technologies such as chemical treatment of leachate.

In designing closure to meet these performance goals, the same alternative procedures for obtaining design and construction standards that were listed for liner systems (Chapter 7) can be considered. These include:

- Use of uniform design criteria.
- Use of categorical design criteria.
- Use of a performance goal approach.

These different approaches are discussed in detail in Chapters 3 and 7 in terms of waste containment, and are briefly summarized in the following paragraphs in terms of closure.

Because of the large variability in mine waste types and mine waste management unit configurations, it is unlikely that uniform design criteria will result in suitable cover systems for all units. The criteria would have to be established for assumed worst-case conditions that would be uneconomic in some cases, infeasible in others, and under some conditions provide a technically inferior solution.

The use of categorical design criteria has merit, provided the categories selected reflect the important differences in the various types of mine wastes, as well as the potential range of site climatic, geologic, and other conditions. As it is difficult to establish a manageable number of categories that still adequately address the wide range of variables that must be taken into account, this approach will invariably incorporate some of the disadvantages of the uniform design criteria approach discussed above. As discussed in Chapter 7, the advantage of the categorical design criteria are their ease of application.

The performance goal approach allows a design to be based on a containment assessment and includes consideration of climate, waste characteristics, waste containment features, and actual site-specific conditions, including geology, geohydrology, and the proximity to, and nature of, ground and surface water bodies. This approach leads to the selection of appropriate closure waste treatment, removal, cover construction or other controls, that when coupled with other site-specific factors listed above, would meet closure objectives and performance goals. It allows closure technologies, or methods, and design standards to be optimally selected for the particular site. As discussed in Chapter 3, this approach has the most potential to yield a cost-effective and environmentally sound closure plan. It requires that more knowledge of the waste management unit and the site be developed than the other approaches, and is more conducive to the introduction of new technology.

8.4.2.2 <u>Practices at Hazardous, Municipal, and Uranium Tailings Waste Management Units</u>

Standards and methods for closing waste sites have been developed for other types of waste disposal units. These include uniform, categorical and performance-based criteria, and are briefly summarized below for hazardous, municipal, and federal uranium mill tailings wastes.

<u>Hazardous Waste Management Units</u>

Performance standards (40 CFR Parts 260-264) and uniform design criteria guidelines (EPA, 1987a) have been developed by EPA for closure of hazardous waste management units. The general performance goals for ponds (surface impoundments),[7] waste piles, and landfills include

---

[7] In the regulations, ponds are referred to as surface impoundments.

providing a cover that minimizes migration of liquids, functions with minimum maintenance, promotes surface drainage, minimizes erosion, accommodates settling, and has a hydraulic conductivity that is equal to or less than the bottom liner system or the natural subsoils underlying the waste management unit. For ponds, the residual wastes can be removed and the pond closed. Alternatively, the free liquids can be removed, the residual wastes stabilized and closure completed. For waste piles, removal of free liquids prior to closure is required, while for landfills the leachate collection system is required to remain operational to prevent the accumulation of free liquids. The uniform criteria guidelines include a detailed description of the type of cover that should be installed at hazardous waste ponds, landfills and piles, including the number, type, and thickness of individual layers. These layers include, from top to bottom, vegetation, soil, filter, drainage, composite infiltration barrier, and potentially other special purpose layers, each with a defined thickness.

This prescriptive uniform design represents one method of constructing a cover which, because of its universal application to all hazardous waste management units, must provide adequate containment for worst case conditions. Such conditions likely include a wet climate, free liquid in the waste, high concentrations of hazardous constituents, and receiving water bodies in proximity to the waste management unit. The EPA, however, does point out that these are guidelines and not regulations and that alternative cover designs may be acceptable depending on site-specific conditions.

State of California requirements for hazardous waste management units are regulated by the Department of Health Services (CCR, Title 22, Division A, Chapter 30) and provide both performance criteria as well as uniform prescriptive design criteria. The major difference when compared to the federal requirements is that the state design criteria are regulations while the federal design criteria are guidelines. However, the prescriptive cover requirements deal only with the cover layer types and do not include layer thicknesses. As for the federal guidelines the state requirements do allow alternative cover designs provided it provides an equivalent level of protection for public health and the environment.

The state requirements for a cover, for ponds (surface impoundments), waste piles, and landfills include the following:

- Function with minimal maintenance
- Accommodate settling

- Include the following layers from top to bottom:
  - Vegetation
  - Topsoil
  - Filter
  - Drainage
  - Compacted soil or geomembrane barrier
  - Foundation

Municipal Waste Management Units

In California, closure of municipal waste management units is regulated by both federal regulations (40 CFR Draft Part 258), and by the California Integrated Waste Management Board (CCR, Title 14, Division 7, *Chapter 3*) and the Regional Water Quality Control Board (CCR, Title 23, Division 3, *Chapter 15*). The performance goals in the federal regulations require minimization of infiltration, maximization of runoff without excessive erosion, and promotion of vegetative growth. Uniform design criteria require 18 inches of soil material with a hydraulic conductivity less than or equal to that of the bottom liner or natural formations and no greater than 10^{-5} cm/sec, and a 6-inch layer for erosion protection and support of vegetation.

The state regulations incorporate similar general requirements, however, provide more detailed minimum uniform criteria such as a 2-foot thick foundation layer, a 1-foot thick infiltration barrier of 10^{-6} cm/sec or 10^{-7} cm/sec hydraulic conductivity and a 1-foot layer of soil to support vegetative growth. The hydraulic conductivity of the cover also needs to be less than that of any bottom liner system or underlying natural geologic materials. Free liquid must also be removed by continued operation of the LCRS.

The regulations allow for an alternative final cover provided it achieves the equivalent performance as one constructed to the prescriptive design criteria.

Uranium Mill Tailings Management Units

The United States Department of Energy (DOE) is mandated by federal law to remediate 24 inactive uranium mill tailings impoundments in 24 states. The remediation work is well advanced on this billion-dollar program. A parallel effort in progress on the part of private industry is intended to ultimately remediate an additional 25 uranium mill tailings impoundments.

As described by Caldwell (1990), the regulations that govern the remediation and closure of both the federal and the private sector facilities are similar. For both, the overriding requirement is to provide closure works that will remain stable for at least 200 years and, to the extent reasonably achievable, for 1,000 years. Covers and other components are to be provided to control the rate of radon flux, inhibit infiltration, limit erosion, and prevent inadvertent human and animal intrusion and dispersion of the wastes. The remediated work must be designed to minimize the need for maintenance, and ongoing active maintenance is not permitted.

On the DOE Uranium Mill Tailings Remedial Action (UMTRA) Project, approximately half of the piles have been or will be relocated to remove them from population centers, riverbanks where flood erosion protection is not feasible, or from areas where they would negatively impact ground water resources. The other half are being remediated in place. Both in place and relocated tailings are being shaped to flatten slopes, enhance runoff control, promote drainage, and provide an aesthetic configuration. A multilayered cover is being placed over the reshaped tailings piles.

The details of the multi-layer cover differ, depending on the site location and the nature of the encapsulated wastes as described by Caldwell (1990 and 1991), DOE (1989), and Portillo (1991). The covers include the following basic elements. The first layer placed over the tailings is compacted clay or silt that controls radon gas emission and limits rainfall infiltration. The next layer is generally a drain of clean sand to expedite shedding of water and to act as a bedding to the overlying layers. At arid sites, the final layer is durable erosion-resistant rock. At other sites, the upper layers consist of a thick zone of soil for frost protection, plus layers of suitable soil and rock that control erosion and facilitate the establishment of vegetation.

8.4.2.3 Relevance of Other Industry Practices to Mine Wastes

The requirements outlined above include a combination of general performance criteria and prescriptive uniform design criteria. While the state design criteria for hazardous and municipal wastes are presented as categorical, i.e., they are provided for each type of waste management unit, the technical cover requirements are actually the same for each type of waste and are therefore effectively uniform criteria.

From a technical perspective, there are differences in the manner in which the various types of wastes are managed. Hazardous wastes require an elaborate prescriptive multilayered cover and municipal waste a specific but less elaborate multilayered cover. While uranium tailings also require a multilayered cover, the performance requirement for some of these layers have to address the escape of radioactivity and are therefore different from those associated with hazardous or municipal waste management unit covers. Furthermore, the cover details are developed based on site-specific conditions.

Considering that the characteristics of the different types of mine wastes described in this book are more diverse than the different types of wastes described above, it becomes even more apparent that there is a need to treat mine wastes using the performance goal approach, which would allow flexibility to develop the necessary cover specifications, or the categorical design criteria approach, where each category applies to a specific mine waste management unit in its site-specific climatic settings.

8.4.3 FACTORS TO BE CONSIDERED FOR CLOSURE DESIGN

As discussed previously, factors that should be considered when designing for closure include:

- Physical and chemical characteristics of the waste.
- Climatic, geologic, geohydrologic, and seismic conditions at the site.
- Proximity of ground or surface water bodies.
- Water balance of the closed waste management unit, i.e., the extent, if any, to which free liquids may continue to accumulate in or migrate through the waste.
- Engineered waste containment provided.
- Performance of the waste management unit during the operating period.

The physical characteristics of the waste generally determine the type of erosion protection required, and the foundation material and construction that is needed, should a cover be necessary. The chemical characteristics of the waste, the amount of rainfall, and the existing waste containment features and liners, determine the need for and requirements of a cover. Low infiltration covers are required when the climate is sufficiently wet to cause free liquids to accumulate in the waste, and where the chemical constituents contained in these free liquids pose a threat to beneficial uses of water. It is usually good practice to design these elements of a cover system to have an effective infiltration rate which is equal to or less than the seepage potential through the liner system or natural geologic formation underlying the waste.(8) This will reduce the potential for quantities of free liquid to accumulate in the waste. The types of cover

(8) This performance goal is frequently stated differently in regulations, i.e., the hydraulic conductivity of the cover should be equal to or less than that of the liner system. The way it is stated in *Mine Waste Management* is more accurate, as it addresses the key issue of water balance and incorporates differences in hydraulic head on the cover and liner systems.

components that should be considered are generally the same as those for liner systems which include clay or geomembrane barrier, and drainage layers (see Chapter 7).

The storm types and patterns in the site area determine the potential for surface erosion from the closed facility. When these are capable of causing surface erosion of the waste and subsequent dispersion in the environment, it will be necessary to incorporate components in the cover design that would prevent this from occurring. These could include revegetation and/or the provision of coarse gravel layers that are erosion resistant.

Long-term integrity should be maintained, and design measures should include appropriate surface water diversion, waste regrading, and erosion protection measures to limit surface erosion. The construction specifications for the various components of a cover should be selected to assure long-term integrity. This requires, for example, the selection of vegetation that can totally adapt to the local soils and climate and gravel or rock material used for erosion protection that is very durable. It also may require a certain amount of regrading of surface slopes of the waste management unit to minimize erosion.

Information about the waste unit performance during operations can be useful. Stability, erosion characteristics, seepage losses, and discharges to surface waters that may have occurred are factors which must be considered when closing a facility.

8.5 CLOSURE TECHNOLOGIES AND DESIGN ELEMENTS

8.5.1 INTRODUCTION

This section describes the technologies that can be used as well as the associated design elements. The following are definitions of the terms used in this section:

- **Closure Plan** - the design and construction details selected to close a waste management unit, in accordance with the closure objectives and requirements of relevant closure regulations. The plan includes a narrative, design drawings, construction specifications, as well as quality assurance, monitoring, and other post-closure maintenance plans.
- **Closure Technologies** - technical approaches that can be used to close the waste management unit. It is useful to distinguish between "passive-care" technologies which, once installed, require very little operation and maintenance, and "active-care" technologies, which require a significant amount of ongoing operation and maintenance. A good example of a passive-care technology is a low-maintenance cover, while typical active-care technology is chemical treatment of a continuous leachate stream.

- **Design Elements** - the detailed components required to implement the technologies selected for a closure plan. These include items such as a soil layer to support vegetation in a cover, the installation of drains to enhance consolidation of a slurry waste, or spray bars to facilitate evaporation of a liquid waste.

Figure 8.1 provides an overview of the technologies and design elements that can be used in the development of a closure plan. Although it may not be exhaustive, it does provide the designer with a starting list of alternatives from which to initiate development of such a plan. The following sections discuss the closure technologies listed in Figure 8.1 which include no action, institutional controls, conditioning or treatment, encapsulation and waste removal for offsite disposal.

8.5.2 NO ACTION

No action may be appropriate when a mine waste management unit, even without closure activities, will not threaten the beneficial uses of surface or ground water. This technology is frequently applicable to mine waste management units that contain "dry wastes" and are located in arid climates. For example, if a waste rock or a rinsed heap leach residue pile will not produce a contaminated leachate or require sediment or dust control, it may be safely left in place and not treated.

8.5.3 INSTITUTIONAL CONTROLS

Institutional controls include measures to restrict access to the waste management unit or any water bodies that may have been contaminated by waste constituents. These controls are:

- Restriction of access, using fencing and/or posting.
- Land use restrictions, using notes on property deeds which prevent certain land uses.
- Regulations or legislation to restrict the use of ground water in areas where beneficial uses have been impaired by the mine waste management unit.

Institutional controls may be appropriate as interim closure technologies that allow time for the permanent closure plan to be implemented, or for remediation of waste constituent releases that may have occurred. An example of where access restriction may be contemplated is the presence of a pool of liquid waste which is evaporated over a period of time prior to the waste management unit being covered. Other instances, where institutional controls may be required as part of

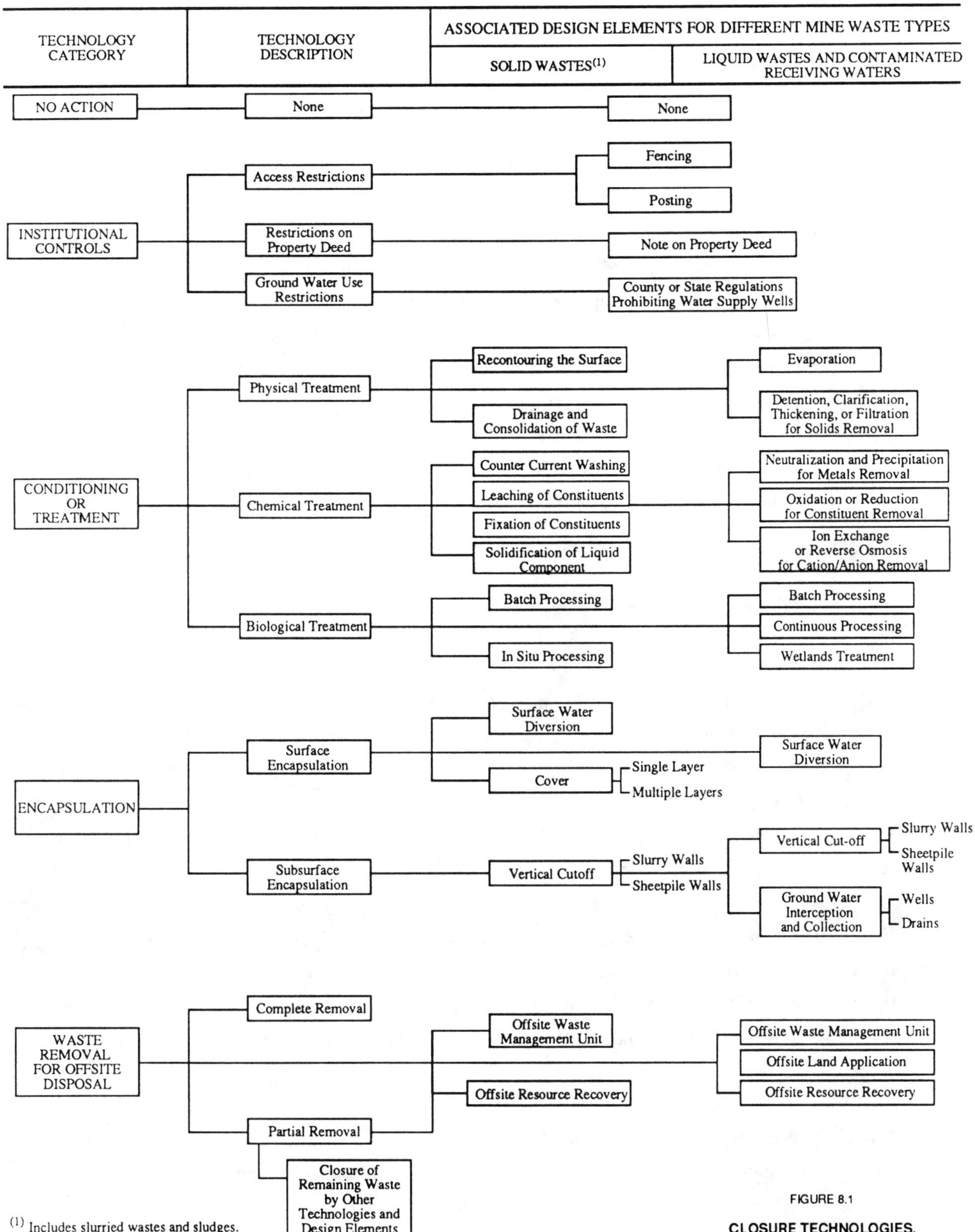

(1) Includes slurried wastes and sludges.

FIGURE 8.1

CLOSURE TECHNOLOGIES, ALTERNATIVES, AND DESIGN ELEMENTS

permanent closure, are the use of deed notes that document the presence of waste management units on the property and restrict future land use that may affect the integrity of the containment systems.

8.5.4 CONDITIONING OR TREATMENT

Conditioning or treatment of the wastes, leachate escaping from the waste, or receiving water that has been affected by waste constituents may be needed to meet closure objectives. It is required when the physical or chemical properties of the waste, leachate, or receiving water must be altered to produce cost-effective closure. Conditioning renders the waste more suitable for closure while treatment reduces constituent concentrations which impair beneficial uses of water. Conditioning is usually physical and can involve grading the surface of the waste to promote drainage, drying tailings to eliminate free liquids or compaction of wastes to produce a suitable cover foundation. In some instances it can also be chemical such as dust control of the surface of solid waste using chemical binders. Treatment can be physical, chemical, or biological. It can range from simple technologies, such as the addition of lime to a waste for pH control to the complex, such as construction of a physical-chemical treatment plant for treating a liquid effluent. The recovery of residual resources, such as metal for example, in the waste also can be considered a treatment technology.

Conditioning or treatment technologies may be applied only once, or they may be required for an extended period of time, or even in perpetuity. Single applications are those that permanently change the characteristics of the waste and include, for example, consolidation of tailings and removal of excess fluids using drains and by loading the surface with a layer of waste rock. Extended treatments, for example, are those that are applied to a continuously generated leachate stream by physical-chemical means.

Extended treatment can be categorized as either "active-care" or "passive-care." Active methods include those involving physical-chemical treatment, and in some cases wetlands treatment, where regular removal of the substrate is required to maintain efficient treatment. In other cases, wetlands treatment may be possible without maintenance requirements and could represent a passive-care technology.

It may be necessary to select the conditioning or treatment technology on the basis of testing, which could include bench scale and laboratory tests, or pilot scale plants or plots. The results of this testing provide information on the effectiveness and costs of various alternatives so that the most appropriate can be selected.

Some treatment technologies that have been used as part of mine waste management unit closure include:

- Spent heap leach piles at mining operations are usually rinsed with water and occasionally treated with an oxidizing agent to reduce the concentrations of reagents and metals in the interstitial water.
- Installation of a physical-chemical treatment plant at a gold mine in a humid climate to treat acid rock drainage (ARD) and tailings seepage for an extended period of time after closure.
- Land application of seepage containing elevated concentrations of cyanide and nitrate back on to the heap leach residue pile from which the seepage is derived. Treatment occurs by processes such as oxidation, volatilization, and biological activity.

8.5.5 ENCAPSULATION

8.5.5.1 Introduction

The objective of encapsulation is to reduce or eliminate the transport of fluids through the waste where these fluids produce a leachate seepage which poses a threat to water quality. It also is intended to prevent wind dispersion of waste particles which could also pose a threat to water quality by direct contact with surface water, or by the leaching of constituents from the redeposited particles. Encapsulation involves the partial or complete isolation of the waste material from the surrounding environment. The key design elements of this technology include: (1) covers which overlie the waste and tie into the underlying natural or engineered containment system, (2) surface water handling methods, such as dikes and channels to prevent runoff contact with the waste and, where necessary, (3) augmentation of the existing containment system by cutoff walls constructed around the facility, or by providing subsurface drainage that prevents ground water from contacting the waste. These are discussed in the following subsections.

8.5.5.2 Covers

Suitable cover types can include the existing surface of the waste, direct vegetation of the waste, a topsoil layer to facilitate the establishment of vegetation or more complex multilayered systems. Figure 8.2 illustrates the alternative cover types applicable to different mine waste management units. These range from utilizing the waste material itself to the comprehensive RCRA (Resource Conservation and Recovery Act) cover recommended by EPA for hazardous waste sites.

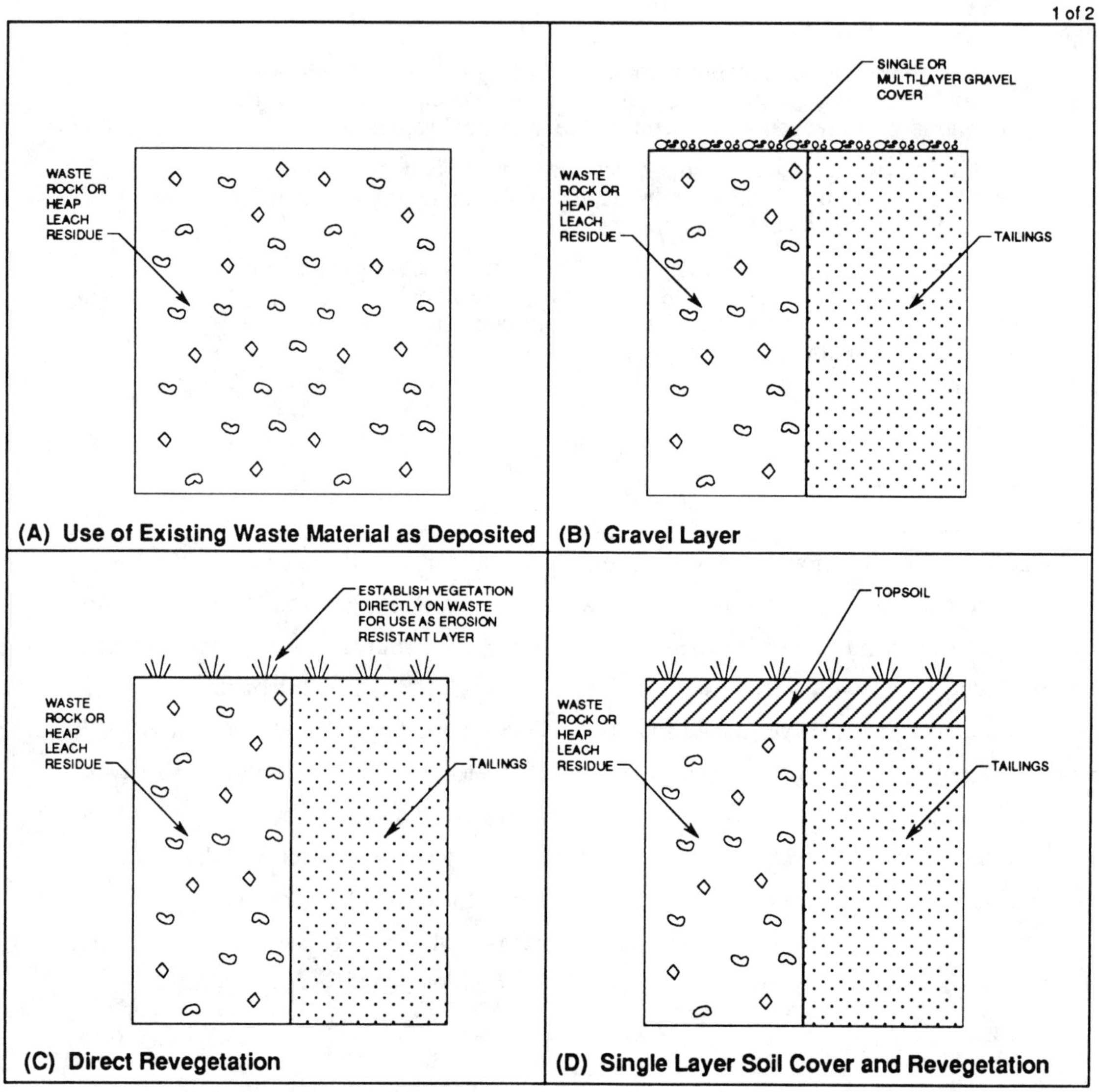

FIGURE 8.2

TYPICAL MINE WASTE MANAGEMENT UNIT COVERS

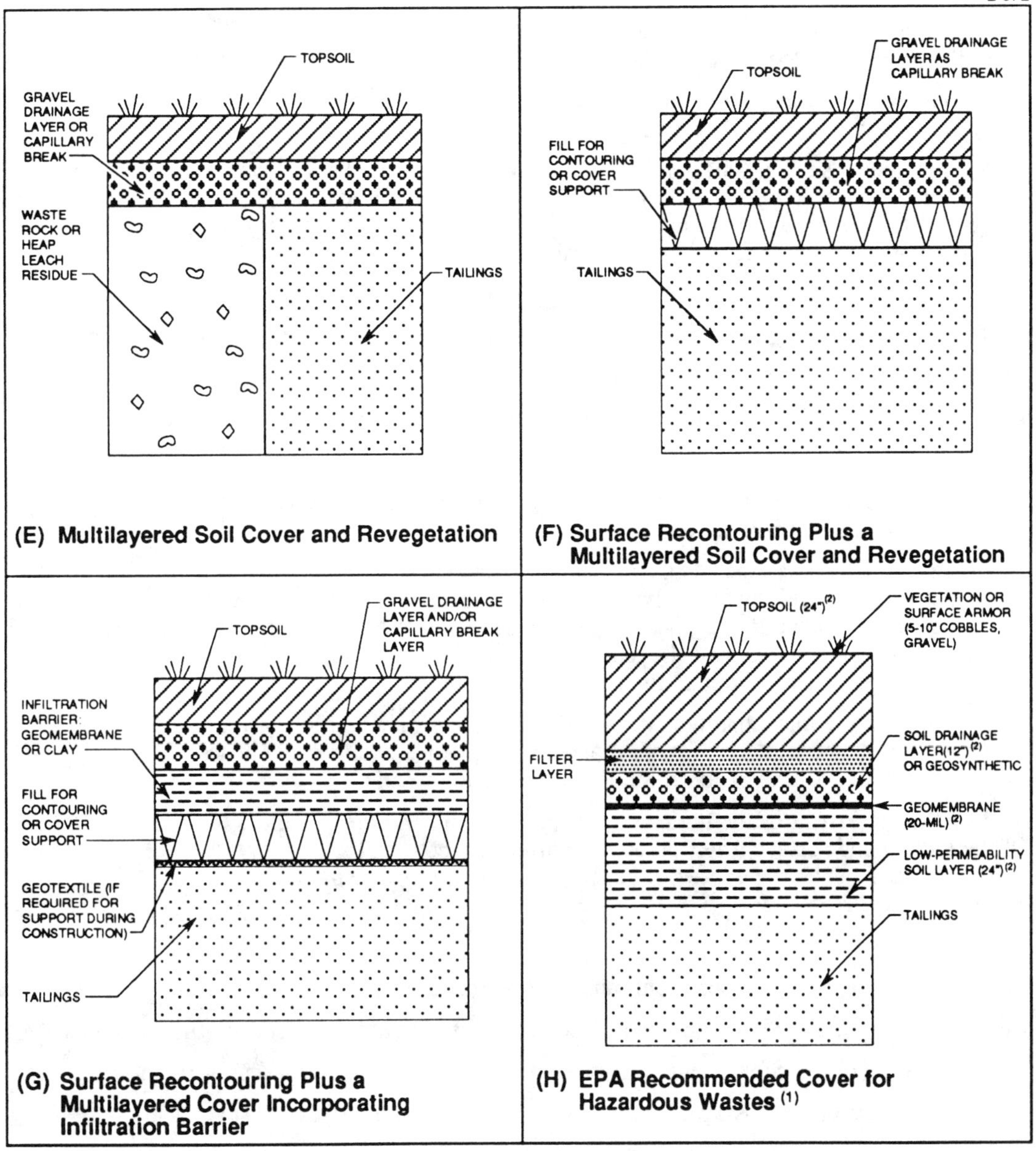

(1) EPA (1989).

(2) Minimum Thickness

FIGURE 8.2

TYPICAL MINE WASTE MANAGEMENT UNIT COVERS

The basic components of multilayered covers (EPA, 1989) are listed below:

- A top layer, consisting of either a vegetated soil layer or armoring to minimize erosion and air transport.
- A drainage layer, that minimizes water infiltration into the waste or the infiltration barrier located under it.
- A one- or two-component infiltration barrier layer, that limits water infiltration into the underlying wastes.
- Special layers, such as biotic barriers to prevent damage to the infiltration barrier by burrowing animals and plant roots, and foundation layers needed to support a cover and/or provide the cover shape necessary for control of surface runoff and internal drainage.

Closure of mine waste management units may not require any components or may only require some or all of the components. The selection and design of the above components must meet the closure objectives and, at the same time, incorporate the site-specific conditions previously listed, i.e.:

- Climatic conditions, that determine the amount of infiltration and the surface water erosion potential.
- Waste characteristics, such as the physical and chemical properties and the shape of the waste surface, that determine the need for a cover and influence the potential quantity and quality of leachate or seepage. The physical characteristics and the shape of the surface also determine the need for any fill and/or foundation layer under the cover.
- The existing containment system and its performance history that determines the need, if any, for additional containment and also can dictate the amount of infiltration control the cover needs to provide.
- The characteristics of the underlying geologic units and the distances to the ground and surface water bodies that affect the potential risks to water quality, and hence the degree of conservatism that must be incorporated into the closure design.

The need for the different cover components and the specifications for these, such as material type, thickness, aerial extent, slope, and method of construction, should be based on the most cost-effective combination of the components that meets the closure objectives and performance goals. In some cases, it may be necessary to undertake a detailed comparison of several alternative combinations or cover system designs that all meet the performance goals. The containment assessment approach outlined in Chapter 3 is directly applicable to cover systems and can be used to assist in the selection of cover components, as well as in the comparison of alternative cover systems.

The sections below and Table 8.2 describe in more detail the design considerations for the various cover components. Some of these have been adapted from the EPA's guidance document for hazardous waste management unit covers (EPA, 1989) to suit the very different conditions found at mine waste management units.

Top Layer

The top layer consists either of a vegetative or armoring layer of gravel-size material (Figure 8.2[A] and 8.2[B]). In some areas, the prevailing climate may inhibit the establishment and maintenance of vegetation, or a planned alternative use of the site may preclude vegetation. In other cases, the surface of the waste may be sufficiently well armored so that no additional treatment is required.

If required, the vegetation component of the top layer should meet the following general specifications:

- Locally adapted perennial plants.
- Resistant to drought and temperature extremes.
- Roots that will not disrupt the infiltration barrier, if present.
- Capable of thriving in low-nutrient soil with minimum nutrient addition.
- Sufficient plant density to minimize cover erosion.
- Capable of surviving and functioning with little or no maintenance.

The selection of plant species is an important consideration. The use of shrubs and trees is usually inappropriate because the root systems extend to a depth that normally would invade either the drainage layer, the infiltration barrier, or the waste.

It is advisable to consult an agronomist, government cooperative extension services, or local university for recommendations of adapted plant varieties and other guidance on local crop cultivation. Several references provide lists of available vegetation and discussions of site-selection criteria (EPA, 1976, 1979, 1983a, 1983c, 1985, and 1987a; Lee, et al., 1984; Thornburg, 1979; and Wright, 1976). These references provide essential information about plant species, seeding rate, time of seeding, and areas of adaptation.

TABLE 8.2

DESIGN GUIDELINES FOR COVER ELEMENTS

LAYER	PURPOSES OF LAYER	LAYER ALTERNATIVES	TYPICAL THICKNESS	GENERAL REQUIREMENTS[1]
TOP LAYER	Minimize Waste Dispersion by Surface Water or Wind Transport.	Granular In Situ Mine Waste	Variable	• Slopes can vary from a minimum of 1-1/2 percent to the angle of repose of the material. • Slopes and material should be stable and provide durable protection against erosion. • Surface contouring to prevent surface runoff concentration in local areas or ponding of water.
		Vegetation	Not Applicable	• Slopes can vary from a minimum of 1-1/2 percent to 50 percent. • Slopes and vegetation should be stable and provide durable protection against erosion. • Vegetation should be: - Persistent. - Drought resistant. - Adaptable to local conditions. - Shallow-rooted.
		Surface Armor	1/2- to 5 feet	• Slopes can vary from 1-1/2 percent to the angle of repose of the waste pile. • 1-inch nominal gravel up to boulders approximately 3 to 4 feet in diameter. • Size and thickness of armoring layer should be based on rainfall intensity and slope.
	Support Vegetation.	Soil	6 to 24 inches	• Slopes can vary from a minimum of 1-1/2 percent to 50 percent. • Required thickness will depend on the vegetation type, end use of reclaimed area, and suitability of the underlying mine waste to partially support vegetative growth.
DRAINAGE LAYER	Minimize Percolation and Damage to Infiltration Barrier. Prevent Upward Capillary Rise of Liquids From the Underlying Waste.	Sand or Gravel	12 to 24 inches	• Slopes range from 1-1/2 percent to a maximum controlled by stability considerations. • Adequate capacity to handle at least five times the anticipated infiltration rate through the layer above. • The layer should include a gravel toe drain or equivalent to direct drainage flows away from the waste management unit. • In some instances, it may be necessary to place a soil filter or geotextile over the drainage layer to prevent clogging by fines.
		Geotextile Geogrid Geonet	Variable (30 to 150 mils)	• Performance should be equivalent to the overlying sand or gravel layer.
INFILTRATION BARRIER LAYER	Minimize Percolation Into the Waste.	Geomembrane	20 to 60 mils	• Slopes range from 1-1/2 percent to a maximum controlled by stability considerations.
		Soil With a Low Hydraulic Conductivity	12 to 36 inches	• Hydraulic conductivity ranging from 10^{-5} cm/sec to 10^{-7} cm/sec depending on site-specific needs. • Should be located below the frost zone.
SPECIAL LAYER	Minimize Damage to Infiltration Barrier.	Biotic Barrier	12 to 24 inches	• Large materials, such as coarse gravel, cobbles, etc.
	Support Cover and Promote Drainage.	Foundation Layers	Minimum 24 inches and up to tens of feet, as required for control of drainage and contouring of surface.	• Frequently can be constructed from mine waste such as centrifuged coarse tailings fractions, spent ore, or waste rock. • These layers may require compaction to assure adequate support of overlying cover layers.

(1) The maximum slope ranges have been increased over those specified by EPA (1989). These flatter minimum slopes have been selected because of the large areas of mine waste management units and because tailings beach angles are frequently in the range of 1-1/2 percent. The steeper maximum slopes have been selected to accommodate the steep slopes typically associated with mine waste piles. Practice has shown that mine waste management units can be successfully closed with this larger slope range.

Where an armoring layer is used, it is recommended that the material possess the following characteristics:

- Capable of remaining in place and minimizing erosion of itself and the underlying soil component during extreme weather events of rainfall and/or wind.
- Contains durable materials that are not likely to weather significantly over an extended period.
- Capable of accommodating settlement of the underlying material without compromising its performance.

A surface armor component of very coarse materials promotes infiltration rather than runoff. Such a layer may, therefore, be more applicable in arid areas or used in combination with an infiltration barrier layer. In those areas, leachate generation due to water infiltration is not usually a concern.

For both vegetated and armored top layers the shape of the top surface should ideally be slightly convex, and uniformly sloped. In non-level terrain, diversion structures should be installed to prevent the run-on of surface water onto the cover. To prevent ponding of rainwater due to irregularities of the surface, the cover should be sloped so as to achieve the slopes indicated on Table 8.2 after allowance for settlement and subsidence (EPA, 1982a). Slopes greater than 5 to 10 percent will usually require special controls to prevent erosion such as armoring or vegetation.

The U.S. Department of Agriculture's Universal Soil Loss Equation (USLE), or the Modified Universal Soil Loss Equation (MUSLE) (USDA, 1980), is often used to evaluate the erosion potential for cover surfaces. As with any modeling approach, the USLE provides only approximate estimates of sediment yield, and should be used in conjunction with engineering judgement. The USLE is particularly useful in comparing the relative sediment yields from different cover alternatives that may be proposed.

Other methods that can aid in the design of erosion resistant covers include the geomorphological and the empirical tractive force methods (DOE, 1989). The geomorphological method utilizes information on erosion resistant slope and particle size relationships obtained from natural drainages in the vicinity of the waste management unit. The tractive force method uses published relationships between surface runoff flow velocities and specified physical properties of the surface materials for stable conditions.

Where revegetation is required to provide erosion protection and/or for aesthetic reasons, it may be possible to directly revegetate the surface of the mine waste. In some instances, this may require chemical and/or biological soil amendments. Where this is not possible, a single- or double-layered soil system needs to be provided to cover the waste (see Figure 8.2[C], [D], and [E]).

The single-layer soil cover simply provides an enhanced medium to provide vegetative growth. The double-layer system incorporates a drainage layer and/or capillary break which primarily provides protection for the vegetation from any chemical constituents contained in the waste. It also provides enhanced drainage in the cover which can reduce the amount of infiltration into the waste. This drainage layer is discussed in the following section.

Recommended guidelines for the soil cover required to support vegetation are as follows:

- For vegetation support, a thickness of 6 to 24 inches. The minimum thickness can be used only if the mine waste underlying the soil layer is not phyto-toxic and can support vegetative growth to some extent. Greater total thickness may be required where maximum frost penetration exceeds this depth, or where greater plant-available water storage is necessary or desirable.
- The soil should be of medium texture to facilitate seed germination and plant root development.
- The layer should receive minimum compaction, plus allow sufficient infiltration to maintain growth through drier periods.

Medium-textured soils, such as loams, have the best overall characteristics for seed germination and plant root system development. Fine-textured soils, such as clays, are often fertile but provide poor drainage which can cause problems such as puddling of water on the surface, or difficulty in initial establishment of plant cover during wet periods. Sandy soils are often a problem due to low water retention and loss of nutrients by leaching. It may be cost-effective to excavate and stockpile topsoil located in the footprints of mine waste management units, mine pits, plant areas, etc. during construction of a mining project. Where only a minimum amount of native topsoil can be saved by stockpiling, the remainder may be made up by selecting local borrow material having appropriate qualities.

<u>Drainage Layer</u>

When required, the drainage layer should be designed to minimize the amount and residence time of water coming into contact with the underlying waste or infiltration barrier, thereby decreasing the potential for leachate generation. Its construction materials and configuration should therefore facilitate rapid and efficient removal of water to an exit drain. The drainage layers also functions as a capillary barrier. The material used requires a grain size distribution sufficiently coarse to prevent the upward migration of fluids by capillary action, thus preventing waste constituents migration to the root zone of a vegetated top layer. Meeting the requirements for drainage usually will satisfy the capillary barrier requirements.

The drainage layer should be designed, constructed, and operated to function without clogging. Where necessary, physical clogging may have to be prevented by incorporating a filter layer of soil or geosynthetic material between the top layer and the drainage layer. The prevention of biological clogging may range from limiting vegetation to shallow-rooted species to the installation of a biotic barrier. In arid locations, the need for, and design of, a drainage layer should be based on consideration of precipitation frequency and intensity, and the sorptive capacity of other soil layers in the cover system. It may be possible, for example, to construct a top layer that will absorb most, if not all, of the precipitation that infiltrates into that layer, thereby eliminating the need for a drainage layer.

The suggested 12-inch minimum thickness of the drainage layer (Table 8.2) allows sufficient cross-sectional area to transport drainage in most situations. In some cases, particularly where relatively flat drainage slopes may be part of the design, drainage layers thicker than 12 inches may be necessary. Infiltration modeling using such tools as the Hydrologic Evaluation of Landfill Performance (HELP) model (Chapter 6, and EPA, 1984a) may aid in the design of drainage layers.

The drainage layer must slope to an exit drain which allows infiltrated water to be efficiently removed. Further information is provided in EPA (1985) and Bureau of Reclamation (1977) publications. Care should be taken in the design to control the velocity of the exiting water, both within and beyond the exit drains, in order to prevent soil loss and destabilization.

Materials used to construct the drainage layer should be durable, clean, or washed or screened prior to construction to remove fines that may promote clogging. To further prevent clogging, it may be necessary that a granular or geosynthetic filter be placed directly over the drainage layer to minimize the migration of fines from the overlying topsoil. If a graded granular filter is used, care should be taken to ensure that the relationship between the grain size distribution in the filter and adjacent material meets specified criteria. Numerous studies have been done to develop and evaluate such "filter" criteria (see Bertram [1940], Karpoff [1955], Thanikachalam [1975]). Although there are numerous variations for filter criteria, those presented by Cedergren (1967) are usually recommended for water drain construction and are used by the EPA (1987) in the design of hazardous waste management unit covers. These criteria, which are summarized in Table 8.3,

TABLE 8.3

FILTER CRITERIA

A. FILTER CRITERIA FOR HIGH FLOW GRADIENT CONDITIONS[(1)(2)]

- $\frac{D_{15}\text{ filter}}{D_{85}\text{ base}} < 5^{(3)}$

- $5 \leq \frac{D_{15}\text{ filter}}{D_{15}\text{ base}} \leq 20$

- $\frac{D_{50}\text{ filter}}{D_{50}\text{ base}} < 25$

B. FILTER CRITERIA FOR GRAVEL ENVELOPE[(4)(5)]

- Drainage layer material must be reasonably well graded and free of substances which could change the hydraulic conductivity with time. To be well graded, the material should have:

  - For drains comprised of gravel sized material: $\frac{D_{60}}{D_{10}} \geq 4$

  - For drains comprised of sand sized material: $\frac{D_{60}}{D_{10}} \geq 6$, and

  - For either granular or sand drains $1 \leq \frac{(D_{20})^2}{D_{10}D_{60}} \leq 3$

- All material should pass the 1-1/2" sieve and not more than 5 percent should pass the No. 50 (U.S.) sieve (0.02 inch). The maximum size for a lower envelope is 3/8" while the smallest size for the upper envelope is the No. 30 (U.S.) sieve (0.033 inch).

- Typical gravel envelopes for various types of base materials are provided in Figure 8.3.

(1) For high flow gradients of 10 to 50.
(2) According to U.S. Department of the Navy (1982), i.e., NAVFAC DM-7.1.
(3) D_{85}, D_{60}, D_{50}, D_{20}, D_{15}, and D_{10} represent the sieve size through which the subscripted percent of the soil would pass.
(4) For low flow gradients of 1 or less.
(5) According to Winger and Ryan, 1971.

accommodate high flow gradients on the order of 10 to 50 in proximity to the drain. However, for cover drains, flow gradients are more likely to be on the order of 1 or less. These conditions are similar to those that occur in agricultural drains for which so called "gravel envelopes" drains are advocated (Winger and Ryan, 1971; U.S. Department of Interior, 1978). The criteria for "gravel envelopes" should, therefore, be considered when designing cover drainage layers. Suitable criteria (Winger and Ryan, 1971) are provided in Table 8.4. Typical examples of such envelopes are included in Figure 8.3.

The authors are not aware of the extensive use of "gravel envelopes" or anything similar for the subsurface drainage of mine waste as covers. However, it is an area which deserves investigation, especially with respect to successful long-term drain performance.

The replacement of granular filter materials by woven or non-woven synthetic materials (geotextiles) is becoming more widespread. Geotextiles are mostly used for economic reasons, i.e., when certain granular materials are not readily available. The characterization of geotextiles as drainage materials is still being undertaken by researchers, and the "filter criteria" which are presently advocated should be considered as tentative. The design criteria can be adapted from those used for designing filters adjacent to perforated pipes, as they are applicable for a filter adjacent to a woven textile (or a woven textile adjacent to natural soil), i.e.:

$$\left(\frac{D_{85} \text{ of fine material adjacent to woven textile}}{\text{opening in woven textile}}\right) \geq 2$$

Where D_{85} is the particle size which is larger than 85 percent of all the particles present.

Marks (1975) produced an extensive study of the use of geotextiles for subsurface drainage. Some of the findings of this study include:

> 'Geotextile filter systems undergo a stabilization; wherein, after a small initial loss of fines, the remaining sand and silt particles interfere with one another and prevent further migration of the remaining fines. The interaction of the soil particles immediately behind the fabric create a filter media within the soil mass. Thus, the effective area of the drain is increased to provide a more natural transition from soil mass to filter. The natural transition greatly reduces the seepage velocities near the boundary between filter and protected soil.
>
> Geotextile filter systems investigated were found to perform as well as conventional aggregate systems under all protected soils considered.
>
> The experimental laboratory tests showed that one major advantage of geotextile filters is the elimination of the serious problem of segregation during placement of conventional aggregate filters.

TABLE 8.4
TYPICAL CLOSURE PLANS

TYPE OF WASTE MANAGEMENT UNIT	WASTE CHARACTERISTICS	POTENTIAL ENVIRONMENTAL ISSUES	TYPICAL CLOSURE TECHNOLOGIES	TYPICAL DESIGN ELEMENTS
Overburden Piles	• Other than turbidity, no potential for water quality impact.	• Runoff turbidity in humid climates.	• Arid Climate: - Institutional Controls	• Limit access to the site.
			• Humid Climate: - Institutional Controls - Encapsulation	• Limit access to the site. • Diversion ditches and revegetation.
Waste Rock and Dry Tailings Piles	• Other than turbidity, no potential for water quality impact.	• Runoff turbidity in humid climates.	• Arid Climate: - Institutional Controls	• Limit access to the site.
			• Humid Climate: - Institutional Controls - Encapsulation	• Limit access to the site. • Diversion ditches and revegetation.
	• Acid generating or containing soluble minerals or salts.	• Runoff turbidity and potential impact of leachate on surface or ground water in humid climates.	• Arid Climate: - Institutional Controls	• Limit access to the site.
			• Humid Climate: - Institutional Controls - Encapsulation - Possible Conditioning or Treatment	• Minimize infiltration and air flow through dump with low permeability cover and surface runoff control. Achieve turbidity control by revegetation. Leachate collection and treatment system may be required, using either passive or active systems. • Where necessary, "compartmentalize" waste rock with acid production potential in low permeability cells within the facility.[(1)] In extreme cases, chemically treat the rock using neutralizing materials such as limestone or slacked lime.
Tailings Impoundments	• Other than turbidity, no potential for water quality impact.	• Runoff turbidity in humid climates. • Disposal of remnant solutions. • Windblown dispersion.	• Humid Climate: - Institutional Controls - Conditioning or Treatment - Encapsulation	• Limit access, remove free liquids using treatment, discharge, land application, or evaporation. Contour surface to promote drainage, construct diversion and flood control ditches as needed. Place soil cover, if needed, and revegetate in humid climates. Place armoring layer in arid climate, if needed for dust control.
	• Ponded and interstitial liquid containing elevated constituent concentrations, with potential for impacting water quality. • Turbidity.	• Leachate or seepage impact on ground or surface water. • Runoff turbidity in humid climates. • Disposal of remnant solutions. • Windblown dispersion. • Surface drainage. • Settlement of Tailings.	• Arid Climate: - Institutional Controls - Conditioning or Treatment - Encapsulation • Humid Climate: - Institutional Controls - Conditioning or Treatment - Encapsulation	• Limit access, remove free-standing solutions from surface by land application, or evaporation. Provide armoring and/or single layer cover if needed. If necessary, grade surface of tailings or surface of cover material to promote runoff; divert surface flows around impoundment. • Limit access, remove free standing solutions from surface by treatment and discharge, land application or evaporation. Construct cover over impoundment to minimize infiltration, provide soil cover and revegetate as appropriate[(2)] Grade surface of tailings or cover to promote runoff as required. Provide leachate collection system as appropriate.

(1) Low-permeability cell boundaries and cover can be constructed with waste materials if sufficient suitable material is available.
(2) Covers will range from single layers to multi-layered systems with drainage layers to reduce infiltration.

TABLE 8.4

TYPICAL CLOSURE PLANS
(Continued)

FACILITY	WASTE CHARACTERIZATION	ENVIRONMENTAL ISSUES	CLOSURE TECHNOLOGIES	CLOSURE DESIGN ELEMENTS
Wastewater Ponds	Waste solutions with a potential to impact water quality. Residual sludges or precipitates which have the potential to produce a leachate with the potential to impact water quality.	Spillage or seepage to surface or ground water.	• Treatment - Encapsulation - Removal	Solutions treated and discharged, removed by land application or by evaporation. Sludges removed and disposed of at suitable facility. Alternatively, sludges treated and disposed of, or reused in beneficiation facility. Pond decommissioned, and liner removed or burial in place with backfilling of pond area.
Sediment Control Ponds	Other than turbidity, no potential for water quality impact.	Turbidity in runoff derived from pond.	• Treatment - Encapsulation - Removal	Solutions drained off and sediment allowed to stabilize. Area graded to prevent ponding and further storage of waters. Surface armored or revegetated if necessary to prevent erosion.

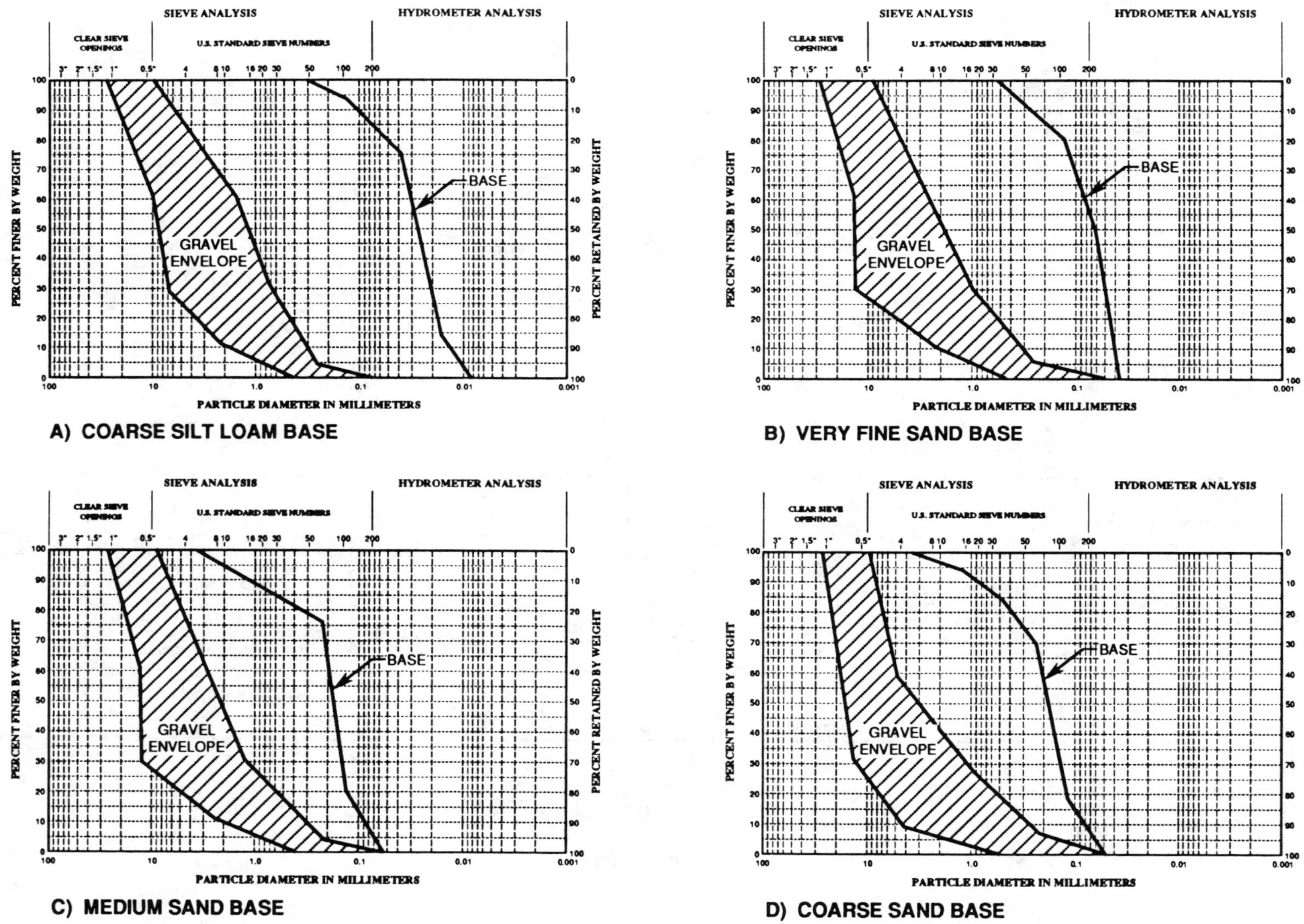

COBBLES	GRAVEL		SAND			SILT AND CLAY FRACTION
	coarse	fine	coarse	medium	fine	

REFERENCE: WINGER AND RYAN (1971).

FIGURE 8.3

EXAMPLES OF GRAVEL ENVELOPES FOR VARIOUS BASE MATERIAL

> Filtration tests produced permeability decreases in aggregate filter systems of about one order of magnitude through internal gradation mechanisms of the aggregate filter system since soil loss did occur during filtration.'

Demery (1980) found that long-term water flow through a soil and geotextile system was controlled by changes in the soil, such as compaction due to flow and clogging due to bacterial growth, regardless of the type or the opening size of filter used. This phenomena was more prevalent when fine soils were located adjacent to the geotextile.

The important conclusion from the above studies is that the percentage of fines in soil layers adjacent to geotextiles should be limited to prevent clogging of the fabric and/or the formation of a filter cake adjacent to it. Therefore, it is, recommended that geotextiles not be used directly against tailings, but that a granular filter material be used between the tailings material and the geotextile. It is recommended that a granular drainage layer adjacent to a geotextile layer should not have more than 2 percent passing the No. 200 sieve.

Other guidelines for the application of geotextiles include:

- The material should have the ability to maintain open pore space and thus transmissivity, under the expected overburden load.
- It should have the ability to conform to settlement of the waste management unit's surface.
- The material should have an adequate useful life.

Geosynthetic drainage materials are manufactured in a variety of configurations which continue to evolve with experience in manufacturing and use. "Geonets" and "geogrids" are drainage components designed to promote rapid flow. They are manufactured as single components that usually must be separated from overlying and underlying soils that could clog them. The separating materials also are geosynthetics in the form of filter fabric. The geogrid, and top and bottom filters (which may also serve as protective bedding and slide-resistant materials), may all be factory-bonded into one unit. These bonded materials, also referred to as "geocomposites," may be applied in one operation as the entire drainage layer. The various forms of geocomposites are well described by Koerner (1989).

Infiltration Barrier Layer

An infiltration barrier is necessary where there is a requirement to reduce the amount of rainfall that can infiltrate a waste management unit. As the objective of this barrier is to avoid the buildup of fluids in the waste, it is important to ensure that hydraulic conductivity of this layer reduces the rate of infiltration to an amount equal to or less than the residual rate of seepage of liquids from the bottom of the waste management unit.

The barrier layer can be provided by either a geomembrane, a compacted soil or clay, or a combination of the two. The characteristics of these materials are described in Chapter 7. A description of the design requirements specifically for a barrier layer are discussed below. It is recommended that this layer be located below the maximum depth of frost penetration.

A geomembrane barrier layer should have the following characteristics:

- It should be of adequate thickness to prevent failure under either potential stress of the post-closure care period, or during construction.
- There should be no surface unevenness, local depressions, or mounding that creates depressions capable of containing or otherwise impeding the rapid flow and drainage of infiltrating water.
- The quality assurance and quality control (QA/QC) during construction should be equivalent to that when constructing geomembrane liner systems (see Chapter 7).
- The geomembrane should be protected by at least 2 feet of overlying material.
- The geomembrane should be in direct contact with the underlying soil or waste, and should be installed on a smoothed surface.
- Bridging or similar stressed conditions in the geomembrane should be avoided by providing slack allowances for temperature-induced shrinkage of the geomembrane during installation and during the period prior to placement of the protective layer or drainage layer. Slack should not be excessive to the extent that folds are created that later may crack.

Requirements for a soil or clay barrier layer are as follows:

- The material should be compacted in layers not to exceed 9 inches in thickness.
- The compacted material must be free of clods, rock, fractured stone, debris, cobbles, rubbish, and roots, etc., that would increase the hydraulic conductivity or serve to promote preferential water flow paths.
- The hydraulic conductivity of the compacted layer should be between 10^{-5} cm/sec and 10^{-1} cm/sec depending on site specific conditions and requirements for infiltration control.

Differential settlement across the cover, which can be significant in the case of a slurried tailings impoundment, may also cause disruptive stresses that should be accounted for in the design of the infiltration barrier. Care should be taken to make allowance for these and other stresses. In some cases, particularly for the installation of covers on slurried tailings impoundments, it may be necessary to first place the foundation layer, allow it to consolidate for a period of a year or more, then install the final cover layers.

In cases where infiltration control is a critical aspect of closure and where there is concern about the constructability of a clay barrier with a low enough hydraulic conductivity, a test fill may be necessary. In construction of the test fill, the soil, equipment, and procedures to be employed in construction of the actual infiltration barrier layer should be used to ensure that design specifications are attainable with the available materials and equipment. In some instances, it may be appropriate to install the test fill directly on the waste to observe what settlement takes place and how compaction can be best achieved.

Other aspects that must be considered in evaluating the design of the soil or clay barrier include desiccation cracking and freeze-thaw cycling. The potential for desiccation of the barrier layer will depend on the physical properties of the clay or soil, design moisture content, local climatology, and moisture content of the underlying waste. The clay content of the soil, the type of clay, and properties such as liquid limit, plasticity, and shrinkage should be used to select a soil with low cracking potential, or in determining placement procedures to reduce cracking potential (EPA, 1979; EPA, 1989).

Compaction of the soil layer wet of optimum is recommended to assure that the lowest hydraulic conductivity is attained with standard Proctor densities (EPA, 1989). To guard against drying in this case, the layer proposed to cover the soil layer should be placed immediately after the soil is compacted. Protection against freeze-thaw cycling can be provided by covering the barrier layer with a soil, gravel or rock layer that is thick enough to prevent frost penetration.

Biotic Barrier

Plant roots or burrowing animals, collectively described as biointruders, may disrupt the integrity of the drainage and infiltration barrier layers. The drainage layer may be especially susceptible to the intrusion of plant roots, which could reduce its capacity. Soil or clay layers could be exposed to both root and animal penetration thus increasing their effective hydraulic conductivity.

Biotic barriers are typically layers of cobbles or coarse gravel that are located between the top layer, and the waste and infiltration barrier layer if present. This biotic barrier layer would directly underlie the soil component of the top layer, perhaps separated from it by a geosynthetic filter layer. In cases where both a biotic barrier and drainage layer are required, a single granular layer can be installed to serve both purposes.

Small-scale field experiments have been conducted to evaluate possible barriers. Experiments with layers of cobbles have been tested in arid and semi-arid conditions, with plants unique to such habitats (Cline, 1979; and Cline et al., 1982). Some research suggests that 3 feet (90 cm) of cobbles, or 6 inches (15 cm) of gravel placed over 30 inches (75 cm) of cobbles, may be effective in stopping root penetration of some deep-rooted plants (DePoorter, 1982), and it may also be effective in stopping the invasion of burrowing animals.

As an alternative to the physical barrier described above, a chemical barrier involving a polymeric herbicide carrier/delivery (PCD) system, used to release herbicide, has been suggested by Cline et al. (1982). It could be installed within a cover, just above the drainage layer to stop the intrusion of roots. The PCD system would contain an herbicide designed to be released slowly over many years. The use of this system may be limited, however, because it may introduce harmful constituents to the cover, and it is not likely to have the longevity of the barrier layer.

Foundation Layer

Cover system designs incorporating drainage layers and infiltration barriers will require foundation layers. Generally the mine waste itself will provide a suitable foundation layer. This is particularly true for waste rock, heap leach residue piles, and tailings facilities that have undergone a successful program of managed deposition or been filled with dry tailings. Most slurried tailings impoundments, however, will have surfaces that are too soft to support a cover system with a foundation layer.

Foundation layers can be constructed by the placement of a layer of granular material, usually waste rock, over the tailings. Where the tailings are extremely soft, it may be necessary to initially place a geotextile on the surface to provide additional support, and to place the foundation material in a lift thickness that is adequate to support the construction equipment without imposing too high a load on the tailings. Experience at the Grey Eagle Mine in Northern California has indicated that this thickness should be between 3 and 4 feet.

In tailings impoundments, a foundation layer frequently is required to provide a mounded surface shape conducive to surface drainage. This can be provided by either placement of granular fill as discussed above, or by regrading the surface of the tailings if it is dry enough and can support the weight of the construction equipment. In some cases it may be possible to deposit slurried tailings, or the courser cycloned tailings fractions to form all or a portion of the necessary mounding.

When selecting the thickness of a foundation layer it is important to know what the ultimate settlement rates are likely to be at various locations in the waste management unit. Potential settlement behavior can be evaluated using standard consolidation theories (Holtz and Kovacs, 1981) or, more accurately, using the finite strain theory (Caldwell, et al., 1984). Application of these theories will require knowledge of the physical characteristics of the material, particularly density and void ratio. In some cases, this may require drilling, sampling, and laboratory testing.

8.5.5.3 Surface Water Management

The objective of surface water management is to protect the waste management unit from erosion damage. Prevention of cover erosion is discussed in Section 8.5.5.2. Prevention of erosion by runoff from adjacent areas is discussed below. Erosion damage due to runoff from areas that drain onto the waste management unit can be prevented by diversions. The design elements that can be utilized for diverting runoff include lined and unlined ditches, as well as dikes and embankments.

Lining of ditches can be accomplished by placement of gravel or rip-rap, or vegetation. Where extremely high velocities are encountered in erodible material, it may be necessary to install materials that are more erosion-resistant, such as grouted rip-rap, soil cement, or gabions. Grouted rip-rap involves injecting a cement grout between the boulders. Soil cement involves adding between 2 and 10 percent by weight of cement to soil before it is compacted to form a ditch liner (Portland Cement Association, 1979). Gabions are prefabricated wire or steel cages into which rip-rap is placed, forming large blocks.

8.5.5.4 Enhancement of Waste Containment

In some cases, it may be necessary to improve the waste containment provided because it did not function as well as originally intended. Consequently, seepage poses a threat to the beneficial uses of water and the proposed cover system, on its own may not alleviate this threat. Waste containment enhancement can be achieved by providing a positive cutoff to seepage, such as a slurry or sheet pile wall, grout curtain, or by preventing contact between shallow ground water and the waste using underdrains and/or ground water extraction wells.

8.5.6 WASTE REMOVAL FOR OFFSITE DISPOSAL

Removal of mine waste and disposal at an alternative site is an extreme measure that is not likely to be economic for an operating mine.

At abandoned mine waste sites, however, this technology has been considered and in one case (DOE, 1990) has been applied. Remediation of the uranium tailings pile near Grand Junction, Colorado, was achieved by moving 4 millions tons of material 15 miles for redisposal at a new site. This was judged to be the most cost-effective method of dealing with the health risks posed by the original tailings impoundment.

8.5.7 CLOSURE DESIGN

Closure design for a mine waste management unit is achieved by selecting appropriate technologies from those shown in Figure 8.1. This selection is best accomplished by considering site- and waste-specific conditions and conducting a waste containment assessment. For any given project, closure requires several technologies or modifications to a given technology.

Also of importance to the selection of the closure method is the project status. Closure design for abandoned and inactive sites, which were not designed to make provisions for final reclamation, must be selected based on practicality and suitability to actual site conditions. In such instances, closure plans are, in effect, "retrofitted" to the site. Data on the facility design, operation, or performance may also be incomplete. It may, therefore, be necessary to undertake a field investigation involving waste and site characterization prior to design for closure.

Active sites generally have sufficient data available to facilitate selection of appropriate closure technologies. Adequate information on design, construction, operation, waste characteristics, and facility performance usually is available. Furthermore, new disposal facilities offer the best opportunities to develop a cost-effective closure plan during the planning phase, and to integrate this plan with the design of the containment system and other operational waste management considerations.

The approach to selection and design of closure systems is discussed in Section 8.4. Table 8.4 summarizes a range of typical closures for various types of mine waste management units. Some important properties that emerge from examination of the typical design elements outlined in Table 8.4 as well as special considerations not included in Table 8.4 are discussed below.

8.5.7.1 Important Perspectives

There is a significant difference in the closure requirements for waste management units in arid climates and those in humid climates. In arid climates there is little potential for leachate formation. Turbidity in the runoff from waste management units is usually not an issue, as it is not significantly different from the runoff from the natural barren ground surrounding the waste management unit. Revegetation is therefore, frequently not necessary to protect beneficial uses of water. Closure requirements can usually be met by limiting access and leaving waste piles as they are. In some cases, tailings impoundments, if they are prone to dusting, may require armoring.

In humid climates, and where the waste has the potential to generate a leachate that poses a threat to the beneficial uses of the waters of the State, more comprehensive closure measures are required. Cover and/or revegetation is usually required for erosion protection and infiltration control. In extreme cases, waste piles and tailings impoundments may require multilayer covers to limit infiltration.

Other closure activities that are used frequently for aesthetic or other reasons, include:

- Recontouring the edge of waste rock or heap leach residue piles to allow some visual blending with the scenery.
- Revegetation of waste management units in arid climates.
- Regrading of heap leach or dried tailings pile slopes to prevent slope raveling.

8.5.7.2 Special Considerations

Slurried tailings impoundments offer special challenges when designing for closure because of the amount of unconsolidated (soft) material left behind which is usually not trafficable, and because the surface of the impoundment is usually shaped like a bowl and collects water. Methods to improve the surface conditions, as well as the shape of the surface, include managed deposition to create a mounded surface, and selectively depositing the courser fraction of cycloned tailings in hollow areas.

As indicated above, the mine waste material itself can frequently be used to construct cover components. Examples include:

- Tailings to construct foundation layers and infiltration barriers.
- Waste rock or heap leach residues to construct armoring layers, drains, and foundation layers.

Finally, the slope stability of covers must also be considered in the design for closure. This is particularly important where infiltration barriers are provided on relatively steep slopes. As outlined in Chapter 7, the materials or interfaces between materials contained within these barriers may have relatively low strengths. Methods for analyzing stability also are described in Chapter 7.

8.5.8 CASE HISTORIES

Brief descriptions of four case histories involving closure of mine waste management units are presented below. These examples show how particular closure technologies were selected and implemented to satisfy the closure objectives. They particularly emphasize the uniqueness of each site and demonstrate the need for a flexible approach to closure planning and design.

8.5.8.1 Case History No. 1 - Closure of an Abandoned Dry Tailings Pile

This case involves a large sidehill deposit of 100,000 tons of dry, gravelly, calcine tailings derived from mercury mining and retorting. The area typically has wet winters, dry summers, and a mean annual precipitation of approximately 33 inches. Although the waste was chemically inert, it provided a source of sediment load into adjacent receiving surface waters.

The goals of the closure plan were to reduce sediment loads in the creek and improve stability of the tailings pile. To achieve this, the tailings slopes were flattened to between 33 and 40 percent, covered with 1 foot of topsoil, and revegetated. Diversion ditches were constructed above the pile to divert runoff. The toe area was stabilized using a rock berm, and a French drain was constructed to provide internal drainage to enhance stability.

8.5.8.2 Case History No. 2 - Closure of an Active Waste Rock Pile

Background

The second case history involves the disposal of over 100 million tons of waste rock. The ore body is largely of sulfides which are highly refractory in nature. A significant portion of waste rock also contains sulfide that has an ARD potential. The sulfide ore is oxidized and treated with hydroxide in the subsequent cyanide leach circuit. The tailings, therefore, have no ARD potential.

Mean annual rainfall in the area is approximately 33 inches, most of it falling in the winter months.

Original ARD Determination

During the planning and design stage of the project, the potential for acid mine drainage was evaluated by testing drill core samples. Seven geochemical rock categories were defined and tested for ARD. Eight percent of the rock was identified as being net acid producing. The test work was interpreted to indicate that, due to this minimal amount of acid-producing rock, it could be commingled with acid- neutralizing rock in the pile, thus providing effective control of ARD.

During these early stages of ARD testing, however, the procedures used were not as conservative as those currently proposed in Chapter 4. Static tests using sulfide sulfur analyses were used to determine the acid generating potential which was then compared directly to the neutralization potential, but without the requirement for the neutralization potential to be at least three times the acid generating potential.

Monitoring of the waste rock pile underdrain, seepage, and surface runoff three years after waste disposal commenced, indicated depressed pH, and sulfide oxidation product enrichment of these waters.

To eliminate the possibility of discharging this water, sumps were constructed to collect underdrain seepage and surface water runoff from the pile. Pumps and pipelines were installed to allow return of the collected water to the beneficiation plant.

Probable Identification

Reassessment of the original test data using the criteria for determining whether waste rock has an ARD potential (outlined in Section 4.3 of Chapter 4) revealed that the amount of acid generating rock could be much higher than the originally estimated 8 percent. Review of actual mine production of the previously identified problem rock type, a volcanic agglomerate, showed that it also constituted a larger percentage than originally anticipated. Also, because of mine production requirements and the limited amount of active dumping locations within the general pile area at any specific time, commingling of the ARD and neutralizing waste was not achieved, and segregated dumping occurred.

One of the first methods employed to better understand waste rock acid leachate chemistry was column leach testing. Samples of four major rock types were gathered and crushed to minus 2 inches. The samples were then loaded into 5-foot tall by 12-inch diameter test columns, and simulated rainwater was applied. Three of the columns produced neutral or slightly alkaline leach solutions, but the fourth, loaded with volcanic agglomerate, produced acidic leachate and much greater levels of dissolved constituents than the other columns. Subsequent column testing identified other acid-producing rock types. The common denominator in all of the acid producers was sulfide enrichment, regardless of basic rock type.

As a result of this testing, waste categorization was simplified to two categories. Low sulfide, nonacid producing waste was categorized as Type A, with high sulfide, and acid producing waste as Type B. With the waste rock separated on a chemical basis, acid mine drainage control planning focussed on the physical properties of the waste type. Type A waste was found to typically consist of serpentinite, unaltered mudstone, and fault gouge of mixed source rock. The important physical attribute of all these rocks was a fine-grain size which could be worked into a clay-like consistency with low hydraulic conductivity.

In contrast, the Type B waste came from mineralized areas hardened by silicification or adularization. This waste broke into large, durable fragments. When placed in a conventional waste dump, the blocky nature of the Type B resulted in a pile highly permeable to both air and water.

Revised Waste Management Plan

A revised waste management plan was developed in which the blocky, acid-generating Type B waste was encapsulated in cells of clayey, nonacid-producing Type A waste. An area to receive waste was first cleared of vegetation and the topsoil was removed and stockpiled. A layer of Type A waste was then placed and spread with a dozer. The thickness of this layer varied with site conditions, but was generally 5 feet.

A 50-foot high lift of Type B waste cell was then placed over the Type A clay. Careful attention was paid to control the gradient of the surface of the waste rock to ensure rapid surface drainage. If areas of the surface did not compact well, clay was placed as a seal over these areas.

The exterior slope of the Type B dump cell was reduced to 40 percent, i.e., flatter than the angle of repose, to facilitate eventual revegetation. Clayey Type A waste was then dumped over this slope to provide a 10- to 15-foot cover. Type A clay was also plug-dumped over the waste cell's surface and dozer spread to form a layer approximately 5 feet thick. Where possible, truck traffic was routed over this surface to enhance compaction. A sufficient thickness of oxidized waste rock (also Type B) was then placed over the clay to ensure a 20 foot minimum total soil thickness over the acid generating material. The Type A material therefore completely encapsulated the Type B rock. This low lift height of 50 feet was selected to increase the frequency of Type A barriers and to reduce the inflow of oxygen (see Figure 4.1 in Chapter 4).

This dumping and spreading sequence was repeated for a second lift to form a 100-foot high dump bench. The outer 150 feet of the surface of the first lift formed a bench that resulted in an overall slope of the pile of 25 percent. This section of dump surface received a second lift of plug-dumped Type A clay to increase its thickness to 10 feet. Careful attention was paid during the clay spreading process to ensure that the bench and slope capping layers were well tied together, forming a continuous cover over the Type B waste.

Finally, soil was spread over the Type A oxidized waste rock to facilitate revegetation. It is believed that the vegetation aids in reducing the oxygen flux into the waste pile.

Operational Aspects

As a result of the above experience and evaluations, the waste rock disposal operations were modified for the ongoing activities. First, waste is classified from visual examination of blasthole cuttings. A geologist inspects each drill pattern and prepares a waste type map. Although the classification is keyed to sulfide mineral observations, periodic lab analyses are run to confirm and calibrate the visual classification. The waste type map is then digitized and merged with the gold assay data. The ore control engineer subsequently draws ore blocks based on gold assays and waste blocks on the remaining areas using the waste type map as a guide. Waste and ore blocks are delineated in the pit with color-coded flags.

The waste pile is designed to have several active dumping faces to accommodate the varying ratios of waste types derived from the pit. The material out of the pit is scheduled for rapid closure of individual Type B cells is in order to minimize exposure time for oxidation and infiltration.

To predict future waste type ratios, a block model was created for the entire body of ore and waste rock. Long-range mine plans are now reviewed not only to check on ore grade but to allow for effective planning of the waste pile construction.

Conclusions

The ARD trends observed appear to have been reversed. However, more observation time will be necessary to determine the ultimate effectiveness of the measures. An important operational consideration is that the combined ore and waste control methods, and the waste cell construction procedures have become routine, demonstrating the operational feasibility of this type of comprehensive waste management.

8.5.8.3 Case History No. 3 - Closure of an Abandoned Tailings Impoundment

This case history involves closure and reclamation of a tailings impoundment containing approximately one million tons of tailings. The impoundment is located in a forested area with an average annual precipitation of 65 inches. The impoundment consists of an earthen embankment behind which wet cyanide tailings were deposited. Due to the steep terrain and the extremely wet climate, managed deposition was not possible until the final stages of deposition, and then only to a limited extent.

The impoundment is unlined, and the foundation material is fractured rock with a hydraulic conductivity of approximately 10^{-5} cm/sec near the surface. A grout curtain and shallow clay-filled cutoff trench were installed under the embankment in order to reduce seepage. A slurry wall was constructed at the downstream toe of the embankment to collect the seepage which was pumped back into the impoundment. This seepage contains not only natural ground water and tailings solutions, but ARD from an old mine adit which had been covered by the embankment.

The tailings solutions contained elevated concentrations of reagents, chemical constituents, and metal values. The tailings remained submerged over most of the project life, and excess water that collected in the impoundment was pumped to a treatment plant and discharged to a land application site. During the last 8 to 12 months of operation, or beaching of tailings was possible along one side of the impoundment.

The primary goals of the closure plan were to: (1) isolate the tailings from runoff infiltration, thereby reducing the hydraulic head and limiting the quantity of interstitial water, and (2) collect and treat the seepage until such time that the water quality of the untreated leachate became acceptable for direct discharge.

The plan consisted of constructing a multilayer, low-permeability cover over the tailings, keying it into the natural ground surrounding the impoundment, and providing surface runoff control. The cover design included a waste rock layer of varying thickness up to 25 feet, a clay layer 3 feet thick, a 1-foot thick gravel capillary barrier over the clay, and a 1-foot thick layer of topsoil.

For economic reasons it was necessary to construct the cover as soon as possible after deposition ceased. To accomplish this, the final stages of tailings were used to create a beach on one side and force the water to a smaller area in the impoundment. This maximized the exposure of tailings and facilitated access for construction. A double layer of geofabric was placed on the tailings surface, forming a stable bed over which a 3-foot layer of waste rock was placed using a small dozer. The remainder of the waste rock foundation layer was then placed according to total thickness designed to compensate for long-term consolidation of the tailings and maintain positive surface drainage over the entire 14-acre surface.

Surface drainage was controlled with perimeter diversion ditches. The main upstream drainage was diverted around the impoundment in a channel cut into rock along the hillside.

Operating data to date show that the closure plan is performing as expected. The tailings have consolidated, and the amount of tailings liquid reporting with the seepage at the downstream toe of the embankment has decreased significantly. Acid mine drainage from the old mine workings is still discharging at the embankment toe and is being treated and discharged to a land application site. Consideration is currently being given to converting the physical-chemical treatment plant to a wetlands treatment system.

8.5.8.4 Case History 4 - Closure of a Heap Leach Pile

This case history deals with the closure of a one million-ton heap leach residue pile. The ore was stacked to a depth of about 125 feet. Detoxification using fresh water rinsing and recirculation of rinse solutions took just under one year.

A number of chemicals and processes were tested and evaluated for detoxification, and fresh water was finally selected as the most appropriate. However, use of fresh water produced rinse solutions which gradually lost their rinse effectiveness due to increased concentrations of the removed metals and cyanides. A proprietary process developed by the mining company was used to treat the water, which contained concentrations of zinc, copper and nickel of approximately 5 mg/L, to near drinking water standards for reapplication to the heap.

An alternative used before development of the treatment process was to dilute the rinse solution with fresh water in a 1 to 5 ratio and use the resultant mixture to irrigate revegetated areas surrounding the pad area. Agricultural water standards were met.

The constituents of concern in the interstitial water of the ore had the following concentrations:

Parameter	Start of Rinsing	After Rinsing
Cyanide (WAD)	30 mg/L	<5.2 mg/L
Copper	100 mg/L	<1 mg/L
Nickel	40 mg/L	<0.5 mg/L

Throughout rinsing, the pH remained at between 9 and 9.5. A total of 69 million gallons was applied to complete the rinsing. Thirty-eight percent was derived from fresh water or rainfall, and the rest was recirculated. Final reclamation of the heap consisted of final contouring, grooming, and revegetation.

8.6 REGULATIONS FOR CLOSURE

8.6.1 GENERAL

The preceding discussion has indicated that:

- The objectives of closure should be the protection of the beneficial uses of water.
- There are numerous closure technologies and design elements which can be used for closing mine waste management units. There is usually a unique combination that provides the best approach to meeting the closure objectives at any waste management unit.

The discussion presented in this chapter demonstrates that it is important to have a flexible regulatory approach governing the selection of closure technologies and designs.

8.6.2 CURRENT CALIFORNIA REGULATIONS

This section provides a summary of the current California regulations and an evaluation of their suitability in light of the discussion on the technical aspects of closure contained in this chapter.

8.6.2.1 Current Regulations

In California, as outlined in Chapter 3, mine waste is categorized as either Group A, B, or C waste. These groups generally represent those that would be judged hazardous (Group A), those that pose a threat by soluble constituents to the waters of the State (Group B), or those that can only impact surface water turbidity (Group C). Prescriptive closure design criteria are established for each category of waste. Engineered alternatives for closing a site are acceptable, provided "equivalent" protection and performance can be demonstrated (see Chapter 3).

Article 7[9], "Mining Waste," of the regulations defines the closure objectives, provides some general closure requirements, and refers to Article 8[9], "Closure and Post-Closure Maintenance" for detailed closure requirements.

The closure objectives in Article 7 are stated as: "New and existing mining waste management units shall be closed so that they no longer pose a threat to water quality. No post-closure land use shall be permitted that might impair the integrity of containment structures." Other requirements contained in Article 7 include:

- There is a required post-closure maintenance period during which monitoring must continue.
- The post-closure maintenance period ends when the Regional Board determines that water quality aspects of reclamation are complete and that the waste no longer poses a threat to water quality.
- The performance standards established for the operating waste management unit continue to apply to the closure and post-closure maintenance period.
- The owner or operator is required to provide financial assurances.
- Vegetation of closed waste management units shall not impair the integrity of the containment systems.
- Irrigation of vegetation shall neither cause nor increase the production of leachate.

The Article 8 requirements apply to all forms of waste, i.e., municipal, industrial, mining, etc. that are governed by the other Articles contained in *Chapter 15*. The specific aspects of Article 8 that are applicable to mining wastes are described below.

---

(9) California Code of Regulations, Title 23, Division 3, *Chapter 15*.

The closure requirement for Group A and Group B "waste piles" are based on the Article 8 landfill closure requirements and include:

- A cover that has the following layers:
  - A 2-foot thick[(10)] foundation layer which must be compacted and can be constructed with waste.
  - A 1-foot thick layer of material with hydraulic conductivity of 10^{-6} cm/s or less.
  - A 1-foot layer of soil to support vegetative growth.
- The routing depth of plants used to vegetate the cover shall not exceed the depth of the soil layer.
- The cover shall be designed to function with the minimum maintenance possible.
- The cover shall be graded to ensure:
  - Slopes of at least 3 percent.[(11)]
  - Elimination of ponding.
  - Erosion protection for slopes greater than 10 percent.
- The waste management unit's leachate collection and removal system will continue to operate for as long as leachate is generated and detected.

The closure requirements for Group A and B tailings impoundments are the same as for the Group A and B waste piles, with the following additional requirements:

- All free liquid remaining at the time of closure shall be removed.

Closure for Group A and B wastewater ponds requires that wastes from waste management units with geomembrane liner systems be removed and redisposed of elsewhere. Where the liner or the outer liner of a double-lined system is constructed of clay, or where containment is provided by natural geologic formation, the unit can be closed with the waste in place, provided:

- All free liquid[(12)] remaining at the time of closure is removed.
- The waste is compacted.
- The moisture content of the waste is less than its moisture holding capacity (i.e., specific retention. See Chapter 6).

---

(10) A lesser thickness can be allowed if it does not affect the structural integrity of the overlying layers.

(11) Lesser slopes may be allowed if an effective system is provided for diverting surface drainage from covered wastes.

(12) Free liquid is defined as liquid which readily separates from the solid portions of the waste under ambient temperature and pressure. Free liquids are not present when a 100 milliliter representative sample of the waste can be completely retained in a standard 400 micron conical filter for 5 minutes without loss of any portion of the waste from the bottom of the filter.

Group C waste management units are required to be closed in a manner which minimizes erosion and the threat to water quality degradation from sedimentation.

8.6.2.2 Evaluation of Current Regulations

The current regulatory requirements largely have been developed for nonmining wastes. Therefore, they do not provide a framework within which cost-effective and site-specific mine waste management unit closure plans can be readily developed. Cases such as waste rock piles with ARD potential, or wet slurried tailings which require very specialized closure design, do not fit the prescriptive categories provided. Furthermore, closure in arid climates frequently demands a different approach than closure in humid climates, due to the limited effect of vegetative cover and the lack of potential for leachate formation.

As a result, many of the current mine waste closure plans are developed through the exemption process outlined in Chapter 3, without the benefit of a decision-making framework or a list of issues that should be addressed.

Incorporation of an alternative, more flexible approach would facilitate development of more site-specific designs and also provide general technical guidelines for developing closure designs and evaluating their effectiveness. Such an approach also would meet the water quality protection goals of closure.

8.6.3 ALTERNATIVE REGULATORY APPROACHES

The following three alternatives represent a range of regulatory approaches which could be used for governing closure of a facility:

- Specification of uniform closure design criteria.
- Establishment of categorical design criteria, based on a range of characteristics and site conditions.
- Selection and design based on performance criteria which are satisfied on a site-specific basis.

These alternative approaches have been discussed in Section 8.4.2 and in Chapter 7. It is apparent that the categorical design criteria approach has some merit, provided it is tailored to mine waste, while the performance criteria approach provides the best opportunity for achieving a sound scientific and cost-effective closure plan.

8.6.4 RECOMMENDED REGULATORY APPROACH

Regulations should allow the planning and design of mine waste management unit closures by either of the following two approaches:

- Categorizing the waste unit characteristics and then designing the cover system in compliance with prescribed regulatory closure requirements based on:
  - Type of facility (e.g., waste pile, or tailings impoundment containing slurried waste).
  - Site and operational factors, including:
    - Potential waste toxicity
    - Water resource values
    - Leachate availability
    - Site factors such as climate and geology
    - Success of management practices
  - Engineered containment provided in operational design.
- Conduct a site-specific waste containment assessment and demonstrate that the waste management unit, with the application of the closure system or other controls, if necessary, does not pose a threat to the beneficial uses of water.

The minimum data to be provided for the first approach should be: (1) waste chemistry, including its acid generation potential, (2) whether the waste management unit is a pile, tailings slurry management unit or impoundment, with data describing the size of the unit and the rate of waste discharge, and (3) how the facility was constructed and operated to maintain integrity of the containment system. The facility then should be placed into a category that provides a conservative list of closure requirements and, if required, specifications for a cover or alternate covers based on the factors listed above. At the applicants choice, additional data concerning water resource values, site climate, hydrogeologic information, and special waste management practices may be provided to allow the facility to be placed into a more appropriate category.

If conditions vary within the waste management unit, the applicant may explain the differences so that the appropriate cover may be used at each location. Otherwise, the system for the most restrictive conditions should be assumed and applied throughout.

Regulations should allow the applicant to select alternative materials as components of the prescribed cover, so long as it can be demonstrated that such alternative materials will not compromise function. Regulations also should provide for engineered alternatives to the categorical specifications, so long as it can be demonstrated that such alternatives would be consistent with applicable performance goals for protection of beneficial uses of water.

If the second approach is pursued, the applicant should provide the following type of data:

- Waste properties, including grain size, density, acid generation potential, and soluble constituent concentrations.
- Ultimate size and configuration of the waste management unit, and anticipated post- closure settlement.
- General geologic characteristics of the area, extending an appropriate distance in all directions from the waste management unit.
- Subsurface conditions at the waste management unit, including types of soil and rock, stratigraphic conditions, and the potential existence of discontinuities (e.g., fractured or faulted zones). Unsaturated zone property data should include natural permeability, natural water content, and estimated field moisture capacity.
- A summary of the results of the monitoring program and an evaluation of the performance of the containment system.
- Depth to ground water and estimated extraction rate potential per individual well and per acre of site area.
- Existing ground water (and surface water, when appropriate) quality, existing and identified beneficial uses, and potential applicable beneficial use objectives.
- The locations of all existing wells or open boreholes to ground water within 1,000 feet of the facility boundary.
- A description of the current and future land use surrounding the waste management unit and the planned land use for the waste management unit.
- A closure and post-closure maintenance plan which includes:
  - As-built drawings of the waste management unit and the monitoring systems.
  - Description of past performance of the containment systems.
  - Description of how the waste management unit will be closed, and the staging of closure, if relevant.
  - Drawings and specifications for the closure activities.
  - Post-closure maintenance and monitoring plan.
  - Post-closure land use plan.
- Where a cover is to be provided as part of closure, a detailed description of the proposed cover systems and maintenance procedures shall be provided, including:
  - The manner in which conditions will vary throughout the waste management unit, and variations in cover systems to be used to accommodate such variability.
  - Components of each type of cover system.

- The manner in which each cover system component will be constructed, including preparation of the waste, removal of free liquids, QA/QC procedures, and the methods used to accommodate settlement of the waste, if necessary.
- Specifications for each cover component, including vegetation or material type, physical and chemical attributes, and design loading conditions.

- Where the cover is to act as an infiltration barrier, calculations showing the performance of the individual cover system components and demonstrating that the integrated system will operate as intended.
- Where enhanced containment is included as part of closure, evaluations of seepage rates and times, based on conservative performance estimates of the enhanced containment system and operational conditions (see Chapter 3). If constituent attenuation is used in this estimate, the method for verifying attenuation potential should be provided (see Chapter 5).
- Where necessary, cover stability calculations should be provided to demonstrate that waste movement which could damage the integrity of the cover system would not occur under either static or dynamic loading conditions.

8.7 BIBLIOGRAPHY

Bertram, G. E. *An Experimental Investigation of Protective Filters.* Harvard Soil Mechanics. Series No. 7. 1940.

Caldwell, J. A. *Cover Design for Uranium Mill Tailings Piles: A Standard for the Future.* Proceedings of the Western Regional Symposium on Mining and Mineral Processing Wastes. University of California, Berkeley. May 30-June 1, 1990.

Caldwell, J. A. *Disposal Cell Design.* In *Deserts as Dumps* edited by Reith and Thompson, University of New Mexico Press. 1991.

Caldwell, J. A., K. Ferguson, R. L. Shiffman, and D. Van Zyl. *Application of Finite Strain Consolidation Theory for Engineering Design and Environmental Planning of Mine Tailings Impoundments, Sedimentation/Consolidation Models - Predictions and Validations.* R. W. Yong and E. C. Townsend (Ed.). American Society of Civil Engineers. 581-606. 1984.

Cedergren, H. R. *Seepage, Drainage and Flow Nets.* New York. John Wiley. 1967.

Cline, J. F. *Biobarriers Used in Shallow-Burial Ground Stabilization.* Pacific Northwest Laboratory Report for U.S. Department of Energy. PNL Report No. 2918. 1979.

Cline, J. F., D. A. Cataldo, F. G. Burton, and W. E. Skiens. "Biobarriers Used in Shallow Burial Ground Stabilization". In Nuclear Technology, 58. August 1982. 150-153.

Demery, P. M. *Research Analysis of Plastic Filters.* Proceedings of the Irrigation and Drainage Specialty Conference. Boise, ID. July 1980. 21.

DePoorter, G. L. *Shallow Land Burial Technology Development - Presentation for the Low-Level Waste Management Program Review Committee.* Los Alamos National Laboratory, Los Alamos, NM. 1982.

Holtz, R. D., W. D. Kovacs. *Introduction to Geotechnical Engineering.* Englewood Cliffs, NJ: Prentice-Hall. 1981.

Holtz, R. D. *Geotextiles for Drainage and Erosion Control at Hazardous Waste Landfills.* EPA 600/2-86-085. NTIS PB 87-129557. Hazardous Waste Engineering Research Laboratory, Cincinnati, OH. 1984.

Karpoff, K. P. *The Use of Laboratory Test to Develop Design Criteria for Protective Filters.* ASTM Proceedings. Vol. 55. 1955. 1183-1198.

Koerner, R. M. *Designing with Geosynthetics.* Second Edition. Englewood Cliffs, NJ: Prentice-Hall. 1989.

Lee, C. R., J. G. Skogerboc, K. Eskew, R. W. Price, N. R. Page, M. Clar, R. Kort, and H. Hopkins. *Restoration of Problem Soil Material at Corps of Engineers Construction Sites.* Instruction Report EL-84-1, United States Army Engineer Waterways Experiment Station, Vicksburg, MS. 1984.

Marks, B. D. *The Behavior of Aggregate and Fabric Filters in Subdrainage Applications.* Department of Civil Engineering. The University of Tennessee, Knoxville. 1975. 149.

McKenzie, C. L. *Heap Leach Detoxification, Carson Hill Gold Mine Leach Unit No. 1.* Preprint No. 26 presented at Northwest Mining Association Annual Meeting, Spokane, WA. 1990.

Nyhan, J. W., W. V. Abeele, B. J. Drennon, W. J. Herrera, E. A. Lopez, G. J. Langhorst, E. A. Stallings, R. D. Walker, and J. L. Martinez. *Development of Technology for the Design of Shallow Land Burial Facilities at Arid Sites.* LA-UR-35-3278. Proceedings of the Seventh Annual Participants' Information Meeting, DOE, Low Level Waste Management Program. 1985.

Oliver, H. J. *Through and Overflow Rockfill Dams - New Design Techniques.* Proceedings of ICS - Rockfill Hydraulic Engineering: Elsevier Publishing, New York, NY. 1967.

Pertusa, M. *Materials To Line or To Cap Disposal Pits for Low-Level Radioactive Wastes.* Geotechnical Engineering Report GR80-7. University of Texas, Austin, TX. 1980.

Portillo, R. *The UMTRA Project.* In *Deserts as Dumps* Reith and Thompson (ed.), University of New Mexico Press. 1991.

Portland Cement Association. *Soil-Cement Construction Handbook.* Engineering Bulletin. 1979.

Robertson, A. Mac G., J. W. Fisher, Dirk van Zyl. *The Production and Handling of Dry Uranium and Other Tailings.* Symposium on Uranium Mill Tailings Management. Colorado State University. Fort Collins, CO. December 9-10, 1982.

Schafer and Associates, Reclamation Research Unit - Montana State University, and CH_2M Hill, *Silver Bow Creek RI/FS. Streambed Tailings and Revegetation Studies - Phase III,* Prepared for the Montana Department of Health and Environmental Sciences, Document SBC-STARS - Phase III - WP-F-RO-110189. November 1, 1989.

Thanikachalam, V. *Studies on the Design of Protective Filters for Earthen Structures.* Ph.D. Thesis. University of Madras, India. 142. 1975.

Thornburg, A. A. *Plant Materials for Use on Surface Mined Lands.* TP-157 and EPA-600/7-79-134. U.S. Department of Agriculture, Soil Conservation Service, Washington, D.C. 1979.

United States Bureau of Reclamation. *Design of Small Dams.* U.S. Government Printing Office, Washington, D.C. 1977.

United States Department of Agriculture. Soil Conservation Service. *Predicting Soil Loss Using the Universal Soil Loss Equation.* 1980.

United States Environmental Protection Agency. *Erosion and Sediment Control; Surface Mining in the Eastern United States, Planning.* Vols. 1 and 2. Report No. EPA-625/3-76-006. 1976.

United States Department of Energy. *Technical Approach Document.* UMTRA-DOE/AL050425.0002 DOE UMTRA Project Office, Albuquerque, NM. 1989.

United States Department of Energy. *UMTRA Project, Chaney Remedial Action Plan.* November 1990.

United States Department of Energy. Uranium Mill Tailings Remedial Action (UMTRA). *Remedial Action Planning and Disposal Cell Design to Comply with the Proposed EPA Standards (40 CFR Part 192).* UMTRA-DOE/AL 400503.0000. January, 1989. 102 and Appendices.

United States Department of the Navy. *Soil Mechanics.* Design Manual 7.1. NAVFAC DM-7.1. May 1982.

United States Environmental Protection Agency. *Evaluating Cover Systems for Solid and Hazardous Waste.* EPA SW-867. NTIS PB 87-154894. Municipal Environmental Research Laboratory, Cincinnati, OH. 1982a.

United States Environmental Protection Agency. *Remedial Action at Waste Disposal Sites.* EPA-625/6-82-006. NTIS PB 82-239054. Municipal Environmental Research Laboratory, Cincinnati, OH. 1982b.

United States Environmental Protection Agency. *Hazardous Waste Land Treatment.* EPA SW-874. NTIS PB 81-182107. Municipal Environmental Research Laboratory, Cincinnati, OH. 1983a.

United States Environmental Protection Agency. *Lining of Waste Impoundment and Disposal Facilities.* EPA SW-870. NTIS PB 86-192796. Municipal Environmental Research Laboratory, Cincinnati, OH. 1983b.

United States Environmental Protection Agency. *Standardized Procedures for Planting Vegetation on Completed Sanitary Landfills.* EPA 600/2-83-055. NTIS PB 83-241018. Municipal Environmental Research Laboratory, Cincinnati, OH. 1983c.

United States Environmental Protection Agency. *The Hydrologic Evaluation of Landfill Performance (HELP) Model.* Vol. 1 and 2, User's Guides for Version 1. EPA/530-SW-84-009 and EPA/530-SW-84-010. NTIS PB 85-100840 and PB 85-100832. Municipal Environmental Research Laboratory, Cincinnati, OH. 1984.

United States Environmental Protection Agency. *Final Covers for Uncontrolled Hazardous Waste Sites.* Report No. EPA-540/2-85-002. NTIS PB 87-119483. Hazardous Waste Engineering Research Laboratory, Cincinnati, OH. 1985.

United States Environmental Protection Agency. *Engineering Guidance for the Design, Construction and Maintenance of Cover Systems for Hazardous Waste.* EPA/600/2-87/039. NTIS PB 87-191656/AS. Hazardous Waste Engineering Research Laboratory, Cincinnati, OH. 1987a.

United States Environmental Protection Agency. *Geosynthetic Design Guidance for Hazardous Waste Landfill Cells and Surface Impoundments.* EPA/600/2-87/097. NTIS PB 88-131263/AS. Hazardous Waste Engineering Research Laboratory, Cincinnati, OH. 1987b.

United States Environmental Protection Agency. *Design and Construction of Covers for Solid Waste Landfills.* EPA-600/2-79-165. NTIS PB 80-100381. Municipal Environmental Research Laboratory, Cincinnati, OH. 1979.

United States Environmental Protection Agency. *Technical Guidance Document: Final Covers on Hazardous Waste Landfills and Surface Impoundments OSWER.* EPA/530-SW-90-047. 39. July 1989.

Winger, R. J. and W. F. Ryan. *Gravel Envelopes for Pipe Drains-Design.* Transactions. American Society of Agricultural Engineers. Vol. 14. No. 3. 1971.

Wright, M. J. (Ed.). *Plant Adaptation to Mineral Stress in Problem Soils.* Cornell University Agricultural Experiment Station. Ithica, NY. 1976.

CONTENTS

CHAPTER 9
HEAP LEACHING

PAGE NO.

LIST OF TABLES/LIST OF FIGURES 455

9.1 Introduction 457
- 9.1.1 Heap Leaching 457
- 9.1.2 Purpose and Scope 457
- 9.1.3 Integration With Other Chapters 459
- 9.1.4 Study Participants 459
- 9.1.5 Sources of Information 459

9.2 Heap Leach Practice 459
- 9.2.1 Description of Heap Leaching 459
  - 9.2.1.1 Mine/Ore Source 459
  - 9.2.1.2 Ore Preparation 460
  - 9.2.1.3 Heap and Pad 460
  - 9.2.1.4 Solution Application/Collection System 461
  - 9.2.1.5 Pregnant Solution Containment 462
  - 9.2.1.6 Metal Recovery Circuit 462
  - 9.2.1.7 Barren Solution Containment 462
- 9.2.2 Alternative Heap Leach Methods 463
  - 9.2.2.1 Reusable Pad Method 463
  - 9.2.2.2 Expanding Pad Method 465
  - 9.2.2.3 Valley Leach Method 467
- 9.2.3 Containment Systems 469
  - 9.2.3.1 Factors Influencing Choice of Liner 470
  - 9.2.3.2 Single Liners 472
  - 9.2.3.3 Double Liners 475
- 9.2.4 Solution Control 476
  - 9.2.4.1 Solution Application 476
  - 9.2.4.2 Collection and Head Control 477
  - 9.2.4.3 Ponds 479

CONTENTS

CHAPTER 9
(Continued)

	PAGE NO.
9.2.4.4 Climatic Considerations	480
9.2.4.5 Water Balance, Rinsing, and Pond Sizing	484
9.2.5 Slope Stability Aspects	487
9.3 Design Approach	488
9.3.1 General	488
9.3.2 Solution Containment Design	489
9.3.3 Solution Control	489
9.3.4 Preparation of Leach Pile for Closure	489
9.3.5 Examples of Project Risk	490
9.3.5.1 HL1	491
9.3.5.2 HL2	491
9.3.5.3 HL3	491
9.3.5.4 HL4	491
9.3.5.5 HL5	491
9.3.5.6 HL6	492
9.3.5.7 HL7	492
9.3.5.8 Assessment of Risks Associated with HL1 to HL7 Examples	492
9.4 Heap Leach Guidelines	492
9.4.1 Motivation	492
9.4.2 Proposed Guidelines	493
9.4.2.1 Information Pertaining to the Proposed Operation	493
9.4.3 Design and Performance Standards	495
9.5 Bibliography	497

CONTENTS

CHAPTER 9
(Continued)

LIST OF TABLES

TABLE NO.	TITLE	PAGE NO.
9.1	Liner Selection Considerations	471
9.2	Suggested Design Event and Criteria for Heap Leach Operations	481

LIST OF FIGURES

FIGURE NO.	TITLE	PAGE NO.
9.1	Schematic of a Heap Leach Operation	458
9.2	Reusable Pad Method Schematic	464
9.3	Expanding Pad Method	466
9.4	Valley Leach Method	468
9.5	Typical Heap Leach Liner Systems	473
9.6	Typical Pressure Emitter Application	478
9.7	Pond Design Considerations	483
9.8	Heap Leach Operation Water Circuits	485

CHAPTER 9

HEAP LEACHING

9.1 INTRODUCTION

9.1.1 HEAP LEACHING

Heap leaching is the term given to the hydrometallurgical process of extracting metals by trickling a solution through a pile or "heap" of ore. The solution reacts with the ore by dissolving metal values, thereby "leaching" the metal out of the ore. The resulting "pregnant" solution is then collected at the base of the heap, drained off to a pregnant solution holding pond, from where it is conveyed to the metal recovery plant. In the plant, metal is extracted, leaving a "barren" solution, which is then conveyed to the "barren" solution pond where it is chemically reconstituted to reuse on the pile. The extracting solutions are, therefore, recirculated. Pregnant solutions of low metal values are frequently recirculated until values are sufficient for efficient metal extraction.

Once the leaching activities have been completed, the heap is rinsed with water or chemicals to complete extraction of residual metal values and to reduce the concentrations of metals and reagents in the pore fluids to acceptable levels. The heap is then closed and reclaimed on the same basis as a mining waste pile.

Heap leaching has been applied to base metals over the past century, while its application to precious metals has become widespread only over the past two decades. The most prevalent heap leaching activity in California involves the use of a weak cyanide solution for the leaching of gold and silver ores.

The general method is illustrated in Figure 9.1. The major components of this method are discussed in more detail in the subsections that follow.

9.1.2 PURPOSE AND SCOPE

Heap leaching involves metals recovery rather than waste management. However, at closure, the heap leach residue or spent ore becomes a waste and falls under the jurisdiction of mine waste management regulations. Due to the final rinsing and/or treatment, the chemical characteristics of the spent ore waste pile pose a lower risk to beneficial uses of the waters than during the operational period.

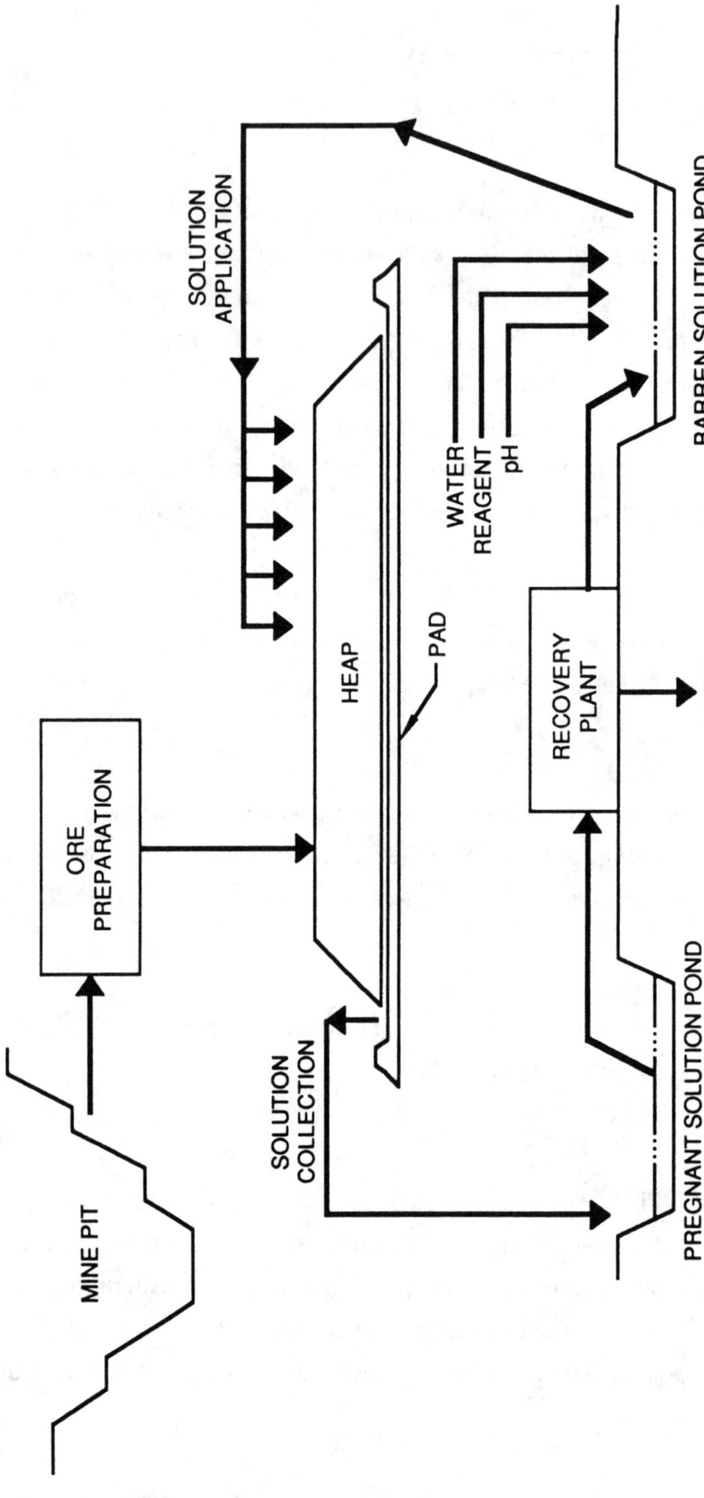

FIGURE 9.1

HEAP LEACH OPERATION

During the operating period (generally years), the operator is obligated to ensure that no adverse water quality impacts occur, and that all leaching solutions are safely contained. With the increased use of heap leaching and the need for containment during operation, a set of guidelines for regulating heap leach operations would be appropriate to assure both adequate containment of fluids during the operating period and compliance with the waste management requirements if the heap is closed in place after leaching ceases. This chapter provides the information required to develop such guidelines.

9.1.3 INTEGRATION WITH OTHER CHAPTERS

This chapter relies on information developed in the following additional chapters of this book:

- Chapter 5 - Attenuation of Chemical Constituents
- Chapter 6 - Climatic Considerations
- Chapter 7 - Liner System Design
- Chapter 10 - Ground Water Monitoring

9.1.4 STUDY PARTICIPANTS

This work was undertaken by the following individuals:

- Robert Dorey, Steffen Robertson & Kirsten (U.S.), Inc.
- Ian P. G. Hutchison, Ph.D., P.E. - Environmental Solutions, Inc.
- Nigel Skirmer, Steffen Robertson & Kirsten (U.S.), Inc.

9.1.5 SOURCES OF INFORMATION

Most of the data sources reviewed for this chapter are listed in Section 9.5, Bibliography.

9.2 HEAP LEACH PRACTICE

9.2.1 DESCRIPTION OF HEAP LEACHING

9.2.1.1 Mine/Ore Source

The material to be leached can be obtained from several sources, such as newly mined ore, previously stockpiled low-grade ore, mineralized waste rock, and/or previously milled tailings. Of these sources, most current heap leach projects involve newly mined ore derived from open-pit mining.

9.2.1.2 Ore Preparation

The metallurgy of the ore dictates how it should be prepared prior to leaching. Generally, ores amenable to leaching are oxidized, although current research is being conducted to develop procedures for conditioning and leaching unoxidized ores. Depending upon the complex in which the metal is held, the ore needs to have the correct particle size to allow solution contact and metal dissolution. Ore pretreatment can range from none (for run-of-mine ore) to crushing only, crushing and agglomeration, or agglomeration only (for fine-grained ore or tailings).

The major goal of ore preparation for heap leaching is to produce an ore fine enough to allow solution contact with the metals, yet permeable and stable enough to allow an adequate percolation rate through the heap. These requirements may be contradictory in some cases, where the optimum leach particle size results in low permeability when the ore is piled on the leach pad.

9.2.1.3 Heap and Pad

Each operation is unique, and the design of the heap and pad requires consideration of a number of factors. These include ore type/origin, metallurgy, leachability, site topography, site geotechnical and geohydrological characteristics, and climate.

Three basic methods of heap and pad construction and operation are currently used:

- Reusable Pad - In this application, the same pad is continually reused; The spent ore is moved from the pad and disposed of in a separate facility.
- Permanent Expanding Heap[(1)] - Here the ore remains permanently on the pad after the leaching is completed. New pads or portions of the same pad are continually constructed to accommodate the additional ore that is mined.
- Valley Leach - In this application, a valley is used to form the leaching area. An earthen embankment usually provides containment at the downgradient end, and the impoundment that is created is filled with ore. The heap and pad are designed to be stable structures which contain both the ore and the leachate. The pad usually is fitted with a liner to contain the solutions.

The pad includes a liner system which is needed principally to contain the leach solutions within the facility. It also acts as a platform on which the heap is built. Ore placement on the heap can be performed by various methods; from end dumping, to placement by front-end loaders, to conveyor

---

(1) This technique is also known as the "dedicated pad method" because each pad is used only for ore initially placed on it, which is ultimately abandoned in place.

stacking. The main criteria in selecting a heap construction method are to limit layering, compaction, and segregation of particles and to prevent damage to the liner system during ore placement.

Often, choices for the heap location are limited, due to constraints such as haul distance, land status, or topography. The foundation for the heap must be capable of withstanding the loads applied, not only in terms of stability, but also in terms of differential settlement across the heap. Also, in the case of valley heaps or sites with steeply sloping ground, the foundation materials and the foundation liner interface must resist sliding.

9.2.1.4 Solution Application/Collection System

Leach solution is conveyed from the barren pond to the heap through a piping system. A pump system is usually required to give sufficient pressure for applying the solution uniformly over the area to be leached. Until recently, the application method was through sprinklers similar to those used for golf course irrigation. Many modern mines now use drip irrigation systems, which provide more uniform coverage and minimize water losses due to evaporation. In rare cases, mines use ponding on the heaps as a means of solution application. Application can be done with sprinklers, wigglers, or emitters. Typical application rates vary from 0.003 to 0.005 gallons per minute per feet^2 (gpm/ft^2),

Application rates in most instances result in a flow of leachate through heap material that is in a partially saturated condition. Normally, this means that the permeability of the ore within the heap must be at least 10^{-4} cm/sec (100 feet/year). The chemical reactions whereby metals are dissolved by the reagents usually require oxygen. The unsaturated condition of the raw ore also allows more oxygen to be available for the dissolution process.

The flow of solution through the heap is under gravity and is essentially vertical from the surface to the base of the heap. Typically, a layer of permeable material (gravel) and/or closely spaced perforated pipes are placed directly above the liner prior to ore loading. These collector pipes provide drainage under the heap and reduce hydrostatic pressure on the liner. This underdrain system can be directly connected to a piping system or a lined collection channel which leads to the pregnant solution pond.

9.2.1.5 Pregnant Solution Containment

The pregnant solution contains the "metal values" and thus, for economic reasons alone, it is imperative to contain this solution. Therefore, liner systems are usually employed for the pad, the ponds, and the solution collection systems. Although not necessary for operations, it is common practice to place the barren solution and pregnant solution ponds adjacent to each other. This confines the large volumes of solutions to one area of the facility and reduces construction and operational costs.

Since the solution ponds are connected directly to the heap(s), drainage from both leachate and direct precipitation flows to the ponds. The design of the ponds, therefore, requires consideration of the application rate of leaching solution, evaporation and precipitation patterns at the site, and the working level of the solution which needs to be maintained for efficient operations.

In the case of heaps designed to retain solution, such as valley leaches, the pregnant solution may be stored within the space between the ore particles. In this instance, the liner and heap earth-work are required to function as water retaining structures. Extraction of solution for recycling is performed by pumping from sumps within the heap or by gravity drainage from low points within the heap.

9.2.1.6 Metal Recovery Circuit

The metal recovery circuit consists of processes to recover the metal from the solution. For gold operations, recovery can be through zinc precipitation (Merrill-Crowe) or through carbon adsorption (van Zyl, Hutchison and Kiel, 1988).

9.2.1.7 Barren Solution Containment

The barren solution pond, like the pregnant solution pond, functions primarily as a containment and storage system. The solution must be contained to facilitate the addition of reagent, as well as for the prevention of impacts to water quality that could result from solution releases. Alternative containments, such as tanks, can also be used, but most heap leach operations employ lined ponds for holding the solution. As discussed in Section 9.2.4.3, double liners with leachate collection and removal systems (LCRS) are frequently used within normal pond operating levels to minimize the potential for both solution losses and environmental impacts to water resources.

9.2.2 ALTERNATIVE HEAP LEACH METHODS

Determination of which of the three basic methods of heap leaching is to be employed at a site (i.e., reusable pad, expanding pad, or valley leach) depends on site-specific conditions, including the ore body type and tonnage, climate, and topography. The following subsections briefly outline key aspects associated with each of these three methods.

9.2.2.1 Reusable Pad Method

The reusable pad method involves the construction of one or a series of durable pads onto which the prepared ore is loaded, leached, rinsed, chemically treated if necessary, then unloaded for disposal in a spent ore pile (see Figure 9.2).

The major considerations involved in the design of a reusable pad are:

- Requires a predictable/short leach cycle.
- Requires an ore of consistent leachability.
- Is particularly suited to situations where there is limited "flat" land.
- Requires a suitable waste residue pile site.
- Requires a durable "high-stress" resistant liner system.
- Can maintain a zero water balance in wetter climates due to smaller pad area.
- May allow for covering of the heap to improve water balance in wet climates (due to the relatively small leach area).
- Requires smaller ponds than other leach methods, due to lesser impact of flood event on the smaller pad area.
- Requires double handling of ore.
- Provides no conditions for ore to "mature" and be re-leached at a later stage to extract residual metal values.

The size of the pads and number of pad segments are dictated both by the rate at which ore is produced and by the length and predictability of the leach cycle required for optimal metal extraction. Generally, leach times are on the order of 60 to 90 days. For a given production rate, the longer the required leach time, the larger the pad area required under the leach in order to accommodate the larger residence time for the leached ore. Scheduling of the leaching cycle for a fixed production also requires that a limit be set for the time the ore can be leached. Where the ore is variable, therefore, not all of it will be at the required recovery within the limit of the fixed leach period.

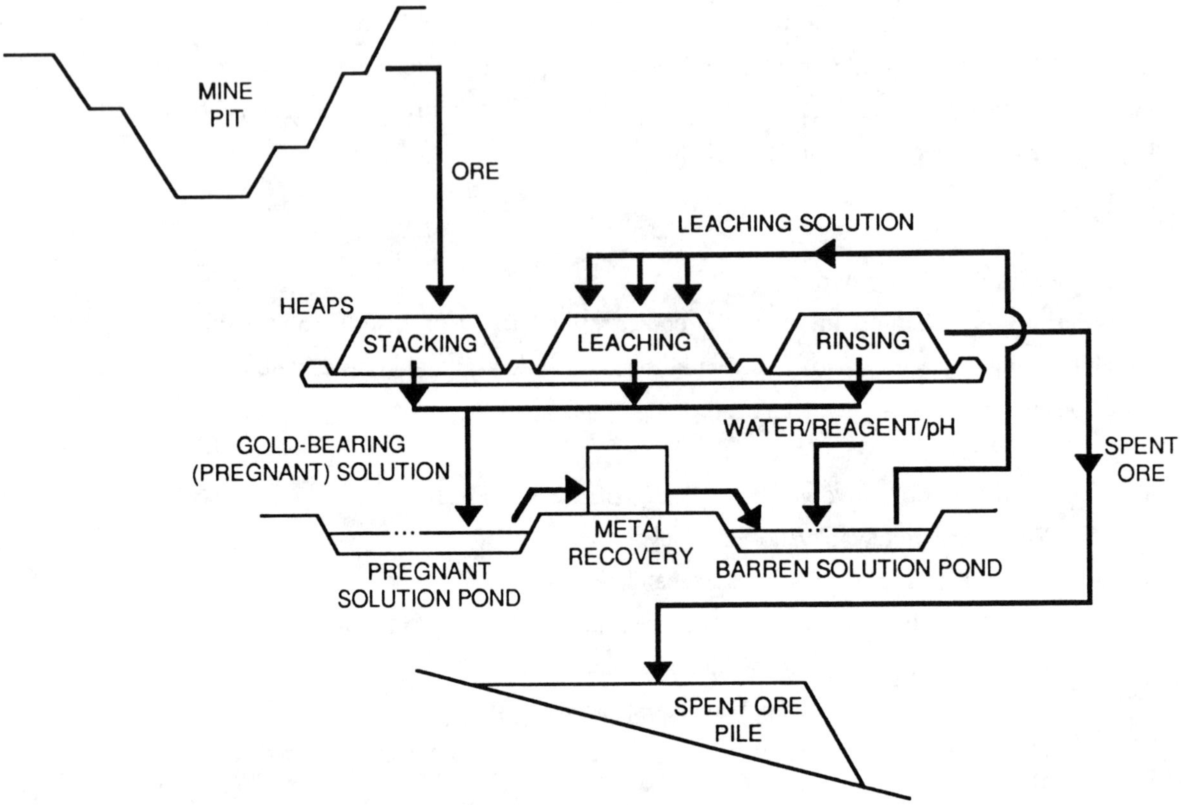

SEQUENCE OF ACTIVITIES:

- **MINE ORE**
- **PREPARE ORE AND STACK ON PAD**
- **LEACH ORE**
- **RINSE/TREAT ORE IF NECESSARY**
- **REMOVE SPENT ORE FROM PAD**
- **DISPOSE OF SPENT ORE IN SEPARATE WASTE PILE**
- **CLOSE PAD AND SPENT ORE PILE**

FIGURE 9.2

REUSABLE PAD METHOD

A drawback of the reusable pad method is this limited flexibility in the leach time. Once ore is removed from the pad, residual values are lost, as there is no opportunity to re-leach the ore or allow it to mature in the heap. The advantage, however, is that the pad costs per ton decrease with the number of times the pad can be used.

Construction of the reusable pad involves a relatively limited area of suitable ground compared to the expanding pad method. The loading and unloading of the pad necessitates a durable liner system capable of tolerating the working stresses involved. Typical liner systems such as asphalt and concrete, in conjunction with synthetic liners, are frequently used. Liner systems are chosen primarily to accommodate the excavation method used to unload the ore. If accurate excavation level control is achievable so that a layer of liner cover (e.g., gravel) is left in place, membrane liners alone could also be considered.

The ore is usually rinsed and/or neutralized to prescribed constituent concentration standards in the residual interstital moisture prior to its transfer from the leach pad to a residue disposal unit that provides containment which is adequate for the residual chemistry of the waste and potential leachate that may be formed. Containment provisions for the residue have to satisfy the appropriate waste management regulations, and will be dependent upon the extent of rinse and/or neutralization prior to removal from the leach pad, the general ore chemistry, and site climatic and hydrogeologic conditions.

9.2.2.2 Expanding Pad Method

The expanding pad method involves the preparation and placement of ore on a continually expanding pad. Following leaching, the ore is left in place, and the pad is expanded (see Figure 9.3). Subsequent re-leaching or washing (rinsing) and treatment, if required, are performed. Additional lifts can be added to the heap, and interim reclamation of rinsed and treated segments can be performed. The major considerations involved in the design of an expanding pad are:

- Requires a large area.
- Requires relatively flat to gently sloping topography.
- Requires a relatively dry climate compared to other methods, in order to maintain zero water balance.
- Needs larger storage capacity than reusable pads for flood events in the ponds.
- Can accommodate variable ore production rates and different leach cycle times for various groups of ore.

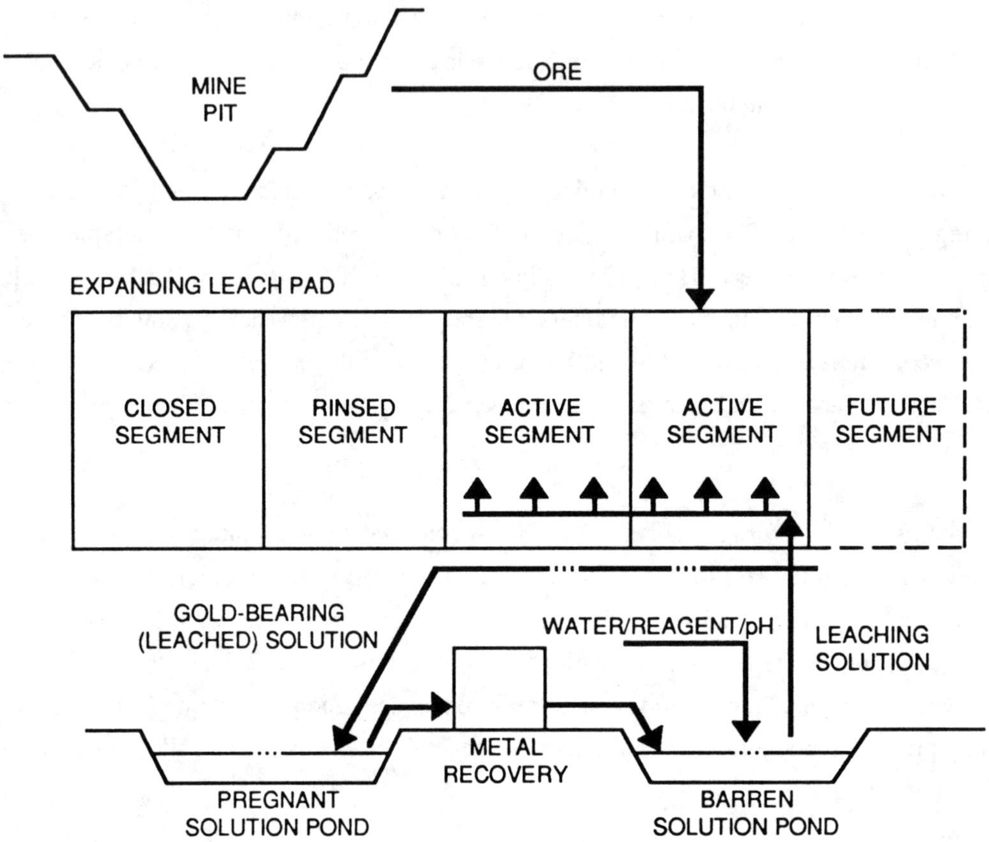

SEQUENCE OF ACTIVITIES:

- **MINE ORE**
- **PREPARE ORE AND STACK ON PAD**
- **LEACH ORE**
- **RINSE/TREAT IF NECESSARY**
- **CLOSE SPENT ORE PILE IN PLACE**

FIGURE 9.3

EXPANDING PAD METHOD

- Requires a relatively simple liner due to limited leach time and "low-stress" environment.
- Liner costs increase with tonnage of ore mined.
- Has low initial capital cost.

The expanding pad method requires relatively large areas of suitable ground on which the pad can be expanded in a staged manner. Typically, ground slope should be less than 10 percent and preferably less than 5 percent.

There are no constraints on the duration of the leach cycle. Ore having variable or long leaching cycles can be accommodated without impacting subsequent ore production. If it is economical, ore can be leached in several cycles before being abandoned.

Liners for the expanding pads are not subject to repeated construction stresses once ore is placed. Generally, membrane liner systems are installed, designed primarily to handle stresses involved during initial ore placement and, if necessary, to handle subsequent lifts. Most expanding pads also include a drainage material and/or perforated pipes immediately above the liner to collect the pregnant solution and to control the hydrostatic pressure. Small, very short-term operations can use natural or amended soil liners. In a few high-risk areas, double liners with a leachate collection and recovery system may be required.

Some expanding pads currently operate with heap heights well in excess of 100 feet. These are generally built in sequential 20- to 40-foot lifts.

9.2.2.3 Valley Leach Method

The valley leach method involves the preparation and placement of ore behind a retaining structure which is usually an earthen embankment. The ore is leached in placed, with subsequent lifts progressing up slope (see Figure 9.4). The majority of the ore remains in contact with the leach solution during operation. At completion of leaching, the ore is rinsed, left in place, and reclaimed.

The major considerations involved in the design of a valley leach facility are:

- Requires a durable ore.
- Can be used in steep terrain.
- Requires reduced storage capacity.

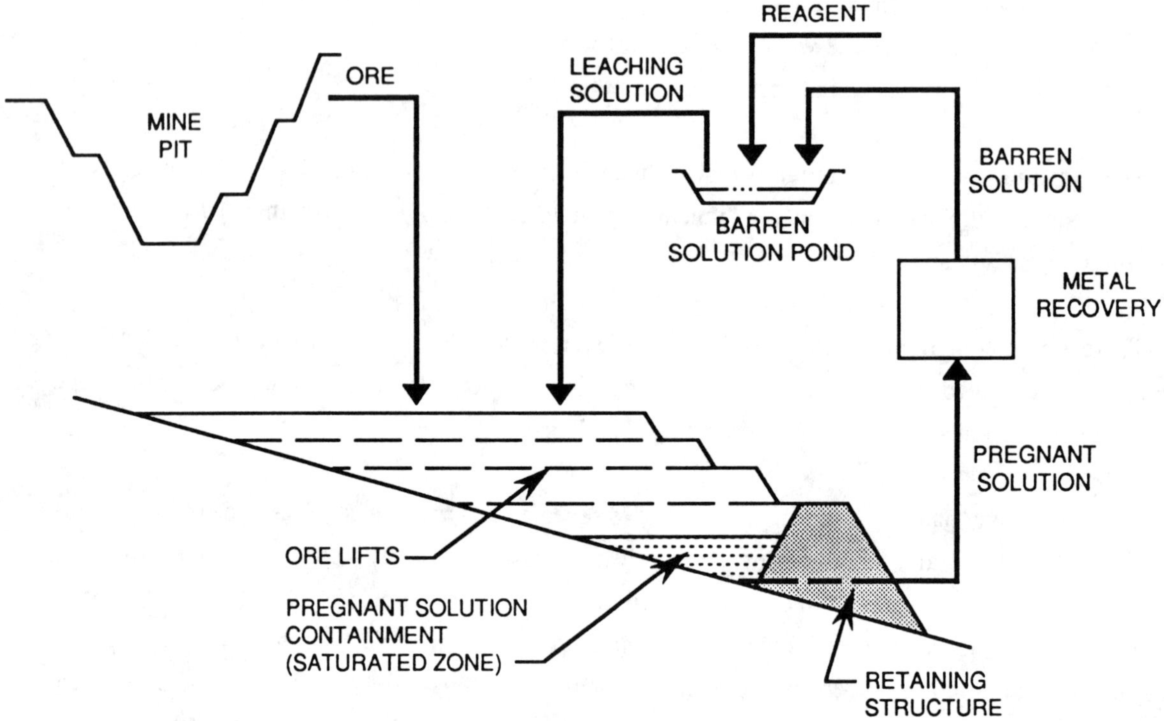

SEQUENCE OF ACTIVITIES:

- **MINE ORE**
- **PREPARE AND STACK IN VALLEY**
- **LEACH**
- **COVER WITH NEW ORE**
- **LEACH**
- **REPEAT PROCESS**
- **RINSE/TREAT IF NECESSARY**
- **LEAVE IN PLACE**
- **RECLAIM**

FIGURE 9.4

VALLEY LEACH METHOD

- Requires a high integrity liner due to the high hydraulic heads.
- Must be placed so as to provide a stable heap, similar to a waste rock dump.
- Requires retaining structure(s).
- Can operate in wide range of climatic conditions.
- Can accommodate extended leach time of up to several years.
- Does not usually require a separate pregnant pond.

The applicability of the valley leach method is dependent upon the durability of the ore and its ability to remain permeable under the stresses of subsequent ore placement. The major advantage of this method is its applicability in steep terrain and its use in a wide range of climatic conditions. The storage capacity of the pore space of the ore is usually used to contain pregnant solution, thus limiting or eliminating the need for a separate pregnant solution pond. However, it is also possible to have a separate pregnant solution pond rather than containing the pregnant solution in the heap.

Typically, the liner system consists of a combination of synthetic and low permeability or amended soil material. At high risk sites, a double-lined system with leachate collection and recovery capabilities may be appropriate. The storage of solution behind the retaining structure requires it to function as a water dam. Careful design of the piping and its placement within the dam is, therefore, critical to the proper functioning of the system.

Operating valley leaches can have heights in excess of 200 feet from toe to crest. Therefore, detailed slope stability evaluations are required when designing such a facility.

9.2.3 CONTAINMENT SYSTEMS

The primary containment systems in a heap leach operation consist of the liner system located under the pad, the solution collection and conveyance system, and the pregnant and barren solution ponds. The design of these systems must be adequate to contain both solutions and precipitation, for economic reasons and for the protection of ground water. Except for liners employed in the reusable pad method, the pad liner system must also meet applicable long-term spent-ore closure requirements.

The following section contains a general discussion of design considerations for various liner systems. The information provided should be reviewed in conjunction with the material provided in Chapter 7 (Liner System Design).

9.2.3.1 Factors Influencing Choice of Liner

The leach pad liners, and the liners in the solution storage ponds, both contain the same liquid. However, the type of liner selected for the leach pad is usually different from that selected for the ponds since the operating conditions are considerably different. Factors influencing the choice of liner are summarized in Table 9.1. It will be seen from the table that, for the leach pad, liner strength is an important factor. The pad is subjected to uniform stresses resulting from the weight of the heap, as well as from local, concentrated stresses caused by the passage of construction equipment, ore loading procedures, or irregularities in pad liner foundation or ore loading conditions. In the case of the pond liners, there are only the uniform hydrostatic stresses from the stored liquid.

Exposure to the elements is another important factor which creates stress for the liner system. In the expanding leach pad method, the ore is placed on the liner and is not moved. Therefore, the liner is exposed only during construction and for a relatively short period while it is being loaded with ore. At the edge of the pad and in solution ponds, however, the liner is continuously exposed to the elements, so more resistant material may be necessary. Finally, the liners must function and maintain their integrity relative to the chemicals to which they will be exposed.

The choice of liner materials primarily consists of geomembranes, soil liners, and amended soil liners. Geomembrane liners can be polyvinyl chloride (PVC), hypalon, high density polyethylene (HDPE), very low density polyethylene (VLDPE), Chevron Industrial Membrane (CIM), chlorinated polyethylene (CPE), and asphalt/hydraulic asphaltic concrete (HAC). Soil and amended soil liners can consist of onsite or local borrow materials, mixtures of both, or bentonite or other soil additive composites.

Factors to be considered are as follows:

- For geomembrane liners:
    - Material type (thickness, strength, durability, cost).
    - Foundation material firmness and smoothness.
    - Cover material protection.
    - Method of placement and seaming.
- For soil liners:
    - Availability of local and imported materials.
    - Soil composition (clay content, chemical stability of clay mineral under action of solution, hydraulic permeability, plasticity).
    - Constructability (liner thickness, preparation, mixing and workability, compaction).

TABLE 9.1

LINER SELECTION CONSIDERATIONS (BASED ON GOLD INDUSTRY PRACTICE)

FACTORS INFLUENCING LINER CHOICE	EXPANDING LEACH PAD	REUSABLE LEACH PAD	LEACH SOLUTION PONDS	LEACH SOLUTION DITCHES	WASTEWATER PONDS
Type of Solution	High pH, Cyanide (CN) Solution	High pH, CN Solution	High pH, CN Solution	High pH, CN Solution	Varies
Physical Loading	Construction and Heap Loading	Heap Material Loading and Removal	Fluctuating Water Level	Moving Fluids	Fluctuating Water Level, Solids Removal
Exposure to Climate	Exposure During Initial Construction	Exposure When Pad Is Empty	Sunlight, Wind, Temperature Extremes	Sunlight, Wind, Temperature Extremes	Sunlight, Wind, Temperature Extremes
Major Design Factor	Seepage Control, Accommodate Construction of Heap	Seepage Control, Accommodate Construction of Heap and Ore Removal	Seepage Control, Resistance to Climate, High Hydraulic Head	Seepage Control, Resistance to Climate and Moving Fluids	Seepage Control, Resistance to Climate, Dry Surface, High Hydraulic Head
Example of Liner Alternatives	Protected HDPE, VLDPE, Hypalon, PVC, Clay	Asphalt, Protected Synthetic, Clay	HDPE, Hypalon, Protected Clay. Often Double-Lined with LCRS.	HDPE, Hypalon, PVC, Asphalt, Concrete	HDPE, Hypalon, Protected Clay. Often Double-Lined with LCRS.

HDPE: High Density Polyethylene
VLDPE = Very Low Density Polyethylene
PVC = Polyvinyl Chloride
LCRS = Leachate Collection and Removal Systems
Reference: Modified from Strachan and van Zyl, 1988.

- Protection from damage during construction.
- Protection from shrinkage.
- Protection from solution flow erosion damage.
- Shrinkage potential.

Liner systems typically involve one of the following:

- Single liners with loading protection.
- Composite single liners.
- Multicomponent liners with leachate control and recovery capabilities.

Liner system selection must be primarily based on seepage potential considering the hydrostatic pressure from the applied leach solution, but must also include considerations such as puncturing, shear strength, durability, slope and stability of the heap, duration of leach and rinse cycles, and local environmental conditions (e.g., climate, hydrogeology). Furthermore, proper installation of the liner is a prerequisite for reliable performance. Experienced contractors, fully supervised by qualified geotechnicians using proper test procedures, are needed to ensure proper construction.

9.2.3.2 Single Liners

Single liners consist of one low permeability membrane only, such as a clay layer, a geomembrane, or a composite of both a geomembrane and low permeability soil (see Figure 9.5). These types of liners can function satisfactorily, provided they are carefully designed and constructed.

Compacted Clay Liners

Clay liners are usually required to have a permeability of between 10^{-6} and 10^{-7} cm/sec (1.0 to 0.1 foot/year). However, to ensure an adequate level of containment at the site, the following important factors need to be considered:

- Characteristics of the clay material.
- Method of placement and compaction.
- Method of liner system protection against loading.
- Method to avoid cracks from drying and/or shrinkage.

In general, increasing the clay content of soil decreases the permeability. Thus, a silty clay may be more desirable than a silty, clayey sand. Clay mineralogy also affects permeability. For example, montmorillonite clays (bentonites) have a lower permeability than kaolinite or illite clays. Also, sodium bentonites are generally an order of magnitude less permeable than calcium or magnesium

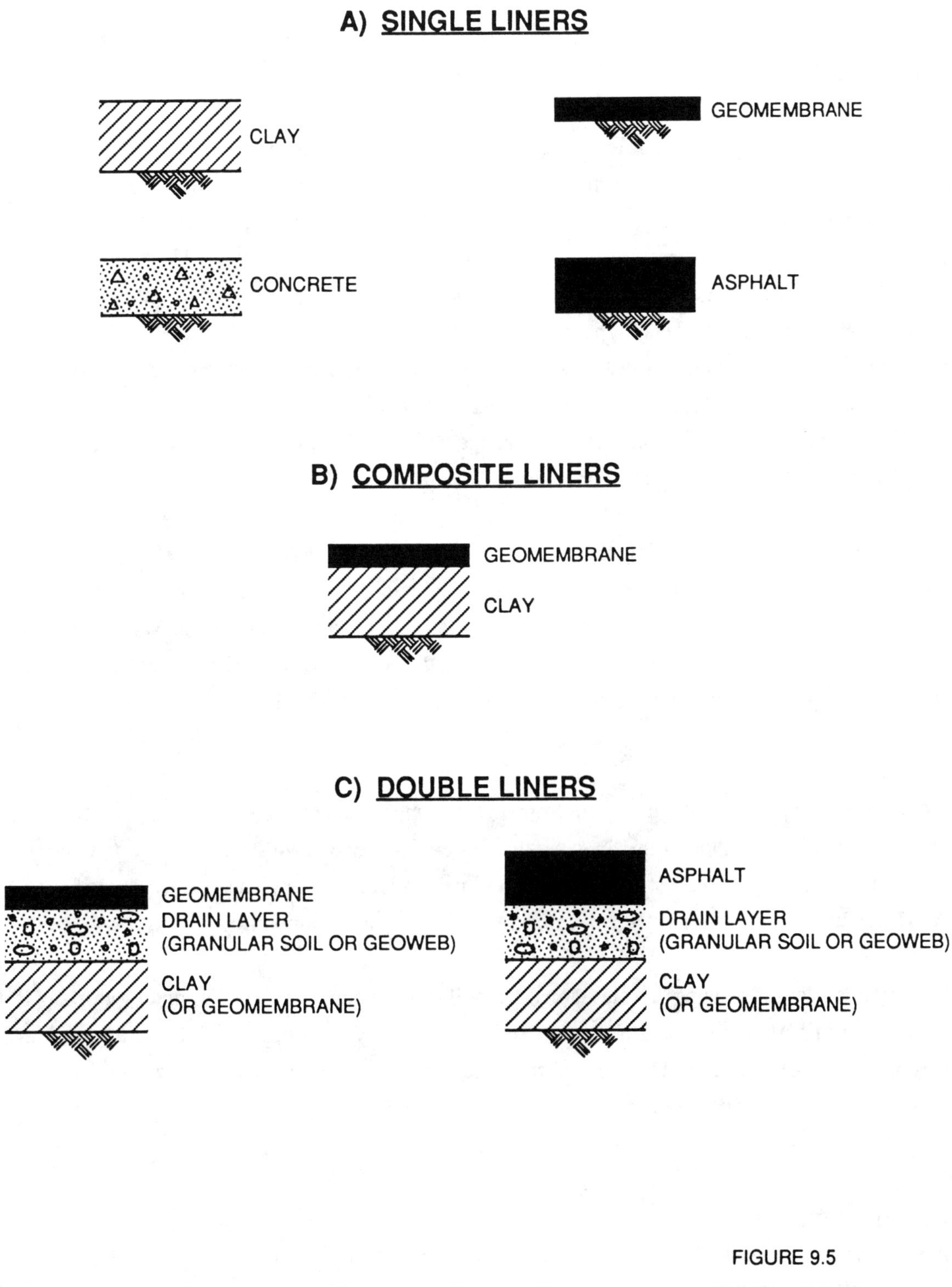

FIGURE 9.5

TYPICAL HEAP LEACH LINER SYSTEMS

bentonites. Permeability is also influenced by the moisture content of the clay liner during placement and compaction; the higher the moisture content, the lower the permeability.

The principal reason for unreliable performance of clay liners is poor or improper construction techniques or cracking due to drying desiccation. Although highly plastic clays have low permeability, if not thoroughly mixed during excavation, conditioning, and placement, they tend to form clumps (or zones), resulting in areas of differential permeabilities. Therefore, special measures are necessary during construction to ensure that uniform conditions are achieved. As a result, proper recognition of clay type, content, and moisture condition is important. Silty clays, rather than highly plastic clays, often make better liner materials because they are easier to compact into uniform, void-free layers. Channel seepage can occur through a clay liner subjected to continuous hydraulic head. Although this seepage would be low, it should be considered in the design of the system (see Chapter 7 for additional discussion).

Desiccation cracking or shrinking of clay can occur within several days of placement, especially during hot, dry periods. Therefore, the compacted clay must be wetted frequently, or covered with a plastic sheet or supplemental soil layer to retain the compacted moisture. Care must also be taken to avoid cracking caused by differential settlement in the foundations. Proper compaction of subsoils avoids settlement cracking, and use of a fine sand cover will prevent drying and shrinkage cracking.

Increases in clay permeability may occur from a geochemical reaction between the clay mineral and the leach solution. Although this is generally more of a concern with certain hazardous waste solutions, compatibility tests should nevertheless be conducted prior to heap leach applications. Also, clay liners must be protected from erosion where there are solution flows.

Similar considerations apply to clay-amended soil liners using native soils mixed with small amounts of imported, powdered bentonite, and to chemically-amended soil liners using sodium carbonate or a petroleum resin. Details of these techniques and further references can be found in Strachan and van Zyl (1988) and Smith and Gierzerski (1987).

Geomembrane Liners

A more frequently used alternative for single liners is synthetic geomembranes (plastic liners). The majority of heap leach liners use these materials, which typically do not react with commonly used leach reagents. Various materials are used, but the most common in heap leach applications is

HDPE. Also, PVC is used successfully but, to avoid deterioration from sunlight exposure, requires more rapid covering than does HDPE. Recently developed VLDPE liners, which exhibit the most attractive qualities of both HDPE and PVC, are anticipated to be used frequently in the future. Other synthetic membrane types include spray-ons such as CIM and HAC which, although costly, are durable and therefore useful for reusable pad liners.

Special considerations when selecting geomembranes are:

- Static and impact loading during construction and heap placement, necessitating proper bedding and cover layers.
- Methods of seaming, such as bonding, welding, and taping.
- Slope stability between the geomembrane and its foundation and/or cover.
- Exposure to elements, leading to ultraviolet degradation and/or thermal shrinkage and resultant seaming problems. A related consideration is the time of year when the liner is installed.
- Cover requirements to prevent puncture by ore placement. Increasing the liner thickness or adding a geotextile cover may be preferable alternatives to sand covers.

Composite Liners

A composite liner is a type of single liner consisting of a geomembrane in combination with a low permeability soil. This system costs considerably more than either a clay or geomembrane liner but, when appropriate, has the advantage of superior seepage control. The potential for leakage from composite liners is much less than with single liners because, for failure to occur, leakage must occur at the same location in both layers. Furthermore, any leakage from the upper/usually synthetic liner is substantially reduced by the presence of a lower permeability (usually clay) layer. Chapter 7 contains details on additional protection provided by composite liner systems.

9.2.3.3 Double Liners

Double liners consist of two low-permeability membranes separated by a drainage layer. An example is an asphalt or geomembrane layer placed over a constructed drainage layer. Performance problems with geomembrane liners are primarily related to flaws in the joints, especially where these are welded in the field. Additional details and examples of potential leakage scenarios are provided in Chapter 7.

Double liners provide a "second line of defense" in the event of excessive leakage (when leakage through the top liner exceeds the capacity of the monitoring/withdrawal system). They are particularly appropriate for impoundments, which must withstand relatively high hydrostatic pressures for long periods or for heap leach pads that have long leach cycles and shallow ground water. Further, double liners have a lower seepage potential than do single liners.

The use of a drainage, or leakage detection, layer between the two low-permeability membranes allows monitoring and collection of leakage. More importantly, it can reduce the hydraulic head on the lower membrane. An important aspect of the design of a double-lined system is incorporating the principal of an acceptable leakage rate through the upper liner. This rate must be low enough that the potential loss from the second, outer liner is sufficiently low that the liner system as a whole performs adequately. Chapters 2 and 7 contain further discussion of the degree of containment required and the performance characteristics of double liner systems.

9.2.4 SOLUTION CONTROL

Solution control covers the various stages of leaching, from initial application to the heap, through collection at its base, to containment in the ponds, together with consideration of natural surface water runoff control.

9.2.4.1 Solution Application

Solution application involves wetting the ore with the required reagent mixed with water. The objective is to achieve maximum contact between the stacked ore and the solution. A general description of application was provided in Section 9.2.1. The following addresses more specific aspects.

It is important to remember that unsaturated flow occurs in most heaps, with the exception of portions of the heap in valley leach operations. The unsaturated flow conditions allow entry of the necessary oxygen. In practice, this means that hydrostatic heads on the liner pad are low, thereby minimizing the potential for leakage.

Occasionally, solution is applied to the surface of the heap by flooding or ponding. This technique is restricted to locations such as desert climates where evaporation losses are high and where excessively high evaporative losses would occur if a fine spray were used. Such flooding cannot be used on very porous ores because of high seepage rates, which would reduce the effectiveness of the leaching solution.

Most solutions are applied by some form of sprinkler (wobbler) or pressure emitter system. Wobblers use an off-center rotary action and apply solution in fairly large drops, thereby minimizing evaporation. Sprinklers are positioned to achieve uniform application rates with attention to coverage of the edges, where extraction can be a problem. Pressure emitters are similar to agricultural drip irrigation systems. Emitters provide continuous dripping with a minimum of drop impact, thus reducing migration of fines and concentrated channel flow in the ore. The ore heap therefore tends to be wetted both laterally and vertically by capillary action. Other significant advantages of pressure emitters are reductions in evaporation losses and ultraviolet degradation, and improved operation in freezing weather. In harsh winter climates, emitter lines are commonly buried 8 to 10 inches below the surface of the ore. Because of these advantages, emitters are becoming the more commonly used system. A typical emitter installation is shown in Figure 9.6.

9.2.4.2 Collection and Head Control

The collection system consists of a series of components that collect solution in the heap and convey it by gravity to the pregnant solution pond. A secondary function of the component within the heap is to reduce the hydrostatic head or zone of saturation above the liner, thereby reducing the potential for leakage. The entire collection system is designed to accommodate the flows of leaching solution runoff, and seepage through the heap during precipitation events.

Typically, the collection system within the heap consists of a drainage zone between the liner and the heap. These collection systems can be constructed using the ore itself, if it is sufficiently permeable, and/or filter materials such as gravel and/or perforated collection pipes. In many operations, these drainage layers are placed before the ore piled onto the heap and they also serve as a protective layer for the liner.

The collection system within the heap is designed to minimize the thickness of any saturation layer above the liner. Its design is based on the slope of the leach pad, the hydraulic conductivity of the ore, and the application rate of the leaching solution. Where the ore's hydraulic conductivity is very high with respect to the application rate, the ore itself may be adequate for drainage. If the ore has a low hydraulic conductivity and has the potential to develop a significant saturated zone above the liner which must be reduced, an extensive collection system including a gravel blanket and drainage pipes may be required.

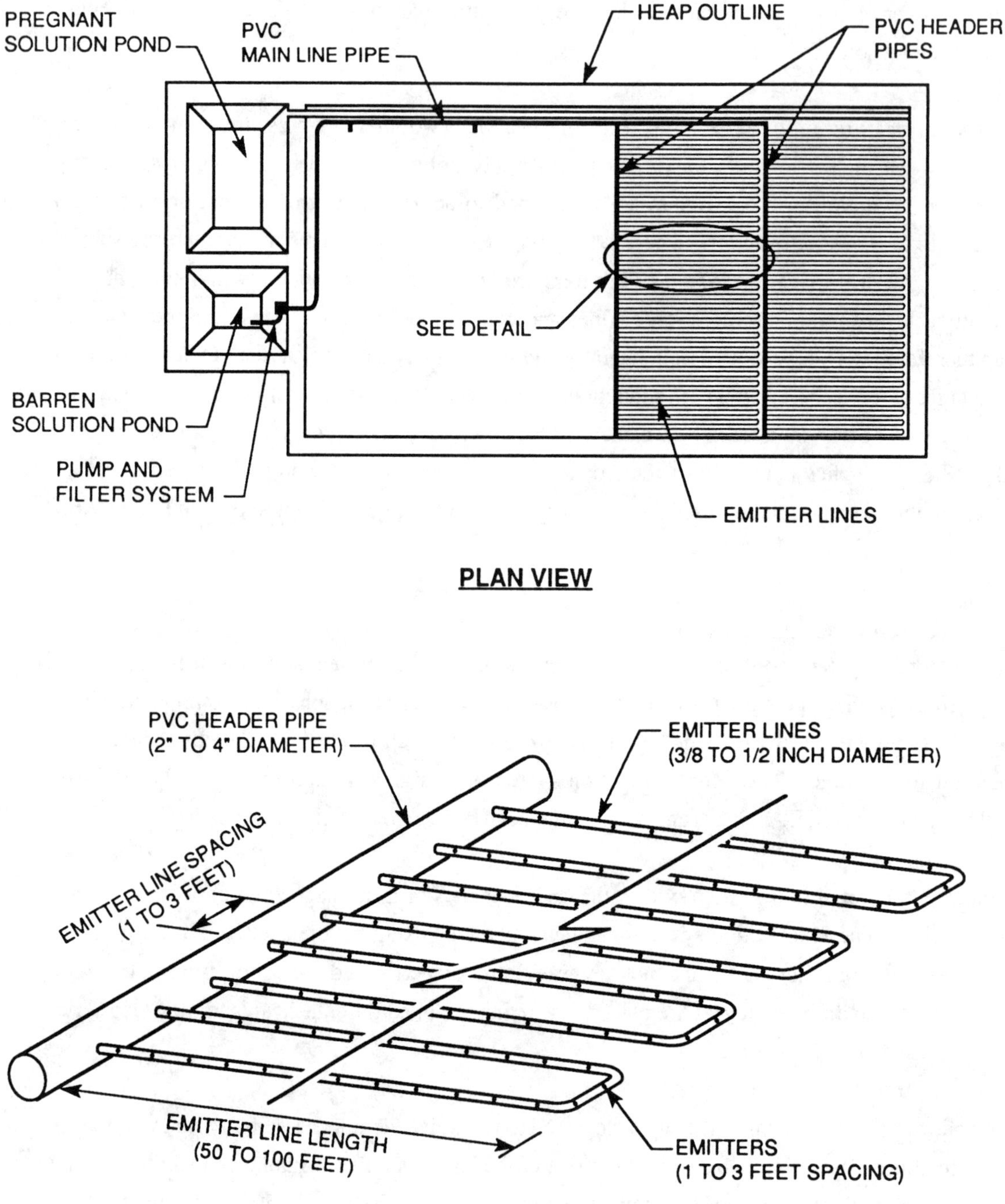

FIGURE 9.6

TYPICAL PRESSURE EMITTER APPLICATION

REFERENCE: O. MUHTADI (1991)

To determine accurately the necessary drain pipe spacing in a heap, the agricultural drain equation is generally sufficient, and results obtained from it are conservative (Duke, 1973). Based upon the anticipated solution application rate and permeability of the ore, this equation can be used to estimate the height of the saturated zone above the liner (McWhorter and Sunada, 1977). The designer can then judge whether or not this height is acceptable and determine the necessity of adding drainage materials or pipes to reduce the height of the saturated zone to acceptable levels.

At some heap leach facilities with excessive fines, the collection system may also act as a filter for the ore, so that small particles of the ore are not conveyed by the solutions into the collection pipes or trenches. As with the liner, the collection system must be chemically compatible with the leach solutions in order to prevent scale or precipitate buildup on pipe openings, and to minimize the breakdown of drain materials by the solutions.

The collection system outside of the heaps must be designed to accommodate solution from both leaching and storm runoff. Solution is usually conveyed in open, lined trenches or in piping above lined ditches that provide secondary containment. The ditches, which also must be capable of conveying storm runoff flows (Section 9.2.4.4), must be designed to convey flows without resulting in erosion, ripping of synthetic liners, or spillage.

9.2.4.3 Ponds

The solution storage ponds must be sized to contain the leach solution from normal leaching operations, storm runoff, the volume of solution that would "drain down" after leaching ceases, and any rinse water that may be added. Factors that need to be considered in the design of ponds include liner selection, pond volume and shape, and inflow/outflow details. Because of the economic value of the stored solutions and the potential for water quality impact, the pond liner should have as low a permeability as possible. For this reason, and because of the relatively small area of the ponds, pond liners are frequently geomembrane. In most instances, a double liner system with leachate collection is used in pond areas that are continuously exposed to leaching solution during normal operations. Single liners are usually adequate for areas that provide only short-term surge capacity.

The side slopes of a pond are dictated by the liner material and the overlying and underlying bedding and protection materials. In order to accommodate equipment traffic and synthetic liner installation, compacted soil and synthetically lined slopes are typically not steeper than 2.5 to 1 or

3 to 1. The liner should be resistant to sunlight, temperature fluctuations, wave action, and wind pressures. The liner should also accommodate inflow through an open ditch or pipe without experiencing damage. Pumping systems, access barges or ramps, and other operational features should be designed so as not to cause leakage or otherwise damage the liner.

The geometry of a pond is dictated by the required capacity, available space, and the ease with which the ground can be excavated. For small to medium size operations, concrete or steel tanks, or similar facilities, have been used to store solution. Larger operations require storage capacity that typically is provided by an excavated, lined pond. In environmentally sensitive areas, consideration has been given to using tanks for solution storage.

9.2.4.4 Climatic Considerations

Climatic factors that need to be considered in the design of solution ponds include the following:

- Short-term, intense rainfall that occurs over a period of minutes to hours. The sizes and configurations of the ponds, ditches, and containment dikes must be designed to accommodate such a rainfall event so that erosion damage and overtopping do not occur.
- Long-term wet or dry periods. Extended wet periods may cause buildup of solution. Extended dry periods cause loss of solution due to evaporation and determine the maximum makeup water requirements. Design must include measures to accommodate either extreme without overtopping of the ponds or damage to exposed line materials.

Local climatic data provide the basis for predicting hydrologic events, and is important in determining pond sizes. For example, the storm event that has a 1 percent probability of occurring or being exceeded in any one year may be used in the design of a specific component. This is commonly referred to as the 1 in 100-year storm event. Suggested design event frequencies for different components of a heap leach facility are given in Table 9.2.

Determination of storm design parameters requires consideration of the critical duration of the design event. The critical duration is that which results in the maximum accumulation of excess water on a pad and pond system during a storm period. Durations shorter than the critical result in less storm water accumulation. For durations that are larger than critical evaporation losses start reducing the accumulated storm water.(2) The critical duration is usually determined by a trial and error process in which durations from several days though several months are evaluated to

(2) For arid climates with high evaporation rates, critical durations are short, generally on the order of days to weeks. For humid climates, the critical duration is longer, i.e., usually months.

TABLE 9.2

SUGGESTED DESIGN EVENT AND CRITERIA FOR HEAP LEACH OPERATIONS[(1) (2)]

FACILITY	TYPE OF DESIGN EVENT
Access Road, Culverts, and Drainage Ditches	1 in 10- to 50-year flood peak.
Drainage Courses and Ditches Outside of Leach Pad Perimeter Berm, Pregnant Pond, and Barren Pond	100-year flood peak.
Internal Freeboard within Leach Pad, Pregnant and Barren Solution Ponds	Maximum of: 1) Average hydrologic conditions plus a short-term, 100-year storm event.[(3)] 2) A longer term 1- in 100-year event over a period of several months or years.[(4)]

(1) Modified from van Zyl, (1988).

(2) Site-specific conditions or regulatory requirements may demand more stringent criteria than those listed in this table.

(3) The duration of a short-term event should be selected by trying durations of, for example, 1, 7, 15, and 30 days and the worst case selected.

(4) The duration should be selected by trying durations of 3, 6, 12, 24, and 36 months and the worst case selected.

determine which yields the maximum storage requirement for the pad and pond system. How this is accomplished is outlined in Table 9.2 and discussed further in Section 9.2.4.5.

The National Oceanic and Atmospheric Administration (NOAA) has regional maps of precipitation that can be used directly for most design purposes. Determining snowmelt parameters is more difficult since there are usually no published maps of regional snowpack. However, the Soil Conservation Service (SCS) and other agencies can often provide information that can be supplemented with local observations and topographic considerations such as exposure and the potential for snow drifting. The snowmelt must be converted to its water equivalent, taking into account variations in snow density.

Depending upon the climate, evaporation can be an important factor in determining water balance, particularly when spray application is used. Spray losses vary greatly, depending on size of droplet, temperature, humidity, and wind. Thus, losses may be both evaporative and windblown. Spraying may have to be reduced during high winds to avoid solution being transported beyond the area of containment. Various techniques are available for estimating evaporation from open water surfaces in ponds and spray losses from heaps (van Zyl, 1988).

For sizing culverts and ditches around the heap leach facility, flood runoff calculations can be made using either the Rational Formula or the SCS unit-graph techniques, with the appropriate storm precipitation as input.

Evaluation of the flood storage capacity of the ponds is relatively straightforward. The appropriate flood volume is obtained by multiplying the total collecting area by the design excess precipitation or snowpack water equivalent. This simplified approach is also applicable to the area within the leach pad containment berm and the pregnant and barren ponds.

Design considerations for the pregnant and barren ponds are shown in Figure 9.7. The ponds must have sufficient storage capacity to contain the following:

- Minimum operating volume sufficient to facilitate operation of the pumps.
- Heap draindown volume.
- Rinsing cycles.
- Normal seasonal fluctuations in water volume (based on average climatic conditions).
- Flood surge (based on the critical design event).

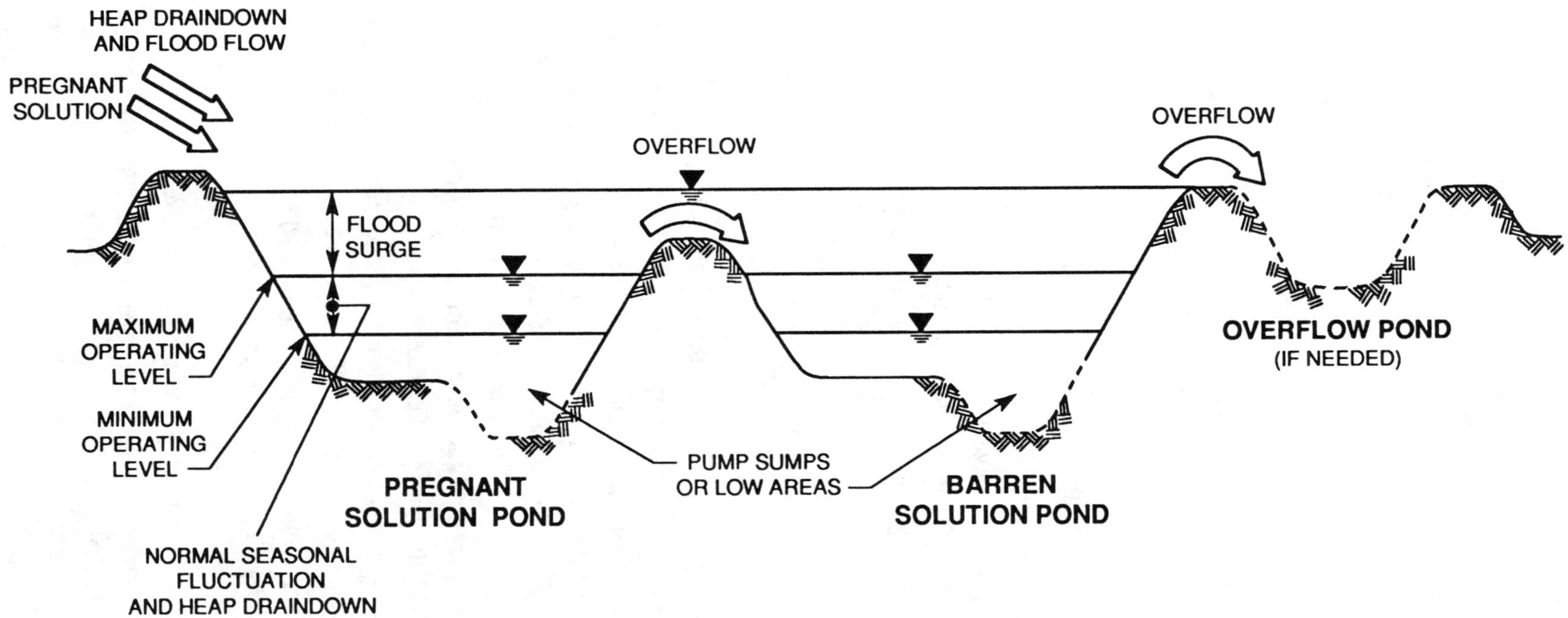

(NOT TO SCALE)

FIGURE 9.7

POND DESIGN CONSIDERATIONS

As shown in Figure 9.7, the pregnant pond capacity can be reduced by allowing it to spill over into the barren pond during the design flood event. In some instances, it may be appropriate to include a third (overflow) pond to eliminate the potential for a surface discharge in environmentally sensitive areas. The liner provided for the overflow pond would not need to be as extensive as that used in areas with continuous ponding.

9.2.4.5 Water Balance, Rinsing, and Pond Sizing

The water balance of a heap leach operation has two main components:

- Recovery circuit.
- Natural water cycle.

The recovery circuit is the relatively constant and predictable flow of recovery fluids through the system. It includes make-up water, reagent addition, heap rinse water, and start-up water requirements. The natural water cycle is superimposed on the recovery circuit and includes time, varying rainfall and/or snowmelt, and evaporation losses. Figure 9.8 illustrates these water circuits.

Water Balance

Throughout a typical operating year, the water circulating through a heap leach operation is stored in different facilities. For example, during start-up, water is added to the system to fill the barren pond and to start the leaching operation. Initially, and after each new ore addition to the heap, much of the leachate that is applied goes into soil moisture storage in the ore. Once the ore is wetted to greater than field capacity (i.e., the ability of the ore to hold water against the force of gravity), the leachate starts to collect around the perimeter of the pond (see also Chapter 6). Evaporation of the leachate spray and the ponds reduces the water in the system, while precipitation on the pads and ponds increases the amount of water in the system. Liquid reagent also adds a small amount to the system water. During temporary or permanent shutdowns, such as in winter in colder climates, some of the leachate collected in the ore will drain out and must be stored in the ponds. Generally, precipitation and make-up water represent the largest water inputs, while spray evaporation and ore wetting result in the greatest water losses.

The key aspect in designing a leach facility is to determine what the overall average water balance is (i.e., whether the system will continuously gain or lose water). Systems that continuously gain, such as those in wet climates or those requiring extensive post-leach rinsing, may require some

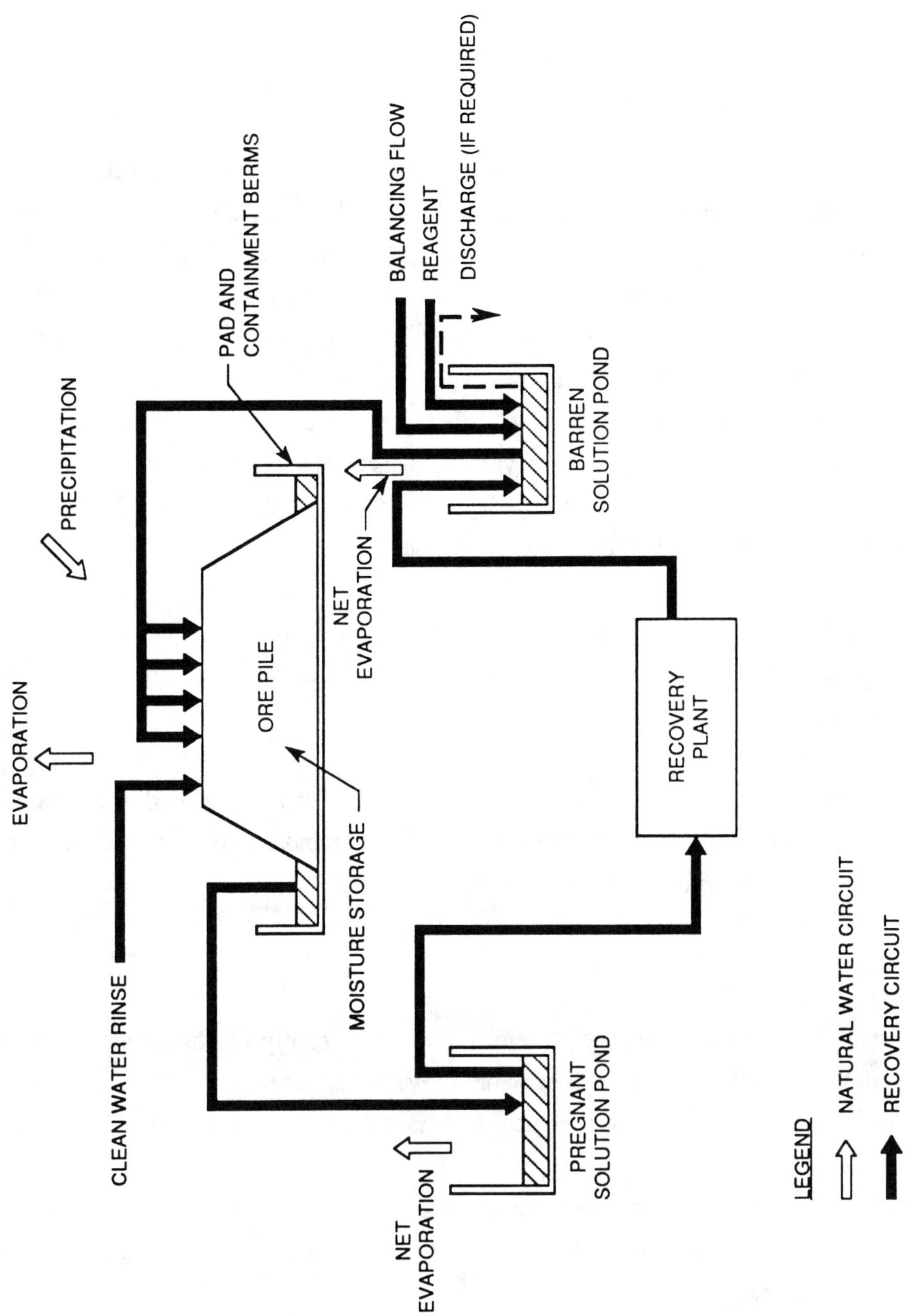

FIGURE 9.8
HEAP LEACH OPERATION WATER CIRCUITS

sort of water disposal system to achieve a balance. This could include evaporation ponds or a surface treatment and discharge system. Systems that continuously lose water, on the other hand, require a constant input of make-up water.

Rinsing

Laboratory tests are usually required to determine the amount of rinse water and/or chemical additives that must be applied to clean the spent ore adequately before disposal. Depending on the residual cyanide or other constituent levels required, the amount of rinse water can vary up to seven or eight pore volume displacements. Where this amount of rinsing is judged to be excessive, it may be necessary to rinse with chemicals, such as hypochlorite or peroxide, in order to enhance cyanide destruction. Other innovative techniques, using recycled rinse water or the introduction of reagent-destroying bacteria, can also be considered. On-off type rinsing has also been found to improve the efficiency of removing reagents from the pile. Rather than rinse the pile on a continuous basis, it is rinsed for a few days. The water is then turned off and the pile is allowed to drain for a few days before rinsing commences again.

Over the years, the leach pile rinse water becomes part of the system water circuit, and a balance must be achieved. When rinsing occurs, the rinse cycle contributes to a short-term increase in pond storage requirements.

An important aspect of rinsing is the sampling and analytical method that is used to determine when the residual intertial pore fluids have reached the prescribed standard. A recommended sampling and analytical protocol is included in Appendix E.

Pond Sizing

It is important that the water storage system, represented mainly by the pregnant and barren solution ponds, be sized to store the total amount of water contained during a relatively rare, wet climatic event. This sizing should also include conditions during winter shutdown, if applicable.

In wetter climates, where the water balance is close to zero, or where water must be discharged to maintain a zero balance, the critical duration is long and may vary from an entire wet season to several wet years in sequence.

In drier climates where, on average, the heap leach operates with a large negative water balance, the critical duration of the design storm event is usually relatively short, varying from 1 to 60 days. It is generally only during these relatively short periods that the precipitation exceeds evaporative losses from the system and that water volumes actually increase.

It must also be considered that the design values developed as outlined above will be subject to inaccuracies. In cases where the calculated balances are close to zero, therefore, development of contingency plans for dealing with excess water should be an important part of the design phase, and the operation must be monitored carefully.

Finally, it is important to note that the way a heap leach system is actually operated can affect the potential for overtopping and, hence, the pond size. This is particularly true in areas where the water balance is close to zero. Operators need to be aware of the maximum normal operating water levels in the ponds and the volumes of draindown and rinse water that the ponds were designed to accept. They must understand that these maximum named levels should not be exceeded except during extreme storm events. Details of how to determine water balance, together with worked examples of pond sizing, are provided in van Zyl (1988).

9.2.5 SLOPE STABILITY ASPECTS

Slope failure on a heap, or along a liner due to sliding, can cause solution to escape and should be prevented, or if allowed, be collected by secondary containment features. Critical cases that usually require close evaluation of stability include: (1) heaps on a weak foundation, (2) heaps on steeply sloping sites, (3) heaps intended to reach a considerable height, and (4) heaps within which buildup of a zone of saturation is expected. The plane along geosynthetic materials in a liner system can also represent a potential surface for sliding. Stability analyses are used to aid the designer in assessing whether or not the selected heap configuration, leach pad, and collection system will be acceptably stable (Strachan and Dorey, 1988).

The evaluation of slope stability consists of selecting a two-dimensional section through the heap, defining the likely failure surface, then calculating the factor of safety along that surface (see Chapter 7). A thorough evaluation of slope stability requires a knowledgeable selection of the most likely failure surfaces, as well as a technically correct and consistent method of calculating the safety factor. In marginal cases, slope stability can be improved by creating setbacks or benches between lifts and providing drainage to reduce pore pressure.

Seismic stability of a heap is also an important consideration (see Chapter 7). Heaps are usually at the angle of repose and the external slopes are, therefore, at a static safety factor of approximately one. Ravelling of these slopes can be anticipated during an earthquake and must be accommodated in the design. When ponds are located at the downstream limit of the pad, the area in which maximum impact of ravelling could occur, the designer must consider the potential for ore to fall into the pond and result in overtopping. Flattening the heap in the pond area or locating the ponds outside of the potential area of ravelling are approaches that have been adopted to alleviate this concern.

The major items affecting stability of a heap are:

- The geometry of the section that is analyzed.
- The shear strength of the materials.
- The pore water pressure conditions in the heap and subsoils.

The geometry and pore pressure conditions can be modified during design to achieve acceptable stability. Shear strength is a function of the materials to be used and is changed primarily by using different materials.

Typical shear strengths of different granular materials, clay, and the interface of various combinations of geomembranes are presented in Chapter 7.

9.3 DESIGN APPROACH

9.3.1 GENERAL

In terms of environmental impact, the two most important considerations when designing a heap leach operation are:

- Provide sufficient solution containment under the pads, ponds, and solution collection and conveyance systems.
- Provide adequate berms around the pads and sufficient pond volume to safely store water that could accumulate in the system under extreme climatic conditions.

Prior to closing a heap leach pile as a waste pile, and after any post-operational rinsing and treatment, the most important considerations are to determine the residual chemical properties of the material and provide measures for control.

A recommended approach for developing heap leach designs is presented in the following sections.

9.3.2 SOLUTION CONTAINMENT DESIGN

The design for solution containment involves selecting appropriate liner systems for the pad or pads, ponds, and the solution collection and conveyance systems between the pads and the ponds. The approach that is recommended when selecting these liner systems is to use the types of containment assessment procedures discussed in Chapter 7. These procedures evaluate the degree to which liquid is contained by a liner system, taking into account the liner system configuration, the physical conditions during construction and operations, chemical characteristics of the solution, the hydrostatic pressure on the liner, the length of time the ore is leached, the nature and characteristics of the underlying geologic formations, and the distance to ground water.

9.3.3 SOLUTION CONTROL

As outlined in the previous sections, it is recommended that a probabilistic approach be adopted to ensure that adequate containment is provided. It is further recommended that the 100-year storm event be considered as a minimum standard. It is necessary that berms, provided around the leach pad area to prevent the influx of natural flood water, be sized to safely accommodate the design storm events. Ditches which connect the leach pad(s) to the ponds must also be sized to safely convey the peak flow from the design storm without overtopping or excessively damaging liner materials.

The ponds must be sized to include the maximum normal operating level, any draindown and rinse water additions, and the storm or snowmelt accumulation during the design precipitation event. If pumping to recirculate the water is used in addition to pond capacity for handling precipitation, assurance is required that power will be available to the pumps even during a very large storm event. Evaluating the critical duration of the design precipitation event will require several water balance evaluations for storms of different storm durations, allowing selection of the duration that provides the maximum pond volume.

9.3.4 PREPARATION OF LEACH PILE FOR CLOSURE

The degree of waste containment that will be required at closure depends in part on the chemical characteristics of the spent ore pile. To a large extent, these characteristics can be controlled by the rinsing and/or chemical treatment of the ore after the leach cycles have been completed.

A design and operational approach to address this issue involves the following:

- During the heap leach design process, conduct laboratory and analytical tests to determine the residual chemical characteristics of the spent ore when subjected to a range of rinsing and/or treatment cycles.
- Based on the above test results, determine the degree of rinsing and/or treatment that will be undertaken in the full scale operation and hence the amount of waste containment that will be required when the piles are closed according to regulatory requirements.
- When designing and constructing the heap leach system, ensure that adequate provision is made for the containment that will be required on closure.
- During operations and after the first portion of the spent ore has been rinsed or treated, sample and test it to determine how the actual chemical properties compare to the predicted properties. If necessary, make field adjustments.
- Prior to closure, conduct sampling and testing on all heap segments to ensure that their chemical characteristics are consistent with the closure plans.

To ensure consistency in the testing of the spent ore, a standard procedure should be developed to evaluate and determine the concentration levels of the constituents of concern. A recommended procedure is included in Appendix E. At the time of this writing, there are two ongoing studies in the State of South Dakota which are likely to provide further valuable guidance in this area. One, conducted by the Department of Water and Natural Resources, is studying the neutralization of spent ore at Brohm Mining Corporation in Deadwood. The second, conducted by the South Dakota School of Mines in Rapid City, is an evaluation of test methods for estimating the neutralization of spent ore.

9.3.5 EXAMPLES OF PROJECT RISK

The following seven generic examples have been formulated to illustrate increasing degrees of risk to water quality. The examples are fictitious, but they do provide comparisons with actual situations. They illustrate the large variation in potential impacts and beneficial uses of water that can be encountered for various types of conditions, as well as the variability in operator experience, operational skills, and financial resources. The range of these variables emphasizes the inappropriateness of using uniform, prescribed measures in regulations that apply to all sites. It is important that there be sufficient flexibility to allow operators and regulators to adjust designs and operating procedures to reflect the level of risk at a particular site.

The examples outlined below (HL1 through HL7) reflect increasing levels of risk.

9.3.5.1 HL1

This facility is situated in a hot desert climate on gently sloping terrain with a deep ground water table. The vadose zone consists of a considerable thickness of layered, silty soil over deep granitic bedrock. The ore is oxidized. The operation is projected to last three years. Leaching of ore at any single location will last for less than one year. Segments of leach pad will be abandoned as leaching of the ore in each location is completed. Abandonment is expected to require only fresh water rinsing and protection from erosion. The nearest human settlement is 15 miles away, and the closest paved highway is 2 miles distant. Operations are year-round by a large U.S. mining company with prior experience in heap leaching.

9.3.5.2 HL2

This facility is similar to HL1 except that it is located at a high elevation in moderately steep terrain with a cold, although dry, winter climate. Operations occur only in the summer months.

9.3.5.3 HL3

This facility is similar to HL1 and HL2 except that the location is in coastal mountains, and the climate is far wetter than the desert locations. Operations are projected to last 7 years, but in the summer months only, since the elevation is high, and winters are cold.

9.3.5.4 HL4

This facility is also in the coastal mountains, but at lower elevations than HL3. It is close to a major river, with overlying silty soils. It is alongside a paved highway, and the nearest human settlement is 2 miles downstream. Year-round operations are expected to last 12 years, and the spent ore is to be thoroughly rinsed prior to abandonment.

9.3.5.5 HL5

This facility is similar to HL3 except that it is situated close to a large lake which happens to be renowned for its fishing and is the location of a major recreational resort. The facility is across the lake from the resort. The ore body is large, consisting of agglomerated tailings from an earlier mining operation, and the mine is planned to operate year-round for 17 years. Precipitation is moderate, but the elevation is high, and winters are cold. Rinsing and some chemical treatment of spent ore will occur.

9.3.5.6 HL6

This facility is a small operation situated high in the mountains, with fairly high precipitation and cold winters. It is a 2-year operation leaching an old, abandoned tailings pile. The facility is located in the floodplain of a stream situated over sand and gravel deposits. The stream supports fish. The location is fairly remote, but a number of dwellings downstream obtain water from ground water wells sunk in the floodplain deposits. It is not planned to rinse the ore since natural precipitation is thought to be adequate to clean the heaps in time. Year-round operation is planned. The mining concern is inexperienced in heap leach operations.

9.3.5.7 HL7

This facility is situated in the wet, coastal mountains close to population centers. It is a very large, but rather low-grade, deposit of sulfide ores. A large scale mine is planned with year-round operations. The facility is planned as a valley leaching scheme over sand and gravel stream deposits and shallow ground water. The stream is tributary to a major fish-bearing river. The company is a large, foreign controlled corporation, with no prior experience in heap leaching.

9.3.5.8 Assessment of Risks Associated with HL1 to HL7 Examples

It is clear from these examples that the threat to beneficial uses of water, and to the environment in general, are very different for the variety of mine settings and operational conditions described above. It is therefore important to take the site-specific conditions into account when designing a heap leach operation, particularly the containment features such as the liner and pond systems. A basis for doing this is to use the containment assessment appraisal which is outlined in detail in Chapter 3.

9.4 HEAP LEACH GUIDELINES

9.4.1 MOTIVATION

As evident from the discussions above, an operating ore heap is subject to different conditions than is the spent ore after closure. Therefore, it is neither appropriate nor necessary to apply waste management unit design standards and requirements during the operating phase. However, guidelines could be developed for compliance with applicable waste discharge limits during the operating phase, with the ultimate waste management requirements applying to the pile left in place on closure.

9.4.2 PROPOSED GUIDELINES

Guidelines should ensure protection of water quality in the vicinity of proposed heap leach operations, and that the ultimate disposal of waste meets the requirements applying to the waste pile left in place on closure. This chapter outlines a proposed set of guidelines which consist basically of two parts. The first specifies site-specific information and results of analyses that the operator should provide. The second outlines the performance standards that must be met when designing and operating a heap leach facility.

These guidelines should apply throughout the operating period of the heap and to the end of any spent ore rinsing and/or treatment activities. Thereafter, the heaps become waste piles regulated as waste management units.

9.4.2.1 Information Pertaining to the Proposed Operation

The following information should be provided:

- Name, location, and mailing address of the facility.
- Name, mailing address, and phone number of the applicant.
- Land ownership status of the facility (federal, state, private, or public).
- The legal structure (corporation, partnership, etc.) and residence of the applicant.
- A topographic site map and aerial photos, if appropriate, extending at least 1 mile beyond the outer limits of the facility site, identifying and showing the location and extent of the following features within one mile of the facility:
    - Wells, springs, wetlands, surface waters, and irrigation ditches.
    - Make-up water source(s).
    - Public and private drinking water source(s).
    - United State Geological Survey (USGS) identified floodplain areas (as shown on USGS sectional quadrangle maps).
    - Service and public roads.
    - Buildings and structures.
- A surface and subsurface description of the site and a characterization of the local hydrogeologic regime, including the unsaturated zone and any significant perched ground water conditions that may be present.
- An operating plan for the facility, including at least the following items:
    - A flow schematic of the recovery system.
    - The characteristics of the solution to be used for the heap leach operation.

- The pond size necessary to cope with all anticipated solutions, plus the critical 100-year extreme precipitation event.
- The liner systems to be used to provide the primary solution containment for the pad(s), solution collection and conveyance areas, and the ponds.
- The methods to be used to place ore onto the liner system and verification that adequate protection from penetration is provided for both dynamic and static loading conditions.
- The maximum hydraulic head to which the liner system may be exposed during operations.
- The maximum length of time each element of the containment system will be exposed to solutions.
- Methods, such as diversion ditches and dikes which will be used to prevent run-on from large storms outside the pad area.
- A Water Management Strategy that describes the water balance, management methods used during normal conditions as well as periods of excess water due to forced and unforced draindowns, rinsing and treatment of the piles, and extreme precipitation and/or snowmelt events.

• A Liner System Containment Assessment, including:
  - An analysis of the amount of leakage which could occur if the liner system: (1) functions as designed, and (2) experiences a worst-case leakage volume, based on the materials used and quality assurance/quality control (QA/QC) procedures to be implemented.
  - An analysis of the potential depth of migration of a seepage front in the event of the worst-case leakage assumption and the resulting potential threat to ground water for the maximum operating period.

• A Monitoring Strategy that describes:
  - The existing surface and ground water quality (baseline).
  - A proposed monitoring system for surface and ground water, and the vadose zone, if practical, that would effectively detect drainage or seepage from the operation.
  - Proposed monitoring for discharges into the unsaturated zone, when appropriate.
  - Proposed monitoring of hydraulic head above liners or containment systems, where appropriate.
  - Proposed monitoring of leachate collection systems and the seepage volumes that indicate a liner system failure.

• A Discharge Response Strategy that describes procedures and methods to be implemented for the abatement and/or remediation of any pollutant that may escape containment and pose a threat to the beneficial uses of the waters.

• Topographic maps and aerial photos, when appropriate.

- Preliminary engineering drawings showing locations and design of those portions of the facility intended to contain, treat, or dispose of solutions. These drawings should be in sufficient detail to allow determination of design capability and should include:
  - Drawings which show surface gradients and the flow of solutions.
  - Predicted flows of runoff and run-on.
  - Design criteria and recovery schematic; leach pad and basin cross sections.
  - Typical details of liner systems for pads, details of pad loading and final heap configuration, solution impoundments and emergency containment impoundments.
  - Leached ore treatment process schematics.
  - Leak detection/monitoring system details.

  The containment portion of the design should be certified by a registered professional engineer.
- Construction and material specifications that meet the design criteria. These should address major construction requirements related to materials of construction, inspection and testing requirements (including liners), and necessary manufacturer certifications. Construction specifications shall include a quality assurance procedure for liner installations and a procedure for leak testing of impoundments.
- A description of the post-operational rinsing and treatment that will be carried out and the resultant chemical property of the waste material.
- A Closure Plan that describes the general procedures and schedule to be implemented at the facility for the reclamation of the heaps, treatment and disposal of solution, and the control and monitoring of discharges and potential discharges for a reasonable period of time. This plan will be based on site-specific conditions and include measures to assure compliance with ultimate waste management unit requirements.

9.4.3 DESIGN AND PERFORMANCE STANDARDS

The following proposed minimum design and performance standards are intended as a baseline for protection of water quality. These standards should apply unless the applicant can demonstrate with more specific information that an alternate to a specific standard will provide equivalent protection.

- Containment Design Criteria - A facility should be designed to manage the maximum expected operating water storage requirements, the draindown of heap moisture on cessation of leaching and the 100-year, critical storm or snowmelt event.
- Pond and Impoundment Design - Ponds and impoundments for containment of solution should be designed for efficient leak detection and provide for leak recovery adequate to protect water quality.
- Liner System Criteria - Liner systems shall be designed to protect the beneficial uses of water. The design should be based on: (1) state-of-the-

art design, construction, and QA/QC procedures, (2) the hydraulic pressures to be applied to the liners, and (3) the length of time the solution will be applied. The basis for the liner system design should be a containment assessment which will incorporate:

- Chemical and physical properties of the material being leached.
- Static and dynamic loading conditions to be applied to the liner system.
- The physical and geochemical properties of the liner system and the natural geologic material between the liner and the ground water table.
- The quality and quantity of ground water available and the potential beneficial uses of this water.

In addition; the liner system should meet the following requirements:

- Have a competent foundation designed to withstand the projected static and dynamic loading and projected differential settlement.
- Be structurally competent at all times and under all potential loading conditions.
- Be chemically compatible with materials contacting the liner system.
- Be designed, with special protection if necessary, to prevent damage during ore loading and unloading.
- Where appropriate, ensure minimal hydraulic head above the liner.
- Where appropriate, incorporate an acceptable leakage through the upper liner of a double-lined system.

In the absence of a design based on a containment assessment the liner system should incorporate as a minimum either a 12-inch thick soil layer of 10-7 cm/sec hydraulic conductivity or a 40-mls thick geomembrane.

- Water Monitoring - A ground water and/or surface water monitoring program should be required for a facility, unless the applicant demonstrates that no reasonable potential exists for impairment of beneficial uses. Monitoring of seepage, moisture, or chemical changes, or soil gas in the unsaturated zone above the ground water, may be required at sites where predictions show that greater than expected leakage from the containment system could cause a significant threat to the beneficial uses of the waters. Also, monitoring of the hydraulic head on liner systems may be required in systems which rely on minimization of head to control leakage. The monitoring program should be dependent on location, design, and operation of the facility, and should be capable of indicating the facility's impact on the surface and/or ground water most likely to be affected by the operation. The monitoring program should be designed to give early detection of any unauthorized discharge.
- Design for Disposal or Closure of Leached Ore Piles - Design for disposal or abandonment of the leached ore should satisfy waste management unit requirements for the characteristics of the material at the time of closure.
- Storage Requirements - Reagents, such as cyanide, shall be stored in lined areas and be separated from substances that are not chemically compatible.

9.5 BIBLIOGRAPHY

Daniel, D.E. *Hydraulic Conductivity Tests for Clay Liners,* in *Geotechnical and Geohydrological Aspects of Waste Management,* eds. D. van Zyl et al. Chelsea MI: Lewis Publishers, Inc. 1987. 15-34.

Deatherage, J.D., J.R. Fahy, and L.A. Hansen. *Sheer Testing of Geomembrane Soil Interfaces. Geotechnical Aspects of Heap Leach Design.* ed. D. van Zyl, SME-AIME. 1987. 45-50.

Duke, H.R. *Drainage Design Based Upon Aeration.* Hydrology Paper No. 61, Colorado State University. 1973.

Hutchison, I. P.G. *Surface Water Control - Water Balance, Introduction to Evaluation, Design and Operation of Precious Metal Heap Leaching Projects,* eds. D. van Zyl, et al. 1988. 203-253.

Leps, T.M. "Review of Shearing Strength of Rockfill". J.S.M. Division, ASCE, 96. (July, 1970).

Martin, J.P., R.M. Koerner and J.E. Whitty. *Experimental Friction Evaluation of Slippage Between Geomembranes, Geotextiles and Soils. Proceedings of International Conference on Geomembranes.* Denver, Industrial Fabrics Association International. 1984. 191-196.

McWhorter, D.B. and D.K. Sunada. *Ground Water Hydrology and Hydraulics.* Fort Collins, Co: Water Resources Publications. 1977.

Muhtadi, Omar. *Personal Communication.* 1991.

Saxena, S.K. and Y.T. Wong. *Frictional Characteristics of a Geomembrane.* Proceedings of International Conference on Geomembranes. Denver. Industrial Fabrics Association International. 1984. 187-190.

Smith, M.E. and G.J. Gierzerski. *Design of Chemically Amended Soil Liners. Geotechnical Aspects of Heap Leach Design.* ed. D. van Zyl, SME-AIME. 1987. 7-14.

Strachan and Dorey. *Geotechnical Studies. Introduction to Evaluation, Design and Operation of Precious Metal Heap Leaching Projects.* eds. D.J.A van Zyl et al. 1988. 152-175. Littleton, Co: Society of Mining Engineers.

Strachan C.L., and D.J.A. van Zyl. *Leach Pads and Liners. Introduction to Evaluation, Design and Operation of Precious Metal Heap Leaching Projects.* eds. D.J.A van Zyl et al. 1988. 176-202. Littleton, Co: Society of Mining Engineers.

van Zyl, D.J.A., I.P.G. Hutchison, and J.E. Kiel. *Introduction to Evaluation, Design and Operation of Precious Metal Heap Leaching Projects.* Littleton, Co: Society of Mining Engineers. 1988.

Vick, S. *Planning, Design, and Analysis of Tailings Dams.* New York: John Wiley. 1983.

CONTENTS

CHAPTER 10
GROUND WATER MONITORING

PAGE NO.

LIST OF TABLES/LIST OF FIGURES 501

10.1 Introduction 503
- 10.1.1 Purpose 503
- 10.1.2 Integration With Other Chapters 503
- 10.1.3 Study Participants 503
- 10.1.4 Sources of Information 503
- 10.1.5 Chapter Organization 504

10.2 Factors To Be Considered When Designing a Monitoring System 504
- 10.2.1 Monitoring System Design Philosophy 504
- 10.2.2 Background and Baseline Definitions 505
- 10.2.3 Waste Characteristics 507
- 10.2.4 Waste Containment and Waste Management Practices 508
- 10.2.5 Environment 510
- 10.2.6 Potential Waste Constituent Migration Pathways 511
  - 10.2.6.1 Potential Seepage Through Containment System 512
  - 10.2.6.2 Seepage Through the Vadose Zone 512
  - 10.2.6.3 Seepage Through the Saturated Zone 518
  - 10.2.6.4 Seepage Pathway Analysis Approach 521
- 10.2.7 Potential Impacts of Waste Constituent Migration 522

10.3 Elements of a Monitoring System 524
- 10.3.1 In-Waste Monitoring 524
- 10.3.2 Vadose Zone Monitoring 526
  - 10.3.2.1 Background 526
  - 10.3.2.2 Types of Systems 527
  - 10.3.2.3 Point Methods 530
  - 10.3.2.4 Non-Point Measurement Techniques 538
- 10.3.3 Ground Water Monitoring 540
  - 10.3.3.1 Drainage Systems 541
  - 10.3.3.2 Single-Screened Monitoring Wells 541
  - 10.3.3.3 Multiple-Screened and Cased Monitoring Wells 544

CONTENTS

CHAPTER 10
(Continued)

PAGE NO.

10.3.3.4 Clustered Monitoring Wells 545
10.3.3.5 Single Wells With Multiple Sampling Points 545
10.3.4 Chemical Constituents for Monitoring 546
10.3.4.1 General 546
10.3.4.2 Selection of Constituents of Leak Detection 546
10.3.4.3 Monitoring Constituent Concentration Limits 548
10.3.5 Frequency of Monitoring 550
10.4 Technical Approach to the Design of a Ground Water Monitoring System 555
10.4.1 The Need for a Ground Water Monitoring System 555
10.4.2 Objectives of a Ground Water Monitoring System 556
10.4.3 Baseline and Background Considerations 556
10.4.4 Selection and Design of In-Waste Monitoring System 557
10.4.5 Selection and Design of a Vadose Zone Monitoring System 558
10.4.5.1 Background Monitoring 559
10.4.5.2 Detection Monitoring 560
10.4.5.3 Frequency of Measurement 560
10.4.6 Selection and Design of a Ground Water Monitoring System 561
10.4.6.1 Background Monitoring Locations 561
10.4.6.2 Downgradient Monitoring Locations 561
10.5 Regulatory Implications 563
10.5.1 Federal 563
10.5.2 State of California 564
10.5.2.1 Background 564
10.5.2.2 General 564
10.5.2.3 Monitoring Programs 565
10.5.2.4 Detection Monitoring 565
10.5.2.5 Verification or Evaluation Monitoring 566
10.5.2.6 Corrective Action Monitoring 566
10.5.2.7 Regulatory Flexibility 567
10.6 Bibliography 568

CONTENTS

CHAPTER 10
(Continued)

LIST OF TABLES

TABLE NO.	TITLE	PAGE NO.
10.1	Summary of Sampling Methods for Vadose Zone Monitoring	528
10.2	Probability of False Readings From Sampling Method	529
10.3	Solute Transport Travel Time Assessment - Alluvial Environment	551
10.4	Solute Transport Travel Time Assessment - Fractured Rock Environment	552

LIST OF FIGURES

FIGURE NO.	TITLE	PAGE NO.
10.1	Summary of Site Conditions Important to the Design of Ground Water Monitoring Systems	506
10.2	Types of Waste Containment	509
10.3	Typical Soil Moisture Characteristics	514
10.4	Potential Seepage Through the Vadose Zone	516
10.5	Potential Seepage Through the Ground Water System	523
10.6	Vadose Zone Monitoring Devices	531
10.7	Ground Water Monitoring Devices	543
10.8	Typical Heap Leach Operation in an Alluvial Environment	553
10.9	Tailings Disposal Facility on Fractured Rock	554

CHAPTER 10

GROUND WATER MONITORING

10.1 INTRODUCTION

10.1.1 PURPOSE

This chapter presents guidelines for the design of monitoring systems for mine waste management units. These guidelines include in-waste, unsaturated or vadose zone, and ground water monitoring.

The monitoring issues addressed herein focus on locating monitoring points in the media of interest, selecting appropriate monitoring equipment, and determining monitoring parameters (i.e., chemical and physical properties) that should be used to provide early detection of waste constituent migration that poses a threat to beneficial uses of ground water. This chapter does not address monitoring system operating procedures or reporting requirements.

10.1.2 INTEGRATION WITH OTHER CHAPTERS

The type of monitoring system suited to a particular mine waste management unit is determined by the facility's design, site geology and geohydrology, climatic factors, and waste characteristics which determine the potential for leachate formation. Chapters 3, 4 and 5 deal with methods that can be used to predict the likely leachate quality of a waste. This information is required when selecting parameters for a monitoring program. Climate conditions under which leachate is likely to form are discussed in Chapter 6.

10.1.3 STUDY PARTICIPANTS

This study was conducted by the following individuals:

- Ian P. G. Hutchison, Ph.D., P.E., Environmental Solutions, Inc.
- Joel Siegel, Morrison Knudsen Engineers
- Adrian Smith, Ph.D., ASCI

10.1.4 SOURCES OF INFORMATION

Through the course of this study, several sources of information were utilized. Interviews with operators of mines were conducted to obtain input on operating monitoring systems. The report entitled *RCRA Groundwater Monitoring Technical Enforcement Guidance Document* (EPA, 1986) was used to reference the current EPA approach to siting monitoring wells. The discussions on vadose zone monitoring techniques relied extensively on *Vadose Zone Monitoring for Hazardous Waste Sites* (Everett et al., 1984). Information regarding mine waste in California was obtained from the report *Mining Waste Study* (U.C. Berkeley, 1988).

Most of the data sources reviewed are listed in Section 10.6, Bibliography.

10.1.5 CHAPTER ORGANIZATION

The remainder of this chapter covers the considerations essential to designing a comprehensive monitoring system:

- Section 10.2 discusses the factors to be considered when designing a monitoring system. These include the waste characteristics, climatic conditions, waste containment measures, depth to ground water, vadose zone characteristics, and ground water quality.
- Section 10.3 deals with the elements of a monitoring system which can include in-waste, vadose zone, and ground water sampling and sensing devices.
- Section 10.4 deals with the selection and design of a monitoring system for the waste, and the unsaturated (vadose) and saturated (ground water) zones.
- Section 10.5 discusses the regulatory implications and provides comments on the recent changes in the monitoring regulations in California.

10.2 FACTORS TO BE CONSIDERED WHEN DESIGNING A MONITORING SYSTEM

10.2.1 MONITORING SYSTEM DESIGN PHILOSOPHY

Fundamental to the design of an effective ground water monitoring system is an understanding of the waste management unit, the underlying geology, seepage migration potential in the vadose zone, and ground water flow conditions and quality, and the most probable and significant pathways of potential migration of waste constituents. The conditions under which waste constituent migration could occur and the potential magnitude of the migration should also be known or estimated. The monitoring system is then designed to provide sampling points along these likely migration pathways at "detection" locations. For comparative purposes, sampling is also conducted at "baseline" points before waste disposal occurs, and "background" points in areas not expected to be unaffected by waste constituents. The degree to which waste constituent migration occurs can then be judged by comparing monitoring data from the "detection" points to the "background" values.

It is also important to know what the potential impacts of any waste constituent migration may be, including the extent of the influences, the potential concentration increases, and the migration time periods. These factors need to be incorporated into the design of the monitoring system.

In summary, the factors which need to be considered in designing the monitoring system include:

- Background and baseline conditions.
- Waste characteristics.
- The degree and nature of the waste containment.
- Disposal environment (climate and local geologic and geohydrologic conditions).
- Potential constituent migration pathways.
- Potential impacts of constituent migration.

In order to illustrate the above concepts, a typical mine waste management unit, and potential waste constituent migration pathways, are shown in Figure 10.1. The containment provided at the waste management unit will determine the nature and quantity of seepage which could occur. Where the liquid portion of the waste is contained by a clay liner or by natural granular geologic formations, general seepage over relatively large areas would be the most likely scenario. Where synthetic liners are used, point seeps at random locations could occur.

As illustrated in Figure 10.1, the facility overlies the vadose zone, which in turn overlies the ground water system. As shown, clay lenses may be present in the vadose zone which form barriers to vertical migration. When covering an extensive area, such clay layers may be termed aquitards. Geologic features such as faults, fractures, folds, etc., also may exist and may influence the rate and direction of fluid migration.

10.2.2 BACKGROUND AND BASELINE DEFINITIONS

One of the most important components of a monitoring system is the establishment of an accurate basis against which the water quality data obtained from monitoring in the vadose and saturated zones can be compared in order to reliably determine whether there has been migration of waste constituents. This basis for comparison is established by obtaining background and baseline water quality data.

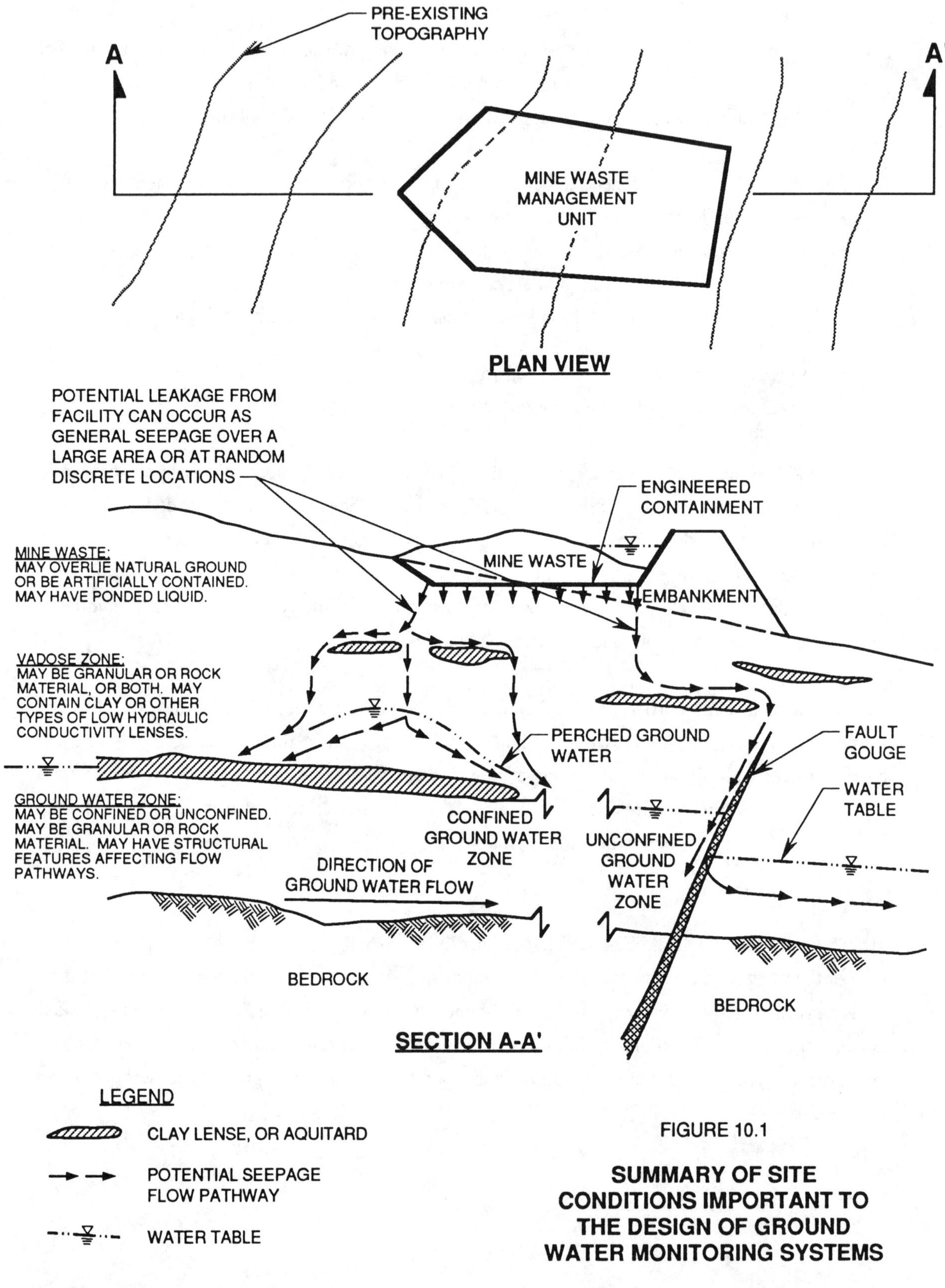

FIGURE 10.1

SUMMARY OF SITE CONDITIONS IMPORTANT TO THE DESIGN OF GROUND WATER MONITORING SYSTEMS

The definitions of background and baseline are as follows:

- Background - Background data describe the quality of ground water in areas which are upgradient from the waste management unit and known to be out of reach of any liquid that could escape from the facility. These data, which are collected both before and during the operation of a waste management unit, are used to define the natural averages and temporal variations in the concentration of constituents.
- Baseline - Baseline data describe the quality of ground water both immediately below and downgradient from the facility prior to any waste disposal. These data, which are collected prior to the operation of a waste management unit, or before there has been sufficient time for potential seepage to migrate to the monitoring points, are used to define the "pre-operational" concentration of constituents and to trace any changes to these during facility operation.

When determining representative background and baseline constituent levels, it must be recognized that instantaneous values can vary seasonally and from year to year. Therefore, it is necessary to collect samples over a sufficiently long period of time so that typical averages and ranges of concentrations can be developed.

Liquids which have seeped from the waste management unit are detected by changes to baseline concentrations which are inconsistent with anticipated natural changes as defined by the background concentrations and any baseline data collected. Also, changes are compared to waste liquid constituents and their concentrations to further determine whether leakage has occurred.

10.2.3 WASTE CHARACTERISTICS

The concentrations and mobility of constituents in the waste largely determine which chemical parameters are of concern when designing a monitoring program. These constituents are not necessarily limited to those inherent to the waste material and the processing reagents added, but may also include reaction products. For example, waste containing pyrites may produce acid rock drainage (ARD) when exposed to water and oxygen, whereas pyrite itself would not be a constituent of concern. (See Chapter 4 for further discussion.)

Some of the primary groups of constituents of concern in California mine waste, as described by the *Mining Waste Study* (U.C. Berkeley, 1988), are:

- Cyanide and thiourea - Used for leaching gold.
- Mercury - Associated with ore deposits containing mercury.

- Heavy metals - Such as copper, zinc, cadmium, and nickel from sulfide deposits, generally associated with ARD. Arsenic, associated with evaporite deposits, and chromium, manganese, and tungsten from oxide deposits.
- Sulfate - Associated with ARD.
- Asbestos - From asbestos mines and mills, and some disturbed serpentine rocks.
- Borates and other dissolved salts - Mainly from evaporite deposits in arid regions.
- Turbidity - From surface disturbance and precipitation of salts, oxides, etc.
- Processing reagents - Including flotation reagents, lixiviants, and solvent extraction reagents.

Additional discussion of the constituents of concern is provided in Chapter 3.

10.2.4 WASTE CONTAINMENT AND WASTE MANAGEMENT PRACTICES

Mine waste which requires special containment can be handled in a number of ways, as summarized in Figure 10.2. These range from natural containment by geologic formations through a range of engineered measures, which include clay, synthetic, or a composite of clay and synthetic liners. Chapter 7 provides details on liner types and their performances.

Waste management practices for reducing the potential for releases of leachate involve removal or prevention of a significant hydraulic head (i.e., ponded fluid) from developing on top of the waste containment system formation. In tailings impoundments, this is accomplished by drying the tailings in the process plant or depositing them in thin layers and allowing them to dry; and/or placing underdrains beneath the deposited tailings. In heap leach and waste rock piles, hydraulic head buildup is prevented by the placement of pipe and/or rock underdrains.

The type of containment system utilized determines what form unexpected releases of waste constituents could take. In unlined systems with ponded water, a release could occur as general seepage over a large area. In naturally lined systems, a release could occur over a large area or be restricted to isolated zones of weakness, such as fractures or shear zones. In synthetically lined systems, widespread releases are not expected; small releases may come from holes or tears which may be randomly located. Chapter 7 contains a discussion on potential leakage rates through liner systems.

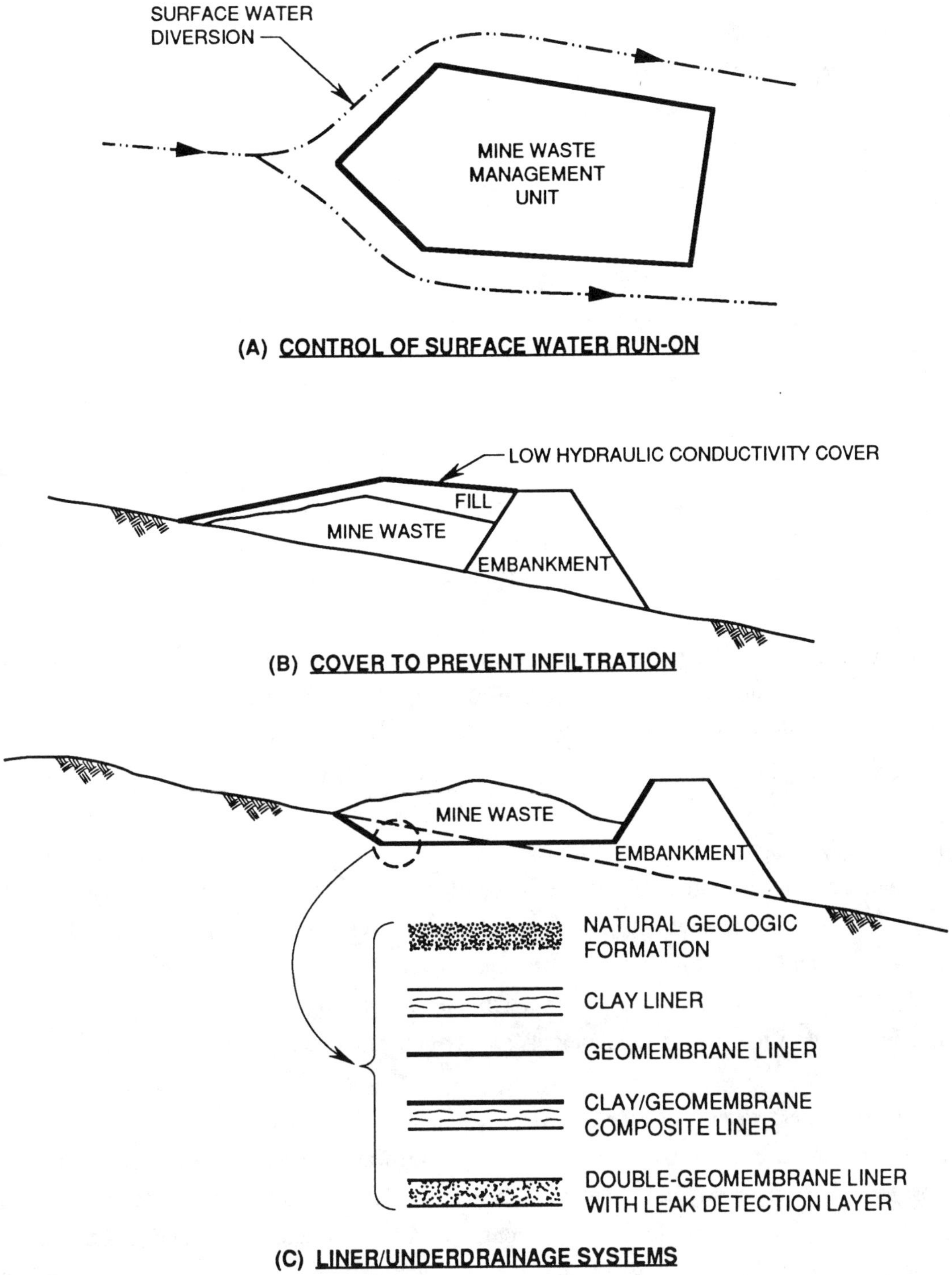

FIGURE 10.2

TYPES OF WASTE CONTAINMENT

10.2.5 ENVIRONMENT

The environment at the waste management unit location can significantly affect the potential for seepage losses to occur and the subsequent pathways it would take. These environmental aspects include:

- Climate.
- Vadose zone characteristics.
- Proximity to ground water.
- Physical properties of geologic formations.
- Hydrogeochemical characteristics of the site.
- Natural ground water flow and quality.

As discussed in Chapter 6, climate is important in determining the potential for a site to generate leachate. In addition, seasonal variations in temperature and precipitation can cause moisture changes in the vadose zone. These variations must be understood and quantified if vadose zone monitoring is to use moisture conditions to detect releases.

The vadose zone characteristics which are relevant to transmitting fluid include permeability, the moisture content versus soil moisture tension relationship, and the natural in situ moisture content. Particularly important is whether the vadose zone is granular soil or a fractured rock mass. This is an important factor in determining likely migration pathways and velocities.

Under conditions of rainfall recharge through the vadose zone, the primary direction of seepage movement is downward, with capillary effects contributing to lateral spreading. Vertical and lateral variations in moisture conditions can also cause lateral, and even vertical migration. Lenses with variable permeabilities and fracture orientations may also cause water to perch and/or spread laterally. The occurrence of perched water and/or lateral spreading can provide an opportunity for placement of one or more wells to monitor vadose zone conditions and potential seepage.

Proximity to ground water is one of the environmental factors which determines the time required for a potential release to reach the water table. If ground water is shallow (near ground surface), vadose zone monitoring may be redundant, with adequate detection capability provided by ground water monitoring. In deep water table conditions, a larger vadose zone may be impacted before a release is detected in the ground water, making vadose zone monitoring more necessary. Depth to ground water is, therefore, an important consideration in determining the need for vadose zone monitoring. Chapter 7 contains further discussion of travel times to ground water.

The hydrogeochemical environment at a waste management site determines the chemical changes experienced by migrating waste constituents. Chapter 4 contains a more detailed discussion of these types of changes that can occur and the constituents that are likely to be affected.

In addition to the chemical changes that can occur in seepage, the background hydrogeochemical characteristics of the ground water may vary in the vicinity of a waste management unit. For example, natural ground water quality may vary spatially. It also may contain elevated levels of constituents which also occur in the waste material. It is, therefore, necessary to clearly understand the hydrogeochemical environment when designing a monitoring system.

The ground water path offers the greatest potential for a liquid waste release to travel beyond the boundaries of the disposal facility. At many mining sites, ground water systems are located in fractured bedrock in which discrete but random flow pathways occur. These aquifers are not as easily characterized as are granular aquifers, and demonstrate that special care must be taken in the selection of monitoring locations. In developing a monitoring program for a ground water system, the flow rates and directions of ground water movement should be defined, so that both background and downgradient monitoring locations can be established.

Some of the interpretations regarding the ground water setting can be obtained from published reports on regional aquifers, geophysical techniques, air photo interpretation, and from spring and seep mapping and sampling. However, information from site exploration activities is usually needed to provide the necessary details. This includes data from drilling, exploration, and water sampling wells. Often at mining projects, exploration and condemnation drilling data can provide useful subsurface data. In most cases, however, supplemental data is required both for purposes of designing the waste management unit and the monitoring system. These data include ongoing observations and sampling, from piezometers completed in a representative number of boreholes. Field permeability determinations, using either upset testing or packer testing, are often also necessary to determine the characteristics of selected geologic formations. Samples of ground water and possibly soil and rock should be collected for geochemical testing.

10.2.6 POTENTIAL WASTE CONSTITUENT MIGRATION PATHWAYS

This section discusses how potential seepage pathways can be developed and the associated seepage flow rates quantified.

10.2.6.1 Potential Seepage Through Containment System

When designing a monitoring system, the worst-case seepage loss scenarios need to be postulated. This can be done by assuming that the containment system's performance is significantly less than the design intent. For example, where clay liners or natural geologic formations are being used for waste containment, it can be assumed that their permeabilities are higher than the design value (by one or two orders of magnitude). Where synthetic liners are concerned, the assumptions that can be made include that a large number of perforations or tears have occurred and that these are much larger than the values used in designing the containment system.

An additional factor is the chemical attenuation of waste constituents as seepage occurs through a containment system, such as a clay liner. These processes can reduce concentrations of certain constituents, and are discussed in detail in Chapter 5. There also may be the potential that seepage through a liner leaches out certain constituents as their solubility is enhanced by the chemical properties of the seepage. Low or high pH liquid may be an example of such seepage.

10.2.6.2 Seepage Through the Vadose Zone

Unsaturated Flow Theory

The first part of this section addresses the theory of seepage flow through the vadose zone. This information provides the background which is helpful in postulating likely seepage pathways from the waste management unit and in evaluating the associated moisture conditions that subsequently influence the types of monitoring devices which are required. The final part discusses practical considerations, including the limitations of the flow theories.

The vadose zone is comprised of unconsolidated granular sediments or consolidated rocks which lie above the water table. The intergranular pore space and the spaces provided by the fracture systems in the rock are generally filled with air, water, and possibly seepage from the waste management unit. Within the vadose zone, localized areas of saturated conditions can occur, such as at perched aquifers and soil horizons that are temporarily wetted during major storms.

When the soil or rock is only partly saturated, the water is at a pressure which is less than atmospheric, and the pore water is referred to as being under a tension or suction. At low suctions (wetter conditions), water is held in the pores mostly by capillarity, a phenomenon which results from a balance of surface tension forces between water, air, and the soil matrix. At very high suctions (drier conditions), the water layers might be only a few tens of molecules thick. Under these conditions, water molecules are strongly adsorbed by electrostatic forces to solid surfaces, particularly clays.

The maximum hydraulic conductivity, k, of a soil occurs at full saturation (zero soil moisture suction) and is referred to as k_s. As the soil dries, the cross-sectional area of fluid required for flow transmission decreases, and the flow paths become more tortuous. As a result, k decreases with decreasing water content. It is not uncommon for the hydraulic conductivity of a soil to decrease by a factor of several thousand to more than a million over a change of water content of only 10 to 40 percent, for example.

Typical soil moisture characteristics curves are presented in Figure 10.3, and also in Figure 6.4 of Chapter 6. These indicate how hydraulic conductivity decreases with a decrease in soil moisture content. Neglecting the differences between wetting and drying phases, there is a unique relationship between soil moisture hydraulic conductivity, and grain size or fracture size by rock material. These curves also highlight that coarser-grained material has a much lower hydraulic conductivity than the fine-grained material at low moisture contents.

For purposes of developing an unsaturated flow equation, it is convenient to express the fluid pressure, P, in terms of a pressure head, ψ, having units of length, so that:

$$\psi = P/\rho g$$

Where P = Fluid pressure (force/unit area)
ρ = Fluid density (mass/unit volume)
g = Acceleration (length/time^2)

Moisture migration occurs due to the difference in total hydraulic head, which is:

$$h = z + \psi$$

Where z = Elevation of the point of interest relative to a common datum (length)

The moisture migration rate can then be determined from Darcy's law for unsaturated flow in the x direction, for example, and can therefore be written as:

$$V = -k\,(\psi)\,\frac{dh}{dx}$$

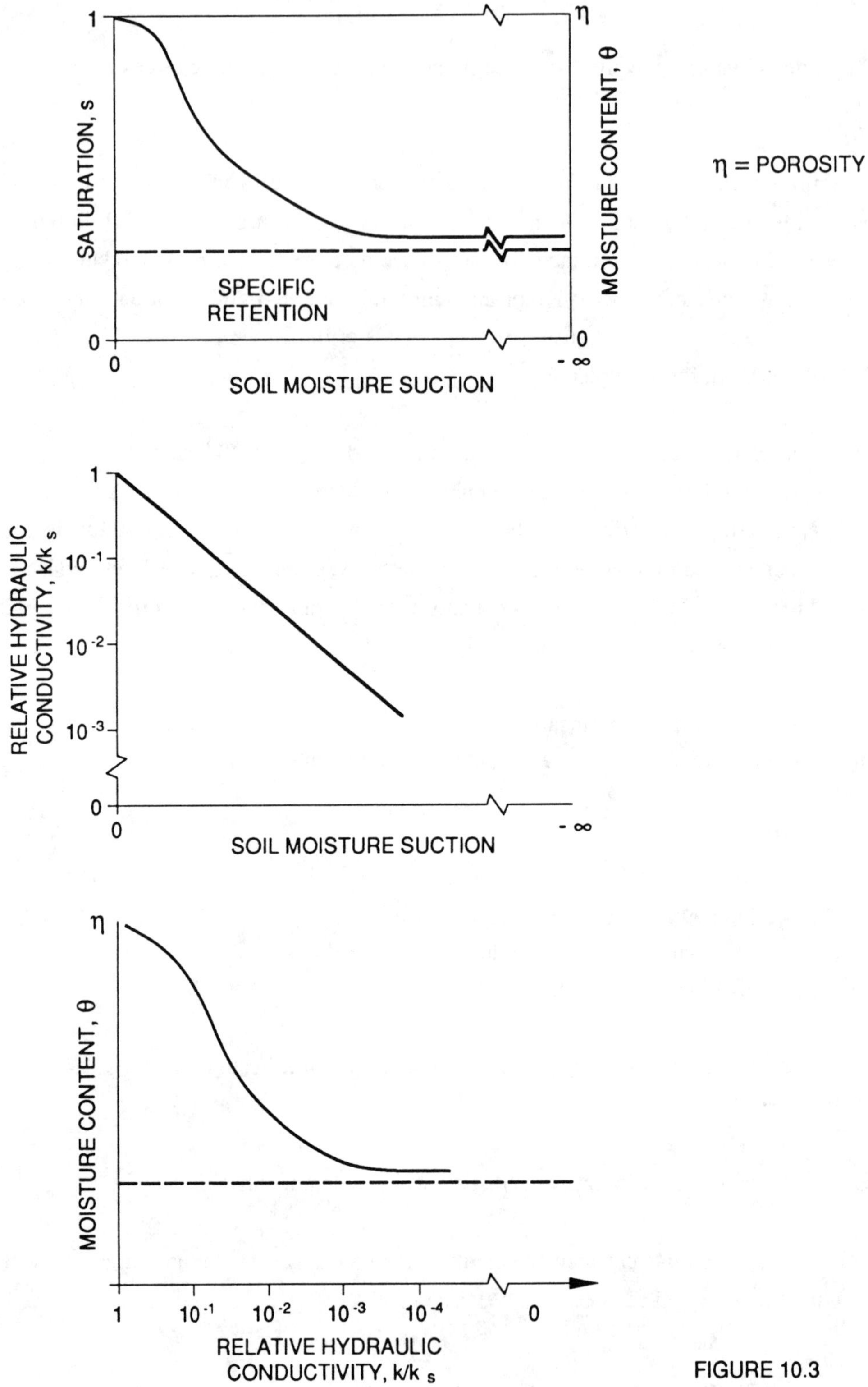

FIGURE 10.3

TYPICAL SOIL MOISTURE CHARACTERISTICS

Where:

V = Darcy's velocity (length/time)
$k(\psi)$ = Unsaturated hydraulic conductivity at pressure head ψ ([length/time])

For areas at approximately the same elevation, the movement of moisture will be determined by the differences in soil moisture suction, with the direction of movement being from the lower suction to the higher suction areas. For granular materials with relatively uniform soil moisture suctions the movement will be controlled by the elevation head and be in a downward direction.

For soils, the nature of the relationship between flow, hydraulic conductivity, soil moisture suction, and moisture content can be evaluated by field sampling and laboratory testing. However, seepage movement through fractured vadose zone materials can be more complex to characterize. Unsaturated flow through fractured rock tends to occur more on a random basis as the migrating fluid particles select those fractures offering the path of least resistance. Fluid may not even enter a major fracture zone if the fractures are open. In such cases, the major fractures act as flow barriers. These fractures are analogous to a very coarse-grained material at low moisture contents which have a very low hydraulic conductivity, as shown in Figure 6.4 in Chapter 6.

For prediction of seepage front movement, required data include in situ moisture content and the hydraulic conductivity-versus-moisture content relationship for the material in the vadose zone (see Figure 10.3). For the latter, several investigators (Laliberte et al., 1966) have related soil characteristics such as grain size distribution to estimate the relationships. McWhorter and Nelson (1979) have proposed an empirical functional relationship between moisture content and seepage flow rate. Mualem (1976) presents a manual on the permeability-moisture-soil suction tension head relationships of many soils. This can be used to estimate the properties of soils which have grain size and texture similar to those listed in the manual. Alternatively, and for more critical applications, laboratory testing may have to be carried out in order to determine the necessary values.

In quantifying these flow conditions, it may be important to initially determine whether the seepage is likely to occur in a saturated or unsaturated condition, as this would indicate the type of instrumentation needed to detect the seepage. This can be determined by comparing the estimated seepage rate with the migration rate in the vadose zone under saturated conditions. If the seepage rate is higher, then saturated flow will occur (Figure 10.4[A]). If not, the flow is likely to be unsaturated (Figure 10.4[B]).

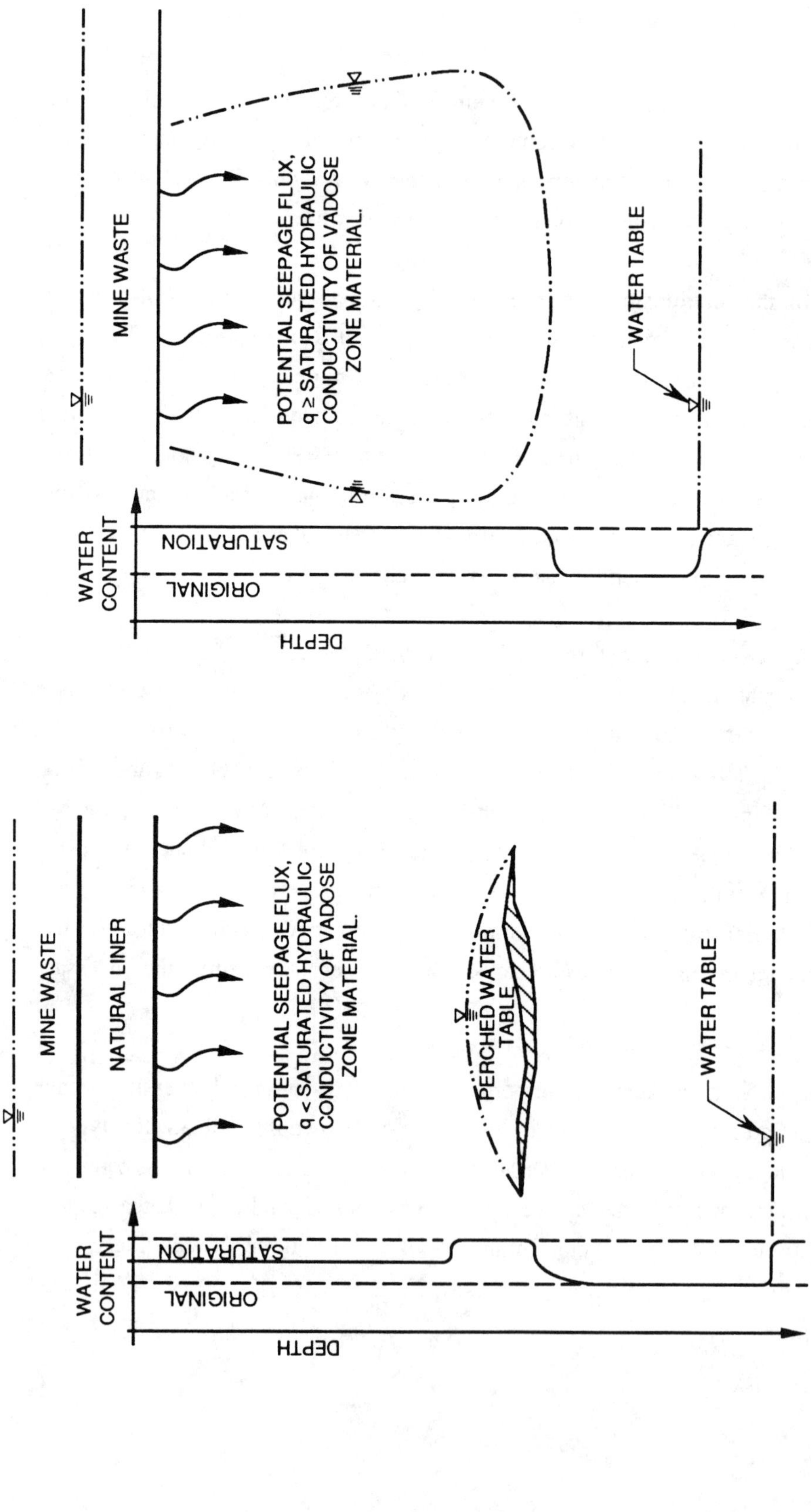

(A) FULLY SATURATED SEEPAGE: MOISTURE CONTENT NOT LIMITED BY RECHARGE. SEEPAGE FLUX AND RATE OF ADVANCE OF WETTING FRONT IS CONTROLLED BY SATURATED HYDRAULIC CONDUCTIVITY OF VADOSE ZONE MATERIAL.

(B) PARTIALLY SATURATED SEEPAGE: MOISTURE CONTENT IS RECHARGE-LIMITED. RATE OF ADVANCE OF WETTING FRONT IS CONTROLLED BY RATE OF SEEPAGE FLUX.

FIGURE 10.4

POTENTIAL SEEPAGE THROUGH THE VADOSE ZONE

Another important consideration when designing a monitoring system is estimating travel time to the water table. The equation for unsaturated flow, developed by McWhorter and Nelson (1979) and discussed in Chapter 7, can be used for this purpose, i.e.:

$$T = \frac{L}{q}\left[(n - \text{Ø}_r)\left(\frac{q}{k}\right)^{\frac{\lambda}{(2+3\lambda)}} + (\text{Ø}_r - \text{Ø}_i)\right]$$

Where:

- T = time for seepage to reach the water table (time)
- L = total length of seepage pathway (length)
- k = saturated hydraulic conductivity of the subsoil (length/time)
- q = seepage flux rate (volume/unit area/unit time)
- n = porosity of subsoil (dimensionless)
- Ø_i = initial volumetric moisture content of vadose zone below the wetting front (dimensionless)
- Ø_r = volumetric moisture content by volume above the wetting front (dimensionless)
- λ = pore size distribution index (dimensionless; typically between 1 and 3)

For saturated conditions, Ø_r is equal to the porosity of the material. For unsaturated conditions, Ø_r is equal to the moisture content at which the vadose zone flow rate equals the seepage loss rate and which is obtained from the soil moisture characteristic curves. As an approximation this value can be set equal to the specific retention value (McWhorter and Nelson, 1979).

Using the above theories, it is possible to determine both the timing of seepage arrival and the expected moisture content of the seepage front. As discussed under liner seepage, liquid migration through the vadose zone also will be subject to attenuation, which reduces the concentration of certain chemical constituents. Chapter 5 contains a more detailed discussion of these processes.

Practical Considerations

Predicting ground water flow is not as mathematically straightforward as indicated by the considerations described above. Hydraulic conductivity is usually variable within aquifer materials which appear to have similar composition. Aquifers may, in fact, be comprised of more than one material type. Sometimes highly permeable preferential pathways of flow occur. There may be either horizontal or vertical hydraulic discontinuity between well locations due to such features as faults, confining layers, or pinching out of geologic features. Therefore, it is extremely important to have a comprehensive understanding of the geology and geohydrology underlying and surrounding the waste management facility.

A further important consideration that emerges from the above discussion is that liquid will not necessarily flow laterally from a partially saturated condition in a geologic formation to a borehole drilled through that formation. Therefore, a monitoring well is not likely to collect any vadose zone seepage of waste constituents. The only time that conventional monitoring wells are suitable is when the seepage loss from a facility causes local saturated conditions to develop adjacent to the well, as occurs in the saturated zone and in any perched water zones that may be present in the vadose zone.

10.2.6.3 Seepage Through the Saturated Zone

This section outlines the theory of saturated ground water flow, which can be used as a basis for developing seepage flow pathways. The mechanisms of contaminant transport in the saturated ground water system are:

- Advection with the fluid flow in the downgradient direction.
- Dilution and dispersion in the ground water.
- Attenuation and release of chemical species due to geochemical reactions.

The flow in the ground water system is given by Darcy's law:

$$q = k \cdot i$$

Where:

q = specific discharge or flux velocity (volume/per unit area/unit time)
k = hydraulic conductivity (permeability) of aquifer (length/unit time)
i = hydraulic gradient within the aquifer (dimensionless)

The migration rate (velocity) of conservative constituents can be evaluated using the following equation:

$$v = \frac{ki}{n}$$

Where:

n = material porosity (dimensionless)

Typical hydraulic conductivity values and porosities for various types of geologic formations are summarized in Figure 6.6. In granular media, or rock which has been subjected to significant pore volume enhancement, such as karst solutioning, porosity values are typically between 20 and 60 percent. However, where flow occurs in fractured rocks and in consolidated cohesive materials that are fractured, such as jointed till, silt or clay, effective porosity is normally much smaller and values on the order of 1 to .001 percent (or 10^{-2} to 10^{-5}) are not unusual (Freeze and Cherry, 1979). In fractured media, therefore, although the hydraulic conductivity may be low, the ground water flow velocities can be large due to the very low effective porosity.

Porosity is always less than one. The linear velocity is always greater than the Darcy velocity, because it accounts for the fact that flow occurs only through the pores of a material rather than through the entire cross-sectional area. As can be seen by the porosity values outlined above, the linear velocity can be orders-of-magnitude greater than the Darcy flux in fractured rock. After the velocity is estimated, the downgradient distance for detection monitoring wells can be selected to ensure timely reporting of seepage.

The above equation describes the advective component of the transport of a constituent. Additional processes that affect the transport and distribution of constituents are dilution and attenuation. Dilution occurs as the seepage flow combines with the regional ground water flow and results in decreased concentrations of the dissolved constituents. Attenuation includes a wide range of physical, chemical and biological processes which are discussed in Chapter 5.

In designing a successful monitoring system, it is important to select monitoring constituents which have concentrations after advection, dilution and attenuation that will not only represent a measurable change in the downgradient concentration, but also will be discernible from background, including the associated natural temporal changes in water quality.

When dealing with ground water flow through fractured rock, two flow components exist. The first represents flow through pores in the intact rock. It is governed by what is called the primary permeability of the rock material. Because primary permeability is often extremely low (10^{-6} to 10^{-9} cm/sec), this flow component often can be ignored.

The second and major flow component occurs through interconnected fractures and open joints in the rock material. This is governed by what is called secondary permeability. On a large scale, flow through fractures can be treated using Darcy's law. The secondary hydraulic conductivity is usually determined by field testing at many points within the rock mass. For purposes of estimating average seepage rates, these values can be averaged using the geometric mean formula as follows (CRC, 1978):

$$k = \sqrt[n]{k_1 \cdot k_2 \ldots\ldots k_i \ldots\ldots k_n}$$

Where:

n = number of measurement points

k_i = hydraulic conductivity for point i (length/unit time)

$\overline{k}$ = average hydraulic conductivity for the formation under consideration (length/unit time)

The above formula is appropriate when the values of hydraulic conductivity are randomly distributed and there is no indication of any interconnected areas of relatively higher conductivity. Where such heterogeneities occur, or where it is known that seepage must pass through alternating layers of higher and lower hydraulic conductivity, different methods of calculating representative hydraulic conductivity values for identified preferential seepage pathways should be used. These can include the following: (1) use of the higher conductivity values in cases where the major fracture zones are known or suspected to be interconnected, (2) use of the above geometric mean formula for selected points all contained within the identified preferential seepage pathway, and (3) use of the harmonic mean formula where the seepage pathway is known to sequentially pass through layers or areas of higher and lower conductivity (CRC, 1978 and Terzaghi and Pack, 1948), i.e.:

$$\overline{k} = \frac{L}{\frac{\ell_1}{k_1} + \ldots\ldots \frac{\ell_i}{k_i} + \ldots\ldots \frac{\ell_m}{k_m}}$$

Where:

L = total length of seepage pathway (length)

m = number of areas or layers of different hydraulic conductivity

l_i = length of each area or layer (length)

k_i = hydraulic conductivity for each area or layer (length/unit time)

10.2.6.4 Seepage Pathway Analysis Approach

A useful technique that can be used to integrate the information presented above in the design of a monitoring system is to complete a seepage pathway assessment using procedures similar to those discussed above. This would provide some indication of the nature of worst case impact and, hence, the degree of monitoring required and the most effective locations for the monitoring devices.

Fundamental to a seepage pathway assessment is a comprehensive geologic and geohydrologic model of the area. This includes a three-dimensional description of the shape of the different geologic formations and their physical properties, such as hydraulic conductivities, porosity, extent and nature of secondary hydraulic conductivity, and an understanding of the spatial variability of these parameters. It also includes a description of the nature and extent of ground water, including vadose zone moisture presence, ground water, recharge and discharge areas, and flow directions, and constituent migration velocities in both the vadose and saturated zones. It will not always be possible to obtain comprehensive and detailed quantification information. As a minimum, however, these data should be sufficient to develop a conceptual model of the conditions. Obviously, the more site-specific the data is, the more focused and, hence, cost-effective the monitoring program can be.

The next step in the seepage pathway assessment is to evaluate worst-case leakage scenarios, such as general seepage over a wide area through a clay liner, point seepage sources through damaged clay or synthetic liners, seepage through containment berms and, possibly, infiltration from spillage areas adjacent to the waste management facility. Factors to be considered in evaluating the potential quantities and rates of leakage include the amount of precipitation, the method of waste placement, and the anticipated areas of free liquid accumulation in the waste.

The next step is to define the potential seepage pathways from the leakage areas to the receiving ground water underlying the waste management facility. These pathways would be based on knowledge of the location and potential quantity of leakage, the geology and geohydrology of the site, and the basic principles of unsaturated and saturated flow. Some assessment of travel time may be possible and useful if there is enough quantitative information on the geohydrologic properties of the geologic formations.

Consideration also needs to be devoted to identifying the key monitoring constituents and the degree to which these concentrations may be altered in the liner, the vadose zone or the saturated ground water. This aspect is discussed further in Section 10.3.4.

Figure 10.5 (A through E) demonstrates several cases of seepage migration for differing hydrogeologic conditions and appropriate monitoring well locations. The concepts shown are taken from EPA (1986). In Figure 10.5 (A), an unconfined water table is located above a bedrock layer. The saturated zone is continuous, and direction and rate of flow are relatively uniform. In Figure 10.5 (B), the ground water zone is confined (under pressure). Although flow in this aquifer may also be uniform, seepage may perch above the confining unit and move laterally.

Figure 10.5 (C) illustrates a site where preliminary drill logs indicated a confining layer of unfractured, continuous clay beneath the site. However, as shown in the figure, the clay layer is not continuous, and the water-bearing unit below the clay layer and the formation above the clay layer are both part of the aquifer. To identify this, field investigations, including measurement of the hydraulic head distribution (particularly with depth), pump tests, and in situ hydraulic conductivity tests can be conducted.

In Figure 10.5 (D), the conditions involve faulting and post-depositional erosion of a low permeability limestone located between two higher permeability sandstones. The fractures in the limestone represent hydraulic communication between the upper unconsolidated sand layer and the sandstone formation below the limestone unit. The aquifer, therefore, includes the unconsolidated sand formation, the limestone formation, and the sandstone formation.

In Figure 10.5 (E), unconsolidated units are underlain by a relatively impermeable bedrock. Borings near the mining waste management facility identified the uppermost aquifer. However, the aquifer unit, which is laterally continuous, is permeable and saturated and can have seepage flow in deeper portions of the aquifer than observed onsite.

10.2.7 POTENTIAL IMPACTS OF WASTE CONSTITUENT MIGRATION

An important factor to be considered when designing a monitoring system is the potential for any waste constituent migration to impair the beneficial uses of the ground water. Generally, where there is an environmentally sensitive natural water body close to the waste management unit, and the seepage quality is such that significant impairment to the uses of that water body would occur if the seepage were to discharge to it, monitoring becomes critical. In these "high risk" situations, an early and reliable seepage detection system should be incorporated.

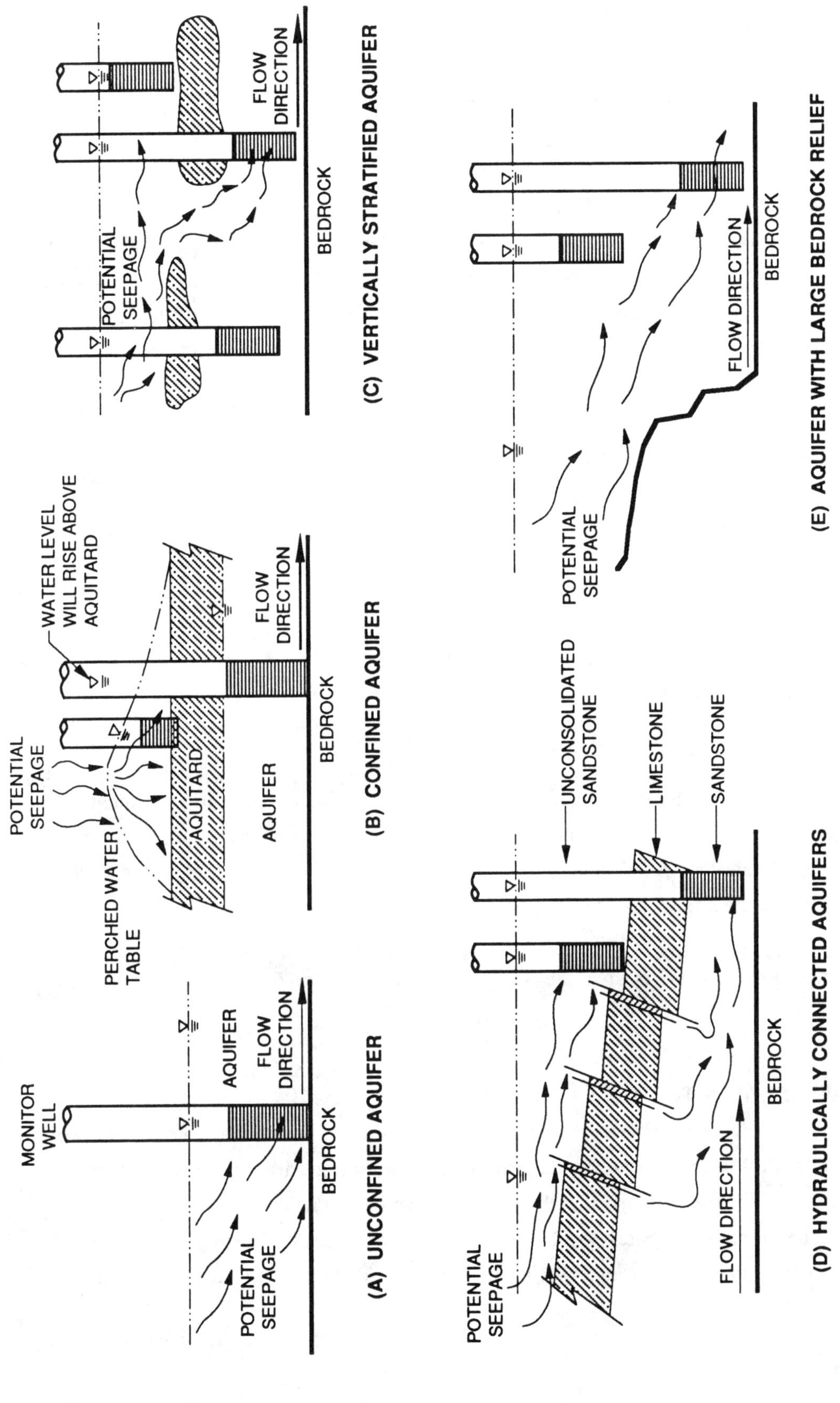

FIGURE 10.5

POTENTIAL SEEPAGE THROUGH THE GROUND WATER SYSTEM

On the other hand, in cases where seepage losses are unlikely to impair beneficial uses, or travel times through the vadose zone are long, there will be less need for early and reliable detection, and ground water monitoring may be all that is necessary.

10.3 ELEMENTS OF A MONITORING SYSTEM

Ground water monitoring systems can be installed to detect seepage losses from a waste management unit, to evaluate a suspected seepage loss for purposes of determining what remedial action is appropriate, or to monitor the performance of a corrective action installed to deal with a seepage problem. This section describes the elements that can be used to develop any one of these monitoring systems.

A ground water monitoring system incorporates some or all of the following key elements:

- In-waste monitoring.
- Vadose zone monitoring.
- Ground water monitoring.

Other important considerations include:

- Location of the point or points of compliance.
- Selection of monitoring parameters.
- Frequency of monitoring.

The point of compliance is usually a point or a plane located at some distance downgradient from the waste management unit. It represents the location or locations at which the operator of the facility must legally meet specified water quality criteria in order to remain in compliance with the relevant laws, regulations and waste discharge permit conditions.

The other important elements outlined above are discussed in more detail in the sections that follow.

10.3.1 IN-WASTE MONITORING

In cases where special waste management practices are required to ensure that no seepage losses occur from a facility, it may be appropriate to install monitoring within the waste and/or the containment system in order to verify that the waste management practices are, in fact, performing as planned. The monitoring system would usually track the moisture content of the waste and, possibly, the containment layers underlying the waste. It also can be used to determine the hydrostatic head on the engineered containment system, or to detect chemical concentrations in the liquids contained in the waste.

For example, in a tailings impoundment where the tailings material is being used as containment, there may be a need to confirm that the thin layer deposition, which is being used to form unsaturated layers of tailings, is effective. In such a case, it may be necessary to install monitoring instruments, or conduct regular sampling and testing, to determine the hydraulic head and moisture conditions in the tailings.

The Jamestown Mine, in Sonora, California represents a case of in-waste monitoring similar to that discussed above. A summary description of the installation and the results of the monitoring program are provided in Appendix F.

When planning in-waste monitoring or sampling systems, it is important to recognize that there is usually a layer of waste material that controls the rate of downward seepage. Therefore, it may be necessary to obtain a profile of soil moisture content and suction through this layer in order to evaluate the magnitude of downward seepage. The types of instruments that could be used to do this are discussed in Section 10.3.2. An alternative approach would involve field sampling, which entails drilling for core samples and subjecting them to laboratory tests to determine density, void ratio, soil moisture content, and soil moisture suction. Selected samples could also be subjected to testing to quantify the relationship of soil moisture content versus unsaturated permeability for the material.

Monitoring the liquid pressure at the base of the waste provides some useful information on the hydrostatic pressure on the containment system below the waste and allows comparison with the design assumptions. This type of monitoring can be achieved using piezometers or pressure transducers (Appendix F).

Monitoring for both the flow and quality of liquid that is collected within the leachate collection and removal systems, provides useful information on the degree to which leachate generation is being controlled, and allows comparison with the design assumptions. Equipment that can be used for this type of monitoring includes flumes and measuring weirs, coupled with either periodic manual level readings or level recording devices which provide continuous readings. If the collected waste is removed by pumping, the seepage collection rate can be measured directly on a totalizer placed in the pipeline.

10.3.2 VADOSE ZONE MONITORING

10.3.2.1 Background

If effective, monitoring of the vadose zone directly beneath and adjacent to the waste management unit can detect releases early, before they reach ground water. With regard to mine waste, there are several applicable generalizations as to the uses and limitations of vadose zone monitoring. First, because of the large areal extent of mine waste disposal facilities, monitoring devices underlying them generally will be remote from their perimeters. Installation of most vadose zone monitoring equipment, therefore, can be very difficult, due to the long distance over which data must be transmitted or samples collected. After placement of the monitoring devices, repair or calibration usually will not be possible.

The type of environment in which vadose zone monitoring is most likely to be successful is where it is used under a facility which is underlain by soil materials (alluvium or colluvium) and where the waste containment is provided by either a clay liner or the natural material itself. The anticipated leakage scenario would involve general seepage over a large area. The system would be most effective if the distances between vadose zone monitoring points are less than the size of anticipated seepage area, and if the equipment installed is sensitive to the anticipated moisture changes in these areas.

Vadose zone monitoring for seepage below a synthetically lined facility, using conventional equipment, is impractical, as leaks are likely to be isolated to a few random locations. As the seepage escapes and moves downward, it gradually spreads laterally (Chapter 7). Therefore, the vadose zone monitoring points would have to be spaced close enough to encounter seepage from any of the random locations and would require a great number of monitoring points. For example, a 100-acre tailings impoundment overlying a loamy sand with a permeability of 10^{-5} cm/sec of soil would require approximately 7,000 instruments at a depth of 10 feet to ensure that any point seepage would be detected (Bumb et al., 1988). Many of the tailings impoundments being constructed today are much larger than 100 acres.

Fewer vadose zone wells would be required, however, if their purpose were to verify that there is not a basic problem with the design concept and that widespread leakage is not occurring. An example is the Mesquite Gold Mine in Imperial County, California, where numerous vadose zone wells are placed around the perimeter of each pad to demonstrate that liner construction and ore loading procedures have not resulted in any systematic failures.

Successful measurement of field parameters, such as water content, is dependent on good contact between the instrument sensor and the surrounding media, good calibration relationships between soil matrix moisture, and the response signal from the sensor. In a fractured bedrock environment, these issues become difficult to achieve. Most vadose zone monitoring techniques are suited for soil (porous) media only, and almost all of the monitoring methods are point source measurements. Installation of a single-point source sampler, such as a suction cup lysimeter, in fractured bedrock is very difficult if not technically impossible, depending on the conditions of the rock.

A further problem with vadose zone monitoring is that it is extremely difficult to determine the background moisture levels under a waste management facility. Just the presence of a large lined area without any leakage will cause a gradual change in the natural moisture conditions immediately below the facility. Differentiating this change, as well as natural seasonal fluctuations in soil moisture, from the impact of a small leak can be very difficult and may provide questionable results.

10.3.2.2 Types of Systems

The basic systems that are used to monitor the vadose zone are direct (sampling) and indirect (non-sampling) point methods, as well as indirect non-point methods. Direct methods collect samples from the vadose zone, whereas indirect methods provide information (such as resistivity or moisture content) from which presence of seepage in the vadose zone can be inferred. Table 10.1 summarizes various direct and indirect methods, their range of applications, and their advantages and disadvantages.

As shown in Table 10.2, both direct and indirect point sampling methods have a probability of yielding false negative readings, i.e., indicating that there is no seepage when a seepage loss has actually occurred. This is because point source measurement techniques monitor only a limited area around each monitoring location. The point direct methods, however, have the advantage that they are unlikely to result in false positive readings, as they involve not only the detection and measurement of soil moisture, but chemical testing, as well. Both the non-point and point indirect methods, on the other hand, have a low probability of producing false positive readings, as they may detect soil moisture changes caused by factors unrelated to seepage loss, such as natural soil moisture fluctuation or responses to recharge events.

TABLE 10.1

SUMMARY OF SAMPLING METHODS FOR VADOSE ZONE MONITORING (MODIFIED FROM EVERETT ET AL., 1984)

SAMPLING METHOD				MEASURED PROPERTY	APPLICABILITY		SPATIAL EXTENT OF MONITORING				DEPTH LIMIT (FEET)			RELIABILITY		DATA COLLECTION SYSTEM			ADVANTAGES (+)/ DISADVANTAGES (-)
					POROUS MEDIA	FRACTURED ROCK	POINT	LINE	LOCALIZED AREA	ENTIRE FACILITY	SURFICIAL	LESS THAN 30	GREATER THAN 30	GOOD	QUESTIONABLE	MANUAL	RECORDING	TRANSMITTING	
POINT METHODS	INDIRECT	TENSIOMETERS	Simple (Bourdon) Type	Soil Moisture Suction	■		◉				◉	◉		◉		◉			+ Good response where soil moisture section is above -0.8 bar.[1]
			Pressure Transducer Type	Soil Moisture Suction	■		◉				◉	◉	◉		◉		◉	◉	- Subject to freezing. Very fragile and easily broken during installation. Tension head does not distinguish between wetting and drying conditions.
			Neutron Moisture Logger	Water Content	■	■	◉	◉			◉	◉	◉	◉		◉	◉		+ Does not disturb soil. Allows continuous vertical profile of moisture. - Difficult to obtain tight seal between access tube and soil. Insensitive under very dry conditions.
			Gama Ray Attenuation	Bulk Density, Water Content	■	■	◉	◉			◉	◉		◉		◉	◉		+ Excellent vertical profiling resolution. High precision and accuracy. - Two access tubes required. Difficult to keep vertical tubes parallel at depth.
			Electrical Resistance Blocks	Electrical Resistivity	■		◉				◉				◉	◉			+ Simple. Easy to install. - Good only at relatively dry conditions. Blocks may deteriorate with time. Resistance changes may also depend on variables other than soil moisture.
			Psychrometer	Relative Humidity	■		◉				◉	◉	◉		◉	◉	◉	◉	+ Good at low moistures, high soil tension conditions. Can be installed at great depth. - Does not distinguish between wetting and drying conditions. Complex, very fragile.
			Vapor Monitoring	Gas Vapors	■	■	◉	◉			◉	◉	◉	◉		◉		◉	+ Allows the vadose zone to be monitored without disrupting the containment system. Could also be used at existing facilities. - Costly to install.
	DIRECT	SOIL AUGER SAMPLING & LABORATORY ANALYSES	Hand	Soil & Pore Water Chemistry, Moisture Content	■		◉	◉			◉	◉		◉		◉			+ Can measure, through lab analysis, geochemical effects of soil on leachate. Great depth range.
			Power		■		◉	◉			◉	◉	◉	◉		◉			- Destructive method. Temporal data cannot be obtained at precisely the same location.
		LYSIMETERS	Suction Cup	Moisture Content, Pore Water Chemistry	■	▪	◉				◉	◉	◉	◉		◉			+ Yields the most representative sample of the direct methods.
			Filter Candle		■	▪	◉				◉				◉	◉			- Subject to clogging. May fail at soil suction below -0.5 bar. Installation depth of filter candle is limited.
			Collection Trenches	Liner Leakage	■	■			◉		◉			◉		◉			+ Provides direct measurement of liner leakage. - Can only sample leakage from a very small area of liner.
NON-POINT METHODS	INDIRECT	GEOPHYSICAL	Resistivity	Resistivity	■	▪			◉	◉	◉	◉	◉	◉		◉			+ Good reproducibility. Non-destructive. Provides accurate readings of bulk vadose zone properties. - Interpretation of soil moisture changes is difficult.
			Time Domain Electromagnetics	Resistivity	■	▪			◉	◉	◉	◉	◉	◉		◉			+ Non-destructive. Provides accurate readings to bulk vadoze zone properties.
			Time Domain Reflectrometry	Dielectric Constant	■	▪			◉	◉	◉	◉	◉	◉		◉			- Interpretation of soil moisture changes is difficult.

[1] 1 bar = 1 atmosphere = 33.9 feet of water.

LEGEND:

Technically Feasible	■
Technical Feasibility Depends on Site-Specific Conditions	▪
Not Technically Feasible	
Applicable	◉
Not Applicable	

TABLE 10.2

PROBABILITY OF FALSE READINGS FROM SAMPLING METHODS

SAMPLING METHOD		POTENTIAL FOR:	
		FALSE NEGATIVE READINGS	FALSE POSITIVE READINGS
POINT	Indirect	⬤	●
	Direct	⬤	•
NON-POINT	Indirect	⬤	●

LEGEND:

Probable	⬤
Low Probability	●
Unlikely	•

The non-point indirect methods also have a probability of yielding false negative results. Even though they generally cover larger areas of the waste management unit than the point sampling method, they may not have the same sensitivity to detect smaller leaks.

10.3.2.3 Point Methods

Tensiometers

Tensiometers are used to measure soil moisture suction pressures in granular material. There are simple (bourdon type) and complex (pressure transducer type) tensiometers which can be used. A simple tensiometer consists of a porous ceramic cup cemented to a rigid plastic tube, a small diameter tubing leading to a manometer or a bourdon gauge, and a filler plug in the rigid plastic tube (Figure 10.6). The internal volume of the system is filled with water, except for the terminal portion of the small diameter tubing filled with mercury. Water moves into or out of the tensiometer system through the pores in the ceramic cup until equilibrium is reached. The mercury level in the manometer will move to adjust to the new water column height. The soil water content can be estimated by using soil moisture characteristics curves.

Tensiometers are simple, providing in-place continuous readings and good reliability. However, at a mining site, installations would have to be limited to perimeter locations. There is no practical way to read the instrument remotely from its point of installation. The units are fragile and can be broken easily. Another limitation is that it is difficult to purge air from the system, especially in dry soils. This can lead to malfunctioning.

Neutron Moisture Logger

Neutron logging consists of lowering a down-hole probe, containing a high energy neutron source and a detector for thermalized neutrons, into a cased borehole (Figure 10.6). Neutron loggers detect the hydrogen content of the soil. A charged particle is emitted during neutron capture and then is detected by a solid or gaseous counter. The count rate then is converted to water content via a calibration curve.

At a mine waste management unit, angle cased holes can be installed at the perimeter to obtain moisture readings under the facility. Also, the probes can be run through horizontal cased holes located under the facility.

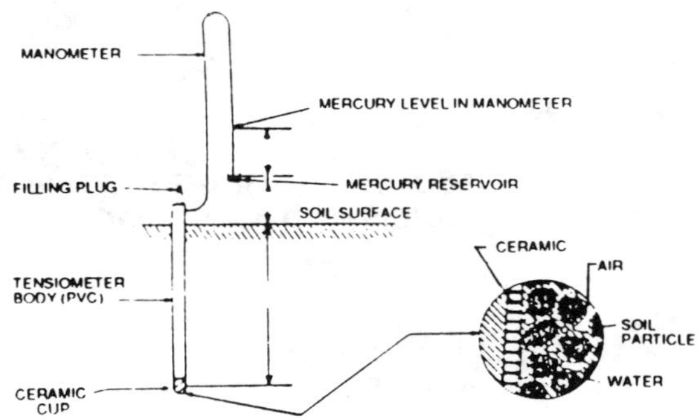

SCHEMATIC REPRESENTATION OF TENSIOMETER AND SECTION THROUGH THE CERAMIC CUP

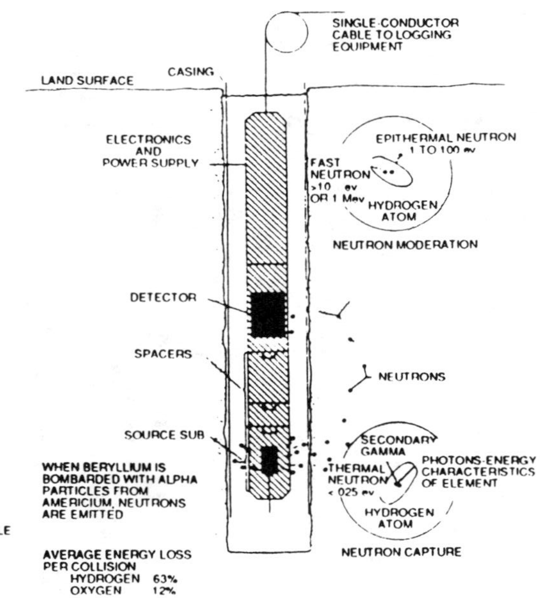

EQUIPMENT AND PRINCIPLES OF NEUTRON MOISTURE LOGGING

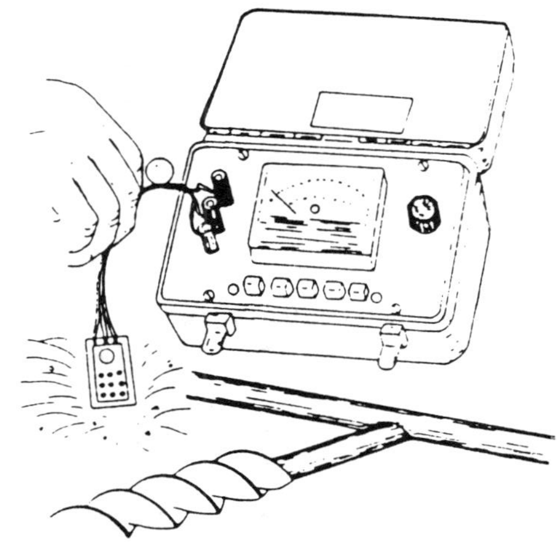

THERMISTOR SOIL CELL, METER, AND HAND AUGER. CELL IS BEING PLACED IN SOIL CAVITY

FIGURE 10.6

VADOSE ZONE MONITORING DEVICES

REFERENCE: EVERETT ET AL. 1984

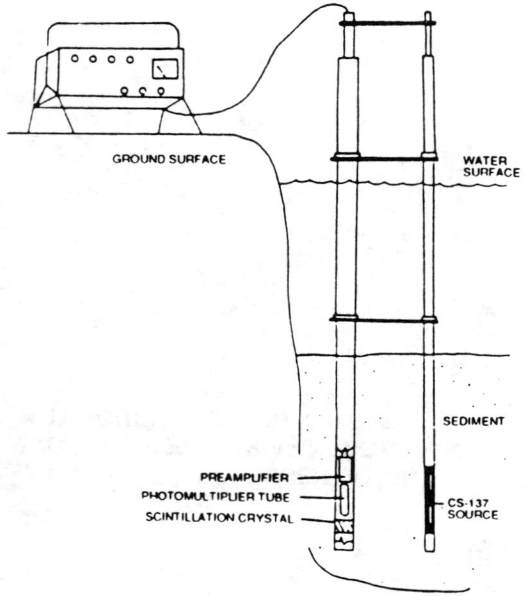

DUAL PROBE USED TO MEASURE WATER CONTENT BY GAMMA RAY ATTENUATION

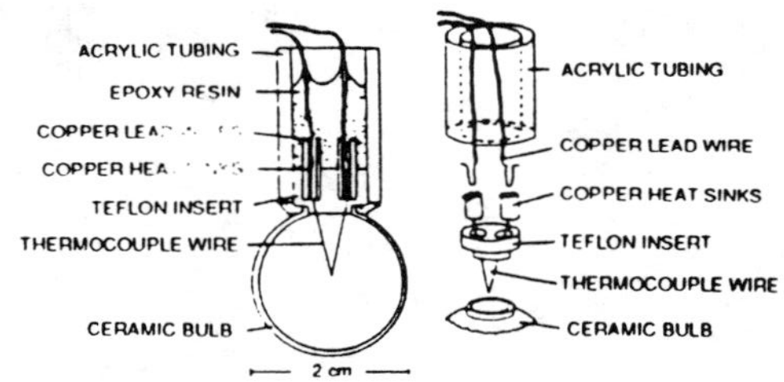

SOIL PSYCHROMETER

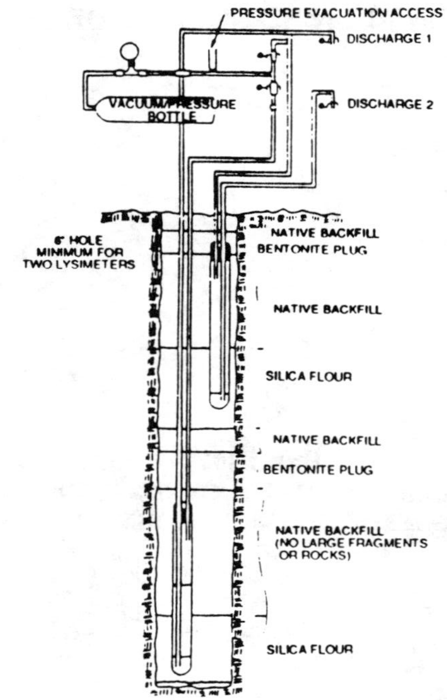

CLUSTERED SUCTION CUP LYSIMETERS IN A COMMON BOREHOLE

REFERENCE: EVERETT ET AL. 1984

FIGURE 10.6

VADOSE ZONE MONITORING DEVICES

Neutron moisture loggers can also be installed in existing wells. A tight fit between the borehole and well casing is necessary to minimize the vertical leakage of water and the subsequent side leakage of neutrons, which would lead to erroneous results.

When using the results of neutron loggers, it also must be recognized that the presence of a high concentration of fast neutron moderators, such as boron, chlorine, and hydrocarbons, may cause erroneous moisture readings.

Gamma Ray Attenuation

Gamma radiation methods can be used to determine bulk density and to estimate water content. There are two techniques of gamma ray attenuation: (1) the two-hole transmission technique, and (2) the scattering technique. The two-hole transmission method entails lowering a dual probe, one probe containing a gamma radiation source and the other containing a radiation detector, into two adjacent boreholes. Parallel holes are drilled to the desired depth and completed with 2-inch aluminum access tubes as shown in Figure 10.6. The count rate is measured through a standard material such as magnesium or aluminum. The transmission method is most suitable for determining bulk density in shallow soils.

The scattering technique consists of lowering a single probe, containing both gamma source and detector separated by shielding, into a borehole to the desired depth. The count rate is measured, divided by the standard count rate, and then dry weight bulk density is calculated. The scattering method requires a higher strength source, but is less accurate than the transmission technique. However, because of the difficulty of installing precisely aligned access tubes for the transmission technique, the scattering technique is more commonly used in the vadose zone. It is particularly suited to rock conditions found at many mining sites. The depth limit is generally accepted to be 30 feet.

Gamma ray attenuation is a nondestructive, in situ measuring technique that is precise and accurate. However, it requires expensive equipment, and the radiation may pose a health hazard. It requires straight, vertical holes for accurate measurements. Instabilities in count rates can be encountered, and changes in the bulk density in swelling or shrinking soils may affect the calibration.

Electrical Resistance Blocks

Electrical resistance blocks consist of electrodes embedded in suitable porous material composed of gypsum, fiberglass, nylon cloth, and plaster of paris. These blocks are used to measure soil-water content or soil-water suction. The water content or negative pressure within the blocks responds to the water content or suction of the soil with which the blocks are in intimate contact. Before installation, these blocks are calibrated in the material from the site in which they are to be installed. Then, a hole slightly larger than the block is dug or drilled to the desired monitoring depth. The block is thoroughly saturated with water and placed in the hole. Loose soil is packed around it, and the wire leads are extended to the ground surface and attached to a resistivity meter. The block is allowed to sit overnight to equilibriate the moisture in the block with the moisture in the surrounding soil.

Electrical resistance blocks are best suited for a general study of soil-water relationships. Their major advantages are that they are inexpensive and can be calibrated for either water content or water suction. There is opportunity for installations both at depth and horizontally inward from a mine waste management unit's perimeter, with measurements taken remotely because the signal is electrical. However, the blocks are difficult to install in coarse soils and consolidated material, and they do not perform well in either coarse, sandy soils or in swelling or shrinking soils because they need intimate contact with the surrounding material. The gypsum blocks eventually dissolve, resulting in an unstable matrix for the sensors. Contact resistance between the porous body and the soil can restrict the exchange of water between the soil and the block. Further, certain types of blocks are insensitive to moisture changes in the wet range and are, therefore, commonly used in conjunction with tensiometers to monitor moisture changes at pressures less than -0.8 atmospheres.

Psychrometers

Psychrometers, more accurately referred to as thermocouple psychrometers (TCP), measure soil-water pressure under very dry conditions, when the tensiometer is ineffective due to air-entry problems. A soil psychrometer measures the relationship between soil moisture suction and the relative humidity of soil-air. Psychrometers consist of a porous bulb chamber to sample relative humidity in the soil, a sensitive thermocouple, a heat sink, a reference electrode, and associated electronic circuitry. There are two types of TCPs. One type is installed in access tubes and consists of psychrometers placed in porous cups at the base of tubing. The second type is called a "sealed-cup" psychrometer in which the thermocouple unit is permanently sealed into a porous enclosure.

Psychrometers have been used effectively at depths varying from a few inches to over 300 feet in a wide range of geologic settings. They are most useful in measuring the soil-water pressure or water potential in very dry soils. The ceramic material in which the thermocouple is embedded must be in equilibrium with the soil moisture. If it is not, the accuracy of the measurement is questionable. In order to facilitate the determination of soil moisture content from the psychrometer reading, characteristic curves must be developed for each soil type in which the units are to be installed.

Good contact between the cup and the surrounding soil may be difficult to achieve with devices installed in tight, well-indurated material and in coarse material. Fluctuations in temperature affect the measurements of the TCP and can result in errors. TCP units are poorly suited for use in acid media due to severe thermocouple wire corrosion problems. Also, access-tube instruments may be susceptible to attack by fungi or bacteria.

Pore Vapor Monitoring

During the past five years, thousands of underground tanks have had vadose monitoring systems installed, consisting of sensors to monitor pore vapors which would be affected by leaking fluids. Such systems only perform for volatile liquids and are most often used for solvents and lighter fractions of petroleum hydrocarbons, such as gasoline. A similar concept can be used to monitor for hydrogen cyanide (HCN) gas in the vadose zone at some mine waste management units to provide early warning of cyanide solution seepage. Such a system has recently been installed beneath a new heap leach facility at the Mesquite Mine in Imperial County, California (Environmental Solutions, Inc., 1990). Verification of the concept's viability for the Mesquite site (e.g., deep ground water) conditions was determined by a combination of field tests conducted by Chatwin (1989) and Gold Fields Mining Corporation, and laboratory simulation tests conducted by Environmental Solutions, Inc.

Although this system is being used for an operating heap leach facility, it could also have application for waste management units such as tailings facilities. The Mesquite system consists of two fans of vadose zone wells, each fan consisting of three wells installed at angles of approximately 10, 20, and 75 degrees from horizontal. Each well has a sealed, perforated collection zone beneath the leach pad's liner system. If seepage of cyanide solution were to occur, its pH would be reduced by buffering in the vadose zone. This would result in the formation of HCN gas at the vadose well collection zone. Gas withdrawn from the wells will be monitored periodically for evidence of HCN gas.

These vadose zone wells also have open ends so that soil samples could be obtained if indications of seepage were observed. In this application, these wells would also serve as direct methods of monitoring.

Auger Sampling and Laboratory Analyses

Tools used for drilling through and sampling of soils include mechanical hand augers and continuous flight power augers. Power augers consist of solid-stem, hollow-stem, and cable-tool drill rigs. Because of the need to collect undisturbed samples, drilling fluids or muds cannot be used. Samplers which can be run downhole to sample at desired depths include split spoon samplers, tube type samplers (California spoons, Shelby tubes, etc.), continuous samplers, and core barrels.

Testing conducted on the samples should include determination of soil moisture. If estimates of seepage migration rates are required, then testing should also include the soil moisture content-hydraulic conductivity relationships. Batch or column leach laboratory testing would be required to determine the chemistry of the seepage.

If the vadose zone consists of fractured rock or a fractured consolidated clay deposit, then it is unlikely that sampling will provide much useful information because of the very low porosity and hence, low availability, of pore water. In all likelihood, it would be impossible to obtain completely undisturbed samples of material. Only in cases where the rock mass is highly fractured and behaves more like a continuous porous medium would there be some merit in contemplating a sampling program. Even then, it often would be difficult to assure that the samples were actually in the fracture zone with seepage flow.

Lysimeters

Wells and open cavities cannot be used to collect pore water or solution flowing in the vadose zone under negative pressure. Therefore, suction samplers are used and they are currently the most widely used field tool for sampling the vadose zone. There are three basic types of suction samplers: ceramic-type (or suction lysimeters), hollow fiber filters, and membrane filters.

Ceramic-type samplers (Figure 10.6) are constructed from ceramic materials and use the same type of cup as tensiometers. These samplers are placed in the soil with the ceramic cups in contact with the pore space of the soil. After the water content of the soil and cup are equilibrated, a vacuum can be applied to the interior of the cup, and a water sample can be pumped to the surface.

Suction cups can be divided into three categories: (1) vacuum-operated soil-water sampler, (2) vacuum-pressured sampler, and (3) vacuum-pressured sampler with check valves (or high pressure vacuum). Vacuum-operated soil-water samplers consist of a ceramic cup attached to the end of a small diameter polyvinyl chloride (PVC) tube (similar to a tensiometer). The PVC tube extends above the ground surface and is capped with a rubber stopper and outlet tubing. A vacuum is applied to the system by a hand pump, and soil-water is pumped into the ceramic cup and then sucked into a collection flask via a small diameter plastic tube inserted into the outlet tubing. Depth limitations of this type of unit are approximately 6 feet below the ground surface.

Vacuum-pressured lysimeters, on the other hand, can extract samples from depths to approximately 25 feet. The design and procedures for these are similar to those described above, except that two copper lines are utilized, and a vacuum-pressure pump supplies air pressure to the system to collect the sample. One disadvantage of this system is that some of the water or solution in the ceramic cup may be forced back into the surrounding media when air pressure is applied.

The high-pressure vacuum lysimeter was designed to eliminate the backflow problem. It consists of upper and lower chambers, isolated except for a connecting tube with a one-way check valve. A sample delivery tube extends from the base of the upper chamber to the surface; it also contains a check valve. A second, shorter, tube is used to deliver vacuum or pressure using nitrogen gas. When a vacuum is applied to the system, it extends to the suction cup through the open one-way check valve; the second check valve is closed at this time. The sample is delivered to the upper chamber which has approximately a 1-quart capacity. After filling the chamber, the vacuum is released, and pressure is applied to the shorter tube. The one-way valve to the cup is closed, and the one-way valve in the delivery tube is opened, thereby forcing a soil-water sample to the surface.

Single lysimeters can be installed in a cavity or clustered with two or more units in a common borehole. There are several factors limiting the operation of suction cup samplers. A good contact between suction cup and soils may be difficult to achieve in coarse-grained soils. Also, solids, bacteria, and chemical precipitates may clog the suction cup. Some of these problems can be alleviated by backfilling the area around the suction cup with silica flour. Further, cups tend to selectively sample from small pores and may not be representative of "average" fluid conditions.

Cellulose-Acetate Hollow Fiber Samples

These operate on a principal similar to the ceramic cup samplers with the exception that a cellulose-acetate hollow fiber is used in place of the cup. These fibers function as molecular sieves. According to Levin and Jackson (1977), the hollow fibers are variable extraction devices, but their application is more suited to laboratory studies.

Membrane Filter Samples

These consist of polycarbonate or cellulose-acetate membrane filters contained in a filter holder to permit applying a vacuum to the system and for delivery of the sample to a bottle. The sampler is placed in a hole together with a glass filter collection sheet. In operation, soil water is drawn into the collector system by capillaries. Subsequently, water flows into the collector sheets toward the glass fiber wicks as a result of the suction applied to the filter holder assembly. The glass prefilter minimized clogging of the membrane filter by the material in the soil liquids. The basic sampling limit can be used to depths of 12 feet.

Seepage Collection Trench

Another concept of vadose zone monitoring is to install a seepage collection trench to capture and sample potential leakage under a small, representative area of a liner. The philosophy behind this technique is that the trench sample leakage from a small area of liner can be extrapolated to the rest of the liner system to determine a total leakage rate. A disadvantage to this method is that it is potentially subject to a high sampling error unless several such collection trenches are constructed.

This type of system has been installed at the American Girl Mine in Imperial County, California. It consists of a 20- x 100-foot lined and backfilled, U-shaped ditch located directly under the primary synthetic liner forming the heap leach pad. It is backfilled with compacted granular native material and includes perforated collection pipes to facilitate collection of seepage for sampling.

10.3.2.4 Non-Point Measurement Techniques

Several innovative techniques have been installed, or are being proposed, for vadose zone monitoring at mine waste management units. What these methods have in common is that they monitor bulk properties of the vadose zone. Therefore, these methods are most suitable for detecting "significant" seepage releases, i.e., those that would cause measurable soil wetting over large areas.

Resistivity Method

An indirect measure that has been used to characterize soil salinity and delineate shallow pollution plumes is electrical resistivity, or its inverse property, conductivity. Most soil minerals are effective insulators (Rhoads and Halvorson, 1977), so electrical conductivity in saline soil is provided by pore water containing dissolved electrolytes. These methods may, therefore, have application to mine wastes where a highly saline waste is involved.

As outlined in Rhoads and Halvorson (1977), the different resistivity techniques that could be used include:

- Four Electrode Method - This method works well in uniform soil and is effective to depths of 5 to 10 feet. Its accuracy is reduced in layered soil strata.
- Electrical Conductivity (EC)-Probe Method - This method consists of a vertical probe that detects changes in soil salinity in discrete depth intervals. The probe is inserted into soil at the bottom of an augured drill hole. It must be calibrated for the different soil types it encounters. It does not work well in rocky, consolidated, or fractured materials.
- Earth Resistivity Method - This method indirectly measures the earth's resistance to an induced current. When used under the proper circumstances, it can be an effective monitoring tool. It has limitations in variable geologic, hydrologic, chemical, and physical settings.

Time Domain Reflectrometry (TDR)

Time domain reflectrometry (TDR) is a method of measuring the dialectric constant (K). Researchers have found K to be strongly dependent upon soil moisture content and weakly dependent upon soil type, density, and salinity when measured in the MHz-GHz range. The TDR method involves propagation of a pulse or electromagnetic wave along a transmission line and measurement of its velocity (vp). Pulse generation and reflection time measurements are made with an off-the-shelf electronic cable testing device or oscilloscope which is coupled to a soil probe via a coaxial cable. The vp can then be calculated and related to soil moisture content.

The TDR method lends itself to a low soil moisture situation; sensitivity decreases as soil moisture increases. This is in contrast to tensiometers, which obtain accurate soil moisture pressures when capillary pressures are greater than -0.8 bars, a relatively moist soil. Furthermore, TDR is not sensitive to factors which influence neutron moisture loggers, such as high soil organic content or the presence of boron or chlorine.

Time Domain Electromagnetic Method

At the Shumake heap leach facility near Mojave, California, a geophysical method known as the time domain electromagnetic method (TDEM) has been installed and is being tried for vadose zone monitoring. The method consists of laying out a square transmitter grid over an area of ground, and pulsing currents on and off over time. A receiver in the middle of the loop measures electromotive forces (EMF) due to induced currents in the ground. The resulting EMF-versus-time curve is then inverted to yield a geoelectric section. The section indicates the resistivity and thickness of horizontally-layered zones. Sequential sets of these data can then be used to estimate whether the subsurface is increasing in moisture content, as characterized by lower electrical resistivity, at certain depth intervals.

The benefit of TDEM is that it requires no disturbance of the ground surface and so can be used in existing facilities. Based on results obtained from a test water flood, it has been shown that saturation of a soil volume 100 feet square and 30 feet deep is readily detected and that moisture changes exceeding 5 percent (by weight) over the same area can be detected. Moreover, leakage of heap leach solution will result in greater decreases in resistivity because of the low conductivity of the leach solution due to elevated dissolved solids levels. This method will not detect small point leaks.

10.3.3 GROUND WATER MONITORING

In selecting and evaluating ground water monitoring systems, it is necessary to distinguish between detection monitoring, which is aimed at identifying whether or not leakage is occurring; and evaluation monitoring, which is aimed at determining the source of the leak, the extent of contamination, and the rate of movement of the contamination.

Different types of monitoring systems are usually required to meet the above two objectives. Detection monitoring generally requires sampling at only one location on each potential seepage pathway. Frequently, one monitoring well can be installed to sample from several potential seepage pathways. Evaluation monitoring, on the other hand, usually requires sampling from individual flow paths and also at several locations along each flow path to facilitate evaluation of contaminant migration rates. Evaluation monitoring systems do not have to operate for as long a period as the detection monitoring systems.

Ground water monitoring systems include the following types of installations:

- Drainage systems.
- Cased, single-screened monitoring wells.
- Cased, multiple-screened monitoring wells.
- Clustered monitoring wells.
- Single wells with multiple sampling points.

A brief description of each type of installation, with advantages and disadvantages, is presented in the following subsections.

10.3.3.1 Drainage Systems

Drainage systems include gravel-filled trenches and buried, perforated pipelines. They can be used for collecting water from shallow perched zones or from shallow water tables underlying waste management units.

The advantage of a drainage system is that they provide for effective detection monitoring and are relatively easy to sample from. However, it is very seldom that the ground water levels underlying a waste management unit is shallow enough to allow this type of system to be installed. There is one instance at the Jamestown Mine in California (Appendix F) where a shallow ground water table exists under a tailings impoundment. Gravel drains have been installed to allow passage of ground water flow without causing a hydraulic head buildup under the liner. These drains also serve as an early detection monitoring system.

10.3.3.2 Single-Screened Monitoring Wells

The use of single-screened monitoring wells is the most common method of monitoring ground water. They can be used to obtain both water level measurements and water quality samples.

Monitoring wells typically include the following features:

- A hole that is drilled and subsequently cased.
- A single screened interval in the casing. The location and extent of the screen corresponds to the vertical zone of ground water to be monitored.
- An annular seal above the screened interval installed to prevent inflow of water from higher ground water zones, or from the surface.
- A surface concrete seal and cap to protect the exposed portion of the well.

In practice, internal well casing diameters of between 2 and 6 inches are used and can be fitted with either dedicated or portable[1] well pumps. The selection of the appropriate diameter and pumps depends on the combined installation and operating costs, and operator preferences. An additional consideration when selecting a well diameter is the ability to develop[2] the well. For example, if wells are relatively deep and need to be drilled with mud, a larger diameter well should be considered. The larger diameter will enhance the efficient movement of water through the screen and also enhance the removal of mud and cuttings from within the casing during development.

Two-inch wells are the most economical to install. However, they can only be fitted with relatively low-capacity positive displacement (PDP) or bladder pumps. They are, therefore, generally only used when fairly long purging and water sampling cycles can be accommodated. Because of the relatively low cost of these types of pumps, they are frequently dedicated to each well.

Four-inch wells can be sampled using either larger PDP or bladder pumps, or dedicated or portable electric submersible pumps. These wells generally can be sampled from greater depths and, where required, the purging sampling cycle can be considerably shorter than for the 2-inch wells. A well diameter greater than 4 inches would only be used in special cases where there was either a need to pump large volumes of water or where the monitoring was combined with a water supply function requiring a larger diameter well.

Typical well completion concepts are shown in Figure 10.7(A) (Campbell and Lehr, 1973). The screen and casing material should be nonreactive to the monitoring constituents, i.e., PVC, is the most widely used, however, stainless steel screen is also used, and it is more durable than PVC although its cost is considerably greater. Both materials are nonreactive to virtually all the constituents of interest for monitoring mine waste management units. Stainless steel screen may be connected to a low-carbon steel-type casing to reduce the materials cost. This is appropriate as long as the sampling method will remove the stagnant water in the casing and replace it several times over with water from the formation, before a sample is collected. If the geologic formations remain stable after drilling, the completion interval can be made an open hole.

The length of the screened interval is usually set at a minimum of 10 to 20 feet. It must be long enough to ensure that at least 10 feet remain in the ground water, even during extreme

---

(1) Dedicated pumps are permanently installed in a well. Portable pumps are lowered into the well only for purposes of sampling.

(2) Well development includes surging and pumping the well immediately after it is constructed in order to remove all fine-grained material that may cause clogging.

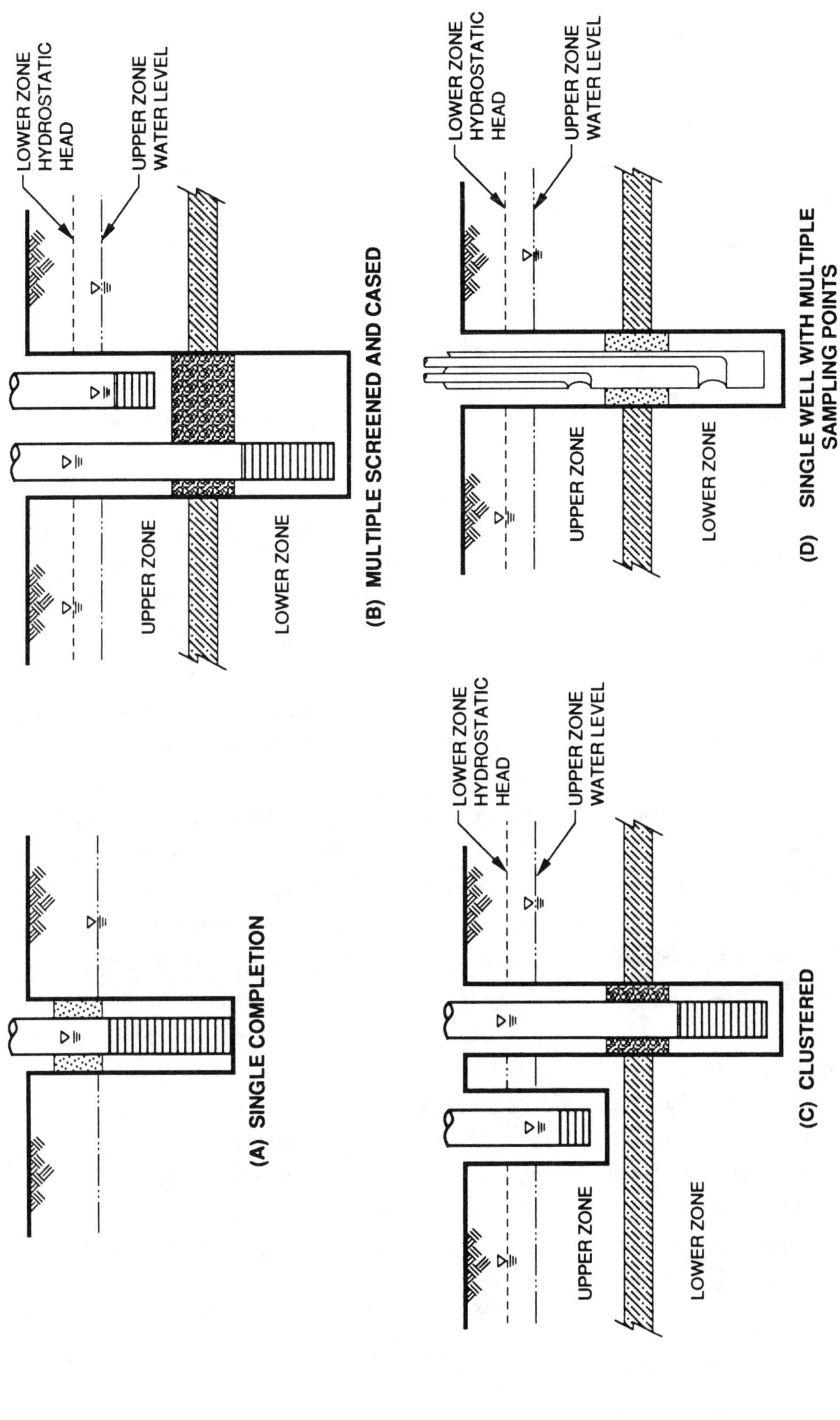

FIGURE 10.7

GROUND WATER MONITORING DEVICES

seasonally low water levels. Where a single completion is to collect and monitor flows from several separate vertical flow zones, the length of the screened interval must obviously be long enough to cover all the zones. An extremely important consideration when installing single completion wells is to make sure that they do not provide conduits for waste constituents to flow into an aquifer that may otherwise be isolated from any seepage sources. Care must also be taken to ensure that chemical constituents of concern are still detectable after being diluted by sampling from multiple seepage pathways and taking into account that only one pathway may contain elevated constituent concentrations.

PVC well casing or screen lengths may be joined by solvent welding of the sections with a slip coupling, a bell and spigot arrangement, or threaded adapters. Solvents in PVC casing are generally of no concern at mining sites as few, if any, organic chemicals are ever of interest.

Sealing materials for monitoring wells include cement grout, bentonite grout, and bentonite pellets or chips. It is preferable to install the bentonite in slurry rather than pellet form, because slurry allows a more continuous and, therefore, complete seal to develop. When using a cement grout seal, a small percent (usually 3 to 5 percent) of powdered bentonite is added to the grout in order to reduce shrinkage when the mix cures. For the same reason, the cement to water ratio is also minimized. In fractured material, cement may enter the completion interval if the gravel pack is not well above the top of the screen. This phenomenon can cause elevated pH in the water samples, so its potential must be recognized when designing and constructing the well.

The single-screened well has one inherent limitation and that is its inability to determine the precise entry point of a waste constituent into the well. In some cases, this may result in excessive dilution and, hence, nondetection of a constituent (Morrison, 1983). However, its main advantages, i.e., straightforward construction procedures, and the ability to detect waste constituents over relatively large depths, result in its widespread and continued use, particularly where detection monitoring is the primary goal.

10.3.3.3 Multiple-Screened and Cased Monitoring Wells

Figure 10.7(B) shows a multiple-screened and cased monitoring well. This type of installation consists of at least two separate casing strings in the same borehole, screened at different vertical intervals and isolated from each other by an impermeable annular sealant. This type of well is used for isolating vertical sections of the ground water zone for constituent detection purposes, or for isolating different ground water zones which would otherwise communicate through the annular space in the absence of a proper seal.

The advantage of this installation is the ability to detect constituents in potentially low concentrations by limiting dilution effects. However, there are also two distinct disadvantages. First, it is extremely difficult to isolate the various screen sections to ensure the representativeness of each well sample at a particular depth. Second, the logistics of drilling the well are complex. A large borehole is required to accommodate the two casing strings, and additional time is required to allow the intermediate seals to set. These costs should be considered when comparing them to those for an equal number of separately drilled wells (see Section 10.3.3.4, Clustered Monitoring Wells).

This type of well is infrequently used, as the risk of cross-flow between different screened intervals is usually too high. Where multiple-screened intervals are required, clustered monitoring wells or single multiport wells are usually used instead.

10.3.3.4 Clustered Monitoring Wells

Clustered monitoring wells (Figure 10.7[C]) serve the same purpose as the multiple-screened and cased wells, but are installed in separate borings at various depths; each as a single-screened well. This installation offers added protection against water mixing between different depth intervals. The main disadvantage is cost, particularly if ground water is deep and/or drilling conditions are difficult.

Clustered wells are usually required in cases where there is more than one aquifer underlying the waste management facility and there is a need for sampling each separately.

10.3.3.5 Single Wells With Multiple Sampling Points

The final option for installation of monitoring wells for sensing different vertical portions of an aquifer(s) is a single well with multiple sampling points. This differs from the other methods in that separate, smaller sampling casings, or tubes, are incorporated in one well casing. Each tube has its own sampling port set at a specific depth. Figure 10.7(D) illustrates the well construction.

The well completion hardware for this type of well system is relatively complex. Some sophisticated systems, such as the one by Westbay Instruments (1990), have been developed.

Individual valved sampling ports in this system are accessed by probes containing sample bottles under vacuum, which are lowered to connect to and open a single valve. The vacuum bottles then retrieve the sample from the specific depth interval (port) of interest by directly drawing water from the geologic formation. Valved ports are isolated by conventional annular sealing grouts or by packers.

Generally, this method will be more costly than clustered wells if relatively few intervals are monitored. However, if a larger number of monitoring intervals are required, this method becomes cost-effective. This technology is usually reserved for instances where very detailed three-dimensional information on ground water concentration distribution is required.

10.3.4 CHEMICAL CONSTITUENTS FOR MONITORING

10.3.4.1 General

When selecting lists of parameters to be sampled and analyzed, it is useful to consider them in two categories. The first, a comprehensive list, includes a complete suite of cations, anions, and metals. Sampling for these constituents would initially occur during the pre-project period to adequately define background and baseline conditions and to develop an understanding of the hydrogeochemical regime of the site. During the operations phase of the waste management unit, it is necessary to sample this comprehensive list of constituents at only a few key locations and on an infrequent basis.

The second category is a short list of leak detection constituents that provide the most reliable indication of any leakage from the waste management unit. Selection of this shorter list of constituents is discussed further in the following subsection.

10.3.4.2 Selection of Constituents for Leak Detection

The following is a list of the basic requirements for the inclusion of constituents in a ground water quality monitoring program:

- It should always be present in the waste stream that is generated.
- It should be present in the waste at a sufficiently high concentration to cause a detectable impact on the receiving water if any seepage occurs.
- The potential impact on the receiving water should be greater than the known or anticipated variation in the background concentration of that constituent.

- The constituent should not be subject to significant secondary reactions or transformations within the unsaturated zone (e.g., attenuation).
- It should not be generated by leaching from natural materials so as to provide a secondary source.
- The resultant concentration level in the ground water should be significantly above the limit of analytical detection.
- It should be soluble in the waste liquid and in the ground water, and consequently be hydrogeochemically mobile.

Constituents which, although site-specific, do not fulfill the above criteria are often of little use as monitoring parameters and may create problems with data interpretation. A discussion of the application of these criteria to waste management at a typical gold mine follows.

For a typical gold mine tailings waste, the following constituents would be present at levels which are detectable. Total dissolved solids (TDS), conductivity, alkalinity, bicarbonate, ammonia, nitrate, cyanide, calcium, sodium, chloride and sulfate. In addition, such species as fluoride, copper, iron, manganese, and zinc may also be present.

Depending on the nature of the ground water, some of the above constituents may not be suitable as monitoring parameters, as they will result in an insufficient contrast with baseline ground water quality, and hence will not be readily detectable if seepage reaches the ground water. For example, if the seepage has an alkalinity of 350 mg/L expressed as calcium carbonate ($CaCO_3$) and the ground water has an average alkalinity of 250 mg/L with a range of 230 to 290 mg/L, a very significant proportion of seepage mixing with the ground water would be necessary to elevate the alkalinity to a level where contrast over the baseline were confirmable.

Some constituents are subject to secondary reactions within the unsaturated zone below a facility, which may attenuate their concentration. This secondary reaction can remove the species from the seepage and, in some cases, release other species which are unrelated to the source itself. Both processes make it difficult to determine the impact of the transport of the constituents on the receiving water. In general, hydrogeochemically nonconservative constituents such as iron, manganese, and copper make poor monitoring parameters.

It is not uncommon to find some constituents that can be leached from geologic materials underlying the waste management unit in response to natural changes in the geochemical controls. For example, a decrease in pH or redox potential can cause an increase in manganese concentrations in seepage through natural systems. Hence, erratic concentrations of manganese can appear in monitoring wells, unrelated to manganese sources, but a function of the variation in the subsite hydrogeochemical regime. Therefore, redox and pH-sensitive constituents are not usually effective monitoring parameters. These include copper, nickel, iron, and manganese.

A further aspect is that the parameters selected for a monitoring program should be soluble, both in the seepage (source) and in the ground water. This is not generally an issue, as most major cations and anions have high solubility products, which is also why they are major species in the ground water. However, the use of some minor species as monitoring parameters is questionable if they have low solubility products and are subject to strong solubility controls.

Applied to waste management units, the above discussion on the criteria for monitoring parameters makes a strong argument for the use of a select range of hydrogeochemically mobile constituents for assessing the potential impacts of seepage.

In conclusion, a short list of constituents for use in leak detection monitoring should be selected as follows:

- Any unique constituent or constituents that are added to the waste stream during processing and which are relatively mobile, geochemically stable and not present in ground water. Cyanide in the gold mining industry is a good example of this type of constituent.
- One to three major cations or anions that occur in elevated concentrations in the waste and are known to behave reasonably conservatively in the environment. Sulfate is usually a suitable candidate.
- One to three metals that are elevated and mobile in the waste and are known not to attenuate significantly in the environment.
- TDS if the waste is highly saline.
- pH if the waste is highly acidic or alkaline.

10.3.4.3 Monitoring Constituent Concentration Limits

An important component of an effective monitoring program design is the selection of realistic concentration limits which are used to determine whether a seepage impact has occurred. Erroneous and nonsignificant data can be produced by setting arbitrarily low ground water standards close to the detection limits and then using these standards to designate the "cutoff" between background or baseline and seepage impact causing "false positive" readings.

The above problem has been recognized for some time by the U.S. Environmental Protection Agency (EPA) and some states when setting limits for National Pollution Discharge Elimination

System (NPDES) permits. Successive analytical method documents by EPA provide values for analytical accuracy at detection limits for constituents, where analytical tests with a number of participating laboratories have been performed. It is well known, however, that accuracy is very low for values approaching detection limits, even though the nominal precision within any particular laboratory is considered reasonable.

The result has been the acceptance by the EPA, and the use by several states, of practical qualification limits or practical concentration limits (PQLs or PCLs). These limits are set so that there is some reasonable assurance that the value established to determine compliance, for example, is set where its exceedance can be described to the impact of seepage of waste constituents.

An example of the derivation of PQLs, as taken from a recently issued NPDES permit for a gold mining project, is provided below. It shows, for a range of metal species, that the PQL is realistically many times the detection limit of the analytical method. The range of the ratios of PQL to detection limit varied from 2.5 to 20.

Constituent	Detection Limit (mg/L)	PQL (mg/L)	Ratio PQL/ Detection Limit
Cadmium	0.0051	0.03	6.0
Mercury	0.0002	0.005	2.5
Lead	0.01	0.075	7.5
Zinc	0.005	0.1	20.0
Copper	0.02	0.1	5.0

The above discussion is also applicable to cyanide, which is usually considered one of the key monitoring parameters for a gold mine waste facility. Analytical detection levels for the three cyanide "groups" which include free, weak acid dissociable (WAD), and total, vary from 0.005 mg/L to 0.01 mg/L, depending on the method, extent and type of cyanide metal complexation, and the laboratory. As has been noted in many publications, for example, the sessions on analytical methods for the Salt Lake City (1982) and the Tucson (1984) cyanide conferences, accuracy and reproducibility at the detection level are not good. Adopting EPA's PQL approach and using the above PQL/detection limits ratios for metals,[3] a reasonable range of PQLs for cyanide would be from about 0.026 mg/L to 0.2 mg/L, presuming that an 0.01 mg/L

(3) Metals analyses are generally more accurate than cyanide analyses, and the PQLs established are therefore conservatively low.

detection limit is reasonable. (It is unlikely that the lower detection limit of 0.005 mg/L is at all reasonable for cyanide.)

A good illustration of the cyanide concentration limit problem is in recent data from a monitoring program at an operating gold mine. Cyanide levels in one well were 0.01 mg/L total cyanide and 0.02 mg/L WAD cyanide, but there were no other constituents present which could be due to seepage from the facility. There was 100 percent variation in these numbers, and the total cyanide, which should be the larger of the two, was only one-half the level of the WAD cyanide. Clearly, it is questionable at this level whether either value was correct, whether cyanide was actually present at all, or whether there was any waste seepage component in the ground water. If a PQL approach were taken, these low concentration values would not be considered.

10.3.5 FREQUENCY OF MONITORING

Typically ground water quality monitoring programs have usually been established with monthly or quarterly sampling frequencies. In general, the net result has been an over-collection of data. Ground water monitoring frequency should be based on a rational approach which takes into account the rate, and variation in the rate of migration of waste constituents in the vadose zone and ground water system, and an evaluation of the impact of such migration on the receiving ground water.

In order to illustrate the above concepts, two examples are considered and are outlined in Tables 10.3 and 10.4, and are illustrated in Figures 10.8 and 10.9. Example 1 (Table 10.3) deals with a heap leach operation located in an alluvial environment with a ground water table 200 feet below the surface. The main area of concern is the protection of a water supply well located approximately 10,000 feet downgradient from the heap leach operation.

The calculations show that, for the assumed seepage rate, travel time through the vadose zone would be 20 years, and through the saturated zone to the water supply well 500 years. A sensitivity analysis could be performed assuming that the site permeabilities were underestimated by an order-of-magnitude. This would yield travel times of 2 and 50 years, respectively.

Based on the above assessment, it is clear that, as far as protecting water supply, monitoring wells around the leach pad coupled with annual or semiannual sampling would provide adequate early warning. However, should vadose zone and ground water remediation be required if excessive

TABLE 10.3

SOLUTE TRANSPORT TRAVEL TIME ASSESSMENT ALLUVIAL ENVIRONMENT

EXAMPLE 1

SETTING: HEAP LEACH OPERATION OVERLYING ALLUVIAL FORMATIONS AS SHOWN IN FIGURE 10.9

SEEPAGE MIGRATION:

(1) THROUGH THE VADOSE ZONE:

Monitoring System Design Leakage Rate	$= 10^{-6}$ cm/sec = 1.0 ft/year
Initial Soil Moisture Content (by volume)	= 10%
Soil Moisture Content (at leakage rate [by volume])	= 20%
Depth to Ground Water	= 200 feet
Travel Time in the Vadose Zone	$= \dfrac{200 \times (20\text{-}10)}{1 \times 100}$
	= 20 years

(2) IN THE GROUND WATER:

Saturated Hydraulic Conductivity	$= 10^{-4}$ cm/sec
	= 100 ft/year
Slope	= 5%
Porosity	= 25%
Solute Transport Velocity	$= \dfrac{100 \times 0.05}{0.30}$
	= 20 ft/year
Travel Time to Closest Water Supply Wells 10,000 feet Downstream	$= \dfrac{10{,}000}{20}$
	= 500 years

TABLE 10.4

SOLUTE TRANSPORT TRAVEL TIME ASSESSMENT FRACTURED ROCK ENVIRONMENT

EXAMPLE 2

SETTING: WET TAILINGS DISPOSAL FACILITY ON UNLINED FRACTURED BEDROCK AS SHOWN IN FIGURE 10.9.

SEEPAGE MIGRATION: Assume saturated flow conditions:

Saturated Hydraulic Conductivity	$= 10^{-6}$ cm/sec
	= 1.0 ft/year
Slope	= 10%
Porosity	$= 10^{-3}$
Solute Transport Velocity	$= \frac{1.0 \times 0.1}{0.001}$
	= 100 ft/year
Travel Time to Monitoring Wells Located 200 Feet from the Edge of the Waste	$= \frac{200 \times 12}{100}$
	= 24 months

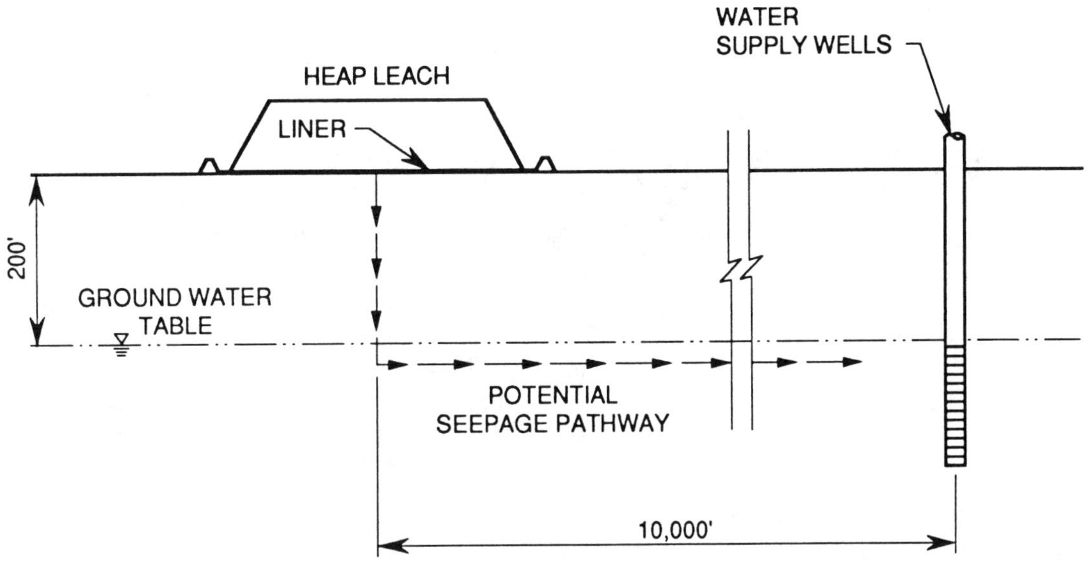

FIGURE 10.8

TYPICAL HEAP LEACH OPERATION IN AN ALLUVIAL ENVIRONMENT

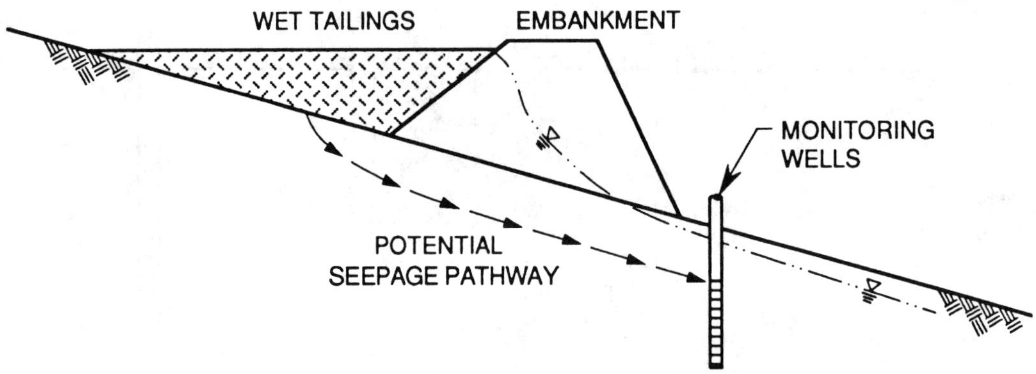

FIGURE 10.9

TAILINGS DISPOSAL FACILITY ON FRACTURED ROCK

seepage occurs, the operator should consider installing early detection monitoring devices in the vadose zone under the heap leach pad. Because of the potential for a relatively short travel time in the vadose zone, it might be appropriate to select a quarterly monitoring period for this system.

Example 2 deals with a wet tailings impoundment located on fractured rock. As shown in Table 10.4, the estimated travel time for any seepage to reach the monitoring wells is 24 months. As before, quarterly or semiannual monitoring should suffice.

The guidelines that can be used to establish monitoring frequency include:

- Establish the relative rate of solute transport in the geologic system.
- Perform simple sensitivity analyses to arrive at a range of solute transport rates.
- Determine the likely concentration impacts to the ground water.
- Establish the seasonal fluctuations in the ground water.
- Set the monitoring frequency to provide the necessary seepage detection response, taking into account the above factors.

Use of the above guidelines will result in a more focused and less voluminous monitoring program, while still providing the necessary detection sensitivity. It will also streamline the data review and regulatory process, as there will be less data, thereby providing the opportunity for greater level of review and evaluation.

10.4 TECHNICAL APPROACH TO THE DESIGN OF A GROUND WATER MONITORING SYSTEM

10.4.1 THE NEED FOR A GROUND WATER MONITORING SYSTEM

Ground water monitoring is required when the intrinsic properties of the waste pose a potential threat to the beneficial uses of the waters of the State. For these types of wastes, some form of containment, either natural, engineered, or a combination of the two, is relied on, and monitoring is required to confirm that this containment is adequate throughout the active life of the waste management unit and for a specified period after closure. The duration of this post-closure period should be sufficient, within practical limits, to allow detection of any contaminants that could potentially migrate out of the mine waste management unit at concentrations that are detectable and which could pose a potential threat to the beneficial uses of ground water. At mine waste management units where any potential leakage would have reached ground water and been detected by the monitoring system prior to closure, it may not be necessary to implement any post-closure monitoring.

The containment assessment approach outlined in Chapter 3 provides a useful tool for determining whether ground water monitoring is required. Evaluation of potential seepage travel times, can be used to assist in determining the duration of the post-closure monitoring period.

10.4.2 OBJECTIVES OF A GROUND WATER MONITORING SYSTEM

The primary objective of a monitoring system usually is to provide early detection of any leakage from a waste management unit. The detection should be early enough to avoid impairment of the beneficial uses of ground water and to prevent the need for any major remediation activities by the operator. The monitoring system need not provide sufficient data for definition of any leakage problems that may occur, as such information can be obtained by monitoring and sampling programs installed after seepage is detected.

Where a liner system is used for containment and there is some uncertainty associated with its performance, a secondary objective of monitoring should be established to verify that the waste unit liner system is generally functioning as designed. This objective could be met by in-waste monitoring as outlined in Section 10.3.1, by vadose zone monitoring close to the liner systems, or by a combination of the two.

10.4.3 BASELINE AND BACKGROUND CONSIDERATIONS

As a prelude to a discussion of monitoring system design, it is important to have a clear understanding of "baseline" and "background" conditions. Baseline constituent concentration values should be represented by a range which considers spatial variation (both areal and depth) in the potential receiving ground water system, as well as by temporal variation (e.g., the "seasonality" caused by ground water flow responses to recharge). Furthermore, the baseline monitoring system should be installed in the same geologic/lithologic and geohydrologic setting as the detection monitoring system, as these factors influence ground water quality.

Background sampling locations must also be chosen with care so as to incorporate, as far as possible, the effects of the natural geologic and geohydrologic setting of the system being monitored. However, it must be recognized that, due to local variability in geology and geochemistry, the variation in water quality in background monitoring wells may not always be the same as the variation in the downgradient detection monitoring wells.

A degree of judgement, therefore, is required to determine whether changes in water quality detected in the downgradient monitoring system are attributable to water quality changes upgradient of the waste management unit or to other conditions. Factors to be considered in such an evaluation involve a comparison of water quality between the upgradient wells and the downgradient detection wells, the timing of upgradient and downgradient changes in water chemistry, the magnitude of such change(s), and the concentrations of the constituents of interest in the waste. In some cases, it may be more informative to compare variations between the quality of water in the individual downgradient detection wells, than to make a generalized comparison between upgradient and downgradient water quality.

10.4.4 SELECTION AND DESIGN OF IN-WASTE MONITORING SYSTEMS

The design of an in-waste monitoring system is dependent on both site-specific and waste-specific conditions. The need for this type of system is predicated on the degree to which onsite waste management and the waste itself are being relied upon to provide containment, the amount of certainty associated with some key aspect of the design or management of the waste management units operation, and the amount of vadose zone or other early warning ground water monitoring that is provided. Monitoring of the hydrostatic head on the liner system represents an example of tracking a key containment design element, as well as a waste management requirement, i.e. for unsaturated waste deposition.

In developing a monitoring system design, the following aspects should be considered:

- Location and type of instruments to be installed within the mine waste in order to monitor changes in moisture content. Suitable equipment for tailings, for example, would include tensiometers or lysimeters. Spacing of the instruments should be such that there are at least several measuring points for each subarea of waste that is deposited in a uniform fashion.
- Location and type of instruments to be installed at the base of the waste in order to detect buildup of hydrostatic pressures. The types of instruments that could be considered include piezometers, or pressure transducers. Spacing considerations would be similar to those for the in-waste monitoring equipment.
- Location of instruments for monitoring flow and quality in the leachate collection and return systems.
- The need for waste sampling using undisturbed sampling methods and analysis instead of, or as a supplement to, the in-waste monitoring.

The frequency of in-waste monitoring is generally higher than for ground water or vadose zone monitoring. Continuous daily or weekly flow readings in the leachate collection systems may be necessary if accurate total drainage volumes are required. Monitoring within and immediately below the waste may be required on a weekly or monthly basis so as to continually evaluate the effects of the waste management practices.

10.4.5 SELECTION AND DESIGN OF A VADOSE ZONE MONITORING SYSTEM

Vadose zone monitoring can be used, provided it is technically and economically feasible. Instances where it is technically feasible include those in which the seepage requiring detection is relatively large and uniformly spread over a wide area, or in cases where spot checks on the performance of a liner system at selected locations, for example, are sufficient.

Examples of conditions where vadose zone monitoring may not be feasible or economic include the detection of minor quantities of seepage, point source leaks in a synthetic liner, and where the facility is underlain by fractured bedrock or otherwise extremely heterogeneous geologic conditions. In such instances, the non-point indirect methods (Table 10.1) may not be able to detect minor quantities of seepage. Monitoring for all potential leaks using the point sampling method (Table 10.1) could require thousands of monitoring points, as could monitoring in extremely heterogeneous geologic formations or fractured rock. Furthermore, it is unlikely that any instruments would operate successfully in fractured bedrock (see Section 10.3.2.1).

The detection instruments have limited useful life, which must be matched, to the extent possible, with the anticipated monitoring life of the facility. The instruments also must be able to operate over the anticipated range of soil moisture conditions under the waste management unit.

Aside from selection of detection instruments, the design of the monitoring system also must consider the extent to which a leak would impact the ground water zone being monitored and the potential consequences of contaminating the ground water. Those situations which represent a more significant risk to ground water could require more comprehensive monitoring systems and possibly incorporate redundancies to assure detection, even if some monitoring or sampling devices fail.

Indirect vadose zone monitoring methods (Table 10.1) only measure soil moisture characteristics and do not sample the water. Judgements on the releases of waste from a waste management unit and their impacts must be made purely on an indirect assessment of seepage quantity. Backup sampling and testing of vadose zone material may have to be included, therefore, as an integral part of the monitoring system. When selecting an indirect monitoring device, the following aspects must be addressed:

- Instrument error must be significantly less than the potential impacts of a seepage loss.
- The natural changes in moisture conditions must be understood or independently monitored.

As is evident from the discussion in Section 10.3.2, innovation in vadose zone monitoring is an ongoing need. Mining companies and regulators should encourage development, and not be hesitant to experiment with techniques and devices that show promise.

Where vadose zone monitoring is infeasible but nevertheless necessary, alternative approaches can be considered. These include, but are not necessarily limited to:

- Additional engineered containment which could reduce the requirement for vadose zone monitoring;
- In-waste monitoring to directly monitor the fluid conditions which could potentially lead to seepage losses;
- Vadose zone sampling at selected locations at key time periods during the active life of the mine waste management unit.

10.4.5.1 Background Monitoring

Background monitoring locations should be selected to be within the same soil type underlying the waste management unit. Monitoring locations need not be a great distance from the facility, as there is usually little potential for seepage to move laterally. The background locations should avoid areas where operational disturbances, such as water spraying for dust control, or excavation, could affect soil moisture.

Ideally, background measurements should define natural soil moisture conditions that would occur in the absence of waste disposal. However, the construction of containment facilities for mine waste disposal can alter vadose zone conditions, even if seepage losses do not occur. It may be necessary, therefore, that a control area be set aside where monitoring could occur under conditions representative of the disposal environment, but in the absence of waste liquids.

Monitoring devices in background areas should be installed at depths similar to those considered for detection monitoring in the disposal area. These depths should be sufficient to prevent diurnal temperature and moisture fluctuations from affecting the instrument readings. Enough devices should be installed to determine average changes in moisture conditions over specific areas.

10.4.5.2 Detection Monitoring

The selection of the location and spacing of detection monitoring devices under containment structures depends largely upon the desired goal in detecting "point" or "area-wide" seepage sources. If point sources are to be detected, the monitoring device should be located under those areas where the point sources are likely to occur. On the other hand, if the point sources are associated with randomly occurring flaws in the liner system, it usually will not be possible to install sufficient point type measurement devices. An alternative approach under these circumstances may be to provide monitoring at only a few representative locations to assure that there is no systematic or reoccurring flaw in the liner system.

Detection of area-wide seepage is generally easier to conduct. Where point measurement devices are used to accomplish this, a sufficient number must be provided to allow evaluation of realistic average changes in moisture conditions. This usually will require a minimum of three to five devices within the area that could be affected by seepage losses. When installing non-point monitoring devices, the key aspect is to ensure that they will be sensitive and accurate enough to reliably detect the moisture changes likely to be caused by the expected seepage loss.

Because of the large size of most mine waste management units, monitoring locations usually will be limited to the perimeter of the facility, especially if the facility is already installed at the time vadose zone monitoring is considered. New units offer an opportunity for installation of monitoring devices immediately below the facility, with remote measurement at the perimeter.

The vertical spacing of measurement devices is also an important consideration. Monitoring in more than one vertical plane can have distinct advantages. For example, the characteristic of the moisture profile at a given time can be used to distinguish whether an elevated moisture at a given depth is due to a continuous seepage source or the passage of a seasonal moisture front.

10.4.5.3 Frequency of Measurement

Vadose zone seepage rates are generally low, and so do not require frequent sampling. However, in order to evaluate seasonal changes, no less than quarterly monitoring should be undertaken.

10.4.6 SELECTION AND DESIGN OF A GROUND WATER MONITORING SYSTEM

As mentioned previously, the design of a ground water monitoring system should be based on the site characteristics and identification of the potential seepage flow pathways. The considerations which go into the design include location of background monitoring points, location of downgradient monitoring points, and whether one or more vertical intervals of the ground water system(s) need to be monitored. The number of monitoring points will increase as the complexity of the natural ground water flow pattern increases. The presence of different geologic structures, variable hydraulic gradients, or different material types within the zone of saturation could necessitate more intensive monitoring than within a homogeneous, uniform gradient ground water system.

The ground water monitoring system should meet the following goals:

- Include background areas within which ground water is not affected by potential seepage of waste constituents.
- Include the potential seepage pathways.
- Detect seepage releases at the earliest possible time.
- Ensure that screened intervals in wells are approximately sized to detect the potential seepage rates.

10.4.6.1 Background Monitoring Locations

Monitoring wells in areas expected to be unaffected by potential seepage flow are important for providing information on the temporal and spatial variations in water quality in the ground water system, against which operational data can be compared. Because of the large areas occupied by mine facilities, several background monitoring points may be necessary. Each geologic formation, which is expected to have unique ground water chemical characteristics, and which is likely to be affected by seepage, should be monitored.

10.4.6.2 Downgradient Monitoring Locations

The horizontal spacing of detection monitoring wells should be based on the need to intercept potential seepage pathways. Generally, these pathways occur as one of the following types: (1) gradually expanding waste constituent concentration plumes in relatively homogeneous, unconsolidated granular or consolidated uniformly fractured formations, (2) discrete pathways in zones of higher hydraulic conductivity, such as residual gravel beds or permeable fault or joint systems, (3) random pathways in interconnected fractures in rock formations, or (4) a combination of the above. Guidelines for locating wells in each of these pathways are discussed below.

The longitudinal and lateral extent of pathways in homogeneous formations can be reasonably well predicted using standard analytical techniques and knowing the extent of the source and the geohydrologic characteristics of the formation. Ideally, the locations of wells should be selected to intersect each potential plume. However, usually it will not always be practical to provide a separate well to cover every potential pathway. In such cases, several monitoring locations may be selected to monitor several typical and also worst-case or most likely seepage pathways. If these typical or worst-case locations were to show seepage, additional wells could be installed subsequently to investigate the total extent of the seepage. In some instances, it also may be appropriate to take into account the degree to which monitoring wells can "capture" surrounding ground water by increasing the volume of water pumped from the well prior to sampling.

Managing cases of higher hydraulic conductivity zones or fault structures is more involved. The site exploration activities must be sufficiently detailed and comprehensive to provide an understanding of the nature and variability of these zones. In unconsolidated formations, this consists of identifying sand or gravel lenses and linear features caused by the historic surface drainage and depositional events. For fractured, consolidated deposits, exploration activities include the type and number of joint sets, details on jointing in each set, and the variation in joint length and spacing underlying the waste management unit. This information can be obtained from a variety of methods, including geologic and joint mapping, core drilling, single well pump testing, and multiple observation well pump testing. Based on the information obtained, the likely extent of seepage migration should be determined and used to select appropriate monitoring well locations. It may be appropriate to stage the construction of the monitoring wells, with the initial wells being logged and pump-tested to provide information for installation of subsequent wells.

The third case mentioned above, i.e., the random seepage pathways, usually involves monitoring for small quantities of seepage in interconnected fracture systems in consolidated geologic formations. There are two basic approaches to dealing with this situation. The adopted approach is likely to depend on the consequences of any undetected seepage that may occur. Where such consequences would not result in significant and immediate threats to human health and the environment a more pragmatic approach should be adopted. This would involve selecting monitoring wells in areas which are judged to represent the more likely seepage pathways, i.e.:

- Formations close to areas of high hydraulic head in the waste management unit.
- Natural low topographic areas which could represent a preferential ground water flow pathway.
- Other areas that may be suspected of containing potential seepage pathways.

It may be appropriate also to install additional monitoring wells in the receiving water bodies immediately upgradient of the locations at which the water is being withdrawn for beneficial use. These wells would provide an additional and independent level of ground water monitoring.

The second approach which is applied in more critical situations, would involve the installation of a monitoring system that has the ability to withdraw water from a contiguous ground water zone that includes all potential seepage migration pathways. This requires installation of sufficient wells that when pumped for sampling develop waste intersecting cones of depression within this contiguous zone. Over time, contained pumping and sampling from such as system will assure that a very large proportion of the ground water containing potential seepage pathways will be sampled. Hydraulic calculations can be performed to develop initial well spacing and required pumping volumes for such a system. A pilot scale field installation and testing program will generally be required to adjust spacing to suit site-specific conditions. There will usually be economic tradeoffs to be made between providing more wells or increasing the time water is pumped from the wells.

10.5 REGULATORY IMPLICATIONS

10.5.1 FEDERAL

Federal regulatory authority over active mine waste management units generally falls under the EPA pursuant to the Clean Water Act and the Resource Conservation and Recovery Act (RCRA). The Comprehensive Environmental Resource Compensation and Liability Act (CERCLA) is applicable to uncontrolled releases of pollutants at waste disposal facilities, and the intent of monitoring under CERCLA is significantly different than detection monitoring. The Clean Water Act (33 U.S.C. 1342) regulates only "point source" discharges to surface waters, and not discharges that may affect only ground water, and is not discussed further. EPA currently is developing a regulatory program for most mine wastes under Subtitle D of RCRA, which is a nonhazardous solid waste program. See Chapter 3 for further details regarding RCRA.

Thus, there are no federal requirements that pertain specifically to ground water monitoring requirements at mine waste management units.

10.5.2 STATE OF CALIFORNIA

10.5.2.1 Background

The State of California recently has adopted revised monitoring requirements applicable to a wide range of waste management units including those for Group A and Group B mine wastes. The approach adopted in these regulations is generally similar to the technical guidelines developed in this chapter. However, they also contain several requirements that may not be consistent with the mine waste containment approach outlined in Chapter 3, or that may not be technically feasible at all mine waste management units. Key elements of these regulations are summarized in the following sections. A discussion of potential concerns with these regulations follows.

10.5.2.2 General

General monitoring requirements are outlined in California Code of Regulations (CCR), Title 23, Division 3, *Chapter 15*, Article 5. The primary goal of these regulations is to detect and locate any leakage from the waste management unit by monitoring the vadose zone (when feasible), surface waters, and ground water adjacent to a mine waste management unit.

Under the Article 5 regulations, the Regional Quality Control Board (Regional Board) establishes the following general requirements and definitions for new or existing mine waste management units:

- Points of compliance are defined as the points at which water quality protection standards are to be applied, and at which monitoring is to be conducted. A point of compliance should be located in the vertical surface located at the hydraulically downgradient limit of the waste management unit and should extend through the uppermost ground water body.
- A compliance period during which monitoring is required is defined as the active life of the waste management unit, plus the closure and a post-closure maintenance period, if necessary. If a discharger is engaged in a corrective action program after the closure period, a post-closure monitoring program is established to ensure that water quality standards have not been exceeded for a period of at least three consecutive years.

10.5.2.3 Monitoring Programs

In accordance with Article 5, dischargers are to conduct a three-phase monitoring program. Initially, a discharger must institute a detection monitoring program to be approved by the Regional Board. If there is "statistically significant evidence of a release," or "significant physical evidence of a release," then the discharger must initiate an evaluation monitoring program. Any remediation that is required also must incorporate a corrective action monitoring program.

10.5.2.4 Detection Monitoring

A discharger must perform vadose (unsaturated) zone, surface water and ground water, monitoring, where practical. The purpose of monitoring the vadose zone directly beneath and adjacent to the waste management unit is to detect releases at the earliest possible opportunity. Monitoring also is to be conducted at designated points of compliance, at additional specified points, and at background points.

A list of "constituents of concern" to which "water quality protection standards" apply must be developed. These include waste constituents, hazardous constituents, and reaction products that are expected to be in, or derived from, waste contained at the disposal facility.

"Water quality protection standards" must be established as a basis for comparing water quality during the facility's active life, closure period, and post-closure maintenance period. "Concentration limits" must be established for constituents of concern for each media, i.e., vadose zone, surface water, and ground water.

Dischargers must monitor for the list of "constituents of concern" periodically to provide adequate water quality monitoring data at a site. The Regional Board also shall specify a shorter list of "monitoring parameters" for more frequent routine sampling, considering the following factors:

- Type, quantity, and concentration of waste constituents.
- Mobility, stability, and persistence of waste constituents and/or their reaction products.
- Detectability of indicator parameters or waste constituents.
- Range of background concentrations of the proposed indicator parameters encountered at the waste management unit.

The discharger must determine the water quality at each monitoring point and background point at least quarterly, including the times of expected highest and lowest elevations of the ground water surface. The discharger also must determine the velocity and direction(s) of ground water flow at these times.

After each sampling period, the discharger must determine if there is a statistically significant increase over water quality protection standards at any monitoring point. If the discharger finds that there is such an increase, the Regional Board must be notified within seven days. The discharger may resample if there is any doubt as to the reliability of the data, and must submit a report to the Regional Board within 90 days. This report should include information on the concentration of all waste constituents in the vadose zone, surface water, and ground water at each monitoring point, and any proposed changes to the monitoring system and/or monitoring frequency, sampling, and analytical and statistical procedures used.

10.5.2.5 Verification or Evaluation Monitoring

If the monitoring parameters, compliance points, and/or compliance period used in detection monitoring are not adequate for evaluation monitoring, the Regional Board may recommend that the system be modified. The sampling procedures which were described under Detection Monitoring (Section 10.5.2.4) are also followed in Evaluation Monitoring. If the discharger or Regional Board determines that water quality protection standards are being exceeded at any monitoring point, the discharger shall notify the Regional Board within seven days, develop all data necessary to design a complete corrective action program and, within 180 days, submit an engineering report to the Regional Board requesting authorization to establish a corrective action program.

10.5.2.6 Corrective Action Monitoring

A discharger required to undertake corrective action to achieve compliance with water quality protection standards shall establish and implement a corrective action monitoring program to demonstrate the effectiveness of the corrective action program. Reports on the effectiveness of the program should be submitted at least semiannually to the Regional Board.

10.5.2.7 Regulatory Flexibility

It is evident from the discussion above that the Article 5 regulations were developed for a wide range of waste management units and consequently contain elements that may be necessary for waste management units requiring the highest degree of waste containment, i.e., units containing liquid wastes with very high concentrations of toxic constituents and that overlie potable ground water resources at shallow depths.

Application of these regulations, therefore, to specific mine waste management units should allow sufficient flexibility for modifications to be made to accommodate the unique properties of mine waste and the site-specific conditions associated with the particular mine waste management unit. These modifications would be dependent on the potential risk to beneficial uses of water posed by the waste and the technical feasibility of the prescriptive requirements contained in Article 5. The specific modifications that should be considered and which relate to the concerns discussed in Section 10.5.2.1 are as follows:

- The amount of monitoring and types of monitoring systems required should, in part, be determined by the potential threat to beneficial uses posed by the waste and the amount of containment that is provided. Waste management units which either pose a low risk or which incorporate several containment features, such as multiple liners and geologic formations with low hydraulic conductivity, should require less intensive monitoring than those that do not have these attributes. This will not only promote a more effective design by allowing the selection of the most appropriate combination of containment features and monitoring elements, but will also provide a positive incentive for the use of sites with lower potential risks.
- Allowance should be made to alter certain monitoring element requirements where they may be technically infeasible. For example, where vadose zone monitoring cannot be effectively accomplished at a large mine waste management unit because it overlies fractured bedrock, alternative approaches such as those outlined in Section 10.4.5 should be considered.
- The location of vadose zone monitoring systems that are installed should be consistent with the containment approach adopted and the monitoring sensors should be located at a depth or distance from the waste at which complete waste containment is achieved. For example, if the containment approach relies on the upper 3 feet of a naturally occurring clay bed to contain liquid wastes, then the vadose zone sensor should be located below this clay bed.

10.6 BIBLIOGRAPHY

Bumb, A. C., C. R. McKee, R. B. Evans, and L. A. Eccles. "Design of Lysimeter Leak Detector Networks for Surface Impoundments and Landfills." Ground Water Monitoring Review 8 (1988): 102-114.

Campbell, M. D., and J. H. Lehr. *Water Well Technology*. New York: McGraw-Hill. 1973.

Chatwin, T.D. *Cyanide Attenuation/Degradation in Soil*. Resource Recovery and Conservation Company. 1989.

CRC Standard Mathematical Tables. 25th Edition, West Palm Beach, Florida: CRC Press. 1978. 503-504.

Cyanide From Mineral Processing. Proceedings of a Workshop. The National Science Foundation, Salt Lake City, UT. February 2-3, 1982.

Cyanide in the Environment. Proceeding of a Conference. Fort Collins, CO: Colorado State University. Tucson, AZ. December 11-14, 1984.

Environmental Solutions, Inc. *Vadose Zone Monitoring Program Vista Leach Pad*. Prepared for California RWQCB - Colorado River Basin. Mesquite Gold Mine. February 6, 1990.

Everett, L. G., L. G. Wilson, and E. W. Hoylman. *Vadose Zone Monitoring for Hazardous Waste Sites*. Park Ridge, N.J.: Noyes Data Corporation. 1984.

Freeze, R. A., and J. A. Cherry. *Groundwater*. Englewood Cliffs, N. J.: Prentice-Hall. 1979.

Laliberte, Brooks and Corey. *Properties of Unsaturated Porous Media*. Hydrology Paper No. 17. Colorado State University. Fort Collins, Colorado. 1966.

Levin, M.J. and D.R. Jackson. "A Comparison of In-Situ Extractors for Sampling Soil Water." Soil Science Society of America Journal 41 (1977): 535-536.

McWhorter, D.B. and J.D. Nelson. Unsaturated Flow Beneath Tailings Impoundments. American Society of Civil Engineers. Geotechnical Division. November 1979.

Morrison, R. D. *Groundwater Monitoring Technology*. Prairie Du Sac, Wisconsin: Timco Manufacturing, Inc. 1983.

Mualem, Yechezkel. *A Catalogue of the Hydraulic Properties of Unsaturated Soils*. Research Project No. 442, Technion Research and Development Foundation, Ltd. Haifa: Technion-Israel Institute of Technology. 1976.

Resource and Conservation Recovery Act of 1976 (PL94-580, October 31, 1976; 42 U.S.C. 6901 et seq.) Subtitle D.

Rhoads, J.D. and A.D. Halvorson. Electrical Conductivity Methods for Detecting and Delineating Saline Seeps and Measuring Salinity in Northern Great Plains Soils. United States Agricultural Research Service, ARS W-42.

Stephens, D. B., and S. P. Neuman. *Analyses of Borehole Infiltration Tests Above the Water Table*. Technical Reports on Natural Resource Systems, No. 35. Tucson: University of Arizona. 1980.

Terzaghi, K., R. Peck. *Soil Mechanics in Engineering Practice*. 1948.

United States Environmental Protection Agency. *Background Documents on Proposed Liner and Leak Detection Rule*. May 1987.

United States Environmental Protection Agency. *RCRA Groundwater Monitoring Technical Enforcement Guidance Document (TEGD)*. OSWER 9950.1. 1986.

University of California at Berkeley. Mining Waste Study Team. *Mining Waste Study, Final Report*. California State Water Resources Control Board. 1988.

Westbay Instruments. MP System product literature, Vancouver, B. C., Canada. 1990.

CONTENTS

CHAPTER 11
ECONOMIC IMPACT EVALUATION

	PAGE NO.
LIST OF TABLES	572
11.1 Introduction	573
11.1.1 Objective	573
11.1.2 Integration With Other Chapters	573
11.1.3 Study Participants	574
11.1.4 Sources of Information	574
11.1.5 Chapter Organization	574
11.2 Regulatory Analysis	575
11.2.1 Background	575
11.2.2 Previously Proposed Regulatory Changes	582
11.3 Cost Analyses	583
11.3.1 Introduction	583
11.3.2 Regulatory Cost Sensitivity	583
11.3.3 Climatic Cost Sensitivity	587
11.4 Conclusions and Recommendations	587
11.5 Bibliography	589

CONTENTS

CHAPTER 11
(Continued)

LIST OF TABLES

TABLE NO.	TITLE	PAGE NO.
11.1	Summary of Some Key Regulatory Implications	576
11.2	Required Mine Waste Containment Under the Current Article 7 Regulations	579
11.3	Required Mine Waste Containment Under the Article 7 Revisions Proposed in June 1988	580
11.4	Required Mine Waste Containment Under the Article 7 Revisions Proposed in January 1989	581
11.5	Regulatory Cost Impact Evaluation - Precious Metal Group B Waste Pile in a Humid Environment Overlying Low Hydraulic Conductivity Bedrock: Total Present Value Costs	584
11.6	Regulatory Cost Impact Evaluation - Precious Metal Group B Tailings Impoundment in a Humid Environment Overlying Low Hydraulic Conductivity Bedrock: Total Present Value Costs	585
11.7	Regulatory Cost Impact Evaluation - Industrial Minerals Group B Waste Pile in a Humid Environment Overlying Low Hydraulic Conductivity Bedrock: Total Present Value Costs	586
11.8	Climatic Impact Evaluation - Precious Metal Group B Tailings Pile: Total Present Value Costs	588

CHAPTER 11

ECONOMIC IMPACT EVALUATION

11.1 INTRODUCTION

11.1.1 OBJECTIVE

As with any regulatory process, the financial impacts of mine waste regulation should be considered. Most water quality impacts cannot be reasonably measured in terms of economic cost alone. However, when an industry is to invest in water quality protection, it is appropriate that the investments are made in the most beneficial manner. As a focal point for discussion, this chapter addresses the costs associated with implementing proposed changes to Article 7 of the California regulations,[(1)] discussed between representatives of the State Water Resources Control Board (State Board) and the Regional Water Quality Control Boards (Regional Board), and representatives of the California Mining Association. When taken in context with the other chapters of this book (especially Chapter 2, Performance Review of Modern Mine Waste Management Units), these proposals can serve as a useful cost-benefit analysis. However, this chapter does not address the benefits of the proposals.

This chapter contains a financial impact evaluation of potential revisions to the design, construction, operating, and closure requirements contained in Article 7. Also discussed is the impact of climate on waste disposal costs. Using the results presented here, both regulators and operators will have some basis for judging how regulatory change affects the economics of mining projects.

The findings in Chapter 2 indicate that modern mining-related activities have had no significant impacts on beneficial uses. The results of the financial analyses presented in this chapter show that modifying some of the basic provisions of Article 7 could have severe cost implications. It is important, therefore, to give careful consideration to any proposed changes, and to complete a thorough analyses of all potential implications of such changes.

11.1.2 INTEGRATION WITH OTHER CHAPTERS

This chapter is different from the others in that it is not part of the discussion on the technology of mine waste management. It is essentially a separate evaluation dealing with the waste management cost implications of a range of technical and regulatory conditions. However, when using the

---

(1) California Code of Regulations, Title 23, Division 3, *Chapter 15*, Article 7.

information presented in the other chapters for the selection and design of waste management facilities, the reader is urged to consider the cost implications outlined in this chapter.

11.1.3 STUDY PARTICIPANTS

This study was conducted by the following individuals:

- Jeff Huggins, Vector Engineering, Inc.
- Ian P. G. Hutchison, Ph.D., P.E., Environmental Solutions, Inc.
- Todd Ramey, Vector Engineering, Inc.
- Mark E. Smith, P.E., Vector Engineering, Inc.

11.1.4 SOURCES OF INFORMATION

Two types of mine waste facilities are evaluated in this chapter: (1) mining tailings impoundments, and (2) mine waste piles. The cost analyses for these types of facilities have been prepared using both site-specific cost information and more general information based on the experience of the authors. The site-specific data were obtained from three California mines; two of which were operational and a third which was in the initial stages of permitting. Some of the data used for the mining tailings impoundment were derived from interviews with engineers and a site visit to the Jamestown Mine, operated by Sonora Gold at Jamestown in Tuolumne County, California. The Dicalite Mine, operated by Grefco Incorporated north of Burney in Shasta County, California, was chosen as the model for industrial minerals waste disposal. The precious metal waste pile of the proposed Spanish Mine operation at Washington in Nevada County, California, was picked as the model for precious metal waste disposal. These sites were chosen on the basis of several factors, including: (1) type of product, (2) size of operation, and (3) location of project. Where data and design criteria were determined to be significantly different from a "typical" operation, some substitutions were made.

11.1.5 CHAPTER ORGANIZATION

Section 11.2 provides a summary of the current technical waste containment and closure requirements and the changes that have been proposed by the State Board staff during the past few years. Cost analyses of the impacts of these changes on three typical waste management units are presented in Section 11.3, which also outlines typical cost changes for waste disposal under different climatic conditions. Conclusions are provided in Section 11.4. Data sources are listed in Section 11.5, Bibliography.

11.2 REGULATORY ANALYSIS

11.2.1 BACKGROUND

The State Board staff has contemplated a wide range of changes to Article 7 over the past few years. Many of these have been discussed with representatives of the mining industry, as well as with the consulting and academic professions. Two draft sets of modified regulations were prepared for comment in both June 1988, and January 1989. Although it is recognized that those two drafts do not necessarily reflect any final modifications that may be implemented, they provide a reasonable basis for discussing and analyzing the cost impact of potential regulatory changes.

Some of the key issues associated with these different sets of regulations are summarized in Table 11.1. The two significant aspects of the proposed revisions are: (1) the method by which waste is characterized, and (2) a requirement for an increased level of redundancy in the containment (liner) systems. The modifications to the waste characterization methods would considerably narrow the application of the Group C,(2) or its equivalent, category. More wastes, therefore, likely would fall within the Group B,(2) or equivalent, category. This factor alone represents the most significant cost to the industry since most current waste dumps are classified as Group C wastes.

Detailed descriptions of the prescriptive containment and closure requirements for each set of regulations are contained in Tables 11.2, 11.3, and 11.4. In developing these tables, the current waste classification system (i.e., A, B, and C categories)(2) was used to identify typical waste facilities on a consistent basis, even though the proposed regulations would no longer incorporate these categories. For the A and B categories, both high and low risk site environments were considered. High risk generally means that the geologic formations are permeable, ground and/or surface water resources are close to the facility, and the climate is wet. Low risk generally means tight geologic formations, poor or limited water resources far removed from the facility, and an arid climate. These two conditions provide the extremes that bracket most "real world" site conditions.

The above-mentioned tables provide a detailed checklist of the prescriptive requirements for waste containment and closure of mine waste management units for each waste category, each waste type (tailings impoundments, waste pile, mine wastewater impoundment), and each site risk category.

---

(2) Refers to California Code of Regulations, Title 23, Division 3, *Chapter 15*, Article 7, for waste classification criteria (Groups A, B, and C which are discussed in Chapter 3).

TABLE 11.1

SUMMARY OF SOME KEY REGULATORY IMPLICATIONS
(All References to Mine Waste Classification Groups Pertain to the Current Regulations)

ASPECT OF MINE WASTE DISPOSAL	CURRENT REGULATIONS	IMPLICATIONS OF THE REVISIONS PROPOSED IN JUNE 1988	IMPLICATIONS OF THE REVISIONS PROPOSED IN JANUARY 1989
Applicability of Requirements	All new waste management facilities. Existing facilities as required by the Regional Board.	All new waste management facilities. Also, at the Regional Board's discretion, all facilities permitted prior to 1984, including historic and abandoned mines.	All new waste management facilities. All existing facilities as required by the Regional Board. All active facilities in regard to surface drainage, monitoring, and closure.
Exemptions From Requirements	Alternatives to prescriptive standards may be allowed (2510[b]),[(1)] provided: (1) prescriptive standard is not feasible, and (2) alternative is consistent with performance goal and provides equivalent protection.	No change.	No change.
Waste Classification	Provides for Groups A, B and C. Classification of waste is based on intrinsic properties of the waste only (with some exceptions). Testing for hazard classification is required.	Provides a more general waste and site characterization involving intrinsic properties of the waste, natural site hydraulic conductivity and attenuative capacity, active life, and climate. Testing for hazard classification, acid mine drainage potential, and deionized water extraction required to determine pollution potential of waste.	Provides a more general waste and site characterization, involving intrinsic properties of the waste, natural site hydraulic conductivity and attenuative capacity, climate, and background soil and water chemical concentrations. Testing for hazard classification, acid mine drainage potential, and soluble pollutants is required. Constituents to be tested for are specified.
Containment Requirements	Prescriptive containment requirements are specified for each waste group and are summarized in Table 11.2.	The A, B, C waste groupings are eliminated. The proposed regulations specify a wide range of containment requirements which are dependent on the waste and site classification. These requirements generally have been increased to provide for double liners and waste underdrain systems for the equivalent Group A and certain equivalent Group B tailings impoundments. Waste underdrains are also required for certain piles. The specified waste characterization could result in less waste falling in the "Group C" (unlined) equivalent category than is currently the case. Table 11.3 summarizes these containment requirements.	The containment requirements are based on a similar approach to the June 1988 proposal, but they have been increased. It is more difficult to use tailings, for example, as the upper liner in an equivalent Group B facility. Also, the bottom liner requirements have been increased to a synthetic plus clay or a natural geologic formation. The specified waste characterization requirement could cause less waste to fall in the "Group C" (unlined) equivalent category than is currently the case. Table 11.4 summarizes these requirements.

(1) Refers to section in CCR, Title 23, Division 3, *Chapter 15*.

TABLE 11.1

SUMMARY OF SOME KEY REGULATORY IMPLICATIONS
(All References to Mine Waste Classification Groups Pertain to the Current Regulations) (Continued)

ASPECT OF MINE WASTE DISPOSAL	CURRENT REGULATIONS	IMPLICATIONS OF THE REVISIONS PROPOSED IN JUNE 1988	IMPLICATIONS OF THE REVISIONS PROPOSED IN JANUARY 1989
Exemptions From Prescriptive Containment Requirements	A Regional Board may exempt a mine waste pile from liner requirements if the discharger can demonstrate that no leachate will form. Also, a Regional Board may exempt Groups A and B facilities from liner requirements or other provisions, provided: (1) there are only minor amounts of ground water present, (2) discharge is in compliance with the water quality control plan, and (3) natural containment prevents vertical, and natural or engineered containment prevents horizontal interconnection with ground water.	Other than the general 2510(b) exemption, no exemptions are allowed. Site attenuation could be taken into account, but only after completion of a rigid field study under supervision of members of the academic profession. However, these stringent requirements for proving a site's natural attenuative characteristics will limit its application.	Engineered alternatives to the prescriptive requirements may be approved, provided they meet the applicable performance goals and provide an equivalent or greater level of performance. Temporary stockpile operations for less than two years can be exempted from some of the requirements. Site attenuation could be taken into account, but only after a rigid field study had been completed and reviewed by at least three academicians. Unlined facilities may only be approved if: (1) waste is composed entirely of unprocessed, oxidized alluvium, or (2) "fate of contaminants" investigation (excluding attenuation) indicates that pollutants will not migrate to waters of the State, or (3) downgradient waters have no beneficial uses, or (4) waste does not contain waste constituent at concentrations greater than background. The above considerations may reduce the numbers of permits granted for unlined facilities.
Monitoring Requirements	Covered under the Article 5 requirements.	No change.	No change.
Exemptions From Monitoring Requirements	Vadose zone monitoring is only required if it is judged to be feasible.	Requirements for a vadose zone monitoring exemption are similar to the current regulations. The relevant site-specific conditions that are to be considered when judging feasibility are defined. No exemptions from vadose zone monitoring are allowed if reliance is being placed on attenuation by natural geologic formations.	Vadose zone monitoring is required to the extent it is technically feasible.
Closure Requirements	These are contained in Article 8, which applies to all landfill types. Table 11.2 summarizes the requirements.	Generally similar to current requirements. Some provisions, however, may require removal of a waste pile located on a synthetic liner before closure. Requirements also incorporate removal of "free liquids" from tailings prior to closure.	Generally, similar to current requirements. Also incorporate removal of "free liquids" from tailings prior to closure.
Exemptions From Closure Requirements	See exemptions from requirements above.	Alternatives may be approved, if they provide equivalent protection to those prescribed.	As for June 1988.
Retrofitting Requirements	None stated.	Some retrofitting is generally implied by applicability of requirements (see above).	As for June 1988.
Remediation Requirements	None stated.	Removal of waste may be required to remediate a containment failure. No closure retrofits required.	None stated.

TABLE 11.1

SUMMARY OF SOME KEY REGULATORY IMPLICATIONS
(All References to Mine Waste Classification Groups Pertain to the Current Regulati
(Continued)

Pa

ASPECT OF MINE WASTE DISPOSAL	CURRENT REGULATIONS	IMPLICATIONS OF THE REVISIONS PROPOSED IN JUNE 1988	IMPLICATIONS OF THE REVISIONS PROPOSED IN JANUARY
Special Provisions for Heap Leach Operations	None provided. During the operating phase, heaps are not considered to be a waste and therefore, not subject to Article 7 regulations.	None provided.	Specifically requires that heap operations receive waste dischi requirements prior to operatin Containment requirements are determined based on the heap characteristics after the "ore pr and neutralization treatment" s Leak detection monitoring requirements are also specified Closure in place may be permi provided no waste is discovere leak detection monitoring syst
Waste Treatment Requirements	None stated.	Treatment or neutralization required when there is an effective method available.	Dewatering, neutralization, or treatment, whenever feasible, i required to minimize the pollu potential.
Waste Containment Contingency Plans	None required.	Required for all new and existing facilities.	Required.
Definition of "Free Liquid"	Liquid which readily separates from the solid portions of waste. Free liquids are not present if a sample can be completely retained on a 400 micron filter for 5 minutes.	Moisture content equal to or greater than the specific retention of the waste.	Not provided.
Definition of Arid Climate	None provided.	Average annual rainfall less than 4.15 inches, plus other specified parameters. Maximum rainfall precludes most of the state from arid definition.	Combination of site and waste conditions such that no leachat be generated. Average annual r shall not exceed 7 inches.

TABLE 11.2

REQUIRED MINE WASTE CONTAINMENT UNDER THE CURRENT ARTICLE 7 REGULATIONS

WASTE MANAGEMENT UNIT	WASTE CATEGORY (1)	SITE CONDITIONS				REQUIRED LINER SYSTEM						REQUIRED CLOSURE PROCEDURES						CONTAIN-MENT/ CLOSURE PROFILE (3)
							UPPER LINER			LOWER LINER					CAP			
		PERENNIAL SURFACE WATER	GROUND WATER	NATURAL CONTAINMENT	CLIMATE	UPPER DRAIN	TAILINGS (2)	CLAY/ SYNTHETIC (2 ft./40 mil.)	LOWER DRAIN	NATURAL CONTAIN-MENT	CLAY (2 ft.)	REMOVE SURFACE LIQUIDS	DEWATER SOLID WASTE	REMOVE RESIDUAL SLUDGES	SOIL COVER (1 foot)	CLAY COVER (1 foot)	FOUND-ATION LAYER (2 feet)	
Tailings Impoundment	A	Close	Shallow, Good Quality	Limited	Wet			●	●		●	●			●	●	●	84T1
	A	None	Deep, Poor Quality	Exceeds 10^7 cm/sec	Dry			●	●		●	●			●	●	●	84T2
	B	Close	Shallow, Good Quality	Limited	Wet		●		●		●	●			●	●	●	84T3
	B	None	Deep, Poor Quality	Exceeds 10^6 cm/sec	Dry		●		●	● (4)		●			●	●	●	84T4
	C	← All Conditions →								●								84T5
Waste Pile	A	Close	Shallow, Good Quality	Limited	Wet				●		●				●	●	(5)	84P1
	A	None	Deep, Poor Quality	Exceeds 10^7 cm/sec	Dry				●	● (4)					●	●	(5)	84P2
	B	Close	Shallow, Good Quality	Limited	Wet				●		●				●	●	(5)	84P3
	B	None	Deep, Poor Quality	Exceeds 10^6 cm/sec	Dry				●	● (4)					●	●	(5)	84P4
	C	← All Conditions →								●								84P5
Surface Impoundment (Wastewater)	A	Close	Shallow, Good Quality	Limited	Wet			●	●		●	●		●				84W1
	A	None	Deep, Poor Quality	Exceeds 10^7 cm/sec	Dry			●	●	● (4)		●		●				84W2
	B	Close	Shallow, Good Quality	Limited	Wet			●	●		●	●		●				84W3
	B	None	Deep, Poor Quality	Exceeds 10^6 cm/sec	Dry			●	●	● (4)		●		●				84W4
	C	← All Conditions →								●								84W5

(1) As required in the current Article 7.
(2) It is assumed that the hydraulic conductivity of the tailings is low enough to constitute the upper liner.
(3) Specified to identify containment and closure profiles used in the cost estimates presented in this chapter.
(4) It is assumed that an exemption to use the natural geologic formation instead of the clay liner has been obtained.
(5) It is assumed that the surface of the waste material is adequately compacted to provide the necessary foundation layer.

Legend: ● Requirements

TABLE 11.3

REQUIRED MINE WASTE CONTAINMENT UNDER THE ARTICLE 7 REVISIONS PROPOSED IN JUNE 1988

WASTE MANAGEMENT UNIT	WASTE CATEGORY (1)	SITE CONDITIONS				CONTAINMENT REFERENCE (2)	REQUIRED LINER SYSTEM						REQUIRED CLOSURE PROCEDURES						CONTAINMENT/ CLOSURE PROFILE (4)
								UPPER LINER			LOWER LINER					CAP			
		PERENNIAL SURFACE WATER	GROUND WATER	NATURAL CONTAINMENT	CLIMATE		UPPER DRAIN	TAILINGS	CLAY/ SYNTHETIC (2 ft./60 mil.)	LOWER DRAIN (2 feet)	NATURAL CONTAINMENT	CLAY (2 feet)	REMOVE SURFACE LIQUIDS	DEWATER SOLID WASTE (3)	REMOVE RESIDUAL SLUDGES	SOIL COVER (1 foot)	CLAY COVER (1 foot)	FOUNDATION LAYER (2 feet)	
Tailings Impoundment	A	Close	Shallow, Good Quality	Limited	Wet	a	◍		●	●		●	●	◍		●	●	●	88T1
	A	None	Deep, Poor Quality	Exceeds 10^{-7} cm/sec	Dry	a	◍		●	●		●	●	◍		●	●	●	88T2
	B	Close	Shallow, Good Quality	Limited	Wet	j	◍	○	◍	●		●	●	◍		●	●	●	88T3
	B	None	Deep, Poor Quality	Exceeds 10^{-6} cm/sec	Dry	j	◍	●		●	● (5)		●	◍		●	●	●	88T4
	C	← All Conditions →				l					●								88T5
Waste Pile	A	Close	Shallow, Good Quality	Limited	Wet	c				●		●				●	●	(6)	88P1
	A	None	Deep, Poor Quality	Exceeds 10^{-7} cm/sec	Dry	c				●	● (5)					●	●	(6)	88P2
	B	Close	Shallow, Good Quality	Limited	Wet	f				●		●				●	●	(6)	88P3
	B	None	Deep, Poor Quality	Exceeds 10^{-6} cm/sec	Dry	g				●	● (5)					●	●	(6)	88P4
	C	← All Conditions →				l					●								88P5
Surface Impoundment (Wastewater)	A	Close	Shallow, Good Quality	Limited	Wet	b			●	●		●	●		●				88W1
	A	None	Deep, Poor Quality	Exceeds 10^{-7} cm/sec	Dry	b			●	●	○	◍	●		●				88W2
	B	Close	Shallow, Good Quality	Limited	Wet	i			●	●		●	●		●				88W3
	B	None	Deep, Poor Quality	Exceeds 10^{-6} cm/sec	Dry	i (5)			●	●	● (5)		●		●				88W4
	C	← All Conditions →				l					●								88W5

(1) The proposed June 1988 revisions did not incorporate these waste categories. They are retained here to allow comparison with Table 11.2 and, as such, reflect the intrinsic pollution potential of the waste material.
(2) Description of containment references contained in the June 1988 proposal.
(3) The requirements for dewatering require removal of liquids to the specific retention of the material.
(4) Specified to identify containment and closure profiles used in the cost estimates presented in this chapter.
(5) It is assumed that an exemption to use the natural geologic formation instead of the clay liner has been obtained.
(6) It is assumed that the surface of the waste material is adequately compacted to provide the necessary foundation layer.

Legend: ● Requirements

◍ Added requirements when compared to current Article 7 regulations (Table 11.2).

○ Deleted requirements when compared to current Article 7 regulations (Table 11.2).

TABLE 11.4

REQUIRED MINE WASTE CONTAINMENT UNDER THE ARTICLE 7 REVISIONS PROPOSED IN JANUARY 1989

WASTE MANAGEMENT UNIT	CURRENT WASTE CATEGORY[1]	SITE CONDITIONS					REQUIRED LINER SYSTEM							REQUIRED CLOSURE PROCEDURES						CONTAINMENT/ CLOSURE PROFILE[3]
								UPPER LINER			LOWER LINER						CAP			
		PERENNIAL SURFACE WATER	GROUND WATER	NATURAL CONTAINMENT	CLIMATE	CONTAINMENT REFERENCE[2]	UPPER DRAIN OR COVER	TAILINGS	SYNTHETIC (60 mil)	LOWER DRAIN	NATURAL CONTAINMENT	SYNTHETIC (60 mil)	CLAY (2 feet)	REMOVE SURFACE LIQUIDS	DEWATER SOLID WASTE	REMOVE RESIDUAL SLUDGES	SOIL COVER (1 foot)	CLAY COVER (1 foot)	FOUNDATION LAYER (2 feet)	
Tailings Impoundment	A	Close	Shallow, Good Quality	Limited	Wet	a	◍		●	●		◍	●	●	◍		●	●	●	89T1
	A	None	Deep, Poor Quality	Exceeds 10^{-7} cm/sec	Dry	c	◍		●	●	●[4]		◍	●	◍		●	●	●	89T2
	B	Close	Shallow, Good Quality	Limited	Wet	g	◍	○	◍	●		◍	●	●	◍		●	●	●	89T3
	B	None	Deep, Poor Quality	Exceeds 10^{-6} cm/sec	Dry	k		○		◍	●[4]	◍		●	◍		●	●	●	89T4
	C	← All Conditions →				q					●	(40 mil.)								89T5
Waste Pile	A	Close	Shallow, Good Quality	Limited	Wet	d	◍		◍	●		●	◍				●	●	[6]	89P1
	A	None	Deep, Poor Quality	Exceeds 10^{-7} cm/sec	Dry	e	◍		◍	●	○	◍	◍[5]				●	●	[6]	89P2
	B	Close	Shallow, Good Quality	Limited	Wet	l				●		●	◍				●	●	[6]	89P3
	B	None	Deep, Poor Quality	Exceeds 10^{-6} cm/sec	Dry	o					●[4]						●	●	[6]	89P4
	C	← All Conditions →				q					●	(40 mil.)								89P5
Mining Impoundment (Wastewater)	A	Close	Shallow, Good Quality	Limited	Wet	b			●	●		●	◍	●		○[7]	●	●	●	89W1
	A	None	Deep, Poor Quality	Exceeds 10^{-7} cm/sec	Dry	b			●	●	●[4]	◍		●		○[7]	●	●	●	89W2
	B	Close	Shallow, Good Quality	Limited	Wet	b			●	●		●	◍	●		○[7]	●	●	●	89W3
	B	None	Deep, Poor Quality	Exceeds 10^{-6} cm/sec	Dry	j			●	●	●[4]	◍		●		○[7]	●	●	●	89W4
	C	← All Conditions →				q					●									89W5

(1) The proposed January 1989 revisions did not incorporate these waste categories. They are retained here to allow comparison with Table 11.2 and, as such, reflect the intrinsic pollution potential of the waste material.
(2) Description of containment references contained in the January 1989 proposal.
(3) Specified to identify containment and closure profiles used in the cost estimates presented in this chapter.
(4) It is assumed that an exemption to use the natural geologic formation instead of the clay liner has been obtained.
(5) 10^{-6} cm/sec requirement.
(6) It is assumed that the surface of the waste material is adequately compacted to provide the necessary foundation layer.
(7) Removal of residual sludge is an alternative to the cap requirements.

Legend:
● Requirements
◍ Added requirements when compared to current Article 7 regulations (Table 11.2).
○ Deleted requirements when compared to current Article 7 regulations (Table 11.2).

The analysis is not meant to represent a rigorous assessment, but merely to point out the general differences in approach that are being contemplated and to provide a basis for the cost sensitivity analyses.

11.2.2 PREVIOUSLY PROPOSED REGULATORY CHANGES

As mentioned above, the changes proposed in June 1988 (CCR, 1988) and January 1989 (CCR, 1989) are summarized in Tables 11.3 and 11.4.

A detailed review of Table 11.3 indicates that additional levels of waste containment over those currently required were contemplated. These included:

- Tailings Impoundments
    - Drainage layers between the waste and the liner system.
    - Additional liners under certain conditions.
    - More emphasis on dewatering of the tailings before closure.
- Waste Piles
    - More reliability for containment was placed on the use of liners than on natural geologic formations. Therefore, there is more emphasis on lining the waste piles.

The emphasis on dewatering requires some clarification. The obvious intent of this requirement is to eliminate the potential for free liquids to migrate from the waste material after closure. However, where tailings are concerned, this poses a significant technical problem. With current technology, the only methods by which the above objective can be achieved are: (1) to vacuum or filter the material and air dry it before, or as, it is placed in the waste management unit, or (2) to physically remove the deposited wet material, spread it in thin layers, and air dry it. These methods obviously would work only in dry climates. All other methods, such as loading and compressing the material and providing it with drainage systems, such as well or wick drains, will merely reduce the moisture content. However, the tailings matrix itself still would be in a saturated or close to saturated condition. Furthermore, the costs of undertaking these types of dewatering efforts can be extremely high.

Table 11.4 indicates the additional changes contemplated by the most recent set of proposed regulations. The more important changes for tailings impoundments, waste piles, and mining impoundments include the use of a clay/geomembrane composite liner system as the lower liner.

11.3 COST ANALYSES

11.3.1 INTRODUCTION

Two basic cost analyses were completed. The first focused on the potential impacts of regulatory change. For this purpose, a typical case involving a 12-million ton Group B waste pile and a 30-million ton tailings impoundment (both from the precious metals sector), and a 5-million ton waste pile (from the industrial minerals sector), were selected. Furthermore, to allow for maximum variability in interpretation of the current and proposed regulations, the facilities were assumed to be located in a humid environment with low hydraulic conductivity geologic formations (10^{-6} to 10^{-7} cm/sec). These conditions actually are typical of much of central and northern California.

11.3.2 REGULATORY COST SENSITIVITY

Three regulatory scenarios shown below were selected to represent a wide range of potential conditions and, hence, cost impacts. They are:

- A reasonable application of the current regulations. This scenario assumed that the approach to waste containment includes the incorporation of well defined site conditions. It is assumed that this approach includes an adequate amount of investigative effort and careful technical consideration by both the design professionals and the regulators. This cost scenario is referred to as the base case.
- A conservative application of the current regulations. This scenario assumes that worst-case assumptions were made through the design and permitting process and that much less effort was devoted to the individual site conditions and optimizing the proposed designs.
- The proposed 1989 regulations were selected as possibly representing an upper envelope of the amount of waste containment that could be required. Furthermore, a conservative application of these regulations is assumed.

Summaries of the waste management costs associated with these three scenarios are presented in Tables 11.5, 11.6, and 11.7. Important conclusions that can be drawn from the information presented in these tables are:

- For precious metal waste rock piles, more stringent containment requirements which involve the introduction of liner systems would dramatically increase the waste management costs. There is an order-of-magnitude spread between the lowest and highest disposal costs evaluated. Furthermore, these costs impact the financial rate of return of the mining project. Cost increases may, in fact, be almost as large as a reasonable profit margin.
- For precious metal tailings impoundments, the proposed requirements increase costs by as much as three times the base case costs.

TABLE 11.5

REGULATORY COST IMPACT EVALUATION PRECIOUS METAL GROUP B WASTE PILE IN A HUMID ENVIRONMENT OVERLYING LOW HYDRAULIC CONDUCTIVITY BEDROCK TOTAL PRESENT VALUE COSTS[(1)]

COST PARAMETERS	REGULATORY SCENARIO		
	CURRENT REGULATIONS		PROPOSED REGULATIONS
	REASONABLE APPLICATION (84P4)[(2)]	CONSERVATIVE APPLICATION (84P3)	ADDITIONAL LEVEL OF REDUNDANCY IN CONTAINMENT (89P3 less the drain layer)
$/Ton of Waste	$0.05	$0.22	$0.50
$/Ounce of Gold	$5.47	$24.80	$55.20
% of Gold Price[(3)]	1.40 %	6.20 %	14.00 %
% of Profit Margin[(4)]	11.00 %	52.00 %	115.00 %
Detailed Cost Table Reference (Appendix G)	G-5	G-6	G-7

(1) Based on 10% interest. Includes construction, operations and maintenance, closure, and post-closure monitoring.
(2) Containment/Closure Profile reference on Tables 11.2 through 11.4.
(3) A gold price of $400/ounce.
(4) Assuming a net profit of 12% on gross sales.

TABLE 11.6

REGULATORY COST IMPACT EVALUATION
PRECIOUS METAL GROUP B TAILINGS IMPOUNDMENT
IN A HUMID ENVIRONMENT
OVERLYING LOW HYDRAULIC CONDUCTIVITY BEDROCK
TOTAL PRESENT VALUE COSTS [1]

COST PARAMETERS	REGULATORY SCENARIO			
	CURRENT REGULATIONS		PROPOSED REGULATIONS ADDITIONAL LEVEL OF REDUNDANCY IN CONTAINMENT	
	REASONABLE APPLICATION (84T4)[2]	CONSERVATIVE APPLICATION (84T3)	WITHOUT TAILINGS DEWATERING (89T3)	WITH TAILINGS DEWATERING (89T3)
$/Ton of Waste	$0.79	$1.20	$2.51	$2.53
$/Ounce of Gold	$7.97	$12.10	$25.32	$25.60
% of Gold Price[3]	2.00%	3.00%	6.30%	6.40%
% of Profit Margin[4]	17.00%	25.00%	53.00%	53.00%
Detailed Cost Table Reference (Appendix G)	G-8	G-9	G-10	G-11

(1) Based on 10% interest. Includes construction, operations and maintenance, closure, and post-closure monitoring.
(2) Containment/Closure Profile reference on Tables 11.2 through 11.4.
(3) A gold price of $400/ounce.
(4) Assuming a net profit of 12% on gross sales.

TABLE 11.7

REGULATORY COST IMPACT EVALUATION INDUSTRIAL MINERALS GROUP B WASTE PILE IN A HUMID ENVIRONMENT OVERLYING LOW HYDRAULIC CONDUCTIVITY BEDROCK TOTAL PRESENT VALUE COSTS [(1)]

COST PARAMETERS	REGULATORY SCENARIO		
	CURRENT REGULATIONS		PROPOSED REGULATIONS
	REASONABLE APPLICATION (84P4)(2)	CONSERVATIVE APPLICATION (84P3)	ADDITIONAL LEVEL OF REDUNDANCY IN CONTAINMENT (89P3)
$/Ton of Waste	$0.10	$0.23	$0.66
$/Ton of Product [(3)]	$0.49	$1.14	$3.31
% of Product Price [(4)]	1.00%	2.00%	7.00%
% of Profit Margin [(5)]	8.00%	19.00%	55.00%
Detailed Cost Table Reference (Appendix G)	G-12	G-13	G-14

(1) Based on 10% interest. Includes construction, operations and maintenance, closure and post-closure monitoring.
(2) Containment/Closure Profile reference on Tables 11.2 through 11.4.
(3) A typical industrial minerals product mixture is assumed.
(4) A product price of $50/ton.
(5) Assuming a net profit of 12% on gross sales.

- For industrial minerals waste rock piles, the effects are similar to those for precious metals. Cost increases up to six times the base case could be incurred. These costs can be impressively high when compared to a reasonable profit margin.

Appendix G contains a list of the assumptions made and the detailed cost estimate calculations used.

11.3.3 CLIMATIC COST SENSITIVITY

This example involves a 30-million ton precious metal tailings impoundment in an arid and wet climate, alternately permitted under the current Article 7 regulations and under the proposed 1989 regulations (Table 11.8). It should be noted that the current regulations do not, in fact, incorporate provisions for arid climate as a form of waste containment. For purposes of this analysis, however, it is assumed that exemptions for certain containment features have been obtained partially because the facility is located in an arid climate.

The potential cost impacts of constructing this facility in a wet climate under both the current and proposed regulations could amount to a 40 to 50 percent increase of the costs in an arid climate. As discussed in Chapter 3, this example demonstrates the economic incentives for mines that are constructed in arid climates.

11.4 CONCLUSIONS AND RECOMMENDATIONS

It is evident from the analyses presented in this chapter that mine waste management costs and the impacts of these costs on the profitability of mining operations are extremely sensitive to both regulatory change and climate. This means that both regulations and climate usually have a direct bearing on whether a mine project will be financially viable and, hence, worth developing. This sensitivity is due mainly to the very large volume of material that a mine generates relative to the volume and value of the commercial product it produces.

The cost sensitivity requires that the regulators, the mining industry, and the design professionals all place an emphasis on optimizing the design of a facility so that it achieves its objectives with the appropriate amount of engineered containment. Erring on either side of the appropriate solution can unnecessarily impact the financial viability of a mining project. Under-engineered facilities are more likely than appropriately-engineered facilities to develop environmental problems that present future cost liabilities.

TABLE 11.8

CLIMATIC IMPACT EVALUATION PRECIOUS METAL GROUP B TAILINGS PILE TOTAL PRESENT VALUE COSTS [1]

COST PARAMETERS	REGULATORY SCENARIO			
	CURRENT REGULATIONS		PROPOSED REGULATIONS (Excluding Tailings Dewatering)	
	LOW RISK ENVIRONMENT/ DRY CLIMATE (84T4)[2]	HIGH RISK ENVIRONMENT/ WET CLIMATE (84T3)[2]	LOW RISK ENVIRONMENT/ DRY CLIMATE (89T4)[2]	HIGH RISK ENVIRONMENT/ WET CLIMATE (89T3)[2]
$/Ton of Waste	$0.79	$1.20	$1.83	$2.51
$/Oz. of Gold	$7.97	$12.10	$18.51	$25.32
% Increases		(152% of low risk environment cost)		(137% of low risk environment cost)
% of Gold Price [3]	2.00%	3.00 %	4.60%	6.30%
% of Profit Margin [4]	17.00%	25.00 %	39.00%	53.00%
Detailed Cost Table Reference (Appendix G)	A-8	A-9	A-15	A-10

(1) Based on 10% interest. Includes construction, operations and maintenance, closure, and post-closure monitoring.
(2) Containment/Closure Profile reference on Tables 11.2 through 11.4.
(3) A gold price of $400/ounce.
(4) Assuming a net profit of 12% on gross sales.

Because of the above considerations, it is recommended that mine waste management regulations at either state or federal levels, be subjected to a waste management cost analysis in order to ensure that both the benefits and the costs are fully evaluated, and that finite financial resources are committed to where the most environmental benefit can be gained.

11.5 BIBLIOGRAPHY

California Code of Regulations (CCR), Title 23, Division 3, *Chapter 15*, Article 7, *Mining Waste Management.* Proposed Draft Regulations. State Water Resources Control Board. June 1988.

California Code of Regulations (CCR), Title 23, Division 3, *Chapter 15, Discharges of Waste to Land.* Register 84, No. 49 - 12-8-84. State Water Resources Control Board. December 8, 1984.

California Code of Regulations (CCR), Title 23, Division 3, *Chapter 15,* Article 7, *Mining* Waste *Management.* Proposed Draft Regulations. State Water Resources Control Board. January 31, 1989.

Grefco, Inc., C. Smith, Burney, CA. *Personal Communication.* 1989.

Panorama Resources Ltd., R. Woodfill, Spanish Mine Project, Washington, CA. *Personal Communication.* 1989.

Sonora Mining Corp., D. Lee and D. Z. Skolasinki, Jamestown, CA. *Personal Communication.* 1989.

APPENDIX A

EDITORS, AUTHORS, AND PEER REVIEWERS

APPENDIX A

EDITORS, AUTHORS, AND PEER REVIEWERS

The following paragraphs provide biographical information on the two editors who were also lead authors on several chapters.

Ian P. G. Hutchison, Ph.D., P.E. (Environmental Solutions, Inc.) is a civil engineer/hydrologist, with 25 years of experience in water resource development, water quality evaluations, waste management, site remediation, and environmental compliance management for mining and other industries. He was contributing author and co-editor of the recently published book on heap leach operations, *Introduction to Evaluation, Design and Operation of Precious Metal Heap Leaching Projects* (Society of Mining Engineers, 1988). Dr. Hutchison has directed the design and construction of mining waste disposal facilities in a large number of the western United States. He has expertise in acid rock drainage control and reclamation of waste disposal facilities, particularly Superfund mining sites. He also has worked extensively on hazardous waste management projects, and provided expert witness testimony. Dr. Hutchison is actively involved in working with agencies in formulating water quality protection policy and mine waste disposal regulations, at both state and federal levels. He is the author of many publications on surface and ground water issues, as well as waste management.

Richard D. Ellison, Ph.D., P.E. (Environmental Solutions, Inc.) is a civil engineer and registered Professional Engineer (P.E.) in 25 states. He has over 30 years of professional experience, specializing in the design and construction of mining and hazardous waste disposal facilities, and earth and rockfill dams. Dr. Ellison has completed designs for state-of-the-art hazardous waste landfills, as well as several mine waste disposal facilities in the western United States. Many of these have incorporated unique features for improving systems for waste containment and reducing costs. He is the author or co-author of more than 35 technical publications and text chapters, and was the first chairman of the Scientific Advisory Committee for the U.S. Environmental Protection Agency (EPA) Hazardous Waste Research Center at Louisiana State University. Dr. Ellison was principal author and editor of the 800-page *Engineering and Design Manual for Coal Refuse Disposal*, the first comprehensive manual for the investigation, permitting, design, and operation of waste disposal facilities.

Other lead authors in alphabetical order, were:

Peter H. Dohms, P.G. (Condor Earth Technologies, Inc.) is a geological engineer with 25 years of experience. He is a specialist in the geologic and economic aspects of gold, copper, and precious metals mining. Mr. Dohms has practiced widely and authored numerous publications on mining-related environmental issues, particularly the relationship between mining waste disposal and ground water quality. He has extensive experience in mining and mine waste disposal in the western United States.

Michael E. Henderson, P.E. (Welsh Engineering) is a civil engineer with more than 15 years of professional experience. He is a recognized expert in tailings dam and earthworks engineering and the design of heap leach operations. He also has expertise in hazardous and mine waste site investigations and remediation, geosynthetic liner design and evaluation, and hydraulic and pipeline engineering. Dr. Henderson has been involved in the engineering and design of waste disposal facilities in several eastern and western states. He has conducted short courses on heap leach design and reclamation for state and federal agencies in the western United States, has been a short course instructor on dam design in conjunction with the University of Nevada, Reno, and has been a review consultant for Nevada, Idaho, and Montana state agencies. Dr. Henderson is a member of the Society for Mining, Metallurgy, and Exploration, the American Society of Civil Engineers, and the American Institute of Mining Engineers. He is a technical reviewer for the U.S. Bureau of Mines' Rock Mechanics and Roof Control Program and is on the International Committee on Large Dams. Dr. Henderson is author or editor of more than 30 reports, papers, and presentations, primarily on the subjects of rock mechanics and hydraulic transport of slurry and solids.

Don Poulter, P.E. (D.P. Engineering, Inc.) is a civil engineer with more than 10 years of professional experience, specializing in mine waste disposal, particularly the design and permitting of tailings facilities. His experience includes waste rock disposal and heap leach pad designs and permitting, and reclamation and closure of mine waste management units. Mr. Poulter has been responsible for the management, design, and construction of more than 15 tailings embankments and impoundments for gold and uranium mines throughout the United States, including the design and construction of a dry tailings deposition system. His experience also includes development of field instrumentation for monitoring systems.

Andy Robertson, Ph.D., P.E., a Principal of Steffen Robertson and Kirsten (B.C.) Inc., is a civil engineer specializing in mine geotechnics. He has over 25 years of experience in the design of mine waste facilities, and has developed a particular interest and expertise in the field of acid rock drainage (ARD). Dr. Robertson has been involved in a large number of key ARD projects in Canada, the United States and internationally. Dr. Robertson was the principal author for the development of the *Draft Acid Rock Drainage Technical Guide* for the British Columbia Task Force on ARD, and has lectured extensively in Canada, the United States, Sweden, the United Kingdom and South Africa on appropriate production, control and reclamation measures. He has provided specialist ARD review and mine decommissioning consulting services to a number of national authorities including the provincial governments of B.C., Saskatchewan and Ontario, and to Norway, Sweden, South Africa, and the United States Bureau of Mines. His participation in mining industry guidance includes guidelines development for the United States Mining Association and participation in ARD prediction and control studies on more than 30 base and precious metal and uranium mines internationally.

Adrian C. Smith, Ph.D. (ASCI, Inc.) is a hydrogeochemical engineer with more than 25 years of professional experience in hydrogeochemistry and mine waste disposal, specializing in contaminant migration and pollution control. His expertise is in tailings impoundments and chemical and solid waste disposal sites, with emphasis on issues involving acid drainage, cyanide, and organic chemicals. Dr. Smith also has prepared environmental assessments of ground water and surface water pollution problems related to waste disposal operations, and has expertise in computer modeling to predict pollution from wastes disposal and interpretation of hydrogeochemical exploration data. Dr. Smith has been involved in mining projects throughout the world, including the United States, Canada, United Kingdom, South Africa, Australia, Brazil, Norway, France, and Greece, and is a Chartered Engineer in the United Kingdom. He has published more that 50 technical papers on waste disposal, water pollution, and environmental control in waste disposal/ hydrogeochemistry and is a Member of the Institute of Mining and Metallurgy (MIMM), Member of the Association of Exploration Geochemists (MAEG), Member of the International Association of Water Pollutant Research (MIAWPR), Member of the Institute of Geologists (MIG), and Companion of the South African Institute of Civil Engineers (SAICE).

Mark E. Smith, P.E. (Vector Engineering, Inc.) is a civil engineer with over 12 years of experience, and an expertise in waste management, slope stability, liner systems, pipeline design, and construction. He is author of more than 10 publications on mine waste management, liners, and heap leach mining and is a member of the International Geosynthetics Society, AIME/Society of Mining Engineers (SME), and the American Society for Testing and Materials (ASTM)

Committees on Geotextiles and Geomembranes, and Waste Disposal. Mr. Smith is also president of the Sierra Nevada Mining and Industry Council, and a director of the Tahoe Area Citizens for Resources and Environment.

Dirk van Zyl, Ph.D., P.E. (EIC, Corporation) is a civil engineer with over 15 years of professional experience. He has expertise in heap leach and tailings impoundment design, construction, and quality assurance for precious metals mining, and in hazardous waste disposal facilities design and contaminated site remediation. Dr. van Zyl is author or co-author of 45 technical publications and research reports, editor of five books and conference proceedings on subjects that include erosion control, tailings management, impoundment design, cyanide chemistry and toxicity, and maximization of heap leaching efficiencies. His affiliations include Society of Mining, Metallurgy and Exploration, Inc., South African Institution of Civil Engineering, and South African Institute of Mining and Metallurgy.

In addition to the above, many others provided technical support for the preparation of the material in the following chapters. These included:

- John P. Barton-Bridges, P.E. (Steffen Robertson and Kirsten [U.S.], Inc.)
- Donald T. Bishop, Ph.D., P.G. (Condor Earth Technologies, Inc.)
- Terry Chatwin, Ph.D. (Resource Recovery Conservation Consultants, Inc.)
- Eric Dille (Steffen Robertson and Kirsten [U.S.], Inc.)
- Thomas M. Donnely, Esq. (Heller, Ehrman, White and McAuliffe)
- Robert Dorey (Stefen Robertson and Kirsten [U.S.], Inc.)
- John Ellis, MICE Chartered Engineer (Environmental Solutions, Inc.)
- Kenneth J. Floom, P.E. (Environmental Solutions, Inc.)
- James Good, Esq. (Gresham, Varner and Savage)
- Jeff Hess, R.G. (Environmental Solutions, Inc.)
- Barry A. Hillman, Ph.D. (Condor Earth Technologies, Inc.)
- Jeff Huggins (Vector Engineering, Inc.)
- Vincent Ip, P.E. (Environmental Solutions, Inc.)
- Gene Nunnelley, P.E. (Condor Earth Technologies, Inc.)
- Marc E. Orman (Vector Engineering, Inc.)
- Kerry Parkinson, P.E. (Environmental Solutions, Inc.)
- Roman Pyrih, Ph.D. (GEOCHEM, Inc.)
- Todd V. Ramey (Vector Engineering, Inc.)
- Lisé Rendina (Environmental Solutions, Inc.)

- Jim Rouse (GEOCHEM, Inc.)
- Donald Shaw (Environmental Solutions, Inc.)
- Joel Siegel (Morrison and Knudsen Environmental Services, Inc.)
- Nigel Skirmer, P.E. (Steffen Robertson and Kirsten [U.S.], Inc.)

Peer reviewers included:

Jack A. Caldwell (Jacobs Engineering) is a civil engineer with 20 years of academic and industry experience in design, construction, and reclamation of mine and industrial waste disposal facilities, including waste rock dumps and tailings impoundments. He has also practiced extensively in many areas of hazardous waste disposal. Mr. Caldwell is author of more than 50 technical articles on geotechnical and ground water engineering, tailings impoundment design and construction, and waste disposal engineering.

John A. Cherry, Ph.D. (Geoflow Ltd.) is a geological engineer with 25 years of professional experience in hydrogeology, specializing in contaminant behavior in ground water, particularly physical and geochemical processes in aquifers and aquitards. He has been a professor at the University of Waterloo for 20 years and Director of the University Consortium Solvents-in-Groundwater Research Program. Dr. Cherry has been involved in the development and commercialization of new equipment for ground water monitoring and is co-holder of several patents. As consultant to various government agencies and corporations in the United States and Canada, he has designed ground water studies and remedial actions for major waste disposal facilities and contamination sites. He is co-author of the 1979 *Groundwater* text and recipient of many awards, including the O.E. Meinzer Award of the Geological Society of America, Horton Award of the American Geophysical Union, and Excellence in Research Award for Liquid and Solid Waste Research from the Ontario Ministry of the Environment. Dr. Cherry was selected as the first Henry Darcy Distinguished Lecturer for the National Water Well Association.

Jean-Pierre Giroud, Ph.D. (GeoSyntec Consultants) is a civil engineer with 25 years of professional experience in geotechnical engineering, with extensive expertise in foundation design, geotextiles, and geomembranes. Since 1970, he has pioneered applications for use of synthetics in landfills and liquid impoundments, such as the protection of geomembranes using geotextiles, double geomembrane liner with granular leakage detection layer, and double geomembrane liner with intermediate synthetic geonet drainage layer. Dr. Giroud is a member of American Society of Civil Engineers, ASTM, and chairman of the Second International Conference on Geotextiles and International Conference on Geomembranes. He has presented courses to the

EPA, Federal Highway Administration, United States Bureau of Reclamation, and other public agencies and universities throughout the United States and Canada. He has also lectured on foundation design, geotextiles, and geomembranes in North and South America, Europe, Africa, the Middle and Far East, and Australia. Dr. Giroud has authored or co-authored more than 200 books and technical papers.

David B. McWhorter, Ph.D. (Chemical and Agricultural Engineering Department, Colorado State University) is an engineer with degrees in petroleum, civil, and agricultural engineering. He has 25 years of academic and industry experience in ground water hydrology, specializing in movement and behavior of injected and in situ fluids in modified environments. Dr. McWhorter has developed widely accepted procedures for estimating water quality changes related to surface coal mines and calculation of seepage rates from lined impoundments. He has served as Chairman of the Western Regional Coordinating Committee for Drainage Research, Advisor for Colorado Water Study (Colorado Department of Natural Resources), and member of the Federal Interagency Committee on Health and Environmental Effects of Advanced Energy Technologies. Honors include Top Exploitation Engineer (Shell Oil Company), Oliver Pennock Distinguished Service Award for Excellence in Teaching, Halliburton Education Foundation Award of Excellence, and A.W. Farrall Engineering Achievement-Young Educator Award (American Society of Agricultural Engineers). Dr. McWhorter is author or co-author of more than 100 articles and presentations, including the 1977 text, *Ground-Water Hydrology and Hydraulics*. He holds a patent for Disposal of Spent Oil Shale and Other Materials.

Stuart D. Miller, Ph.D. (Stuart Miller & Associates) is an environmental chemist with 15 years of professional experience, with expertise in mine waste management and rehabilitation, geochemistry, water quality, and soil contamination and rehabilitation. Dr. Miller has conducted mine waste geochemistry investigations for more than 80 mining projects in Australia, Thailand, Indonesia, Malaysia, China, Solomon Islands, Papua-New Guinea, New Zealand, and the United States. Dr. Miller has been an invited member of Environmental Program Advisory Committee of Australian Nuclear Science and Technology Organization (ANSTO), Technical Advisor to the Mine Environment Neutral Drainage (MEND) Technical Committee of Environment, Canada, and Assessor for the National Association of Testing Authorities (NATA), Australia. He has written numerous publications on mine waste geochemistry and management, and mine environmental management and rehabilitation.

Herbert C. Osborne (H.C. Osborne & Associates) is a metallurgical engineer with 30 years of experience in metallurgy, surface and underground precious metals mining, and coal mining. He has considerable experience in the design, start-up, and operation of heavy media and gravity circuits, conventional cyanide plants, and precious metal flotation plants.

APPENDIX B

CALIFORNIA STATE AGENCIES PRIMARILY RESPONSIBLE FOR REGULATING MINE WASTE DISPOSAL

APPENDIX B

CALIFORNIA STATE AGENCIES PRIMARILY RESPONSIBLE FOR REGULATING MINE WASTE DISPOSAL

CALIFORNIA STATE BOARD

STATE WATER RESOURCES
CONTROL BOARD
901 P Street
Sacramento, California 95814
(916) 322-3133

CALIFORNIA REGIONAL WATER QUALITY CONTROL BOARDS

1. NORTH COAST REGION
   1440 Guerneville Road
   Santa Rosa, California 95403
   (707) 576-2220

2. SAN FRANCISCO BAY REGION
   2101 Webster Street, Room 500
   Oakland, California 94612
   (415) 464-1255

3. CENTRAL COAST REGION
   81 Higuera Street, Suite 200
   San Luis Obispo, California 93401-5414
   (805) 549-3147

4. LOS ANGELES REGION
   101 Centre Plaza Drive
   Monterey Park, California 91754
   (213) 266-7500

5. CENTRAL VALLEY REGION
   3443 Routier Road
   Sacramento, California 95827-3098
   (916) 361-5600

   - Fresno Branch Office
     3614 East Ashlan Avenue
     Fresno, California 96002
     (209) 445-5116

   - Redding Branch Office
     415 Knollcrest Drive, Suite 100
     Redding, California 96002
     (916) 224-4845

6. LAHONTAN REGION
   2092 Lake Tahoe Boulevard
   Suite 2
   South Lake Tahoe, California 96150
   (916) 544-3481

   - Victorville Branch Office
     15428 Civic Drive
     Victorville, California 92392
     (619) 241-6583

7. COLORADO RIVER BASIN REGION
   73-271 Highway 111, Suite 21
   Palm Desert, California 92260
   (619) 346-7491

8. SANTA ANA REGION
   2010 Iowa Avenue, Suite 100
   Riverside, California 92507
   (714) 782-4130

9. SAN DIEGO REGION
   9771 Clairemont Mesa Boulevard
   Suite B
   San Diego, California 92124-1331
   (619) 265-5114

APPENDIX C

TAILINGS IMPOUNDMENT, CONTAINMENT ASSESSMENT CONCEPTUAL MODEL, PARAMETERS AND EQUATIONS

TABLE OF CONTENTS

APPENDIX C
TAILINGS IMPOUNDMENT, CONTAINMENT ASSESSMENT CONCEPTUAL MODEL, PARAMETERS AND EQUATIONS

	PAGE NO.
LIST OF TABLES/LIST OF FIGURES	608
C.1 Introduction	609
C.2 Parameters and Material Properties	609
C.3 Equations Used	609
C.4 Seepage Calculations for Tailings Impoundment	613
C.4.1 Case A - Wet Tailings Deposition in an Unlined Impoundment	613
C.4.1.1 Case A1 - Natural Formation of Fracture Bedrock	613
C.4.1.2 Case A2 - Natural Formation of Clay Soil	614
C.4.1.3 Case A3 - Natural Formation of Weathered Bedrock or Granular Soil	614
C.4.2 Case B - Wet Tailings Deposition in a Double-Lined Impoundment	615
C.4.2.1 Case B1 - Double Clay Liner With a Natural Formation of Fractured Bedrock	615
C.4.2.2 Case B2 - Double Geomembrane Liner With a Natural Formation of Fractured Bedrock	616
C.4.2.3 Case B3 - Double Clay Liner With a Natural Formation of Weathered Bedrock or Granular Soil	617
C.4.2.4 Case B4 - Double Geomembrane Liner With a Natural Formation of Weathered Bedrock or Granular Soil	617
C.4.3 Case C - Managed Tailings Deposition in a Single-Lined Impoundment	618
C.5 Reference	618

CONTENTS

APPENDIX C
(Continued)

LIST OF TABLES

TABLE NO.	TITLE	PAGE NO.
C.1	Symbols of Properties	611
C.2	Numerical Values of Properties Used in Example Calculations	612

LIST OF FIGURES

FIGURE NO.	TITLE	PAGE NO.
C.1	Conceptual Model Tailings Impoundment	610

APPENDIX C

TAILINGS IMPOUNDMENT, CONTAINMENT ASSESSMENT CONCEPTUAL MODEL, PARAMETERS AND EQUATIONS

C.1 INTRODUCTION

1. As shown in Figure C.1, the impoundment used in these examples is located 2.5 miles upgradient of a potential ground water resource that contains water of suitable quality and quantity for municipal use.

2. Three different tailings impoundment cases have been analyzed. Case A is wet (or slurry) disposal in a 100-acre unlined facility; Case B is wet disposal in a double-lined impoundment; and Case C is managed tailings deposition in a single-lined impoundment. Cases A and B calculations are provided in this appendix. Case C calculations are similar to those for Case B and are thus not repeated in this appendix.

3. In order to ascertain the impact of site geologic conditions, three alternative foundation conditions were used to evaluate Case A and two alternative foundation conditions were used to evaluate Case B. One foundation type assumes that the site is underlain by fractured bedrock formations with a low hydraulic conductivity and porosity, i.e., 10^{-6} cm/sec and 0.1 percent, respectively. The next foundation condition considers a low hydraulic conductivity, 10^{-6} cm/sec clay formation, with a porosity of 20 percent. The final foundation condition assumes weathered bedrock or granular (soil) formations with a moderate hydraulic conductivity and porosity, i.e., 10^{-4} cm/sec and 20 percent, respectively.

C.2 PARAMETERS AND MATERIAL PROPERTIES

1. Table C.1 defines the symbols used in the examples for the various material properties. Table C.2 provides the numerical values assigned to these properties.

C.3 EQUATIONS USED

- Darcy's seepage equation (Freeze and Cherry, 1979):
  $Q = .6\ kia$

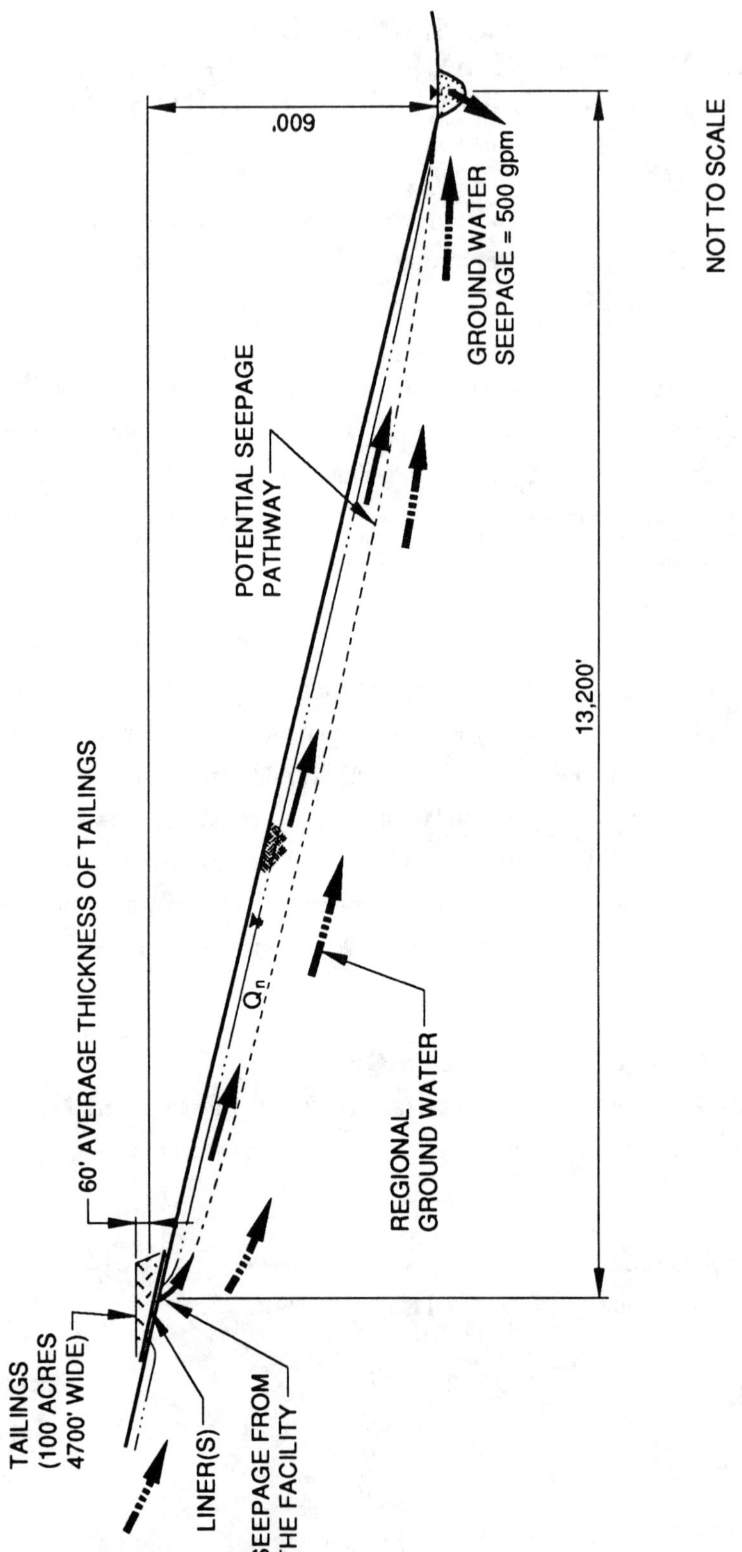

FIGURE C.1
CONCEPTUAL MODEL
TAILINGS IMPOUNDMENT

TABLE C.1

SYMBOLS OF PROPERTIES

MATERIAL	PROPERTY SYMBOL							
	TOTAL SEEPAGE RATE (gpm)	UNIT AREA SEEPAGE RATE (gpm/acre)	HYDRAULIC CONDUCTIVITY (ft/year)	HYDRAULIC HEAD (ft)	SEEPAGE PATH LENGTH (ft)	AREA OF SEEPAGE (acres)	VELOCITY OF SEEPAGE (ft/year)	POROSITY (%)
Tailings	Q_t	--	k_t	H_t	L_t	a_t	--	--
Liner	$Q_{\ell 1}$ $Q_{\ell 2}$	$q_{\ell 1}$ $q_{\ell 2}$	k_ℓ	H_ℓ	L_ℓ	a_ℓ	--	--
Natural Formations	Q_n	--	k_n	H_n	L_n	a_n	V_n	n_n
Leachate Collection and Recovery System (LCRS)	Q_{LCRS}	--	k_{LCRS}	H_{LCRS}	L_{LCRS}	a_{LCRS}	--	--

--: Parameter not used in calculations.

TABLE C.2

NUMERICAL VALUES OF PROPERTIES USED IN EXAMPLE CALCULATION

MATERIAL	PROPERTY VALUE				
	HYDRAULIC CONDUCTIVITY cm/sec (ft/yr)	HYDRAULIC HEAD (ft)	SEEPAGE PATH LENGTH (ft)	AREA OF SEEPAGE OR LEAKAGE (acres)	POROSITY (%)
Tailings	10^{-6} (1)	60	60	100	--
Clay Liners					
• Top	10^{-7} (0.1)	63	3	100	--
• Bottom	10^{-7} (0.1)	3.2	3	100	--
Fractured Bedrock Natural Formation	10^{-6} (1)	660-Case A 600-Case B	13,200	5.5	0.1
Clay Soil Natural Formation	10^{-6} (1)	660	13,200	5.5	20
Weathered Bedrock/ Granular Soil Natural Formations	10^{-4} (100)	660	13,200	5.5	20
LCRS	10^{-1} (100,000)	(1)	(1)	0.1	--

(1) The hydraulic gradient is the controlling parameter and is equal to the slope of the LCRS, i.e., 0.045.

--: Parameter value not used in calculations.

Where:

Q = total seepage rate (gpm)
.6 = conversion factor
k = hydraulic conductivity (feet/year)
i = hydraulic gradient (dimensionless)
a = seepage cross-sectional area (acres)

- Ground water seepage velocity (Freeze and Cherry, 1979):

$$V = \frac{ki}{n}$$

Where:

V = seepage velocity (ft/year)
n = effective porosity (dimensionless)

C.4 SEEPAGE CALCULATIONS FOR TAILINGS IMPOUNDMENT

C.4.1 CASE A - WET TAILINGS DEPOSITION IN AN UNLINED IMPOUNDMENT

TAILINGS IMPOUNDMENT DETAIL

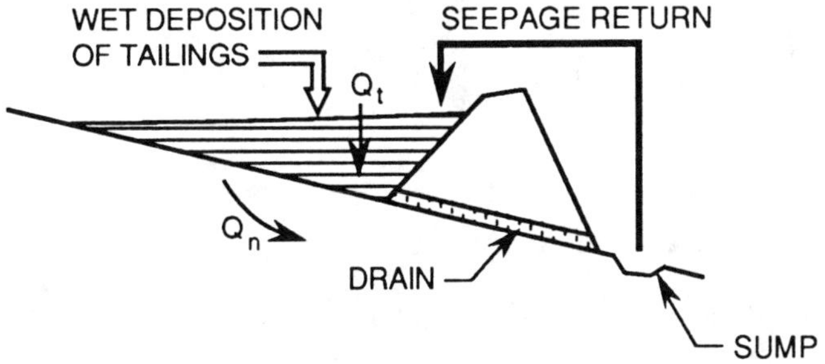

UNLINED FACILITY

C.4.1.1 Case A1 - Natural Formation of Fractured Bedrock

- Determine average seepage through tailings:

$$Q_t = k_t \frac{H_t}{L_t} a_t = (1) \frac{60}{60} (100)\ 0.6 = 62 \text{ gpm} \quad \text{(Eq. 1)}$$

- Determine seepage through natural formation:

$$Q_n = k_n \frac{H_n}{L_n} A_n = (1) \frac{660}{13{,}200} (5.5)\ 0.6 = 0.17 \text{ gpm} \quad \text{(Eq. 2)}$$

- As Q_n is significantly less than Q_t, seepage to sump of 61.83 gpm (62 - .17) is to be expected.

- Determine travel time (T) to ground water resource:

$$V_n = \frac{k_n}{n_n}\frac{H_n}{L_n} = \frac{1}{0.001}\frac{660}{13{,}200} = 50 \text{ ft/yr} \qquad \text{(Eq. 3)}$$

$$T = \frac{13{,}200}{50} = 260 \text{ years} \qquad \text{(Eq. 4)}$$

C.4.1.2 Case A2 - Natural Formation of Clay Soil

- Seepage through tailings: Q_t= 62 gpm (Eq. 1).
- Determine seepage through natural formation:

$$Q_n = K_n \frac{H_n}{L_n} a_n = (1)\frac{660}{13{,}200}(5.5)\ 0.6 = 0.17 \text{ gpm} \qquad \text{(Eq. 5)}$$

- As Q_n is significantly less than Q_t, seepage to sump of 61.83 gpm is to be expected.
- Determine travel time (T) to ground water resource:

$$V_n = \frac{K_n}{n_n}\frac{H_n}{L_n} = \frac{1}{0.2}\frac{660}{13{,}200} = 0.25 \text{ ft/yr} \qquad \text{(Eq. 6)}$$

$$T = \frac{13{,}200}{0.25} = 53{,}000 \text{ years} \qquad \text{(Eq. 7)}$$

C.4.1.3 Case A3 - Natural Formation of Weathered Bedrock or Granular Soil

- Seepage through tailings: Q_t= 62 gpm (Eq. 1).
- Determine flow through natural formation:

$$Q_n = K_n \frac{H_n}{L_n} A_n = (100)\frac{660}{13{,}200}(5.5)\ (0.6) = 17 \text{ gpm} \qquad \text{(Eq. 8)}$$

- As Q_n is less than Q_t, seepage to the sump of 45 gpm (62 - 17) is to be expected.
- Determine travel time (T) to ground water resource:

$$V_n = \frac{K_n}{n_n}\frac{H_n}{L_n} = \frac{100}{0.2}\frac{660}{13{,}200} = 25 \text{ ft/yr} \qquad \text{(Eq. 9)}$$

$$T = \frac{13{,}200}{25} = 530 \text{ years} \qquad \text{(Eq. 10)}$$

C.4.2 CASE B - WET TAILINGS DEPOSITION IN A DOUBLE-LINED IMPOUNDMENT

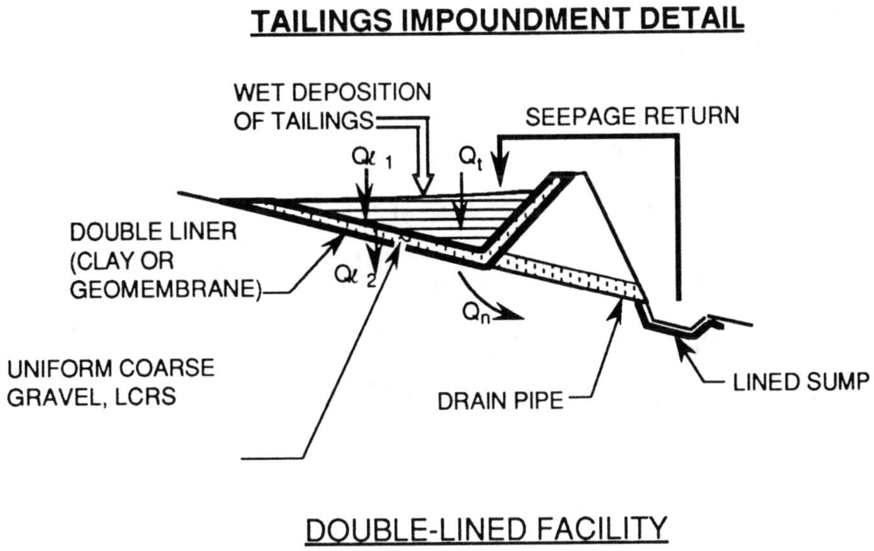

C.4.2.1 Case B1 - Double Clay Liner With a Natural Formation of Fractured Bedrock

- Seepage through tailings: Q_t= 62 gpm (Eq. 1).
- Determine leakage through top clay liner:

  $$Q_{\ell 1} = K_{\ell 1} \frac{H_{\ell 1}}{L_{\ell 1}} a_{\ell 1} = 0.1 \frac{63}{3} (100)\ 0.6 = 126 \text{ gpm} \qquad \text{(Eq. 11)}$$

  Since $Q_t < Q_{\ell 1}$, $Q_{\ell 1} = 62$ gpm.

- Check capacity of LCRS which discharges to drain pipe.

  $$Q_{LCRS} = k_{LCRS} i_{LCRS}\, a_{LCRS} = 100{,}000\ (0.045)\ (0.1)\ (0.6) = 270 \text{ gpm} \qquad \text{(Eq. 12)}$$

  Since $Q_{LCRS} > Q_{\ell 1}$, $Q_{LCRS} = 62$ gpm.

- Determine maximum head H_{LCRS} on bottom liner:

  Step 1 - Calculate average flow area:

  $$a_{LCRS} = \frac{Q_{LCRS}}{k_{LCRS}\, i_{LCRS}} = \frac{62}{100{,}000\ (0.045)\ 0.6} = 0.02 \text{ Acres} \qquad \text{(Eq. 13)}$$

  Step 2 - Calculate corresponding flow depth:

  D_{LCRS} = .02 x 43,560/4,700

  = 0.2 feet

  Where 4,700 feet is the average width of the impoundment.

Step 3 - Calculate total head:

$H_{\ell 2}$ = Liner thickness + D_{LCRS}

= 3 + 0.2

= 3.2 feet.

- Determine leakage through bottom clay liner:

$$Q_{\ell 2} = k_{\ell 2} \frac{H_{\ell 2}}{L_{\ell 2}} a_{\ell 2} = 0.1 \frac{3.2}{3} (100)\ 0.6 = 6.4 \text{ gpm} \qquad \text{(Eq. 14)}$$

- Determine seepage through natural formation:

$$Q_n = K_n \frac{H_n}{L_n} a_n = 1 \frac{600}{13{,}200} (5.5)\ 0.6 = 0.15 \text{ gpm} \qquad \text{(Eq. 15)}$$

Since $Q_n < Q_{\ell 2}$, $Q_{\ell 2} = Q_n = 0.15$ gpm and excess liquid will be removed by the LCRS.

- Determine travel time (T) to ground water:

$$V_n = \frac{K_n}{n_n} \frac{H_n}{L_n} = \frac{1}{0.001} \frac{600}{13{,}200} = 45 \text{ ft/yr} \qquad \text{(Eq. 16)}$$

$$T = \frac{13{,}200}{45} = 293 \text{ years} \qquad \text{(Eq. 17)}$$

C.4.2.2 Case B2 - Double Geomembrane Liner With a Natural Formation of Fractured Bedrock

- Seepage through tailings: Q_t = 62 gpm (Eq. 1)
- Determine leakage through top geomembrane liner:

$$q_{\ell 1} = 0.21\ a^{0.1} k_n^{0.74} H^{0.9} \text{ (see Chapter 7)} \qquad \text{(Eq. 18)}$$

Where:

$q_{\ell 1}$ = unit area leakage rate, m^3/sec/acre
a = area of hole, m^2, = 3.2 x 10^{-6} m^2
k_n = hydraulic conductivity of soil below geomembrane liner, m/sec
= 10^{-3} m/sec
H = hydraulic head on the liner (m)

For:

q = .62 gpm/acre (62/100)

= 3.9 x 10^{-5} m^3/s/acre

H = 0.09 m

= 3.5 inches

Since H is very small, seepage through tailings will control and $Q_{\ell 1}$ = 62 gpm.

- Flow capacity of LCRS, Q_{LCRS}, = 270 gpm (Eq. 12)

  Since $Q_{LCRS} > Q_{\ell 1}$, Q_{LCRS} = 62 gpm (Eq. 19)
  and head on lower geomembrane liner is 0.2 feet (Eq. 13).

- Determine flow through bottom geomembrane liner from Chapter 7 (Table 7.7) (empirical) for $k_n = 10^{-6}$ cm/sec.

  $Q_{\ell 2/acre} \approx .2$ gpd

  $\approx 1.4 \times 10^{-4}$ gpm

  $Q_{\ell 2} = 0.01$ gpm

- Seepage capacity through natural formation, $Q_n = 0.15$ gpm (Eq. 15)

  Since $Q_{\ell 2} < Q_n$, $Q_n = 0.01$ gpm.

- Travel time to ground water = 293 years. (Eqs. 16 and 17)

C.4.2.3 Case B3 - Double Clay Liner With a Natural Formation of Weathered Bedrock or Granular Soil

- Seepage through tailings and upper liner (Chapter 7)

$$Q_{\ell 1} = k_{\ell 1}\left(\frac{L_t + H_\ell}{H_\ell + H_t\dfrac{k_{\ell 1}}{k_t}}\right) a_t\ .62$$

$$= .1\left(\frac{60 + 3}{3 + 60\left(\frac{.1}{1}\right)}\right)(100)\ (.62) = 43 \text{ gpm}$$

- Flow capacity of LCRS, Q_{LCRS} = 270 gpm (Eq. 12) and, since $Q_{LCRS} > Q_{\ell 1}$, Q_{LCRS} = 43 gpm

- Seepage capacity of bottom clay liner, $Q_{\ell 2}$ = 6.4 gpm (Eq. 14)

- Determine seepage through natural formation:

$$Q_n = k_n \frac{H_n}{L_n} a_n = 100\,\frac{600}{13{,}200}\,(5.5)\ 0.6 = 15.5 \text{ gpm} \qquad \text{(Eq. 20)}$$

  Since $Q_n > Q_{\ell 2}$, Q_n = 6.4 gpm and excess liquid will be removed through the LCRS.

- Determine travel time (T) to ground water:

$$V_n = \frac{k_n H_n}{n_n L_n} = \frac{100}{0.2}\,\frac{600}{13{,}200} = 23 \text{ ft/yr} \qquad \text{(Eq. 21)}$$

$$T = \frac{13{,}200}{23} = 574 \text{ years}$$

C.4.2.4 Case B4 - Double Geomembrane Liner With a Natural Formation of Weathered Bedrock or Granular Soil

- Q_{LCRS} = 62 gpm (Eq. 19)
- Determine seepage through bottom geomembrane liner from Chapter 7 (Table 7.7) (empirical) for $k_n = 10^{-4}$ cm/sec:

  $= q_{\ell 2} \simeq 2.8 \times 10^{-3}$ gpm/acre

  $Q_{\ell 2} = 0.28$ gpm
- Flow capacity through natural formation, Q_n = 15.5 gpm (Eq. 20) and since $Q_n > Q_{\ell 2}$, Q_n = 0.28 gpm.
- Travel time (T) to ground water = 574 years (Eq. 21).

C.4.3 CASE C - MANAGED TAILINGS DEPOSITION IN A SINGLE-LINED IMPOUNDMENT

- For these cases, the tailings act as the upper liner and the drainage layer between the tailings and the lower liner as the LCRS. As the seepage rates into the LCRS for Case B are controlled by the tailings, the hydraulic conditions in the LCRS are similar, and hence, the resultant seepage rates from the waste management unit would be the same as those calculated for Case B.

C.5 REFERENCES

Freeze, R.A. and J.A. Cherry. *Groundwater.* Englewood Cliffs, NJ: Prentice-Hall. 1979.

APPENDIX D

HELP MODEL RUN RESULTS

APPENDIX D

SUMMARY OF HELP MODEL RUN RESULTS

CASE	PROFILE	RUN NO.	YEAR	ANNUAL PRECIPITATION (inches)	ANNUAL PERCOLATION (inches)	PERCENT
I	A/Waste Rock	1	1989	35.2	17.8	50
I	B/Waste Rock	2	1989	35.2	7.6	22
I	A/Tailings	3	1989	35.2	0.0	0
I	A/Waste Rock	4	1989	32.3	19.5	60
			1990	32.3	19.87	62
			1991	32.3	19.87	62
			1992	32.3	19.88	62
			1993	32.3	19.86	62
I	B/Waste Rock	5	1989	32.3	9.06	28
			1990	32.3	8.48	26
			1991	32.3	8.48	26
			1992	32.3	8.47	26
			1993	32.3	8.47	26
I	A/Tailings	6	1989	32.3	0.0	0
II	A/Waste Rock	7	1980	15.7	8.32	53
			1981	7.8	3.48	45
			1982	14.7	6.62	45
			1983	22.9	10.48	46
			1984	5.72	2.07	36
II	B/Waste Rock	8	1980	15.7	6.03	38
			1981	7.8	1.21	15
			1982	14.7	2.16	15
			1983	22.9	4.63	20
			1984	5.72	.34	6
II	A/Tailings	9	1980	15.7	.00	0
			1981	7.8	.00	0
			1982	14.7	.00	0
			1983	22.9	.00	0
			1984	5.7	.00	0
III	A/Waste Rock	10	1981	2.1	.01	0
			1982	6.1	.00	0
			1983	9.4	1.86	20
			1984	5.8	.77	13
			1985	2.3	.18	8
III	B/Waste Rock	11	1981	2.1	.00	0
			1982	6.1	.00	0
			1983	9.4	.00	0
			1984	5.8	.00	0
			1985	2.3	.00	0
			1986	4.0	.00	0

A = Nonvegetated
B = Vegetated

APPENDIX E

PROTOCOL FOR DETERMINING NATURALLY LEACHABLE RESIDUALS

APPENDIX E

PROTOCOL FOR DETERMINING NATURALLY LEACHABLE RESIDUALS

This appendix contains a description of a procedure for the assessment of leachable residual species from spent ore on heap leach pads in the natural environment. It can be used to assess compliance with prescribed unloading or decommissioning criteria for spent ore.

The basic procedure involves a "batch" type test in which a known volume of simulated rainwater is mixed with a known mass of the spent ore in order to determine constituent concentrations that could be expected, should leachate escape the waste pile and migrate to a receiving water body.

The dilution ratio used in this test must be related to the criteria used to assess compliance. For example, if the criteria are based on drinking water quality (maximum contaminant level [MCL]), then a 20 to 1 through 16 to 1 liquid to solid ratio would be applicable, in order to maintain consistency with the methodology and rationale developed by the U.S. Environmental Protection Agency (EPA) for Method 1312 and its predecessor, Method 1310, the EP Toxicity Test. Other recognized tests that could be used include the American Society for Testing and Materials (ASTM) D3987-85 "Shake Extraction of Solid Waste With Water" which also has a 20 to 1 liquid to solid ratio.

A known mass of material to be tested is then extracted in a "zero-headspace" vessel for a period of 24 hours using an artificially-simulated rainwater. Specific criteria are:

- Sample Preparation:

  Based on the field moisture content of the material, the undried sample is used in the test, as drying of the sample will artificially volatilize constituents such as cyanide. (The final results can be corrected for moisture content if reporting of waste constituents as a fraction of the dry sample weight is required.)

  The weight/mass of the sample is determined by selecting the appropriate liquid to solid ratio to be used in the test, and ensuring the combined liquid and solid completely fill the test extraction vessel.

- Extraction Fluid:

  The extraction fluid, as simulated rainwater, can be either: (1) deionized water saturated with carbon dioxide and its pH value adjusted to pH 5.0 ±0.1 using sulphuric acid, or (2) using EPA Method 1312, Solution No. 2, with a pH value of ±5.0. (As these extraction fluids contain dissolved sulfates and EPA Method 1312, Solution 2, also contains nitrates, the determination of the resultant concentration in the leachate of these anions will be inexact.)

- Extraction Procedure:

  The extraction, using a rolling bottle or end-on-end extractor, should be continued for 24 hours, without the addition or removal of solutions.

- Preparation for Analysis:

  The equilibrated extract solution should be filtered at 0.45 microns. For cyanide analyses, a portion of the equilibrated solution should be immediately preserved with sodium hydroxide pellets to keep the pH value at ±12. For metal analyses the solution should be used as is and the analyses conducted within a 48-hour period.

The advantages of this test protocol are:

- The use of a zero-headspace vessel allows no vapor phase loss of volatiles such as cyanide and allows for secondary reactions in the system.
- A leach solution that approximates the rainfall pH of 5.0 in the western United States is used in the test as being a reasonable simulation of the leaching that would occur under field conditions.

APPENDIX F

SONORA MINING CORPORATION
JAMESTOWN MINE
DESCRIPTION AND SUMMARY OF PERFORMANCE OF THE TAILINGS MANAGEMENT FACILITY

APPENDIX F

SONORA MINING CORPORATION
JAMESTOWN MINE
DESCRIPTION AND SUMMARY OF PERFORMANCE OF TAILINGS MANAGEMENT FACILITY

F.1 INTRODUCTION

Sonora Mining Corporation constructed the Jamestown Mine, a 4,000-7,000 tons per day (tpd) gold mine and beneficiation facility, during 1986. The project is located 1 mile southwest of Jamestown in Tuolumne County, California.

Beneficiation of ore includes crushing, grinding, and physical separation by flotation of the gold concentrate in a mill. The concentrate is subsequently shipped offsite for final extraction.

The tailings from the flotation circuit are pumped to the tailings management unit where they are deposited onto sloping beaches in thin layers to provide maximum solid/liquid separation and to promote the drainage, consolidation and drying of the tailings. Drainage water from the base of the tailings is removed by the Leachate Collection and Recovery System (LCRS). Surface water that is collected from the surface of the tailings is continuously decanted by the Supernatant Collection System (SCS), which includes a sloping filter on the upstream face of the embankment. The water collected by both systems is conveyed to the Tailings Water Retention Pond (TWRP) through outlet pipes beneath the tailings embankment foundation. The water is then pumped to the mill for reuse (see Figure F-1).

The tailings management unit's containment features include an outer liner and a ground water spine drainage system beneath the outer liner for the tailings storage facility, and a double liner and a ground water monitoring drain for the process water retention pond.

F.2 DESIGN OBJECTIVES

The objectives of the design of the tailings waste management unit are to protect the regional ground water and surface water flows, to minimize any seepage discharge from the tailings during the operational phase of the mine, to eliminate seepage discharge in the long-term, and to form an unsaturated, stable tailings waste pile.

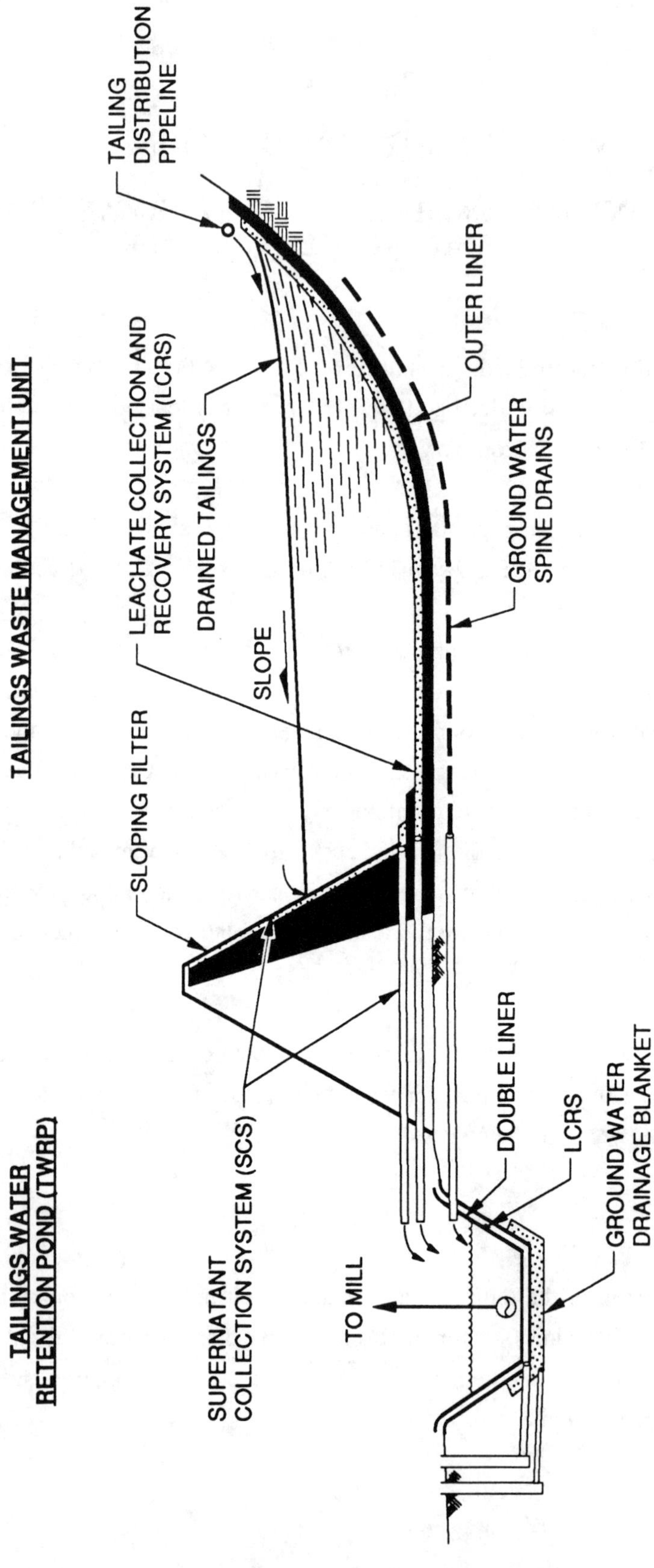

REFERENCE: KNIGHT AND PIESOLD LTD.

FIGURE F.1

SCHEMATIC ARRANGEMENT OF TAILINGS MANAGEMENT SYSTEM

In order to achieve these objectives the design requires that all free-draining liquid be removed from the tailings and recycled to the mill, and that the tailings are allowed to drain and consolidate at all times. In contrast to containment in a conventional, fully saturated tailings pond, the tailings storage facility continuously removes all excess water that would be available for seepage, by decanting, under-drainage, and evaporation, thus greatly reducing the potential for any seepage loss.

The principal requirements of the design are as follows:

- Permanent, secure and total confinement of all solid waste materials in a hydraulically placed waste pile.
- Control, collection, and removal of free-draining liquids from the waste pile during operation for recycling.
- Temporary storage of free-draining liquids in a separate retention pond prior to recycling.
- The inclusion of monitoring systems for key features of the facility in order to ensure that performance goals are achieved.

F.3 DESIGN FEATURES

The features of the tailings management unit are shown schematically on Figure F.1 and include the following:

- A system of ground water drains called spine drains, located beneath the tailings management unit, both to act as a monitoring system, and to achieve separation of ground water from the base of the tailings.
- A clay layer above the spine drains and underlying the tailings management unit to serve as an outer liner.
- A filter blanket and network of perforated pipes on top of the clay forming an under-drainage system, or LCRS, to remove seepage from the tailings and minimize the head buildup on the outer liner.
- An embankment constructed of waste rock and overburden to retain the tailings solids within the facility. This embankment includes a sloping filter and drainage network, or SCS, on the upstream face to allow for continuous removal of all supernatant water and storm runoff from the surface of the deposited tailings.
- Outlet pipe works embedded in concrete under the embankment, to conduct water collected in the SCS and the LCRS to the TWRP.
- A double lined TWRP, fitted with a high density polyethylene and clay double liner, LCRS, and a ground water drainage blanket beneath the outer liner.

Deposition of tailings occurs along the perimeter of the tailings management unit to produce a tailings beach sloping towards its center. From there, surface water discharges into the SCS. Tailings are deposited in a controlled rotational manner to promote drainage and drying, and to increase the density and hence capacity of the tailings management unit. This depositional method is also referred to as subaerial deposition.

F.4 PHYSICAL MONITORING SYSTEM

The monitoring systems for the various components of the tailings management unit includes:

Tailings Management Unit Feature	Monitoring Parameters	Monitoring System
• Ground Water Spine Drains	Ground water level, beneath outer liner	7 Piezometers
• Tailings Management Unit LCRS	Water Level within LCRS and on the outer liner.	7 Piezometers
• Ground water and LCRS Flows	Flow rate for drains	2 Flow Meters
• Supernatant Drains	Water level within drains	2 Piezometers
• Tailings Mass	Water level within the tailings	6 Piezometers
• TWRP	Water level within pond	2 Piezometers
• Ground Water Drains	Ground water level under the TWRP	2 Piezometers
• TWRP LCRS	Leakage from upper liner of the TWRP	1 Flow Meter
• Embankment Foundation and Drains	Ground water level under the embankment	4 Piezometers

The physical monitoring equipment is a series of vibrating wire piezometers and continuously operating flow meters.

F.5 CHEMICAL MONITORING SYSTEMS

Waste Discharge Requirements have been issued by California Regional Water Quality Control Board and require monitoring potential impacts to the regional ground water. These include requirements ongoing monthly monitoring of 22 wells around the perimeter of the tailings management unit and monthly monitoring of the following system components in the tailings management unit and TWRP:

- Supernatant Drains
- Waste Management Unit LCRS Drains
- Ground Water Spine Drains
- TWRP LCRS Drains
- TWRP Ground Water Spine Drains

F.6 MONITORING RESULTS

Since the discharge of tailings commenced in January of 1987, the piezometer systems and flow meters have demonstrated that the Supernatant, LCRS and Ground Water Spine Drains have operated effectively and removed excess water from the tailings management unit. As a result, there has been no hydraulic head buildup on the outer liner, the embankment, and the tailing water pond outer liner. The piezometers have also demonstrated that the tailings mass is draining and consolidating as predicted. In summary, the tailings management unit is meeting its design objectives.

Water quality monitoring of the tailings management unit, TWRP ground water spine drains, and the 22 monitoring wells has demonstrated that there has been no impact on ground water quality from (Table F.1). Within the system, the process water is of high quality as demonstrated by the July 1989 monitoring results in Table F.1, which are typical of the data collected to data (1990).

Of the milling reagents used, only Pine Oil was found to be persistent enough to be analyzed for in its pure form. Carbon disulfide and cresylic acid were found to be persistent breakdown products of Xanthate and Reagent S-5688, respectively which are used in floatation. The reagents or their breakdown products, however, are rarely detected due to dilution and decomposition by exposure to ultraviolet light in the waste management unit. As indicated in Table F.1, the reagents are typically not detected in the Supernatant or LCRS drain systems.

Boron, barium, manganese and nickel are the metals being monitored and generally are present at or near the level of detection. In the past, arsenic, beryllium, cadmium, chromium, copper, iron, lead, mercury, selenium, vanadium, and zinc were routinely analyzed for, but rarely were any of these metals present at concentrations greater than their detection limits. As a result they were dropped from the monitoring program.

TABLE F.1

SONORA MINING CORPORATION MONTHLY SUMMARY - JULY 1989

SITE NAME	DATE	TEMP (C) FIELD (DEG. C)	DEPTH TO WATER (FEET)	DEPTH WSB (FEET)	pH FIELD	pH	B.C. FIELD CONDUCTIVITY (μmho/cm)	TDS (mg/L)	TOC (mg/L)	CALCIUM (mg/L)	MAGNESIUM (mg/L)	SODIUM (mg/L)	POTASSIUM (mg/L)	CHLORIDE (mg/L)	SULFATE (mg/L)	NITRATE (mg/L)	AMMONIA (mg/L)	CARBONATE (mg/L)	BICARBONATE (mg/L)	BORON (mg/L)	BARIUM (mg/L)	MAGANESE (mg/L)	NICKEL (mg/L)	THIOUREA (Cyaminide) (mg/L)	XANTHATES (Carbon Disulfide) (mg/L)	PINE OIL (mg/L)	S-5688 (M.O.P.-Cryselic Acid) (mg/L)
TMULCRS	07-17	20.7	N/A	N/A	7.47	7.40	1,551	1,205	5.4	56.0	87.9	195.0	26.0	89.6	530.0	0.9	0.15	0.00	314.00	0.38	-0.10	1.00	0.02	N/A	-0.01	-0.02	-0.02
TMUSUPER	07-17	24.7	N/A	N/A	7.88	7.50	1,664	1,280	5.8	41.0	34.0	220.0	40.0	112.0	580.0	2.2	0.04	0.00	280.00	0.58	-0.10	0.05	0.02	N/A	-0.01	-0.02	-0.02
TMUGND	07-17	20.5	N/A	N/A	6.55	6.70	806	475	2.2	50.0	73.0	14.0	1.1	16.6	60.0	2.7	0.02	0.00	448.00	-0.10	0.15	-0.01	-0.01	N/A	-0.01	-0.02	-0.02
TWRPLCRS	07-17	19.8	N/A	N/A	7.05	7.80	865	720	3.6	67.0	72.0	40.0	6.2	53.8	215.0	28.8	0.03	0.00	272.00	0.37	0.11	0.04	-0.01	N/A	-0.01	-0.02	-0.02
TWRPGND	07-17	20.7	N/A	N/A	7.47	7.10	819	635	2.2	67.0	59.0	24.0	1.0	56.3	182.0	44.7	0.03	0.00	193.00	0.52	0.14	-0.01	-0.02	N/A	-0.01	-0.02	-0.02
MW1	07-24	19.7	14.89	1,409.75	7.12	7.30	785	530	-1.0	67.0	68.0	16.0	1.1	18.8	112.0	8.0	-0.02	0.00	488.60	0.12	-0.10	0.02	-0.01	N/A	-0.01	-0.02	-0.02
MW3	07-21	20.8	23.05	1,397.82	7.08	7.50	780	525	-1.0	55.0	77.0	6.0	0.5	10.6	136.0	1.8	-0.02	0.00	398.00	0.23	0.10	0.12	-0.01	N/A	-0.01	-0.02	-0.02
MW4	07-12	20.8	36.89	1,356.72	7.87	7.80	627	395	1.3	40.0	48.0	15.0	0.8	12.0	100.0	-0.4	0.04	0.00	257.90	1.10	-0.10	-0.01	0.04	N/A	-0.01	-0.02	-0.02
MW6	07-11	19.0	23.35	1,407.74	7.25	7.10	697	465	3.1	100.0	16.0	22.0	0.5	5.3	130.0	2.2	0.06	0.00	262.90	0.95	-0.10	0.12	-0.01	N/A	-0.01	-0.02	-0.02
MW7	07-14	19.3	14.54	1,598.15	8.35	7.30	415	240	-1.0	36.0	30.0	10.0	0.5	6.7	23.0	4.4	0.04	0.00	238.00	-0.10	-0.10	-0.01	-0.01	N/A	-0.01	-0.02	-0.02
MW8	07-11	19.0	24.80	1,386.12	6.65	6.90	953	595	2.0	136.0	28.0	33.0	0.5	15.2	130.0	1.3	0.04	0.00	448.08	1.60	-0.10	0.20	-0.01	N/A	-0.01	-0.02	-0.02
MW9	07-11	18.3	21.31	1,357.03	7.27	7.80	654	415	2.5	88.0	21.0	16.0	0.5	6.0	132.0	0.4	0.08	0.00	242.00	1.10	-0.10	0.04	-0.01	N/A	-0.01	-0.02	-0.02
MW10	07-12	18.0	28.52	1,350.01	7.41	6.90	837	565	1.5	80.0	40.0	23.0	0.5	74.0	144.0	7.1	0.04	0.00	188.00	0.27	-0.10	0.05	-0.01	N/A	-0.01	-0.02	-0.02
MW11	07-17	18.4	43.04	1,376.96	6.48	7.50	847	540	1.3	51.0	67.0	26.0	0.5	78.6	170.0	12.0	-0.02	0.00	198.00	0.51	-0.14	-0.01	-0.01	N/A	-0.01	-0.02	-0.02
MW12	07-12	20.3	13.01	1,398.42	7.75	7.50	809	500	1.4	62.0	71.0	6.0	0.5	14.2	120.0	-0.4	0.03	0.00	371.80	0.25	-0.10	0.09	0.02	N/A	-0.01	-0.02	-0.02
MW13	07-21	20.6	18.66	1,416.09	7.57	7.70	893	598	-1.0	53.0	64.0	65.0	4.4	17.4	116.0	-0.4	0.46	0.00	481.80	1.00	0.13	0.20	-0.01	N/A	-0.01	-0.02	-0.02
MW14	07-21	20.4	18.29	1,418.03	7.20	7.30	846	545	-1.0	41.0	92.0	13.0	0.4	17.4	116.0	2.7	-0.02	0.00	436.90	0.15	-0.12	-0.01	-0.01	N/A	-0.01	-0.02	-0.02
MW15	07-17	19.3	18.75	1,371.01	6.68	7.00	780	595	2.1	48.0	67.0	17.0	0.4	68.0	150.0	12.8	-0.02	0.00	211.00	0.58	0.16	-0.01	-0.01	N/A	-0.01	-0.02	-0.02
MW16	07-17	19.7	35.75	1,352.18	7.44	8.00	585	385	-1.0	32.0	66.0	6.0	0.5	5.7	86.0	-0.4	0.05	0.00	318.00	0.40	-0.10	0.01	-0.01	N/A	-0.01	-0.02	-0.02
MW17	07-20	23.0	9.72	1,378.67	7.24	7.60	1,335	365	-1.0	73.0	140.0	21.0	0.9	44.6	320.0	92.6	-0.02	0.00	392.00	0.16	0.16	0.01	0.02	N/A	-0.01	-0.02	-0.02
MW18	07-26	20.2	18.05	1,381.51	7.25	7.50	740	500	1.2	78.0	55.0	6.0	0.5	9.9	147.0	-0.4	-0.02	0.00	324.00	0.13	-0.10	0.12	-0.01	N/A	-0.01	-0.02	-0.02
MW19	07-24	27.7	14.70	1,431.71	7.56	7.60	541	310	-1.0	46.0	42.0	11.0	0.7	3.9	12.0	-0.4	-0.02	0.00	357.00	0.13	-0.10	0.12	0.01	N/A	-0.01	-0.02	-0.02
MW20	07-26	19.3	77.55	1,432.45	7.32	7.50	528	340	1.1	32.0	54.0	5.0	0.5	2.8	60.0	2.7	-0.02	0.00	295.00	-0.10	-0.10	-0.01	-0.01	N/A	-0.01	-0.02	-0.02
MW21	07-14	20.4	99.47	1,529.66	7.64	7.90	722	450	-1.0	33.0	90.0	3.0	0.5	2.1	94.0	-0.4	0.04	0.00	430.00	0.16	-0.10	0.02	0.02	N/A	-0.01	-0.02	-0.02
MW22	07-14	21.3	37.32	1,577.50	7.52	7.80	291	195	-1.0	11.0	30.0	5.0	3.2	1.8	6.0	4.0	0.05	0.00	185.80	-0.10	-0.10	-0.01	0.01	N/A	-0.01	-0.02	-0.02

DESCRIPTION OF WELL ABBREVIATIONS:

TMULCRS - Tailings Management Unit LCRS
TMUSUPER - Tailings Management Unit Supernatant Drains
TMUGND - Tailings Management Unit Ground Water Drains
TWRPLCRS - Tailings Water Retention Pond Leachate Collection and Recovery System Drains
TWRPGND - Tailings Water Retention Pond Ground Water Drains
MW - Ground Water Monitoring Well

F.7 SUMMARY

The piezometers and flow meters have demonstrated that the various components of the tailings management units are functioning properly and meeting the design objectives. Further, the quality of the process water within the tailings management unit is quite high and the monitoring system has demonstrated that operation of the tailings facility has had no effect on the regional ground water.

APPENDIX G

WASTE MANAGEMENT COST CALCULATIONS

LIST OF TABLES
WASTE MANAGEMENT COST CALCULATIONS

TABLE NO.	TITLE	PAGE NO.
G.1	Precious Metals Mine 'X' Mine Production Data	640
G.2	Precious Metals Mine 'Y' Mine Production Data	641
G.3	Industrial Minerals Mine Mine Production Data	642
G.4	Unit Cost for Elements in Waste Management Unit	643
G.5	Precious Metals Mine 'X' Cost Analysis Group B Waste Rock Pile (84P4)	644
G.6	Precious Metals Mine 'X' Cost Analysis Group B Waste Rock Pile (84P3)	645
G.7	Precious Metals Mine 'X' Cost Analysis Group B Waste Rock Pile (89P3)	646
G.8	Precious Metals Mine 'Y' Cost Analysis Group B Tailings Impoundment (84T4)	647
G.9	Precious Metals Mine 'Y' Cost Analysis Group B Tailings Impoundment (84T3)	648
G.10	Precious Metals Mine 'Y' Cost Analysis Group B Tailings Impoundment (89T3) (Without Tailings Dewatering)	649
G.11	Precious Metals Mine 'Y' Cost Analysis Group B Tailings Impoundment (89T3) (Without Tailings Dewatering)	650
G.12	Industrial Minerals Mine Cost Analysis Group B Waste Pile (84P4)	651
G.13	Industrial Minerals Mine Cost Analysis Group B Waste Pile (84P3)	652
G.14	Industrial Minerals Mine Cost Analysis Group B Waste Pile (89P3)	653
G.15	Precious Metals Mine 'Y' Cost Analysis Group B Tailings Impoundment (89T4) (Without Tailings Dewatering)	654

TABLE G.1

PRECIOUS METALS MINE 'X'
MINE PRODUCTION DATA

DESCRIPTION	UNIT	TOTAL
Orebody Size	Tons	2,500,000
Daily Ore Production	Tons/day	2,000
Daily Waste Production	Tons/day	9,600
Total Tons Mined	Tons/day	11,600
Stripping Ratio (Waste to Ore)	NA	4.80
Working Days Per Year	Days	250
Mine Life	Years	5
Ore Grade	Ounces/Ton	0.054
Percent Recovery	%	80
Total Metal Recovered	Ounces	108,000
Total Waste	Tons	12,000,000
Total Ore Mined	Tons	14,500,000

NA: Not Applicable.

TABLE G.2

PRECIOUS METALS MINE 'Y' MINE PRODUCTION DATA

DESCRIPTION	UNIT	TOTAL
Orebody Size	Tons	30,000,000
Daily Ore Production	Tons/day	4,808
Daily Waste Production	Tons/day	19,230
Total Tons Mined	Tons/day	24,038
Stripping Ratio (Waste to Ore)	NA	4.00
Working Days Per Year	Days	312
Mine Life	Years	20
Ore Grade	Ounces/Ton	0.11
Percent Recovery	%	90
Total Metal Recovered	Ounces	108,000
Total Waste	Tons	119,995,200
Total Ore Mined	Tons	149,997,120

NA: Not Applicable.

TABLE G.3

INDUSTRIAL MINERALS MINE MINE PRODUCTION DATA

DESCRIPTION	UNIT	TOTAL
Orebody Size	Tons	1,000,000
Daily Ore Production	Tons/day	250
Daily Waste Production	Tons/day	1,250
Total Tons Mined	Tons/day	1,500
Stripping Ratio (Waste to Ore)	NA	5.00
Working Days Per Year	Days	200
Mine Life	Years	20
Percent Recovery	%	100
Total Metal Recovered	Ounces	1,000,000
Total Waste	Tons	5,000,000
Total Ore Mined	Tons	6,000,000

NA: Not Applicable.

TABLE G.4

UNIT COST FOR ELEMENTS IN WASTE MANAGEMENT UNIT

TYPE	THICKNESS	($/Cu.Yd.)	($/Sq.Ft.)
Clay Liner	2 feet	3.50	0.25
Clay Liner	1 foot	3.50	0.12
HDPE Liner	40-mil	NA	0.50
HDPE Liner	60-mil	NA	0.55
Sand Drain Layer	1 foot	12.00	0.45
Rock Drain Layer	2 feet	7.50	0.55
Granular Foundation	2 feet	1.50	0.10
Soil Cover	1 foot	2.00	0.07
Geotextile Layer	NA	NA	0.10
Geogrid	NA	NA	0.20
Finish Grading and Foundation Layer	1 foot	1.50	0.05

HDPE: High Density Polyethylene
NA: Not Applicable

TABLE G.5

PRECIOUS METALS MINE 'X' COST ANALYSIS GROUP B WASTE ROCK PILE (84P4)[1]

CATEGORY	DESCRIPTION	UNIT	QUANTITY	UNIT COST ($)	TOTAL COST ($)
Construction	Waste Pile				
	Clearing and Grubbing	Acres	50	1,000.00	50,000.00
	Finish Grading and Liner Bedding	SF	0	0.00	0.00
	Lower Liner	SF	0	0.00	0.00
	Lower Drain	SF	0	0.00	0.00
	Upper Liner	SF	0	0.00	0.00
	Upper Drain	SF	0	0.00	0.00
	Diversion Ditches	LF	3,000	20.00	60,000.00
	Monitoring Wells	Each	5	7,500.00	37,500.00
	Vadose Zone Monitoring	Each	0	0.00	0.00
	Subtotal Waste Pile				147,500.00
	Siltation Pond				
	Clearing and Grubbing	Acres	15	1,000.00	15,000.00
	Cut, Fill, and Rough Grading	CY	50,000	2.50	125,000.00
	Finish Grading and Liner Bedding	SF	0	0.00	0.00
	Lower Liner	SF	0	0.00	0.00
	Lower Drain	SF	0	0.00	0.00
	Upper Liner	SF	0	0.00	0.00
	Pumping Unit	Each	0	0.00	0.00
	Monitoring Wells	Each	2	7,500.00	15,000.00
	Vadose Zone Monitoring	Each	0	0.00	0.00
	Subtotal Siltation Pond				155,000.00
	Engineering, Permitting, and QA/QC[2]	NA	20%	302,500.00	60,500.00
	Total Capital Cost				363,000.00
	Dollars Per Ton of Waste	Tons	12,000,000	NA	0.03
	Dollars Per Ounce of Product	Ounces	108,000	NA	3.36
Annual Operation and Maintenance	Pumping Cost	KW/HR	0	0.00	0.00
	Ground Water Monitoring	Sample	28	500.00	14,000.00
	Vadose Zone Monitoring	Sample	0	0.00	0.00
	Labor	HRS/YR	0	0.00	0.00
	Vehicle	Year	0	0.00	0.00
	Total Annual Cost				14,000.00
	Total Present Value Over Life of Mine	Years	5	10%	53,071.00
Closure	Clay Cover[3]	SF	0	0.00	0.00
	Geotextile Layer	SF	0	0.00	0.00
	Soil Cover	SF	2,178,000	0.07	152,460.00
	Revegetate Soil Cover	Acres	50	1,500.00	75,000.00
	Total Closure Cost				227,460.00
	Present Value	Years	5	10%	141,235.00
Post-Closure Monitoring	Total Monitoring Cost (Per Year)				14,000.00
	Present Value	Years	5	10%	32,953.00
Total	Total Present Value				590,259.00
	Dollars Per Ton of Waste	Tons	12,000,000	NA	0.05
	Dollars Per Ounce of Product	Ounces	108,000	NA	5.47

(1) Containment/Closure Profile listed on Tables 11.2 through 11.4.
(2) Expressed as percent of total construction cost.
(3) It is assumed an alternative closure method is approved.

SF = Square Foot
CY = Cubic Yard
LF = Linear Foot
LS = Lump Sum

KW/HR = Kilowatts per hour
HRS/YR = Hours per year
NA = Not Applicable

TABLE G.6

PRECIOUS METALS MINE 'X' COST ANALYSIS GROUP B WASTE ROCK PILE (84P3)

CATEGORY	DESCRIPTION	UNIT	QUANTITY	UNIT COST ($)	TOTAL COST ($)
Construction	Waste Pile				
	Clearing and Grubbing	Acres	50	1,000.00	50,000.00
	Finish Grading and Liner Bedding	SF	2,178,000	0.05	108,900.00
	Lower Liner	SF	2,178,000	0.25	544,500.00
	Lower Drain	SF	0	0.00	0.00
	Upper Liner	SF	0	0.00	0.00
	Upper Drain	SF	0	0.00	0.00
	Diversion Ditches	LF	3,000	20.00	60,000.00
	Monitoring Wells	Each	5	7,500.00	37,500.00
	Vadose Zone Monitoring	Each	0	0.00	0.00
	Subtotal Waste Pile				800,900.00
	Siltation Pond[1]				
	Clearing and Grubbing	Acres	15	1,000.00	15,000.00
	Cut, Fill, and Rough Grading	CY	50,000	2.50	125,000.00
	Finish Grading and Liner Bedding	SF	653,400	0.05	32,670.00
	Lower Liner	SF	653,400	0.25	163,350.00
	Lower Drain	SF	653,400	0.45	294,030.00
	Upper Liner	SF	653,400	0.50	326,700.00
	Pumping Unit	Each	0	0.00	0.00
	Monitoring Wells	Each	2	7,500.00	15,000.00
	Vadose Zone Monitoring[2]	Each	2	5,000.00	10,000.00
	Subtotal Siltation Pond				981,750.00
	Engineering, Permitting, and QA/QC[3]	NA	20%	1,782,650.00	356,530.00
	Total Capital Cost				2,139,180.00
	Dollars Per Ton of Waste	Tons	12,000,000	NA	0.18
	Dollars Per Ounce of Product	Ounces	108,000	NA	19.81
Annual Operation and Maintenance	Pumping Cost	KW/HR	0	0.00	0.00
	Ground Water Monitoring	Sample	28	500.00	14,000.00
	Vadose Zone Monitoring	Sample	24	100.00	2,400.00
	Labor	HRS/YR	0	0.00	0.00
	Vehicle	Year	0	0.00	0.00
	Total Annual Cost				16,400.00
	Total Present Value Over Life of Mine	Years	5	10%	62,169.00
Closure	Clay Cover	SF	2,178,000	0.12	261,360.00
	Geotextile Layer	SF	2,178,000	0.10	217,800.00
	Soil Cover	SF	2,178,000	0.07	152,460.00
	Revegetate Soil Cover	Acres	50	1,500.00	75,000.00
	Total Closure Cost				706,620.00
	Present Value	Years	5	10%	438,755.00
Post-Closure Monitoring	Total Monitoring Cost (Per Year)				16,400.00
	Present Value	Years	5	10%	38,602.00
Total	Total Present Value				2,678,706.00
	Dollars Per Ton of Waste	Tons	12,000,000	NA	0.22
	Dollars Per Ounce of Product	Ounces	108,000	NA	24.80

(1) 84W3 Containment/Closure Profile (Table 11.2).
(2) Collection sumps.
(3) Expressed as percent of total construction cost.

SF = Square Foot
CY = Cubic Yard
LF = Linear Foot
LS = Lump Sum

KW/HR = Kilowatts per hour
HRS/YR = Hours per year
NA = Not Applicable

TABLE G.7

PRECIOUS METALS MINE 'X' COST ANALYSIS GROUP B WASTE ROCK PILE (89P3)

CATEGORY	DESCRIPTION	UNIT	QUANTITY	UNIT COST ($)	TOTAL COST ($)
Construction	Waste Pile				
	Clearing and Grubbing	Acres	50	1,000.00	50,000.00
	Finish Grading and Liner Bedding	SF	2,178,000	0.05	108,900.00
	Lower Liner[1]	SF	2,178,000	0.80	1,742,400.00
	Lower Drain	SF	2,178,000	0.55	1,197,900.00
	Upper Liner	SF	0	0.00	0.00
	Upper Drain	SF	0	0.00	0.00
	Diversion Ditches	LF	3,000	20.00	60,000.00
	Monitoring Wells	Each	5	7,500.00	37,500.00
	Vadose Zone Monitoring	Each	0	0.00	0.00
	Subtotal Waste Pile				3,196,700.00
	Siltation Pond[2]				
	Clearing and Grubbing	Acres	15	1,000.00	15,000.00
	Cut, Fill, and Rough Grading	CY	50,000	2.50	125,000.00
	Finish Grading and Liner Bedding	SF	653,400	0.05	32,670.00
	Lower Liner[1]	SF	653,400	0.80	522,720.00
	Lower Drain	SF	653,400	0.45	294,030.00
	Upper Liner	SF	653,400	0.55	359,370.00
	Pumping Unit	Each	0	0.00	0.00
	Monitoring Wells	Each	2	7,500.00	15,000.00
	Vadose Zone Monitoring[3]	Each	2	5,000.00	10,000.00
	Subtotal Siltation Pond				1,373,790.00
	Engineering, Permitting, and QA/QC[4]	NA	20%	4,570,490.00	914,098.00
	Total Capital Cost				5,484,588.00
	Dollars Per Ton of Waste	Tons	12,000,000	NA	0.46
	Dollars Per Ounce of Product	Ounces	108,000	NA	50.78
Annual Operation and Maintenance	Pumping Cost	KW/HR	0	0.00	0.00
	Ground Water Monitoring	Sample	28	500.00	14,000.00
	Vadose Zone Monitoring	Sample	24	100.00	2,400.00
	Labor	HRS/YR	0	0.00	0.00
	Vehicle	Year	0	0.00	0.00
	Total Annual Cost				16,400.00
	Total Present Value Over Life of Mine	Years	5	10%	62,169.00
Closure	Clay Cover	SF	2,178,000	0.12	261,360.00
	Geotextile Layer	SF	2,178,000	0.10	217,800.00
	Soil Cover	SF	2,178,000	0.07	152,460.00
	Revegetate Soil Cover	Acres	50	1,500.00	75,000.00
	Total Closure Cost				706,620.00
	Present Value	Years	5	10%	438,755.00
Post-Closure Monitoring	Total Monitoring Cost (Per Year)				16,400.00
	Present Value	Years	5	10%	62,169.00
Total	Total Present Value				6,047,681.00
	Dollars Per Ton of Waste	Tons	12,000,000	NA	0.50
	Dollars Per Ounce of Product	Ounces	108,000	NA	56.00

(1) Includes clay and synthetic liners.
(2) 89W3 Containment/Closure Profile (Table 11.4).
(3) Collection sumps.
(4) Expressed as percent of total construction cost.

SF = Square Foot
CY = Cubic Yard
LF = Linear Foot
LS = Lump Sum

KW/HR = Kilowatts per hour
HRS/YR = Hours per year
NA = Not Applicable

TABLE G.8

PRECIOUS METALS MINE 'Y' COST ANALYSIS GROUP B TAILINGS IMPOUNDMENT (84T4)

CATEGORY	DESCRIPTION	UNIT	QUANTITY	UNIT COST ($)	TOTAL COST ($)
Construction	Impoundment[(1)]				
	Clearing and Grubbing	Acres	500	1,000.00	500,000.00
	Finish Grading and Liner Bedding	SF	0	0.00	0.00
	Lower Liner	SF	0	0.00	0.00
	Lower Drain	SF	0	0.00	0.00
	Upper Liner	SF	0	0.00	0.00
	Upper Drain	SF	0	0.00	0.00
	Diversion Ditches	LF	10,000	20.00	200,000.00
	Tailings Distribution Pipeline	LF	15,000	25.00	375,000.00
	Monitoring Wells	Each	20	7,500.00	150,000.00
	Vadose Zone Monitoring	Each	0	0.00	0.00
	Subtotal Impoundment				1,225,000.00
	Embankment				
	Clearing and Grubbing	Acres	50	1,000.00	50,000.00
	Foundation Excavation	CY	1,000,000	0.50	500,000.00
	Foundation Preparation	SF	2,178,000	0.15	326,700.00
	Embankment Fill	CY	6,000,000	2.50	15,000,000.00
	Subtotal Embankment				15,876,700.00
	Leachate Collection Pond[(2)]				
	Clearing and Grubbing	Acres	15	1,000.00	15,000.00
	Cut, Fill, and Rough Grading	CY	50,000	2.50	125,000.00
	Finish Grading and Liner Bedding	SF	653,400	0.05	32,670.00
	Lower Liner	SF	0	0.00	0.00
	Lower Drain	SF	653,400	0.45	294,030.00
	Upper Liner	SF	653,400	0.50	326,700.00
	Pumping Unit	Each	2	10,000.00	20,000.00
	Monitoring Wells	Each	2	7,500.00	15,000.00
	Vadose Zone Monitoring[(3)]	Each	2	5,000.00	10,000.00
	Subtotal Leachate Collection Pond				838,400.00
	Engineering, Permitting, and QA/QC[(4)]	NA	15%	17,940,100.00	2,691,015.00
	Total Capital Cost				20,631,115.00
	Dollars Per Ton of Waste	Tons	30,000,000	NA	0.69
	Dollars Per Ounce of Product	Ounces	2,970,000	NA	6.95
Annual Operation and Maintenance	Pumping Cost	KW/HR	300,000	0.08	24,000.00
	Ground Water Monitoring	Sample	88	500.00	44,000.00
	Vadose Zone Monitoring	Sample	24	100.00	2,400.00
	Labor	HRS/YR	3,100	12.00	37,200.00
	Vehicle	Year	1	7,500.00	7,500.00
	Tailings Distribution Pipeline Replacement	LS	20%		75,000.00
	Total Annual Cost				190,100.00
	Total Present Value Over Life of Mine	Years	20	10%	1,618,428.00
Closure	Remove Surface Liquids	LS	1	50,000.00	50,000.00
	Dewater Tailings	LF	0	0.00	0.00
	Remove Residual Sludges	CY	0	0.00	0.00
	Foundation Layer	SF	21,780,000	0.10	2,178,000.00
	Clay Cover	SF	21,780,000	0.12	2,613,600.00
	Geotextile	SF	21,780,000	0.10	2,178,000.00
	Soil Cover	SF	21,780,000	0.07	1,524,600.00
	Revegetate Soil Cover	Acres	500	1,500.00	750,000.00
	Total Closure Cost				9,294,200.00
	Present Value	Years	20	10%	1,381,524.00
Post-Closure Monitoring	Total Monitoring Cost (Per Year)				46,400.00
	Present Value	Years	5	10%	25,865.00
Total	Total Present Value				23,656,932.00
	Dollars Per Ton of Waste	Tons	30,000,000	NA	0.79
	Dollars Per Ounce of Product	Ounces	2,970,000	NA	7.97

(1) An exemption from the prescriptive liner requirements based on low-permeability geological formation and absence of significant ground water resource.
(2) 84W4 Containment/Closure Profile (Table 11.2).
(3) Collection sumps.
(4) Expressed as percent of total construction cost.

SF = Square Foot
CY = Cubic Yard
LF = Linear Foot
LS = Lump Sum
KW/HR = Kilowatts per hour
HRS/YR = Hours per year
NA = Not Applicable

TABLE G.9

PRECIOUS METALS MINE 'Y' COST ANALYSIS GROUP B TAILINGS IMPOUNDMENT (84T3)

CATEGORY	DESCRIPTION	UNIT	QUANTITY	UNIT COST ($)	TOTAL COST ($)
Construction	Impoundment				
	Clearing and Grubbing	Acres	500	1,000.00	500,000.00
	Finish Grading and Liner Bedding	SF	21,780,000	0.05	1,089,000.00
	Lower Liner	SF	21,780,000	0.25	5,445,000.00
	Lower Drain[(1)]	SF	21,780,000	0.55	11,979,000.00
	Upper Liner	SF	0	0.00	0.00
	Upper Drain	SF	0	0.00	0.00
	Diversion Ditches	LF	10,000	20.00	200,000.00
	Tailings Distribution Pipeline	LF	15,000	25.00	375,000.00
	Monitoring Wells	Each	20	7,500.00	150,000.00
	Vadose Zone Monitoring	Each	0	0.00	0.00
	Subtotal Impoundment				19,738,000.00
	Embankment				
	Clearing and Grubbing	Acres	50	1,000.00	50,000.00
	Foundation Preparation	SF	2,178,000	0.15	326,700.00
	Embankment Fill	CY	5,000,000	1.50	7,500,000.00
	Subtotal Embankment				7,876,700.00
	Leachate Collection Pond[(2)]				
	Clearing and Grubbing	Acres	15	1,000.00	15,000.00
	Cut, Fill, and Rough Grading	CY	50,000	2.50	125,000.00
	Finish Grading and Liner Bedding	SF	653,400	0.05	32,670.00
	Lower Liner	SF	653,400	0.25	163,350.00
	Lower Drain	SF	653,400	0.45	294,030.00
	Upper Liner	SF	653,400	0.50	326,700.00
	Pumping Unit	Each	2	10,000.00	20,000.00
	Monitoring Wells	Each	2	7,500.00	15,000.00
	Vadose Zone Monitoring[(3)]	Each	2	5,000.00	10,000.00
	Subtotal Leachate Collection Pond				1,001,750.00
	Engineering, Permitting, and QA/QC[(4)]	NA	15%	28,616,450.00	4,292,468.00
	Total Capital Cost				32,908,918.00
	Dollars Per Ton of Waste	Tons	30,000,000	NA	1.10
	Dollars Per Ounce of Product	Ounces	2,970,000	NA	11.08
Annual Operation and Maintenance	Pumping Cost	KW/HR	300,000	0.08	24,000.00
	Ground Water Monitoring	Sample	88	500.00	44,000.00
	Vadose Zone Monitoring	Sample	24	100.00	2,400.00
	Labor	HRS/YR	3,100	12.00	37,200.00
	Vehicle	Year	1	7,500.00	7,500.00
	Tailings Distribution Pipeline Replacement	LS	20%	NA	75,000.00
	Total Annual Cost				190,100.00
	Total Present Value Over Life of Mine	Years	20	10%	1,618,428.00
Closure	Remove Surface Liquids	LS	1	50,000.00	50,000.00
	Dewater Tailings	LF	0	0.00	0.00
	Remove Residual Sludges	CY	0	0.00	0.00
	Foundation Layer	SF	21,780,000	0.10	2,178,000.00
	Clay Cover	SF	21,780,000	0.12	2,613,600.00
	Geotextile	SF	21,780,000	0.10	2,178,000.00
	Soil Cover	SF	21,780,000	0.07	1,524,600.00
	Revegetate Soil Cover	Acres	500	1,500.00	750,000.00
	Total Closure Cost				9,294,200.00
	Present Value	Years	20	10%	1,381,524.00
Post-Closure Monitoring	Total Monitoring Cost (Per Year)				46,400.00
	Present Value	Years	5	10%	26,145.00
Total	Total Present Value				35,935,015.00
	Dollars Per Ton of Waste	Tons	30,000,000	NA	1.20
	Dollars Per Ounce of Product	Ounces	2,970,000	NA	12.10

(1) Includes sand and geotextile layers.
(2) 84W3 Containment/Closure Profile (Table 11.2).
(3) Collection sumps.
(4) Expressed as percent of total construction cost.

SF = Square Foot
CY = Cubic Yard
LF = Linear Foot
LS = Lump Sum

KW/HR = Kilowatts per hour
HRS/YR = Hours per year
NA = Not Applicable

TABLE G.10

PRECIOUS METALS MINE 'Y' COST ANALYSIS GROUP B TAILINGS IMPOUNDMENT (89T3)

(Without Tailings Dewatering)

CATEGORY	DESCRIPTION	UNIT	QUANTITY	UNIT COST ($)	TOTAL COST ($)
Construction	Impoundment				
	Clearing and Grubbing	Acres	500	1,000.00	500,000.00
	Finish Grading and Liner Bedding	SF	21,780,000	0.05	1,089,000.00
	Lower Liner(1)	SF	21,780,000	0.80	17,424,000.00
	Lower Drain	SF	21,780,000	0.45	9,801,000.00
	Upper Liner	SF	21,780,000	0.55	11,979,000.00
	Upper Drain(2)	SF	21,780,000	0.55	11,979,000.00
	Diversion Ditches	LF	10,000	20.00	200,000.00
	Tailings Distribution Pipeline	LF	15,000	25.00	375,000.00
	Monitoring Wells	Each	20	7,500.00	150,000.00
	Vadose Zone Monitoring	Each	0	0.00	0.00
	Subtotal Impoundment				53,497,000.00
	Embankment				
	Clearing and Grubbing	Acres	50	1,000.00	50,000.00
	Foundation Preparation	SF	2,178,000	0.15	326,700.00
	Embankment Fill	CY	5,000,000	1.50	7,500,000.00
	Subtotal Embankment				7,876,700.00
	Leachate Collection Pond				
	Clearing and Grubbing	Acres	15	1,000.00	15,000.00
	Cut, Fill, and Rough Grading	CY	50,000	2.50	125,000.00
	Finish Grading and Liner Bedding	SF	653,400	0.05	32,670.00
	Lower Liner(1)	SF	653,400	0.80	522,720.00
	Lower Drain	SF	653,400	0.45	294,030.00
	Upper Liner	SF	653,400	0.50	359,370.00
	Pumping Unit	Each	2	10,000.00	20,000.00
	Monitoring Wells	Each	2	7,500.00	15,000.00
	Vadose Zone Monitoring(3)	Each	2	5,000.00	10,000.00
	Subtotal Leachate Collection Pond				1,393,790.00
	Engineering, Permitting, and QA/QC(4)	NA	15%	62,767,490.00	9,415,124.00
	Total Capital Cost				72,182,614.00
	Dollars Per Ton of Waste	Tons	30,000,000	NA	2.41
	Dollars Per Ounce of Product	Ounces	2,970,000	NA	24.30
Annual Operation and Maintenance	Pumping Cost	KW/HR	300,000	0.08	24,000.00
	Ground Water Monitoring	Sample	88	500.00	44,000.00
	Vadose Zone Monitoring	Sample	24	100.00	2,400.00
	Labor	HRS/YR	3,100	12.00	37,200.00
	Vehicle	Year	1	7,500.00	7,500.00
	Tailings Distribution Pipeline Replacement	LS	20%	NA	75,000.00
	Total Annual Cost				190,100.00
	Total Present Value Over Life of Mine	Years	20	10%	1,618,428.00
Closure	Remove Surface Liquids	LS	1	50,000.00	50,000.00
	Dewater Tailings	LF	0	0.00	0.00
	Remove Residual Sludges	CY	0	0.00	0.00
	Foundation Layer	SF	21,780,000	0.10	2,178,000.00
	Clay Cover	SF	21,780,000	0.12	2,613,600.00
	Geotextile	SF	21,780,000	0.10	2,178,000.00
	Soil Cover	SF	21,780,000	0.07	1,524,600.00
	Revegetate Soil Cover	Acres	500	1,500.00	750,000.00
	Total Closure Cost				9,294,200.00
	Present Value	Years	20	10%	1,381,524.00
Post-Closure Monitoring	Total Monitoring Cost (Per Year)				46,400.00
	Present Value	Years	5	10%	26,145.00
Total	Total Present Value				75,208,711.00
	Dollars Per Ton of Waste	Tons	30,000,000	NA	2.51
	Dollars Per Ounce of Product	Ounces	2,970,000	NA	25.32

(1) Includes clay and synthetic liners.
(2) Includes sand and geotextile layers.
(3) Collection sumps.
(4) Expressed as percent of total construction cost.

SF = Square Foot
CY = Cubic Yard
LF = Linear Foot
LS = Lump Sum

KW/HR = Kilowatts per hour
HRS/YR = Hours per year
NA = Not Applicable

TABLE G.11

PRECIOUS METALS MINE 'Y' COST ANALYSIS GROUP B TAILINGS IMPOUNDMENT (89T3)

(With Tailings Dewatering)

CATEGORY	DESCRIPTION	UNIT	QUANTITY	UNIT COST ($)	TOTAL COST ($)
Construction	Impoundment				
	Clearing and Grubbing	Acres	500	1,000.00	500,000.00
	Finish Grading and Liner Bedding	SF	21,780,000	0.05	1,089,000.00
	Lower Liner(1)	SF	21,780,000	0.80	17,424,000.00
	Lower Drain	SF	21,780,000	0.45	9,801,000.00
	Upper Liner	SF	21,780,000	0.55	11,979,000.00
	Upper Drain(2)	SF	21,780,000	0.55	11,979,000.00
	Diversion Ditches	LF	10,000	20.00	200,000.00
	Tailings Distribution Pipeline	LF	15,000	25.00	375,000.00
	Monitoring Wells	Each	20	7,500.00	150,000.00
	Vadose Zone Monitoring	Each	0	0.00	0.00
	Subtotal Impoundment				53,497,000.00
	Embankment				
	Clearing and Grubbing	Acres	50	1,000.00	50,000.00
	Foundation Preparation	SF	2,178,000	0.15	326,700.00
	Embankment Fill	CY	5,000,000	1.50	7,500,000.00
	Subtotal Embankment				7,876,700.00
	Leachate Collection Pond				
	Clearing and Grubbing	Acres	15	1,000.00	15,000.00
	Cut, Fill, and Rough Grading	CY	50,000	2.50	125,000.00
	Finish Grading and Liner Bedding	SF	653,400	0.05	32,670.00
	Lower Liner(1)	SF	653,400	0.80	522,720.00
	Lower Drain	SF	653,400	0.45	294,030.00
	Upper Liner	SF	653,400	0.55	359,370.00
	Pumping Unit	Each	2	10,000.00	20,000.00
	Monitoring Wells	Each	2	7,500.00	15,000.00
	Vadose Zone Monitoring(3)	Each	2	5,000.00	10,000.00
	Subtotal Leachate Collection Pond				1,393,790.00
	Engineering, Permitting, and QA/QC(4)	NA	15%	62,767,490.00	9,415,124.00
	Total Capital Cost				72,182,614.00
	Dollars Per Ton of Waste	Tons	30,000,000	NA	2.41
	Dollars Per Ounce of Product	Ounces	2,970,000	NA	24.30
Annual Operation and Maintenance	Pumping Cost	KW/HR	300,000	0.08	24,000.00
	Ground Water Monitoring	Sample	88	500.00	44,000.00
	Vadose Zone Monitoring	Sample	24	100.00	2,400.00
	Labor	HRS/YR	3,100	12.00	37,200.00
	Vehicle	Year	1	7,500.00	7,500.00
	Tailings Distribution Pipeline Replacement	LS	20%	NA	75,000.00
	Total Annual Cost				190,100.00
	Total Present Value Over Life of Mine	Years	20	10%	1,618,428.00
Closure	Remove Surface Liquids	LS	1	50,000.00	50,000.00
	Dewater Tailings(5)	LF	5,450,000	1.00	5,450,000.00
	Remove Residual Sludges	CY	0	0.00	0.00
	Foundation Layer	SF	21,780,000	0.10	2,178,000.00
	Clay Cover	SF	21,780,000	0.12	2,613,600.00
	Geotextile	SF	21,780,000	0.10	2,178,000.00
	Soil Cover	SF	21,780,000	0.07	1,524,600.00
	Revegetate Soil Cover	Acres	500	1,500.00	750,000.00
	Total Closure Cost				14,744,200.00
	Present Value	Years	20	10%	2,191,631.00
Post-Closure Monitoring	Total Monitoring Cost (Per Year)				46,400.00
	Present Value	Years	5	10%	26,145.00
Total	Total Present Value				76,018,819.00
	Dollars Per Ton of Waste	Tons	30,000,000	NA	2.53
	Dollars Per Ounce of Product	Ounces	2,970,000	NA	25.60

(1) Includes clay and synthetic liners.
(2) Includes sand and geotextile layers.
(3) Collection sump.
(4) Expressed as percent of total construction cost.
(5) Based on wick drains spaced at 10-foot center over 50 percent of the area, installed to an average depth of 50 feet.

SF = Square Foot
CY = Cubic Yard
LF = Linear Foot
LS = Lump Sum

KW/HR = Kilowatts per hour
HRS/YR = Hours per year
NA = Not Applicable

TABLE G.12

INDUSTRIAL MINERALS MINE COST ANALYSIS GROUP B WASTE PILE (84P4)

CATEGORY	DESCRIPTION	UNIT	QUANTITY	UNIT COST ($)	TOTAL COST ($)
Construction	Waste Pile[1]				
	Clearing and Grubbing	Acres	35	1,000.00	35,000.00
	Finish Grading and Liner Bedding	SF	0	0.00	0.00
	Lower Liner	SF	0	0.00	0.00
	Lower Drain	SF	0	0.00	0.00
	Upper Liner	SF	0	0.00	0.00
	Upper Drain	SF	0	0.00	0.00
	Diversion Ditches	LF	2,500	20.00	50,000.00
	Monitoring Wells	Each	5	7,500.00	37,500.00
	Vadose Zone Monitoring	Each	0	0.00	0.00
	Subtotal Waste Pile				122,500.00
	Siltation Pond[2]				
	Clearing and Grubbing	Acres	1	1,000.00	1,000.00
	Cut, Fill, and Rough Grading	CY	50,000	2.50	125,000.00
	Finish Grading and Liner Bedding	SF	0	0.00	0.00
	Lower Liner	SF	0	0.00	0.00
	Lower Drain	SF	0	0.00	0.00
	Upper Liner	SF	0	0.00	0.00
	Pumping Unit	Each	0	0.00	0.00
	Monitoring Wells	Each	2	7,500.00	15,000.00
	Vadose Zone Monitoring[2]	Each	2	5,000.00	10,000.00
	Subtotal Siltation Pond				151,000.00
	Engineering, Permitting, and QA/QC [3]	NA	20%	273,500.00	54,700.00
	Total Capital Cost				328,200.00
	Dollars Per Ton of Waste	Tons	5,000,000	NA	0.07
	Dollars Per Ton of Product	Tons	1,000,000	NA	0.33
Annual Operation and Maintenance	Pumping Cost	KW/HR	44,529	0.08	3,562.00
	Ground Water Monitoring	Sample	20	500.00	10,000.00
	Vadose Zone Monitoring	Sample	24	100.00	2,400.00
	Labor	HRS/YR	0	0.00	0.00
	Vehicle	Year	0	0.00	0.00
	Total Annual Cost				15,962.00
	Total Present Value Over Life of Mine	Years	20	10%	135,896.00
Closure	Clay Cover	SF	0	0.00	0.00
	Geotextile	SF	0	0.00	0.00
	Soil Cover	SF	1,524,600	0.07	106,722.00
	Revegetate Soil Cover	Acres	35	1,500.00	52,500.00
	Total Closure Cost				159,222.00
	Present Value	Years	20	10%	23,667.00
Post Closure Monitoring	Total Monitoring Cost (Per Year)				12,400.00
	Present Value	Years	5	10%	6,987.00
Total	Total Present Value				494,751.00
	Dollars Per Ton of Waste	Tons	5,000,000	NA	0.10
	Dollars Per Ton of Product	Tons	1,000,000	NA	0.49

(1) An exemption from the prescriptive liner requirements based on low permeability geological formation.
(2) Collection sumps.
(3) Expressed as percent of total construction cost.

SF = Square Foot
CY = Cubic Yard
LF = Linear Foot
LS = Lump Sum

KW/HR = Kilowatts per hour
HRS/YR = Hours per year
NA = Not Applicable

TABLE G.13

INDUSTRIAL MINERALS MINE COST ANALYSIS GROUP B WASTE PILE (84P3)

CATEGORY	DESCRIPTION	UNIT	QUANTITY	UNIT COST ($)	TOTAL COST ($)
Construction	Waste Pile				
	Clearing and Grubbing	Acres	35	1,000.00	35,000.00
	Finish Grading and Liner Bedding	SF	1,524,600	0.05	76,230.00
	Lower Liner	SF	1,524,600	0.25	381,150.00
	Lower Drain	SF	0	0.00	0.00
	Upper Liner	SF	0	0.00	0.00
	Upper Drain	SF	0	0.00	0.00
	Diversion Ditches	LF	2,500	20.00	50,000.00
	Monitoring Wells	Each	5	7,500.00	37,500.00
	Vadose Zone Monitoring	Each	0	0.00	0.00
	Subtotal Waste Pile				579,880.00
	Siltation Pond[1]				
	Clearing and Grubbing	Acres	1	1,000.00	1,000.00
	Cut, Fill, and Rough Grading	CY	50,000	2.50	125,000.00
	Finish Grading and Liner Bedding	SF	43,560	0.05	2,178.00
	Lower Liner	SF	43,560	0.25	10,890.00
	Lower Drain	SF	43,560	0.45	19,602.00
	Upper Liner	SF	43,560	0.55	23,958.00
	Pumping Unit	Each	2	10,000.00	20,000.00
	Monitoring Wells	Each	2	7,500.00	15,000.00
	Vadose Zone Monitoring[2]	Each	2	5,000.00	10,000.00
	Subtotal Siltation Pond				227,628.00
	Engineering, Permitting, and QA/QC[3]	NA	20%	807,508.00	161,502.00
	Total Capital Cost				969,010.00
	Dollars Per Ton of Waste	Tons	5,000,000	NA	0.19
	Dollars Per Ton of Product	Tons	1,000,000	NA	0.97
Annual Operation and Maintenance	Pumping Cost	KW/HR	44,529	0.08	3,562.00
	Ground Water Monitoring	Sample	20	500.00	10,000.00
	Vadose Zone Monitoring	Sample	24	100.00	2,400.00
	Labor	HRS/YR	0	0.00	0.00
	Vehicle	Year	0	0.00	0.00
	Total Annual Cost				15,962.00
	Total Present Value Over Life of Mine	Years	20	10%	135,896.00
Closure	Clay Cover	SF	0	0.00	0.00
	Geotextile	SF	0	0.00	0.00
	Soil Cover	SF	1,524,600	0.07	106,722.00
	Revegetate Soil Cover	Acres	35	1,500.00	52,500.00
	Total Closure Cost				159,222.00
	Present Value	Years	20	10%	23,667.00
Post-Closure Monitoring	Total Monitoring Cost (Per Year)				12,400.00
	Present Value	Years	5	10%	6,987.00
Total	Total Present Value				1,135,560.00
	Dollars Per Ton of Waste	Tons	5,000,000	NA	0.23
	Dollars Per Ton of Product	Tons	1,000,000	NA	1.14

(1) 84W3 Containment/Closure Profile (Table 11.2).
(2) Collection sumps.
(3) Expressed as percent of total construction cost.

SF = Square Foot
CY = Cubic Yard
LF = Linear Foot
LS = Lump Sum

KW/HR = Kilowatts per hour
HRS/YR = Hours per year
NA = Not Applicable

TABLE G.14

INDUSTRIAL MINERALS MINE COST ANALYSIS GROUP B WASTE PILE (89P3)

CATEGORY	DESCRIPTION	UNIT	QUANTITY	UNIT COST ($)	TOTAL COST ($)
Construction	Waste Pile				
	Clearing and Grubbing	Acres	35	1,000.00	35,000.00
	Finish Grading and Liner Bedding	SF	1,524,600	0.05	76,230.00
	Lower Liner(1)	SF	1,524,600	0.80	1,219,680.00
	Lower Drain	SF	1,524,600	0.55	838,530.00
	Upper Liner	SF	0	0.00	0.00
	Upper Drain	SF	0	0.00	0.00
	Diversion Ditches	LF	6,000	20.00	120,000.00
	Monitoring Wells	Each	5	7,500.00	37,500.00
	Vadose Zone Monitoring	Each	0	0.00	0.00
	Subtotal Waste Pile				2,326,940.00
	Siltation Pond(2)				
	Clearing and Grubbing	Acres	1	1,000.00	1,000.00
	Cut, Fill, and Rough Grading	CY	50,000	2.50	125,000.00
	Finish Grading and Liner Bedding	SF	43,560	0.05	2,178.00
	Lower Liner(1)	SF	43,560	0.80	34,848.00
	Lower Drain	SF	43,560	0.45	19,602.00
	Upper Liner	SF	43,560	0.55	23,958.00
	Pumping Unit	Each	2	10,000.00	20,000.00
	Monitoring Wells	Each	2	7,500.00	15,000.00
	Vadose Zone Monitoring	Each	2	5,000.00	10,000.00
	Subtotal Siltation Pond				251,586.00
	Engineering, Permitting, and QA/QC(3)	NA	20%	2,578,526.00	515,705.00
	Total Capital Cost				3,094,231.00
	Dollars Per Ton of Waste	Tons	5,000,000	NA	0.62
	Dollars Per Ton of Product	Tons	1,000,000	NA	3.09
Annual Operation and Maintenance	Pumping Cost	KW/HR	44,529	0.08	3,562.00
	Ground Water Monitoring	Sample	20	500.00	10,000.00
	Vadose Zone Monitoring	Sample	24	100.00	2,400.00
	Labor	HRS/YR	0	0.00	0.00
	Vehicle	Year	0	0.00	0.00
	Total Annual Cost				15,962.00
	Total Present Value Over Life of Mine	Years	20	10%	135,896.00
Closure	Clay Cover	SF	1,524,600	0.12	182,952.00
	Geotextile	SF	1,524,600	0.10	152,460.00
	Soil Cover	SF	1,524,600	0.07	106,722.00
	Revegetate Soil Cover	Acres	35	1,500.00	52,500.00
	Total Closure Cost				494,634.00
	Present Value	Years	20	10%	73,524.00
Post-Closure Monitoring	Total Monitoring Cost (Per Year)				12,400.00
	Present Value	Years	5	10%	6,987.00
Total	Total Present Value				3,310,639.00
	Dollars Per Ton of Waste	Tons	5,000,000	NA	0.66
	Dollars Per Ton of Product	Tons	1,000,000	NA	3.31

(1) Includes clay and synthetic liners.
(2) 89W3 Containment/Closure Profile (Table 11.4).
(3) Expressed as percent of total construction cost.

SF = Square Foot
CY = Cubic Yard
LF = Linear Foot
LS = Lump Sum

KW/HR = Kilowatts per hour
HRS/YR = Hours per year
NA = Not Applicable

TABLE G.15

PRECIOUS METALS MINE 'Y' COST ANALYSIS
GROUP B TAILINGS IMPOUNDMENT (89T4)
(Without Tailings Dewatering)

CATEGORY	DESCRIPTION	UNIT	QUANTITY	UNIT COST ($)	TOTAL COST ($)
Construction	Impoundment				
	Clearing and Grubbing	Acres	500	1,000.00	500,000.00
	Finish Grading and Liner Bedding	SF	21,780,000	0.05	1,089,000.00
	Lower Liner[1]	SF	21,780,000	0.55	11,979,000.00
	Lower Drain	SF	21,780,000	0.45	9,801,000.00
	Upper Liner	SF	0	0.00	0.00
	Upper Drain[2]	SF	21,780,000	0.55	11,979,000.00
	Diversion Ditches	LF	10,000	20.00	200,000.00
	Tailings Distribution Pipeline	LF	15,000	25.00	375,000.00
	Monitoring Wells	Each	20	7,500.00	150,000.00
	Vadose Zone Monitoring	Each	0	0.00	0.00
	Subtotal Impoundment				36,073,000.00
	Embankment				
	Clearing and Grubbing	Acres	50	1,000.00	50,000.00
	Foundation Preparation	SF	2,178,000	0.15	326,700.00
	Embankment Fill	CY	5,000,000	1.50	7,500,000.00
	Subtotal Embankment				7,876,700.00
	Leachate Collection Pond				
	Clearing and Grubbing	Acres	15	1,000.00	15,000.00
	Cut, Fill, and Rough Grading	CY	50,000	2.50	125,000.00
	Finish Grading and Liner Bedding	SF	653,400	0.05	32,670.00
	Lower Liner[1]	SF	653,400	0.55	359,370.00
	Lower Drain	SF	653,400	0.45	294,030.00
	Upper Liner	SF	653,400	0.55	359,370.00
	Pumping Unit	Each	2	10,000.00	20,000.00
	Monitoring Wells	Each	2	7,500.00	15,000.00
	Vadose Zone Monitoring[3]	Each	2	5,000.00	10,000.00
	Subtotal Leachate Collection Pond				1,230,440.00
	Engineering, Permitting, and QA/QC[4]	NA	15%	45,180,140.00	6,777,021.00
	Total Capital Cost				51,957,161.00
	Dollars Per Ton of Waste	Tons	30,000,000	NA	1.73
	Dollars Per Ton of Product	Tons	2,970,000	NA	17.49
Annual Operation and Maintenance	Pumping Cost	KW/HR	300,000	0.08	24,000.00
	Ground Water Monitoring	Sample	88	500.00	44,000.00
	Vadose Zone Monitoring	Sample	24	100.00	2,400.00
	Labor	HRS/YR	3,100	12.00	37,200.00
	Vehicle	Year	1	7,500.00	7,500.00
	Tailings Distribution Pipeline Replacement	LS	20%		75,000.00
	Total Annual Cost				190,100.00
	Total Present Value Over Life of Mine	Years	20	10%	1,618,428.00
Closure	Remove Surface Liquids	LS	1	50,000.00	50,000.00
	Dewater Tailings	LF	0	0.00	0.00
	Remove Residual Sludges	CY	0	0.00	0.00
	Foundation Layer	SF	21,780,000	0.10	2,178,000.00
	Clay Cover	SF	21,780,000	0.12	2,613,600.00
	Geotextile	SF	21,780,000	0.10	2,178,000.00
	Soil Cover	SF	21,780,000	0.07	1,524,600.00
	Revegetate Soil Cover	Acres	500	1,500.00	$750,000.00
	Total Closure Cost				9,294,200.00
	Present Value	Years	20	10%	1,381,524.00
Post-Closure Monitoring	Total Monitoring Cost (Per Year)				46,400.00
	Present Value	Years	5	10%	26,145.00
Total	Total Present Value				54,983,258.00
	Dollars Per Ton of Waste	Tons	30,000,000	NA	1.83
	Dollars Per Ton of Product	Tons	2,970,000	NA	18.51

(1) Includes sand and geotextile layers.
(2) Collection sumps.
(3) Expressed as percent of total construction cost.
(4) 89W4 Containment/Closure Profile (Table 11.4).

SF = Square Foot
CY = Cubic Yard
LF = Linear Foot
LS = Lump Sum

KW/HR = Kilowatts per hour
HRS/YR = Hours per year
NA = Not Applicable

ISLAND JEWELS

THE NATURAL HISTORY OF WESTERN AUSTRALIA'S ISLANDS

IAN ABBOTT & ANDREW A BURBIDGE

BookReality
Helping Writers Become Independent Authors

Contents

INTRODUCING THE AUTHORS iii

CHAPTER 1: INTRODUCTION 3

Acknowledgements 9

CHAPTER 2: TROPICAL CAYS 17

CHAPTER 3: KIMBERLEY ISLANDS 39

CHAPTER 4: NORTH-WEST INSHORE ISLANDS 67

CHAPTER 5: NORTH-WEST OFFSHORE ISLANDS 93

CHAPTER 6: ISLANDS OF GUTHARRAGUDA (SHARK BAY) 115

CHAPTER 7: MID-WEST INSHORE ISLANDS 139

CHAPTER 8: HOUTMAN ABROLHOS 157

CHAPTER 9: PERTH REGION 181

CHAPTER 10: SOUTH-WEST CORNER ISLANDS 211

CHAPTER 11: ARCHIPELAGO OF THE RECHERCHE 239

CHAPTER 12: SIMILARITIES AND DIFFERENCES BETWEEN REGIONS IN THE CHARACTERISTICS OF THEIR ISLANDS 268

CHAPTER 13: NATURE CONSERVATION ON WESTERN AUSTRALIAN ISLANDS 296

APPENDIX 1. COMMON AND SCIENTIFIC NAMES OF MAMMALS, BIRDS AND WELL-KNOWN REPTILES MENTIONED IN THIS BOOK 317

APPENDIX 2. UNFAMILIAR WORDS EXPLAINED 325

APPENDIX 3. FURTHER READING 330

INDEX 354

INTRODUCING THE AUTHORS

Ian Abbott I became aware of islands during primary school in Sydney when I read RL Stevenson's novel *Treasure Island*. For Book Week in 4th class, I wrote and illustrated a book which was modelled on *Treasure Island*.

I did not make footfall on any island until 1966, when a party of 2nd Year University of Sydney undergraduate students visited Heron Island for one week. This small tropical cay is within the Great Barrier Reef and has a tourist resort and research station. We walked on the exposed reef, admired the marine life, and on land observed the black noddy terns nesting in *Pisonia* trees. Green turtles coming ashore to lay their eggs were observed closely. At dusk, wedge-tailed shearwaters returned to their burrows, and we became familiar with their cacophonous nocturnal calls. Dr Mary Besly, our university lecturer and guide, made sure that we did not overlook the more subtle aspects of the island ecosystem.

It was not until 1969 that my interest in island ecology became serious. I moved to Melbourne and commenced research for a PhD degree at Monash University. My research proposal was to study the ecology and evolution of the land bird faunas of Tasmania and three large islands in Bass Strait, and compare these with mainland Victoria. My main supervisor, Dr Douglas Dorward, was keen on comparative studies of bird species. My proposal was defended at a seminar, accepted, and I commenced three years of field work. My research program was greatly influenced by Robert MacArthur and Edward Wilson's book (1967) on theoretical aspects of island biotas, David Lack's books on island bird ecology, and publications of Peter Grant and Allen Keast. Seven major papers resulted from this research.

In 1972 I was awarded a post-doctoral fellowship to work with Peter Grant, based at McGill University, Montréal, Canada. He had studied for his PhD the ecology of land birds on several islands of Mexico. I studied the feeding ecology of Darwin's ground-finches on islands in the Galápagos. The outcome was seven major publications.

In 1974 I was awarded a post-doctoral fellowship at The University of Western Australia. For three years I studied the ecology of silvereyes on islands and coastal mainland sites of south-west Western Australia (WA). Camped out on islands for 1-2 weeks, I had much spare time after I had completed my morning mist-netting and leg-banding duties. I therefore interested myself in vegetation, flora, seabirds, and reptiles. My plant specimens were submitted to botanists at the WA Herbarium, developing my skill in identifying and correctly naming species. Lizards were captured alive and donated to the collection of the WA Museum. Fifteen papers resulted from this period.

Through the courtesy of Prof AR (Bert) Main, I became aware of a card index of plant species recorded during the 1950s by Glen Storr on many of the stacks around Rottnest Island. This provided a baseline against which I could compare my species lists on the same islets, and measure species turnover and thus draw conclusions about the stability of island floras. The importance of the impact of nesting seabirds, a topic previously researched in eastern Australia by Mary Gillham, on island vegetation and floristic composition provided another point of comparison. This program was extended to all islets in the Perth metropolitan region.

In 1979 I gained my first permanent employment, as a public servant scientist. This did not involve island ecology. It was not until 2000 that I recommenced my interest in this discipline. My research was necessarily opportunistic. I published papers advocating an accelerated but cautious program of translocating mammal species to islands, studying change in vegetation and floristic composition on Carnac Island over the long-term, recording the natural history of Saddle Island, surveying the history of changes in human use of WA islands, documenting the potentially-catastrophic consequence of misidentification of the rat species found on Woody Island, analysing the causes of species declines on North Island (New Zealand) following settlement by Europeans, and contributing to a database of all known occurrence of mammal species on WA islands.

With partial retirement in 2008, I commenced a monographic review of the ecology of the islands of WA. This, co-authored with Allan Wills, was published in 2016. This book builds on the knowledge accrued by me on islands since 1966, and has provided a most interesting opportunity to synthesise my knowledge with information available in the scientific literature, with a sharp focus on the islands of WA.

Andrew A Burbidge My interest in islands was kindled by reading a 1961 paper by Professor Bert Main (who later became my PhD supervisor at The University of Western Australia) on the occurrence of kangaroos and wallabies on WA islands and its climatic and ecological implications, and his 1971 paper, with a student, on the conservation of kangaroos and wallabies in reserves in WA, which compared reserves of remnant vegetation (islands in a sea of farmland) with islands surrounded by ocean. During my PhD studies I accompanied colleague Jack Kinnear, who was researching marsupial nutrition, to East Wallabi Island in the Houtman Abrolhos. My role was to run after and catch tammar wallabies so Jack could collect samples from them, before they were released. I also assisted another scientist who was trapping bush rats.

In 1969 I visited Barrow Island for the first time and have been there many times since. At that time an oilfield was being developed on an important nature reserve. More recently government approved the Gorgon Gas project, which constructed a major liquefied natural gas plant there. I have advised on quarantine from the start of the Gorgon project's planning and still help monitor the implementation of the company's Quarantine Management Plan.

After commencing work with the state Department of Fisheries and Fauna in 1968, and being aware of the value of islands to nature conservation, I first documented the occurrence of mammals on WA islands and worked to ensured that important ones were reserved for nature conservation. I then started a card index documenting seabird breeding on WA islands, initially based on information in a book published in 1971, again working to ensure that the important ones were protected. The card index developed into a computer database, which has since been used by state authorities planning responses to oil spills. Wanting to improve knowledge of seabirds breeding on islands, with colleagues I started surveying important seabird islands, starting with the Houtman Abrolhos, where colleagues and I visited c. 150 islands in 1991-93. We also set up transects to monitor numbers of the lesser noddy, which, in WA, nests only in small areas of white mangroves on three Houtman Abrolhos islands. I have returned there several times to help count nests on these transects.

One job I was given early in my career was to eradicate rabbits from Carnac Island, which is located close to the Perth coast, and with help from pest control experts in another government department, they were eradicated. This led to a series of invasive species eradications on WA islands, starting with black rats on Bedout Island, off the Pilbara coast, an important seabird island. Later I led or was involved with eradications of goats, black rats, and house mice on several islands, culminating in the eradication of black rats and feral cats on the 100 or so islands and islets in the Montebello Islands. Eradication of introduced predators can allow seabirds to start breeding again, as has happened on Rat Island in the Houtman Abrolhos; however, mammals need help. One project I led translocated endangered mala (or rufous hare-wallabies) to Trimouille Island in the Montebellos, a significant logistical exercise, as they came from the Tanami Desert in the Northern Territory, a straight line distance of c. 1700 km. The translocation involved the wallabies being captured in the middle of the night, driven in four-wheel drive vehicles to an airstrip, a long flight in a fixed-wing aircraft that refuelled in the middle of the desert, and finally out to the island in a helicopter. This could not have occurred without input from colleagues in the Northern Territory, and the support of an oil and gas company that provided the helicopter.

Part of my duties were biological survey and conservation reserve selection, and that allowed me involvement, often with my colleague Norman McKenzie, in major surveys of Kimberley Islands in 1971, 1972 and 1973, and again in the 1980s and early 2000s. Working with local Aboriginal people on the island surveys led to me advising them on conservation on Kimberley native title land, which includes many islands.

Since retirement, I have continued my interest in islands and have visited Lord Howe Island, Christmas Island, the Cocos (Keeling) Islands, the Galápagos and some islands in Alaska.

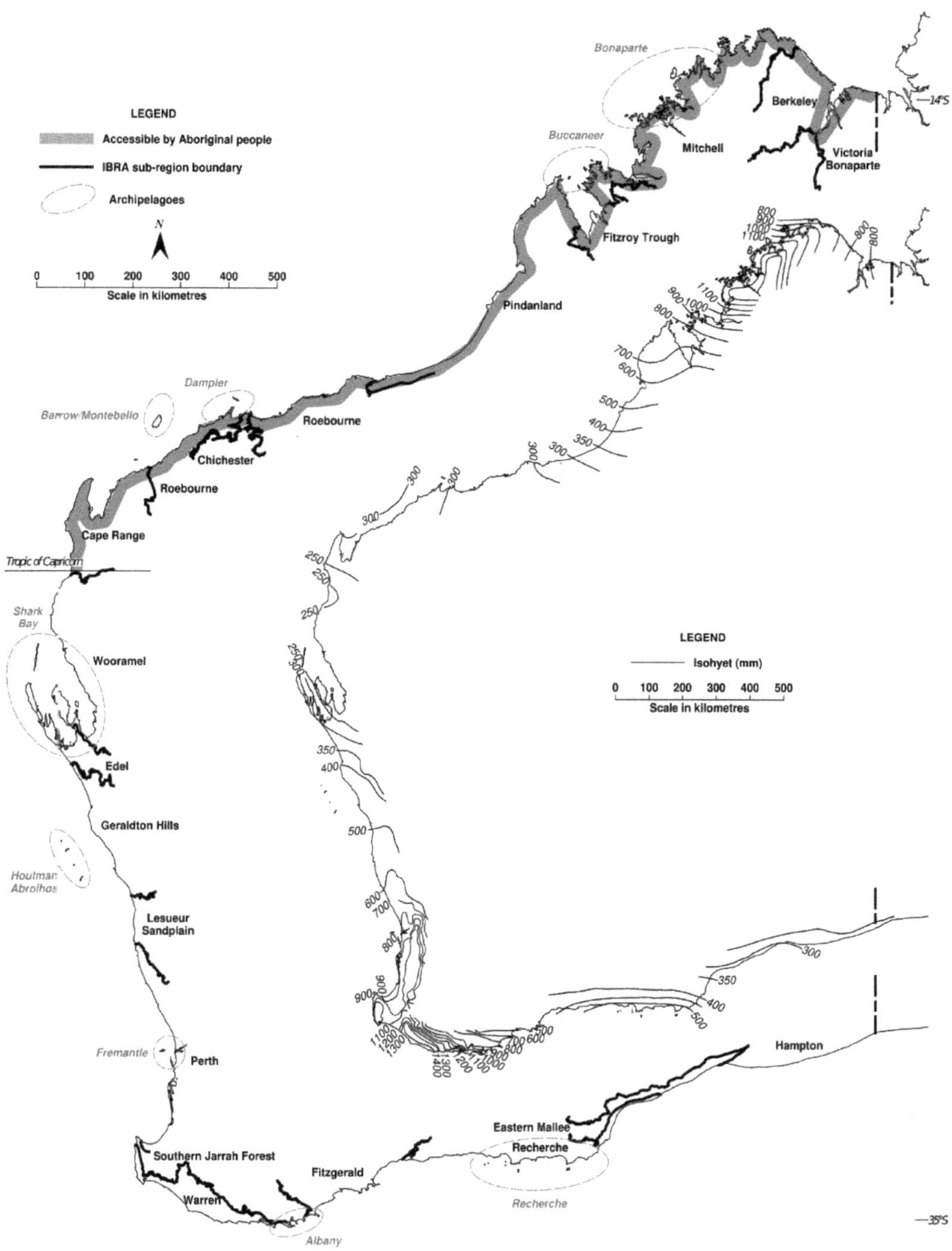

Map 1.1. Coastline of Western Australia, showing where Aboriginal people visited offshore islands, boundaries of coastal bioregions, location of major archipelagoes, and (inset) mean annual rainfall along the coastline.

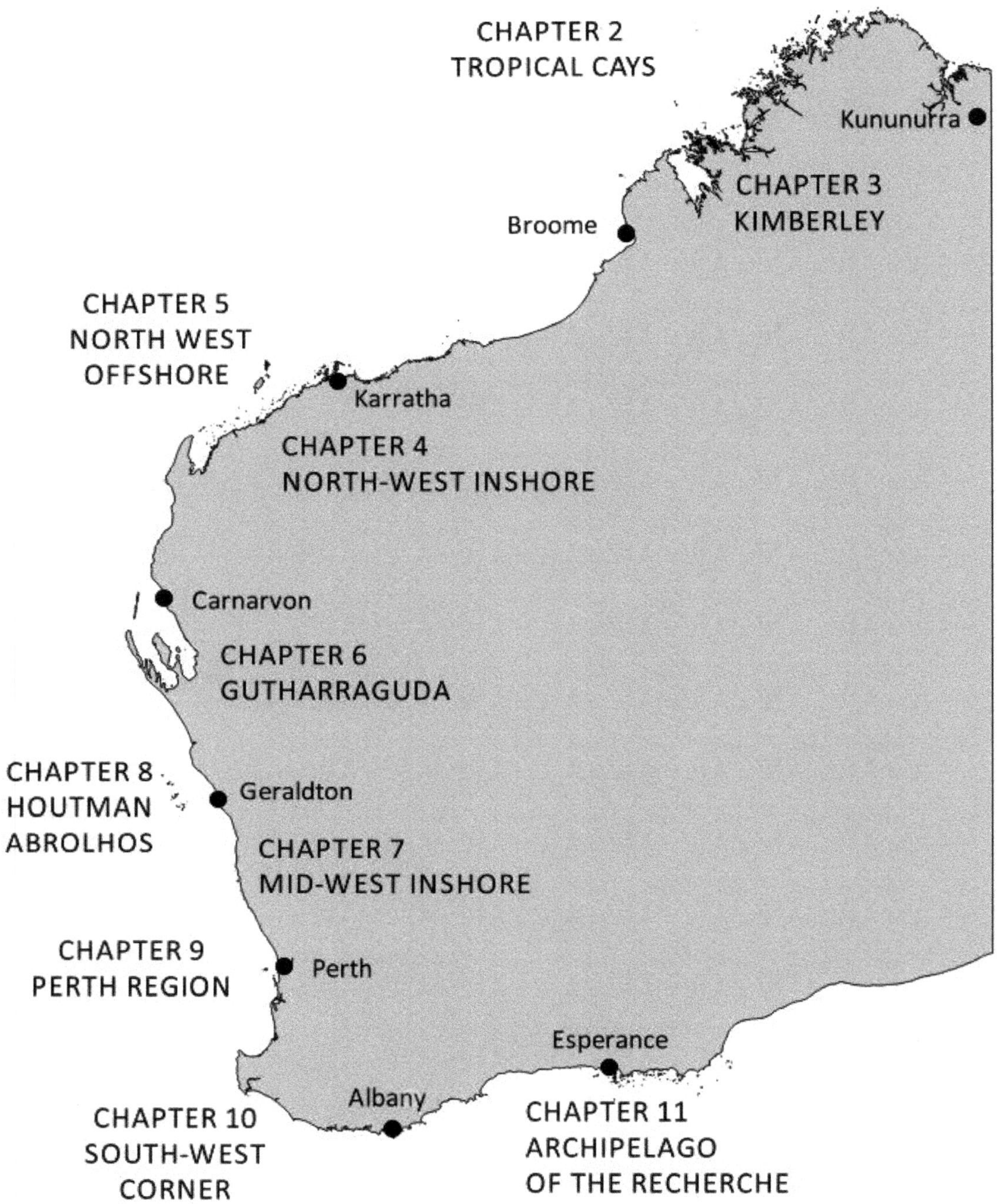

Map 1.2. Map of Western Australia, showing the link to each chapter of this book.

CHAPTER 1: INTRODUCTION

We have written this book to share our knowledge of, and enthusiasm for, the islands of Western Australia (WA). The State of Western Australia, despite it governing a landmass one-third of the area of Australia, is remote from where most Australians live, and indeed is remote from most parts of the globe. Perth shares with Honolulu the distinction of being the most isolated capital city on Earth. Tourists who travel to Australia from Europe or North America mostly visit the famous landmarks on the side of Australia opposite to WA.

Only two of Western Australia's islands are familiar to tourists from within the state, elsewhere in Australia, or outside Australia. The best known is Rottnest Island because of its proximity to Perth and recently-acquired global fame from 'quokka selfies' taken by 'celebrity' visitors. The next most known island is Penguin Island (just south of Perth), where little penguins and sea lions can be viewed. Barrow Island has become familiar to the many workers building and operating the oil field and liquefied gas plant, but it is not a tourist destination. Apart from these few, very few visitors from outside WA would be familiar with its more than 3500 islands. This situation is slowly changing with tourism authorities marketing some other island destinations, including Dirk Hartog Island and the Houtman Abrolhos archipelago. However, most people who live in WA or have the opportunity to visit will never make landfall on any island other than Rottnest.

This book is intended to provide a vicarious experience of doing so. We have both studied WA's islands periodically for 50 years, having visited several hundred islands and researched aspects of some in great detail. We have published the results of our studies in the appropriate science and natural history journals. Having retired, we think it now appropriate to distil the essence of what we have learned and present this knowledge and understanding for those with an interest in island life. We combine this with information gathered by many other scientists and natural historians.

In addition, books about Australian islands published in Australia and books published about Earth's islands usually overlook or minimise the national and global importance of WA islands (Table 1.1). However, the situation with books about Australian islands is improving. Books with a global focus continue to pay scant regard. This is another reason for our writing this book.

Western Australia was once a remote colony of the British Empire. It was slow to develop because most colonists from Britain preferred to settle in New South Wales or South Australia. WA's infertile soils made farming a doubtful proposition for many, and the distance from markets often made trading an unprofitable enterprise. The Mediterranean-type climate of the State's south-west, with little rainfall between October and April, resulted in low density stocking of sheep and cattle. This, together with the small area of most islands, had a great benefit for island life. Very few WA islands were farmed or used for pastoral activities, and the footprint of European settlers on them was light and intermittent. This fortunate circumstance resulted in almost all of WA's islands retaining their full complement of terrestrial mammal, seal, seabird, land bird, reptile, frog, invertebrate, and plant species that were present during the period from 1829 to the 1880s.

What makes islands interesting?

From the 1400s to the 1800s, Portugal, Spain, Britain, the Netherlands and France sought to increase their sovereignty over lands outside Europe. Often the first landfalls were on islands, such as the Canary Islands, West Indies, and islands in Southeast Asia or the Pacific Ocean. Navigators became aware of the presence of different species of plants and animals on these islands, as well as local peoples. This aroused the interest of scholars in Lisbon, Madrid, London, and Paris and led to naturalists travelling to islands. So began the collection of specimens and their return to museums and herbaria in Europe, followed by scientific description and naming of the species discovered. At this stage, scholarly interest in the floras, faunas, and indigenous people was largely observational.

This changed in the 1830s and 1850s with the visits of Charles Darwin and Alfred Wallace to many islands. Both independently attempted to explain why so many different species occur on islands in archipelagoes. Their observations and reasoning led to a significant breakthrough in understanding, with the formulation of the concept of evolution of species via the process of natural selection. Isolation of populations originally of the same species for thousands to millions of years resulted in their divergence and the formation of new species. By visiting many islands, Darwin, Wallace and their successors could compare and contrast island life on islands differing in rainfall, vegetation, area, elevation, and distance from the mainland or from other islands.

It was also realised that the origin of islands was an important factor in explaining variation in the numbers and kinds of species present. Oceanic islands have never been joined to continents and are usually millions of years old. The species present on them either came from elsewhere or evolved on the island. In contrast, continental islands were once part of the adjacent continent, and have existed only since sea levels rose after the last ice age ended some 15 000 to 7000 years ago. They are therefore much younger than oceanic islands and seldom have species that evolved on them. The defining characteristic of life on continental islands is that it has resulted from extinction of some species that were present when sea level rose.

What makes continental islands interesting?

There are many more continental islands on Earth than oceanic islands and almost all islands within the State of Western Australia are continental. Because most are visible from the coast of continents, they are familiar and may have been unjustly regarded as ordinary, commonplace, and unremarkable. This is an undeserved reputation!

Some continental islands provide the most convincing examples of natural extinction of populations of species during recent millennia, i.e. in the absence of humans. Subfossils (unfossilised animal bones no older than a few hundred years or, in sheltered locations, a few thousand years) found in caves or in the surface soil provide strong evidence of the impact of millennia of isolation on island populations of bird and mammal species. Factors implicated in this degradation of ecosystem integrity are numerous, and include small population size, inbreeding reducing the genetic fitness of isolated populations, loss of apex predators and consequent change in the 'balance' between species, competition between species for scarce resources, large nesting populations of seabirds fertilising the soil and thereby benefiting some plant species and disfavouring others, and increased exposure to salt-bearing winds following encirclement of the newly isolated landmass by the ocean. Subfossils can also reveal changes since European settlement, for example on Dirk Hartog and Faure Islands where several mammal species disappeared, with their bones remaining in caves, after pastoral stations were established and cats became feral.

Continental islands also offer a sobering outlook for species conservation in mainland reserves and national parks. In Australia such areas are no more than 120 years old. Many are now completely isolated because the surrounding vegetation has been removed or greatly modified in order to create farms, pastoral properties, suburbs, mines, and plantations of introduced tree species. It is most unlikely, based on the situation observed on islands, that these mainland remnants will retain all of the species present at the time these remnants were set aside for the conservation of nature. The study of islands may help predict and manage change.

What makes Western Australian islands interesting?

WA's islands lack the high levels of biodiversity present on the large tropical islands of Java, New Guinea, Sumatra and Borneo, on the ancient islands of New Zealand, New Caledonia and Madagascar, or on oceanic archipelagoes (including the Indonesian, Galápagos, Canary, Seychelles, and Hawaiian islands). WA's islands are young and there has been insufficient time for many endemic species and subspecies to evolve.

WA's islands do, however, offer much more biodiversity than Greenland, the Arctic islands of Canada, islands of the subantarctic, or the British islands, despite many of these islands being large. Most of these islands have experienced extensive glaciation during the last ice age and parts remain ice-bound for much of the year, as well as having long periods of human impact. WA has not experienced glaciation for hundreds of millions of years.

As previously stated, most of WA's islands are continental in nature and are the high hills and plateaux of a drowned landscape. However, a few islands are cays, the result of deposition of sand or coral rock on rocky reefs.

Western Australia has ten archipelagoes, not all named as such: Bonaparte, Buccaneer, and Dampier Archipelagoes, Montebello and Lowendal Islands, three groups in the Houtman Abrolhos, and the Archipelago of the Recherche (Map 1.1). Like the world famous Galápagos and Hawaiian archipelagoes, some WA islands provide the opportunity for tourists and scientists to compare and contrast the biota of individual islands. For example, Rottnest Island has the dugite, a species of snake, whereas nearby Garden Island has the tiger snake and carpet python. Each also has a different species of wallaby (quokka on the former, tammar wallaby on the latter).

The many islands of south-west WA are remarkable because they were not accessed by First Nations people, who colonised Australia at least some 50 millennia before the islands were formed. The Aboriginal people south of the Dampier Archipelago presumably lost the ability, or no longer needed, to construct watercraft. This circumstance is very unusual for continental islands elsewhere in the world. In Australia, it is shared with islands of South Australia and those between Tasmania and Victoria. These larger islands are renowned for their dense populations of wallabies and other mammal species. This differs from the islands of north-west WA, which Aboriginal people visited by means of rafts. Many of these islands lack wallabies, and if they do occur, only small wallaby species are present and they occur only on larger islands. The smallest of these is Wollaston Island (708 ha).

Our experience with islands

We have visited, and in some cases, lived on for a week or more, many islands. International highlights include the Galápagos, Britain, Shetland, Manhattan, Sjælland (Denmark), Bintan (Indonesia), Vancouver Island and the Queen Charlotte Islands in Canada, some Alaskan islands, and North and South Islands (New Zealand). Australian islands (excluding WA) that we have visited or where we have researched include Norfolk, Lord Howe, Tasmania; Curtis, Deal, Flinders and King islands in Bass Strait; Kangaroo (South Australia); Heron (Queensland); Phillip (Victoria); and Christmas Island and the Cocos Keeling Islands (Commonwealth Territories). The results of scientific studies have been published on the avifaunas of Tasmania, Bass Strait islands, and Kangaroo Island.

Our observations made on the floras and faunas of these islands provide useful points of comparison with our own studies of the biota of WA islands. Each of us has landed on several hundred WA islands, including all of the archipelagoes. We are of course not the first scientists to research aspects of the ecology and natural history of WA islands. Published reports by others provide historical perspective and complement our studies.

Islands – their awesome and sublime nature

Approaching, and then landing upon, an island often evokes feelings of awe, mystery, and curiosity (Plates 1.1-1.9). Before a first visit the emotive response is: What will I discover? Will there be species not commonly found on other islands? Will species expected to be present instead be absent? A quick look for animal tracks on beaches often reveals which mammals might occur there. Will there be unusual communities of species, unexpected vegetation types and habitats, and large populations of breeding seabirds? These questions even arise with second and subsequent visits.

Boat owners are also attracted to islands but not usually to study their natural history. The opportunity to relax and picnic on a beach in sight of the mainland coast, with few, or no other, people present is highly prized, especially if the beach is protected from prevailing winds. Sometimes the presence of sea lions sleeping on the beach evokes a sense of wonder about the natural world and inspires admiration of such remarkable species. Such situations provide authentic learning opportunities for children.

We think it no coincidence that islands have often provided fictional settings for high drama in human interactions, as well as emphasising the necessity for marooned people to demonstrate adaptability, resourcefulness, and resilience. Famous examples include *The Tempest* (William Shakespeare, c. 1610), *Robinson Crusoe* (Daniel Defoe, 1719), *The Swiss Family Robinson* (Johann Wyss, 1812), *The Coral Island* (Robert Ballantyne, 1857), *Treasure Island* (Robert Stevenson, 1883), *The Secret Island* (Enid Blyton, 1938), *And Then There Were None* (Agatha Christie, 1939), *Lord of the Flies* (William Golding, 1954), and *Dirt Music* (Tim Winton, 2001).

Adaptability and flexibility have also confronted populations of plant and animal species isolated from populations on the adjoining mainland. More often than not, however, species have fixed behavioural and ecological responses. It is improbable that seabirds attempting to establish new breeding populations on newly formed large islands could succeed in the presence of mammalian predators. Instead seabirds are more likely to nest successfully on small islands, on which predators cannot persist.

This book

We have written this book with three broad objectives in mind.

First, the islands of WA are a magnificent natural heritage dating from the conclusion of the last ice age. Their natural history deserves to be better known, understood, conserved, and appreciated.

Second, continental islands offer a gloomy prognosis of the long-term future of many national parks and nature reserves on mainland Australia. Left as they are, with minimal intervention, areas of remnant natural vegetation present on the mainland set aside for nature conservation will degrade and continue to lose species. Islands tell us what the flora and fauna of these mainland reserves may resemble within several thousand years. Scientifically-researched actions taken now and continued into the future should ensure that this process of local extinction can be slowed, and in some cases minimised and even prevented.

Third, islands continue to offer significant protection from destructive processes currently underway on mainland Australia. The most significant are:

- Exclusion of exotic predators, particularly the cat, fox, black rat, and house mouse.
- Exclusion of the cane toad, which poisons some native predator species (although toads may arrive on some Kimberley islands during floods).
- Exclusion of vectors of plant and animal disease, e.g. *Phytophthora*, viruses.
- Mitigation of the impact of changes in climate. Island environments are buffered from increasing temperature by the ocean surrounding them. In the future islands may provide sanctuary for species sensitive to prolonged or extreme heat.
- Reduced risk of wildfire. Those WA islands that were not visited or occupied by Aboriginal people should have experienced less fire than those that were visited. Such islands are likely to have provided safe haven for species sensitive to particular combinations of fire frequency, intensity, and season of burning.

Structure of this book

We recognise that a broad subject such as the islands of WA can be written about from many perspectives. This book could have been organised by focusing each chapter on groups of species, such as seabirds, land birds, mammals, reptiles, frogs, or trees and other plants. We could have recognised themes defined by insular characteristics (area, degree of isolation, elevation, degree of exposure to gales, annual rainfall, geology, whether part of an archipelago). Or we could have focused on a few representative islands that have been well studied.

Another alternative would have been to structure the book in terms of actual and potential threats to island life. These include tourism, farming of crops, grazing and browsing by sheep, goats or cattle, the introduction of predators detrimental to island species, and the frequency of occurrence of cyclones and tsunamis.

Instead we decided to examine the nature of island life by geography, starting from the tropical north of WA (Tropical Cays and Kimberley region), continuing south along the north-west coast (Pilbara region), farther southward along the west coast (Shark Bay, mid-west, south-west), and then east along the south coast (Map 1.2).

We have used common names for mammals, birds and well-known reptiles, and provided scientific names for less well-known reptiles, frogs and plants. Common names we have used are shown in Appendix 1 together with their scientific names. Technical terms have also been avoided as much as possible, and those we found necessary to use are defined in Appendix 2 - Unfamiliar terms explained. We have kept referencing to a minimum. For economy of space we have preferenced, where available, reviews as these provide hundreds of references to the extensive literature that has been published about the islands of WA. References we have consulted are shown in Appendix 3 - Further reading.

Acknowledgements
We thank the following for their assistance: Kim Doohan (anthropological advice for Chapter 3); Alex Errington PSM and Anne Greig (land tenure information); Alex George AM and Elizabeth Rippey (botanical information); Cheryl Lohr (Chapter 4); Keith Morris (Chapter 13); and Lisa Wright and Nicole Wrexford (Librarians in the Department of Biodiversity, Conservation and Attractions [DBCA]). Nic Dunlop, Ron Johnstone, and Denis Saunders kindly drew our attention to several errors in the first printed edition (3 November 2022).

Most of the images in this book are from our visits to WA islands, many during the 1970s.

Although we used high quality cameras then available, images are usually inferior to those that can be taken with digital cameras that only came into general use in the 1990s. Our photographs, all from the 1970s and 1980s, were taken on colour film and processed into slides. For this book, these had to be professionally scanned with a high quality digital scanner.

We thank Rowan Clarke (Chapter 2), David Pearson (Chapter 7), and Chris Surman (Chapter 8) for kindly allowing us to use images from their research undertaken on islands along the WA coastline. Lucy Clausen (DBCA) allowed us to include an image used in Chapter 13. Lauren Cabrera (DBCA) permitted use of images from the Bert and Babs Wells Collection, held by DBCA. Sally Treasure facilitated access, and Peter Nicholas scanned the slides.

Table 1.1. Relevant books that have come to our attention about islands, and their coverage of Western Australian islands.

Year of publication	WA islands mentioned	No. (%) of pages with reference to WA islands	Total No. of text pages in the book	Reference
Australian scope				
1965	Rottnest, Houtman Abrolhos, Montebello Islands, Cockatoo	16 (8%)	203	Beatty 1965
1970	Rottnest, Houtman Abrolhos, Dirk Hartog, Barrow, Buccaneer Archipelago, Cockatoo	12 (9%)	133	Baglin & Mullins 1970
1986	Barrow, Archipelago of the Recherche, Houtman Abrolhos, Bernier, Dorre, Dirk Hartog, Montebello Islands, Rottnest	27 (17%)	159	Dutton 1986
1997	Pelsaert, Bernier & Dorre, Dampier Archipelago, Bigge, Mondrain	61 (22%)	281	Chester & McGregor 1997
2019	Buccaneer Archipelago, Lacepede, Montebello, Barrow, Dirk Hartog, Houtman Abrolhos, Rottnest, Archipelago of the Recherche	33 (15%)	222	Wheeler 2019
2019	Rottnest, Houtman Abrolhos, Lacepede, Browse, Ashmore	46 (20%)	233	Taylor 2019
2019	Rottnest, Dampier Archipelago, Archipelago of the Recherche, Dirk Hartog, Penguin, Houtman Abrolhos, Buccaneer Archipelago	56 (24%)	232	Anon. 2019
Global scope				
1965		0	407	Carlquist 1965
1969		0	221	Berrill & Berrill 1969
1976		0	139	Sparks 1976
1980		0	?	Amos 1980
1994		0	153	Stevenson & Talbot
2009	Rottnest	3 (0.3%)	982	Gillespie & Clague 2009
2012	Rowley Shoals, Houtman Abrolhos	2 (0.7%)	306	Fischer 2012

Plate 1.1. Safe landing on Chatham Island via dinghy. May 1975. Credit I Abbott.

Plate 1.2. Safe landing on Wilson Island via helicopter. April 1993. Credit: AA Burbidge.

Plate 1.3. Safe landing on Woody Island via wharf. February 1976. Credit: I Abbott.

Plate 1.4. Safe landing on Bedout Island via LARC by a navigation aids team. June 1982. Credit: AA Burbidge.

Plate 1.5. Campsite of a biological survey team on Salisbury Island. March 1982. Credit: AA Burbidge.

Plate. 1.6. Campsite of a biologist on Carnac Island. 1970s. Credit: I Abbott.

Plate 1.7. Scientist at work: Radio-tracking translocated mala, Trimouille Island, Montebello Islands. August 1998. Credit: AA Burbidge.

Plate 1.8. Scientist at work: A rock parrot captured temporarily in a mist-net, Woody Island. February 1976. Credit: I Abbott.

Plate 1.9. Awesome grandeur: The sheer cliffs on the southern side of Chatham Island, 186 m high. May 1975. Credit: I Abbott.

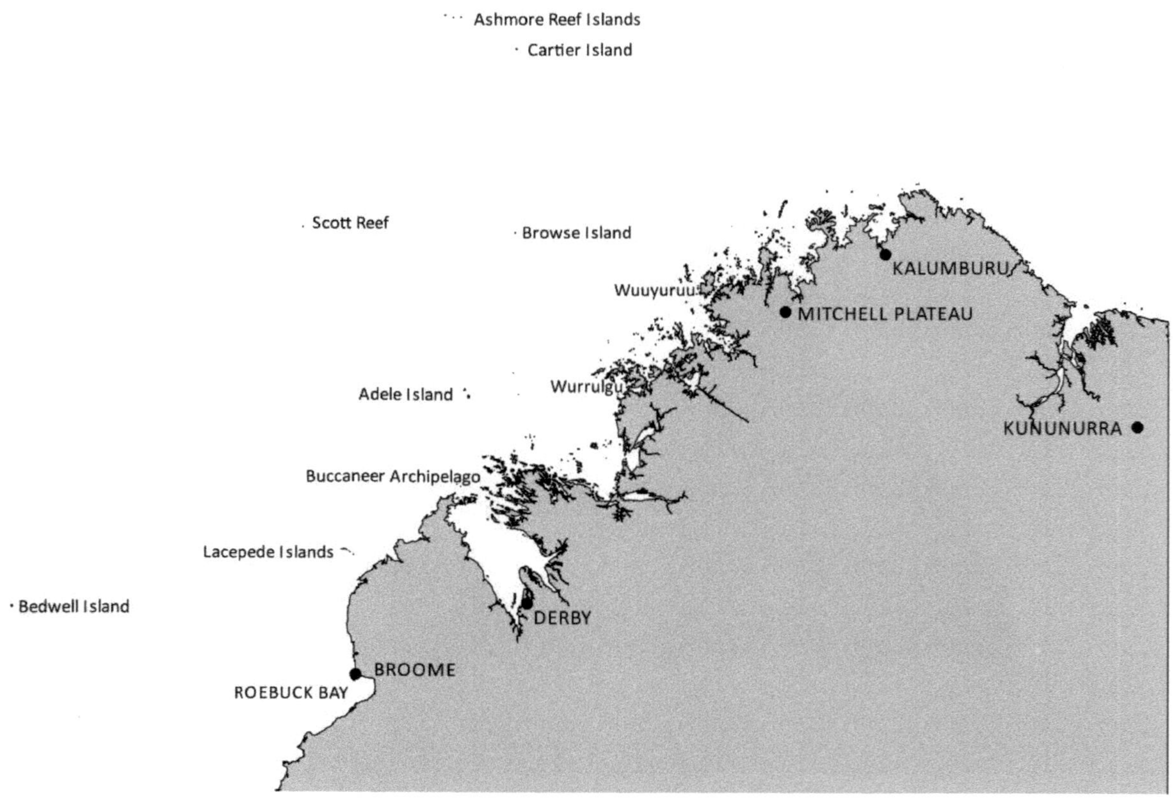

Map 2.1. Tropical cays of Western Australia.

CHAPTER 2: TROPICAL CAYS

'As we approached [Adele Island], the number of birds seen was much greater than had ever before been observed by us in this region, and as we were wading across the reef we could see them hovering over the island in a perfect cloud, the air being filled with their harsh cries'. James Walker, 1891 visit of HMS *Penguin.*

INTRODUCTION

There are c. 24 sand cays north of latitude 20° S (Table 2.1). The reason for the uncertainty is that a few reefs (Mermaid Reef, Seringapatam Reef, Hibernia Reef) appear to have sandbanks which are exposed only at low tides. Little Turtle Island is awash at HW spring tides. Although seabirds may alight on these pieces of land, there is no possibility of breeding. Therefore we do not regard ephemeral sandbanks as cays.

Coral cays are islands that have been formed by ocean currents depositing coral sand and fragments on rocky reefs or coral reefs. Some are atolls, ring-shaped islands built on the rim of the crater of submerged volcanoes, enclosing a lagoon. They are restricted to warm (> 18° C) ocean waters in the tropics. Many cays are small (<5 ha) and most are of low elevation (<5 m). Some are the most remote islands in north-western Australia, with the Ashmore islands closer to Roti (Indonesia) and possessing an informal Indonesian name, Pulau Pasir.

Cartier Island and the cays situated on Ashmore Reef are Australian Government territories; the remainder are in Western Australian Government waters (Map 2.1). Bathymetric charts reveal that the cays on Rowley Shoals, Ashmore Reef, and Scott Reef, as well as Browse Island and its reef, emerge from deep waters (200-500 m).

European discovery and naming. Almost all of the cays were named before the colony of Western Australia (WA) was founded in 1829. Surprisingly, none were discovered as a result of shipwreck. William Dampier sighted Browse Island in 1699, but did not name it. Cartier was the first island to be discovered and named (1800). French navigators named Lacepede and Bedout islands in 1801, and Adele, North Turtle, and Lesueur islands in 1803. Peter Heywood RN in HMS *Vulcan* discovered Scott Reef in 1801 but its cays were left unnamed until 1909. Ashmore Reef was first sighted in 1811 by Samuel Ashmore, captain of the *Hibernia*. Phillip King RN named Bedwell and Jones islands during his 1818-19 voyage. Browse Island was named in 1838 after the captain of the *Lynher*, which landed the first British explorer (George Grey) on the Kimberley coast. Other cays have descriptive names (Sand, Sandy) or are named because of close proximity to a coastal feature on the mainland (Stewart). No cays were visited by Europeans until after they settled the north-western mainland coast at Cossack in 1863, followed soon after by small settlements at Port Hedland, Broome, and elsewhere. Britain annexed Ashmore Reef in 1878 and Cartier Island in 1909, but ceded them to Australia in 1934.

Biological discovery. Knowledge of the natural history of the cays began to accrue in May 1891, with the leisurely visit to Adele Island by James Walker RN on HMS *Penguin.* The following year he published a paper on the birds observed there. He recorded 10 seabird species (four breeding), and noted that one species was shy and another was indifferent to human presence. In April 1901 naturalist John Tunney visited Bedout Island and found six seabird species breeding. There was then a protracted hiatus, with observations on the seabirds first published for Cartier, Ashmore, Browse, Lacepede, and Scott Reef islands in 1952, and for North Turtle Island in 1975.

It can be difficult to visit cays – they are far from the nearest port; anchorage is unreliable; timing has to align with high tide in order to facilitate access; and the time available for inspection on the cay is constrained by the tide. We therefore sympathise with Dom Serventy's plight on his visit to Adele Island in 1949: It took him 90 minutes each way to wade across the reef at low tide, leaving him only 60 minutes to observe the avifauna on only part of this large island.

BIOLOGICAL CHARACTERISTICS

Tropical cays are some of the most simplified ecosystems on Earth. Most are scantily vegetated with cosmopolitan plant species dispersed by oceanic currents (Plate 2.1, 2.2). For example, the Ashmore cays (total area 55 ha) have only 23 plant species, with grasses, herbs, and creepers predominant. Trees are absent. Few cays provide sufficient area of habitat to sustain viable populations of native land bird, reptile, and mammal species. In 1977, one of us (IA) used a sweep net to collect invertebrates from *Spinifex longifolius*-dominated vegetation on Adele, Bedout, and Lesueur islands. This technique yielded many species including numerous insects, possibly resulting from the absence (or low numbers) of insectivorous land birds and lizards. Even larger predators, such as eagles and other raptors, are absent from most islands, despite prey species (seabirds) being abundant.

These cays have never been part of mainland Australia. Their biota has either flown there, been carried passively by wind, or has arrived as flotsam via ocean currents. No amphibians or native mammals have established. Only one species of land bird, the tawny grassbird, is resident (on Adele and Lesueur islands) and only one lizard species, the skink *Ctenotus fallens*, has been recorded (on Bedout Island). Although an unidentified (probably brown) quail and a buff-banded rail have been recorded on Adele and Browse Islands respectively, neither has established. This rail is present on some of the Ashmore islands. Most of the cays are renowned for the great diversity of seabird species that breed there (Table 2.2), often in prodigious numbers (discussed below).

Notable fauna. Seven cays have an outstanding number of seabird species breeding: East Ashmore (15 species); Middle Ashmore (14); Adele, Bedout and West Lacepede (10); Middle Lacepede (8); and West Ashmore (5). No one cay has all 21 species breeding (Plates 2.3-2.9). The wedge-tailed shearwater breeds only on West Ashmore, and both tropicbird species breed on the three large islands on Ashmore Reef, with the red-tailed tropic bird breeding also on Bedwell Island (Table 2.2). Breeding populations of the three booby and two frigatebird species occur in WA only on these cays. However, most tern species, and the shearwater, pelican, cormorant, and silver gull also breed on many of the inshore islands of WA. Eastern reef egrets also occur in large numbers on many cays. Breeding raptors are, however, few and limited in their occurrence: White-bellied sea-eagle on Bedout, North Turtle, and East Lacepede islands, and eastern osprey on Lacepede and Lesueur islands. Letter-winged kites have twice been recorded on Adele Island, where they subsist on Pacific rats.

In >100 years of study, the land bird fauna of Adele Island has remained unchanged, with the same species (tawny grassbird) present. Such lack of species turnover is unusual by global standards. Nor have West Ashmore, Browse or Bedout islands been colonised by any land bird species.

Turtles come ashore on cays, often in large numbers, and lay their eggs above the reach of high tide. For example, the crew of HMS *Beagle* in 1840 brought 30 green turtles aboard from Little Turtle Island in four hours; 'the number that could have been here obtained was enormous'.

Accidental fauna. Species of land bird that migrate annually to and from Australia, as well as vagrants from the Australian mainland and Asia, have been recorded on some of these cays. Their presence is temporary; none breed there, and they are usually recorded in very small numbers. Examples include black-eared cuckoo, little grassbird, and red-backed button-quail (West Ashmore Island), sacred kingfisher (Cartier, East Lacepede Islands), mangrove kingfisher (Bedout Island), nankeen kestrel (Adele, Browse Islands), brown falcon (Adele Island), boobook owl (Adele Island), rainbow bee-eater (Cartier Island), fairy martin (West Ashmore, Cartier, Browse Islands), pied honeyeater (West Ashmore Island), Australian pipit (Bedout Island), and magpie-lark (Adele, Browse Islands).

In December 1997 the Arctic warbler was recorded on one of the Lacepede Islands. The lesser coucal was seen on West Ashmore in 2005. Two migratory Asian land bird species, the Asian brown flycatcher and the island monarch, were first recorded (as vagrants) on

West Ashmore Island in 2009. Up to 2019 a further 32 species have been observed there. These records have no ecological significance as they represent stray birds.

BIOLOGICAL ATTRIBUTES OF PARTICULAR ISLANDS

These cays with wondrous numbers of seabirds, many breeding seabird species, and large colonies of seabirds have attracted the attention of many biologists. Some of their research demonstrates the dynamic nature of these populations.

Ashmore cays The best-studied of all the tropical cays of north-western Australia, Ashmore cays have been visited by experts on 17 occasions between 1998 and 2010. Both species of tropicbirds are irregular in their nesting on the three largest cays, with red-tailed tropicbird more common than the white-tailed (e.g. 1-17 vs 1-2 nests on West Island). On Middle and East Islands, the numbers of other seabird species are considerable, with the following maxima having been counted: Sooty tern (40 000); brown noddy (31 000); brown booby (5000); greater crested tern (2800); lesser frigatebird (2500); black noddy (450); red-footed booby (220); and great frigatebird (65). However, there is remarkable variation in numbers of birds, and their nests, from year to year.

Box 2.1. Rohan Clarke (1973-)

Rohan Clarke first visited Ashmore Reef in the year 2000 as part of a birdwatching expedition, and has been fortunate to return regularly. As an ecologist at Monash University he leads a research program focused on documenting population trends and ecology of breeding seabirds, and migrant shorebirds and land birds that visit this system. This is a team effort, with expeditions including expert staff, PhD students, and skilled citizen scientists. Collectively these efforts have resulted in >20 scientific publications and a similar number of management reports to agencies and industry. Because research at Ashmore Reef has an applied focus, outputs have extended to whole-of-island management, marine park recommendations, and invasion ecology of Asian house geckoes and tropical fire ants. These ants threaten the terrestrial biodiversity values of the Ashmore Reef system. Rohan also works on monitoring and conservation-based programs on the Cocos Keeling Islands in the Indian Ocean and Norfolk Island in the Pacific Ocean.

Adele Island Located c. 100 km NNE of Cape Leveque, Adele Island is a low, flat, sandy cay situated on a large reef. Low, often inundated, areas support grasslands of saltwater couch *Sporobolus virginicus*, while slightly more elevated soils have a dense cover of *Spinifex longifolius*.

Adele Island has the greatest population of breeding seabirds of any Kimberley island: Red-footed booby (first reported and breeding recorded in 1990, 14 pairs); brown booby (up to 7500 pairs); masked booby (up to 400 pairs); great frigatebirds (up to 300 pairs) and lesser frigatebirds (up to 3000 pairs); Caspian tern (up to 140 pairs); and, in the past, lesser crested tern. As well, other birds that depend on the ocean for food breed there: pied cormorant, Australian pelican, and eastern reef egret (up to 300 birds).

Adele Island is a major nesting site for two seabird species: Brown booby and lesser frigatebird. It is the only WA island where red-footed boobies and great frigatebirds breed.

Brown boobies have a wide distribution on and near tropical islands worldwide. Western Australian nesting colonies of brown boobies in the Kimberley, plus the one on Bedout Island in the Pilbara, are probably the largest in the world, totalling c. 10 000 pairs.

Bedout Island This island, like the Ashmore cays, demonstrates the dynamic nature of breeding populations of seabirds, as it has been visited frequently by ornithologists and experienced birdwatchers. Appreciable declines in numbers of the four most common species (masked booby, brown booby, lesser frigatebird, brown noddy) occurred between 1972 and 1978 (based on counts made in May).

The masked booby is a pantropical seabird species, with six subspecies recognised. The Bedout Island population belongs to a subspecies shared with Adele and North Keeling Islands. However, limited evidence is suggestive of genetic distinctiveness and may indicate that the Bedout Island population does not interbreed with other island populations.

Lacepede Islands Located 120 km north of Broome and 20 to 35 km offshore from the Dampier Peninsula, they are internationally recognised as an 'Important Bird Area' and are a major breeding site of sea turtles. All four islands are part of the Lacepede Islands Nature Reserve, which extends to low-water mark. Seabirds have been recorded breeding on all cays, but the major colonies are on Middle and West Islands. Here there are large colonies of brown booby and lesser frigatebirds. A 1982 visit counted 7370 occupied nests of the brown booby on West Island and 10 300 on Middle Island. There were also 5700 occupied nests of the lesser frigatebird on West Island, but none on Middle Island. In October 2010 c. 30 000 roseate terns were seen on West Island.

Black rats established on the Lacepede Islands in the late 1800s, probably from pearling luggers wrecked during a cyclone. They would have had a major impact on smaller seabirds such as terns and noddies. In 1986, an eradication project led by Dr Robert Prince, while researching sea turtles, removed this threat.

Bedwell Island This cay is located within Clerke Reef, the middle of the three Rowley Shoals c. 310 km west of Broome. Clerke Reef was named by Phillip King in HMC *Mermaid* in 1818 after Captain Clerke, a whaler, who first reported the feature. Bedwell Island, named by King after the *Mermaid*'s master's mate, is best known to ornithologists as a nesting site of the red-tailed tropicbird, which nests there in low numbers (up to 10 pairs). There are two records of the white-tailed tropicbird nesting, in 1993 and 2003. Other birds use the island as a resting place. Dead green turtle hatchlings were observed in 1982, but there is no evidence of nesting since then. Most visitors go to the Rowley Shoals to dive on the coral reefs and fish in designated places, as the group is a marine park.

Browse Island A coral rock and sand cay of c. 25 ha, of which c. 9 ha is vegetated, Browse Island is located on a large coral reef 220 km north of Koolan Island and c. 175 km from the nearest mainland coast. A light tower is situated there, as is a helipad used by the oil and gas industry. In 1699 William Dampier recorded that it was 'inhabited only by boobies and man-of-war birds [frigatebirds]'.

Browse Island once held large deposits of guano but these were mined out between 1870 and 1890, resulting in a highly-disturbed surface. A visit in June 1972 reported that there was sparse vegetation behind the beaches, mainly of the creeper *Ipomoea brasiliensis* and a few scattered low shrubs. Reef egrets were preparing to breed with nests being built in a solitary mangrove tree, but no seabirds were nesting. There were many hundreds of sea turtle nests on the beach. Introduced house mice can be abundant. In June 1977 c. 300 greater crested terns were nesting. Brown boobies (200 birds) and 12 bridled terns were recorded in June 1972, with one chick of the latter species found. This is yet another illustration of the dynamic nature and resilience of the seabird fauna.

FACTORS LEADING TO CHANGE IN THE BIOTA

Natural factors. Tropical cays do not exist in a benign physical environment. They are subject at least once each year to extreme winds and wave surges generated by tropical cyclones. The tidal range is 6-10 m.

Extra high (spring) tides can flood low lying parts of cays, as was observed on West Lacepede Island in May 2012 when nests and eggs of brown boobies were affected. The generation of severe wave action by the combination of low air pressure (during passage of a tropical cyclone) and tidal surge combined with a high tide can result in low cays being washed over, with dire consequences for vegetation and populations of seabirds. Tsunamis have doubtless helped impoverish cays of their biota. These massive waves are extremely powerful when land and adjacent reefs are encountered, and would be physically destructive to cays. In time, they would rebuild as a result of deposition of sand and coral fragments, and their previous biota would recolonise from other cays. Most of the seabird species breed all year, and losses from wind and waves can soon be made good.

Global warming has initiated small rises in sea level in the oceans. These are expected to continue, resulting in erosion of cays (reducing the area available as breeding sites for seabirds), more frequent over-washing by waves (impacting on nesting success), and ultimately in the submergence of cays, as the coral reefs on which they are situated may not keep pace with rising sea levels due to increasing ocean acidity. Coral reefs should also be affected by more frequent bleaching events. The long-term future of the seabird fauna in the waters of north-western WA is therefore uncertain.

The major cause of natural biotic disturbance on cays is the roosting and nesting of large numbers of seabirds. This has transferred, during centuries, huge quantities of nitrate and phosphate from fish populations to island soils, and has helped shape the floristic composition and vegetation structure. Predation by exotic rodents, often in plague numbers, can eventually cause breeding of smaller seabirds to cease. This then disrupts the transfer of nutrients from ocean to land, with knock-on effects to plant growth and population size of invertebrates.

Nesting turtles dig large craters in which to deposit their eggs. With large numbers coming ashore, as on Cartier Island and Sandy Island (Scott Reef), much of the surface substrate is regularly disturbed. This may reduce the extent of vegetation.

Human factors. Cays have been impacted to varying extent by Aboriginal people, Asians, and Europeans.

Aboriginal use. Aboriginal people living on the Kimberley coast made simple but buoyant rafts (Chapter 3). These watercraft were seen in c. 1916 as far offshore as 80 km north of Cape Bougainville. It seems improbable that islands beyond this range (Ashmore, Cartier, Browse, Cunningham, Bedwell, Adele) were visited. Bedout and North Turtle islands lie off that part of the Pilbara coast where Aboriginal people did not possess watercraft, and so were unvisited.

Inshore islands were visited for food, including turtle meat and eggs, and seabird meat and eggs. Such visits were necessarily seasonal and less likely to take place during the period when cyclones occur (October to March). These resources were probably consumed on the cay as rafts had limited storage capacity. Dom Serventy speculated that Aborigines may have been responsible for extirpation of some populations of breeding seabirds on inshore islands.

Indonesian and European voyagers. Remoteness and isolation should confer a high degree of protection from human disturbance. Sadly, this has not been so on these cays. Few are pristine.

From the 1700s, Makassan fleets of prahus (boats) annually visited the islands and bays of north-western Australia, collecting trepang. In 1803 the French navigator Nicolas Baudin came across c. 25 prahus near Cassini Island, 30 km from the Kimberley coast. Crew of HMS *Beagle* found a broken jar during their 1840 visit to Little Turtle Island. These remote cays supplied food for visitors from Asia in the form of turtles, seabirds and their eggs. No information about the extent of their depredations became available until 1949, when Dom Serventy visited many of the cays and found explicit evidence of wildlife poaching. He saw two prahus near Middle Ashmore Island. In September 1986 c. 25 prahus were present at Ashmore Reef. Between 1999 and 2001 female green turtles on Browse Island were killed by Indonesian fishers for their eggs, with the animals left to die.

These waters are now patrolled by the Royal Australian Navy and Australian Border Force vessels and aircraft, and killing of wildlife is thereby much reduced. Indonesian fishers may have previously exterminated populations of the red-footed booby on Middle and East Ashmore islands. This species has only been definitively recorded there since 1995. Nesting first occurred in 2002-2003.

In 1872 North Turtle and Bedout islands were leased for one year, for pastoral purposes. It is improbable, however, that livestock were ever landed, as these cays lack freshwater supplies.

Millennia of breeding on cays by large populations of seabirds has resulted in extensive deposits of their faeces (guano). In 1876 the government leased Browse and Lacepede islands for guano mining for a period of 10 years. Guano was mined (under WA Government permit) and mostly exported for use as agricultural fertiliser, usually until the resource was mined out (Plate 2.10). The largest quantities were extracted from the Lacepede Islands (37 226 tons), Browse Island (23 177 tons), and Jones, Lesueur, and Stewart islands (collectively 2577 tons). However, some of the guano resource was removed illegally before the government became aware of this practice. Even after 1873 the government exercised no direct supervision and thus did not always obtain the appropriate royalty payment.

In 1876 the government discovered that guano was being illegally removed from the Lacepede Islands. Americans were involved, and the government seized the vessel. This resulted in an international diplomatic incident. A resident commissioner was then dispatched, and he estimated the amount of guano available as c. 5000 tons (East Island), 40 000 tons (Middle Island), and 15 000 tons (West Island). Removal of guano involved many men living on islands, e.g. 27 men on the Lacepede islands in February 1877, and

c. 40 Chinese employed on Jones Island in 1884. In 1878 an American was said to be in 'possession' of Ashmore Reef. Guano mining of the Lacepede cays ceased in 1904.

Adele Island seems to have escaped guano mining. James Walker in 1891 reported 'a small patch of guano' near its south-eastern part. A guano lease was granted for Bedout Island in 1928, but was cancelled in 1930 before any mining took place. Little is known about guano extraction from the Ashmore islands, but deposits were depleted by 1891.

Other evidence of human visits on some cays is the presence of graves on East Ashmore, Cartier, Lacepede, Bedout, and North Turtle islands. Several islands (Cunningham, Adele, Browse, East Lacepede, Bedout, Lesueur) have automatic navigation aids (lights exhibited from a tower). We should be thankful that no lighthouses were constructed, as resident staff would have eaten seabird and turtle eggs, and possibly introduced pet dogs or cats. Until recently, these islands were serviced by an amphibious vehicle (LARC) that could at high tide pass over the surrounding reef and drive to the light tower. These regular visits are unlikely to have inadvertently transported any exotic species such as house mice or black rats. During the 1939-45 war, Cartier Island was used for bombing practice and was cleared of unexploded ordinance in 1986.

Following European settlement in 1863, its gradual expansion aided by the establishment of ports along the mainland coast, and increased local shipping, mariners took advantage of natural resources offered by cays. In April 1863 pelican eggs were collected from North Turtle Island, and in July 1863 'a score or two' of young pelicans were killed for their oil on the same island. Green turtles were also taken. In 1900 a company was granted a lease for capture of turtles on the Lacepede Islands, but it is unclear to what extent this occurred. At this time there was criticism in the press of large numbers of turtle being killed (on the Lacepede Islands?) solely for their eggs. In 1899 a Japanese man was said to have killed 1200 turtles for their shells, leaving the meat to waste.

In 1884 the Lacepede Islands were used by pearlers as a depot for abducted Aboriginal people who were forced to dive for pearl shell.

The predictable outcome of human presence in the region was the arrival of foreign species – black rats, Pacific rats, house mice, tropical fire ants, and various plant species. The black rats and house mice presumably came with the guano diggers (probably via off-loading of cargo from vessels) and from guano-loading vessels and pearling luggers being driven ashore during cyclones. A tropical cyclone in 1877 wrecked six ships that were anchored at the Lacepede Islands, waiting to load guano. Much wreckage was blown ashore. In March 1878 two guano-loading vessels were wrecked at Browse Island by a tropical cyclone, and in December another cyclone caused a vessel to run ashore there. Guano extraction sometimes entailed the construction of a tramway and jetty (as on the Lacepede Islands and

Jones Island), thus further facilitating the potential introduction of rodents. Usually a horse was used to convey guano to the jetty via a tramway; this would have necessitated the import of hay, increasing the chance of introducing weed species.

The local extirpation of breeding populations of sooty terns and brown noddies from Bedout Island may be attributable to black rats. These rodents were eradicated in 1981. Cats, pigs, domestic fowls, house mice, and black rats were also introduced to Browse Island. All except the mice are now absent. Black rats were eliminated from the Lacepede islands in 1986. Black rats are also present on East and Middle Ashmore islands, but were eradicated on West Ashmore Island in the 1980s. House mice occur on Middle and East Ashmore islands. The Pacific rat was probably introduced to Adele Island by Indonesian seafarers. It was first recorded in 1891, but was probably there many years before. Attempts to eradicate it began in 2004 but have so far failed.

Four species of exotic plants are known from the Ashmore cays, with two species established, and another two (corn, coconut palm) not yet naturalised. The fire ant population on the Ashmore cays is being baited, with the aim of eradicating this species (which impairs the hatching and fledging success of seabirds).

The Asian house gecko *Hemidactylus frenatus* was first noted on West Ashmore Island in 1990 but is suspected to have been introduced well before then because it is commensal with humans in Asia.

Most islands have accrued quantities of jetsam – bottles, fishing nets, cans, and plastic objects. One and a half days spent on Browse Island in August 2018 resulted in the retrieval of 7 kg of such rubbish.

A major risk to seabirds is spillage of oil from ships and offshore oil fields, and the discharge of sewage from tourist and other vessels. In 2009 a 75-day oil and gas leak in the Montara oil field in the Timor Sea resulted in an extensive oil slick. A 'small number' of terns and boobies were affected but the impact of the spill on the cay fauna does not appear to have been adequately studied.

Island management. The Australian Government has set a high standard with its regular surveillance of the biota of the Ashmore cays and Cartier Island since 2002. The WA Government actively monitored seabirds on Adele and Bedout islands during the 1970s-80s but has discontinued this. Visits by ornithologists and birdwatching tours have occurred intermittently since. Two surveys every third year seems a desirable frequency, given the expense involved in travel.

Cays at Rowley Shoals are included in a management plan, which has now expired. None of the other cays have specific management or monitoring plans. However, baiting on Ashmore Reef cays is taking place to protect seabirds from the tropical fire ant.

Sources of information

Introduction: Australia Pilot 1972; Berry 1993; British Admiralty 1880; Fairbridge 1948; Henderson & Henderson 1998; Royal Australian Navy 1966a, 1966b.

Biological characteristics: Abbott 1982; Carter *et al.* 2010, 2011; Carter & Swann 2011, Clarke *et al.* 2009, 2017; Coate *et al.* 1994; Fuller & Burbidge 1998; Johnstone & Storr 1998; Kenneally 1993; Milton 2005; Serventy 1952; Swann 1998; Tunney 1902; Walker 1892.

Biological attributes of particular islands: Abbott 1979, 1982; Burbidge & Fuller 1996; Burbidge *et al.* 1996; Bush & Lodge 1977; Coate 1977; Coate *et al.* 1994; Fuller & Burbidge 1998ab; Kingsley *et al.* 2020; Moro *et al.* 2019.

Factors leading to change in the biota. Natural factors: Abbott & Wills 2016; BOM 2021; Collis 2013.

Factors leading to change in the biota. Human factors: Basedow 1918; Bridge 2020; Brockman 1987; Coate & Coate 1986, 2000; Coghlan 1885; Collis 2013; Hunt 1986; newspapers; Somaweera *et al.* 2020; Stokes 1846; Stuart 1923; Taylor 1987.

Island management: DEC & MPRA 2007; Mott *et al.* 2020; Woodward 1917.

Table 2.1. Synopsis of physical and biological attributes of tropical cays.

Island	Area (ha)	Elevation (m)	Distance from mainland WA (km)	Presence of vegetation	No. breeding seabird species
West Ashmore	28	2.5	335	yes	5
Middle Ashmore	13	3	325	yes	14
East Ashmore	13	3	320	yes	15
Splittgerber	c. 1.5	?	320	yes	1
Adele	218	3	85	yes	10
Browse	25	6	175	yes	2
Sandy Lacepede	6	4	25	yes	1
West Lacepede	82	7	33	yes	10
Middle Lacepede	63	2	30	yes	8
East Lacepede	<0.5	2	20	yes	2
Bedout	24	4	40	yes	10
North Turtle	50	10	13	yes	3
Cunningham		4	300	no	0
Bedwell		2	300	no	1
Cartier	1.7	1	?350	no	1
Sandy (Cape Talbot)	15	3	0.8	yes	?
Sand (Long Reef): 2 cays	1 (both)	6	24 (both)	?	3
Sand (Jamieson Reef)	0.05	2.7	33	no?	0?
Sandy (NW) (Scott Reef)	5.4	?	250	no	2
Jones	17	3	26	yes	1
Stewart (2 cays)	6 (W); 21 (E)	6	6 (W); 5 (E)	yes	1
Lesueur	77	3	9	yes	1

Table 2.2. Seabird and turtle species breeding on tropical cays.
Abbreviations used:
For Ashmore island names: W, West; M, Middle; E, East; S, Splittgerber;
For Lacepede island names: W, West; M, Middle; E, East, S, Sandy.
X = present as a breeding species

Species	Ashmore	Adele	Bedout	Lacepede	North Turtle	Cartier	Browse	Bedwell	Sand (Long Reef)	Jones	Scott Reef	Stewart	Lesueur
Red-tailed tropicbird	W M E							X					
White-tailed tropicbird	W M E												
Wedge-tailed shearwater	W												
Australian pelican		X		W M	X				X				
Lesser frigatebird	M E	X	X	W M									
Great frigatebird	M E	X											
Red-footed booby	W M E	X											
Brown booby	W M E S	X	X	W M S							X		
Masked booby	M E	X	X										
Great pied cormorant		X		M	X				X				

Species	Ashmore	Adele	Bedout	Lacepede	North Turtle	Cartier	Browse	Bedwell	Sand (Long Reef)	Jones	Scott Reef	Stewart	Lesueur
Brown noddy	M E		X	W M E							X		
Lesser noddy	M E												
Black noddy	M E												
Silver gull		X	X	W M						X		X	
Sooty tern	M E		X	M E									
Bridled tern	M E			W			X						X
Fairy tern				W									
Caspian tern		X	X	W	X								
Roseate tern	M E		X	M									
Lesser crested tern	E	X	X	W M									
Greater crested tern	M E		X	W		X	X		X				
Total	5 14 15 1	10	10	10 8 2 1	3	1	2	1	3	1	2	1	1
Green turtle	W	X	X	X	X	X	X				X		
Flatback turtle				X									
Hawksbill turtle			?X		?X								

Omitted (no data): Sand Island (Jamieson Reef); Sandy Island (Cape Talbot).

Plate 2.1. Aerial view of Bedout Island, showing vegetation dominated by beach spinifex. June 1982. Credit: AA Burbidge.

Plate 2.2. Splittgerber Cay, Ashmore Reef. Credit: Rohan Clarke, Monash University.

Plate 2.3. Masked Booby, Bedout Island. June 1982. Credit: AA Burbidge.

Plate 2.4. Brown Booby, Adele Island. June 1972. Credit: AA Burbidge.

Plate 2.5. Lesser Frigatebird, Adele Island. June 1972. Credit: AA Burbidge.

Plate 2.6. White-tailed tropicbird, Ashmore Reef.
Credit: Rohan Clarke, Monash University.

Plate 2.7. Great frigatebird, Ashmore Reef. Credit: Rohan Clarke, Monash University.

Plate 2.8. Red-footed boobies and great frigate birds sitting on their nests, Ashmore Reef. Credit: Rohan Clarke, Monash University.

Plate 2.9. Large number of seabirds, Ashmore Reef. Credit: Rohan Clarke, Monash University.

Plate 2.10. The legacy of disturbance caused by mining for guano, Browse Island. June 1972. Credit: AA Burbidge.

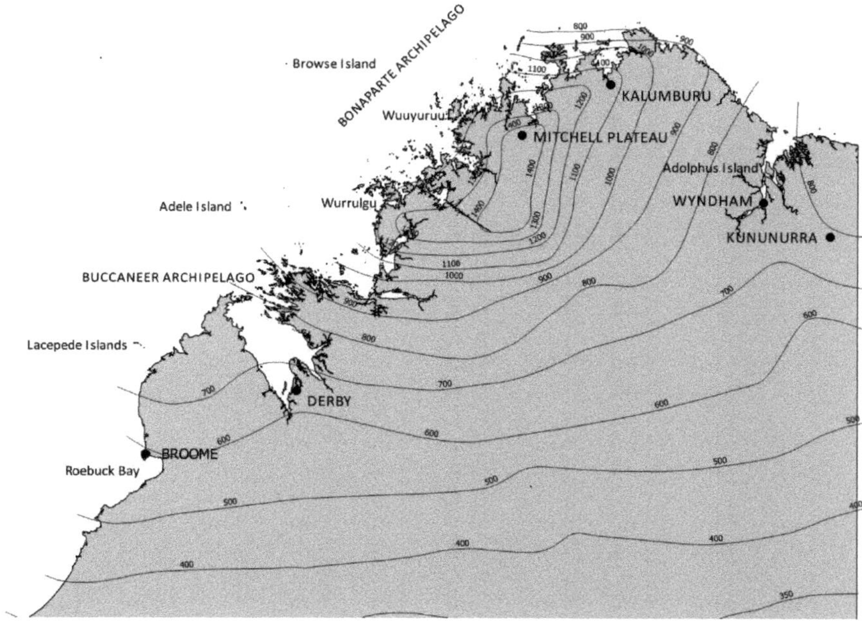

Map 3.1. Location of Kimberley region and its archipelagoes and major islands. Also shown are isohyets (mean annual rainfall in millimetres).

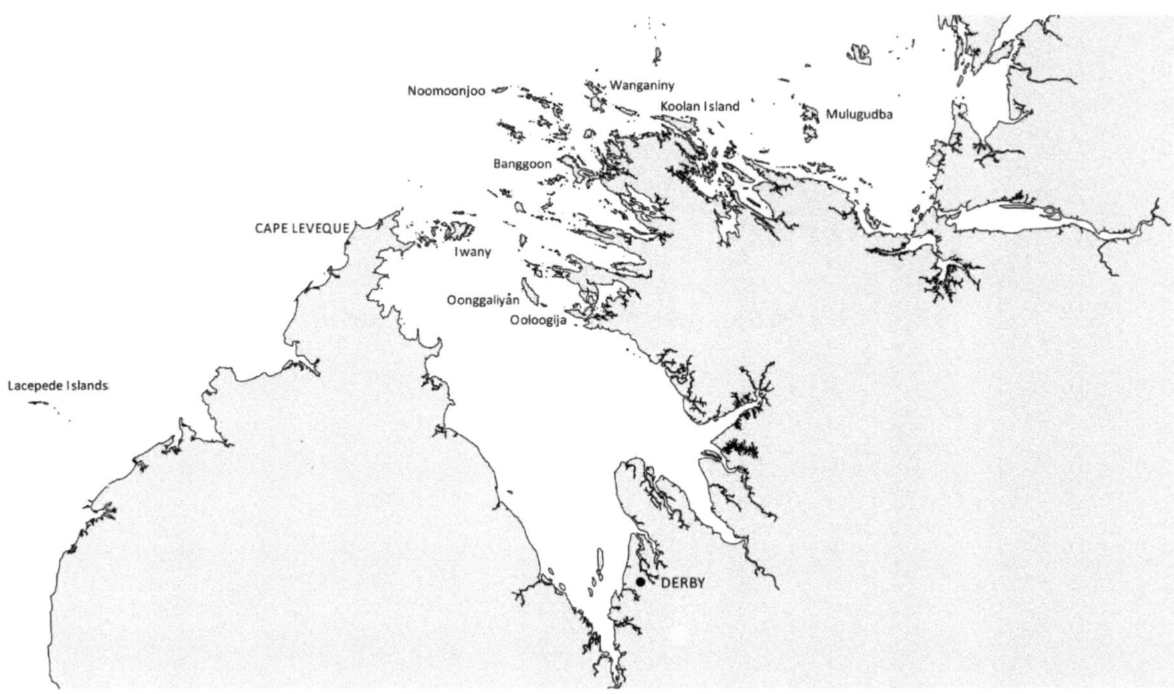

Map 3.2. The Buccaneer Archipelago, Kimberley region.

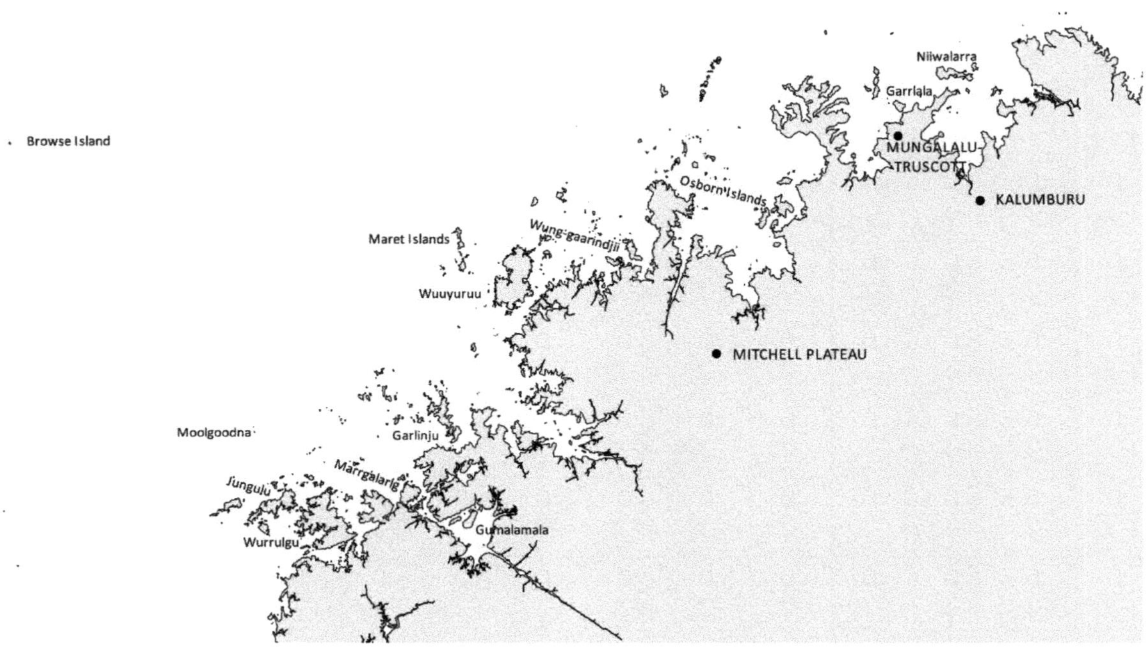

Map 3.3. The Bonaparte Archipelago, Kimberley region.

CHAPTER 3: KIMBERLEY ISLANDS

One island is a greedy Wandjina named Nyimbaragooj (Degerando Island), *while the other is a generous Wandjina named Nyilanggooj* (Jarumu Island). *The two islands are these two women Wandjina who were fishing. One was greedy and did not give to the other. The other one was generous and did share. You have to respect them both and talk to the generous one so that she will give you fish if you go fishing in that area. But the ungenerous one will not give, even if you ask respectfully. She competes with you for fish.*

Janet Oobagooma, who lived on Wurrulgu (Augustus Island) and the adjacent mainland as a child.

When we see smoke on the islands we would make fire too and they would come here and visit.

Janet Oobagooma

The 'canoe people' who moved between the Montgomery Islands and the coast during the early 1900s were family groups ...

Blundell *et al.* (2017)

INTRODUCTION

Table 3.1. Kimberley islands by size category

>1000 ha	20
100 ha to 1000 ha	145
20 ha to 100 ha	258
10 ha to 20 ha	177
1 ha to 10 ha	908
<1 ha	1125
Total	**2633**

Geography. Because the Kimberley coast is largely high, rugged, ancient sandstones and because it is a 'drowned coastline', the Kimberley has more islands than any other equivalent part of Western Australia (WA), or, for that matter, Australia as a whole. There are c. 8300 islands in Australia, including external territories; Western Australia has more than 3500 of them, of which 2633 occur in the Kimberley (32% of all Australian islands; more than 70% of WA islands). Most Kimberley islands are small, but 165 have an area greater than 100 ha. Of these, 20 are larger than 1000 ha, while 258 have an area of between 20 ha and 100 ha.

Apart from a few cays well offshore (Chapter 2), Kimberley islands lie close to the mainland coast and were joined to the mainland during and after the last glacial maximum of around 20 000 years ago when sea levels were c. 100-120 m below today's levels. As the climate warmed, sea levels gradually rose and most Kimberley islands were separated from the mainland by about 8000 to 10 000 years ago. The extreme tides along the Kimberley coast, up to 12 m, would have scoured channels once they developed, but the presence of hard sandstones and volcanic rocks would have limited erosion. A few inshore islands are connected to the mainland at low tide.

Although many islands have not yet had their maximum elevation measured, there are at least 74 islands with their highest point in the range 50-99 m, 29 islands in the range 100-149 m), nine (150-199 m), three (200-249 m), and one (250-299 m). The four tallest are Saint Andrew (288 m), Boongaree (239 m), Adolphus (220 m), and Middle Osborne (216 m). The two biggest islands are not particularly high: Augustus (178 m) and Bigge (138 m).

All the sandstone Kimberley Islands are rugged with typically steep topography. Most of the larger sandstone islands rise well above 50 m above sea level, with the highest being Gumalamala (288 m) and Wunandarra (239 m). Many of the smaller sandstone islands can reach up to 50 m or more above sea level. Wharton Sandstone islands, such as Umpatayi, often have undulating plateaux. The volcanic Niiwalarra has a maximum height of 192 m, while the depositional islands found close to shore (e.g. Mary Island and an unnamed island in King Sound) are only 1-3 m high.

Discovery and naming. The numerous islands and surrounding waters of the north-west Kimberley have been used and occupied by Aboriginal people for tens of thousands of years, both when they were joined to the mainland during the period of lower sea levels during the last 'ice age', and since separation after sea levels rose. After separation, access to islands was via rafts (kalwa/kalum often pronounced as gaalwa by Bardi, and galam by Worrorra, and called biyal-biyal by Mayala) typically made up of two overlapping raft platforms constructed from logs cut from the trunks of the kapok mangrove *Camptostemon schultzii*, and some other species, and joined with wooden nails and/or vines (Plate 3.1), although single log rafts (swimming rafts) were also used. A wooden paddle made of cypress pine was used to propel the raft but long distance journeys relied on tides and currents to carry the raft. Saltwater people are proud of their detailed knowledge and particular skills that allow them to navigate the massive tides and often treacherous coastal regions. In the north Kimberley, dugout canoes have also been used for sea travel over the last few hundred years. Evidence suggests that these craft were initially obtained from Makassans (present day Indonesians) by Kwini, Gaambera, Wunambal and Dambimangari people, who later learned to make them themselves, with the skill spreading southwards. People often took pet dingoes on rafts and in canoes when going to islands, mainly to aid hunting, and also to sacrifice if a large saltwater crocodile threatened. On the east Kimberley coast, logs were used to support swimmers (swimming logs) and transported people as far offshore as Pelican Island, 7 km from the mainland.

Kimberley coastal Aboriginal people were treated badly by European settlers. People used canoes (the term 'canoe' may have been applied to rafts shaped like a canoe as well as dug-out canoes) and rafts to island-hop to hunt and to avoid being killed by aalmara (white people, including police), before moving to the mission at Kunmunya to escape persecution.

Use and occupation by Aboriginal people has included active and passive management of valuable resources. For the Traditional Owners (TOs) these islands are places that represent complex entities of cultural and ecological values. For instance, islands are rarely known by a single name. Instead, they have layered meanings of cultural significance. Complex relationships of caretaking and naming have continued into the present, despite significant geographic and mobility constraints. The islands are components of Aboriginal cultural landscapes and seascapes that are of enormous significance to TOs. As such, islands are elements of the complex enculturated space that Kimberley TOs call their 'saltwater country'.

The islands are now included in several contiguous native title determinations that recognise private property rights of Aboriginal peoples in the islands and the seas. Tenure includes exclusive and non-exclusive possession native title, and covers areas that under state law include Aboriginal Reserves and unallocated Crown land.

Table 3.2 Native title on Kimberley islands

Native Title Determinations: Islands	**Native Title Group**
Wyndham to Napier Broome Bay, including Niiwalarra (Sir Graham Moore) and adjacent islands	Balanggarra
Anjo Peninsula to Cape Wellington	Uunguu
Cape Wellington to Wotjalum, including Koolan and Cockatoo Islands	Dambimangari
Buccaneer Archipelago, excluding Iwany (Sunday) Island	Mayala People
Iwany (Sunday) and adjacent islands	Bardi and Jawi
Valentine Island (adjacent to Dampier Peninsula)	Bindunbur
Mary Islands and un-named islands in the south of King Sound	Boorroola Moorrool Moorrool

Balanggarra Native Title stretches from Napier Broome Bay to Cambridge Gulf and Wyndham town. TOs associated with the Belaa (also known as Kwini) and Yeiji language groups and the Argna, Arawari, Gular and Guragona language groups, among others, are connected to this country with a distinctive set of local laws and customs. The Wanjina Wunggurr native title community covers most of the North Kimberley with their overall territory consisting of the Wilinggin, Uunguu and Dambimangari native title determination areas. Members of the Wanjina Wunggurr community share a law and culture derived from Wanjina and Wunggurr ancestral beings. Within the overall Wanjina Wunggurr community, the Wunambal and Gaambera language groups and languages are associated with the Uunguu determination area, while the Worrorra language group and language are associated with the Dambimangari (now often spelled Dambeemangaddee) determination area. The Bardi and Jawi native title community includes the Bardi language group from the tip of the Dampier Peninsula and the Jawi from Iwany and other neighbouring islands. The Mayala determination is associated with the Ugarrang/Unggarranggi and Umiide/Umiida languages and the determination area is connected to their island country on the basis of two distinct cultural traditions, rather than one. Some are connected on the basis of the Wanjina Wunggurr culture, while others are connected on the basis of beliefs that are more similar to those of Bardi and Jawi culture.

The north-west Kimberley is famed for its traditional rock art and stone arrangements. They occur all through country, including on islands. Traditional owners are concerned that their knowledge about images in their country is being lost and misrepresented. To them the art and stone arrangements are part of an ancient, interconnected record of cultural knowledge–they are part of the country itself and *Lalai* (dreaming), and they have been there since time immemorial. Traditional owners are now recording information about sites. The books *We are coming to see you* and *Barddabardda Wodjenangorddee: We're telling all of you* are examples. These books were created mainly for future generations of Wanjina Wunggurr traditional owners and also for other people who are interested in the Wanjina Wunggurr culture and the meaning of the art and stone arrangements.

European history. Early European navigators such as Abel Tasman (1644), William Dampier (1699), Nicolas Baudin (1801), and Matthew Flinders (1802) kept well out to sea as the extreme tides and murky seas, combined with sailing ships that were hard to manoeuvre in light winds, led them to put safety above inshore exploration. It was left to Phillip Parker King in HMC *Mermaid* in 1819 and 1820, and in HMSV *Bathurst* in 1821, to prepare the first Admiralty charts of the coastline and islands. Many European island names derive from Baudin and King. King's journals make interesting reading as the coast he saw differs little from the coast as seen today.

Naming. Kimberley islands have had traditional names for thousands of years. Some are now being adopted officially by nomenclature authorities and it is likely that more will be adopted in future. Dual naming may be a sensible first step. Where possible, this chapter uses traditional names rather than the 'European' names applied by explorers and settlers. Most Kimberley islands lie within named archipelagos. The largest of these are the Bonaparte and Buccaneer Archipelagos, but there are several smaller island groups, including the Osborn Islands in Admiralty Gulf, Nganguru Archipelago in Vansittart Bay and the Niiwalarra north of Kalumburu. Some named island groups lie within archipelagos: Ninemba, Heywood, Garlinju and Institut Islands are part of the Bonaparte Archipelago, while the Buccaneer Archipelago is divided into North East, South East, West and South West Groups. It does not include islands south of Bedford Islands and Hidden Island, including High, Iwany (Sunday), Mermaid, Oonggaliyan (Long) and nearby islands and islands in southern end of King Sound.

Many offshore Kimberley Islands were named by the French Baudin Expedition of 1801-1803, while other, often further inshore, islands were named by Phillip Parker King in HMC *Mermaid* in 1819 and 1820, and HMSV *Bathurst* in 1821. In most cases islands were named after now little-known French and English officials and have little relevance to modern Australia and the trend to recognise First Nation's knowledge and culture is important. Hopefully, the traditional names of many islands, and other places, will receive official recognition.

The Kimberley's Institut islands were named by Nicolas Baudin in 1801 to celebrate the founding of the *L'Institut de France* in 1795. Many of the islands in this group were named after members of the Institute, e.g. Condillac, Corneille, Descartes, Lafontaine, Lamarck, Laplace, Lavoisier, Montesquieu and Racine, who were notable scientists and intellectuals. However, few WA islands overall are named after scientists. In the Kimberley, Storr Island (indigenous name Yuwadan) was named after Glen Milton Storr (1921-1990), ornithologist and herpetologist at the Western Australian Museum, who named many WA reptiles, including Kimberley species, and Glauert Island was named after Ludwig Glauert, a taxonomist and former Director of the WA Museum. These names may be superseded by Aboriginal names, used for thousands of years.

Table 3.3. Aboriginal and place names for Kimberley islands mentioned in this chapter.

Aboriginal name	Place name†	Language	Native Title Claim
Banggoon	Hidden Island	Mayala	Mayala
Garlinju	Coronation Islands	Wunambal	Uunguu
Garrlala	Mary Island	Gaambera	Uunguu
Gumalamala/Ngarlangkarnanya	Saint Andrew Island	Worrorra	Dambimangari
Iwany	Sunday Island	Jawi/Jaawi	Bardi and Jawi
Jirritmirii/Jirridmiri	Adolphus	Forrest River	Balanggarra
Marrgalarlg	Uwins Island	Worrorra	Dambimangari
Miya	Byam Martin Island	Worrorra	Dambimangari
Moolgoocna	Booby (White) Island	Worrorra	Dambimangari
Mulugudba/ Mooloogoob	Kingfisher Island	Umiida/Umiide	Dambimangari
Nganguru	Eclipse Islands	Gaambera	Uunguu
Ngurraali	Middle Osborn Island	Gaambera	Uunguu
Niiwalarra	Sir Graham Moore	Kwini/Belaa	Balanggarra
Ninemba/Ninaamba	Champagny Island	Winjarrumi	Dambimangari
Noomoonjoo	Caffarelli Island	Oowini/Bardi	Mayala
Ooloogija	Lachlan Island	Oowini/Bardi	Mayala
Oonggaliyan	Long Island	Oowini/Bardi	Mayala
Umpatayi	South West Osborn	Gaambera	Uunguu
Wanganiny	Bathurst Island	Oowini/Bardi	Mayala
Wunandarra	Boongaree Island	Wunambal	Uunguu
Wunggaarindjii/Wunggaarindjii	Katers Island	Wunambal	Uunguu
Wurrulgu/Wurroolu	Augustus Island	Worrorra	Dambimangari
Wuuyuruu	Bigge Island	Wunambal	Uunguu
Yuduwala	unnamed (west of	Worrorra	Dambimangari
Yuwadon	Storr Island	Worrorra	Dambimangari

† place names from the Gazetteer of Australia

PHYSICAL CHARACTERISTICS

Geology and geomorphology. A Kimberley island's geology and rainfall affects its appearance and vegetation, and often which terrestrial animals are present. All of the Kimberley coast between Cambridge Gulf and King Sound, including the islands in the northern part of King Sound, is part of the Kimberley Craton, an ancient, stable portion of the Earth's crust where the basement rock crops out at the surface. West and south of King Sound is the Dampier Peninsula, which is really a higher-rainfall part of the Great Sandy Desert. Apart from a few inshore islands of mangroves, only the Lacepede Islands occur off the Dampier Peninsula, and they are sand cays. There are no islands south of Broome until the Pilbara.

Most Kimberley Craton islands are continental islands: parts of mainland Australia separated by rising sea levels during the past few thousand years. Most are composed of sandstone, but a few are of volcanic origin (Plates 3.2, 3.3). Two sandstones dominate: Wunaamin Miliwundi (formerly King Leopold) Sandstone and Wharton Sandstone, but a few islands in the far northeast are Pentecost Sandstone. The rock of the volcanic islands is called Carson Volcanics. All these rocks are c. 1800 million years old. Wunaamin Miliwundi Sandstone is the oldest, Carson Volcanics (mostly basalt) overlies Wunaamin Miliwundi Sandstone and Wharton Sandstone overlies Carson Volcanics rocks. Hart Dolerite, slightly younger again, intrudes other rocks in a few places, such as on Wuuyuruu. On a few small islands, such as the Maret and Montalivet Islands, Tertiary bauxite, similar to that at Mitchell Plateau, overlies Carson Volcanics, forming plateaux. A small area of laterite plateau also occurs on the eastern end of Niiwalarra. Many islands of the Buccaneer Archipelago reflect the folded rocks of the Yampi/Talbot area which contain rock types from the Kimberley Basin and other metamorphic rocks and granites found in in the Yampi fold belt of the southern Kimberley. Much younger alluvial deposits occur, mainly sand and coastal mud. Island coastlines are mostly rocky. Unlike the adjacent mainland, however, some islands have sandy beaches but, in most protected bays, mud and mangroves dominate.

Climate. The north and north-west Kimberley have two distinct seasons: the 'wet', lasting roughly from November to April, and the 'dry' from May to September or October. The start of the wet season varies from year to year, mainly depending on when a monsoon trough develops. There are few weather stations near the Kimberley coast and only one on an island—Troughton Island, which is located 17 km north of Bougainville Peninsula and has significant gaps in its weather data. The existence of weather stations at Wyndham, Kalumburu, Truscott, Derby and Broome has allowed development of rainfall isohyets (Map 3.1) suggesting that the mean annual rainfall on Kimberly continental islands varies from 800 mm in the southern Buccaneer Archipelago and on north-eastern islands towards Wyndham, to 1200 mm for larger islands such as Wurrulgu and Wuuyuruu. However, due to the lack of long-term stations in the higher-rainfall area, these data must be considered preliminary. Kuri Bay, located on the coast near Wurrulgu, with weather data available from 1962 to 2010, had a mean annual rainfall of 1449 mm (range 749 to 2081 mm), suggesting that rainfall on nearby islands is actually higher than Map 3.1 shows. A weather station at Mitchell Plateau, operated between 1970 and 1986, had a mean annual rainfall of 1532 mm, varying from 1093 to 2157 mm, but it was located at a higher elevation than nearby islands. Rainfall is a major determinant of island vegetation and animals. It seems that parts of the north-west Kimberley coast and adjacent islands receive among the highest rainfall of anywhere in WA with mean annual rainfall of over 1400 mm, while in some years more than 2000 mm may fall.

Temperatures in this tropical area are mild-to-hot throughout the year, with the 'wet' season having high humidity and the 'dry' season being a little cooler and with lower humidity. The range of temperatures on the islands, surrounded by ocean, is much smaller than on the adjacent mainland, where minima at Mitchell Plateau can be around 10°C in June and maxima as much as 41°C in January. Troughton Island has a mean annual maximum of 31.1°C (range 30.1 to 32.6°C) and the temperatures on some offshore islands would be somewhat similar but those close to the coast probably have a greater range.

BIOLOGICAL CHARACTERISTICS

Biological discovery. Aboriginal people had, and continue to have, a wealth of knowledge about the plants and animals of their country, as well as management of it, passed down from generation to generation. Traditional names are now being recorded by some groups. However, almost nothing was known scientifically about the flora and fauna of Kimberley islands prior to the 1970s. Before that there had been some limited recording and collecting of museum specimens at mainland missions such as Wotjulum on the Yampi Peninsula, which operated between 1934 and 1937, Kunmunya, which operated between 1913 and 1951, Pago (from 1908, moved to Kalumburu in 1932) and Oombulgurri (Forrest River) from 1913 to 1968. The only mission on an island was at Iwany in King Sound, which operated between 1899 and 1934. One natural history publication came from this period, a list of birds by R. Söderberg published as part of the results of the Mjöberg Swedish Scientific Expedition of 1910-1913. John Warham, who visited Cockatoo Island in 1956, published a bird list from there. Intriguingly, Warham reported a 'rabbit-rat' (*Mesembriomys* sp. or *Conilurus* sp.) on Cockatoo Island, which has not been reported since.

In 1971, biologists from the then Department of Fisheries and Wildlife and the WA Museum were attached to the 5th Field Survey Squadron of the Royal Australian Survey Corps, Australian Army, which was doing the field work for the preparation of 1:100 000 topographic maps of the Kimberley coast. The biologists were able to make brief visits to 15 islands, noting basic information about them and collecting animal specimens for the Museum. Access to the islands was by helicopter. Most islands were visited for a short time, from a few minutes up to three days depending on the work being undertaken by the army surveyors.

Then in 1972 the first boat-based strategic surveys were undertaken by scientists from the Department of Fisheries and Wildlife, the WA Museum and the WA Herbarium. Based on the 45-foot wooden Fisheries patrol vessel *Dampier*, two teams visited 15 islands over five weeks, including the remote Adele Island. This research was continued in 1973, this time on the Fisheries research vessel *Flinders*, visiting nine islands not examined in 1972, and further examining one visited the previous year. During these two years, as well as documenting which vertebrate animals occurred on islands and collecting specimens for the Museum, plant specimens were collected for the first time and lodged in the WA Herbarium. The resulting publication included papers on environment, mammals, birds, and amphibians and reptiles, and concluded with recommendations about protection of the more biologically important islands. After that, Kimberley biological survey work shifted to the mainland, to Prince Regent Nature Reserve (now a national park) in 1974 and Drysdale River National Park in 1975.

In May and July 1978, one of us (IA), while on the lighthouse tender MV *Cape Don*, visited 22 tropical islands, including several in the Kimberley, recording land and seabirds.

The next detailed survey was in 1982, when Department of Fisheries and Wildlife research scientists and Museum and Herbarium staff, based on a Fisheries patrol vessel, conducted a biological survey of selected islands in the Buccaneer Archipelago and nearby islands in King Sound. Botanical collections were lodged in the WA Herbarium and zoological collections were lodged in the WA Museum. Islands examined were Wanganiny (Bathurst and Irvine Islands), Noomoonjoo, Conilurus, East Sunday, Gibbings, Banggoon, King Hall, Oonggaliyan, Ooloogija, Macleay, Sir Frederick, Iwany and East Sunday. Results from this work were included in the publications from the 2007-2010 surveys.

Box 3.1 Norman L McKenzie AM (1945 -)

Norm visited Kimberley islands in 1971, 1972, 1973 and 1982 as part of a Department of Fisheries and Wildlife team recording and collecting animals and plants. As leader of the state government's biogeography research group, Norm was instrumental in setting up biological surveys state-wide that were site-based and repeatable. Among many other projects, Norm led further surveys of Kimberley islands in 1987-1989 during the Kimberley rainforests survey of the late 1980s, which included Traditional Owners. During the more detailed Kimberley Islands Biological Survey of 2007-2010, he further developed working relationships with Traditional Owners, with them participating fully in field surveys. In 2012, he became a member of the Uunguu Evaluation and Monitoring Committee, first as a representative of the Department of Biodiversity Conservation and Attractions and its predecessors, and later as an independent scientist. His contributions to knowledge of Kimberley islands and to working with Traditional Owners have been without parallel.

Between 1982 and the 2000s there were further scientific visits to Kimberley islands, with some published accounts. Between 2002 and 2005, scientists from the WA Museum and other organisations visited 35 Kimberley islands, 27 of which were examined for the first time, recording mammals and reptiles.

During the Kimberley rainforest surveys of 1987 to 1989 rainforest patches on Wurrulgu, Wuuyuruu, Wunandarra, Glauert, Ngurraali, Gumalamala and Umpatayi Islands were examined. Natural history data were collected during The Western Australian Naturalists' Club's ecotours led by Kevin Coate and Kevin Kenneally, particularly on Kampamantiya (Naturalists Island, 185 ha) in Prince Frederick Harbour. Adele Island and the Lacepede Islands continue to be visited by birding groups, the Lacepedes also by Kimberley coastal cruise companies.

At Koolan Island, long a site where iron ore has been mined and exported, various people have recorded its animals and plants; these were compiled and published in 1995.

In 2006-2007 biological survey work commissioned by INPEX Operations Australia as part of their proposal to build a liquefied natural gas plant on the Maret Islands (since abandoned) resulted in publication of detailed information on the vegetation, flora, terrestrial and underground fauna of North and South Maret Islands and adjacent small islands, mainly those with lateritic plateaux.

Between 2007 and 2010 a major Kimberley Islands Biological Survey took place. Teams of biologists, working with Traditional Owners and Aboriginal rangers, visited mostly large islands during both dry and wet seasons. Islands surveyed were Jirritmirii/Jirridmiri, Wurrulgu, Wuuyuruu, Wunandarra, Miya, Coronation, Jungulu, Banggoon, Katers, Mulugudba Ooloogija, Oonggaliyan, Ngurraali, Yuduwala (west of Yuwadan), Manjalaaly, Niiwalarra, Gumalamala, Yuwadan, Iwany, Marrgalarlg, unnamed (15.9367°S, 124.4590°E) and Wulalam. These comprehensive surveys resulted in a number of scientific publications collated in 'Biodiversity values on selected Kimberley islands, Australia' *Records of the Western Australian Museum* Supplement No. 81.

Even after all this scientific work, we know relatively little about the animals and plants on most Kimberley Islands, particularly the smaller ones, where in most cases nothing has been recorded. Even for comparatively well-surveyed islands, species lists are rarely complete. On Wurrulgu, which had been examined by wildlife biologists several times, a camera trap survey by Dambimangari Rangers and WWF Australia in 2013 added one mammal species—the scaly-tailed possum. That more research is needed and more islands need surveying is demonstrated by the Uunguu Rangers' examination of Jar Island (154 ha) in Vansittart Bay. Using traditional knowledge and camera traps, they identified the echidna and the golden-backed tree-rat, two terrestrial mammal species previously unrecorded for this island.

Vegetation and flora. The vegetation of Kimberley islands is a microcosm of that of the adjacent mainland, reflecting recent separation, closeness to the mainland and similar geology and soils (Plates 3.2-3.8). The main vegetation types vary with geology. Sandstone islands have mostly sparse vegetation with few trees. Native fig *Ficus platypoda* and emu apple *Owenia vernicosa* grow in rock fissures, and shrubs such as Kimberley heather *Calytrix exstipulata* can be common in places. Spinifex (*Triodia* spp.) is widespread. Volcanic red soils have low woodlands of *Eucalyptus* spp. over grasslands. Steep volcanic slopes support shrublands. Rainforest patches have a variety of plant species that do not commonly occur elsewhere, including *Albizia lebbeck*, *Eugenia grandis*, and lianas such as *Tinospora smilacina*.

Differences from the mainland are related to habitats not present on islands, particularly wetlands such as lakes and alluvial floodplains. Only the largest islands have permanent creek pools that contain aquatic plants such as water lilies and are fringed by riparian vegetation, including paperbark trees (*Melaleuca*), *Terminalia* species and screw pine (*Pandanus*). The flora is also similar. The Kimberley islands biological survey of 2007 to 2010 found only six plant species not known from the mainland and that may be due to insufficient survey on the adjacent mainland.

Fauna. Because of their high rainfall and variety of habitats, including woodlands and rainforest patches, Kimberley islands, more so than WA islands further south, support a greater variety of parrots, cockatoos, honeyeaters, rainforest specialists such as orange-footed scrubfowl and rainbow pittas, goannas and other lizards, and invertebrates including land snails.

Mammals. While several north-west Kimberley mammals have declined and become restricted to high rainfall, rugged sandstone areas or become very rare, no extinctions have been documented — the only part of mainland Australia where this is the case. Nineteen species of non-flying mammals have been recorded on Kimberley islands (Plates 3.9-3.12) and 28 species of bats (three species of flying foxes and 25 species of insectivorous bats). The list of islands with non-flying mammals is incomplete, as relatively few islands have been subjected to detailed trapping and some species are difficult to detect. The list of bats is near-complete for the larger islands because species can be identified using modern bat detectors that record their high frequency echolocation calls over several nights.

The number of non-flying mammal species detected on Kimberley islands has a relationship with island area (Figure 3.1). Of interest is that Wurrulgu (Augustus Island) has many more species than Wuuyuruu (Bigge Island), even though area and mean annual rainfall are similar. Wurrulgu is composed mainly of Wharton Sandstone and has a more complex topography and semi-permanent watercourses, while Wuuyuruu is composed mainly of Wunaamin Miliwundi Sandstone and has a simpler topography. However, Wunandarra (Boongaree Island), also composed mainly of Wunaamin Miliwundi Sandstone, has the same number of non-flying mammals as Wuuyuruu, so the differences cannot be easily explained

Dingoes. Aboriginal people took dingoes to Kimberley islands, partly to aid hunting and partly as sacrificial animals should large saltwater crocodiles threaten their rafts. In 1971, one of us (AAB) visited Ngurraali Island in Admiralty Gulf. He soon noticed dog tracks and, at night in his tent, he was serenaded by the howling of dingoes. In 2007 and 2008, when Ngurraali was visited by scientists taking part in the Kimberley Islands Biological Survey, dingoes were no longer present. Dingoes have been recorded on other Kimberley islands: Jirritmirii (Adolphus) Island by Phillip Parker King in 1819 (not present in 2009), Wollaston Island in 1972 (not revisited since), East Purrungku Island in 2004 (not revisited), Marrgalarlg (Uwins) Island in 2003 and 2008, and Wurrulgu (Augustus) Island in 1972 (not recorded in 2008). It appears that dingoes established on some islands during visits by Aboriginal people, but in some cases they have failed to persist in the medium- to long-term.

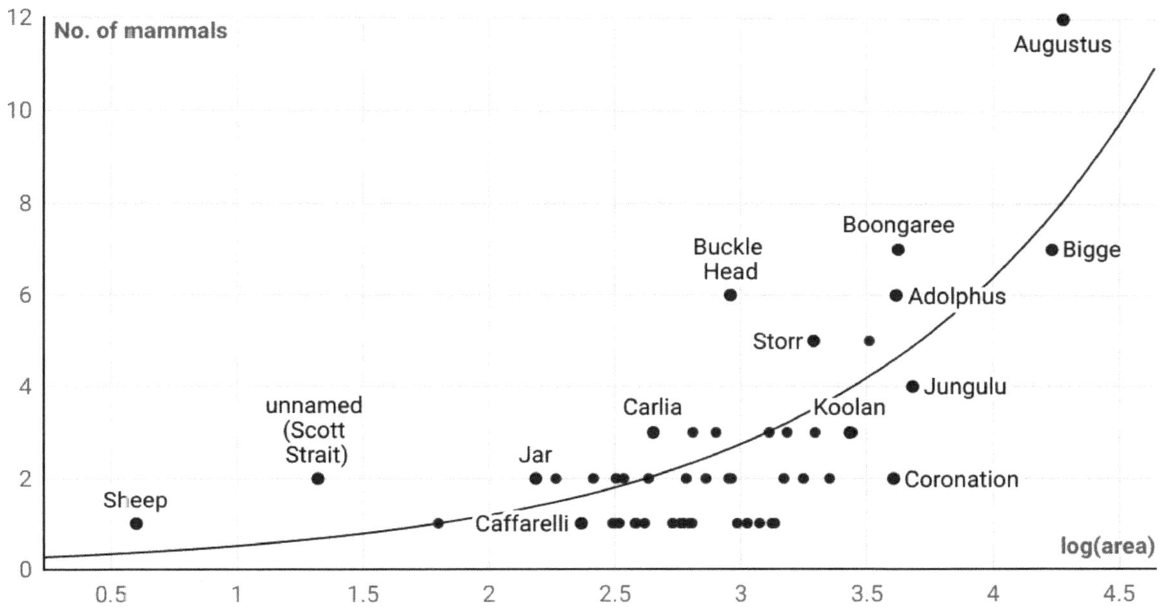

Figure 3.1. Relationship between island area and number of non-flying mammals detected on Kimberley islands, with exponential trendline.

It is noteworthy that an experimental introduction of dingoes to Pelorus Island in Queensland resulted in the reduction of the goat population to one or two male goats within two years. It is interesting to speculate that dingoes left on Kimberley islands may have eradicated larger mammals that once occurred on them; no mammals larger than small rock-wallabies have been detected on Kimberley islands, with the exception of agile wallabies on two islands, one of which is so close to the mainland that they may have reinvaded. Common larger mammal species on the adjacent mainland, such as euros and short-eared rock-wallabies, are notably absent.

Land birds. Because of their closeness to the mainland, a subset of mainland birds occurs on Kimberley continental islands. The lack of significant wetlands means few wetland birds, but other groups are well represented. During the Kimberley Islands Biological Survey a total of 157 species was recorded from 24 islands visited. The inclusion of other records increased the total number of bird species known from these islands to 179, with the number of species for individual islands ranging from 22 to 95 species. Only one threatened bird, the northern subspecies of the masked owl, is known to occur on the islands. Rainforest patch specialists include rose-crowned fruit-dove, Torresian imperial-pigeon, emerald dove, rainbow pitta, spangled drongo, yellow oriole, shining flycatcher and orange-footed scrubfowl. The islands also have many eucalypt woodland species (e.g. white-bellied cuckoo-shrike, red-tailed black cockatoo, sulphur-crested cockatoo, blue-winged kookaburra and white-winged triller) and rugged sandstone species (e.g. white-quilled rock pigeon, sandstone shrike-thrush and little woodswallow).

Seabirds. Brown boobies nest on Moolgoodna (Booby Island). It was formerly known as White Island, so named because the seabird droppings have painted white the Carson Volcanics rocks (Plate 3.13), which are black, but the place name was changed to avoid confusion with another Kimberley White Island, located in Admiralty Gulf. The breeding colony on Moolgoodna comprises up to 2000 pairs. Brown boobies have a wide distribution on and near tropical islands world-wide. A 1987 paper suggested that the Western Australian nesting colonies of brown boobies in the Kimberley, plus the one on Bedout Island in the Pilbara, are probably the largest in the world, totalling c. 10 000 pairs. About 500 pairs of greater crested terns have also been recorded breeding on Moolgoodna.

Other Kimberley islands that have records of seabird nesting colonies include Aunt (greater crested tern), Bird (greater crested and roseate terns), Jones (silver gull), Mulgudna (Caspian and greater crested terns), Myres (Caspian tern), North West and South West Twin (roseate tern), Randall (greater crested, lesser crested and roseate tern), Sandy (Long Reef) (greater crested tern), Sandy (Scott Reef) brown noddy, Iwany (greater crested and bridled terns), Troughton (little tern), and Warn (bridled tern). That seabird colonies can move from island to island is shown by records from Low Rocks and Sterna Islands in Admiralty Gulf. Three species of terns—crested, lesser crested and roseate—were recorded breeding on Low Rocks in the 1990s, but in 2008 had shifted their colonies to Sterna Island.

Other birds that depend on the ocean when breeding on islands have been recorded on Kimberley islands. Of particular note is the breeding colony of Australian pelicans on Pelican Island in Joseph Bonaparte Gulf. Great pied cormorants also breed on some Kimberley islands.

Reptiles. During the Kimberley islands biological survey, 79 species of terrestrial reptiles were recorded (Plates 3.14-3.15). The number of species on the 24 islands examined ranged from 14 to 32, with the nine islands having the greater species richness being generally larger, having higher rainfall, and a more rugged topography. Many species detected were north Kimberley endemics and ten were known only from islands, although that may be due to a lack of sufficient surveys on the adjacent mainland. When records from other island surveys were added, a total of 87 terrestrial reptile species are known from Kimberley islands.

As well as terrestrial reptiles, the saltwater (or estuarine) crocodile occurs on and adjacent to Kimberley islands, but is not known to breed on them. Numbers of large crocodiles along the Kimberley coast have been slowly increasing since they were protected in the early 1970s and large, dangerous crocodiles are now common.

Sea turtles nest on several Kimberley islands. Six of the world's seven species of sea turtles occur in Kimberley waters however only four species are known to nest here – green, flatback, hawksbill and olive ridley turtles. Because of the lack of suitable beaches on almost all mainland Kimberley north of the Dampier Peninsula, most nesting takes place on islands. Nearly every sandy beach on Kimberley islands shows some evidence of nesting, but many are of only a few individuals. Nesting occurs during both summer and winter. Major rookeries occur on the Lacepede Islands where both green turtles and flatback turtles nest. Islands with medium-to-high density nesting are Maret, Cassini and Parry Islands. More than 20 islands have medium-low density nests. Flatbacks nest on at least 20 Kimberley islands, with most beaches on these islands having medium-to-low density rookeries. Hawksbills nest on Nganguru, while olive ridley turtles nest on Joonoolgoo (Darcy Island). There are no records of loggerhead turtles nesting in the Kimberley.

Frogs. The Kimberley islands biological survey was the first scientific expedition to visit islands during the wet season. A total of 23 frog species were detected, ranging from 0–10 per island with the largest islands generally having the greatest number of species: Wuuyuruu (ten species), Wurrulgu (nine species) and Wunandarra (eight species). These three are also in the high rainfall area. The Kimberley region is home to over 40 species of frogs, but as the islands do not have significant wetlands it is unsurprising that, while Kimberley islands have more frog species than islands elsewhere in WA, a relatively low number of species occur on them.

Land snails. Land snails in the family Camaenidae occur in a wide variety of habitats in eastern Asia and Australia. Numerous species occur in northern Australia, especially in rainforest patches. Many Kimberley species have very small ranges, often being restricted to only a single or a very few rainforest patches. Of the 89 camaenid species detected during the Kimberley Islands Biological Survey, 73 were endemic to the islands, many to a single island. Of the 24 islands sampled, in only four cases was the same species recorded on two islands and only a single species was recorded on three islands. Some islands had endemic genera. The survey also demonstrated that higher-rainfall islands had more camaenid species. Land snails from other families also occur on the islands, but do not show the same pattern of highly restricted geographic ranges.

Subterranean fauna. Very little is known about animals that are restricted to underground habitats on Kimberley islands. The 2006-2007 survey of the Maret and adjacent islands revealed no stygofauna, but did find several species of troglofauna, all of which appeared to be undescribed species.

BIOLOGICAL ATTRIBUTES OF PARTICULAR ISLANDS

Niiwalarra Islands. These comprise Niiwalarra (Sir Graham Moore Island), Neawangu (Scorpion Island), Kim Islands and the three Geranium Islands and are situated at the northern end of Napier Broome Bay and north of Anjo Peninsula. They are within the Balanggarra Native Title Determination and are proposed to become the Niiwalarra Islands National Park, jointly vested in and managed by the Balanggarra Aboriginal Corporation and the Parks and Wildlife Commission. Niiwalarra has a diverse geology of sandstone, laterite and sand dunes resulting in a varied vegetation, wetlands, a rich bird fauna, and turtle nesting beaches. It is an area of cultural importance to the Kwini People and people lived there until well into the 20th Century. During the latter part of World War II, United States Air Force personnel constructed a radar and LORAN station, the concrete foundations of which can still be seen, and RAAF personnel constructed a radar station giving early warning of any Japanese aircraft approaching the Truscott Airfield or Kalumburu. Mammals recorded there are *Waambu minya* (rakali or water-rat), grassland melomys, delicate mouse, and red-cheeked dunnart, with the last being the only known island population of this species.

Pigs were released there during World War II and more were released later by the Kalumburu Mission. Despite an attempt at eradication because of the risk of exotic diseases establishing, pigs were still present in 2007, but now appear to be absent.

The Osborn Islands - different geology can lead to different biota. The Osborn Islands in Admiralty Gulf, all with a mean annual rainfall of at least 1200 mm, show how different geology can affect which species are present. The larger islands, north to south, are Borda Island, composed of rugged Wunaamin Miliwundi Sandstone, Ngurraali Island that looks very different as it is made of basalt, and Umpatayi which is also basalt, but with Wharton Sandstone overlying it, and leading to the occurrence of two sandstone plateaux and cliffs. Carlia Island is similar to Umpatayi but smaller.

Non-flying mammals and reptiles were briefly examined on Borda Island (728 ha) in 1972. Two mammals species were detected, nabarlek and Kimberley rock-rat. Ngurraali (2252 ha) has been examined in more detail, and eight bat species are known, but although there are some small rainforest patches, the northern blossom bat has not been detected there. Only two non-flying mammals have been found, short-beaked echidna and Kimberley rock-rat. On Umpatayi (1326 ha), which has large rainforest patches below the cliffs of the Wharton Sandstone plateaux, a flying-fox and the northern blossom bat were seen. Only two non-flying mammals have been found there, the Kimberley rock-rat and the rakali or water-rat. However, the large rainforest patch supports a variety of rainforest birds, including the rose-crowned fruit-dove, emerald dove, rainbow pitta, Torresian imperial-pigeon and orange-footed scrubfowl.

On the smaller Carlia Island (450 ha), three species of non-flying mammals were detected during a brief 1972 visit: the northern quoll, the golden-backed tree-rat and the Kimberley rock-rat.

To some extent, these differences in mammal occurrences reflect chance; which species were present when the island separated from the mainland and whether they survived. Habitat availability also affects which species survive, with the rugged Wunaamin Miliwundi Sandstone Borda Island the only one in the Osborn Islands with a rock-wallaby.

Wurrulgu. The largest island off the Kimberley coast (18 929 ha) Wurrulgu is also the largest Wharton Sandstone island. It is the third largest WA island after Dirk Hartog and Barrow Islands. Twelve species of terrestrial mammals have been recorded, although one record, of the golden-backed tree-rat, is based on a sighting only. Notable terrestrial mammals include the nabarlek, golden bandicoot and northern quoll, all listed as threatened species, and the scaly-tailed possum. Thirteen species of bats have been detected, including the ghost bat. Twenty-seven species of reptiles and eight species of frogs have been recorded.

Wuuyuruu. At 17 128 ha, Wuuyuruu is the largest mainly Wunaamin Miliwundi Sandstone island, with limited areas of Hart Dolerite and sand. It is the fourth largest island in WA. Due to its large size, but also to its geology and high rainfall (c. 1200 mm per year), it has a rich fauna. Seven species of terrestrial mammals and 11 species of bats are known to occur there. The terrestrial species include the short-beaked echidna, scaly-tailed possum, monjon and Kimberley rock-rat. Thirty-one species of reptiles have been recorded. On no island have both species of small rock-wallaby (monjon and nabarlek) been recorded. However, interestingly, in 2003, Dr Tony Start picked up two old weathered skulls, possibly subfossil, behind mangroves on Wuuyuruu Island and they turned out to be from a nabarlek–so at some time in the past both species may have occurred on the island, or perhaps the skulls date from when sea levels were lower and the island was part of the mainland.

Garrlala. Garrlala (918 ha) is the only large Kimberley island not mostly comprised of sandstone or volcanic rocks. While it has some exposed Wunaamin Miliwundi Sandstone, most surfaces are sandy. It is one of only two islands in the Kimberley with a population of agile wallabies, possibly reflecting its sandy soils and suitable forage. However, it is close to the mainland and it is possible the wallabies occasionally swim or 'wade' across at extreme low tides. There is anecdotal evidence from Aboriginal informants of wallabies swimming to islands. The only other non-flying mammal known from Garrlala Island is the grassland melomys. Only 14 species of reptiles have been found there, the lowest number of any island surveyed during the Kimberley Islands Biological Survey. Frogs have not been surveyed, as the island has not been visited by scientists during the wet season.

Marrgalarlg. Lying at the mouth of St George Basin and the Prince Regent River, rapid tidal currents pass Marrgalarlg, situated as it is in an area with the largest tides in the Kimberley. From a conservation point of view, it has significant populations of northern quoll, golden bandicoots and golden-backed tree-rat, as well as a population of short-beaked echidna. Rakali (water-rats) have also been recorded there. Of interest is a small population of dingoes, first noted in 2003 and still present in 2008. Nineteen species of reptiles have been recorded.

Koolan Island. Due to it having had an iron ore mine and a town, Koolan Island is the most thoroughly inventoried island, biologically, in the Kimberley. A week-long flora survey during the 1993 wet season, supported by herbarium records and earlier work, revealed 282 plant taxa (species, subspecies and varieties) occurring in five major vegetation units. The most widespread vegetation community is *Eucalyptus miniata - Corymbia confertiflora* open woodlands, generally over *Triodia* hummock grassland. Large mangrove communities occur in two sheltered bays south of the settlement. In addition, much of the coastline of the island supports a narrow mangrove fringe. Overall, the flora and vegetation is very similar to that of the adjacent mainland and nearby Cockatoo Island.

Three non-flying mammal species, the northern quoll, a planigale, and common rock-rat, occur on Koolan Island (as well as feral goats, possibly eradicated), along with 17 species of bats. Dogs and guinea pigs were kept as pets but did not establish. Thirty-five species of reptiles, a single species of frog, and 116 species of birds have been recorded. Some of the invertebrate fauna has also been investigated, with three earthworm, eight land snail, two scorpion, one centipede, nine spider, two ant, and 34 butterfly species present. The WA Museum holds an unsorted collection of moths and beetles.

Buccaneer Archipelago. The 1982 biological survey of some of the islands in the archipelago revealed a rich flora and fauna. Non-flying mammals known from the islands include northern quoll (Sir Frederick, Chambers and Hidden Islands), nabarlek (Hidden and Long Islands), golden bandicoot (Lachlan Island), golden-backed tree-rat (Conilurus, Chambers, Hidden and Lachlan Islands), grassland melomys (Sunday, Lachlan and Long Islands), and rakali (Hidden and Long Islands). Several species of bats were recorded.

FACTORS LEADING TO CHANGE IN THE BIOTA

Non-indigenous species. Apart from pigs that were released on Niiwalarra (above), black rats on the Lacepede Islands (both now no longer present), and Pacific rats on Adele Island (still present), no non-indigenous animals have been detected on Kimberley Islands. Black rats once occurred on Iwany Island around the mission, but searches in 2009 and 2012 did not detect them and they probably no longer occur there. Similarly, weeds are not common on Kimberley islands. A 2016 study found records of weeds on 52 Kimberley islands, with only Koolan (82), and Cockatoo (36) islands, both with mines and towns, having more than a few. Sunday Island, which was occupied by a mission, has 28 recorded weed species. The most pervasive weed is the stinking passionflower *Passiflora foetida*, spread by birds, that invades rainforest patches and some other vegetation types. Biological control is currently being researched.

Australia has the worst record for mammal extinctions of any country, due mainly to the introduction of red foxes and cats. Foxes do not occur in the north-west Kimberley, but feral cats do. The occurrence of mammals on islands where there are no cats reduces the risk of extinction for threatened species such as northern quoll, nabarlek, monjon, pakooma, golden bandicoot, and ghost bat. Other Kimberley endemic mammals that occur on islands are the scaly-tailed possum, monjon and Kimberley rock-rat.

Cane Toads. The northern quoll has declined from much of its range in northern Australia and, more recently, many populations have been eliminated, poisoned by eating the introduced cane toad *Rhinella marina*. Quolls occur on 17 Kimberley islands. However, there is the possibility that cane toads may be carried onto some islands during wet season floods, as has occurred in the Northern Territory, especially islands near the mouth of major rivers and close inshore, such as Marrgalarlg at the mouth of the St George Basin, into which the Prince Regent River drains, and other islands close inshore. A study of northern quoll on Koolan Island revealed that it has an abundant population, but both sexes are smaller than their mainland counterparts. Quolls were recorded moving over 4 km on this 2712 ha island.

Kimberley islands harbour unique taxa and reptile assemblages, as well as numerous species known to be vulnerable to cane toad toxins and, as such, are important conservation refuges. A risk to Kimberley island reptiles, as is the case with the northern quoll, is the possible invasion by the poisonous cane toad that may get swept out to sea by mainland river systems following heavy rainfall, something that has happened in the Northern Territory.

The cane toad is known to affect predatory species that eat frogs and other small animals, such as quolls, snakes, monitors and large skinks. Reptile species on islands that may be affected are sand goanna, Mertens' water monitor *Varanus mertensi* and black-palmed water monitor *Varanus glebopalma*. The northern blue-tongued skink *Tiliqua scincoides intermedia* is particularly susceptible to toads as are some elapid snakes such as the Kimberley death adder *Acanthophis cryptamydros*, greater black whipsnake *Demansia papuensis* and Kimberley mulga snake *Pseudechis weigeli*. Jirritmirii, Wunandarra, Gumalamala, Marrgalarlg and Yuwadan/Yeewadan Islands all lie either close to, or in the mouths of, major tropical river systems that may wash toads onto islands during flooding from high rainfall.

The impact of cane toads, should they establish on islands, on native frogs, is unclear. The ecological impact of introduced cane toads on native frogs has been an area of much conjecture, complicated by the difficulties in assessing changes in frog populations. In a review of cane toad impacts on Australian fauna, Professor Ric Shine of Macquarie University concluded that toads may affect native frogs in a range of negative ways: through competition for food or shelter in larval (tadpole) and adult stages; via lethal ingestion of the toad's toxic eggs, tadpoles or sub-adults by native frogs; or the transfer of parasites from toads to native frogs. However, it is thought that the rocky nature of mesic environments on Kimberley islands is unlikely to favour extensive breeding by cane toads and the hot dry seasons are likely to limit their population growth, therefore reducing levels of competition with native taxa.

Human factors. Tourism, mostly boat-based, is the main human use of Kimberley islands, and only a very few are visited; however, some sites, such as one on Wuuyuruu, have many hundreds of visitors each year. With islands covered by native title, Aboriginal Corporations have developed a visitor pass system that both promotes respectful experiences by visitors and raises funds to support Aboriginal Rangers and island management. The Uunguu Healthy Country Plan states 'We are happy for people to visit our country, share our knowledge, but it has to be done in the right way. That's why we have put together the Uunguu Visitor Management Plan with the Uunguu Visitor Pass as a one stop permission system for visitors to travel to approved places in Wunambal Gaambera country and have a learning and respectful experience.'

Ecological restoration. Eradication of black rats on the Lacepede Islands and pigs on Niiwalarra has promoted ecological integrity of these islands. No extinctions of any native species have been recorded on any Kimberley island.

Koolan Island has been mined for iron ore since the 1960s and much of this island has been stripped of vegetation. At least 43 plant and 10 animal species were introduced to Koolan. The plants included 11 grasses, rubber vine *Cryptostegia grandiflora*, poinciana *Delonix regia* and buffel grass *Cenchrus ciliaris*; the last three are recognised as invasive environmental weeds. The introduced animals include Singapore ant *Monomorium destructor*, an earthworm, a slug, a land snail, a cockroach and four mammals but, of the mammals, only the goat has managed to survive. For restoration and management planning on Kimberley islands, it will be instructive to see which exotics manage to persist now that the town on Koolan, with its watered gardens, has gone.

Island management. The main management issues on islands are associated with fire and visitors. Islands are managed by Aboriginal Rangers working within Healthy Country Plans. Support for rangers comes from funds raised from carbon credit schemes associated with fire management, visitor passes, and from the Commonwealth and state governments.

Historically, islands were burned frequently for a range of reasons including, looking after country (land management), hunting, and smoke signals. Burning on islands was noted by early explorers. After Aboriginal people moved to settlements on the mainland, mostly in the 1900s, fires became less frequent but more intense due to fuel build up.

Traditional Owners had, as is the case throughout Australia, an intimate knowledge of the relationship between fire and country on islands, including its animals. Sylvester Mangolamara, when visiting Garrlala (Mary Island) during the Kimberley Islands Biological Survey said 'Nobody burned that place for a long long time and it's really scrubby, need to burn it to clear that place to make kangaroos healthy, burn it bit by bit, if we make one big fire what will happen to the animals? We need to let them smell that fire and move into another area'. More recently, early dry season burning, usually ignited from aircraft, is reducing the occurrence of large, hot, late, dry season fires on larger islands and on the adjacent mainland.

Kimberley islands are situated within marine parks: the North Kimberley Marine Park, North Lalang-garram Marine Park, Lalang-garram / Camden Sound Marine Park, Lalang-garram / Horizontal Falls Marine Park, the proposed Maiyalam Marine Park and the proposed Bardi Jawi Marine Park. All are, or will be, jointly managed with Native Title holders.

In recent times, TOs have undertaken natural and cultural resource management projects and programs aimed at identifying and managing the values and resources in their traditional country, including islands. Healthy Country Plans now guide management of native title areas, including islands.

Sources of information

General: Abbott & Wills 2016; Burbidge & McKenzie 1978.

Aboriginal people, past and present: Akerman K 2015; Blundell *et al.* 2017; Crawford 1968, 2001; Karadada *et al.* 2011; Mangolamara *et al.* 2018; Vigilante *et al.* 2013; Wunambal Gaambera Aboriginal Corporation 2010.

European exploration: Baudin 1974; King 1826.

Scientific exploration and natural history: Gibson & McKenzie 2012a; McKenzie *et al.* 1995.

Vegetation and flora: Henson *et al.* 2014; Keighery *et al.* 1995; Lyons *et al.* 2014; McKenzie *et al.* 1991.

Weeds: Lohr *et al.* 2016.

Mammals: Gibson & McKenzie 2012b; Lamont *et al.* 2014; McKenzie & Bullen 2012; Spencer *et al.* 2017.

Birds: Pearson *et al.* 2013; Lamont *et al.* 2014.

Reptiles: Cowan & Doughty 2013; How *et al.* 2006; Lamont *et al.* 2014; Palmer *et al.* 2013.

Frogs: Doughty *et al.* 2012.

Land snails: Gibson & Kohler 2012.

Troglofauna: Humphreys 2014.

Plate 3.1. Double log-raft built and used in the past by Aboriginal people along parts of the Kimberley coast. Credit: Museums Victoria.

Plate 3.2. Landscape, hummock grassland, and sandstone rocks, western coast of Wuuyuruu (Bigge Island). August 2007. Credit: AA Burbidge.

Plate 3.3. Carson Volcanics rocks, Ngurraali (Middle Osborn Island). August 1971. Credit: AA Burbidge.

Plate 3.4. Wunaamin Miliwundi (formerly King Leopold) Sandstone and vegetation, Wuuyuruu (Bigge Island).
The tree is emu apple *Owenia vernicosa*. June 1972. Credit: AA Burbidge.

Plate 3.5. Mangal on east coast of Wunandarra (Boongaree Island). August 2007. Credit: AA Burbidge.

Plate 3.6. Hummock grassland, Borda Island. June 1972. Credit: AA Burbidge.

Plate 3.7. Wharton Sandstone and rainforest patch, Umpatayi (South West Osborn Island). June 1972. Credit: AA Burbidge.

Plate 3.8. Vegetation with palms and hummock grassland, Niiwalarra (Sir Graham Moore Island). August 1971. Credit: AA Burbidge.

Plate 3.9. Northern quoll. Credit: Bert & Babs Wells collection, courtesy of the Department of Biodiversity, Conservation and Attractions.

Plate 3.10. Djini (pale field-rat), Wunandarra (Boongaree Island). August 2007. Credit: AA Burbidge.

Plate 3.11. Golden bandicoot, Wuuyuruu Island. June 2003. Credit: AA Burbidge.

Plate 3.12. Monjon, the smallest species of rock-wallaby, Bigge Island. 2007. Credit: D Pearson, Department of Biodiversity, Conservation and Attractions.

Plate 3.13. Long-term impact of breeding seabirds (brown booby) on Booby Island (formerly known as White Island). The white mass is guano overlying the black Carson Volcanics rocks. September 1986. Credit: AA Burbidge.

Plate 3.14. . North-west prickly gecko *Heteronotia planiceps*, Wuuyuruu (Bigge Island). August 2007. Credit: AA Burbidge.

Plate 3.15. Kimberley death-adder, Wunandarra (Boongaree Island). August 2007. Credit: AA Burbidge.

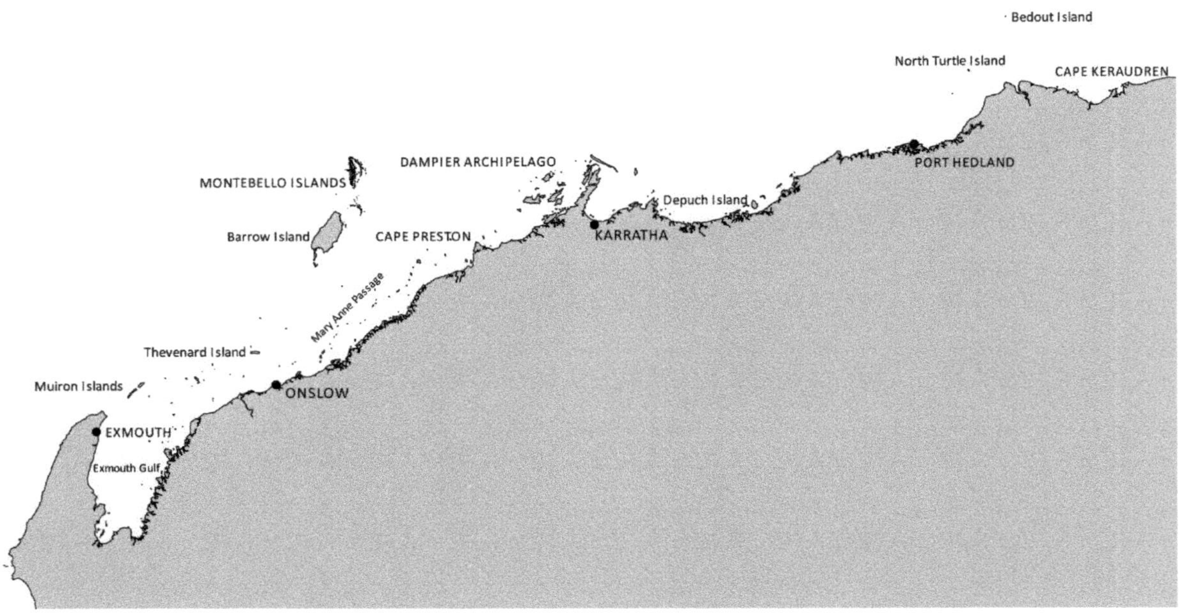

Map 4.1. Near-shore islands of the north-west and Pilbara coasts.

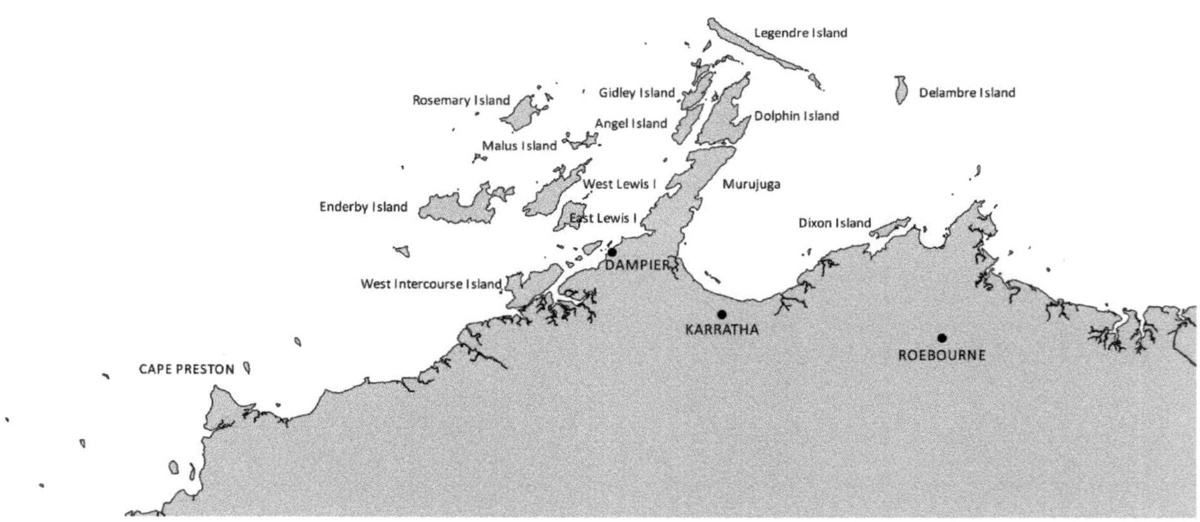

Map 4.2. The Dampier Archipelago, Pilbara region.

CHAPTER 4: NORTH-WEST INSHORE ISLANDS

'... the Dampier Archipelago contains, arguably, the greatest concentration of petroglyphs in the world'.
Ken Mulvaney

INTRODUCTION

The representation of islands along the 1200 km of coastal waters between Roebuck Bay in the north and Exmouth Gulf in the south is not geographically uniform. The north-eastern sector south to Port Hedland has only 17 small islands on mudflats and two tropical cays, Bedout and North Turtle Islands (discussed in Chapter 2). From Port Hedland to Point Samson (200 km) there are just 15 islands. Then in the next 80 km occur the 81 islands comprising the Dampier Archipelago. The remaining 200 km of coastline south-west to Exmouth Gulf is studded with 99, mostly small, low, and sandy, islands (Maps 4.1, 4.2).

Some inshore islands between Exmouth Gulf and Cape Preston are difficult to delineate and challenge the definition of a continental island. West Intercourse Island (2254 ha) is only 200 m from the mainland, separated by a channel. Hope Island (667 ha) is nearly joined to the mainland by mangal, being separated by a channel only 15 m wide. Tent Island (1415 ha) is joined by extensive mangal to the mainland coast. Such links very likely act as a barrier to the movement of plant and mammal species to these islands, though not to invertebrate and land bird species that live in the foliage of mangrove trees. Exceptionally low tides have, however, facilitated access of foxes to these islands. There is thus some ambiguity about the degree of isolation.

The northern limit of Dampier Archipelago is well defined, but its western and eastern boundaries are not geographically clear. For convenience, the western and eastern boundaries are here respectively designated as Cape Preston and Cape Lambert.

Physical attributes and environmental settings. These islands usually experience little rainfall between April and December and high temperatures from October to May. Between December and April tropical cyclones can bring strong winds and much rain. If they coincide with high tides, low islands risk extensive inundation, and intensified wave action causes erosion of soil. Tsunamis have also reached these islands, possibly from large meteorites crashing into the ocean, though not in recent times. This is evidenced by blocks of rock weighing up to 600 tonnes resting 8 m above sea level on Legendre and Dolphin islands in Dampier Archipelago. The presence of lithified oysters in these blocks has been used to date this tsunami at c. 3000 years ago. This wave washed right over Legendre Island (31 m in elevation).

Few islands outside the Dampier Archipelago are composed entirely of rock. They are sand islands, having been formed by deposition of sand onto rocky reefs, presumably after the ocean attained its current sea level at c. 5000 years ago (Plates 4.1, 4.2). The islands comprising Dampier Archipelago have a diverse array of ancient igneous rocks (basalt, granophyre, gabbro, granite) and Pleistocene aeolianite (Plate 4.3). All have, to some degree, areas of sand, alluvium, and mud flats.

Most sand islands are of low maximum elevation (<10-15 m), in contrast to most rocky islands (>20-30 m). The tallest sand islands are North Muiron (19 m) and South Muiron (18 m), and the tallest rocky islands are Depuch (158 m), Murujuga (127 m), Dolphin (125 m), and Enderby (96 m).

The approximate time of formation of the islands can be inferred from hydrographic charts. At the height of the last glacial maximum c. 20 000 years ago, sea levels were c. 100 m lower than at present. As global temperature rose, the ice began to melt and the water released gradually inundated the coastline. The higher hills were isolated from the new mainland and became islands. About 8000 years ago, the future Rosemary, Malus islands, and adjacent islets of the future Dampier Archipelago were part of one large island, the only island yet formed. By c. 6000 years ago, six island groups had formed. Once sea level stabilised the archipelago included nearly 20 islands more than at present. Wave refraction around the outer islands combined with shallow waters inshore led to the deposition of sand and the joining of these islets to form larger islands. The result of these processes is best demonstrated by satellite imagery of the eastern side of Enderby Island and the Lewis and Malus groups.

Many of the inshore islands between Roebuck Bay and Exmouth Gulf are so close to the mainland that they partially join to it at low tide. Examples include Finucane, Downes, Weerdee, the Forestier Group, Jarman, Dixon, Picard, Preston, Thringa, Carey, Weld, Yammadery, and Direction Islands. Some mainland bluffs (e.g. Cape Keraudren and Cape Preston) were once islands but have been captured by deposition of sand.

The largest island, formerly known as Dampier Island, was originally connected at low water to the mainland. The tidal range is 4.3 m. It is now recognised as a peninsula (Murujuga, formerly known as Burrup Peninsula), following the development of saline ponds at its southern end and the construction of a road and railway to the port of Dampier.

Sandy islands most remote from the mainland include Airlie (28 km), Bessieres (21 km), Thevenard (21 km), North Sandy (20 km), Serrurier (19 km), and South Muiron (17 km). In the Dampier Archipelago, Rosemary (21 km), and Delambre (17 km) are most distant from the mainland.

Island area. Islands with small area are very numerous compared with islands with area >20 ha (Table 4.1). Islands greater than 100 ha in area are more plentiful between Exmouth Gulf and Cape Preston (17 islands), and in Dampier Archipelago (17), than between Cape Lambert and Roebuck Bay (4).

Table 4.1. The number of north-west inshore islands, classified by area.
*mangal excluded.

Area (ha)	Exmouth Gulf to Cape Preston	Dampier Archipelago	Cape Lambert to Roebuck Bay
0.1-0.9	2	18	3
1.0-4.9	20	17	11
5.0-9.9	15	6	5
10.0-19.9	13	9	4
20.0-29.9	13 (NNW Tent, Fly, Flat, Locker, Ashburton, Direction, Great Sandy, North Sandy, unnamed #1, #2, S Carey, Steamboat, Angle)	4 (Quartermaine, Lady Nora, Whittaker, Collier Rock #5)	1 (W Moore)
30.0-39.9	7 (unnamed x2, N Dugong, WNW Tent, Y, Airlie, Round)	4 (Tozer, Collier Rocks #6 & #7, E Mid Intercourse)	1 (Bezout)
40.0-49.9	3 (NW Doole, unnamed, Fortescue)	3 (NE Regnard, Keast, W Mid Intercourse)	
50.0-59.9	3 (Simpson, Burnside, Bessieres)	1 (Goodwyn)	1 (Reef)
60.0-69.9	2 (unnamed, Long)	2 (Cape Bruguieres, SW Regnard)	
70.0-79.9	2 (Roberts, NW Yammadery)		2 (Weerdee, Sable)
80.0-89.9	1 (South Mangrove)		
90.0-99.9	1 (S Tent)		
100.0-199.9	6 (S Hope, NE Burnside, E Tent, Middle Mangrove, North Mangrove, Potter)	2 (Eaglehawk, Hauy)	1 (Ronsard)
200.0-299.9	2 (Doole, Serrurier)	2 (E Intercourse, N Gidley)	

Area (ha)	Exmouth Gulf to Cape Preston	Dampier Archipelago	Cape Lambert to Roebuck Bay
300.0-399.9	3 (Coolgra Point, Yammadery, Sholl)	2 (Malus, Delambre)	1 (Downes)
400.0-499.9	2 (South Muiron, North Muiron)	1 (Dixon)	
500.0-599.9	1 (near Yammadery)		
600.0-699.9	2 (Thevenard, Hope*)		1 (Finucane)
700.0-799.9			
800.0-899.9		2 (Angel, Gidley)	
900.0-999.9			
1000.0-1999.9	1 (Tent)	2 (Rosemary, Legendre)	1 (Depuch)
2000.0-2999.9		1 (W Intercourse)	
3000.0-3999.9		3 (Enderby, Dolphin, Lewis)	
4000.0-4999.9			
5000.0-5999.9			
6000.0-6999.9			
7000.0-7999.9			
8000.0-8999.9			
9000.0-9999.9			
10 000-10 999			
11 000-11 999		1 (Murujuga)	

Aboriginal history. The coast between Roebuck Bay and Exmouth Gulf supported 12 groups of indigenous people. Two of these (Yaburara and Mardudunera) are known to have possessed watercraft and thus could readily visit or occupy islands in Dampier Archipelago. The Ngaluma lacked watercraft but could wade across to Depuch Island.

The watercraft (*yowarda*) in use at Dampier Archipelago was a buoyant, round log c. 2 m long and 30 cm thick, with the bow slightly bent upwards. The Aboriginal sat astride with the feet placed on 2-3 pegs and propelled the watercraft by paddling with the hands. The simplicity of its construction, its functionality, and ease of handling impressed European visitors, resulting in many descriptions of the contrivance. The navigator Phillip King in 1818 depicted the watercraft in use. The island people inhabited many of the islands and were present until at least 1880. On the islands, they probably hunted breeding seabirds, sea turtles, terrestrial reptiles, and mammals. Some of the islands, especially Murujuga, and Dolphin, Angel, Gidley, Enderby, and Depuch islands, remain sites of cultural importance, evidenced by the numerous petroglyphs and middens.

Many animals are depicted in the rock art, and some can be identified to species. Examples are the echidna, dugong, and dingo. Also included are the thylacine (Plate 4.4) and Tasmanian devil, both of which became extinct on mainland Australia 3000-4000 years ago.

European history. None of the islands were known to Europeans until the renowned privateer William Dampier visited in 1699 in the *Roebuck*. His landfall, which he named Rosemary Island, was actually (the later named) East Lewis Island. On this island he observed the little corella, found evidence of human occupation, and collected specimens of two plant species that are now preserved in the Oxford University Herbarium. The leaves of one species, *Olearia axillaris*, have an odour of rosemary when crushed. The other plant species is *Swainsona formosa*, unjustly known today as Sturt's desert pea. The genus name *Willdampia*, commemorating Dampier's collection, has been published but is not widely accepted by botanists.

In 1801 came the French (Nicolas Baudin, *Le Géographe*), who steered clear of much of the coast and thus named islands mostly distant from the mainland: Rivoli; Muiron; Serrurier; Rosily; Thevenard; Bessieres; Malus; Legendre; Haüy; Delambre; Bezout; Picard; Forestier Group (comprising Depuch and Sable); and Ronsard. Landfall was made only on Malus and Depuch islands, though not by any of the scientists on the vessel. Baudin named the archipelago after Dampier.

The islands of Dampier Archipelago remained little known to Europeans until they were charted in 1818 by Phillip King on HMC *Mermaid*. He cruised within the archipelago and named Enderby, Goodwyn, Gidley, Lewis (which he mapped as one island), and Intercourse islands. Most of the numerous low and sandy islands in Exmouth Gulf east to Cape Preston were left unvisited and unnamed. The ship's botanist, Allan Cunningham, unflatteringly described these as an 'archipelago of sterile sandy islands'. He did record that *Acacia coriacea* was frequent on the islands in Exmouth Gulf, occurring as a tree 3-4 m tall.

In Dampier Archipelago, Cunningham collected specimens of 36 plant species, mostly on Enderby Island with a few on Intercourse and Malus Islands. He noted 'White Cockatoos' (little corellas) and abundant 'pigmy Kangaroo' (the Jerdamurra or Jerdanmurra, its Aboriginal name as recorded later by John Durlacher, and known now to zoologists as Rothschild's rock-wallaby).

Sailors aboard the *Mermaid* were the first Europeans to meet a Dampier Archipelago Aboriginal, when a man was forcibly brought aboard. This was only the second time that it was realised by Europeans that Aboriginal languages in western Australia differed from those spoken in eastern Australia. The Sydney Aboriginal man Boongaree, who was on board, attempted to interview him but was unable to understand his words and thus obtain any information about local resources. After being offered food and provided with gifts, the detainee was released. Subsequent interactions with Aboriginal people proved more amicable, resulting in King naming an island after this friendly social intercourse.

The final visit by European navigators took place c. 20 years before the mainland coast was settled by colonists from Perth. In 1840 John Clements Wickham and John Lort Stokes on HMS *Beagle* (the same ship that five years previously had transported Charles Darwin on his voyage around the world) visited Depuch Island. They observed extensive sand-banks connecting the islands of the Forestier Group to the mainland and noted that at low tide it is possible to walk across them. No Aboriginal watercraft were seen, but their huts and an 'immense number' of rock art images were found on the island. Wickham sketched some of these and published them in 1842 in a London geographical journal. A 'small kind of kangaroo' (black-footed rock-wallaby) was shot and the painted finch was collected and later described from this type specimen by John Gould as a species new to science. In the early 1960s, proposals to develop a deep-water iron ore export port at Depuch Island led to a 1962 survey by the Western Australian Museum of its engravings, flora, and fauna. Fortunately, the port development did not proceed.

It was not until the 1880s and 1890s that most of the smaller islands received a European name. This resulted from colonial and British hydrographic surveys of inshore waters following the establishment of pastoral stations and the coastal towns of Onslow, Roebourne, and Port Hedland. Some of the more insignificant islands in Dampier Archipelago were not named, however, until the 1960s.

BIOLOGICAL CHARACTERISTICS

Island vegetation and plant species. It seems that comprehensive botanical collecting on these islands did not take place until 1962 and 1970. On the low sandy islands, hummock (*Triodia*) grassland, heath, and shrubland predominate. Mangroves (as thickets) are commoner on islands close to the coast. Not all sandy islands are alike in their floras. For example, on Thevenard Island *Acacia coriacea* is the dominant shrub species, whereas on other islands *A. bivenosa* is. However, both species are known from >50 islands.

In Dampier Archipelago igneous rocks predominate, resulting in greater variety of vegetation types. Small trees (kurrajong, five species of fig) and grassland occur on rock piles and outcrops. Slopes support the most extensive type of vegetation, hummock grassland (Plates 4.5, 4.6). Drainage lines have a diverse flora of shrubs and herbs, and run-on areas (flats) are temporarily waterlogged after rain and contain many grass and herb species. Beaches and fore-dunes are dominated by *Spinifex longifolius*, creepers, and Sturt's desert pea. The inland sand dunes support *Acacia coriacea* and *A. bivenosa* scrub. Rocky shores and bays are lined by mangroves (*Avicennia marina, Rhizophora stylosa*). Eucalypts (*Eucalyptus microtheca, E. patellaris, E. victrix, E. xerothermica*) are limited in distribution to Dolphin Island (three species) and West Lewis and Enderby (one species). Nearly 300 plant species have been recorded in Dampier Archipelago, with the two largest islands the most diverse floristically (Dolphin Island, 157 species; Enderby Island, 117 species).

Sand islands lack species of fig, and mangroves are of limited occurrence (particularly on offshore islands) compared to islands in the Dampier Archipelago (Plate 4.7). Widespread plant species include *Rhagodia eremaea, R. latifolia, R. preissii, Cynanchum viminale, Scaevola cunninghamii, Sesuvium portulacastrum, Spinifex longifolius, Threlkeldia diffusa, Triodia epactia and T. pungens.*

Animal species. The fauna of these islands was first given serious attention in 1918 by Frederick Whitlock. He briefly visited 13 islands *en route* to Barrow Island, recording details about the shorebirds, breeding colonies of seabirds, and raptors seen. No information about land birds was published. Next came visits to Dampier Archipelago by David Ride and Glenn Storr in 1962 and Andrew Burbidge and Robert Prince in 1970 (vertebrates), and to Depuch Island by Storr in 1962. Although all of the islands between Roebuck Bay and Exmouth Gulf have since been visited by biologists, they are deserving of more attention.

Only two frog species have been recorded on these islands – *Cyclorana maini* on Angel, Dolphin, and Finucane Islands, and *Litoria rubella* on Dolphin Island. Only the much larger Murujuga 'Island' has both frog species.

Reptile species are well represented on the islands, with 41 species recorded on the sandy islands between Exmouth Gulf and Cape Preston. About 37 species occur within Dampier Archipelago, including many species of geckoes, goannas, skinks, and snakes. As would be expected, the largest island (Murujuga) has the most speciose reptile fauna, with 61 species. Some species of lizard, including *Ctenotus saxatilis*, *Heteronotia binoei*, *Lerista bipes*, *Lophognathus horneri*, and *Menetia greyii*, occur on many sand and rock islands. Others have more restricted occurrence, including *Lerista jacksoni*, *L. nevinae* and *L. verhmens*.

Four species of sea turtle (green, flatback, hawksbill, and loggerhead) come ashore on islands with sandy beaches to lay their eggs.

Seabirds breed on many islands, including six species of tern (greater crested, Caspian, fairy, bridled, lesser crested, roseate), and the ubiquitous silver gull. The only species in this region that constructs and breeds in a burrow is the wedge-tailed shearwater. It has been reported breeding on many sandy islands between Muiron and Bezout Islands.

Many shorebird species migrate annually to Australia from their Northern Hemisphere breeding sites to these islands. The grey-tailed tattler occurs in globally-significant numbers on some of these islands. Shorebirds resident in Australia also breed on these islands, for example the pied and sooty oystercatchers. Several islands are a stronghold for the beach stone-curlew.

Three species of marine raptor (eastern osprey, white-bellied sea-eagle, brahminy kite) are present, constructing their bulky nests on the ground. These birds of prey mostly take fish from the surrounding inshore waters, but the sea-eagle will take seabirds, and in the Dampier Archipelago, mammals.

Few species of land bird occur on the small sandy islands, in contrast to the larger, rocky islands of Dampier Archipelago. This is doubtless linked to the larger area and more varied vegetation of the latter. Within Dampier Archipelago, ten raptor species have been recorded, with four species breeding. The bar-shouldered dove is the most widespread pigeon or dove species and the little corella is the most ubiquitous cockatoo species. The welcome swallow, Australasian pipit, willie wagtail, singing honeyeater, brown honeyeater, pied butcherbird, white-breasted woodswallow, and Torresian crow occur on many islands.

Extensive areas of mangrove forest and thickets in Dampier Archipelago support nine bird species restricted (or almost so) to this vegetation type: bar-shouldered dove, collared kingfisher, mangrove robin, mangrove whistler, white-breasted whistler, mangrove fantail, dusky gerygone, Australian yellow white-eye, and white-breasted woodswallow. However, three of these species (the dove, white-eye, and woodswallow) also occur on small sandy islands outside the archipelago lacking mangroves and forage in *Acacia* thickets or in the air.

Many interesting island occurrences of land bird species have been recorded. These include the little button-quail (on 9 islands), galah (8), yellow-throated miner (8), spinifexbird (8), white-winged fairy-wren (7), budgerigar (6), bush stone-curlew (5), white-plumed honeyeater (4), grey-headed honeyeater (4), spinifex pigeon (3), purple-backed fairy-wren (3), and crested bellbird (1). It is not yet clear which of these occurrences represent vagrants; this cannot be ascertained until more visits by naturalists and biologists have taken place.

Sandy islands generally lack native mammal species, but if one is present, apart from Thevenard Island, it is the djini/pale field rat (Table 4.2), which no longer occurs on the adjacent mainland, where it is thought to have been eradicated by the much more aggressive introduced black rat. The next species best-represented on the islands is surprising - a large kangaroo, known to local Aboriginal people (and recorded by John Durlacher) as the *Pickoordar* or *Chickoodar*, and to zoologists as *Osphranter robustus*, the euro. This species is a capable swimmer and thus has been recorded on several of the sandy islands close to the mainland coast. The second largest species (Rothschild's rock-wallaby) is well-represented on islands. It occurs on the three largest islands in Dampier Archipelago and another large island (Plate 4.8).

The 'island' of Murujuga supports the most diverse mammal fauna (15 species). Nearby Dolphin, at about one quarter of the area, has only six species. The smallest islands (11-51 ha) in the archipelago with only one mammal species are Keast and Kendrew islands, both with the same rodent species (djini).

Distributional patterns. A detailed study of variation between islands in their reptilian fauna has been undertaken on ten islands in Dampier Archipelago, ranging in area from 0.9 ha to 140 ha. Nineteen species were found, with 1-9 species per island. Environmental attributes were measured on each island. The number of reptile species occurring on each island was highly correlated with island area, degree of isolation, and distance from the nearest land (whether mainland or another island).

Much of the information presented in the remainder of this section has been sourced from a database (comprising 57 000 rows!) painstakingly compiled by Dr Cheryl Lohr of all species known to occur on the islands examined in this chapter.

Box 4.1. Cheryl Lohr (1983-)

Cheryl is a research scientist with an eclectic experience in wildlife ecology and invasive species management. Her interest in biology began on the family farm in the WA wheatbelt when she witnessed the response of feral rabbits to the implementation of biocontrol. During her research for the BSc degree in Natural and Agricultural Science at

The University of Western Australia she worked with Dr David Algar to further development of feral cat baits. During her studies for the MSc degree at UWA she researched the use of contraceptives to manage over-abundant native wildlife at the Perth Zoo.
Cheryl obtained the degree of PhD at the University of Hawaii at Manoa. This research focused on the human dimensions, cultural aspects, and cost-benefit analysis of managing introduced animals. In 2012, Cheryl joined the Department of Biodiversity, Conservation and Attractions, and became involved in the development of island biosecurity models and systematic conservation planning for the Pilbara Islands. These types of models require large quantities of biological data, much of which was sourced from archival material. This task took 18 months working full time, and the database is now used frequently by the DBCA's Pilbara Region.

Goannas have an interesting distribution on these islands. Eight species occur, four of which have very restricted occurrence: *Varanus bushi* on Finucane Island; *V. brevicauda* on Finucane and Potter Islands; *V. eremius* on Finucane and Murujuga; and *V. pilbarensis* on Murujuga, Dolphin and West Intercourse Islands. The other four species are more widespread: *V. gouldii* (on 15 islands); *V. panoptes* (9); *V. acanthurus* (8); and *V. tristis* (7). The larger islands have the most goanna species: Murujuga (6 species); Finucane (5); and Dolphin (4). Most islands (17) have just one species and another eight have two species. The question arises as to why certain combinations of two species occur on some islands and not on others, and why so many islands have only one species, which is not always the same species. Interestingly, several sand islands have one goanna species present. Does this hint at goannas being able swimmers or capable of hitching a ride on vegetation washed out to sea when coastal rivers flood? Alternatively, are sand islands older than 5000 years and actually are fragments of the former coastal plain? These topics deserve detailed ecological and molecular study.

Patterns of the distribution of species among islands are also well illustrated by the native mammal fauna. (With plants, island lists of species are either not available or based on limited collecting; with land birds most visits are infrequent, making it difficult to distinguish vagrant species from breeding resident and seasonal visitor species.) We assembled and published the first comprehensive list of mammal species on Australian islands in 1979. Since then we have confined our attention to WA islands. The information presented in Table 4.1 is extracted from a database published in 2017 and updated and maintained by us.

Cardinal principles of island biogeography are that large islands support more species than small islands, that large species occur only on large islands, and that predator (flesh-eating) species cannot persist on small islands. In archipelagoes, particularly those formerly connected to the mainland, there should be a regular pattern of species occurrence in terms of island area and degree of remoteness of islands from the mainland. Islands created by gradually rising sea level should exhibit an orderly sequence of species becoming extinct, thereby generating a regular pattern.

The data assembled for the mammal faunas of the islands between Roebuck Bay and Exmouth Gulf are only partially consistent with predictions from these principles. First, there is no regular pattern! This is particularly apparent for the nine largest islands (>1000 ha in area), and is suggestive of the importance of other factors. Some species may exclude related species through competition when resources become scarce (e.g. food items such as seeds, insects, and herbage). Could the chequerboard distribution of the two rock-wallaby species be the outcome of interspecific competition? For some other species, their habitat requirements (in relation to vegetation structure, shelter, refuges) may not have been adequately met.

Some absences may be the result of bad luck. When a population is reduced to small numbers, an incident such as an intense fire, long absence of any fire, or passage of an intense cyclone, may result in local extinction from the island. Could the absence of rock-wallabies from West Lewis and Legendre islands be the result of hunting by Aboriginal people who lived on these islands?

Species that can swim short distances may be able to reach islands close to the mainland or large islands and live on small islands for brief periods. The euro, a very large kangaroo (up to 55 kg in body mass), requires a large home range, and so it appears that only Dampier and Dolphin Islands are large enough to support a viable population. It is interesting to note that the euro has not been recorded on Enderby, Rosemary, Legendre and other large islands more distant from the large inshore islands. Most of the islands on which the euro is present are close to the mainland and periodically joined to it. However, its occurrence on Potter Island, 3 km offshore, is difficult to explain.

All islands of area <40 ha that have mammals have only one native mammal species present, but this is not always the same species of rodent. The djini (pale field-rat) occurs on both small and large islands, but is not always present on islands connected to the mainland. Two other rodent species appear to have persisted on larger islands, with the djoori (common rock-rat) on seven islands with area >900 ha and on two other very small islands, and the mingkiri (sandy inland mouse) on eight islands larger than 300 ha.

Finucane Island demonstrates the advantage of proximity to the mainland, thereby facilitating (repeated) dispersal to the island. Six mammal species is a large number for an island with area of 550 ha.

We have presented the rather complex Table 4.1 and discussed it at some length to illustrate that, even for a well-known group of animals, there is still much to be discovered about the ecology of these island populations.

FACTORS LEADING TO CHANGE IN THE BIOTA

Natural factors. The occurrence of tsunamis, cyclones (extreme wind speed and wave action), heatwaves (prolonged very high temperatures over successive days), high rainfall during a short period (local flooding), and drought (no rain during successive months extending over several years) should have influenced which species persisted following the creation of these islands by rising sea levels. The historical occurrence of these factors from the earliest years of European settlement up to 1927 has been documented in great detail by the Bureau of Meteorology.

The impact of these extreme events on the biota is, however, yet to be studied on these islands. Droughts, floods, cyclones, and heatwaves have recurred frequently in the past 150 years of European settlement, and presumably before then. Populations of flora and fauna on the islands should have experienced bottlenecks under these conditions, perhaps putting those species with small populations at risk of local extinction. The cumulative impact of all of these extreme events during many millennia on the biota of these islands is yet to be studied.

Human factors: Aboriginal impacts. Aboriginal people residing on islands, or regularly visiting (by means of watercraft or swimming) the smaller islands in Dampier Archipelago, as well as visiting those islands that could be reached by wading may have had significant impact on some populations of island species. This was probably in the form of hunting animal species that were easiest to obtain, namely eggs of seabirds and turtles and the meat of turtles ashore to lay eggs. It seems unlikely that their watercraft were suitable for transporting a dingo (as was the case for Kimberley region islands), and none of the early navigators reported the presence of the dingo in Dampier Archipelago. The French found evidence of their presence on Depuch Island, and it is likely that dingoes occurred on Murujuga. Dingoes could have swum (particularly at low tide) the short distance from Murujuga to adjacent Dolphin, Gidley, and Angel islands, as well as to West Intercourse Island from the mainland.

Aboriginal burning on islands, and when, how often, and at what scale it took place remains unknown. Fire would have initiated ecological succession of plant communities but should not have permanently altered the island flora.

Aboriginal occupation and its impacts declined and virtually ceased following epidemics of smallpox and measles in the 1860s and 1880s.

Human factors: European impacts. In 1863 'hills of whale bones', from shore-based whaling, were observed on 'the islands' of Dampier Archipelago. It seems that this refers to West Lewis Island. This record precedes that of 1870 on Malus Island, where a trypot remained until the 1970s.

During 1875-76, Thevenard, Delambre, and Rosemary islands were leased for 'turtling'. No other islands relevant to this chapter were then in use, some 13 years after Europeans settled the region.

West Lewis Island was occupied by John Durlacher after 1876. Sheep were kept on this island, but for how long is unknown. By 1883 a substantial house had been built. Sheep pastured well on hummock grassland, and costs were minimal as islands required no fencing to keep dingoes out. It does not seem that any other islands were used to graze livestock. Freshwater is lacking, and the nutritionally-poor vegetation would not have permitted high-density stocking. The vast area of the mainland presented a more economic means of depasturing sheep.

Construction of lighthouses or automatic light towers took place between 1888 and 1959 on ten islands, and entailed limited destruction of vegetation. Although the only (formerly) manned lighthouse (on Jarman Island) is close to the mainland, it is likely that goats were kept there from time to time, perhaps to supply milk and meat. If so, browsing of palatable plant species should have reduced their population size.

Three islands were used for the detention of Aboriginal people. Bezout Island was used as a leprosarium from 1911 to 1931. In 1874, women on Enderby and Delambre islands were sold to the highest bidder as sexual slaves.

The earliest type of physical destruction resulted from guano mining in the 1800s. This activity was fortunately restricted to three islands, namely Thevenard, Great Sandy, and Airlie islands. Airstrips (with consequent destruction of vegetation and reduction of animal habitat) have been constructed on Rosemary (now abandoned), Sholl, Direction, and Thevenard islands. There is also a helipad on Airlie and Thevenard islands.

The greatest physical destruction to islands occurred with the industrialisation of the Pilbara region, with Mistaken, East Intercourse, and Finucane islands used for stockpiling iron ore or salt prior to export. Iron ore was first exported from Dampier in 1966 and salt production commenced in 1967. Part of Murujuga has been an industrial hub for liquefied natural gas processing since 1978 and fertiliser production since 2006. This has resulted in damage to petroglyphs on Murujuga and presumably other islands nearby. Decommissioning of oil facilities on Thevenard and Airlie islands (Plate 4.9) provides an opportunity to restore and enhance the value of these islands for nature conservation.

Arguably, the worst impact of European settlement has been the cessation of isolation through the construction of a bridge, causeway, road or railway from the mainland. Finucane, Murujuga, East Intercourse, East Mid Intercourse, Mistaken, and Preston islands have experienced this indignity.

The Mackerel Islands tourist resort occupies part of Thevenard Island (Plate 4.9). Many other islands have been used for unsupervised tourism (camping out), usually centred on recreational fishing, including Delambre Island, nine islands in Dampier Archipelago, and seven islands between Cape Preston and Exmouth Gulf. This activity risks pet dogs being brought ashore and intruding into nesting seabird colonies and also disturbing shorebirds. Pet dogs have been reported on Angel, Dolphin, and Thevenard islands. Another risk to island biota is when visitors use cooking fires inappropriately. In 2003 visitors started a fire that burnt most of Angel Island, and in 2013 a fire occurred on Tent Island.

Due to their low relief, many of the cays will be affected by rising sea levels caused by anthropogenic climate change. Not only will this reduce their size but it will erode beaches used by sea turtles for nesting, and rising sea and beach sand temperatures will affect the sex ratio of hatchlings (see Chapter 13).

Introduced species. Any disturbance by Europeans creates opportunities for exotic plant and animal species to reach an island (seeds on clothing, seeds blown by wind, stowaway animals in boxes) and then gain a foothold (e.g. in areas cleared for gardens and paths). Some have proliferated on many islands: Buffel grass and kapok bush are known from >50 and > 30 islands respectively. These species pre-empt space that should be occupied by native plant species. Once cured, the biomass of buffel grass serves to provide a ready source of fuel for a wildfire.

Islands with the most weed species are: Murujuga (19 species); Thevenard (18); Rosemary (9); Legendre (8); Angel (7); Dolphin, Enderby (6); and Airlie, North Muiron, West Intercourse, West Lewis South, South Muiron (5).

Buffel grass is well established on the old airstrip on Rosemary Island; Kapok Bush occurs on some sandplains, and the prickly pear did occur on Enderby Island but was eradicated by departmental officers in 1985. Prickly pear is well established on East and West Lewis Islands and has been subject to control attempts by the former Agriculture Protection Board. Spraying with the herbicide 245-T in 1979 was unsuccessful; however biological control trials with *Cochineal* insects (1984-1986) were successful. In October 1986 Cochineal insects were distributed to all Prickly Pear infestations on East and West Lewis Islands.

Palms (*Cocos nucifera*) and a Tamarisk tree have been planted on the North West Game Fishing Club lease on Rosemary Island. A palm *Phoenix dactylifera* has been planted near the well site on Rosemary Island. *Salsola kali,* which is common on many islands, is now regarded as an introduced species.

Black rats were once present on many vessels, especially pearling luggers, and have been recorded on 15 islands. They have only been eradicated on Bezout Island. Black rats have a well-documented history from many islands around the world of preying on eggs of seabirds and hatchling turtles. Rats are capable swimmers, but it seems likely that they gained access when pearling luggers were driven ashore during cyclones or were careened on island shores.

The red fox established on the north-western coast in the 1920s, having travelled overland after being released in Victoria in the 1870s. Foxes easily crossed to Murujuga and, being strong swimmers over short distances, reached the islands adjacent. They also reached many of the islands temporarily tied to the mainland. Foxes have been recorded on 22 islands. Aerial deployment of poison baits in the 1980s in Dampier Archipelago eradicated them from nine islands, but this is seldom permanent as they re-invade.

The feral cat has been reported from 14 islands. Cats can swim but seldom choose to do so. They have crossed to many of the tied islands and appear to have been released on several of the more isolated islands, including one cat on each of Thevenard and Serrurier islands where they no longer occur. Eradication took place on four islands in Dampier Archipelago in the 1980s. Even one cat on an island could during its life annihilate a small population of prey species unable to seek refuge (e.g. nesting seabirds, land birds with limited powers of flight, and small mammal species).

The remaining non-native mammal species of potential concern is the house mouse. When in large numbers, mice can make it very difficult for nesting seabirds to raise their young. This species is known from ten islands, mostly inshore and close to the mainland coast. House mice were introduced accidentally to Thevenard Island in 1986. It is impractical to bait this island because a native rodent species, the kerakenga, is also present. The only solution would be to capture a small population, maintain them off site, bait the island and kill all house mice (and the remaining native species), and then repatriate the native species. This would be an expensive operation, and there is no guarantee that house mice would not be inadvertently re-introduced, as there is a tourist camp on the island.

To conclude this account of introduced mammals, Finucane Island has at one time had rabbits, horses, and cattle; West and East Lewis had sheep; and Jarman Island had goats and a monkey (species unknown).

Extinctions of island populations. The red fox caused the extinction of the black-footed rock-wallaby on Depuch Island between 1962 and 1982. The dingo has also disappeared from Murujuga and Depuch Island, probably as a result of the widespread use of strychnine poison on the adjacent mainland from the 1860s. The ngadji disappeared from Murujuga sometime before 1994. No extinctions of island populations of plant, bird, or reptile species have yet been ascertained.

Translocations of species. Islands can be suitable places for establishing insurance populations of threatened species. This is especially so when the factors threatening a species are absent, for example introduced predators (foxes, cats, rats), inappropriate fire regimes, and introduced herbivores. So far, four mammal species have been released on islands on which they had not previously occurred. The first successful translocation was of Rothschild's rock-wallaby to West Lewis Island from Enderby Island in the 1980s. In 1996 the kerakenga (northern short-tailed mouse) was introduced from Thevenard Island to Serrurier Island. Then the golden bandicoot, involving 92 animals collected on Barrow Island, was released on Doole Island in Exmouth Gulf in 2010-11. The earlier release, involving the same islands, was of 224 djoongari (a species of native rodent) in 1993-2001 from Bernier Island (via Perth Zoo). It failed by 2005, and it is not obvious why the bandicoot thrived and the rodent did not.

Table 4.1 also exemplifies an important point about ensuring more effective nature conservation on islands. No one island has retained all of the 23 mammal species known to occur on the islands between Roebuck Bay and Exmouth Gulf. Even the largest, Murujuga, is missing eight species. If the view is taken that Murujuga should not be regarded as an island, then at least seven islands are required in order to conserve insular populations of all 23 species. This point demonstrates the value of comprehensive biological survey before islands are used for purposes other than nature conservation. The most precautionary (default) position should be to reserve all islands for nature conservation.

Island management plans. These documents are the foundation of strategic and operational actions that attempt to balance conflicting human activity so that the conservation value of islands is not diminished. Three plans have been published, in 1990, 1999, and 2020. All use zoning to minimise conflict. Recreational activity is prohibited on a few islands (no access at any time), limited to coastal parts of some islands (camping permitted within 100 m of high-water mark), or allowed anywhere on some islands (day trips, with or without camping). Dogs are prohibited.

Box. 4.2. Keith Morris (1953-)

Keith was a volunteer for the then Department of Fisheries and Wildlife (F&W) before obtaining a position in 1982 as Reserves Management Officer based in Karratha. There his main responsibilities were island survey and management, especially for the Dampier Archipelago, authoring the 1990-2000 management plan for this group. His duties expanded to other Pilbara islands, leading teams that eradicated black rats on the southern end of Barrow Island and on the nearby Double, Middle, Boodie, and Pasco Islands. In 1985 he also led a team that eradicated feral goats on Bernier Island in Shark Bay.

After transferring to the Science Division of the Department of Conservation and Land Management (into which the wildlife part of F&W was transferred) and moving to Perth in 1988 he continued involvement in island research and management. He led translocations of djoongari to Doole Island (Exmouth Gulf) and North West Island (Montebello Islands), before being part of the team that translocated golden bandicoot, boodie, spectacled hare-wallaby and brushtail possum from Barrow Island, where land was being cleared to construct a liquefied natural gas plant, to the Montebello Islands, Matuwa (ex-Lorna Glen pastoral lease), and Cape Range National Park.

More recently, just prior to his retirement in 2020, he planned and commenced mammal translocations to Dirk Hartog Island National Park as part of one of Australia's most ambitious fauna reconstruction projects. As well as his research and conservation actions associated with islands, Keith has also researched a wide range of projects aimed at conserving threatened mainland mammal species.

The Pilbara islands database provides a valuable record of information about the occurrence of species. However, many of the records are old and how the islands were examined by naturalists and biologists is unknown. We recommend a biological survey of all islands consistent with current best practice. A plan produced in 2015 as to how to prioritise weed management on the islands needs to be implemented.

Sources of information

Introduction: Australia Pilot 1972; Locate 2021.
Physical attributes and environmental settings: BOM 2021; Foley 1957; Hickman & Gibson 1982; Hickman & Smithies 2001; Hickman & Strong 2003; Locate 2021; Playford *et al.* 2013; Royal Australian Navy 1967, 1968, 1974a, 1979, 1983, 2019; Van De Graaff *et al.* 1982; Williams 1968.
Island area: Locate 2021.
Aboriginal history: Donaldson 2009; Durlacher 2013; Green 2020; Mulvaney 2009, 2015; Orchard & Orchard 2018; Tindale 1974; Wickham 1842.
European history: Bridge 2020; George 1971, 1990, 1999, 2018; Marchant 1988; Murray & Hercock 2008; Orchard & Orchard 2018, 2020; Stokes 1846; Wickham 1842.
Island vegetation and plant species: Burbidge & Prince 1972, Lohr 2017; Start & McKenzie 1992.
Animal species: Abbott 1982; Burbidge & Prince 1972; Durlacher 2013; Johnstone 1990; Johnstone & Storr 1998, 2004; Kendrick 2007; Lohr 2017; Ride & Neumann 1964; Start & McKenzie 1992; Whitlock 1919.
Distributional patterns: Burbidge & Abbott 2017; Connell 1983; Lohr 2017.
Natural factors: Foley 1957; Hunt 1929; Playford *et al.* 2013.
Human factors. Aboriginal impacts: Abbott 1980; Abbott & Wills 2016; Green 2020.
Human factors. European impacts: Abbott 2006; Abbott & Wills 2016; CALM 1990; Gibb 2010.
Introduced species: Burbidge & Abbott 2017; Burbidge & Prince 1972; Lohr *et al.* 2016.
Extinctions of island populations: Burbidge & Abbott 2017.
Translocations of species: Burbidge & Abbott 2017.
Island management plans: CALM 1990; DBCA 2020; Lohr 2017; Lohr *et al.* 2015; NPCA 1999.

Species codes: (in the following table):
Rt *Rattus tunneyi*;
Oro *Osphranter robustus*;
Za *Zyzomys argurus*;
Ph *Pseudomys hermannsburgensis*;
Pn *Pseudomys nanus*;
Hc *Hydromys chrysogaster*;
Dr *Dasykaluta rosamondae*;
Pr *Petrogale rothschildi*;
Ta *Tachyglossus aculeatus*;
Dh *Dasyurus hallucatus*;
Plg *Planigale* sp.;
Pdl *Pseudomys delicatulus*;
Pm *Pseudantechinus macdonnellensis*;
Ll *Leggadina lakedownensis*;
Sm *Sminthopsis macroura*;
Sy *Sminthopsis youngsoni*;
Pds *Pseudomys desertor*;
Pla *Petrogale lateralis*;
Pw *Pseudantechinus woolleyae*;
Nt *Ningaui timealeyi*;
Oru *Osphranter rufus*;
Pc *Pseudomys chapmani*;
Na *Notomys alexis*.

Table 4.2. List of all islands in this region known to have at least one native mammal species, arranged by decreasing island area (rows) and decreasing insular occurrence of species (columns).
*island temporarily joined to the mainland at low tide or permanently by a causeway or bridge.

Island	Area (ha)	Species																							Total no of species
		Rt	Oro	Za	Ph	Pn	Hc	Dr	Pr	Ta	Dh	Plg	Pdl	Pm	Ll	Sm	Sy	Pds	Pla	Pw	Nt	Oru	Pc	Na	
Murujuga*	11804		X	X	X		X	X	X	X	X	X	X	X						X	X	X	X		15
Dolphin	3300	X	X	X					X		X	X													6
Enderby	3241	X		X	X				X																4
W Intercourse*	2367		X																						1
Tent*	2015		X										X			X									3
W Lewis	1982	X		X																					2
Legendre	1364	X		X	X					X															4
Rosemary	1122			X	X		X		X																4
Depuch	1121						X												X						2
E Lewis	993	X																							1
Angel	926	X		X																					2
Gidley	856	X	X																						2
Hope*	850		X		X			X																	3
Thevenard	589														X										1
Finucane*	550		X		X			X									X	X						X	5
Dixon*	512		X		X					X															3
Malus	349	X																							1
Downes*	315		X		X																				2
Sholl	302					X																			1
E Intercourse*	250		X																						1
Potter	160		X			X		X																	3
M Mangrove	110	X																							1

Island	Area (ha)	Species																							
		Rt	Oro	Za	Ph	Pn	Hc	Dr	Pr	Ta	Dh	Plg	Pdl	Pm	Ll	Sm	Sy	Pds	Pla	Pw	Nt	Oru	Pc	Na	Total no of species
Haüy	109	X																							1
Roberts*	83	X																							1
Collier Rocks	80	X																							1
Carey*	72		X																						1
SW Regnard	65	X				X																			2
Burnside*	58		X																						1
N Mangrove	57	X																							1
Simpson	57	X	X																						2
Keast	51	X																							1
Fortescue	41	X				X																			2
Tozer*	37		X																						1
NE Regnard	36						X																		1
Cohen	32						X																		1
W Moore	32	X																							1
Ashburton	27	X																							1
Islet NE of W Lewis	23	X																							1
Weld	23	X																							1
Angle	21			X																					1
Kendrew	11			X																					1
Thringa*	8					X																			1
'Little Roberts'	5	X																							1
Total no of islands		22	15	9	8	5	5	4	4	3	2	2	2	1	1	1	1	1	1	1	1	1	1	1	

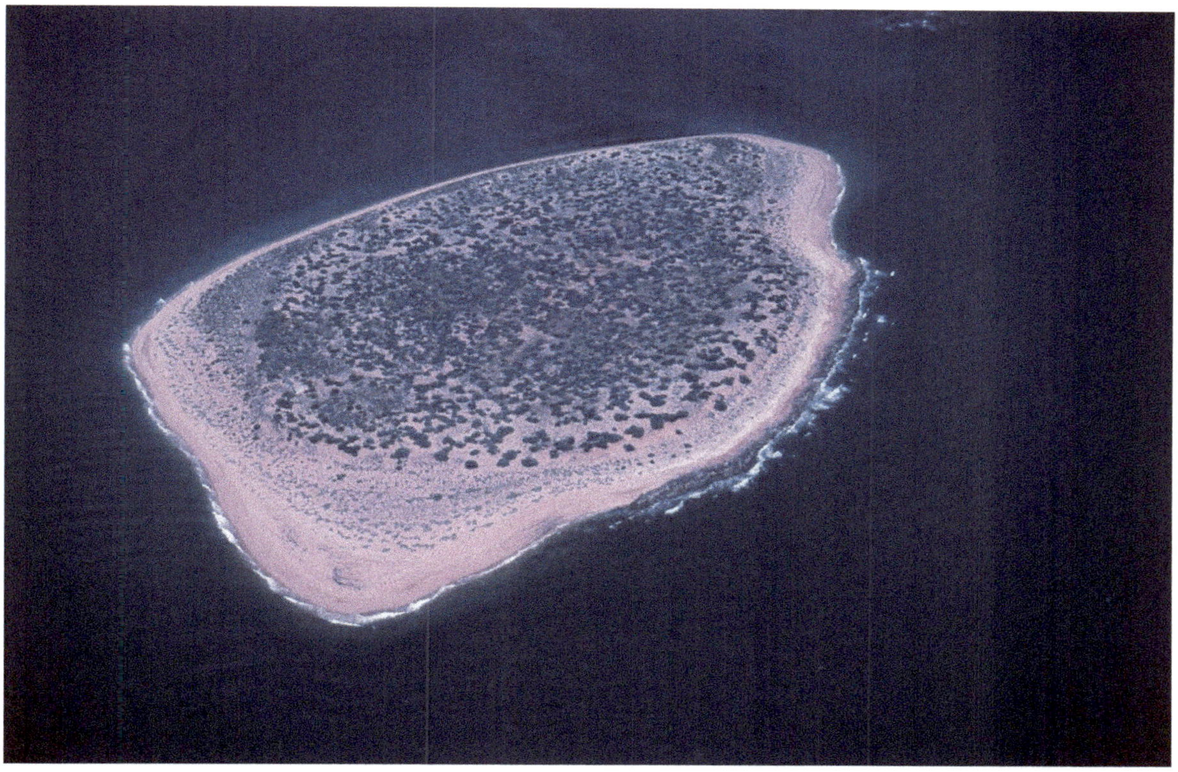

Plate 4.1. Aerial view of Direction Island, a sand cay. February 1981. Credit: AA Burbidge.

Plate 4.2. Aerial view of Airlie Island, showing vegetation and light tower. February 1981. Credit: AA Burbidge.

Plate 4.3. Enderby Island, Dampier Archipelago. June 1970. Credit: AA Burbidge.

Plate 4.4. Rocky gully, Angel Island, Dampier Archipelago. June 1970. Credit: AA Burbidge.

Plate 4.5. Petroglyph presumed to be of a thylacine, Angel Island, Dampier Archipelago. June 1970. Credit: AA Burbidge.

Plate 4.6. Granophyre rocks and hummock grassland, Angel Island, Dampier Archipelago. Credit: AA Burbidge.

Plate 4.7. Vegetation on Thevenard Island. July 1979. Credit: AA Burbidge.

Plate 4.8. Rothschild's rock-wallaby, Enderby Island, Dampier Archipelago. Credit: Bert & Babs Wells collection, courtesy of the Department of Biodiversity, Conservation and Attractions.

Plate 4.9. Oil base and tourist resort, Thevenard Island. February 1981. Credit: AA Burbidge.

Map 5.1. Islands offshore from the Pilbara coast.

CHAPTER 5: NORTH-WEST OFFSHORE ISLANDS

According to a recent review and evaluation conducted by a panel of experts in charge of global weather and climate extremes within the WMO Commission for Climatology, the new record wind gust, not related to tornados, registered to date is 408 kilometres per hour. This record occurred during Tropical Cyclone Olivia on 10 April 1996 on Barrow Island, Australia. The previous record was of 372 kilometres per hour, registered in April 1934 across the summit of Mount Washington, New Hampshire (USA).

World Meteorological Organization, 22 January 2010.

INTRODUCTION

Between 55 and 80 km from the Pilbara coast lie three groups of islands: Barrow and nearby islands; the Lowendal Islands; and the Montebello Islands. At the time Aboriginal people arrived in Australia, c. 65 000 years ago, sea levels were around 90 m lower than today's, and the three island groups were joined to each other and to the mainland. Noala people occupied or visited all three groups until c. 9000-6500 years ago, when the sea level rose, and the islands probably became too far offshore for easy access.

These islands have low, unreliable rainfall that occurs mainly in mid-winter or from summer tropical cyclones. Tropical cyclones pass over or near the islands almost every year, but not always close enough to bring substantial rainfall. The only relatively long-term weather data are from Barrow Island Airport and the former Barrow Island weather station located in the centre of the island. Mean-annual rainfall at the airport station (2000 to 2020) was 275 mm but was highly variable (84–640 mm) and at the central station (1970 to 1991; destroyed by a cyclone) it was 318 mm (137–747 mm). Temperatures are warm to hot, but summer maxima are cooler than the adjacent mainland, as they are ameliorated by surrounding seas, with mean monthly maxima at the airport station ranging from 24°C (July) to 33.5°C (February and March). Temperatures are higher near the centre of the island with recordings of over 40°C.

BARROW GROUP ISLANDS

INTRODUCTION

Barrow Island (23 569 ha) is the second-largest island in Western Australia (WA); only Dirk Hartog Island is larger. Just to the south of Barrow are Middle (361 ha) and Boodie (115 ha) Islands, and the 2.5 ha Pasco Island. Off its east coast are North and South Double Islands (12 ha and 20 ha) and some islets.

European history, discovery and naming. Phillip Parker King in HMC *Mermaid* named Barrow Island in 1818, after Sir John Barrow, second Secretary to the Admiralty and founder of the Royal Geographical Society. Pasco Island was probably named after Frederick Pasco RN by JW Combe RN during the 1899 survey of Mary Anne Passage (the channel between Barrow Island and the mainland). Double Islands were first mentioned as 'Double Islet' in the voyage of the *Tien Tsin*, a barque that traded along the WA coast, published in the *Perth Gazette* in January 1865. Middle Island was in common usage among oil workers on Barrow Island before being approved in 1972. Boodie Island's name was suggested in 1973 by Department of Fisheries and Wildlife staff, after Harry Butler and Andrew Burbidge discovered that boodies, a potoroid marsupial, occurred there; prior to its naming locals thought that it was Pasco Island.

Geology. Composed largely of Tertiary limestone, Barrow Island is slightly undulating dome-shaped, with a maximum elevation of 62 m, and has mostly red, shallow, eluvial soils as well as slightly deeper alluvial soils. There areas of exposed rock, especially towards the western side of the island, where there are low cliffs (Plate 5.1). Near the coast, particularly towards the south end, as well as on Boodie and Middle Islands, coastal low dunes and flats of white sand occur. Barrow Island was connected to the mainland by a land bridge until c. 7400 years ago, when rising sea levels cut it off.

Biological discovery. The first natural history observations were made by John Lort Stokes and the crew of HMS *Beagle* in 1840. The naturalist John Tunney visited in 1900 and made collections that were lodged in the WA Museum. Frederick Lawson Whitlock visited in 1917 and 1918 on behalf of Henry Luke White, mainly to study the birds and collect eggs. Some members of the British atomic weapons test team working at the Montebello Islands visited in 1952. In the 1950s, the island was visited by personnel associated with the testing of the first British nuclear weapon and was, for a while, considered as a potential site for these tests. By this time the value of Barrow Island to mammal conservation had been recognised and it was proposed by English politicians that the native fauna could be herded to the southern end of the island, a fence constructed across it (a distance of approximately 10 km) and the atomic weapons detonated at the northernmost point! Fortunately this option was not pursued and the nuclear weapon tests were carried out in the Montebello Islands, 40 km to the north.

Then, in 1958, Dominic Serventy and Jock Marshall visited and provided a comprehensive description of the island and its fauna, although, strangely, they did not report boodies, which are abundant. Later, extensive and detailed natural history observations and collections were by WH (Harry) Butler in 1964 and subsequent years.

Reservation. In 1908, Barrow Island was declared a reserve, mainly because of what was then considered to be a distinct species of kangaroo (now a subspecies, the Barrow Island euro) and the endemic black-and-white fairy-wren. The original purpose was 'Protection of Flora and Fauna' and is now 'Conservation of Flora and Fauna'. It was proclaimed Class A in

1910 and is one Western Australia's oldest nature conservation reserves. However, it was not vested in a statutory body until 1985. The 1965 report by the Western Australian Sub-

> THE PERMANENT RESERVES ACT, 1899.
>
> *Reserve* 11648, *Barrow Island.*
>
> Department of Lands and Surveys,
> 7608/07. Perth, 18th February, 1910.
>
> HIS Excellency the Governor has been pleased, under "The Permanent Reserves Act, 1899," to classify Reserve 11648, Barrow Island, as Class "A" for the protection of "Flora and Fauna."
>
> R. CECIL CLIFTON,
> Under Secretary for Lands.

Declaration of Barrow Island reserve as Class A. *Government Gazette* 18 February 1910, page 537.

Committee of the Australian Academy of Science Committee on National Parks recommended its vesting in a nature conservation statutory body, but this recommendation was not acted on until well after the 1975 Environmental Protection Authority (EPA) report to government 'Conservation Reserves for Western Australia', which also recommended vesting. The Academy of Science report also recommended that no wharf or jetty be constructed to minimise the threat of introduction of non-indigenous rodents, a recommendation that was ignored when oil was found. Later, after the EPA report was acted on, all the adjacent islands were reserved. Barrow Island is reserved to low water mark. The Boodie, Double and Middle Islands Nature Reserve was declared in 1984.

BIOLOGICAL CHARACTERISTICS

Vegetation and flora. Barrow's vegetation is largely *Triodia* spinifex hummock grassland (Plate 5.2), with scattered emergent native figs *Ficus platypoda* and other *Ficus* spp. and weeping pittosporum *Pittosporum phillyreoides*. Near the west coast, there are a few small patches of a low mallee *Eucalyptus xerothermica*, one tallish group of which is known locally as 'the valley of the giants'. There are also some patches of yulbah (or bats-wing coral tree *Erythrina vespertilio*). White sands are dominated by beach spinifex with occasional large shrubs of *Acacia coriacea*. Geologically and biologically Barrow Island is related to North West Cape and Cape Range, not the Pilbara mainland, and is part of the Carnarvon Basin bioregion.

At least 377 species of native plants have been recorded on Barrow Island Nature Reserve. While typical of the flora of parts of the Pilbara region, the flora has affinities with that of Cape Range. White mangroves occur as small stands on the east and south coasts.

Mammals. Biologically, Barrow is best known for its mammals (Plates 5.3-5.7). Eleven species of terrestrial non-flying mammals occur there as well as three species of insectivorous bats. While some Kimberley islands have more mammal species, Barrow has more non-flying species (11) than any other WA island except the Kimberley's Augustus Island (12), which has much higher rainfall. Five of Barrow's species are listed as threatened species or subspecies and four subspecies are endemic.

Table 5.1. Terrestrial, non-flying mammal species of Barrow Island

Mammal species/subspecies	Listed as threatened?	Endemic to Barrow Island?	Occurrence elsewhere	Comments
Barrow Island euro	yes	yes	-	endemic subspecies
Black-footed rock-wallaby	yes	no	Cape Range, south west WA	
Barrow Island spectacled hare-wallaby	yes	yes	subspecies *leichardti* occurs in northern Australia	endemic subspecies
Barrow Island boodie	yes	yes	boodies once occurred across vast areas of inland Australia, but became extinct on the mainland	reintroduced to Boodie Island; introduced to Alpha Island, Montebellos
Barrow Island golden bandicoot	yes	yes, including Middle Island	mainland subspecies was once widespread; remains only in the north Kimberley and other islands	subspecies not accepted by all taxonomists; introduced to Hermite Island, Montebellos
common brushtail possum (south-western subspecies)	no	no	South-west and west of WA	
fat-tailed false antechinus	no	no	widespread in arid areas	
common planigale	no	no	widespread in northern Australia	
rakali (water-rat)	no	no	widespread	occurs in both fresh and marine waters, on Barrow Island

Mammal species/subspecies	Listed as threatened?	Endemic to Barrow Island?	Occurrence elsewhere	Comments
				lives around sheltered coastlines
djoorri (common rock-rat)	no	no	widespread in northern Australia	
moolpoo (western chestnut mouse)	no	no	widespread in northern Australia	subspecies *ferculinus*, endemic to Barrow Island, no longer accepted as valid

One species, the black-flanked rock-wallaby, demonstrates what happens when small populations are isolated for a long time. The rock-wallabies occur in small areas of suitable habitat along and near Barrow's west coast and counts suggest fewer than 150–200 animals occur there. Genetics research has shown that the Barrow Island population of has extremely low levels of genetic variation and suffers from inbreeding depression (reduced fecundity, skewed sex ratio, increased levels of fluctuating asymmetry) and a small effective population size. Nevertheless, this population has survived a long period of isolation (c. 1600 generations).

Four species of bats have been recorded on Barrow Island: white-striped mastiff-bat, common sheath-tailed bat, northern mastiff-bat and Finlayson's cave-bat. As shown in Table 4.1, the golden bandicoot also occurs on Middle Island and the boodie was reintroduced to Boodie Island.

Birds. About 120 species of birds have been reported from Barrow Island, including seabirds, shorebirds, and 48 species of land birds. Of the land birds, only 15 are thought to be permanent residents, six are regular migrants, and 27 are irregular visitors. Barrow Island is internationally recognised by BirdLife International as an 'Important Bird Area' mainly because of its extensive intertidal areas. Those in the south and south-east parts of the island, including Bandicoot Bay, support large numbers of shorebirds. The island has an endemic bird subspecies, the Barrow Island white-winged fairy-wren (also known as the Barrow Island black-and-white fairy-wren), which has been translocated to Hermite Island, Montebellos. It is listed as Vulnerable but considered Least Concern in *The Action Plan for Australian Birds 2020.* Tangling melaleuca *Melaleuca cardiophylla* shrubs provide key habitat for this species.

The Double Islands are considered a regionally significant site for breeding wedge-tailed shearwaters, which also nest on Boodie and Middle Islands. Ground-nesting seabirds do not breed on Barrow Island as the abundant golden bandicoots would predate their eggs.

From a modern biodiversity standpoint, Barrow Island is best known for its mammals and underground animals. However, it was also the black-and-white fairy-wren that led to its reservation. In 2001, Steve Pruett-Jones from the University of Chicago and Erin O'Donnell from the University of Adelaide visited Barrow Island to study its birds. They estimated the total numbers of the more common birds occurring there; estimates of the three most abundant species were 24 623 spinifexbirds, 7519 black-and-white fairy-wrens and 3920 singing honeyeaters.

Reptiles and frogs. Barrow Island has a rich reptile fauna, with 43 species recorded: dragon lizards (3 species), legless lizards (5), geckoes (5), skinks (19), blind snakes (3), monitors (or goannas, 3) and snakes (5). The Barrow Island leopard skink, is endemic. A single species of frog, Main's frog *Cyclorana maini*, occurs on the island, spending most time underground and only emerging after heavy rain to breed in claypans and seasonal watercourses.

Major sea turtle nesting rookeries occur on the island (Plate 5.8). Mainly on its western beaches, large numbers of the green and flatback turtles nest, while lesser numbers of hawksbill and loggerhead turtles also nest on Barrow and on some adjacent islands. Satellite tracking has shown that some Barrow Island nesters will, outside the breeding season, swim large distances to the north into the Timor Sea. Records show that 63 turtles were taken from Barrow Island in 1840, and in 1864, Captain Jarman of the *Tien Tsin* and his passengers conducted a turtle hunt on the east coast of Barrow Island where 'the bays were swarming with them'. Fortunately, Barrow Island was excluded from licences issued for turtle 'fishing' in the period from 1910 to 1973, although enforcement of the 'fishery' was minimal.

Invertebrates. A comprehensive survey of terrestrial invertebrates was made prior to the construction of the liquefied natural gas plant. It revealed c. 2480 indigenous species, but also found 25 species that were non-indigenous and another seven that are possibly non-

WH (Harry) Butler AO, CBE (1930-2015)

Naturalist Harry Butler first visited Barrow Island in 1964 as the oil field was being developed, and became the conservation consultant to the oil company soon afterwards. In 1964-65 he made early collections for the WA Museum and recorded information about Barrow Island's animals, making many 'first records' for the island. From then onwards he educated the workers, oversaw rehabilitation of disturbed areas, and set up the first quarantine system. Harry specialised in working at the interface of industry and conservation. He helped many scientists and government workers who visited the island gain a better understanding of it. After the Gorgon Gas Project was approved by government in 2009, Harry continued to be a champion for the island, visiting to educate the thousands of workers building the LNG plant.

Harry's contribution to Barrow Island and other conservation issues led to Chevron Australia funding the 'Harry Butler Institute' within Murdoch University, which promotes research at the intersection of business and biodiversity conservation.

indigenous, all associated with disturbed areas, and most likely to have resulted from the construction and management of the oilfield.

The subterranean terrestrial and aquatic species that inhabit Barrow's extensive karst systems are world-renowned. The anchialine system (sometimes referred to as subterranean estuaries), with freshwater overlying and interacting with seawater, protects a unique array of short-range endemic species: the many animals that dwell only in the cool, dark caves and in groundwater. Such species are termed stygofauna if they are aquatic (stygobionts if they can only survive in dark underground waters), and troglofauna if living outside water (troglobionts if they can only survive in completely dark caves). Many of the invertebrate stygofauna have their closest relatives occupying similar habitats on opposite sides of the North Atlantic, primarily in the Caribbean and Canary Islands. There are also stygobitic and troglobitic vertebrates: the endemic Barrow Island cave gudgeon *Milyeringa veritas*, an endemic species of subterranean, very long blind snake *Anilios longissimus* (the first truly troglobitic reptile to be described globally) and a subterranean eel, which has been not been described. By 2013, 82 animal species had been recorded living underground on Barrow Island.

FACTORS LEADING TO CHANGE IN THE BIOTA

Barrow Island provides an example of what some of mainland arid Australia, particularly *Triodia* (spinifex)-dominated areas, looked like before European settlement and the introduction of red foxes, feral cats, European rabbits, and other pests. It is free of introduced vertebrates, but does have a small number of non-indigenous invertebrate species.

Non-indigenous species. Barrow Island is one of the largest land masses in the world without introduced rodents; however, Barrow and adjacent islands have not always been rat and house mouse free. Black rats were introduced to the Double Islands, and Middle and Boodie Islands, well before the oil field was developed, probably from pearling luggers. Eradication was achieved in 1983-1985. In 1985, rats were discovered on Boomerang Island (5 ha) which is joined to Barrow at low tide. Fortunately they had not spread to the main island and eradication was quickly achieved. How long they had been there is a mystery. In July 1990, a black rat was trapped near the southern tip of Barrow Island and subsequent survey revealed that they occurred on white sand surfaces but, fortunately, not on the major rocky areas to the north. Eradication was achieved using bait stations that excluded all native mammals except the two species of native rodents, which recolonised the area after the rats were eradicated and the bait stations were removed. House mice were discovered in warehouses in 1994, 1995 and 1996, but were eradicated and did not establish. Constant vigilance is required to ensure that introduced rodents do not get to Barrow Island and, if they do arrive, they are detected quickly, and eradicated with minimal effects on the native fauna.

One condition of government approval of the Gorgon Gas Plant was for Chevron Australia, as operator, to develop a new quarantine management system (QMS) aimed at preventing the establishment of any non-indigenous species. Implementation of the QMS is monitored by an independent Quarantine Expert Panel (QEP) that advises the company and the state Minister for Environment. The Barrow Island QMS is now arguably the best island quarantine system in the world. One of us (AAB) first visited Barrow Island in 1969 and has been there many times since. AAB assisted with the development of the QMS and currently (2022) chairs the QEP. To date, only one non-indigenous species, a jumping spider, has established and is believed to pose minimal threat to the island's biodiversity. Importantly the island remains free of black rats and house mice, one of the largest areas in the world with this enviable status.

There are few weeds on Barrow Island, with only 22 species having been recorded. Chevron has active weed eradication programs for buffel grass *Cenchrus ciliaris* and kapok bush *Aerva javanica*, both of which are now very restricted in area. Most other weeds are either eradicated or monitored with removal occurring when located. Spiked malvastrum *Malvastrum americanum* and speedy weed *Flaveria trinervia* are considered naturalised and eradication is no longer possible.

In addition to buffel grass, seven weed species have been recorded on Middle, Boodie and Pasco Islands: five on Middle Island, three on Pasco Island and two on Boodie Island.

Most of the spinifex hummock grassland has not been burned for many decades. That, along with the lack of introduced mammals, has meant that Barrow Island is a haven for wildlife. Visitors to Barrow are amazed at how abundant the native mammals are. Workers entering dining areas can have golden bandicoots run indoors around their feet. Outside, spectacled hare-wallabies, possums and boodies forage near the accommodation areas. And when driving the roads (which have a low speed limit and a lower one at night) animals are often seen on and near the roads. It is not only mammals that are frequently sighted; perenties, Australia's largest lizard and the island's top predator, behave as though they own the island, taking little notice of vehicles or people (Plate 5.9).

Island management. A management plan for Barrow group nature reserves was approved in 2015. The plan covers the Barrow Island Nature Reserve (23 483 ha) and the Boodie, Double and Middle Islands Nature Reserve (586.7 ha). Parks and Wildlife staff are based on Barrow Island and work closely with staff from Chevron Australia, the operator of the Gorgon Gas Plant and the oil field.

LOWENDAL ISLANDS

INTRODUCTION

Fifty-one islands, islets, and rocks make up this archipelago, which is located north east of Barrow Island. Like Barrow, it is an oil and gas base, this time for an offshore field, and the islands are protected in the Lowendal Islands Nature Reserve.

European discovery, history and naming. The islands were named in 1801 by Nicolas Baudin, sailing in *Le Géographe*, after Ulric Frédéric Woldemar, Comte de Lowendahl, a member of the French Academy of Science. Until relatively recently, most of the individual islands were not named.

An oil and gas hub was constructed on Varanus Island in 1987 by the Bond Corporation and has undergone ownership changes since - initially to Apache Energy and later to Santos Limited.

Biological discovery. When the largest island in the Lowendals was proposed for an oil and natural gas base, Harry Butler and Andrew Burbidge visited the islands in 1985, made some early animal and plant observations and scientific collections, and proposed plant and animal names for the larger islands, that were later adopted: Varanus (83 ha), Bridled (27 ha), and Abutilon (27 ha). 'Beacon Island' (2.9 ha, east of Bridled Island) is not yet formally named. Varanus Island was named after the spiny-tailed goanna, Bridled Island after the bridled tern, and Abutilon Island after a species of *Abutilon*.

Reservation. The Lowendal Islands Nature Reserve was declared in March 1976; it is not Class A.

BIOLOGICAL CHARACTERISTICS

Birds. The Lowendal Islands are recognised by BirdLife International as an Important Bird Area because of the large breeding colony of wedge-tailed shearwaters that occurs on Varanus Island. Studies have shown minimal impact of the oil and gas facilities on them. Caspian, greater crested, and bridled terns also breed on Varanus Island. Wedge-tailed shearwaters and the eponymous bridled tern also nest on Bridled Island along with the silver gull. Both greater and lesser crested terns, as well as silver gulls, have been recorded breeding on 'Beacon Island', located to the east of Bridled Island.

Reptiles. As well as the eponymous *Varanus acanthurus* and some smaller lizards, Varanus Island has a significant hawksbill turtle nesting rookery. A smaller rookery occurs on Beacon Island. Globally, the IUCN Red List considers hawksbill turtles to be Critically Endangered and Australian populations are one of very few remaining abundant. Satellite tracking has shown that female hawksbill turtles nesting on Varanus Island feed between 50 km and 450 km from their nesting beaches.

FACTORS LEADING TO CHANGE IN THE BIOTA

Non-indigenous species. In 1993, house mice were detected on Varanus Island, apparently arriving in food containers sent to the oil company, naturally extending their range to the adjacent Bridled and 'Beacon' Islands. After initial attempts to eradicate failed, eradication was finally achieved in 1997.

Island management. Day-to-day management of the Lowendal Islands Nature Reserve is mainly administered by the oil company and their consultants with only occasional visits by Parks and Wildlife Service staff.

MONTEBELLO ISLANDS

INTRODUCTION

To older Australians the Montebello Islands are infamous because, in 1952 and 1956, the British tested three nuclear weapons there. Now they are a conservation park and the surrounding waters are a marine park. They are located c. 20 km north of Barrow Island and about 80 km north-west of Cape Preston on the adjacent mainland.

The Montebellos (Plate 5.10) consist of c. 180 islands, islets and rocks, the largest being Hermite (1022 ha), Trimouille (520 ha), North West (135 ha), Alpha (117 ha) and Bluebell (65 ha). The total area of all islands is c. 2200 ha.

European discovery, history and naming. The first European visitors were Thomas Bright and survivors of the English vessel *Tryal* (or *Tryall*), wrecked on Tryal Rocks, north west of the Montebellos in May 1622 - the first known shipwreck in Australian waters. Brooke Island is named after the *Tryal*'s captain John Brooke, but Thomas Bright was the real hero, managing to save many of the crew, camping for a short time on North West Island, and then sailing in a longboat to Batavia. The next visitors were the crews of *Le Naturaliste* and *Le Géographe* in 1801, commanded by Nicolas Baudin. The Montebello Islands were named by Baudin in 1801 after Jean Lannes, a French general who was with Napoleon Bonaparte when the French defeated the Austrians in 1800 at the battle of Montebello. Bonaparte made Lannes the Duke of Montebello. (Montebello is Italian for 'beautiful mountain', a most inappropriate term for these flat, arid islands.) Early maps and reports used 'Monte Bello' as the name for the islands, but this was later corrected. Baudin also named Trimouille and Hermite Islands, the former in honour of the La Trimouille family and the latter after one or other of two French naval officers of that name.

HMS *Beagle* visited in 1840 and the crew, wanting fresh meat, shot spectacled hare-wallabies on Trimouille Island. Whether they proved palatable is unknown as, in 1912, Paul Montague (see box) stated 'the flesh was by no means good to eat, tasting very strong; this was the only instance we found wallaby at all unpalatable'. There were few visits for some time after that, mainly by pearling luggers (some of which are known to have been wrecked there) that were probably the source of the black rats and the cats. Montague wrote that there were regular visits by pearlers during his time at the Montebellos. He attributed the presence of black rats to a schooner that was wrecked 12 years before his visit.

Before the 1950s very few islands had names. In 1951, Royal Navy surveys prior to the British nuclear weapons tests of 1952 and 1956, named several islands after European plants (including Bluebell, Primrose, Crocus, Marigold, Foxglove, Pansy, Jonquil, Kingcup, Daisy, Ivy, Buttercup and Dahlia). They also named points after British Prime Ministers (Churchill, Balfour, Disraeli, Gladstone, Palmerston and Atlee), bays after drinks (Claret, Whiskey, Champagne, Chianti, Chartreuse, Stout, Vermouth, Hock, Cocoa, etc.) and channels after physicists (Stephenson, Kelvin, Wheatstone, Faraday and Bunsen). Karangi Island was named after a Royal Australian Navy vessel that supported the tests. Following the Royal Navy's decision to name islands after English flowers, the black rat and feral cat eradication team in the 1990s proposed naming many un-named islands after Australian flowering plants: Hakea, Mulga, Acacia, Karri, Waratah, Boab, Bloodwood, Solanum, Banksia, Spinifex, Beaufortia, and many more. These names were accepted by nomenclature authorities.

Later natural history observations were made by Keith Sheard (CSIRO) in 1950, Frank Hill (a member of the nuclear weapons test team) in 1952, Dominic Serventy and Jock Marshall in 1958, Harry Butler in 1966 and Andrew Burbidge (with other Fisheries and Wildlife staff) in 1970, and in 1971 with Harry Butler.

Reservation. After the nuclear weapons tests of 1952 and 1956, the Commonwealth government declared the islands and surrounding seas a prohibited area, although this was not enforced—at least one person visited and collected scrap metal, probably very radioactive, and sold it. The Montebellos were returned to state control in 1992 and declared a conservation park.

BIOLOGICAL CHARACTERISTICS

Mammals. Montague found mummified remains of golden bandicoots on Hermite Island in 1912 and reported 'The cats have evidently been responsible for its extermination.' He found spectacled hare-wallabies on Hermite (but not on Trimouille) and stated 'The cats which have been introduced to Hermite Island appear to be breeding rapidly; wherever introduced they soon become exceedingly shy and wary, and grow to a large size. They will, no doubt, in a few years time (sic) have accounted for the wallabies, as they have for the bandicoots. If they cannot kill a full grown wallaby—though I am inclined think they do—they make short work of the young ones.' Montague's prediction proved to be correct.

A single species of bat, Finlayson's cave-bat, was reported by Montague and later visitors. In addition to the mammals mentioned above, tracks of rakali (or water-rats) were reported on Trimouille Island by Harry Butler in the 1970s, and later visitors reported tracks on

Hermite Island in 1971 and Alpha Island in 1983. The species has not been recorded since and may have succumbed to predation by feral cats.

Birds. By 2000, 70 species of birds had been recorded in the Montebello Islands. Some are resident, but most are vagrants or migrants. Two, possibly three, resident species have disappeared from the islands -- the spinifexbird, collared kingfisher and, possibly, the black-and-white fairy-wren. The last of these was recorded by Keith Sheard of CSIRO in 1950, but not by Montague in 1912, Hill in 1952, or later observers. Sheard also reported the subspecies on Sholl Island, but no one else has seen it there. It seems possible that Sheard mistook the species for another.

The Montebellos are a significant seabird breeding place. Wedge-tailed shearwater has been recorded nesting on 10 islands, silver gull on five islands, Caspian tern on 14 islands, lesser crested tern (one island), greater crested tern (5), roseate tern (8), fairy tern (2), and bridled tern (4). Other birds that depend on the ocean for their food also breed in the Montebellos: white-bellied sea-eagle, eastern osprey, pied cormorant, eastern reef egret, beach stone-curlew, and pied and sooty oystercatchers.

About 12-15 pairs of beach stone-curlews breed on the beaches, and the islands' rocky shores are a stronghold for the northern sooty oystercatcher.

Reptiles. Twenty-one terrestrial reptile species have been recorded. Gould's goannas are common. One subspecies of legless lizard, *Aprasia rostrata rostrata*, was considered to be endemic to the Montebellos, but taxonomic research has shown that the subspecies is not valid and the species is relatively widespread on the mainland.

The Montebellos are a major sea turtle nesting place, with green and hawksbill turtles having large rookeries and flatbacks and loggerheads also nesting there. Satellite tracking has revealed that turtles of all four species swim long distances, south to Barrow Island, east to the adjacent mainland coast, and north into the Timor Sea. PD Montague reported that, in 1910, a Mr Louis Williams, known locally as 'Balmy Louis' camped on Trimouille Island 'for several years, hunting hawksbill turtle, which formerly came up on to the beaches to breed in considerable numbers, though now they are nearly extinct in that neighbourhood.' The legal taking of sea turtles that occurred in WA up to 1973 included the Montebello Islands. Records do not disclose the precise number taken from the Montebellos, but it was probably several thousands.

FACTORS LEADING TO CHANGE IN THE BIOTA

Non-indigenous species. Black rats were reported on many islands, including very small ones, by Montague in 1912 and he attributed their presence to a pearling lugger wrecked c. 20 years before his visit (i.e. c. 1890). Because of their extensive distribution in 1910 it is possible that the rats established earlier than 1890, as pearling vessels were active in the area from the 1860s. It is not known how the cats established, but pearling luggers seem the most likely source. House mice were first reported in 1983, but are likely to have been present well before that. It is unclear whether they remained after the rats were eradicated.

Weeds are prevalent, particularly buffel grass and kapok bush. It is unknown when these weeds were introduced, but they were undoubtedly spread by the activities of the nuclear weapons test teams who disturbed many parts of the islands.

The nuclear weapons tests. After World War II the United States of America stopped cooperating with the British concerning nuclear weapons development, due in part to risk of English spies passing secrets to the Soviet Union (which indeed happened). The British decided to develop their own weapons, led by scientists who had been involved in the Manhattan Project. The Montebello Islands were chosen as the site for the first test. When the British Prime Minister of the time, Winston Churchill was asked in the House of Commons whether the wild life of the islands might be affected he replied that investigations had been made. He said that the investigating officer had reported only 'some lizards, two eagles, and something that looked like a canary sitting on a perch'. Mr Churchill felt able, therefore, to give an assurance that no serious damage to wild life could be done by the weapons tested there. Clearly, either the investigating officer was incompetent or his advice was ignored.

At the time the Russians did not possess missiles capable of carrying nuclear weapons and there was concern that they might bring one into a port in a ship. So the first, 1952 test, was of a device in a ship. The 'Hurricane' test was the detonation of a 25 kilotonne plutonium-based device 2.7 m below the water line in HMS *Plym*, an obsolescent frigate, moored c. 400 m off Cocoa Beach, Trimouille Island. The blast on 3 October 1952 vaporised most of the ship and lifted thousands of tonnes of contaminated salt water and sediment into the air, much of it landing on Trimouille Island, nearby islands and the adjacent ocean. It left a saucer-shaped crater on the seabed 6 m deep and 300 m across. The British returned in 1956 and detonated two further devices, one of 15-20 kilotonnes on the northern end of Trimouille and one of 98 kilotonnes on Alpha Island. Both were on top of towers to simulate an air drop (Plate 5.11). The Alpha Island nuclear weapon, stated at the time to be 60 kilotonnes, was the largest tested anywhere in Australia. However, the actual yield was 98 kilotonnes. Since the test yield broke an assurance made personally by Prime Minister Anthony Eden of the United Kingdom to Prime Minister Robert Menzies of Australia that the yield would not exceed 2.5 times that of the 1952 test (thus c. 62 kilotonnes), the true yield was concealed until 1984.

Although now much lower than in the past, areas of slightly elevated radiation remain (Figure 5.1) and visitors should not stay in elevated radiation zones for more than a few hours. They also should not disturb soil or touch any metal objects anywhere on Trimouille or Alpha Islands, as Plutonium remains in the soil. While natural history records were kept by people associated with the tests, there was no systematic biological survey and without the work by PD Montague in 1912, we would know very little about the islands' biota prior to the tests.

One significant sea turtle nesting place is the north-east beaches of Trimouille Island and visitors after the nuclear weapons tests reported hundreds of dead turtles on its beaches. Despite these deaths, turtles still nest there, but presumably in much lower numbers. Research by the Australian Nuclear Science and Technology Organisation in 2015-2017 examined local samples including turtle skin, bones, and hatchlings as well as eggshells, sea sediments, diet items and beach sands. Radionuclide levels in turtle hatchlings and adults were relatively low; however, they would have been high for c. 10 years after the tests. Unbelievably, the turtle 'fishery' continued to operate after the nuclear weapons tests. It takes from 30 to 50 years for a green turtle hatchling to reach sexual maturity, so recruitment into the adult population may now be occurring.

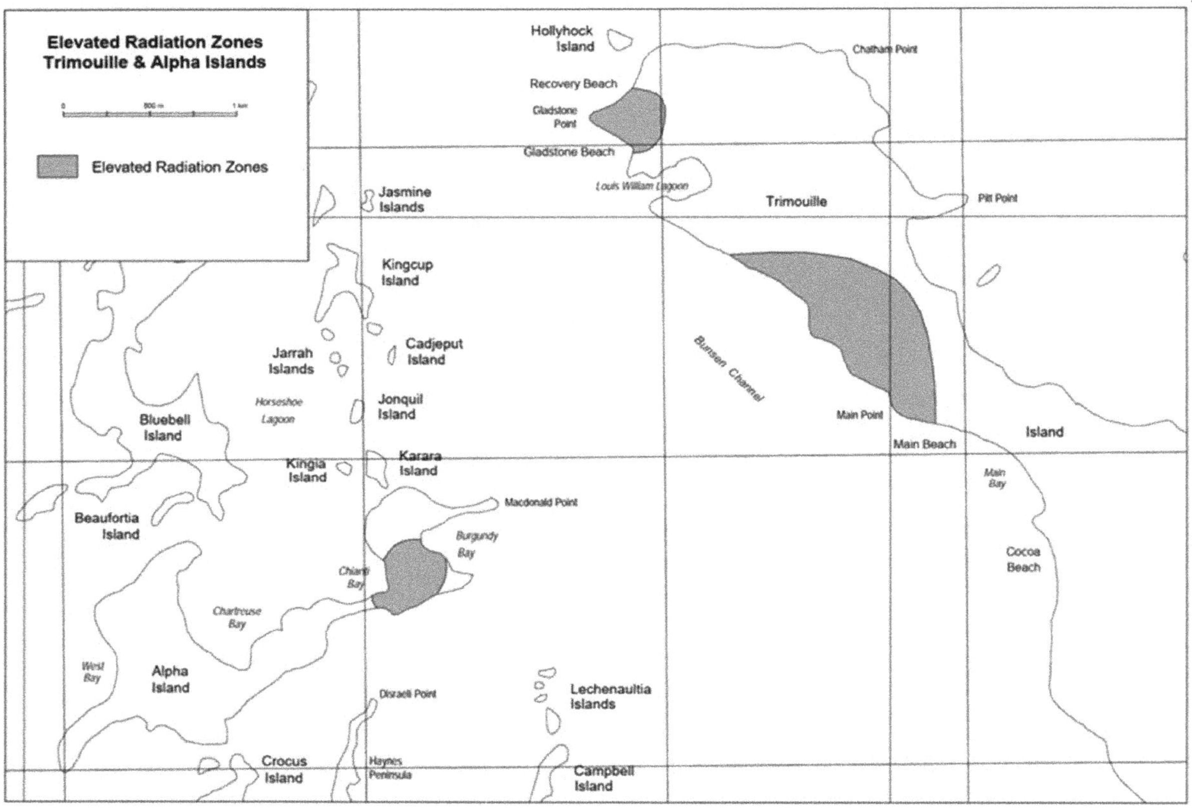

Figure 5.1. Elevated radiation zones on Trimouille and Alpha Islands.

Ecological restoration. 'Montebello Renewal' commenced in 1994 with a biological survey of the islands, partly aimed at recording which islands had cats and rats, but also looking at distribution of possible non-target animals (those that might be affected by rodenticide). Rats were found on almost every island visited. Cats were present only on Hermite Island, although they had been reported on Trimouille Island in 1970 and Bluebell Island in 1985. The Commonwealth government provided funding for the eradication of rats and cats and reintroduction of locally extinct animals, but the quantity of funds was inadequate to achieve this aim and had to be augmented by the state conservation department. In 1995 a trial rat eradication was conducted on one island, later named Renewal Island, and then, in 1996 a major project, carried out mainly by volunteers, mostly from the Department of Conservation and Land Management who volunteered their leave time, laid rodenticide bait stations on a 50 m grid on all the larger islands, with further intensive baiting on all smaller islands. Rats were present even on islands as small as 4 ha. Hand baiting in bait stations designed to allow rats to enter but exclude possible non-target animals such as brown quail, bar-shouldered doves and raptors, was chosen to minimise impact on other species.

Monitoring revealed that rats remained on four islands. The smaller ones were rebaited, but Hermite was too large to rebait by hand. Research during and after the 1996 baiting revealed that there were no significant non-target effects, so aerial baiting of Hermite and all western islands, using a spreader bucket slung beneath a helicopter navigated via GPS, was conducted in 1999 and 2001, and achieved eradication.

Feral cat eradication commenced in 1999 with c. 1100 specially-developed baits being laid throughout Hermite Island from a helicopter. Four cats survived the baiting and three were trapped soon afterwards. The fourth, however, eluded the trappers for six weeks before being trapped and euthanased. Later searches revealed that no cats remained on the island.

With the islands free of introduced predators, planning for translocations commenced. Because the Montebellos had been seriously disturbed by the weapons tests and by the introduction of feral predators, the islands were considered ideal for translocations of threatened species, particularly those threatened by feral cats and red foxes on the mainland. The first introduction was of the highly-endangered mala (mainland subspecies of the rufous hare-wallaby), which at the time was classified as 'Extinct in the Wild'. Thirty (20 females, ten males) were flown from a semi-captive colony in the Tanami Desert in the Northern Territory to Trimouille Island in June 2002. Although this species had not occurred on Trimouille, the introduction was approved due to its highly-threatened status and that spectacled hare-wallabies were to be reintroduced to Hermite Island. The Montebellos provided a unique opportunity for animal introductions due to the considerable changes wrought on the environment by the nuclear weapons tests. The introduction was successful and mala are now widespread on Trimouille. The next translocation was an introduction of djoongari (or Shark Bay mouse), to North West Island in 1999, releasing 59 animals bred at Perth Zoo. Djoongari were once widespread on the mainland but became restricted to a single island, Bernier Island in Shark Bay. Like the mala, djoongari did not occur in the Montebellos. It is now abundant.

When governments approved the development of the Gorgon Gas Project's liquefied natural gas plant on Barrow Island, an offset condition funded the removal of some species of animals from the site of the plant and their translocation. Spectacled hare-wallabies (111 in 2010), golden bandicoots (165 in 2010), black-and-white fairy-wrens (37 in 2010-11) and spinifexbirds (47 in 2010–11) were translocated to Hermite Island and boodies (40 in 2010-2011) to Alpha Island. Boodies had not been recorded from the Montebellos, but their introduction provided an insurance population for the Barrow Island population, which differs genetically from those on Bernier and Dorre Islands. All these animals came from Barrow Island, many from the area to be cleared for the construction of the Gorgon Liquefied Natural Gas plant, and all translocations have been successful.

In 2004, the islands were surrounded by the Montebello Islands Marine Park of c. 58 000 ha. The conservation park and marine park are now visited by fishers, surfers, and tour groups.

Island management. While severely damaged in the past by the introduction of cats and black rats, and by nuclear weapons testing, the Montebellos are now a valuable part of the state's island nature conservation estate. This is particularly so since radiation levels have fallen, the rats and cats have been eradicated, and threatened mammals and birds have been reintroduced and introduced. The surrounding marine parks adds to the area's biodiversity conservation values. Due to the islands' remoteness, visits by Parks and Wildlife staff from Karratha are irregular.

Sources of information

History and island names: Murray & Hercock 2008.
General: Smith *et al.* 2006.
Archaeology: Veth 1993, Veth *et al.* 2017, Ward *et al.* 2022.
Geology: Veth 1993, Department of Parks and Wildlife 2015.
Vegetation and flora: Burbidge 1971, Moro & Lagdon 2013.
Fauna general: Abbott & Wills 2016, Burbidge *et al.* 2000, Montague 1913, 1914, Moro & Lagdon 2013, Serventy & Marshall 1964.
Subterranean fauna: Humphreys *et al.* 2013.
Mammals: Abbott & Burbidge 2017, Algar *et al.* 2002, Burbidge 2008, Langford & Burbidge 2001, Morris 2002.
Birds: Pruett-Jones & O'Donnell 2004.
Reptiles: Kendrick 2003, Halkyard 2014.
Invertebrates: Majer *et al.* 2013.
Biosecurity: Scott *et al.* 2017.
Nuclear tests and radiation: Anon. 1952, Cathcart 1994, Johansen *et al.* 2019, Kendrick 2003.

Plate 5.1. 'John Wayne country', Barrow Island, with limestone cliffs and hummock grassland. October 2003. Credit: AA Burbidge.

Plate 5.2. Landscape view of hummock grassland, Barrow Island. August 2005. Credit: AA Burbidge.

Plate 5.3. Oil well and euros, Barrow Island. August 2005. Credit: AA Burbidge.

Plate 5.4. Black-footed rock-wallaby and limestone, Barrow Island. August 2005. Credit: AA Burbidge.

Plate 5.5. Spectacled hare-wallaby, Barrow Island. February 1981. Credit: AA Burbidge.

Plate 5.6. Boodie, Barrow Island. February 1981. Credit: AA Burbidge.

Plate 5.7. Barrow Island golden bandicoot. February 1981. Credit: AA Burbidge.

Plate 5.8. Green turtle, Barrow Island. April 1985. Credit: AA Burbidge.

Plate 5.9. Perentie, Barrow Island. October 2006. Credit: AA Burbidge.

Plate 5.10. Aerial view of part of the Montebello Islands, showing southern end of Hermite Island in foreground. February 1981. Credit: AA Burbidge.

Plate 5.11. Plinth on Alpha Island, the site of the third atomic weapon test in the Montebello Islands, in 1956. May 1994. Credit: AA Burbidge.

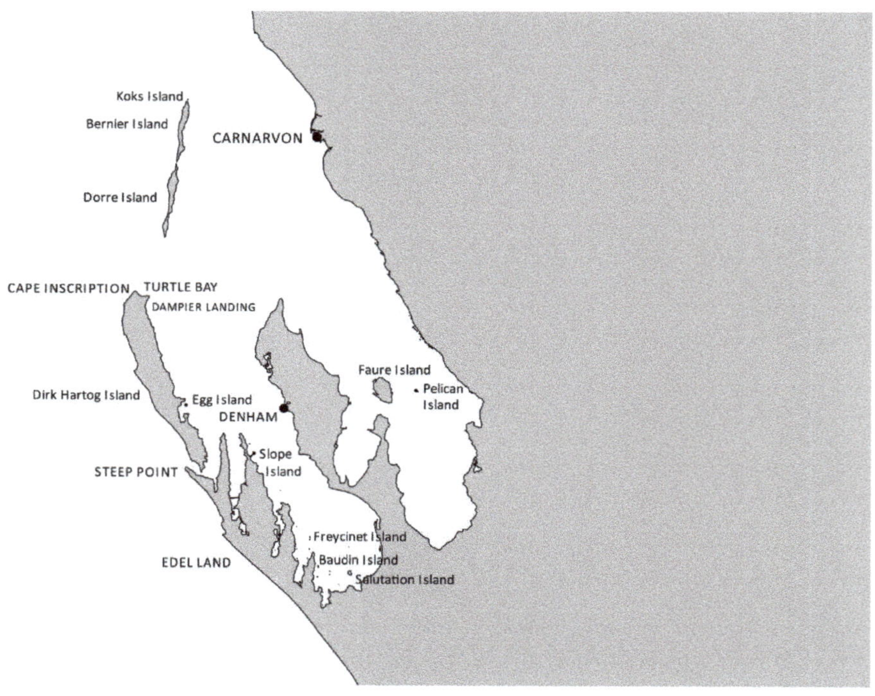

Map 6.1. Islands of Gutharraguda (Shark Bay).

CHAPTER 6: ISLANDS OF GUTHARRAGUDA (SHARK BAY)

INTRODUCTION

1616, DEN 25 OCTOBER IS HIER AENGECOMEN HET SCHIP D EENDRACHT VAN AMSTERDAM, DEN OPPERKOPMAN GILLIS MIBAIS VAN LVICK SCHIPPER DIRCK HATICHS VAN AMSTERDAM DE 27 DITO TE SEIL GEGHM NA BANTAM DEN ONDERCOOPMAN JAN STINS OPPERSTVIERMAN PIETER DOEKES VAN BIL Ao 1616.

1616, on the 25th October, arrived here the ship Eendracht of Amsterdam; the upper merchant, Gilles Mibais of Luyck; Captain Dirk Hartog of Amsterdam; the 27th ditto set sail for Bantam; undermerchant Jan Stein, upper steersman, Pieter Doekes from Bil, A[nn]o 1616

The original words and an English translation of the pewter plate left on Dirk Hartog Island in 1616. The plate is now in the Rijksmuseum, Amsterdam, The Netherlands, having been removed in 1697 by Willem de Vlamingh, who replaced it with one of his own. Vlamingh's plate, which has the words on the faded Dirk Hartog plate in full, is in the Western Australian Museum–Shipwrecks Gallery, in Fremantle.

World Heritage. Gutharraguda (Shark Bay) is a World Heritage Property, recognised under several criteria, including under Criterion X: 'Containing the most important and significant natural habitats for in-situ conservation of biological diversity, including those containing threatened species of outstanding universal value from the point of view of science or conservation.' 'Most important and significant natural habitats ... containing threatened species ...' included two Shark Bay islands that are of outstanding nature conservation significance: Bernier and Dorre, with Dirk Hartog and Faure Islands becoming significant habitats since the removal of sheep, goats and feral cats and the reintroduction and introduction of threatened mammals. Gutharraguda has several smaller islands that complement these larger ones, with seabird nesting colonies and other valuable features.

Traditional ownership. Gutharraguda is the traditional country of three Aboriginal language groups: Malgana, Nhanda and Yingkarta. Malgana is the traditional language of the people of central Shark Bay. Gutharraguda is the Malgana name for Shark Bay, which means 'two bays' or 'two waters' and the Malgana name for Dirk Hartog Island is Wirruwana.

Table 6.1. Shark Bay islands by size category

>1000 ha	4
100 ha to 1000 ha	1
20 ha to 100 ha	0
10 ha to 20 ha	1
<10 ha	36
Total	**42**

Geography. Four large islands and many small ones occur within Shark Bay. In order of size they are Dirk Hartog (62 000 ha), Faure (5800 ha), Dorre (c. 4920 ha), and Bernier (c. 4800 ha). Next in size are Salutation (125 ha), and Baudin (19 ha). Other islands and islets are smaller than 10 ha.

Climate. The climate of Gutharraguda is transitional, with both winter and summer rainfall. Rainfall is very low and variable and falls mostly in the winter between May and August; however, summer cyclones and storms can bring occasional high rainfall. The nearest long-term rainfall stations to the islands are Denham, having data from 1893 to the present, with some gaps, and Carnarvon. Mean annual rainfall at Denham is only 221 mm, varying from 78 mm to 520 mm, with a median rainfall of 202 mm, reflecting the occasional very high rainfall years. Carnarvon Post Office data are from 1883 to 1950 and Carnarvon Airport data are from 1945 to 2021. Similar to Denham, the mean annual rainfall at Carnarvon is 220 mm, varying from 68 to 619 mm. Dirk Hartog homestead had data from 1893 to 1948 and recorded a mean annual rainfall of 313 mm. The mean maximum temperature at Denham is 26.8°C, varying from 21.8°C in July to 31.8°C in February, while the mean minimum temperature is 19.5°C (12.7°C in July to 22.9°C in February). The highest temperature recorded at Denham was 47.0°C in February 1991. Temperature extremes on the islands would be ameliorated by the adjacent ocean.

Geology, and bioregions. Geologically, Gutharraguda is within the Carnarvon Basin. The western islands are composed mostly of the late-Pleistocene and Holocene Tamala aeolianite (or Tamala limestone), which occurs along much of the west coast of WA from Gutharraguda to near Albany. The western side of the three larger western islands is mostly bounded by steep cliffs varying in height from 1-2 m to c. 80 m near Herald Heights on Dirk Hartog Island (Plate 6.1). There is a sandy beach near Dirk Hartog's north-eastern tip, Turtle Bay. The larger islands' eastern, protected sides have a few mostly low cliffs with beaches between. Sand is the only soil, derived from the limestone or from the ocean. Sand blowouts occur on Dirk Hartog Island and Bernier Island.

Dirk Hartog, Bernier, Dorre, Koks, Salutation and Baudin Islands are part of the Yalgoo bioregion, while Faure Island lies within the Carnarvon bioregion. The Yalgoo bioregion is transitional, with biotic elements from both the south-west and the arid zone.

Large rocks thrown up by one or more mega-tsunamis, probably caused by meteorites crashing into the Indian Ocean, are located along the west coast of the western islands. One rock on Dirk Hartog Island, known as 'The Block', is situated beside the west coast, 6.5 km southwest of Cape Inscription (Plate 6.2). This calcrete block measures 10.5 m x 8 m x 4 m, and is estimated to weigh at least 700 t. It was derived from the top of the shoreline cliff and thrown by a mega-tsunami c. 15 m above sea level and 120 m behind the cliff. There are many smaller blocks along Dirk Hartog's west coast, and below the Cape Inscription cliffs, and also on Bernier, Dorre, and Koks Islands. Where the coastal cliffs are lower than c. 15 m, blocks of the dislodged calcrete have been carried above the cliff as far inland as 300 m, reaching maximum elevations of c. 25 m. Where a cliff is higher than 15 m, blocks are seen to be lying on or at the base of the slope. Large parts of these islands are higher than 35 m above sea level, the expected runup height reached by the tsunamis. Dating of molluscs attached to the rocks suggests that the mega-tsunami or tsunamis occurred several thousand years ago, with the most recent c. 3500 years ago.

BERNIER AND DORRE ISLANDS

Geography. Bernier and Dorre, together with the small Koks Island (2.9 ha), situated c. 370 m north of Bernier Island and only 280 m long, form the north-western margin of Gutharraguda. The combined area of Bernier and Dorre Islands is 9720 ha, with Dorre slightly the larger. The islands were separated from the mainland c. 8000 years ago as sea levels rose after the last glacial maximum. However, it is thought that the two islands were joined until 3000-5000 years ago. There is a 500 m gap between the two islands. The two larger islands are relatively low, rising to heights of between 40 and 50 m above sea level.

Discovery and naming. Aboriginal people did not occupy these islands at the time of European settlement, although they would have occupied them when they were joined to the mainland. From 1908 to 1918, Bernier and Dorre Islands were used as lock hospitals for the isolation and treatment of Aboriginal peoples from WA believed to be suffering from venereal diseases, tuberculosis and leprosy. Initially, in 1908, the women were held on Bernier Island and the men on Dorre Island, but this arrangement was reversed in mid-1910. Daisy Bates, who visited in 1910, famously described the islands as 'tombs of the living dead'. It is not known how many people were transported there, but more than 700 patients were admitted to the hospitals, of whom more than 200 died. In 1986 the lock hospitals were registered as protected areas under the WA *Aboriginal Heritage Act 1972*. The islands' heritage significance was also recognised with their inclusion on the *Register of the National Estate* in 1987.

Dorre Island was named Dorre Eylandt by Dirk Hartog in October 1616, meaning 'dry or barren island'. William Dampier in HMS *Roebuck* landed on Bernier Island on 6 August 1699. Bernier Island was named by the French Baudin expedition of 1801-03 after Pierre François Bernier (1779-1803), the astronomer aboard *Le Géographe*. Members of the Baudin Expedition probably landed on both islands, but details of their collections were never written up due to the early deaths of the expedition's biologists. Explorer George Grey and party landed on Bernier Island in February 1839 before their ill-fated attempt to row to Perth in two whaleboats.

BIOLOGICAL CHARACTERISTICS

Biological discovery. The islands' importance for mammals was recognised early in WA's history. WA Museum collector John Tunney visited Bernier and Dorre in March and April 1896, and for two weeks in February 1899. Details of his visits are unknown as his letters to the Museum have disappeared, apparently stolen, and his field diaries were destroyed after his death by his family on his instructions. Guy Shortridge, a noted early collector of mammals and birds for the British Museum, spent five weeks on Bernier Island in June and July 1906. He wrote 'The two wallabies and *Bettongia* swarmed on the island. In the case of *Lagostrophus* I have never seen any animal not even rabbits (considering the small amount of area) in such numbers. It has been a particularly dry season and they were very thin.' and 'The bandicoot's skull I found on Bernier Island will probably be this kind [Shark Bay bandicoot]. I took a lot of trouble in trying to get a specimen of it, but haven't yet got one. There is a man on the island looking after a flock of c. 60 sheep - he had five cats there and on a small island like Bernier they have perhaps thinned them out.' Very fortunately, the cats did not breed and become feral.

Otto Lipfert, the WA Museum's taxidermist, collected on the islands in August and September 1910, and an account of the mammals and birds was published. He reported that the wallabies were 'not at all numerous', but that he was able to 'get 12 good specimens' of the Shark Bay bandicoot. Most of his observations were about birds. At about the same time GF Hill visited Bernier Island, but no published account of his visit exists. EL Grant Watson visited Bernier between 1910 and 1911 with an anthropological expedition, and reptiles he collected are in the WA Museum. He reported 'A great many mice wallaby and bandicoots'. Many of the early visitors to Bernier Island took advantage of the pastoral lease and the lock hospitals as they provided transport and limited accommodation.

From 14 to 27 July 1959, a WA Museum and WA Herbarium expedition visited both islands to make a detailed record of the islands' flora and fauna. This resulted in a comprehensive account of the islands' physical environment, botany, mammals, birds, and reptiles.

Vegetation and flora. Both Bernier and Dorre islands have extensive areas of Acacia shrubland (*Acacia ligulata, A. sclerosperma, A. tetragonophylla*) over low shrubs and chenopods. On the west coast there is dwarf scrub or open heath of southern diplolaena *Diplolaena dampieri*, thick-leaved fan-flower *Scaevola crassifolia* and stiff westringia *Westringia rigida*. On Bernier Island there are areas of open hummock grassland of the spinifex *Triodia plurinervata* with mallees (*Eucalyptus oraria, E. obtusiflora* subsp. *dongarraensis*) and shrubs of *Thryptomene baeckeacea*, *Acanthocarpus preissii* and *Beyeria cinerea*. Both islands, but particularly Bernier, have areas of un-vegetated, mobile white sand (blowouts). The mobile dunes on both Islands were possibly caused or exacerbated by browsing by introduced goats, and on Dirk Hartog, also sheep.

Dorre Island has had extensive fires in the past. In 1860, the Julius Brockman camped on Dorre Island and his diary gives a graphic account of fire on the island – 'The island was high with spinifex, which had never been burnt ... There were thousands of wallaby. The wind was blowing a gale from the south and when I lighted a fire to camp, it swept the island bare for twenty two miles.' Later he noted 'I walked about the island, which was now bare as a sand patch and what were not burnt of the animals would have no food except seaweed until rain fell again.' Further devastating fires were reported on Dorre Island in 1908 and 1973 and the widespread destruction led to concerns for the survival of the island's maning (banded hare-wallabies). There are no records of large fires on Bernier Island.

Mammals. The two islands have the only remaining natural populations of the Shark Bay bandicoot (Plate 6.3), maning (Plate 6.4), and djoongari (Shark Bay mouse, Bernier Island only), all of which once occurred on the adjacent mainland. Both islands have populations of the rufous hare-wallaby, which became 'extinct in the wild' on the mainland, and the boodie (burrowing bettong), which, while becoming extinct on the mainland, also survived on Barrow Island. The Shark Bay islands' populations of the rufous hare-wallaby and boodie are endemic subspecies. Both Bernier and Dorre have populations of the ash-grey mouse. A specimen of the desert mouse was collected on Bernier Island in 1910 and is now in the National Museum of Victoria, but the species has not been located since. Two species of bat, the lesser long-eared bat and Finlayson's cave-bat have been recorded on the islands.

Research has shown that numbers of hare-wallabies, boodies and bandicoots on the islands vary considerably with rainfall. Combined estimates of minimum population size for both islands in 1991-92 were from c. 4000 Shark Bay bandicoots to c. 10 000 maning. However, a drought in the late 1980s reduced populations of the bandicoots by up to 75%, boodies by 65%, and rufous hare-wallabies by up to 60%. Concern has been expressed about the future of these mammal populations as winter rainfall declines, and temperatures increase, due to climate change. The two island populations of the rufous hare-wallabies are genetically different from the remnant mainland population (mala) that last occurred in the wild in the Tanami Desert, Northern Territory, and became 'extinct in the wild', surviving only in captivity, before translocation. The two island populations do not differ sufficiently from each other to be considered different subspecies, with the name *Lagorchestes hirsutus bernieri* being applied to both.

Birds. One interesting bird species present on the islands is the rufous fieldwren. A subspecies, the Dorre rufous fieldwren, restricted to Bernier and Dorre Islands, was described in 1912 on the flimsiest of grounds and remains (2020) listed as Vulnerable. During a study of Dirk Hartog Island bird subspecies, morphometric and genetic analyses revealed that Bernier and Dorre field-wrens do not differ significantly from mainland birds and the authors recommended that the subspecies should not be recognised and that it should be removed from the list of threatened species. *The Action Plan for Australian Birds 2020* considered the subspecies to be Least Concern, but commented that genetic evidence suggests it should not be recognised. Neither population is under threat as long as the islands remain undisturbed and free of invasive species such as black rats and cats.

Reptiles. Twenty-eight reptile species have been recorded on Bernier Island and 18 on Dorre Island. The difference between the two lists probably indicates differing survey effort. None of the reptiles is of conservation significance as all are widespread in the region.

FACTORS LEADING TO CHANGE IN THE BIOTA

George Baston camped on Bernier Island in the 1880s to cut sandalwood, a lucrative endeavour in those days. Later he leased Bernier Island as a pastoral station and built a small cottage for his wife and five children. The family returned to Carnarvon in the early 1900s. His cottage was used by the lock hospitals.

Non-indigenous species. Goats were introduced to both islands during the short period when they were pastoral leases prior to the setting up of the lock hospitals. They did not survive on Dorre Island, where a skeleton of an old male was located in 1969, but they persisted on Bernier Island where they caused extensive damage to the vegetation, resulting in degradation and erosion. A 2019 study, based on satellite observations, showed that grazing from feral goats on Bernier Island had significant impacts on vegetation cover levels. The study showed that the area of sand dunes on Bernier Island, following the eradication of the goats, declined at a near steady rate, from an area of 693 ha in 1973 to 488 ha in 2018. Vegetation cover levels in 2018 on both islands were among the lowest observed in the Landsat record (1988 to 2018), being likely due to a period of extremely low rainfall, showing what may happen as winter rainfall declines in the south-west of WA.

As mentioned in 'Biological discovery' George Baston kept sheep and cats on Bernier Island in the early 1900s, but neither become feral.

Reservation. In 1907, Dorre Island was protected under the *Game Act 1892* as a 'reserve for native game', which prohibited shooting. However, this reservation was rescinded in 1909, being reinstated after the lock hospitals were abandoned. Bernier Island was protected under the Game Act in 1919. Bernier and Dorre Islands were not reserved under the Land Act, for the purpose 'Conservation of fauna' (later amended to 'Conservation of flora and fauna') until 1957, at the same time being proclaimed Class A. Koks Island was reserved for 'Conservation of flora and fauna' in 1976 and proclaimed Class A in 1989. Both reserves are vested in the Conservation and Parks Commission and managed by the Parks and Wildlife Service.

Ghurkhas and goats. Goats were introduced to Bernier Island in 1899, when about six, including a single male, were taken to what was then a pastoral lease. Numbers increased and several hundred were present by the 1960s, when it was clear that they were causing significant degradation of the vegetation and major soil erosion. Ground shooting resulted in many being culled with perhaps less than 50 left by 1966. In 1970, the commanding officer of the 2nd King Edward VII's Ghurkha Rifles, then stationed in Singapore, requested permission to visit Bernier Island by helicopter from a Royal Navy helicopter carrier that was to travel along the Western Australian coast, as he was a keen naturalist and wanted to see banded hare-wallabies. This proposal developed into one where the goats might be controlled or eradicated by shooting from helicopters. Unfortunately, this did not eventuate, as the helicopter carrier was diverted to assist with serious flooding in East Pakistan (now Bangladesh). However, c. 500 Ghurkhas did arrived to carry out exercises with Australian defence force soldiers, and a request was received that some of them might visit Bernier Island to shoot goats. On Monday 21 June 1971, a party of 15 soldiers accompanied by two members of the SAS Regiment and two Departmental staff, one of whom (AAB) is a co-author of this book, were taken to the island from Carnarvon on a prawn trawler. Over the next several days, the Gurkhas shot 22 goats, four of which were eaten at evening dinner. One day the trawler took 12 soldiers to the north end of the island and they walked southwards in a line across the island, each person walking c. 23 km. It became clear that ground shooting would never achieve eradication as goats were able to hide in thick scrub or in caves on the west coast. Eradication was finally achieved in 1984 by a helicopter team of an experienced pilot and marksman, who had controlled large feral animals elsewhere in WA, supported by Departmental staff who provided a camp, food and fuel. While the team was on the island they found, partly hidden on a beach, a large cache of marijuana, which they duly reported to the police.

DIRK HARTOG ISLAND (WIRRUWANA)

INTRODUCTION

Geography. Dirk Hartog (c. 62 000 ha) is the largest island in WA, being 79 km long and 11.5 km wide at its widest point. Cape Ransonnet, the southern tip, is separated from the mainland at Edel Land by South Passage, a gap of only c. 1.5 km. West Point (and some coast south of it) is the most westerly point of Australia, apart from Indian Ocean territories.

Discovery and naming. Dirk Hartog Island, while not occupied by Aboriginal people at the time of European settlement, is part of the Malgana peoples' traditional lands. Aboriginal people were employed as workers on the Dirk Hartog Island pastoral lease, but for much of the time in the past they were either paid nothing or had their wages retained by the government.

Shark Bay was visited by several early European ships and explorers. The earliest recorded visit to Dirk Hartog Island was by Dirk Hartog in the *Eendracht*, who left a plate recording his visit on Cape Inscription. Willem de Vlamingh in the *Geelvinck*, anchored at Turtle Bay in 1697, removed the plate and replaced it with one of his own. (See epigraph at start of this chapter.) Vlamingh was the first to publish an account of Shark Bay's natural history.

William Dampier in HMS *Roebuck* visited in 1699, with the crew landing on Dirk Hartog Island at a site now known as Dampier Landing. The crew spent four days there and Dampier described the vegetation, noting that 'of Trees or Shrubs there are divers sorts; but none above ten Foot high ...' and 'There were also beside some Plants, Herbs, and tall Flowers, growing on the Ground, that were sweet and beautiful, and for the most part unlike any I had seen elsewhere.' He also reported animals that 'were a sort of Racoons, different from those in the West Indies, chiefly as to their legs; for these have very short Fore-Legs; but go jumping upon them as others do.' It has been assumed that this description was of the maning (banded hare-wallaby). Allan Cunningham's description of a small kangaroo (see below) possibly confirms that assumption.

On 30 March 1772, François de Saint-Aloüarn in *Gros Ventre* landed at Turtle Bay. There he claimed possession of the western half of New Holland for King Louis XVI of France. To support the claim he buried two bottles containing statements of proclamation written on parchment. Two silver coins were placed on the top of each bottle, which were sealed by lead caps. The coins, lead seals and one of the bottles were discovered by archaeologists in 1998 and are now on display in the Western Australian Maritime Museum.

In 1801 Nicolas Baudin's corvettes, *Le Géographe* and *Le Naturaliste*, separately spent 70 days in Shark Bay with the ships' companies exploring, mapping and naming many features and making extensive zoological collections. In 1803, Baudin returned with *Le Géographe* and *Le Casuarina*, a vessel purchased at Port Jackson (Sydney) in 1802. François Péron, a trained zoologist, described zoological observations from the expedition and named new species, including the maning, which he named *Kangurus fasciatus*, noting 'The striped kangaroo breed in great numbers on the three islands of Bernier, Dorre and Dirk Hartighs, but we could not discover any of them on any part of the continent or on the other islands that we successfully explored.' However, Péron's death at the age of 35, and the earlier deaths of many of his colleagues, prevented a complete account of the expedition's biological achievements.

Louis de Freycinet, a senior officer in the Baudin expedition, returned in 1818 in the *L'Uranie*. Crew members went to Dirk Hartog Island to recover de Vlamingh's plate, which was delivered to the Royal Academy of Inscriptions and Elegant Literature in Paris. There it stayed until 1947, when the French Government presented it to the people of Australia. The plate is now in the Western Australian Maritime Museum. The two naturalists on board *L'Uranie*, Jean René Constant Quoy and Joseph Paul Gaimard, named several new species including the Shark Bay bandicoot, boodie, and quokka (the quokka was collected at King George Sound).

Phillip Parker King in HMSV *Bathurst* landed at Cape Inscription in January 1822 and Allan Cunningham, the brig's botanist, collected plants on Dirk Hartog Island. Cunningham also reported 'I started a black-back Opossum, a small Kangaroo from beneath the *Acaciae*' (possibly a maning or banded hare-wallaby) and 'We also roused a 'hair seal' (Australian sea lion, well north of its usual range). There were many later visits after the Swan River Colony was established in 1829. In the 1870s and 1880s Dirk Hartog Island was used as a campsite for pearlers, and, in 1878 stone was mined for ballast.

BIOLOGICAL CHARACTERISTICS

Biological discovery. Early explorers including William Dampier, the Baudin expedition and Phillip Parker King with Alan Cunningham, recorded plants and animals. More recently, visits by ornithologists and students from Wesley College made observations, but mainly of birds. The first comprehensive biological survey took place in September 1972 by Andrew Burbidge and Alex George, supported by Trevor Evans, invited and facilitated by Sir Thomas Wardle, who was the pastoral lessee at the time. Since then, scientists associated with the 'Return to 1616' project have continued to add biological information.

Vegetation and flora. The island's vegetation is mostly open heath, low shrubland and hummock grassland (Plates 6.5, 6.6). There are extensive areas of white sand dunes with no or sparse vegetation. A feature of the island (and nearby peninsulas) are 'birridas', the Malgana name for claypans with gypsum floors up to 5 m thick. They are sparsely vegetated with low shrubs. One, near the homestead, is used as an airstrip. Shark Bay is the site of the first authenticated botanical collections made by Europeans in Australia. Dampier landed on Dirk Hartog Island in 1699 and collected plants there. Twenty-four of his collections are in the Fielding-Druce Herbarium, University of Oxford, in England, including the oldest known collection of a *Beaufortia, B. sprengelioides*. Fortunately, Dampier rescued his specimens when the *Roebuck* was wrecked at Ascension Island in March 1701. The dominant plants on

Dirk Hartog are *Acacia ligulata*, *Thryptomene dampieri*, and the spinifex *Triodia plurinervata*. Other common tall shrubs include *Acacia coriacea*, *Pittosporum phillyraeoides*, *Diplolaena grandiflora*, and *Alyogyne cuneiforms*. There are areas of low kwongkan (heath), mainly towards the island's northern end, with species more typical of the south-west of WA. The vegetation has suffered from over-grazing by sheep and goats, but is now slowly recovering (see 'Return to 1616' below). The September 1972 survey, preceded by a high-rainfall winter, collected 269 species of plants in flower, with more having been added since bringing the total flora to around 300 species. Some southwestern plant species occur here at their northern limit, including two species of orchids.

There are no records of extensive bushfires on Dirk Hartog Island. However, on 9 December 1850, Frank Helpman recorded 'As we neared Cape Inscription, the land was quite obscured by dense smoke from the bush of Dirk Hartog's Island being on fire' and on 15 December (when a little farther south) 'Dense smoke over the land.' The 1972 visit did not detect evidence of bushfire.

Mammals. Recent surveys and identification of subfossil remains in caves show that the island once had 13 or 15 species of native mammals, including eight marsupials (Table 6.1).

Table 6.1. Original mammal fauna of Dirk Hartog Island (DHI)

Common name	Comments
mulgara	Extinct on DHI
chuditch (western quoll)	Extinct on DHI; listed threatened species
dibbler	Extinct on DHI; listed threatened species
little long-tailed dunnart	Persists on DHI
Shark Bay bandicoot	Extinct on DHI; listed threatened species
boodie	Extinct on DHI; listed threatened species
maning (banded hare-wallaby)	No subfossils located but recorded by William Dampier in 1699 and reported by François Peron in 1807; listed threatened species
rufous hare-wallaby	No subfossils located; but occurrence possible; listed threatened species
Finlayson's cave-bat	Persists on DHI
lesser long-eared bat	Persists on DHI
wopilkara (greater stick-nest rat)	Extinct on DHI; listed threatened species
noodji (ash-grey mouse)	Persists on DHI
Shark Bay mouse	Extinct on DHI; listed threatened species
sandy inland mouse	Persists on DHI
heath mouse	Extinct on DHI; listed threatened species

Of the 13 or 15 original native mammal species, only five survived. The local extinction of the others has been attributed to the introduction of feral cats, which arrived on the island with pastoralists in the 1800s. Degradation of the vegetation by goats and sheep would have contributed to the local extinctions. The pattern of small terrestrial mammals and bats persisting, while medium-sized mammals, those with a body weight between 35 g and 5.5 kg (the critical weight range), disappear is typical of most of Australia, except on islands with no cats or foxes.

Alexander Baynes (1944–)

Dr Alex Baynes has been the leading WA expert on locating and identifying subfossils for many decades. Subfossils — the remains, usually bones, of dead animals that are not fossilised — are typically found in caves or other shelters, where they have generally been accumulated by predators, but also in lag deposits left when sand dunes have been deflated by wind. They may be hundreds or even thousands of years old. Identifying mammal subfossils, which requires meticulous examination of bones and teeth, can show which species occurred in an area prior to European settlement. In Australia this has been particularly important in documenting the distributions and abundances of mammal species prior to the establishment of feral cats and red foxes. Alex's data have shown that many of the extinct and threatened Australian mammals once had very much wider mainland distributions, revealing the extent of the loss since settlement.

At Shark Bay, Alex has identified which mammals occurred on Dirk Hartog (Wirruwana) and Faure Islands, as part of a broader study of the pre-European mammal faunas of the peninsulas, islands and adjacent mainland begun in 1970. Without this information, conservation managers would not have known which mammals to reintroduce to islands once introduced predators are eliminated, as has been done on Faure and is underway on Wirruwana.

Birds. More than 70 species of birds, including land and shorebirds, have been recorded on Dirk Hartog Island. Most are widespread species. Three subspecies, listed as threatened, are endemic to Dirk Hartog Island: the Dirk Hartog rufous fieldwren, the Dirk Hartog southern emu-wren, and the Dirk Hartog Island white-winged fairy-wren. All subspecies are considered to be Least Concern in *The Action Plan for Australian Birds 2020*; however, genetic evidence suggests that they are not valid. The widespread variegated fairy-wren is also present on the island. Research in 2015 revealed that there were c. 37 000 southern emu-wrens, 24 000 white-winged fairy-wrens, 38 000 variegated fairy-wrens and 20 000 rufous fieldwrens on the island. Early bird observers described Dirk Hartog as 'an island of wrens'. However, 2015 was preceded by high-rainfall year and counts in 2016, preceded by a dryer year, showed lower numbers. DNA research indicates that some of these subspecies are possibly invalid as they are not genetically significantly different from mainland birds of the same species.

The Dirk Hartog Island western grass-wren is considered Extinct having not been recorded for more than 100 years. The reason for the extinction are unclear but may be similar to those causing the extinction of many mammal species: predation by feral cats and unremitting grazing by sheep and goats.

Few seabirds have been recorded breeding on Dirk Hartog, probably due partly to feral cats and mice but also due to seabirds preferring to nest on small islands. Silver gulls, Caspian terns, fairy terns, and bridled terns have been reported nesting in small numbers. There is a major nesting colony, estimated at c. 2500 breeding pairs in 1920 and 1972, of great pied cormorants at Quoin Bluff South on the east coast (Plate 6.7).

Reptiles. Thirty species of reptiles have been recorded on Dirk Hartog Island: seven geckoes, three legless lizards, four dragon lizards, ten skinks, one python, and four elapid snakes. Most are widespread species, but the skink *Ctenotus youngsoni* is restricted to Dirk Hartog Island and the adjacent Edel Land.

Turtle Bay, at the northeast corner of Dirk Hartog Island, is a significant loggerhead turtle nesting rookery. The loggerhead, the second-largest sea turtle, has a wide distribution in tropical and subtropical waters but has been over-exploited in most of the world and is listed as Endangered. Turtle Bay has the largest loggerhead turtle breeding colony in Australia, with more than 1000 females nesting in some years.

FACTORS LEADING TO CHANGE IN THE BIOTA

Non-indigenous species. Sheep and goats, introduced by pastoralists, have grazed and degraded the vegetation since the middle of the 19th Century (Plate 6.8). Cats arrived with pastoralists and became feral. As well as sheep, goats and cats, during the long period of pastoralism, several other non-indigenous mammals were introduced to Dirk Hartog Island. At times, the station used horses and camels, and cattle were there for a period. House mice became established, presumably having arrived in food containers. Captain Denham of HMS *Herald* released pigeons (rock doves) in 1858, presumably for food for shipwrecked mariners. Rock doves occur now but whether they are derived from this early release is unknown.

Island management – 'Return to 1616'. 'Return to 1616' is a major project, funded by the Gorgon Barrow Island Net Conservation Benefits Fund and implemented by the Parks and Wildlife Service, Department of Biodiversity, Conservation and Attractions, aimed at ridding Dirk Hartog Island National Park of feral mammals and reintroducing locally extinct ones (Table 6.1). Removal of sheep commenced in 2007 with the pastoral lessee exporting c. 4000 animals by barge. Teams of Parks and Wildlife Service staff then trapped and culled the few remaining sheep and most of the goats between 2008 and 2013, with the eradication of the feral goats aided by 'Judas goats'– goats released with radio transmitter collars that joined remaining groups, thus allowing a helicopter to find them. By November 2017 a total of 5185 sheep and 11 133 goats had been eliminated from Dirk Hartog Island, with the last of the goats being the most challenging to locate. The final seven Judas goats, all of them females with radio collars, were shot in November 2017, having performed their task. Satellite and ground monitoring in 2019/2020 showed that the island had experienced a rapid improvement in vegetation cover since the removal of sheep and goats.

During this period introduced plant species were surveyed, with some species subsequently targeted for eradication.

The most difficult task was feral cat eradication. In 2014, a fence was constructed to divide the island into two sections to make monitoring more efficient. The 1.8 m high fence was made of rabbit netting with an overhang at the top and three electric wires. A gate allowed vehicles to continue to travel north-south along the main track.

Aerial baiting for cats occurred in May 2015 and May 2016 and intensive monitoring took place both before and after baiting. Three types of traps were used: automated camera traps, soft leg-hold traps and cage traps. Sand pads were used to detect tracks. Leg-hold traps used scent, and visible and audible lures, to attract feral cats, as food lures are more likely to attract other animal species. Cage traps were baited with fried chicken. Trapped cats were euthanased.

The last feral cat detected south of the fence was in August 2015, and the last in the north was in October 2016. Checks south of the fence were made using detector dogs during the winters of 2015 and 2016. Monitoring for feral cats north of the fence began in 2015 and detector dogs checked the northern part of the island in winter 2017. This was followed by two years of systematic 20-day searches covering the whole island four times a year. The major effort led to the confident declaration in October 2018 that feral cats had been eradicated from Dirk Hartog Island, a major achievement of international significance.

The next stage of the project was fauna reconstruction, starting in August 2017 with the trial release of maning and rufous hare-wallabies near the southern end of the island. The wallabies were transported by helicopter from Bernier and Dorre Islands. With the success of the trial and declaration of the island as cat-free, the first full-scale translocation of hare-wallabies involved the release of 90 maning and 50 rufous hare-wallabies from Bernier and Dorre Islands Nature Reserve onto Dirk Hartog Island in October 2018, with another 49 rufous hare-wallabies the following year. Monitoring has shown survivorship to be very high, followed by breeding. In August/September 2019, an additional 50 rufous hare-wallabies were translocated from Bernier and Dorre Islands, along with 72 Shark Bay bandicoots. A further 27 bandicoots were translocated in September 2020. In October 2019, 26 dibblers from a captive-breeding program at Perth Zoo were released and another 31 in October 2020. Post-release monitoring was conducted using radio-tracking, live-capture trapping and remote cameras. However, follow-up monitoring was suspended during part of 2020, due to the COVID-19 pandemic and damage to infrastructure by Cyclone Seroja. Later monitoring centred on scat analysis.

In 2021 djoongari, wopilkara, and Shark Bay mice were released; at the time of writing (February 2022) it was too early to say whether these translocations will succeed, but past translocations indicate they will. Further translocations are planned, including the western grass-wren. As well as reintroductions, it is proposed to introduce two additional threatened mammals, the woylie and the desert mouse, both of which are known to have occurred in the Shark Bay area.

Other animals that were predated by cats, but persisted on Dirk Hartog Island in low numbers, such as the sandy inland mouse, ash-grey mouse, knob-tailed gecko and bush stone-curlew, are now thriving.

Dr David Algar was the driving force behind the eradication of feral cats from Dirk Hartog Island and led the team that developed the Eradicat® bait, now used in Western Australia for feral cat control. The removal of feral cats from Dirk Hartog Island was the largest successful feral cat eradication campaign for an island in the world. Feral cats took a horrendous toll on the wildlife of Dirk Hartog Island, which had lost 10 of its 13 ground-dwelling mammal species due to feral cat predation. After the last feral cat was removed from the island in October 2016, 46 000 km of monitoring and 114 684 camera trap nights have confirmed success, which was independently confirmed in 2018 by Island Conservation (an international NGO with the mission to prevent extinctions on islands). This outstanding achievement meant the island could become a refuge for some of Western Australia's most iconic threatened animals including hare-wallabies, boodies, woylies, dibblers, chuditch and Shark Bay bandicoots. In December 2020, Dave was recognised with a national Froggatt Award, an award that recognises outstanding achievements in Australia's fight against environmental weeds, diseases and pest animals.

You can keep up-to-date with Dirk Hartog Island National Park Ecological Restoration Project by reading 'Wirruwana News' at https://www.sharkbay.org/news/.

Reservation. Part of Dirk Hartog Island became a pastoral lease in 1868, and the lease was expanded to almost the entire island in 1873. In 2009, the pastoral lease was cancelled with most of the island becoming a national park. The original homestead is now a privately-run ecotourism lodge. Small areas of freehold land occur at Sunday Island Bay, and the northern tip of the island is managed by the Shire of Shark Bay. Campsites have been established at several locations within the national park.

FAURE ISLAND

INTRODUCTION

Geography. Faure Island (5800 ha) is located in Disappointment Reach, the main eastern Bay of Gutharraguda. It and Pelican Island are situated on the Faure Sill, a sandbar overlying sandstone that crosses the eastern gulf of Gutharraguda from Peron Peninsula to the mainland. The Sill limits water movement southwards into Hamelin Pool, where hypersaline conditions support stromatolites.

Discovery and naming. Faure Island was named by Nicolas Baudin in 1801 after Pierre Faure, the geographer aboard *Le Naturaliste*. There is a 6 km separation from the mainland; however, this can be reduced to 2 km at extreme low tides. Most of the island, was a pastoral lease that ran sheep and, later, angora goats, from 1873 until 1999, when the lease was purchased by Paruna Sanctuary Ltd, now the Australian Wildlife Conservancy (AWC), and named the Faure Island Wildlife Sanctuary, with the intention of removing threats to the island's wildlife and translocating threatened mammals to it.

PHYSICAL CHARACTERISTICS

Geology and climate. The island is composed of the Peron Sandstone, a soft, poorly consolidated, orange-red aeolianite sandstone, which is exposed around much of the island's edge. Orange-red sand, derived from the sandstone, is the main soil, with areas of white sand on coastal fore-dunes and tidal flats. The maximum height above sea level is 33 m. As in the rest of the region, the climate is dry with most of the annual average 205 mm of rain (data from Hamelin Pool) falling in winter.

BIOLOGICAL CHARACTERISTICS

Vegetation and flora. The island has five major plant communities: Acacia shrublands, mallee shrublands, spinifex grasslands, samphire and saltbush shrublands, and white mangrove woodlands, which occur along the coast. Surveys have recorded 142 native plant species and 21 introduced species of vascular plants. The vegetation is mostly Acacia thicket, dominated by *Acacia ligulata* and *A. rostellifera*. The flora is a subset of that of the adjacent mainland. Differences between Faure and the western islands are probably due to differing geology and soils.

Mammals. Examination of subfossil deposits showed that the island's original, pre-feline, mammal fauna comprised the Shark Bay bandicoot, woylie, djini, and Shark Bay mouse. Since declaring the island feral-free (apart from the house mouse), AWC has successfully established populations of four nationally threatened mammals: boodies, maning, Shark Bay bandicoots and Shark Bay mice. Woylies were also translocated but failed to establish, with the failure of the woylie reintroduction being attributed by AWC to predation by the island's population of sand goannas.

Birds. Ninety-seven species of birds had been recorded by 2005, with others since. Great pied cormorants nest in large numbers in the mangroves. All land birds are typical of the Shark Bay area. The extensive mud flats provide important habitat for migratory waders (shorebirds).

Reptiles. Ground vertebrate surveys in 2000 and 2002 found 12 species of skinks, five of geckoes, two of dragon lizards, the bungarra, a python and one species of venomous snake. More species have been detected since with the total now at 36. None is endemic to the island.

FACTORS LEADING TO CHANGE IN THE BIOTA

Human impact and non-indigenous species. The long period of pastoralism led to significant changes on the island. The island was free of rabbits and foxes, but had horses, goats, feral cats and house mice, and, in earlier times, sheep. The horses, sheep, and goats were removed and the feral cats were eradicated by scientific staff from the Wildlife Research Centre, then part of the Department of Environment and Conservation.

Island management. The island is managed by the Australian Wildlife Conservancy.

OTHER SHARK BAY ISLANDS

Salutation Island (125 ha) has an introduced population of wopilkara. This stick-nest rat became extinct on mainland Australia, surviving only on the Franklin Islands, South Australia. Prior to Salutation Island becoming a nature reserve, it was occasionally used as a ram paddock. The conservation introduction, in 1990, was successful and wopilkara are now common there, providing an insurance should something happen to the original population and being the source for the translocation to Dirk Hartog Island.

Eight species of seabirds breed on at least 40 Shark Bay islands and islets (Plate 6.9). Another six species recorded breeding on Shark Bay islands depend on the sea for their food: Australian pelican, eastern reef-egret, eastern osprey, white-bellied sea-eagle, pied oystercatcher and sooty oystercatcher. Data from a 1997 study are provided in Table 6.2.

Table 6.2. Seabirds recorded nesting on Shark Bay islands in 1997.

Species	No. of breeding pairs	No. of breeding islands
wedge-tailed shearwater	580	5
great pied cormorant	3430	7
Pacific gull	16	7
silver gull	c. 1000	26
Caspian tern	11	11
greater crested tern	441	5
roseate tern	230	2
fairy tern	3	1

The lesser crested tern has been recorded breeding on Meade Island, but was not seen during the 1997 survey.

Shark Bay is renowned for the large numbers of great pied cormorants that live there, with a probable total of several thousand birds. As well as the large colony on Dirk Hartog Island of up to 2500 nests, in 1997 there were c. 1500 nests on Freycinet Island, and 1150 nests on Pelican Island. Another large colony exists on Faure Island. However, great pied cormorants may shift nesting colonies from island to island in different years. Shipping channel markers are favourite perching places and are quickly crowned with a thick layer of guano. The first land-based industry in Shark Bay was guano mining. It began in 1850 on Egg Island (Plate 6.10), off the east coast of Dirk Hartog Island, and progressed to 13 other islands in the Bay; these small islands now have a 'moonscape' appearance. In 1850, to protect the industry from outsiders, the colonial government established a garrison at Quoin Bluff South on Dirk Hartog Island and later at Cape Heirisson.

The colony of Australian pelicans on Pelican Island, near Faure Island in Disappointment Reach, the eastern large bay of Shark Bay, is one of c. 17 islands where pelican nesting has been recorded in WA. Between 50 and 100 pairs have been seen nesting there. In the mid- to late-1960s, Slope Island was joined to the mainland to develop a port for Shark Bay Salt. Wedge-tailed shearwaters once bred there, with 150-200 pairs recorded nesting in 1965, but the port development has destroyed their breeding habitat.

The western spiny-tailed skink occurs on Baudin Island (19 ha) where it shelters under limestone slabs (Plate 6.11). The island was named after the leader of the French Expedition of that name by Captain Henry Mangles Denham in HMS *Herald* in 1858. The Baudin Island skink population was originally described as an endemic subspecies, the Baudin Island spiny-tailed skink *Egernia stokesii aethiops*. Currently (2022) the Baudin Island spiny-tailed skink is listed as the western spiny-tailed skink *Egernia stokesii badia*, a Vulnerable subspecies that also occurs on the Shark Bay mainland, as well as in the wheatbelt and Murchison, but morphological and genetic research as to the distinctiveness of the Baudin Island population is contradictory and requires further study.

Sources of information

World Heritage: https://www.sharkbay.org/world-heritage/shark-bay-world-heritage-values/, http://www.environment.gov.au/heritage/about/world-heritage

Traditional ownership: https://www.sharkbay.org/culture-history/aboriginal-heritage/aboriginal-languages/

European history: Abbott 2006, Christensen 2008.

Naming of islands: Murray & Hercock 2008.

Climate: BOM 2021, Wyrwoll *et al.* 2000.

Geology and soils: Playford *et al.* 2013, Playford 2014.

Lock hospitals: www.lockhospital.com.au, Bates 1944.

European history: George 1999, https://www.sharkbay.org/culture-history/

Vegetation & Flora: Burbidge & George 1978, Keighery & Muir 2008, Milentis 2022, Ride *et al.* 1962.

Mammals: Lipfert 1912, Anon 1971, Baynes 1990, 2008, Burbidge & George 1978, Eldridge *et al.* 2018, Orchard & Orchard 2018, Ride *et al.* 1962.

Birds: Black *et al.* 2021, AH Burbidge *et al.* 2021a, 2021b, Burbidge & Fuller 2000, Dell & Cherriman 2008, 2021c, 2021d, Storr 1990.

Reptiles: Cowan 2018, Storr & Harold 1978.

Return to 1616: https://www.sharkbay.org/restoration/dirk-hartog-island-return-1616/, Algar *et al.* 2019, DBCA Dirk Hartog Island National Park ecological restoration project annual reports.

Plate 6.1. Western cliffs of Wirruwana (Dirk Hartog Island). September 1972. Credit: AA Burbidge.

Plate 6.2. The block, deposited by a tsunami, Wirruwana. November 2015. Credit: AA Burbidge.

Plate 6.3. Shark Bay bandicoot, Dorre Island. Credit: Bert & Babs Wells collection, courtesy of the Department of Biodiversity, Conservation and Attractions.

Plate 6.4. Maning (banded hare-wallaby), Dorre Island. Credit: Bert & Babs Wells collection, courtesy of the Department of Biodiversity, Conservation and Attractions.

Plate 6.5. Eastern bay on Wirruwana. November 2015. Credit: AA Burbidge.

Plate 6.6. Cape Inscription lighthouse, northern end of Wirruwana. November 2015. Credit: AA Burbidge.

Plate 6.7. Great pied cormorant breeding colony, Quoin Bluff South, Wirruwana. September 1972. Credit: AA Burbidge.

Plate 6.8. Vegetation overgrazed by sheep and goats near a water source, Wirruwana. November 2015. Credit: AA Burbidge.

Plate 6.9. Eagle Island (area = 0.45 ha). May 1997. Credit: AA Burbidge.

Plate 6.10. Egg Island (area = 0.4 ha). May 1997. Credit: AA Burbidge.

Plate 6.11. Baudin Island spiny-tailed skink. May 1997. Credit: AA Burbidge.

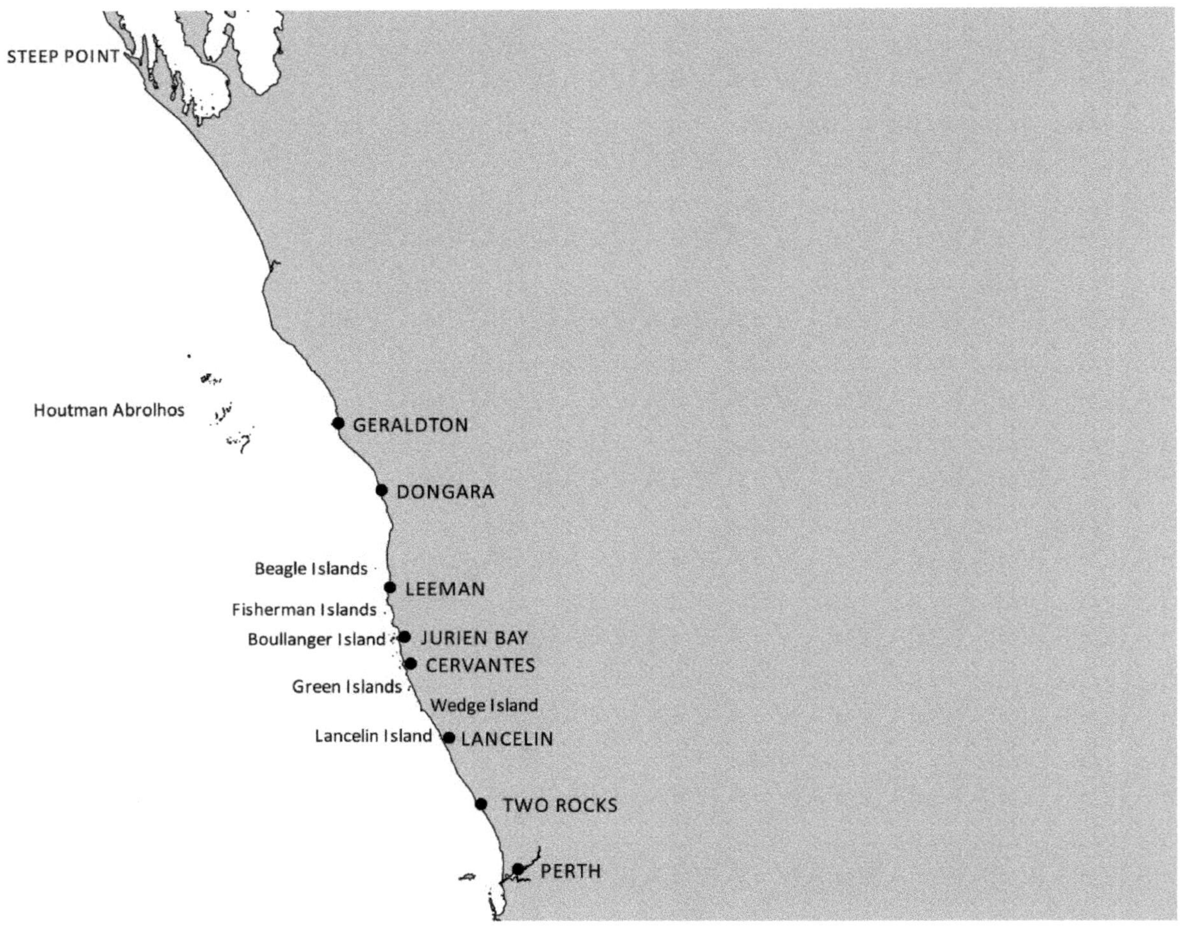

Map 7.1. Inshore islands of the Mid-west coast.

CHAPTER 7: MID-WEST INSHORE ISLANDS

'Today [1.12.1959] *was the day of a giant discovery, but how great I did not realise until I got to the Royal Society* [of Victoria] *Symposium in Melbourne a week later ...* [On North Fisherman Island] *We assumed they were hair seals* [Australian sea lions], *which were uncommon so far N but for which there was 1 authenticated record, the skull collected by Glen* [Storr] *on the Wallabi's* [sic] *on our Abrolhos trip. It seem there might be a question however – they might be fur seals, tho' this was 800 miles from their nearest known breeding station on the Recherche ... At the symposium the following week, where the learned were gathered to discuss evolution, my discovery took the mammalogists by storm. Dom* [Serventy] *& Dr. Ride, director of the Perth Museum were partic^y* [particularly] *excited ... Eventually controversy waxed so fierce ... Fortunately I had black & white & coloured photos in plenty, also teeth & bones, & was able to produce all these...but with no degree of concord ... Dr. Fleming of N.Z.* [New Zealand] *thought they were hair seals. I was inclined to be with him, finding them less "chubby" than the fur seals of N.Z., pulling out on sand instead of rocks & having fairly prominent light manes in the case of the bulls. At the time of writing no verdict has been given so I retain an open mind on the subject – basking in the glory of an accidental find unknown to either fishermen or scientists. The size of the colony & state of the much modified veg.* [vegetation] *indicated fairly long occup^n* [occupation] *and illustrated the scarcity of observers'.* Mary Gillham, Diary.

INTRODUCTION

The 640 km of coastline between Steep Point (the south-western part of Shark Bay) and Two Rocks (opposite Wreck Point near Yanchep) in the south has two distinct groups of islands. These comprise the archipelago known as Houtman Abrolhos (Chapter 8), 60-80 km offshore, and an inshore chain of nearly 40 islands (mostly islets). Much of the coastline from Steep Point to the south of Dongara, some 400 km, is devoid of islands. Between Green Islands and Two Rocks (100 km) there are only six islands. Thus, the majority of the 40 islands occur along the 100 km stretch of coastline between Beagle Island and Green Island (Map 7.1). This sector is promoted by tourism authorities as the Turquoise Coast, centred on the towns of Leeman, Jurien and Cervantes.

Most of these islands are small (<5 ha) and close to the mainland (<5 km). All lie within the 5 m bathymetric line, and it has been estimated that Lancelin and Boullanger islands were isolated from the mainland 6000 and 4000 years ago respectively. The largest islands (Plates 7.1, 7.2) are Boullanger (34 ha), Escape (11 ha), and Lancelin (7.6 ha). The remotest are the Beagle Islands (9 km offshore).

All islands are above-water remnants of aeolianite substrates, and the larger islands consist of sand-blown dunes overlaying the aeolianite base. These geological features align with the coastal limestone and Safety Bay sand prevalent on the adjacent mainland coast. Elevation of these islands seldom exceeds 10 m, with the highest islands being Wedge (19 m), Lancelin (17 m), and Favorite (14 m). Soils are shallow over aeolianite plateaux but deeper where sand dunes exist. North Boullanger Island (2.4 ha) is joined by a tombolo to the larger Boullanger Island (Plate 7.3), and during summer Wedge Island is joined to the mainland by a tombolo.

The climate of these islands is Mediterranean, with mild, wet winters and a long hot, dry period (December to March). Official meteorological records from Jurien and Lancelin indicate that mean maximum temperature varies from c. 19°C in July/August to c. 30°C in February, and mean minimum temperature ranges from c. 9°C in August to c. 18°C in February. The high specific heat of the ocean should slightly increase the minimum temperature and decrease the maximum temperature on the islands. Median annual rainfall is 598 mm (Jurien) and 637 mm at Lancelin, with most rainfall recorded in June (107 mm, Jurien; 155 mm, Lancelin).

Naming of islands. Despite several Dutch navigators sailing near the islands in 1658 and in the 1690s, none were named by them. About a century later, the French expedition under Baudin named Boullanger Island, Lancelin Island, and Ronsard Rocks. Surveys in the colonial era named the Beagle Islands (after HMS *Beagle*) and Snag, Favorite, Fisherman, Green, Escape, Wedge, and Edward islands. Cervantes Island was named after a ship wrecked nearby in 1844. Most of the other islands and many reefs occur close to the mainland coast, resulting in ships avoiding these waters. As a result, these islands were left unnamed until the 1960s when naturalist Julian Ford began studying the fauna of the islands. He needed to refer to island names in his publications and thus proposed that naturalists be honoured, resulting in nine names being officially accepted. These naturalists, all with interests in fauna, are KG Buller, J Drummond, OH Lipfert, AW Milligan, CLE Orton, PT Sandland, W Webb, FL Whitlock, and HM Whittell.

BIOLOGICAL CHARACTERISTICS

Vegetation. On the smallest islands shrubland of nitre bush *Nitraria billardierei* is often the only type of vegetation. Herbfields occur on some, but are now comprised mostly of exotic plant species. Herbfields are most extensive on Escape, South Cervantes, and Lancelin islands. Heath is present on the deeper sands, and is most extensive on Favorite, Boullanger, Whitlock, Escape, North Cervantes, North and South Green, and Wedge islands. Only on the largest island (Boullanger) does Acacia shrubland occur (Plate 7.1).

On East Beagle Island *Nitraria* has a tree-like habit, caused by the activities of the large sea lion population on this small (1.5 ha) island. The impression is of a miniature woodland. On Sandland Island the *Nitraria* plants approach the tree-like habit resembling that on East Beagle Island as a result of the smaller sea lion population.

Flora. As expected, the exposure of these islands to salt-laden winds, especially during gales, has resulted in all plant species being capable of tolerating wind-borne salt and high levels of sodium in the soil. Conspicuous native species include the herbs *Carpobrotus virescens, Calandrinia calyptrata, Cotula cotuloides, Senecio pinnatifolius, Threlkeldia diffusa, the* grasses *Bromus arenarius, Spinifex longifolius,* and *Sporobolus virginicus,* and the shrubs *Atriplex cinerea, Enchylaena tomentosa, Lepidium foliosum, Malva preissiana, Rhagodia baccata, Roepera billardierei, Salsola australis, Wilsonia humilis,* and *Frankenia pauciflora*. Taller shrubs (0.5-1 m or sometimes 2 m in height) comprise *Atriplex isatidea, Myoporum insulare, Nitraria billardierei, Olearia axillaris,* and *Scaevola crassifolia.*

Species typical of heath include the low trees *Acacia cyclops* and *Acacia rostellifera* (2 m or more in height), the herbs *Acanthocarpus preissii, Conostylis candicans,* and *Cassytha racemosa,* the grasses *Austrostipa flavescens* and *Poa poiformis,* and the shrubs *Eremophila glabra, Exocarpos sparteus, Pimelea gilgiana, Spyridium globulosum,* and *Templetonia retusa.*

The flora of these islands has been studied since 1959, with most knowledge accrued during the 1970s and 1980s. The number of native plant species recorded is 73. About one third of the native flora belongs to three families: Asteraceae, Chenopodiaceae, and Poaceae. The size of the native flora on each island varies from 1-43 species, but not in the expected sequence from smallest to largest island. For example, consider just the six islands with the greatest number of native plant species (38-43 species). Four of these have an area between 2 and 11 ha. The largest island (24 ha) has 38 species, the same number as on the island of only 6 ha. This confronting anomaly does suggest that other factors override area. Differences between islands in habitat variety, substrate (sand, aeolianite), elevation, distance offshore, and nesting seabirds would seem to be significant determinants.

Reptiles. This component of the fauna comprises 17 species (13 species of skink, 4 species of gecko). No snakes occur. As expected, larger islands usually have more species than smaller islands, with the largest number of species recorded on Boullanger and Lancelin islands (7 species). But exceptions occur: The small South Cervantes Island (0.9 ha) has 6 species, one more than the larger Escape Island (10.5 ha) and the same as nearby North Cervantes Island (3.2 ha). Seventeen of the 36 islands visited have no reptiles present, with 0.1 ha (South Essex Island) the smallest island on which a lizard species has been recorded.

The most widespread species are the gecko *Christinus marmoratus* and the skinks *Egernia multiscutata* and *Morethia obscura*, all on seven, but not necessarily the same, islands. Five species occur only on one island: *Ctenotus lancelini* and *Cyclodomorphus celatus* on Lancelin Island, *Egernia pulchra* on Favorite Island, *Lerista elegans* on North Fisherman Island, and *Lerista praepedita* on South Cervantes Island.

Such distributions have resulted in notable differences in species composition from island to island: Only Boullanger, Escape, and Favourite islands have two of the three *Egernia* species; North Cervantes Island has three of the four gecko species; and both South Cervantes and Boullanger islands have two gecko species.

King's skink is a large lizard and occurs on six islands, including four as small as 0.9-3.4 ha. These islands have populations of two shearwater species, the burrows of which are used by the skink. This lizard consumes the eggs of seabirds, and this may have enabled it to persist on small islets. On the adjacent mainland coast, King's skink appears to be relatively scarce, whereas it is plentiful on the islands.

The status of the Lancelin Island skink (Plate 7.4) changed between its discovery in 1957 and detailed surveys conducted in 1991-1992. This resulted in a Recovery Plan in 2000, but improving the status of this species on Lancelin Island is hampered by lack of precise knowledge as to the cause(s) of decline. A recent study identified this species as 19th out of the 20 Australian snake and lizard species most likely to become extinct in the wild by 2040, and the only reptile species on an Australian continental island to be at risk.

Land birds. Our knowledge of this component of the avifauna is based largely on Julian Ford's observations, made from 1957 to 1964. The rock parrot occurs commonly on at least 12 islands, and the nankeen kestrel and welcome swallow are also common breeding species, having been recorded on at least 16 islands.

Box 7.1. Julian Ford (1932-1987)

Julian trained as a chemist but his main hobby was natural history. Between 1957 and 1964 he made 44 visits to the islands between Dongara and Lancelin, and meticulously documented the vegetation and reptile and bird faunas of these previously overlooked islands. He was appointed Honorary Associate of the Western Australian Museum in 1963, which now holds the 3000 specimens he collected during his three decades of field research. Although attaining the rank of Senior Lecturer in chemistry at the Western Australian Institute of Technology (now Curtin University), his real research interest concerned the distribution of bird species and their hybridisation and speciation, and the identification of geographic barriers, for which he earned the PhD degree in 1983 from The University of Western Australia. Through his numerous publications, he made many notable contributions to ornithological knowledge across Australia.

The small size of many of these islands precludes the development of residential populations of most land bird species. It is more likely that these species move between island and mainland, and from island to island. Ecologists term this situation a metapopulation. Other species are vagrant but may breed occasionally on some islands. Thus the buff-banded landrail has bred at least once on Lancelin Island and five other species (little grassbird, silvereye, singing honeyeater, Australian pipit, western (Australian) raven have bred on a few islands.

During 1986, a barn owl was resident on Boullanger Island and, being a nocturnal forager, largely subsisted on house mice, whereas the nankeen kestrel (a diurnal forager) fed on reptiles and insects. The transient presence of sacred kingfishers on migration may also exert some pressure on lizard populations, but not to the extent that nesting or resident nankeen kestrels would.

Seabirds. Three species (wedge-tailed shearwater, white-faced storm-petrel, and little shearwater) breed in burrows on 24, 14, and 12 islands respectively. All three species show high fidelity to their breeding islets. Eleven seabird species nest on the surface of islands; these comprise two cormorant, two gull, and seven tern species. Those species nesting on many islands include the great pied cormorant (26 islands), silver gull (24), bridled tern (22), Caspian tern (21), greater crested tern (15), Pacific gull (15), and roseate tern (14). This last species in the 1960s bred south only to the Green Islands. It was first recorded breeding on Lancelin Island in 1975. The sooty tern and brown noddy have also recently begun nesting on Lancelin Island.

Two fish-eating raptors – the eastern osprey and white-bellied sea-eagle - have been recorded nesting on 29 and 5 islands respectively. In the 1970s a spotted harrier was observed eating bridled terns.

Land mammals. Given the small area of these islands, it is not surprising that only two native species occur. Again, as expected, only the largest island (Boullanger) and its close satellite (Whitlock) – a combined area of c. 30 ha - have the dibbler. The grey-bellied dunnart occurs only on Boullanger Island. The discovery of the dibbler (Plate 7.5) in 1985 by Phillip Fuller was remarkable. The dibbler in colonial times had a wide geographical range in mainland south-west WA until c. 1900 when an epizootic swept through the region and feral cats became widespread. So rapid was its decrease in abundance and contraction in distribution that it was feared extinct until its rediscovery in 1967 on the south coast. Boullanger and Whitlock Islands are its only known insular locations. Both mammal species remain abundant there.

Marine mammals. The Australian sea lion breeds in large numbers on East Beagle (c. 150 animals), North Fisherman (c. 100), and Buller (c. 60) Islands. These are the only breeding islands on the west coast of Australia, apart from islands in the Houtman Abrolhos. Another 18 islands are used as resting ('haul-out') sites. The 4-5 month pupping season occurs every 17-18 months and is not determined by the time of year. Thus, for example, the three consecutive cycles from 1989 to 1991 took place in the following periods: December/January 1987 – May/June 1988; June/July 1989 – November 1989; and November/December 1990 – April 1991. The number of pups was 79 on East Beagle Island (March 1991), 63 on North Fisherman Island (November 1989), and 39 on Buller Island (November 1989).

Female sea lions show very high fidelity to their breeding islands. The populations on East Beagle and North Fisherman islands have genetic characteristics similar to those on the islands of Houtman Abrolhos. The genetic features of the Buller Island population is inexplicably distinctive from these.

Species turnover. This is a term used by biogeographers to describe change in the composition of the biota of islands between surveys. Species can disperse to an island from the mainland or other islands ('immigration') and others can disappear from an island ('extinction'). It is most realistic to document such turnover with plant species on small islands: The plants are firmly rooted in the soil and a thorough and careful search of a small island in spring to early summer makes it nearly impossible to overlook species as most flower at these times.

The most comprehensive survey of plants was performed in 1985-1986 (Greg Keighery, Jeni Alford, Vanda Longman), enabling astute comparison with detailed lists made in 1957-1964 (Julian Ford), 1959 (Mary Gillham), one year (unspecified) in the 1960s (Eric Lindgren), 1976 (Ian Abbott), and 1998 (Tony Start). The native component of the flora of most of the islands is dynamic, with the result of many local extinction and immigration events evident, particularly on 23 islets. No change in floral composition was found on ten islands.

With reptiles, only two surveys have been performed. The second one is a list of species from a 1985 survey. The first one is based on Julian Ford's research. One additional gecko species was recorded on the second, but not the first, survey. Two additional species of skink appear to have been recorded on the second survey. It is suspected that some of this change has resulted from differences in taxonomy.

With land birds, their small population size on these small islands makes it difficult to distinguish vagrants from residential populations. At this stage, it seems unwise to infer whether species turnover occurs.

The composition of the native mammal component of the fauna has not changed in the past 30 years.

Speciation. Given the small area of these islands, their closeness to the adjacent mainland, and the relatively short period since they were isolated from the mainland (up to 6000 years ago), the absence of endemic species and subspecies is unsurprising. It would be very unlikely for the land bird species that occur on these islands, but possible for the less vagile species (invertebrates, reptiles, mammals).

The Lancelin Island skink was first collected in 1961 and later described (in 1969) by Julian Ford as being restricted to this island. Searches of the adjacent mainland coast during the period 1993-1995 resulted in only one individual being sighted, immediately opposite the island where it may have been accidentally transported from the island in equipment. The record is not specimen based. Does this signify that the species should no longer be considered as an island endemic species? We think not. The grey-bellied dunnart on Whitlock and Boullanger islands was described in 1999 as an endemic subspecies (*boullangerensis*), based on a genetic analysis published the year before. This was later reconsidered and rejected as invalid.

FACTORS LEADING TO CHANGE IN THE BIOTA

Natural factors: Although there have been no specific studies, it is likely that drought would reduce the population size of plants and animals living on these small islands. Both short- and long-term changes in the direction, flow, and intensity of the warm-water Leeuwin Current would also impact on the abundance and availability of small marine animals on which seabirds feed, and thereby influence the extent and success of their breeding in some years.

Salt spray deposited, particularly during gales, kills plant tissue. This limits the height of vegetation (except where sheltered). This process may have caused the elimination of many plant species soon after the islands were separated from the mainland coast by rising sea levels. The potency of salt is apparent when the flora and vegetation of small and large islands, and flat and topographically-diversified islands are compared.

On small islands, large populations of sea lions breed among the vegetation and form pathways that result in soil compaction. This hampers germination of native plant species.

Large breeding colonies of seabirds have a significant influence on island ecology, both beneficial and detrimental. Their faeces (dominated by uric acid) are rich in nitrogen, phosphorus, and potassium salts. For example, Whitlock Island (with large populations of seabirds) has 7-20, 9 and 6 times as much nitrate, phosphate, and potassium salts respectively compared with nearby Boullanger Island (which has fewer seabirds). Over a period of hundreds to a few thousand years, faecal deposits from surface-nesting seabirds become substantial and form hardened guano. This fertilises the vegetation and by favouring some plant species and disadvantaging others (particularly sclerophylls) helps shape the floristic composition of these islands.

This variation in the tolerance of native plant species to fertilising and trampling by colonial seabirds results in roosting and nesting areas with a flora different from areas on the same island lacking these activities. Mary Gillham, visiting from Britain, noticed that the Mediterranean climate experienced by these WA islands renders the guano more toxic than on British islands. This is because the guano is not leached away throughout the year, as a result of rainfall being concentrated in the winter months. Sclerophyllous plant species are replaced by succulents and herbs, perennials by annuals, and native species by exotic species.

Box 7.2. Mary Gillham MBE (1921-2013)

Mary served in the Women's Land Army in Britain during the war of 1939-45. This made her eligible for free university education, gaining her degrees in agriculture and botany at the University of Wales. In 1953 she completed her PhD degree with a thesis analysing the impact of seabirds on the vegetation and flora of several Welsh islands. This may have been the first study to reveal the dynamic nature of island plant assemblages. She then published several influential papers exploring themes concerning the impact of grazing, manuring, trampling, and burrowing by animals on the plant ecology of islands. Between 1957 and 1960 she extended her research to islands of the southern hemisphere: New Zealand 1957; Victoria, Tasmania (including Macquarie Island), and Queensland 1958-60; Western Australia 1959; and South Africa 1960. Between 1961 and 1988 she lectured in adult education in Wales. She wrote many books about her island observations and experiences, mostly after retirement, based on her diaries, all of which are now archived in Cardiff.

The manuring effects of seabird colonies should also increase the population size of invertebrates feeding on the vegetation, leaf litter, and in the upper layer of the soil, made up of naturally-infertile sand. The dibbler illustrates the beneficial impact of these changes on an insectivorous species positioned high in the food chain. On Whitlock Island, intermittently linked by a 300 m long sandbar to the larger Boullanger Island, many males live for 2-3 years. On Boullanger Island very few males survive beyond two years and all males die off in winter and spring. These differences between islands are attributable to the greater density of seabirds on Whitlock Island (370 shearwater burrows/ha and 400 terns on its 5.4 ha), compared with only c. 200 storm-petrel burrows/ha on the 26 ha Boullanger Island. Dibbler populations are twice as dense on Whitlock as on Boullanger Island.

The absence of lizards on Edward, Snag, and Webb Islands was attributed by Julian Ford to the continual deposition of guano by nesting great pied cormorants. He also suggested that the burrows created by shearwaters and storm-petrels serve to create additional habitat for lizards.

In conclusion, high densities of nesting and roosting seabirds on islands have profound impacts on the ecological identity of islands and their transfer of marine nutrients to the island environment ramify through the entire insular food web. Seabirds therefore deserve the appellation of 'ecosystem engineers'.

Another significant ecosystem engineer is the Australian sea lion. These large and heavy animals on the Beagle Islands restrict nesting opportunities for the ground-nesting fairy tern. Similarly, the massive weight of sea lions moving across an island forms a hard crust on the soil surface of their paths, and may have reduced the carrying capacity for lizards. The result is fewer lizard species and decreased abundance. On East Beagle Island, Julian Ford found only one individual lizard; a later survey found no lizard species present. Other islands with breeding populations of sea lions have only one or two lizard species present.

Human factors: None of the Aboriginal groups (Amangu, Yued) living along the coastline with adjacent islands constructed watercraft. However, given the proximity of the islands to the mainland (some are only 100 m offshore), their visibility, and the intervening shallow water, and presuming an interest in wading at low tide and an ability to swim the short distance involved, Aboriginal visits and impact on the fauna should not be dismissed. Because documentation has not so far been found, we decline to speculate further.

The first known incursion by Europeans was by Abraham Leeman in 1658. He was Upper steersman on the Dutch vessel *Waeckende Boey* sent to search for survivors of an earlier shipwreck. A change of weather resulted in Leeman's inshore surveying party being abandoned. Before sailing in an open boat for Java, his party of 13 sailors provisioned at Outer Rock, Wedge Island, Lancelin Island, and Boullanger Island. This included the harvesting of 'sea celery' and the killing of gulls for food, and some Australian sea lions for their meat and skins. These brief visits probably had only minor impact on island ecology.

In contrast, the impact of European settlers is well documented. The major factor is the colonisation of the island by 48 exotic plant species. How this happened is not perfectly understood. None of the islands has been settled or farmed. In 1951, Lancelin was only an anchorage and no buildings were present. By 1959 shacks had been constructed by crayfishers. In 1959 there was no settlement at Jurien Bay, only a few holiday shacks, and the area was merely a collecting centre for rock lobsters. There were no shacks in 1959 on the mainland opposite the Fisherman Islands. Very little of the adjacent mainland coast has been farmed, but there are now small towns at Jurien and Lancelin and until recent times there were many illegal cottages close to beaches (now removed except at Grey).

It is most unlikely that plant species were directly planted on any of the islands. Despite many of the islands having at least one beach, which attract people with small boats for picnicking, it seems implausible that this is an effective means of introducing plant species. Furthermore, at one least exotic plant species occurs on the 13 islands without a beach.

The major and most likely sources of exotic plant species are from seeds windblown from gardens, roadsides, and waste areas (rubbish dumps) on the nearby mainland coast, from seeds ingested by silver gulls feeding in disturbed areas, and from silvereyes which eat berries and drupes and later void the inedible seeds.

Most of the exotic species are grasses (15 species), Caryophyllaceae (6 species), and Brassicaceae (5 species). All these have small seeds that could be dispersed by wind or ingested by gulls. The species likely to be dispersed by silvereyes are the edible berries of African boxthorn (Solanaceae) and black berry nightshade (Solanaceae).

The extent of infestation of the islands by exotic plant species is astounding. The proportion of exotic species in their floras ranges from 0 (3 islands) to 63% (Webb, Edward islands), with five islands having more than half of the flora comprised of exotic species. The most widespread of these species are: *Mesembryanthemum crystallinum* (iceplant, Aizoaceae) occurring on 30 islands; *Sonchus oleraceus* (common sow thistle, Asteraceae) on 26 islands; *Hordeum leporinum* (barley grass, Poaceae) on 13 islands; *Cakile maritima* (sea rocket, Brassicaceae) on nine islands; and *Avena barbata* (bearded oat, Poaceae) on eight islands. These species usurp space that would otherwise be available for occupation by native plant species.

In contrast, it is surprising that the nettle *Urtica urens*, a species which thrives in highly-fertilised soils, has been recorded only on two islands (South Cervantes, Tern).

Exotic mammal species are represented by three species. House mice are present on Boullanger and Whitlock islands. Foxes reached a small 'island' just north of North Head, Jurien Bay, between 1928 and 1930, and so wedge-tailed shearwaters no longer breed there. Foxes crossed to Wedge Island in 1960 and its small population of wedge-tailed shearwaters became extinct. In October 1961, Julian Ford found several bridled terns there that had been killed by foxes.

Visitors to Lancelin Island have been observed with their pet dogs, but it is not known how well-controlled these dogs were during these visits. It is illegal to bring a dog onto any island because they are nature reserves.

There are no records of domestic cats or rabbits on any of the islands. During 1980 one western grey kangaroo was present on Boullanger Island, no doubt having been released by a visitor, but it was absent in 2000.

Numerous other threats to native biota on the islands are evident, with most being inadequately documented: habitat destruction caused by vehicle access to Wedge and Escape islands, the construction of a light tower on Escape Island in 1930; bombing practice on Target Rock and Lancelin Island; occasional fire (an osprey nest on Drummond Rock was burnt in 1961); entanglement of seabirds and sea lions feeding at sea in discarded fishing gear; removal of topsoil (Snag Island); shooting of a sea lion on East Beagle Island; and trampling of seabird burrows and disturbance of surface-nesting and -roosting seabirds by visitors (minimised on Lancelin Island by the construction of a boardwalk).

Ecological restoration. Removal of exotic species is the first priority in reversing ecological damage to islands. It would be desirable to eliminate house mice from Boullanger and Whitlock islands, as they compete to some extent with the two native mammal species present and their high densities probably impact on breeding seabirds. However, deploying toxic baits would also eliminate the dibblers, as they are known to eat baits. The other native mammal species (grey-bellied dunnart) does not eat baits. The only feasible option is to temporarily remove a large sample of dibblers, hold them at Perth Zoo, then deploy poison baits, and once all mice are killed repatriate the dibblers to the islands.

With plant species, only African boxthorn has so far been eliminated from the four islands on which it once occurred. It is not feasible to eliminate any of the other weed species, especially given the ease with which they would recolonise.

The other strategy available for improving the conservation status of threatened island populations is to use some islands as a place to establish additional (insurance) populations. This has been done successfully with the dibbler and the Lancelin Island skink.

In 1998 dibblers were translocated to Escape Island. These animals were derived from animals captured in 1997 on Whitlock and Boullanger islands, transferred to Perth Zoo, and bred there. In 2002-3, 133 Lancelin Island skinks were translocated from individuals bred since 1995 at Perth Zoo to Favorite Island (Plate 7.6) in Jurien Bay, and the species appears to have established there.

Island management. A Management Plan was published in 2004. It declared that human access to ten islands is to be limited, and is prohibited to another 26 islands. The Plan makes provision for ongoing monitoring to detect loss of native vegetation and invasion of environmental weeds (such as tree mallow), as well as removal of African boxthorn (since accomplished, mostly by volunteers). Because the islands lack a designated ranger and are infrequently patrolled, most of the remaining action items are difficult to monitor and enforce compliance with the objectives specified in the Plan.

Sources of information

Introduction: Abbott 1978; Australia Pilot 1972; BOM 2021; Ford 1963; Keighery *et al.* 2002; Lowry 1974.
Naming of islands: Ford 1963; Murray & Hercock 2008.
Vegetation: Ford 1963; Keighery *et al.* 2002.
Flora: Abbott 1980; FloraBase website; Gillham Diary; Keighery *et al.* 2002.
Reptiles: DEC/CCWA 2004; Ford 1963, 1965a; Geyle *et al.* 2021; Pearson & Jones 2000.
Birds: Abbott unpubl.; Dickman *et al.* 1991; Dunlop 1979; Ford 1965b; R Johnstone pers. comm.; Johnstone & Storr 1998, 2004; Sandland 1931; Storr unpubl.
Land mammals: Bencini *et al.* 2001; Bennison *et al.* 2016; Fuller & Burbidge 1987; Miller *et al.* 2003; Mills & Bencini 2000; Wolfe *et al.* 2004.
Marine mammals: Campbell 2005; Campbell *et al.* 2008; Gales *et al.* 1992, 1994; Gillham 1961; Gillham Diary.
Species turnover: Abbott 1981; Abbott & Wills 2016; DEC/CCWA 2004; Ford 1963, 1965a; Keighery *et al.* 2002.
Speciation: Peacock & Jones 2000; Start *et al.* 2006.
Factors driving change in the biota: Natural factors: Bencini *et al.* 2001; Ford 1965a, 1965b; Gillham 1961; Miller *et al.* 2003; Wolfe *et al.* 2004.
Factors driving change in the biota: Human factors: Bennison *et al.* 2016; Burbidge & Abbott database; Gillham Diary; Moor & Ferguson 1979; Tindale 1974; Tyler & Ross 1977.
Ecological restoration: Bennison *et al.* 2016; Moro 2002; Hartley & Pearson 2009; Pearson & Jones 2000.
Island management: DEC/CCWA 2004.

Plate 7.1. Lancelin Island. View from ocean. December 1985. Credit: AA Burbidge.

Plate 7.2. Wedge Island. View from mainland coast. November 1992. Credit: AA Burbidge.

Plate 7.3. Boullanger Island. View of beach and tombolo. December 1985. Credit: AA Burbidge.

Plate 7.4. Lancelin Island skink. August 2004. Credit: D Pearson, Department of Biodiversity, Conservation and Attractions

Plate 7.5. Dibbler. Credit: Bert & Babs Wells collection, courtesy of the Department of Biodiversity, Conservation and Attractions.

Plate 7.6. Favorite Island; mainland in background. March 2016. Credit: D Pearson, Department of Biodiversity, Conservation and Attractions.

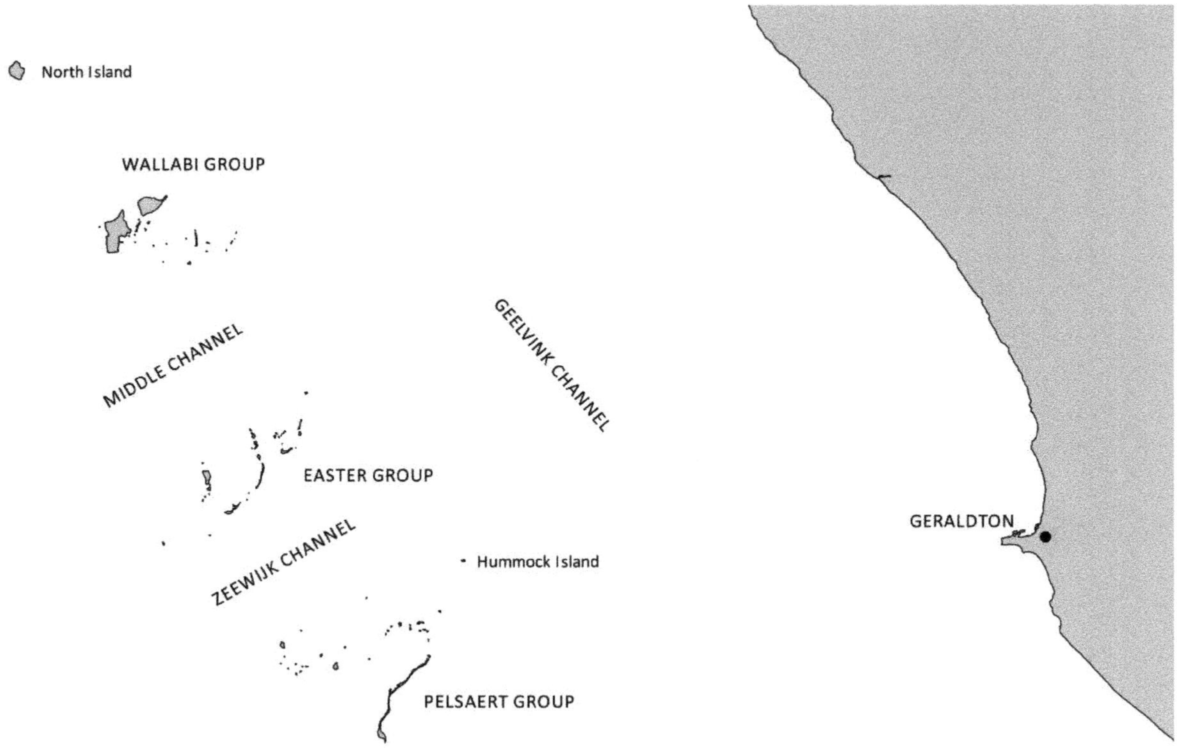

Map 8.1. The Houtman Abrolhos.

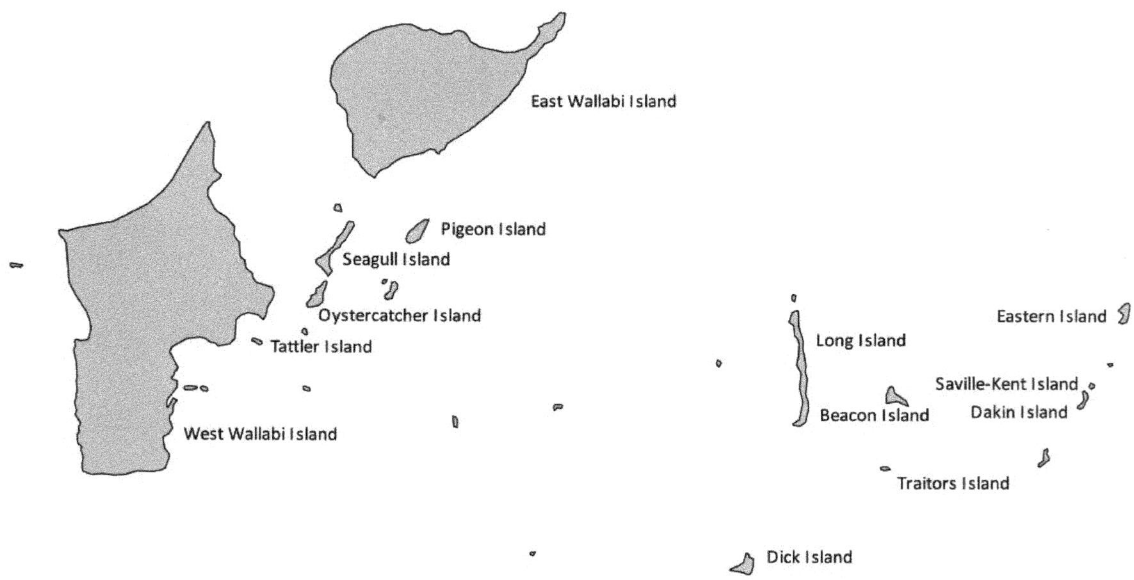

Map 8.2. Wallabi Group, Houtman Abrolhos.

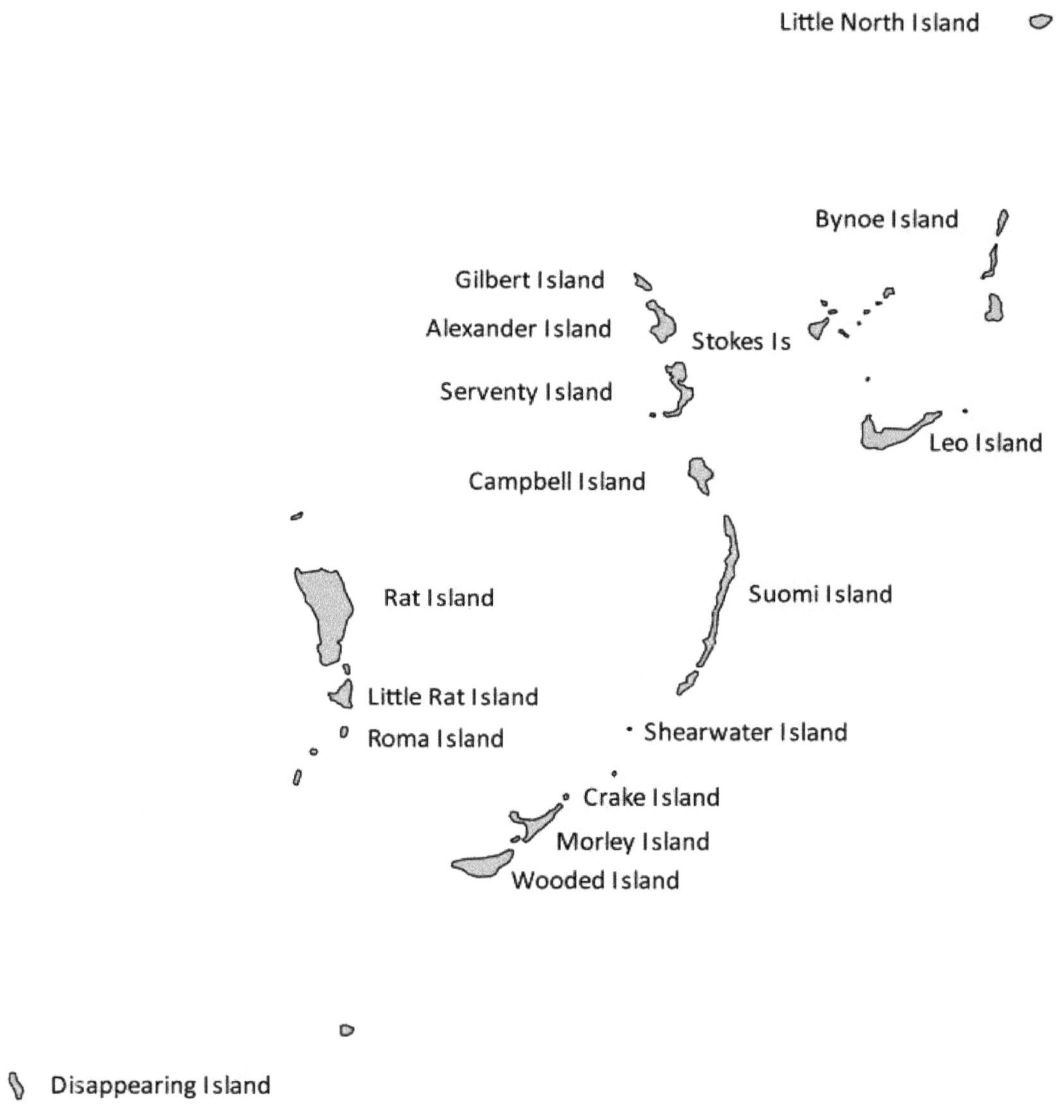

Map 8.3. Easter Group, Houtman Abrolhos.

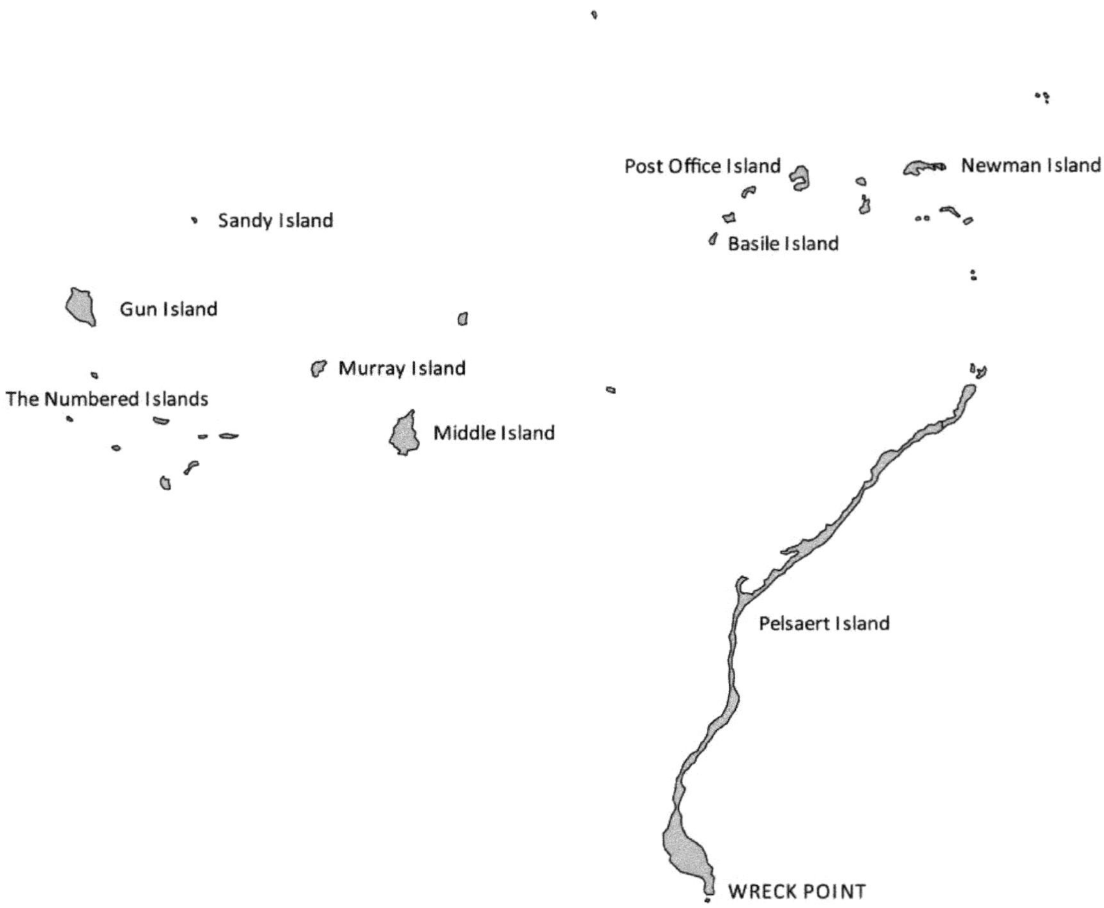

Map. 8.4. Pelsaert Group, Houtman Abrolhos.

CHAPTER 8: HOUTMAN ABROLHOS

All the internal parts of this portion of the island was dense thickets of samphire, &c, in which amazing numbers of birds were breeding, particularly the common brown noddy (Anous stolidus)*; these birds were in one spot covering several acres in extent, in such prodigious numbers, and their nests of seaweed so closely built to each other, that it was no easy matter to avoid crushing birds or eggs at every step; ...*

John Gilbert in a letter to the Editor of *The Inquirer* describing a visit to Pelsaert Island in January 1843.

In the afternoon we stroll over the island. What a wonderful revelation. Why, the whole place is actually alive with birds, some breeding on every bush, some breeding beneath the bushes, and some breeding beneath the bushes underground. ... Locomotion over the ground is rendered extremely insecure, owing to its honey-combed nature caused by countless burrows of petrels, or so-called mutton birds.

Archibald J Campbell in the *Australasian* newspaper describing a visit to Rat Island in 1890.

INTRODUCTION

Geography. The Houtman Abrolhos lie between 65 km and 75 km off the Western Australian (WA) coast near Geraldton. The 210 islands, islets, and rocks[1] are divided into three groups: Wallabi, Easter and Pelsaert.

Table 8.1. Islands in the Houtman Abrolhos National Park by size category; c. 19 small islands occupied by commercial fishers are not included in the national park.

Group	Wallabi	Easter	Pelsaert
100 ha to 1000 ha	3	0	1
20 ha to 100 ha	0	2	1
10 ha to 20 ha	0	5	1
1 ha to 10 ha	8	11	9
<1 ha	47	58	45
Total	**58**	**76**	**57**

The Wallabi Group includes the three largest islands, composed of aeolianite and sand: West Wallabi (587 ha), East Wallabi (307 ha), and North (176 ha), as well as many small coralline islands. The three larger islands are continental in origin, having once been attached to the mainland until isolated by rising sea levels by 8000 to 6000 years ago, after the world's oceans rose

[1] There have been several estimates of the number of islands, islets and rocks in the Houtman Abrolhos. The draft National Park Management Plan stated that there are 210 islands in the Abrolhos. Table 8.1, from Appendix 1 in the draft plan, includes 191 islands. Small islands may disappear or appear after storms.

following the end of the last glacial maximum. The highest point of all the islands is on East Wallabi Island, just 14 m above sea level.

Easter and Pelsaert Groups are comprised of predominantly low islands. These are remnants of reefs or collections of coral rock accumulated on reefs by waves, or a combination of both. Sand has accumulated on some islands, allowing burrow-nesting seabirds to breed there.

Discovery and naming. Aboriginal people would have occupied the islands when they were joined to the mainland, but have not visited since the islands were isolated from the mainland 6000–10 000 years ago. There is only limited archaeological evidence of Aboriginal occupation. A stone artefact made from Eocene fossiliferous chert that was probably derived from outcrops located on the now-submerged continental shelf and once quarried by Aboriginal stone tool makers, was found on Beacon Island during WA Museum excavations relating to the *Batavia* mutiny. Aboriginal people of the adjacent mainland possessed no watercraft, and so had no means to access the islands.

Dutch mariners using the Brouwer Route, a 17th-Century route used by ships sailing from the Cape of Good Hope, South Africa, to the Dutch East Indies (now Indonesia) were the first Europeans to visit the Abrolhos. The route took ships south from the Cape (34°S) into the Roaring Forties, then east across the Indian Ocean, before turning northeast for Java. Thus it took advantage of the strong westerly winds for which the Roaring Forties are named, greatly increasing sailing speed, but if the turn north came too late it brought ships close to the WA coast.

The islands were named 'Abrolhos' by the Dutch navigator Frederick de Houtman in the *Dortrecht*, accompanied by the *Amsterdam*, in July 1619. The name has for many years been linked to the Portuguese expression 'abre olhos', meaning 'open your eyes'. Now the name is thought to have come from the Portuguese word 'abrolhos', which, in 1616, meant 'spiked obstructions'. Houtman may have named WA's Abrolhos after the Arquipélago de Abrolhos off the coast of Brazil. He warned 'One should stay clear of this shoal, for it lies most treacherously for ships that want to call in at this land.'

The most infamous ship that encountered the islands was the *Batavia*, the then flagship of the Dutch United East India Company (VOC - *Vereenigde Oostindische Compagnie*), which was wrecked, early on 4 June 1629, on Morning Reef in the Wallabi Group. It is Australia's second oldest known shipwreck. Most crew and passengers survived and camped on the nearby 2.5 ha Beacon Island. Commander Francisco Pelsaert was a merchant, not a seaman. The ship's captain or skipper, Adrian Jacobsz, was responsible for navigation. After the wreck, Pelsaert and 46 crew went to the mainland in one of the ship's boats to seek fresh water. When they found none, they sailed to Batavia (now Jakarta, Indonesia), taking 33 days to get there, to organise a rescue. During Pelsaert's absence, mutiny on Beacon Island led by Jeronimus Cornelisz, Batavia's under-merchant, resulted in a massacre of c. 125 men, women and children. A group of soldiers, led by Weibbie Hayes, was tricked by Cornelisz into going 6 km away to West Wallabi Island under the pretence of searching for water. The soldiers were actually better off, as they soon found a rock hole with fresh water as well as food (tammar wallabies and other animals). Hayes and his men constructed a walled shelter (Plate 8.1), the first European-built structure in Australia. The mutineers raided twice, and attempted to kill the soldiers, but were driven off. Cornelisz then attempted to lure the soldiers into a trap, but they captured him and five accomplices. Fourteen days later, the rescue ship, the *Sardam*, arrived and the soldiers warned Pelsaert that the mutineers meant to take over his ship and become pirates, leading to the mutineers being captured as they attempted to board the *Sardam*. Some of the mutineers, including Cornelisz, were executed on Long Island. Two were marooned on the mainland near the mouth of the Murchison River. Of the 316 people aboard the *Batavia* when it was wrecked, only 116 survived.

The *Batavia*'s wreck was once thought to be in the Pelsaert Group, as wreckage was sighted on a reef there by the crew of the *Zeewijk* in 1727 and by HMS *Beagle's* crew in 1840. The *Beagle*'s captain, John Clements Wickham, named the southern group of islands 'Pelsaert Group' reflecting this belief. However, historical research by Henrietta Drake-Brockman, based on a translation by ED Drok of Pelsaert's journals in Old Dutch, revealed that the wreck was likely in the Wallabi Group, where it was eventually located in 1963. During the wreck's excavation, part of the hull was uncovered. It was carefully raised and, after a number of years of conservation treatment, the remains were rebuilt in the Shipwreck Galleries in the WA Maritime Museum in Fremantle. The hull is the centrepiece of the *Batavia* Gallery.

On 9 June 1727, another Dutch ship, the *Zeewijk*, was wrecked on Half Moon Reef in the Pelsaert Group. The ship's crew had, despite warnings from Houtman and others and instructions from the VOC, decided to 'steer ENE, if there is an opportunity and if it is feasible, to call at the land of Eendracht' (Eendracht was named after Dirk Hartog's ship, see Chapter 6). The survivors camped on what was later named Gun Island, where they subsisted by eating seabirds, sea lions and food salvaged from the wrecked ship. Drinking water was obtained from rock holes filled by rain and from 'wells' on Middle and Murray Islands, as well by catching rain in canvas. A rescue group of 11 of the fittest survivors and First Mate Pieter Langeweg set off on 10 July for Batavia in the ship's longboat, but were never heard of again. From the timbers salvaged from of the *Zeewijk*, the remaining survivors, once they decided rescue was unlikely, built a small ship, the *Sloepie*, in which 82 out of the initial crew of 208 were able to reach their original destination of Batavia on 30 April 1728. The *Zeewjik* survivors reported seeing wreckage near the wreck of their vessel.

During British Admiralty surveys of the north-west coast of WA in 1840, Commander John Clements Wickham, Lieutenant John Lort Stokes and the crew of HMS *Beagle* discovered a brass gun of about three pounds calibre and an iron swivel gun on which paint still adhered, as well as numerous other maritime artefacts signifying European occupation on an island in the Pelsaert Group.

Wickham named the island Gun Island and the passage between the Easter and Pelsaert Groups, Zeewijk Channel. They also reported finding wreckage, including 'the beams of a large vessel' (according to Stokes) or 'some remains of large timber, evidently a beam of a ship, through it an iron bolt of considerable dimensions' (according to Midshipman Crawford Pasco), at the southern end of an island in the southern group (later named Pelsaert Island) and this led them to mistakenly identify the wreckage as being associated with the *Batavia*. They named the group of islands where they discovered wreckage after Pelsaert, the bay where they found the wreckage, Batavia Roads, and the point at the southern end of Pelsaert Island, Wreck Point. Guano miners based on the islands in the late 1800s also found material from wrecks on both Gun and Pelsaert Islands.

The site of the wreck of *Zeewijk* was discovered in 1968. Between 1976 and 1979 there were three WA Museum expeditions to the *Zeewijk* site but extensive in-water research did not locate any evidence of a second wreck. Further Museum research, including aerial magnetometer surveys in 2008 and 2014, did not reveal any additional targets that may have been an early shipwreck. Several lines of evidence let to a conclusion that the wreckage found on Pelsaert Island was flotsam, and not from a wreck that had occurred locally.

The wreckage seen by the crew of the *Zeewijk* could possibly have been from one of three Dutch vessels that disappeared after leaving the Cape of Good Hope for Batavia: the *Ridderschap Van Holland*, which left the Cape in February 1694, the *Fortuin* in 1724 or the *Aagtekerke* in 1726. However, the *Ridderschap Van Holland* was eventually found to have been wrecked at Madagascar, leaving two possible early Dutch ships that may have contributed wreckage. Wreckage seen by the crew of HMS *Beagle* may have been from the *Flying Foam* that went missing during a gale on 10 March, 1812 and most likely sank in deep water south of the Abrolhos.

Due to the treacherous nature of the area, with low islands and, in earlier times, poor navigation aids, there were several other early shipwrecks at the Abrolhos. These include the 268-tonne barque *Ocean Queen* which, on the night of 13 September 1842, struck the Pelsaert Group. The 18 crew members, including the captain, landed on Pelsaert Island where they stayed for 10 days. They then set out for Fremantle in the ship's boats, but due to strong winds, they landed 130 km north of the Fremantle settlement and walked the remaining distance. On 10 April, 1851, *en route* to Fremantle from Batavia and Singapore, the 28-tonne schooner *Venus* struck one of the reefs surrounding the Pelsaert Group. The crew survived on Middle Island for six months until 15 October when three of them rowed to the mainland. The others were rescued a week later. The 19-tonne schooner *Preston* ran ashore on Pelsaert Island on the morning of 20 September 1855. Anchors were used to try to drag the vessel off the shore, but this failed. Its cargo was unloaded and the vessel was abandoned on 2 October. The crew then left for Port Gregory in the ship's boats. About 60 vessels have been lost in the Abrolhos, including, in 1879, the steel ship *Ben Ledi* on Pelsaert

Island near Big Lagoon, and, in 1908, the steel steamer *Windsor* on Half Moon Reef. Its boiler is now a landmark there.

PHYSICAL CHARACTERISTICS

Geology. Two channels reaching depths of approximately 40 m separate the northern Wallabi Group from the central Easter Group (Middle Channel) and the Easter Group from the southern Pelsaert Group (Zeewijk Channel). Geelvinck Channel separates the Abrolhos from the mainland.

The Houtman Abrolhos islands have been sculpted by falling and rising sea levels, ocean swells, and the warm southward-flowing Leeuwin Current. Each of the three Abrolhos groups consists of a central coral-reef platform laid down c. 125 000 years ago during the last interglacial, on which more recent reefs developed. The smaller islands consist of remnant reef platforms from periods of higher sea level or from loose or consolidated coral rock. Many islands are only 2–3 m above sea level, less at spring high tides. Some small coral cays are only c. 1 m above sea level and are partly or mostly submerged during storm-driven high tides.

A notable feature of some islands is ridge 'waves' (Plate 8.2). These are ridges of coral rock from adjacent reefs, thrown up by winter storm swells and from tropical cyclonic waves, particularly when local sea levels rose for a short time from c. 6000 to 4000 years ago. They give the appearance of long waves on land and occur on dozens of islands in the archipelago. They are particularly noticeable on Long, Serventy, and Pelsaert Islands. Dating has shown that they were formed between 6000 and 500 years ago.

The Houtman Abrolhos has been, and continues to be, greatly influenced by the warm southerly-flowing Leeuwin Current. Because of it, the coral reefs on which the islands are built are the southernmost in the Indian Ocean, are one of the highest latitude reef systems in the world and are the only extensive coral reefs so far south on the western margin of any continent. On other continental west coasts, cool currents predominate. Further, the Leeuwin Current greatly influences marine food chains on which seabirds and other animals depend.

Climate. The Abrolhos islands have a Mediterranean climate with hot, dry summers and cool, moist winters; however, temperatures are milder than those on the adjacent mainland. An automatic weather station on North Island has operated since 2001. Annual mean rainfall there is 270 mm, with most rain falling from May to August. Rainfall on the southern islands may be marginally higher. Mean minimum and maximum temperatures on North Island range from 9.3°C to 19.5°C in July and from 19.1°C to 32.4°C in February. The Abrolhos is a very windy place. During summer a high pressure ridge lies to the south, causing persistent winds from about November to April, and known locally as 'southerlies'. These blow from the south-east or south-west, at speeds exceeding 17 knots (31 km/h) almost half the time. During autumn and winter, the ridge moves northwards, increasing the atmospheric pressure over the islands resulting in highly-variable winds. Winter tends to be the time of both the strongest gales from cold fronts and the most frequent periods of calm. Tropical cyclones have passed near the islands about once every three years.

Naming. Geelvinck Channel was possibly named by Vlamingh in 1696 after his ship when searching for the missing *Ridderschap Van Holland*, with the channel being given the same name by PP King in HMSV *Bathurst* in 1822. Apart from those named during exploration, mentioned above, many more recent island names reflect the variety of people who have visited and worked there. Some are named for fishers or their vessels: Suomi (the Finnish name for Finland) after a fishing boat, Basile, Morley and Newman after local fishing families, Leo after Leo Antillas, a local fisherman, and Roma because many local fishers were of Italian descent. Others are named after naturalists and scientists who have recorded aspects of the Abrolhos' natural history: Alexander, Bynoe, Campbell, Dakin, Gibson, Gilbert, Saville-Kent, Serventy and Stokes. Oystercatcher, Pigeon, Shearwater, Tattler and Crake Islands are named after birds. This is not an exhaustive list of the 192 islands, as many remain unnamed.

BIOLOGICAL CHARACTERISTICS

Because of their isolated species, varying in distribution from island to island, and especially because of the abundant seabirds, the Houtman Abrolhos has long fascinated biologists and naturalists.

Biological discovery. John Gilbert visited Pelsaert, East Wallabi, West Wallabi and other islands in January-March 1843 on behalf of John Gould. AJ Campbell visited Rat and Pelsaert Islands in December 1889. His observations and some of those of GK Beddoes (then manager of the Rat Island guano works) were published by Campbell in 1890 and 1901. Then came visits by CP Conigrave in 1897, R Helms in 1898, R Hall in October 1899, and CG Gibson in November 1907. Under the leadership of WJ Dakin, the Percy Sladen Trust expeditions to the Abrolhos in 1913 and October-November 1915 produced the first detailed accounts of the islands and their fauna. These early visitors were followed by visits by Vincent Serventy, John Warham and Glen Storr. Aquinas College in Perth organised five expeditions to the Abrolhos from 1964 to 1970 and St Patrick's College in Geraldton conducted a single expedition in the 1960s. More recently, Andrew Burbidge, Phil Fuller and Randall Owens conducted island-wide surveys of seabird populations in the 1980s and Chris Surman and Lisa Nicholson conducted a seabird nesting survey of all islands in 2006. The Abrolhos islands' vegetation and flora was extensively described in 2001 by Judith Harvey and others. Paramount among recent scientists is Dr Christopher Surman, who has visited the islands, mainly Pelsaert Group, every year for over 30 years studying the seabirds and other aspects of the environment.

Vegetation and flora. The vascular flora of the Houtman Abrolhos has been thoroughly surveyed and vegetation maps and species lists have been published for 119 islands. As at 2001, these lists totalled 239 species from 68 families, including 95 introduced species, the latter mostly on islands with settlements. East Wallabi Island has 124 species and West Wallabi has 97 species. *Eucalyptus oraria* and *Acacia didyma* occur only on East Wallabi Island. Sandy and shell grit areas have pure stands of grey saltbush *Atriplex cinerea* or saltbush mixed with other low plant species. Pavement limestone areas are sparsely vegetated, while beach dunes on the Wallabi Islands have a flora similar to that of the adjacent mainland coastal dunes. Loose coral rock has sparse or no vegetation. Stands of white mangroves *Avicennia marina* exist on 33 islands on protected shores or in land-locked tidal lagoons but only occur as substantial areas on Wooded, Morley and Alexander Islands in Easter Group and on Pelsaert, Post Office, Burnett, Mangrove and Newman Islands in the Pelsaert Group. Pelsaert Island, which has several relatively large stands (for the Abrolhos), has two areas of mangroves of c. 1.7 and 1.0 ha. The other two islands where lesser noddies nest also have relatively large areas: Morley with 0.9 ha and Wooded with 0.54 ha.

There is no evidence of bushfires on most islands. However, a fire burnt a large part of North Island in 1935 causing soil erosion, with areas of bare rock and sparse vegetation evident today.

Mammals. Two native terrestrial mammals occur on the Wallabi Islands, the tammar wallaby (Plate 8.3) and the bush rat. Tammars were the first Australian marsupial seen by Europeans, the crew of the *Batavia*. They occur on both East and West Wallabi Islands and are relatively common, sheltering under low shrubs during the day and feeding at night. JL Stokes, of HMS *Beagle*, reported that in 1840 'on one day, within four hours, I shot 36, and that between three guns we killed 76 ... which gave rise to the name of Slaughter Point ...'. Tammar numbers are kept in check by two indigenous predators, the white-bellied sea-eagle and the carpet python. As is the case with other mammals on small islands, the tammars on the Wallabi Islands have low genetic diversity and high levels of inbreeding. The bush rat occurred on both islands, but has not been located on East Wallabi Island since 1967 and is likely locally extinct. The cause of this extinction is unknown but it may be due to a combination of the drying climate and poor nutritional status, as West Wallabi Island has large populations of breeding seabirds that bring in guano (fertiliser) from the ocean, while East Wallabi only has a few seabirds that seldom nest. The Wallabi Islands bush rat was originally described as a separate species, *Rattus glauerti*, later thought to be a subspecies, but it is not recognised as such today.

The Abrolhos population of the Australian sea lion (Plate 8.4) is the most northerly breeding population of this rare species, which is restricted to Australia. The Abrolhos population is small, perhaps c. 100 individuals, but appears to be slowly increasing. There were once many more sea lions at the Abrolhos. The shipwrecked crews of the *Zeewijk* (1727) and the *Venus* (1851) both killed them for food and it has been estimated that the crew of the *Zeewijk* killed at least 153. John Gilbert stated in 1843 'all the islands of the group were thickly inhabited by seals', so numbers seem to have reduced a lot since then. William Whitecar, on board an American whaling vessel visiting the Abrolhos in the 1850s while waiting to hunt humpback whales wrote 'We soon found a novel and exciting sport in the destruction of seals, which occur here in great numbers.' Australian sea lions stay in the area where they were born, so there is limited possibility of more arriving from further south. The Australian sea lion is listed as Vulnerable in WA and Endangered under Commonwealth legislation. The IUCN Red List lists the species as 'Endangered and decreasing'.

Reptiles. After many years of study, 25 species of reptiles have been collected on the islands of the Houtman Abrolhos. The highest number of species occurs on East and West Wallabi Islands with 19 species on each. Three species are known only from West Wallabi and one only from East Wallabi – the legless lizard *Aprasia clairae*, which is rare on the mainland. Carpet pythons occur on both islands, but are now very rare on East Wallabi, probably due to the disappearance of bush rats.

No reptile species is restricted to the islands. However, two reptile subspecies are endemic: the Abrolhos dwarf bearded dragon *Pogona minor minima* and the Abrolhos spiny-tailed skink *Egernia stokesii stokesii* (Plate 8.5) named after Stokes of HMS *Beagle*. The spiny-tailed skink subspecies occurred on eight islands. The crew of the *Beagle* kept several as pets, one of which survived the ship's return to England. Its extinction on Rat Island has been attributed to feral cats and is the only known reptile extinction anywhere in the Abrolhos.

A study between 2002 and 2012 found that species distribution is highly correlated with geomorphological composition of the islands, with two species groups occurring on the aeolianite islands and the others occurring on rocky islands and coral cays.

No frogs occur on the Abrolhos. Two early European museum specimens that are stated to have been collected there are now thought to have been incorrectly labelled.

Land birds. While many land bird species have been sighted at the Abrolhos, only a few are considered to be resident. Silvereyes occur on many islands and appear greener than mainland silvereyes. Other resident birds include the buff-banded rail and spotless crake. Of particular note are the brush bronzewing and the painted button-quail. Brush bronzewings are relatively common on the Wallabi Islands but uncommon on the adjacent mainland north of Perth. The Houtman Abrolhos painted button-quail, as its name suggests, is a subspecies that is restricted to the Abrolhos, occurring in the Wallabi Group on East and West Wallabi, Seagull and Pigeon Islands. It is listed as Vulnerable in both legislative lists and *The Action Plan for Australian Birds 2020*. It previously occurred on North Island but recent surveys have failed to locate it, with its disappearance linked to the introduction of tammar wallabies. Camera trap surveys on East Wallabi Island in March 2018-June 2020, revealed low rates of detection. Most occur in the sandy peripheral areas of the island where there are shallow sands or dunes.

Two raptors are common, the white-bellied sea-eagle and the eastern osprey. White-bellied sea-eagles are particularly abundant in Wallabi Group, where they have access to year-round food – tammar wallabies (throughout the year), wedge-tailed shearwaters (August to

May), and little shearwaters (April to December). Examination of sea-eagle nests has shown that shearwaters are the main food source. On West Wallabi Island on 5 December 1992, Andrew Burbidge and Phillip Fuller counted 18 sea-eagles soaring over the island at the one time. The 1991-1993 island-wide survey (see below) counted 15 breeding pairs, nesting on 14 islands.

Ospreys build large nests from sticks and other flotsam and jetsam, with one island being named Post Office Island due to the tall nests there sitting atop stone piles (erected as an early 20th Century navigational aid) resembling old-fashioned pillar post boxes. One nest on Wooded Island was measured to be 1.7 m high. Ospreys also nest on structures built by lobster fishers. Some fishers build platforms for ospreys to nest on, in the hope that they will not nest on lobster pot stacks. The 1991-1993 survey counted 47 pairs in the Abrolhos. In October 1975, 13 nests were counted on Pelsaert Island, three of which had eggs or young birds.

Shorebirds are relatively abundant. Many migratory waders visit the islands, feeding on the inter-tidal zone. The ruddy turnstone and grey-tailed tattler range across reef and coral shorelines whilst bar-tailed godwits and great knots aggregate at the few intertidal sand bars. Four resident shorebird species nest there: pied and sooty oystercatchers, red-capped plover, and eastern reef egret.

Seabirds. Houtman Abrolhos seabirds are famed among ornithologists and bird watchers both because of the very high numbers of breeding species and of birds that breed there (Plates 8.6-8.10). Thirteen species of seabirds breed on at least 122 islands, with well over half a million nesting pairs being present in some years. Breeding species comprise a mixture of tropical and temperate birds. White-faced storm-petrels and little shearwaters do not breed further north. The brown noddy and sooty tern, both tropical species, did not nest farther south, but have recently extended their breeding range southward 275 km to Lancelin Island, the brown noddy in 1992, and the sooty tern in 1998, presumably as a result of increasing sea temperatures. Since the 1960s, off WA, sea temperatures have risen by c. 0.5°C above the long-term average. While this may not sound like much, it can have a significant impact on the distribution and physiology of marine species.

During the last week of November and the first week of December 1991 (Easter Group), 1992 (Wallabi Group) and 1993 (Pelsaert Group), all islands were visited and numbers of breeding pairs counted or, where numbers were too high to count, estimated (Table 8.2). In December 2006, another survey, of all three groups, was conducted.

Table 8.2. Summary of breeding seabirds recorded at the Houtman Abrolhos, November-December 1991-1993 and December 2006 (from Fuller, Burbidge & Owens 1994, Surman & Nicholson 2009).

Species	No of pairs (December 1991, 1992, 1993)	No of islands (December 1991, 1992, 1993)	No of pairs (December 2006)	No of islands (December 2006)
Wedge-tailed shearwater	c. 1 117 800	12	644 883	7
Little shearwater	c. 30 600	26	60 356	46
White-faced storm-petrel	c. 4230	14	17 995	13
Great pied cormorant	c. 1180	9	2234	8
Little pied cormorant*	11	1	-	-
Silver gull	162	17	389	25
Pacific gull	51	39	152	63
Caspian tern	81	9	70	22
Roseate tern	3441	19	4210	19
Sooty tern	c. 260 300	9	105 407	8
Bridled tern	c. 7050	66	2274	90
Fairy tern	481	14	547	13
Greater crested tern	2159	9	2928	10
Brown noddy	c. 132 000	1	121 320	1
Houtman Abrolhos lesser noddy	c. 48 900	3	34 103	3
14 species	**c. 1 680 500**	**122**	**962 765**	**-****

*nesting noted in mangroves on Wooded Island in December 1993 but not during the 1992 survey

** total of 148 islands recorded with breeding birds but these included records of eastern reef egret, eastern osprey, white-bellied sea-eagle and oystercatchers that, while they depend on the ocean for their food, are not strictly seabirds

The red-tailed tropicbird has been recorded nesting in the Abrolhos, but not during the above counts.

The figures in Table 8.2, while valuable, need clarification. Firstly, in 1991-93, the numbers of the three species of burrowing seabirds were estimated by counting or calculating the number of burrows judged to be occupied. This assumption can result in over-estimates compared with examining burrows with a 'burrow-scope' (a miniature camera mounted on the tip of a long flexible probe) to see if a burrow is actually occupied, a time-consuming exercise. The 2006 survey did use a burrow-scope and, while a sample of burrows was examined to produce estimates of wedge-tailed shearwaters, is likely more accurate than the 1990s survey. During the 2006 survey little shearwaters were absent, so the estimate was of the number of burrows. More breeding locations were detected than in the 1990s. More white-faced storm-petrel burrows were also located than in the 1990s.

Some of the differences between the two estimates may be due to varying 'participation' as the proportion of birds that nest in a given year can vary due to differing food availability.

Some species, such as the gulls, breed at other times or may have finished breeding during the December visits and roseate terns are known to nest in autumn as well as early summer. No doubt there have been changes since these data were collected. Easter Group sooty tern colonies are known to nest on different islands in different years and participation (actual number of nesting pairs compared with the number of pairs present) varies from year to year, depending on food availability.

Box 8.1 Dr Christopher Surman (1968-). Chris has studied seabirds and other marine life at the Houtman Abrolhos for 30 years. His long-term scientific research into the Vulnerable lesser noddy is world-class. He pioneered the use of micro GPS units to reveal where lesser noddies went to feed, discovering the incredible journeys that breeding adults make to beyond the edge of the continental shelf (Figure 8.2). His photographs have appeared in numerous magazines and on websites, including *BBC Wildlife*, *Canadian Wildlife Magazine* and *Landscope*. He has published many scientific papers and presented his work at conferences and workshops around the world. In 2019, Chris published his book *Houtman Abrolhos. A natural history*, a comprehensive treatise that includes many of his stunning photographs. Working in a remote location has meant that many people and organisations have helped, but Chris has funded much of his research himself.

Of all the islands where seabirds nest, Pelsaert has the largest number of nesting seabirds and the largest number of breeding species. Twelve seabird species totalling more than one million birds nest there. It is the only island in the Abrolhos where brown noddies nest.

Two species of seabirds that nest in the Abrolhos are listed as threatened: the fairy tern and the lesser noddy. The fairy tern is listed as Vulnerable under WA and Commonwealth legislation. The WA population is distributed over an extensive stretch of coastline (c. 2500 km) from the Pilbara to the eastern south coast. The spring/summer-breeding population winters, primarily, on islands in the Wallabi and Easter Groups of the Houtman Abrolhos, before migrating to breeding areas as far south as Point Malcolm on the eastern south WA coast and as far north as the Ningaloo coast, near Exmouth. In WA, unlike in eastern Australia, most fairy terns nest on islands, with the Abrolhos islands being a major breeding place and also an important non-breeding feeding and roosting place. Because of this, the WA population of the fairy tern is less threatened with extinction than the eastern one. In late March or early April, in all years from 2015 to 2019, night-roosting flocks of 500 to 2000 fairy terns were observed on the airstrip on Rat Island. On 4 June 2019, fairy terns were counted during the twilight arriving at the North Island airstrip with 4100 individuals estimated to be wintering there. Abrolhos fairy terns nest mainly during late spring and summer, while the small Pilbara population nests in winter.

Figure 8.1. Estimated number of breeding pairs of lesser noddies in the Houtman Abrolhos, 1989 to 2014. From Surman *et al.* (2016).

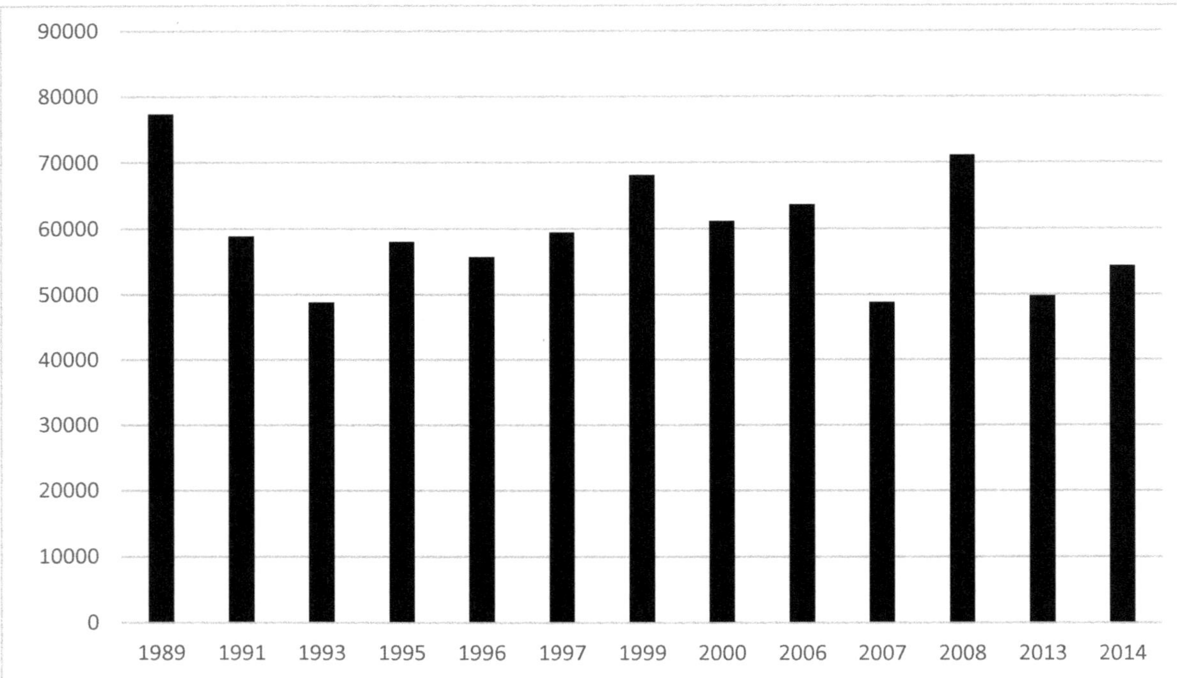

The lesser noddy breeds on either side of the Indian Ocean. In the western Indian Ocean, breeding occurs in Seychelles, Mascarene and Agalega Islands (Mauritius), Maldives, and the Chagos Archipelago. The Houtman Abrolhos is the only place where it breeds in the eastern Indian Ocean and this population has been described as a separate subspecies, the Houtman Abrolhos lesser noddy, which is listed as Vulnerable. A few lesser noddies have been recorded nesting on the Ashmore Reef islands (see Chapter 2), but which subspecies is unknown. In the Abrolhos lesser noddies nest on branches of white (or grey) mangroves on Pelsaert, Wooded and Morley Islands. Numbers have been monitored on Pelsaert Island since 1986, and on all three breeding islands since 1989, by counting the number of nesting pairs along fixed transects and their biology has been researched since 1991. Numbers of nesting pairs from 1986 to 2014 (counts were not conducted every year between 1986 and 2006) varied from c. 77 000 in 1986 to c. 48 000 in 1993 and 2007.

Total numbers at the Abrolhos have been relatively stable but in some years estimated numbers nesting have been low. Research has shown that participation (numbers of pairs that breed) and timing of breeding varies with food availability and that tends to vary with the strength of the Leeuwin Current. Years with poor food availability, mainly El Niño years, can also lead to the death of chicks from starvation.

Lesser noddies, which weigh only 97–120 g, fly very long distances to feed. Tracking using 1.6 g GPS tags, temporarily taped to the bird's tail feathers, showed that adults, when raising chicks, commute an average of 35 km offshore to reach their feeding grounds. However, when incubating an egg, they may extend their foraging range to at least 140 km south-west from their nest and may spend as long as 15 hours searching for food (Figure

8.2). Tracking data show that, to raise a single chick, each lesser noddy parent would have flown at least 2800 km to and from their foraging areas, often against 30 knot (55 km/h) southerly winds on the way out.

FACTORS LEADING TO CHANGE IN THE BIOTA

Non-indigenous species. Past and current threats to the Houtman Abrohlos' biodiversity are associated with introduced species and human use. Future threats include increased visitation and climate change.

On 11 April 1840, Lieutenant Stokes of HMS *Beagle*, reported 'The centre island [of the Easter Group] we named Rat Island, from the quantity of that vermin with which it was infested'. This 56 ha island once had a very large breeding aggregation of seabirds, probably the largest aggregations anywhere in the eastern Indian Ocean. In 1889, AJ Campbell estimated that the mixed colony of brown noddies and sooty terns held 1 452 000 birds or 726 000 pairs. The island also had very large numbers of nesting wedge-tailed shearwaters, meaning that Rat Island once had more nesting seabirds than Pelsaert Island does today. The Rat Island seabird colonies were extirpated by the late 1930s through the combined effects of guano mining, the introduction of cats, and egg collecting by fishers. The black rats, apparently present before 1840, seem not to have had a major effect, although it is known that they have impacted seabird colonies elsewhere. Possibly, the very large number of breeding seabirds, present for only part of the year, meant that rat numbers, while increasing during the seabird nesting season, collapsed after the birds left.

R Helms in 1898, stated 'These rats, however, may have been the native species *Rattus fuscipes*, as Rat Island, the largest island in the Easter Group of Houtman Abrolhos seems to have escaped shipwrecks of Dutch vessels. ... Moreover, Rat Island is a continental island (previously connected with the mainland) and because of the large number of seabirds breeding there, the native species of rat could have persisted in large numbers.' This hypothesis suggests that black rats may not have been introduced until commercial fishing began in 1848, or guano extraction commenced in 1884, or even from a visit by an American whaler (active along the west coast from the 1840s). There is, however, no evidence that the rats were originally the southern bush rat. Between c. 5500 until after 4200 years ago the sea level at the Abrolhos was c. 1 m higher than at present. Rat Island is only c. 2–3 m above high water mark, less at extreme spring tides, so it is possible that the sea may have washed over Rat Island during cyclonic storm surges during the period of higher sea levels. If so, it is unlikely that a native rodent would have survived there after the sea levels rose isolating the Abrolhos from the mainland; however, the spiny-tailed skink did survive this period, which makes the submergence of Rat Island less likely. Spiny-tailed skinks occur on several low islands and may have been distributed by sea-eagles, which may carry prey items for a long time. If the rats on Rat Island in 1840 were black rats, it remains a mystery as to how they established.

The guano miners recorded that the rats were so extraordinarily numerous in the 1880s that they would even run over the table when the men were having their meals and, unless provisions were kept in tanks, nothing was safe from them. This led them to introduce cats, which, along with removal of all topsoil, caused the local extinction of the largest seabird breeding colonies in the eastern Indian Ocean, as well as the Rat Island population of the spiny-tailed skink. The eradication of black rats from Rat Island in 1991, with the last cat dying in around 2000, has allowed seabirds to start nesting there again. Nesting seabirds were first seen again in 1999, with 21 fairy terns nesting on a beach at the northern end - something that would not have been possible when black rats and cats were present. Since 2003, numbers of nesting seabirds have increased greatly. In 2007, 300 pairs of fairy terns nested. By 2012, eight species were nesting, with sooty terns and bridled terns being the most numerous. In December 2012, there were more than 72 000 pairs of sooty terns nesting. However, numbers have varied since then, as sooty terns also nested on other islands in Easter Group, and colonies move between islands. The establishment of a colony of white-faced storm-petrels, sometime after 2008, was particularly remarkable as these small, burrow-nesting seabirds are very vulnerable to black rat predation and would not have been able to persist on Rat Island, even prior to the introduction of the cats. Burrow nesters such as the wedge-tailed shearwater and storm-petrel are unlikely to return in significant numbers as guano mining removed almost all soil.

Weeds are widespread. Ninety-five species have been recorded, mostly on islands with settlements. Some species were introduced by fishers and their families as garden plants. Of major significance in relation to seabirds is African boxthorn *Lycium ferocissimum*, which has been recorded on Wooded Island in Easter Group and on 14 islands in Pelsaert Group. Its long thorns are a significant danger to birds. It has been eradicated from most islands, including Pelsaert Island. However, surveillance is required as it occurs on the adjacent mainland and its seeds float. *Mesembryanthemum crystallinum* (iceplant) is widespread and can form near-complete ground cover in places, inhibiting burrowing by seabirds.

Non-indigenous mammals have been introduced to the islands, either deliberately or by accident. In November-December 1991, Andrew Burbidge and Phil Fuller of the Department of Conservation and Land Management, Randall Owens from the Department of Fisheries, and Ken Johnson from the Conservation Commission of the Northern Territory, undertook a successful program to eradicate black rats from Rat Island and adjacent small islands. They were unaware at the time that house mice also occurred on Rat Island. Black rats have also been eradicated from Pigeon Island and house mice from Beacon Island. House mice still occur on Rat Island, North Island and some other smaller islands. Cat eradication was conducted from 1991 on Rat Island, with the last surviving cat dying around 2000. European rabbits were introduced to several islands. They did not survive on North, Rat, Gun, Middle and Pelsaert Islands, and were eradicated on Leo, Morley and Wooded Islands. Goats were released on East Wallabi Island in 1888 and fortunately did not survive. Horses, goats, sheep and dogs were introduced to Rat Island during guano mining days and did not survive the departure of the guano miners.

Tammar wallabies were introduced to North Island from the Wallabi Islands by lobster fishers, firstly in 1928, which did not result in establishment, and then in 1985, which did. Without natural predators they rapidly increased in numbers resulting in vegetation and environmental degradation, including the disappearance of the local population of the threatened Houtman Abrolhos painted button-quail. Attempts to eradicate them are under way, with 800 animals being culled in 2008 and more in 2019/20. Further work is aimed at eliminating tammars from North Island and reintroducing button-quail.

Mining. During the 19th century many islands were mined for guano. Guano mining began in earnest on Rat Island in 1885 and wound up in 1915. The predominantly Malay workers dug and swept up the guano-enriched soil, levering out the surface limestone in order to get access to the material. Low embankments were constructed for trolley lines that carried the excavated material back to the shipping stockpiles at the north-eastern end of the Island before the product was loaded onto vessels via trolley lines that ran out onto a stone jetty, which is still visible, with a more recent jetty built over it. Many other islands were also mined, with over 48 000 tonnes of guano shipped between 1884 and 1896. Mining recommenced between 1943 and 1945 during World War II, mainly on Pelsaert Island. Remains of that era can be seen on several islands, including Rat and Pelsaert Islands.

Guano mining had a major impact on 16 islands, with topsoil and vegetation being removed from islands, mainly in the Pelsaert and Easter Groups. Mined islands now have little soil and much exposed rock, often piled into heaps.

BP Petroleum Development Australia drilled a stratigraphic well on Gun Island in 1968, but it did not detect commercial hydrocarbons and it was subsequently abandoned.

Human impact. The western rock lobster (still known locally by its earlier name 'crayfish' or 'crays') industry started in 1907, when Italian fishers camped on Rat Island and later on Pelsaert Island. A small canning factory operated on West Wallabi Island during the 1930s. The industry expanded during World War II as a result of a decision by the Defence Foodstuffs Administration in 1941 to supply canned lobster to canteens for the armed forces. With further development of the industry, the Houtman Abrolhos became a base for fishers and 'camps' were constructed on many of the smaller islands, as well as on West Wallabi. The western rock lobster was Australia's most valuable fishery until the government of China prevented import in 2020. In the past, fishers and their families lived on the 22 occupied islands for several months each year. Some 'camps' (small dwellings and associated infrastructure) are elaborate. Some larger islands had a nursing post, a school, and a social club. Recently, due to live export of lobsters and bigger, faster boats, few fishers live on islands. There are airstrips on North Island, East Wallabi Island and Rat Island, but none in the Pelsaert Group. A proposal by lobster fishers to construct an airstrip on Pelsaert

Island in the early 1980s was refused after the then Department of Fisheries and Wildlife published a report showing the island's significance to seabirds.

The first attempt to develop tourism used buildings on Pelsaert Island abandoned by the World War II phosphate miners. It was not successful. Tourism today is based mainly on vessels from Geraldton and aircraft that conduct scenic flights and land on East Wallabi Island. There is a research base on Rat Island.

Climate change. Climate change is a current and looming threat to the Houtman Abrolhos. As the sea level continues to rise, many islands will disappear, with low coral cays going first. At least one island, named Sandy Island (although actually composed of coral shingle), was washed away in the 1980s. More recently, Faole Island in the Pelsaert Group was inundated and the camp was destroyed during a strong winter low pressure system. Fishers' 'camps' on West Wallabi Island have also been damaged during high tides and strong winds. Marine heatwaves, already becoming more frequent, can lead to coral bleaching and even coral death, and greatly affect food availability for seabirds, resulting in cessation of breeding or death of chicks. The coral reefs that protect the islands may increase in height as the sea level rises. However, about 30% of CO_2 in the atmosphere resulting from the burning of fossil fuels and land use changes dissolves in the ocean, making it more acid. Lower pH causes carbonate ions to be relatively less abundant which can make it difficult for animals that use calcium carbonate to lay down shells, or, in the case of corals, structures, so it is possible that the corals building the Abrolhos' coral reefs will slowly die.

In recent years, mangroves have died, including those used by lesser noddies for nesting. Some deaths, particularly on Wooded Island, are due to guano from nesting great pied cormorants. However, some may be due to drought, while it is possible that some are due to lesser noddy guano build up not being washed into the ocean by the reducing rainfall. Increasingly lower winter rainfall due to ongoing climate change may exacerbate this trend. Climate change will also affect the terrestrial flora and fauna, with reduced winter rainfall possibly contributing to the disappearance of the southern bush rats from East Wallabi Island. What effects lower winter rainfall and higher air temperatures will have on the Abrolhos' plants and animals is difficult to predict, but some change is inevitable. In the medium- to-longer term, global climate change (the climate emergency), leading to increasing air and sea temperatures, marine heatwaves, increased ocean acidity and rising sea levels, and possible changes in the Leeuwin Current, are likely to greatly affect the islands and its animals.

Reservation. The first islands reserved for protection of native game under the *Game Act 1892* was the Houtman Abrolhos in 1898, many years after guano extraction began. This forbade shooting and the taking of seabird eggs, but did not prevent the destruction of their nesting habitat. For a long period the Abrolhos islands were included in Class A reserve 20253, first proclaimed in 1929, initially for the purpose 'Public Recreation and Tourist Resort', and later for 'Conservation of Flora and Fauna, Tourism, and purposes associated with the Fishing Industry'. The reserve was vested initially in local people, then the Superintendent of Fisheries, then in Trustees of the Houtman Abrolhos and later in the Minister for Fisheries. In July 2019, the Houtman Abrolhos National Park was declared. It includes all islands except some small islands and parts of larger islands where fishers have 'camps', which remain in Reserve 20253, along with the area between high and low water marks.

Ecological restoration. The eradication of cats and black rats from Rat Island and the eradication of rabbits from several islands have greatly reduced the impact of invasive species. The eradication of introduced tammar wallabies from North Island, once completed, should allow the vegetation to recover, and, in the longer term, permit the reintroduction of the Abrolhos painted button-quail. Ongoing removal of African boxthorn minimises its impact on seabirds.

Island management. With the Houtman Abrolhos recently declared a national park and with increasing visitor numbers, good management and sufficient management resources are necessary to ensure that the islands remain a valuable biodiversity conservation area. Human impact on seabird nesting colonies can be significant if not properly managed.

A draft management plan for the national park was released for public comment in 2021.

Sources of information

General: Abbott & Wills 2016.
Geology: Collins *et al.* 1993, Scheffers *et al.* 2012, Teichart 1947, Wyrwoll *et al.* 1995.
History: Abbott 2006, 2008, 2011, Drake-Brockman 1963, Edwards 1966, 1970, Green 2018.
Island names: Murray & Hercock 2008.
Vegetation and flora: Harvey *et al.* 2001.
Mammals: Abbott 2008, Burbidge & Abbott 2017, Whitecar 1864.
Birds: Barnes *et al.* 2021, Burbidge & Fuller 1989, Davis *et al.* 2021b, Dunlop & Greenwell 2020, Dunlop *et al.* 2015, Fuller & Burbidge 1981, Fuller *et al.* 1994, Gilbert 1843, Greenwell 2021, Storr *et al.* 1986, Surman 2019, Surman & Nicholson 2009, Surman *et al.* 2016, 2017 .
Reptiles: How *et al.* 2004, 2020, Maryan *et al.* 2009.
Coral reef decline: Davis *et al.* 2021a.

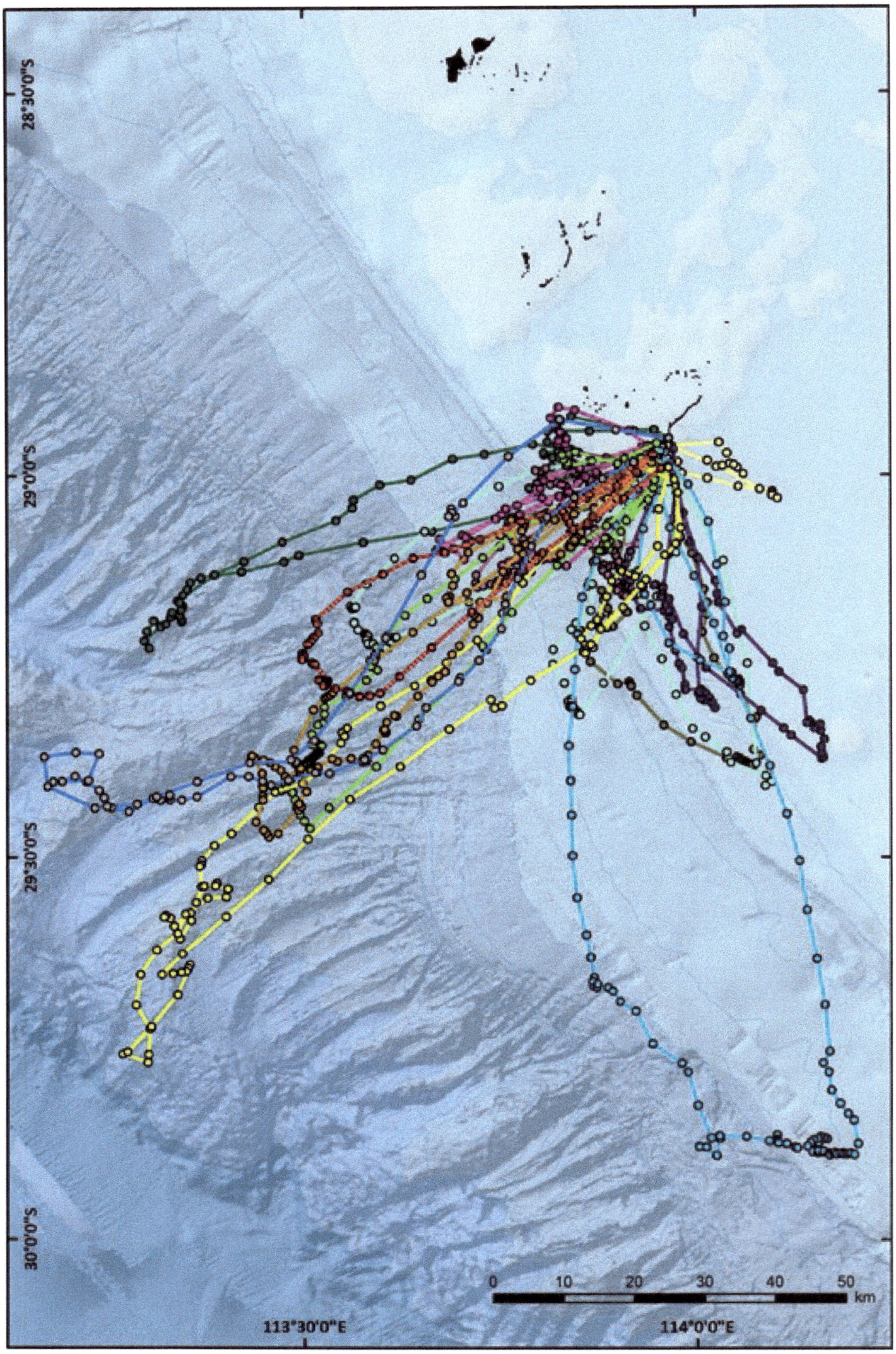

Figure 8.2. The foraging tracks of 12 individual lesser noddies nesting at Pelsaert Island in November 2014, reproduced with permission from Surman *et al.* (2017).

Plate 8.1. Remains of Webbie Hayes' fort (1629), West Wallabi Island. The oldest surviving European building in Australia. December 2017. Credit: AA Burbidge.

Plate 8.2. Geomorphology and vegetation of Pelsaert Island, looking south. December 2016. Credit: AA Burbidge.

Plate 8.3. Tammar wallaby, West Wallabi Island. December 2017. Credit: AA Burbidge.

Plate 8.4. Australian sea lion, Pelsaert Island. December 2016. Credit: AA Burbidge.

Plate 8.5. Abrolhos spiny-tailed skink *Egernia stokesii stokesii*, West Wallabi Island. December 2017. Credit: AA Burbidge.

Plate 8.6. Lesser noddy with chick, on nest in mangal, Pelsaert Island. December 2015. Credit: AA Burbidge.

Plate 8.7. Brown noddy, Pelsaert Island. December 2016. Credit: AA Burbidge.

Plate 8.8. Sooty tern, Pelsaert Island. December 2016. Credit: AA Burbidge.

Plate 8.9. Colony of greater crested terns, Pelsaert Island. December 2016. Credit: AA Burbidge.

Plate 8.10. Roseate tern, nesting colony, Houtman Abrolhos. November 2016. Credit: AA Burbidge.

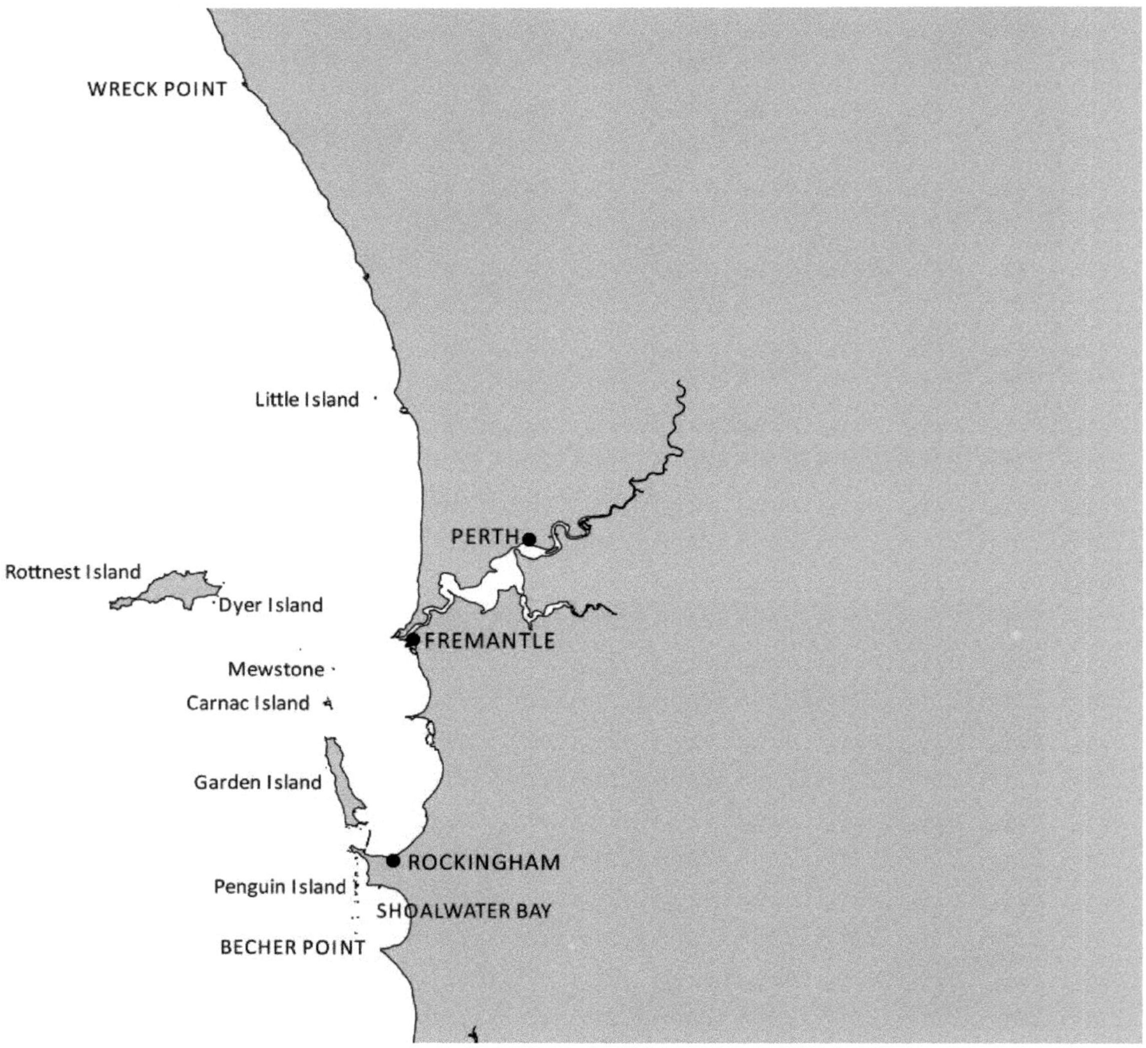

Map 9.1. Islands of Perth region.

CHAPTER 9: PERTH REGION

'*... the island* [Rottnest] *was well wooded ... they had seen two seals and a bush cat as well as many droppings of other animals, but had observed none*'. From the journal of the *Waeckende Boey*, March 1658. (Source: Zuiderbaan 1977).

'*... he with his crew had carefully surveyed the island* [Rottnest] *but they did not find any animals except bushrats, some of which they had shot and brought with them, they also saw a lot of trees with an agreeable smell...[I sent some people ashore] to cut down some trees for fire-wood which is very plentiful here*'... From the journal of the *Geelvinck*, December 1696. (Source: Robert 1972).

'*The ground is covered with little or no soil, chiefly white sand, and rocky, in my opinion unsuitable for cultivation. There are very few Birds here* [Rottnest Island] *, and no Animals, only a kind of rat as big as a common cat, whose dung is found in abundance all over the Island; also very few Seals*'. From the journal of the *Nijptangh*, December 1696. (Source: Robert 1972).

INTRODUCTION

There are 121 islands in the metropolitan area of Perth, the capital city of Western Australia (WA). For the purposes of this chapter, the coastline under consideration lies between Wreck Point (near Yanchep) and Becher Point (south of Rockingham) – a distance of 95 km (Map 9.1).

Most are small islets (termed stacks), formed by erosion of headlands from frequent pounding by waves during several millennia. The four largest islands are Rottnest (1900 ha, 45 m maximum elevation), Garden (1200 ha, 55 m), Carnac (16 ha, 18 m), and Penguin (12 ha, 18 m). Owing to the presence of salt lakes (Plate 9.1), the actual area of land on Rottnest is 1700 ha. Rottnest Island is 18 km from the mainland, Garden Island is 2 km from the mainland, Carnac Island is 3.5 km north of Garden Island, and Penguin Island is only 700 m offshore. Most other islands are no larger than 100 m^2, and many are situated close to the larger islands. Nearly all islets around Rottnest can be easily reached by wading across reefs.

Coming from the north, the first island encountered is Little Island (2 m high), c. 2 km from the mainland near the Perth suburb of Hillarys. Next is Trigg Island (5 m), joined by a short tombolo to the mainland beach. Arthur Head (at the mouth of the Swan River at Fremantle) and Cape Peron (near Rockingham) are former islands joined by natural forces to the mainland coast c. 2000 years ago. Rottnest Island has 47 islets around its shoreline. Between Rottnest and Carnac Islands are situated three vegetated Straggler Rocks (tallest 12 m) and the Mewstone (16 m). Carnac Island has 12 small islets nearby and Garden Island has only eight islets adjacent. South of Cape Peron there are 31 islands in Shoalwater Bay and Warnbro Sound, with Penguin Island the largest and Seal Island (1 ha) the next largest. These islands lie along a north-south line which terminates at The Sisters.

Discovery and naming. The offshore islands of Rottnest, Garden, Carnac and Mewstone are all clearly visible from the mainland, as are the inshore islands situated in Shoalwater Bay and Warnbro Sound. Aboriginal people living on the adjacent mainland probably had names for all of them, but only those of three islands were recorded by European settlers: Wadjemup (Rottnest), Meeandip (Garden), and Ngooloomayup (Carnac). Aboriginal people in the Perth region could not swim and none possessed watercraft, resulting in the islands being unvisited by humans for many millennia. Penguin Island may have been waded to at very low tide.

The first Europeans to visit any of the islands were the Dutch. Samuel Volkersen on the *Waeckende Boey* discovered and landed on Rottnest Island in 1658 but did not name it. He was in search of the *Vergulde Draeck*, wrecked in 1656. The second visit was in 1696 by Willem de Vlamingh on the *Geelvink*, with the *Nijptangh*. These ships were in search of another Dutch ship, the *Ridderschap van Holland*, which failed to arrive at Batavia after sailing from the Cape of Good Hope in 1694 (Chapter 8). De Vlamingh charted a large island, which he named T'Eijlandt Rottenest (Rats' nest, after the quokka).

The next European visitors were the French. *Le Géographe* (Nicolas Baudin) and *Le Naturaliste* (Jacques Hamelin) were at Rottnest in June 1801 and *Le Casuarina* (Louis de Freycinet) in March 1803. The Dutch name, Rottnest, was retained (with a slight change of spelling), following convention, and its eastern and western capes were named for the first time. The other islands were given various informal names, but the final names bestowed were Buache (now known as Garden Island) and Berthollet (Carnac). Rottnest, Buache, and Berthollet Islands as a group were named Les Îles Louis-Napoléon. The Shoalwater Bay islands were collectively called Le Grand Chariot.

The penultimate official visit to Rottnest Island before British settlement at Swan River was by Phillip King RN in January 1822 in HMSV *Bathurst*. James Stirling RN, aboard HMS *Success*, visited both Buache and Rottnest Islands in April 1827. Having grown up in the era of war between England and France, he declined to endorse the French names of the islands, replacing them with Garden Island and Pulo Carnac. Carnac Island was named after an officer on the *Success*. Pulo is the Malay word for island, and was soon discarded. When Stirling returned as Lieutenant-Governor in June 1829 he briefly made Garden Island the site of the first settlement in WA.

The names of the other islands are obviously descriptive: Mewstone (gulls); Penguin (little penguin); Shag (great pied cormorant); Seal (Australian sea lion); and Bird (unspecified seabirds).

PHYSICAL CHARACTERISTICS

Hydrography. Rottnest Island is the oldest island in the region, having been the first to become isolated when the sea level rose following melting of the polar icecaps. When the 7 m bathymetric line was breached c. 7000 years ago, the island ceased to be connected to the mainland. At the 5 m line, all of the smaller satellite islands close to Rottnest, Garden, Carnac, and the Shoalwater Bay islands remained part of either the larger island or mainland. Erosional features at various heights above current sea level indicate that further rises in sea level of nearly 3 m occurred c. 5000 years ago, implying that many small islets were likely to have been submerged or nearly so. Around this time, the western peninsula of Rottnest was an island. There were also eight islands situated in the extensive bay in the north-eastern sector of Rottnest. These were gradually incorporated into the lake system of Rottnest.

Geology, geomorphology, and soil. All of the islands consist of aeolianite (Tamala limestone) up to 20 m thick (Plate 9.2), laid down during the Quaternary era. Only the four largest islands have Holocene sand dunes that were deposited by strong winds. On the two largest islands these dunes reach elevations of 47 m (Rottnest) and 55 m (Garden). Rottnest Island has one unique feature: four large hypersaline lakes with total area of 200 ha, comprising some 10% of the island area. During the Holocene period these lakes were arms of the sea.

Charles Fremantle thought the name 'Garden' Island a misnomer on account of the 'very sandy' soil observed by him in 1829. Its chief merit was firewood 'in great abundance'; otherwise the island was 'fit for no other use'. George Moore visited the island in November 1830 and commented on the 'unpromising appearance of the soil which seemed to be little else than white sand'. He was surprised by the 'tolerably good' production of garden vegetables. In December 1833 Charles von Hügel recorded that it would 'be hard to imagine an uglier, more barren spot' than Garden Island.

Climate. These islands share the Mediterranean-type climate of the south-west mainland of WA. Summers (November – March) are long, dry, and hot: <12 mm (median) of rain falls each month, and the mean daily maximum temperature is c. 25°C. Most rainfall occurs during the period May – August (>100 mm median each month), and the daily minimum temperature is c. 13°C between May and October.

Official data for Rottnest Island, Garden Island, Kwinana, Fremantle, and Swanbourne (coastal weather stations) permit an assessment of similarities and differences between the islands and mainland coast. The high specific heat of the ocean moderates island temperatures. The average maximum temperature on Rottnest Island is c. 2°C lower, and the average minimum temperature is c. 2°C higher than at Swanbourne. The difference between Garden Island and Kwinana is much less (c. 0.5°C), probably because of the proximity of the island to the mainland.

Both islands receive much less rain than the mainland coast: Average annual rainfall is 167 mm less on Rottnest Island and 148 mm less on Garden Island. In conclusion, these islands have a narrower range of daily temperature but are drier than the adjacent mainland coast.

BIOLOGICAL CHARACTERISTICS

Biological discovery. It does not appear that the Dutch and French visits yielded much useful information about the flora and fauna of the islands. The botanist Allan Cunningham, aboard HMSV *Bathurst* in 1822, spent one productive day ashore on Rottnest Island. This yielded specimens of 17 plant species and valuable notes about the vegetation. The German botanist C von Hügel collected plants on Garden and Carnac Islands in 1833, finding two species on the latter 'here and nowhere else'. The naturalists James Drummond, John Gilbert, and Ludwig Preiss collected specimens on Rottnest Island in August 1839. Gilbert also visited Garden Island in August and December 1839. Preiss collected on Carnac Island in November 1839.

Preiss collected c. 50 plant species on Rottnest. A parrot collected by Gilbert only on Rottnest Island in December 1839 was described the next year by John Gould as belonging to a species new to science, the rock parrot. It has the unusual habitat of nesting in holes in aeolianite cliffs. Gilbert obtained specimens of the tammar wallaby on Garden Island and the fairy tern on Rottnest, but missed the red-capped robin on Rottnest Island. About this time William Clark visited the four largest islands and published in a newspaper information of a general nature about some conspicuous species.

There was then a lengthy hiatus until Archibald Campbell visited Rottnest Island in November 1889. He identified the shearwater species breeding at the west end and collected and scientifically described the eggs of the rock parrot. Frederick Whitlock visited in November 1903 and published the first list of bird species for Rottnest Island. The other main island received little attention until Wilfred Alexander recorded bird species on Garden Island in 1919.

Plant lists for were first collated by William McArthur in 1952 (for Rottnest, Garden, and Carnac Islands) and by Glen Storr in 1961 (for Penguin Island). Seven reptile and one frog species were collected on Rottnest (the only island examined) when two German scientists (Wilhelm Michaelsen and Robert Hartmeyer) visited WA in 1905, and most of the invertebrate species collected there proved to be new to science and were later named by experts.

Frequent visits by many scientists and naturalists since the 1930s have led to a sound understanding of the flora and vertebrate fauna of all islands in Perth region. Invertebrates and fungi, however, have yet to achieve the same level of attention and knowledge.

BOX 9.1. Glen Milton Storr (1921-1990)

Glen began his working life in Adelaide as a land surveyor with his hobby being natural history, publishing his first paper at the late age of 26 years. His first association with islands was in 1952 when he assisted Dr Dom Serventy with his shearwater research on Fisher Island, Bass Strait. He settled in Perth WA in 1953, soon published his first paper on WA birds, and began study at UWA, finishing there with a PhD in 1960.

Storr commenced his research on Rottnest Island in 1955 and published important papers on its flora, vegetation, and birds between 1959 and 1965, all incidental to his main focus on the nutrition of the quokka. He also published on the flora of the Shoalwater Bay islands.

He kept meticulous records on his 62 trips to Rottnest Island from 1953 to 1962. His acuity of observation when coupled with his intellect produced several significant conceptual advances in understanding the basis of ecological change on the island.

Between 1955 and 1965 he visited >50 WA islands, and during his career published 24 papers about island species. His scientific reputation rests, however, on his many valid descriptions of WA reptile species and subspecies new to science, and his comprehensive lists of bird species of Queensland, Northern Territory, and the eight regions of WA. These were based on exceptional effort in locating and reading nearly all sources of information then available in scientific journals, books, diaries, and records made by amateur naturalists. He was a polymath with a remarkable ability to synthesise facts into a biogeographical framework of great value in assessing when and where populations of birds have changed in their distribution since European settlement.

One species in particular has been extremely well studied – the famous quokka, a medium-sized marsupial species of great antiquity and endemic to south-west WA. These studies resulted from the appointment of Horace (Harry) Waring as professor of zoology at The University of Western Australia in 1948. Coming from England, he was keen to study marsupials. He discovered that there were few of these species suited to laboratory study still surviving in the metropolitan region. On Rottnest, the quokka was abundant, and the island was easily accessible by ferry from Fremantle. In 1953 a field station was established in disused wartime buildings and so began an era of intensive education and research. Many PhD theses and articles in scientific journals were produced about all aspects of the biology of the quokka, as well as about the ecology of intertidal invertebrates.

Original vegetation and its floristic composition. The only safe anchorage at Rottnest Island for sailing ships was at its eastern end. Consequentially this was the portion usually inspected by the Dutch, French, and English visitors in the 170 years before British settlement. The eastern third of island is mostly well-protected from gales and the deposition of salt spray, and there its vegetation was described by the French in 1801 as 'completely timbered', with the trees 'generally tall'. This is low forest and woodland c. 8 m tall. In 1822 Allan Cunningham noted that the hills surrounding the salt lakes were 'thickly clothed' with pine. In October 1829 Thomas Wilson landed on the south side of Rottnest Island and walked to its centre (Plate 9.3). He recorded sand hills, 'many of which are entirely destitute of any kind of herbage; in the valleys are some stunted trees and shrubs, and a very little grass. It appeared astonishing to us, that Vlaming could speak in raptures of this island, which we found so miserably barren'. In 1839 John Gilbert wrote that the island was 'thickly wooded to the water[']s edge'. By the 1880s much of the interior of the island was dominated by wattle thickets (closed scrub) up to 8 m in height. The vegetation of headlands, which are fully exposed to salt-laden winds, is comprised of salt-tolerant herbs and shrubs.

Garden Island, with its north/south orientation, provides a larger area than Rottnest on its sheltered eastern side, with the result that its low forest/woodland is more extensive. The botanist Charles Fraser visited in April 1827 and recorded that it was 'thickly stocked with Cypress' (Plate 9.4) and 'extensive thickets' of tea tree. In April 1829 Charles Fremantle three times tried to walk over the island 'but from the thickness of the trees and underwood it was impossible to move'; on another occasion it took about one hour to travel c. 1 km. William Breton visited the island in 1829 and wrote that it 'was so densely covered with trees and underwood as to be in many places quite impervious'. Thomas Wilson landed in October 1829 at the southern end with the intent of crossing the island – 'but, from the thickly interwoven underwood, our progress being painful and slow, the attempt was abandoned'. Alexander Collie in 1829 wrote of the trees, thick underwood, and low shrubs 'which are penetrated with some difficulty'. The difficulty of penetrating its vegetation was also experienced by von Hügel in 1833: 'It is densely covered with shrubs tightly entangled with Clematis and a species of *Viscum*. It sometimes takes an hour to cover a few hundred yards'.

Carnac and Penguin Islands consist of low shrubs on aeolianite headlands and wattle scrub on sand dunes where these are topographically sheltered from excessive salt spray. The map in de Freycinet's 1803 field notebook has Carnac Island named as Île Pelée, literally meaning 'bald', doubtless referring to its low vegetation and the lack of trees. The published account uses the term 'barren'.

Three tree species dominate the low forest on Rottnest Island: Rottnest Island pine (6 m), Rottnest teatree (8 m), and weeping pittosporum (3 m). The 'pine' and teatree occur also on Garden Island, whereas the pittosporum is known also (not as a tree) on Garden and Penguin Islands.

The number of vegetation communities on the larger islands is 15 (Garden Island), 12 (Rottnest Island), and 9 (Carnac and Penguin Islands). On the small islets, six well-defined habitat types were recognised by Glen Storr:

(1) Honeycombed rock with soil (if present) restricted to cracks and depressions, with the native plant species *Carpobrotus virescens*, *Nitraria billardierei*, and *Threlkeldia diffusa*.
(2) Presence of talus slope (accumulated rock debris and sand weathered locally from aeolianite). Species as for (1); also *Apium prostratum*, *Calandrinia calyptrata*, *Tetragonia implexicoma*, and *Enchylaena tomentosa*.
(3) Level or gently sloping rock with a thin mantle of soil. Characteristic species: *Carpobrotus virescens*, *Frankenia pauciflora*, *Lepidium foliosum*, *Malva preissiana*, and *Rhagodia baccata*.
(4) Raised beaches on the eastern parts of islets, with *Spinifex longifolius*.
(5) Windward slopes of dunes, with *Alyxia buxifolia*, *Conostylis candicans*, and *Spyridium globulosum*.
(6) Sheltered eastern slopes of dunes, with *Acacia rostellifera*, and *Olearia axillaris*.

The number of habitat types increases with islet area. Most islets (107) have only habitat type (1), seven islands have habitats 1-3 (Plate 9.5), one (Joan) has habitats 1 and 2, and Phillip Rock and two others have habitats 1 and 3. According to this classification, Carnac and Seal Islands have habitat types 1-4 and 6; only Penguin Island has all six habitat types.

Size of the island floras. The first complete list of plant species on Rottnest Island was compiled in 1956 from surveys by the WA Herbarium, with 114 native species recorded. Storr's list of 124 native plant species, published in 1962, was based on several years of study during the 1950s. Re-examination of all records in 1981 yielded 105 plant species. The most recent list, assembled by Elizabeth Rippey and others, comprises 101 native species. The number of native plant species on the remaining large islands is 84 (Garden Island), 43 (Carnac Island), and 40 (Penguin Island).

As expected, the smaller islets (the majority) have few plant species. Those smaller than c. 4600 m^2 and with maximum elevation <3.3 m are unvegetated (no plant species present).

Islet area and elevation jointly determine much of the intensity of exposure to strong wind and sea spray. Small islets sheltered by another, larger, island are protected from extensive wave action, and as a consequence have a greater variety of habitats and larger floras than islets with more exposure. For example, Green Island near Rottnest has 20 species and Middle Shag Island in Shoalwater Bay has 23 species.

BOX 9.2. Elizabeth Rippey (1943-)
After growing up in Zimbabwe, Elizabeth settled in Perth in 1981 and later became interested in coastal plants. This resulted in a publication of a book (co-authored with her botanical-artist mother) on the plant species of the coast and islands of Perth region. A revised and expanded edition was published in 2004.

She attained the degree of PhD in 2003 at UWA, based on her study of the dynamics of vegetation of some of the islands of Perth region. Elizabeth has (with associates) compiled herbaria (with vouchered specimens) for Rottnest Island, Garden Island, and the Shoalwater Bay islands. Her papers in scientific journals document in detail the changes in the flora of the Shoalwater Bay islands, provide a detailed appraisal of changes in the flora of Rottnest Island, and assess the impact of fire and browsing of plant species by quokkas on the vegetation of Rottnest Island.

Terrestrial invertebrates. The expedition conducted by Michaelsen and Hartmeyer in 1905 was the first to document the species of arthropods and earthworms occurring on any WA island – in this case, Rottnest Island. Their collections were worked on by taxonomists, and many species new to science were recognised. Since then additional species of invertebrates have been named, including ants, dragonflies, lacewings, ticks, and spiders. Butterflies, however, have been sampled the most comprehensively, with 15 and 13 native species recorded on Rottnest and Garden Islands respectively.

Frogs. Rottnest Island is the only island with amphibians (three native species). The other islands lack swamps.

Reptiles. These (geckoes, skinks, snakes) are well represented: Rottnest: 16 native species; Garden 14 species; Carnac 4 species. The largest lizard species is King's skink, which is much commoner on the larger islands than the adjacent mainland coast. This difference may be attributed to the abundance of seabird eggs on these islands, and their absence on the mainland coast. This species preys on seabird eggs and thus occurs on several small islands, including Dyer and Penguin Islands.

In 1977 Paul Coster studied the lizards present on many of the larger islands, and found that they were most abundant on Rottnest and Garden islands, followed by Penguin, Green, and

Parrakeet Islands. One individual marbled gecko was found in September 1977 on the first small (450 m^2) islet south of Penguin Island – how did it get there?

On his visit to Rottnest Island in November 1889, Archibald Campbell stated that c. 160 dugites had so far been killed in spring. Aboriginal prisoners received half a stick of tobacco for every snake killed. The carpet python on Garden Island must have been common in 1829, as both Charles Fremantle and Augustus Gilbert recorded it. This population exhibits striking sexual dimorphism in maximum length (male 159 mm, female 231 mm) and body mass (male 1.2 kg, female 5.2 kg). As a consequence the mammalian portion of their diet differs: Only females can eat tammar wallaby, whereas males eat the (introduced) house mouse. Both sexes also capture and eat lizards and birds.

Garden Island also has a population of tiger snakes, as does Carnac Island. It is uncertain if the latter population is indigenous or introduced. Some naturalists regard the placidity of the snakes there as evidence of indigeneity. The only record of snakes is from newspaper reports published in 1878 and 1882. The weight of negative evidence, however, is not suggestive that the occurrence of tiger snakes is natural. Charles Fraser in April 1827 and Charles Fremantle in May 1829 visited Carnac Island and recorded the species encountered. In June 1829 the ship *Parmelia* (conveying settlers from Britain) went aground and was damaged when it attempted to enter Cockburn Sound by traversing the unsurveyed passage between Carnac and Garden Islands. As an emergency measure the 28 female passengers were landed on Carnac Island for five days. William Clark spent a fortnight there in 1831, detained by a winter gale, and lived on fish and penguins. George Moore and Charles von Hügel visited the island in February and December 1833 respectively. Rabbits were released on the island in 1842 and shooters visited for sporting purposes. Whaling (1838-9) and quarantine stations (1892-8) also operated briefly. In 1909, 30 naturalists spent one night and a day on the island. No snakes were reported on any of the above occasions. In 1981 information became available that 70-100 snakes were released in c. 1929 by Rocky Vane, who exhibited venomous snakes to the public in Perth. The origin of the Carnac Island population of tiger snakes - indigenous or introduced - remains unresolved at present.

Land birds. Fifteen species originally occurred as breeding members of the land bird fauna of Rottnest Island: one species each of quail, kestrel, pigeon, parrot, cuckoo, swallow, pipit, robin, scrubwren, honeyeater, chat, silvereye, and raven, and two species of whistler. In 1801 the French recorded 'partridges [probably the brush bronzewing] and crows'.

The original status of the land bird species of Garden Island is imperfectly known but was probably similar to that of Rottnest. The small islands of Carnac and Penguin have been too much disturbed since the 1830s to infer with confidence the identity of the species breeding there then, other than the silvereye. In 1829 Charles Fremantle shot 'a bird like a partridge' on Carnac Island and saw two brace of quail on Garden Island, and Augustus Gilbert observed a pigeon, kingfisher, and small parrot on Garden Island in the same year.

Silvereyes move in flocks, and will visit islands to feed on the small soft fruits of *Rhagodia baccata*, *Threlkeldia diffusa*, and *Nitraria billardierei*, species that occur on small islets. Silvereyes also glean foliage for small invertebrates. Carnac Island was used during the 1970s for a study of the abundance of silvereyes throughout the year. Most birds were captured in January, April/May, and November. Nesting occurred. Some birds were re-trapped on the island up to 15 months after being leg-banded, but it could not be established if these birds had moved to the mainland and returned. A similar study on Rottnest Island and Bold Park on the adjacent mainland has demonstrated movement between these localities. On Rottnest Island, mark and recapture studies have revealed that some live for 6-8 years.

Marine raptors. The eastern osprey is the only species present on the islands and islets; the white-bellied sea-eagle has never been recorded as breeding there. Neither species is regarded as a seabird even though they subsist largely on fish. The sea-eagle is also capable of preying on quokkas and tammars. In 1839 the osprey was, according to John Gilbert, 'most common' in Perth region on Rottnest Island. In 1889 it was reported as preying on poultry there. In the early 1960s only three pairs nested on Rottnest Island. During the period 1977-1984, four islets were used for nesting, but only three of them were in use in 1977. There are now 14 nesting sites of the osprey, with only 4-5 nests used each year. Garden Island had at least one active nest in the 1970s; in the 1990s there were 2-3 nests in use.

Shorebirds and waterbirds. In 1839, curlew sandpipers, sharp-tailed sandpipers, and ruddy turnstones were at times abundant and Australian shelducks, Pacific black ducks, and banded stilts ('snipe', Plate 9.6) were reported in large numbers during the 1830s on Rottnest Island, which has extensive areas of habitat suited to their foraging. Most species are non-breeding visitors. Only the shelduck, pied and sooty oystercatchers, red-capped plover, and red-necked avocet regularly breed there.

Bi-annual surveys of trans-equatorial waders on Rottnest Island from 1999 to 2020 have revealed considerable variation in their abundance from year to year. Two species (red-necked stint, sanderling) ostensibly increased, nine species (bar-tailed godwit, grey-tailed tattler, curlew sandpiper, ruddy turnstone, greater and lesser sandpiper, grey plover, common sandpiper, terek sandpiper) decreased, and three species (whimbrel, red-necked phalarope, common greenshank) varied little. Declines have resulted from loss of habitat between Australia and the northern hemisphere breeding grounds of these species.

Seabirds. Twelve species breed on these islands. The species that attracted most comment from the early visitors was the wedge-tailed shearwater, which breeds on Carnac, Rottnest, and some satellite islets. De Vlamingh in 1696 found the soil on Rottnest to be perforated in 'every direction with burrows', and Cunningham in 1822 recorded that the soil to be 'perforated in every direction as well perpendicularly and horizontally into long Burrows'. The agent remains unidentified, but may refer to a (now extinct) population of this species at the eastern end of the island. Perhaps these were eaten by Aboriginal prisoners.

In contrast, John Gilbert made no comment about their occurrence on Rottnest Island in 1839, despite visiting the west end. It was not until 1889 that the species was verified, when the ornithologist Archibald Campbell visited the west end. Since the 1950s, the total number of burrows there has increased from 590 (in 1956) to 2630 (in 1982), and then to 5865 (in 1993) and 11 745 (in 2002).

On Carnac Island, Charles Fraser in 1827 recorded 'mutton birds', Charles Fremantle in 1829 recorded great pied cormorants and little penguins, George Moore in 1833 recorded 'mutton birds' and their burrows, and Charles von Hügel later in the same year also observed the burrows and eggs of 'a species of seagull' [wedge-tailed shearwater].

The little penguin is known to breed on Penguin (c. 700 pairs), Garden (number unknown), and Carnac (c. 80 pairs) islands – the northernmost breeding occurrence on the west coast of Australia. Other breeding seabird species are the little shearwater (uncommon on Penguin I.), white-faced storm-petrel (uncommon on Bird Island and formerly breeding on Seal Island), great pied cormorant (abundant on small islands around Rottnest Island, Carnac Island, and islands in Shoalwater Bay south to The Sisters), silver gull (many islands, with the largest colonies on Carnac (Plate 9.7), Seal, and Penguin Islands), Pacific gull (uncommon on many islands in the 1830s, but now only a visitor), Caspian tern (uncommon on Carnac and Bird Islands), greater crested tern (common on Carnac and Seal Islands), bridled tern (now on many islands, but first recorded in the region in 1889 on Phillip Rock by Archibald Campbell), and fairy tern. Australian pelicans and fairy terns first bred on Penguin Island in 1998 and 2019 respectively, and have continued to do so. In addition, colonial surface-nesting species often vacate their nesting areas and move to a different part of the island or to another island. In 1921 a pair of red-tailed tropicbirds successfully bred for the first time on Rottnest Island.

Land mammals. Only two native species occur: the quokka on Rottnest Island (Plate 9.8), and the tammar on Garden Island. The quokka was the second marsupial species to become known (in 1658) to Europeans. It has since become the natural icon of Rottnest Island, internationally known through selfies taken of it by celebrities. The Dutch in the 1600s found the animal to be common, judged by its faecal pellets. The French in 1801 found it 'very numerous'. Much of our scientific knowledge of the species has, however, been accrued only during the past 70 years. Environmental conditions on the island have changed considerably since the 1800s, and so it is difficult to ascertain the quokka's ecology before Europeans intruded.

Its home range is c. 1.9 ha, though there is 23% overlap in home range between individuals. The average population size is c. 8000, but there is marked seasonal variation (least in summer [5000], highest during the rest of the year [12 000]). On Rottnest, quokkas breed only once a year, in summer-autumn. They browse vegetation mainly at night and spend the day resting and sleeping in the shade under vegetation. Rottnest quokkas are slightly smaller than those on the mainland and have much reduced genetic variation, expressed as low heterozygosity.

The tammar wallaby was the first marsupial species to become known (in 1629) to Europeans, but that was on islands in Houtman Abrolhos (see Chapter 8). There are estimated to be c. 5000 tammars on Garden Island. Charles Fremantle in May 1829 was the first to record on Garden Island a small 'kangaroo...not larger than a hare'; Augustus Gilbert observed a 'kangaroo rat' there in June 1829; and in June and August 1829 Alexander Collie found 'a small species of Kangaroo' in 'very great numbers'. He examined the pouch of several females and collected evidence from other naval officers that the young 'crept among the fur of the mother towards her belly and towards the opening of the abdominal pouch...its progress...was about as expeditious as that of a snail'. This was the first published account of the mode of reproduction characteristic of marsupials.

In 1801 the French also observed on Rottnest Island a second species of mammal, 'the size of a very large rat'. There is no subsequent record of it, and presumably it refers to a juvenile quokka.

Marine mammals. Seals were seen on the three largest islands between 1658 and 1829. Rottnest Island: 'two seals' (1658); 'a very few' (1696); 'very great numbers ... on the various sandy beaches' (1801); 'seen on the Rocks, & several were observed in the shoal water in the extensive sweep of sandy beach' (1822); Garden Island: 'covered with a great many seals' which are 'present in far greater numbers than on Rottnest Island' (1801); at least 34 (1829); Carnac Island: abundant (1827); at least 13 (1829).

Seals 'sometimes go quite a long way into the forests' (1801) and had formed well-defined paths on Rottnest Island (1822), were present in the shore vegetation on Garden Island (1829), and occurred all over Carnac Island (1829). This behaviour is typical of the Australian sea lion and not the long-nosed fur seal (which the French identified on Rottnest Island in 1801 based on black pups). It is likely that sea lions bred on all three islands, but this was specifically described only on Rottnest Island (1801). The total population before Europeans arrived is estimated at 400-500 animals. In 2014 fur seals colonised Cathedral Rocks near

Rottnest Island. This population fluctuates between 10-80 individuals but breeding has not yet been observed.

Speciation. Having been isolated for only c. 7000 years, Rottnest and Garden Islands are not expected to have evolved any distinct species, endemic to one or both islands. One species on Rottnest Island, the singing honeyeater, was named as a new species in 1911 on the basis of its large size (c. 20% heavier than on the mainland) and duskier coloration. This is no longer accepted as valid. Subspecies have been described for six reptile species on Rottnest Island, but only two of these are currently recognised (dugite and bobtail).

Biogeographic curiosities. Why should the two largest islands differ so markedly in the composition of their floras and faunas? They have similar-sized floras but c. 39 native plant species are found on Rottnest but not Garden Island and c. 25 are found on Garden Island but not Rottnest. Striking examples include *Acacia cochlearis*, *Exocarpos sparteus*, *Hardenbergia comptoniana* and *Melaleuca huegelii* only on Garden Island, and *Acacia littorea*, *Spyridium globulosum*, and *Templetonia retusa* only on Rottnest Island. (The occurrence of several samphire and sedge species only on Rottnest is probably linked to the occurrence of the salt lakes.) There are five species of butterflies only on Garden Island and seven species only on Rottnest Island. Garden Island has five reptile species not on Rottnest, and there are seven species on Rottnest that are absent from Garden Island. Rottnest Island has 14 land bird species in comparison to the 13 species on Garden Island, which lacks the red-capped robin, rufous whistler, and spotted scrubwren. Garden Island also lacks the wedge-tailed shearwater and breeding populations of the great pied cormorant. Garden Island has the grey fantail, willie wagtail and grey butcherbird, which are absent as breeding species from Rottnest Island.

Species turnover. Because of their proximity to Perth, most of these islands have been visited many times by scientists, resulting in exceptional understanding of the degree of natural change in the species composition of the native biota.

Native plant species. A reasonable sample of the small islets near Rottnest Island (28 islets) were visited in 1975, 1977, and 1978. During the period 1975-77, immigrations of species predominated, and this was reversed in the period 1977-78. Most of these local extinctions and immigrations involved annual species. On three islands both extinctions and immigrations occurred. The species composition remained the same on 13 islets (28% of the sample studied).

Native reptile species. The skink *Egernia napoleonis* has not been recorded on Rottnest Island since 1986, and a detailed survey in 2002 failed to find it. It is unusual for reptiles to become extinct on Australian islands.

Native land bird species. On Rottnest Island the brush bronzewing and a species of quail ('in large coveys') were last reported in the period 1870s-1920s and in 1831, respectively. Nine species have immigrated or if previously a seasonal visitor have become breeding species: banded plover (c. 1934); western gerygone (c. 1950); fantailed cuckoo (1960s); rainbow bee-eater (1977); sacred kingfisher (1980s); tree martin (c. 1983); galah (c. 1998); painted button-quail (c. 2002); and brown goshawk (2011). On Garden Island a quail species was last reported in 1829. Garden Island has gained six species, namely the banded plover, rainbow bee-eater, and Australian magpie in the 1970s, painted button-quail (1985), and magpie-lark and black-faced cuckoo-shrike in 1996. The buff-banded rail (treated here as a land bird) has been present on Penguin Island since 1977 and breeds there. It first bred on Rottnest Island in c. 2002.

Native aquatic species. The nankeen night heron bred on Bird Island, but become locally extinct there in the period 1900-1920s. This species was recorded breeding in caverns on Carnac Island in December 1839, but this is no longer the case. The Australian pelican began breeding on Penguin Island in 1998.

Seabird species. In c. 2009 the little penguin began breeding in a rockwall at the naval base on Garden Island. There has been a natural expansion in the geographic range of the bridled and roseate terns, with the former first recorded on Rottnest Island in 1889, and the latter on Rottnest Island and islands in Shoalwater Bay in recent decades. During the 1800s the Pacific gull occurred along the coast (and the islands) of the metropolitan region 'but nowhere in any considerable numbers', according to John Gilbert in 1839. It is now recorded only as a vagrant.

Native mammal species. There has been no change with the two marsupial species. The Australian sea lion no longer breeds on any island in the region; all individuals (mostly on Carnac, Seal, and Little Islands) are males and travel to some of the islands of the inner mid-west to breed (Chapter 7).

FACTORS LEADING TO CHANGE IN THE BIOTA

Natural factors. Studies of pollen deposits on Rottnest Island (the only WA island so far examined in this respect) have revealed that species typical of tuart woodland, including tuart, jarrah, marri, *Allocasuarina/Casuarina* spp., *Agonis flexuosa*, *Banksia* spp., *Macrozamia fraseri*, and *Xanthorrhoea preissii*, occurred there c. 7000 years ago. Around 5000 years ago the island became drier. Exposure to salt-bearing winds from all sides intensified with complete separation from the mainland, killing foliage of susceptible species, and increasing chloride levels in soil and thus gradually modifying plant species composition and decreasing the height of vegetation. On exposed sites on Rottnest Island, individual tea trees have been bent towards the north by these winds.

The lack of mammal predators (such as chuditch) would have facilitated establishment of breeding populations of seabirds, which then altered the chemistry of soil. The acidic guano of nesting seabirds is unfavourable to the persistence of sclerophyllous vegetation (which requires low levels of phosphates and nitrates) and advantageous to coprophilous plant species. Succulent and annual species thrived, with many such species benefiting from the combination of high levels of chloride, phosphate, and nitrate in soil.

Another significant factor was the absence of Aboriginal people. These islands therefore did not experience the fire regime they applied to the adjacent mainland coast. This included deployment of fire in the warmer months, in patches of various sizes, with patches burnt every 3-10 years. The only source of fire on the islands was from lightning strikes. It is thought that the natural interval between fires on these islands may have been as long as a century, and when they occurred they would have been intense and burnt a large area in a short time. In April 1829 Charles Fremantle reported that Garden Island appeared to have 'been burnt all over, as there is a quantity of young wood [*Callitris preissii*] coming up'. A study of growth rings of this species in the 1960s revealed two age groups, 50-60 years and 100 years, suggestive of wildfire in the 1910s and 1860s respectively.

The Leeuwin Current is a natural phenomenon that is known to influence the abundance of plankton and fish along the coast of WA. During years when the current runs strongly it brings warm tropical water low in nutrients, plankton and fish become less common, and this impacts on the population of seabirds. The population of little penguins breeding on Penguin Island has been well studied. The timing and strength of the Leeuwin Current affects surface sea temperature, and both factors correlate with breeding success (number of chicks produced, first laying date, length of the breeding season, and the proportion of eggs that hatch and chicks which fledge).

Human factors. The arrival of European visitors and the settlers was the beginning of transformational change on the islands. There were both direct and indirect impacts, with some having immediate effect and others more protracted.

Fauna. The first immediate impact was the shooting of sea lions by Dutch, French, and English navigators between 1658 and 1829, including 'large numbers' slaughtered on Garden Island in 1801 for their well-tasting fat. In April-May 1829, 34 were killed there and another 13 were killed on Carnac Island for their oil. An 1831 map of Rottnest has a lake named 'Sealer's Lake' (now Lake Bagdad), suggestive of seal-hunting there in the 1820s. Sea lions have returned to Carnac Island, but no longer breed there or on any other island in the region.

Shooting of fauna was a popular activity during the 1800s. In 1827, skins of 'immense numbers' of quokka were obtained on Rottnest Island. These made 'excellent fur pockets or rugs'. The state governor holidayed on Rottnest Island from the 1850s and indulged in shooting quail, pigeons, banded stilts, ducks, and quokkas.

In 1827 James Stirling released one cow, two ewes in lamb, and three goats on Garden Island. None survived to the next visit in 1829. The careening of the damaged *Parmelia* in 1829 at Garden Island provided the opportunity for any house mice or black rats aboard to make landfall – the first possible arrival of exotic rodents on a WA island. Only the house mouse has persisted there. Careening Bay was the main harbour for Perth until the one at Fremantle was constructed in 1897. Garden Island served as a temporary official settlement (the first free settlement in Australia) for several months in 1829, and unspecified numbers of cows, goats, sheep, horses, rabbits, sows, geese, turkeys, ducks, fowls, and at least one canary, dog, and cat were landed there but apparently not permanently.

Rottnest Island has a long history of introduced animals from the 1870s until the 1930s: rock dove, California quail, helmet guineafowl, peafowl, ring-necked pheasant and even malleefowl. Only the pheasant has persisted as a breeding species. The peafowl has been reduced to a few males. Turkeys were introduced in 1946 to Garden Island but were shot out in 1958. The spotted dove, laughing dove, and rock dove, released around Perth in c. 1900, had self-introduced in the 1930s and 1950s to some of the islands. Wallabies (species unknown) were liberated on Penguin Island in 1918. As late as 1958, a retired farmer put 150 sheep briefly on Garden Island.

The earthworm *Microscolex dubius*, introduced into south-west WA in the 1800s, was first found on Rottnest Island in 1905. The Mediterranean snail was introduced (surely accidentally) to Rottnest Island in the 1920s. The area occupied increased from 69 ha in 1936, reaching its maximum in 1978 (687 ha). Since then its distribution has declined.

Rabbits were released on Carnac Island in 1842 and persisted until c. 1897, and were re-introduced in 1965 and eradicated in 1969. Their precise impact on palatable native plant species is unknown, but they did interfere with little penguin breeding burrows. Rabbits were also put on Penguin Island in c. 1933 but became extinct in c. 1952, possibly from myxomatosis. House cats were present on Penguin Island in 1986 but were removed when leases for holiday dwellings were cancelled.

Local extinctions and establishment of native bird species on these islands are as itemised previously in this chapter. The rock parrot population on Rottnest Island is perilously close to extinction. In 2012 only seven birds were detected, down from flocks of many dozens in the 1930s. This may be attributable to trapping of birds in the past.

Flora. In 1827 the botanist Charles Fraser sowed seeds of garden plants and bananas on Garden Island, and Charles Fremantle in May 1829 planted grape vine cuttings and seeds of mustard, cress, and lettuce there, but was not confident that they would thrive. None persisted. The most remarkable change in flora on all islands is, however, the establishment of exotic plant species characteristic of Britain and southern Africa. On Rottnest Island many were deliberately introduced (in gardens and farmland), but most were self-introduced. Rottnest Island now has some 80 introduced plant species. Originally it had one native annual grass species; now it has 28 exotic grass species. The flora of the islands in Shoalwater Bay have been severely degraded, with decline of native perennial plant species and replacement by exotic species. The proportion of exotic plant species present is as follows: Bird Island (79%); Seal Island (53%); Penguin Island (51%); and Middle Shag Island (50%).

Silver gulls are mostly responsible for introducing seeds of exotic annual plants to islands; these seeds are ejected from the crop in viable form onto soil already cleared of vegetation by trampling and manuring for their reception. The gull population has increased greatly since 1829, particularly on Carnac, Seal, and Penguin Islands as a result of birds crossing to and from the mainland, where they preferentially feed around rubbish dumps and other areas with exotic plant species. In the 1940s-50s only small colonies were present on Penguin Island, subsequently increasing to c. 1200 pairs (1978) and 3 000 pairs (1992).

Change in predation and fire regimes. Hunting became frequent on Rottnest Island from 1839 when the island became a prison for male Aborigines, who were allowed to hunt for fresh meat (quokka) on Sundays. At this time c. 70% of the island was dominated by wattle scrub, and there was little ground cover and hence little food for quokkas. The quokka population was therefore likely to have been low. The governor established a summer retreat in 1848 and during later decades led hunting parties for sport; this was a new form of predation. The prison closed in 1903, the island became available for tourism, and commercial ferry services commenced. In 1917 Rottnest Island advertised 'splendid shooting' for tourists. However, reduction in predation pressure during the 1920s is thought to have led to a rapid increase in quokka numbers by the 1930s.

Fire also became more frequent, probably from negligence. Wattle stands become senescent after c. 20-30 years, resulting in much dead wood and thus extensive fuel awaiting ignition. In 1909 a 'large' fire started near the main lighthouse but could not be controlled. In 1955 a wildfire incinerated 730 ha. As the quokka population recovered in numbers, they ate all wattle seedlings and suckers. Unpalatable species, such as *Acanthocarpus preissii*, increased in abundance, leading to an increase in low dense heath that now covers c. 30% of the island. In 1997 another wildfire occurred, with 90 ha burnt, and this provided another opportunity for study of the impact of quokkas on regenerating vegetation.

A bushfire on Penguin Island in 1918 destroyed penguin nesting sites. Wildfire on Garden Island in January 1956 burnt c. 70% of the island.

Climatic change. Increasing sea surface temperatures are likely to impact on breeding success of some seabird species. Little penguins tend to produce fewer fledglings per pair in years when water temperature is high for the three-month period before breeding in February March. High wind speed during breeding also reduces breeding success. On land little penguins become stressed once ambient temperature exceeds 30°C.

Habitat destruction. Settlement of Rottnest Island in the 1830s initiated clearing of native vegetation for farming, roads, firewood, and buildings, though at a small scale. There were once eight fresh-water swamps; all but three were excavated during the 1970s as a source of material to construct roads. Now only one (Barker Swamp) remains. New tourism settlements in the 1970s further destroyed vegetation. In 1907 the southern portion of Garden Island was advertised for sale as 'the only freehold island subdivision' in WA, but this did not proceed. It was not until the establishment of a naval base on Garden Island from 1973 that major habitat destruction occurred, with c. 30% of the original vegetation affected.

Ecological restoration. This does not include the planting of exotic shade trees, a form of landscape 'beautification' which began in the 1880s. Tuarts were planted on Rottnest Island in the 1930s-40s in a misguided attempt to create woodland, oblivious to the three tree species genuinely native to the island. The Forests Department advised about afforestation in the 1970s, but few of the tree species recommended were of local provenance. This policy has changed in recent decades with the planting of local species in fenced blocks so as to exclude quokka browsing. These blocks are visible in satellite images (Locate 2021). However, only 64 ha of restored and natural vegetation occur.

Another misguided attempt at restoration involved the planting in the 1980s of the exotic marram grass on Penguin Island. The objective was to stabilise dunes, and this was successful. On Garden Island >50 000 native plants have been established to stabilise soil and enhance wildlife habitat. Arum lily was first recorded on this island in 1978; in 1997 it was reported as 'the gravest weed threat' there. It forms dense monocultures 1 m tall, and its seed is spread mainly by silvereyes. About $1M were spent spraying the lily between 1989 and 1996. It seems likely that biological control will be necessary.

Progress has been made on Penguin Island with removing weeds, but replanting with the native *Rhagodia baccata* was hampered when silver gulls pulled out the root stock. It was found that covering the planted area with mesh and *Rhagodia* branches with ripe fruit was superior to direct seeding. There are also plans to remove the aggressive weed *Malva arborea* from the Shoalwater Bay islands and re-establish the native species *Malva preissiana*.

The installation of roof tiles on Duck Rock (near Rottnest Island) to provide nest-shelters for rock parrots succeeded. During the period 2015-17 13 nestlings fledged. It does not seem that the eradication of feral cats on Rottnest Island in 2002 has increased the population of rock parrots on the main island. There were only four animals, two of each sex, present in 2002.

Island management. Rottnest, Garden, and Carnac Islands have separate management plans. A third plan includes all islands in Shoalwater Bay. All but that for Rottnest Island have expired and have not been replaced. These plans specify the aims and means of limiting impact by visitors on the island biota without impairing their experience of the scenic beauties and other attractions. This is of course a delicate balance. Garden Island is a Commonwealth of Australia-owned island, and has its own management to minimise the impact of workers at the naval base on the natural values of the island. A State government ranger is also based at the island.

It is legitimate to view tourism as a potential threat to some island species and island values (Plates 9.9-9.11). In 2017 nearly 750 000 people visited Rottnest Island. It is evident from the current management plan that Rottnest Island is managed primarily for visitors and not for its biota. On Garden Island the naval authorities only provide the public with limited access – by boat during the day to northern parts (no camping), and no access via the causeway linking the island to the mainland. Nevertheless, 72% of the island remains accessible to the public. Garden Island is unique in the sense that the naval base has 500 resident sailors; no pets or potted plants are permitted.

Quokkas have benefited from tourism on Rottnest Island. Their body condition is better in the settled sector as more food is available there, reproductive success is improved, and the population (15 animals/ha) is five times more dense than in woodland. Buff-banded rails and King's skinks forage in the picnic area on Penguin Island in the presence of tourists (c. 138 000 visitors in 2014-15). Since 1995 this island has operated a penguin-viewing facility (Plate 9.12), with adult birds that have been rehabilitated from illness and injury or chicks that were abandoned but unable to be released into the wild.

On Penguin Island, the pre-existing sand walking track was replaced in 2010 by a boardwalk 180 m in length and with two lookouts (Plate 9.13). Further erosion of the sand dunes should thus be prevented. This small island also has areas of nesting seabirds to which visitors are denied entry. In order to limit damage to dune vegetation, a seal-viewing platform near Cathedral Rocks, Rottnest Island, was erected in 2014.

Monitoring of species is a task that is not always well-performed. In the case of Penguin Island, however, the response to the arrival of black rats in 2011 was exemplary. It was soon noticed that the rodents were killing lizards and seabirds, resulting in the entire island being baited in 2013. This successfully eradicated the rats, and the lizards and seabirds have since regained their previous abundance.

Sources of Information

Introduction: Abbott 1977a; Australian Pilot 1972; Locate 2021.
Discovery and naming: Appleyard & Manford 1980; Murray & Hercock 2008; Robert 1972; Stirling 1827; Zuiderbaan 1977.
Hydrography: Department of Planning & Infrastructure 2002; Department of Transport WA 1995; Royal Australian Navy 1967a.
Geology, geomorphology, and soils: Playford *et al.* 1977; Searle *et al.* 1988; Storr 1961; Wilde & Low 1980.
Climate: BOM 2021.
Biological discovery: Fisher 1992; J Gilbert undated; Hallermann 2020; Lehmann 1844-47; Michaelsen & Hartmeyer 1907-1930; Orchard & Orchard 2020; Péron & de Freycinet 1824; Robert 1972; Zuiderbaan 1977.
Original vegetation and its floristic composition: Abbott & Wills 2016; Clark 1994; Cottesloe 1979; Marchant & Abbott 1981; Orchard & Orchard 2018; Rippey & Rowland 1995; Rippey *et al.* 2002; Wilson 1835.
Size of the island floras: Abbott 1977a; Abbott & Black 1980; McArthur 1957; E Rippey pers. comm.; Rippey *et al.* 2003; Storr 1961, 1962.
Terrestrial invertebrates: Abbott & Wills 2016.
Frogs: Abbott & Wills 2016.
Reptiles: Abbott & Wills 2016; Abjornsson 1910; Anon. 1882; Appleyard & Manford 1980; Brooker *et al.*2006; Cameron 2006; Campbell 1890; Clark 1842; Clark 1994; Fraser 1830; Pearson *et al.* 2002; Pearson & Bonnet 2007; Péron & de Freycinet 1824.
Land birds: Abbott & Wills 2016; Brooker *et al.* 1995; Campbell 1890; Clark 1842; AH Gilbert 1829; Péron & de Freycinet 1824; Sansom & Blythman 2015; Saunders & de Rebeira 1993; Storr 1964-5.
Marine raptors: Abbott 1977b; Campbell 1890; Holsworth 1965; Singor 2017; Storr 1964-5.
Shorebirds and waterbirds: Cameron 2006; Campbell 1890; Clark 1842; Clark 1994; Mather 2020; Storr 1964-5.
Seabirds: Bancroft *et al.* 2004; Bell 2021; Cameron 2006; Campbell 1890; Clark 1842; Clark 1994; Fraser 1830; Greenwell *et al.* 2020; Rippey *et al.* 2002; Robert 1972; Storr 1964-5.
Land mammals: Collie 1830; De Barros & Coughran 2017; AH Gilbert 1829; Main *et al.* 1959; Orchard & Orchard 2018; Phillips 2016; Waring 1956.
Marine mammals: Campbell 2005; Clark 1842; Cottesloe 1979; Fraser 1830; AH Gilbert 1829; Péron & de Freycinet 1824; Stirling 1827; R Gabbitus (Rottnest Island Authority) pers. comm.
Speciation: Abbott & Wills 2016.
Biogeographic curiosities: Abbott & Wills 2016; Williams 1997.
Species turnover: Abbott 1977a; Abbott & Black 1980; Abbott & Wills 2016; Abbott *et al.* 2000;
Factors leading to change in the biota. Natural factors: I Abbott per. obs.; Backhouse 1993; Cannell *et al.* 2012; Churchill1960; Storr 1963.
Factors leading to change in the biota. Human factors: Abbott *et al.*2000; Errington 2012; Fraser 1830; Hercock 1998; Johnson & Colombelli-Négrel 2021; Joske *et al.* 1997; Pen & Green 1983; Rippey & Hobbs 2003; Rippey *et al.* 1998; Storr *et al.* 1959.
Ecological restoration: Algar 2011; Edmiston & White 1974; Labbé *et al.* 2017; Sansom *et al.* 2019.
Island management: Department of Environment and Conservation 2007; Department of Defence 1993; Rottnest Island Authority 2020.

Plate 9.1. Aerial view of the lakes (and airstrip) situated on the eastern part of Rottnest Island. Credit: I Abbott.

Plate 9.2. Rottnest Island, showing aeolianite bedding at Fish Hook Bay. Credit: I Abbott.

Plate 9.3. Typical vegetation of Rottnest Island. Foreground, wattle scrub; planted tuart and moonah along roads. December 2003. Credit: AA Burbidge.

Plate 9.4. Roadside view of *Callitris preissii* forest on Garden Island. Credit: I Abbott.

Plate 9.5. Green Island, a well-vegetated aeolianite stack 6 m high and 0.12 ha in area, and sheltered from ocean swell by Rottnest Island. Credit: I Abbott.

Plate 9.6. Banded stilts. June 2019. Credit: AA Burbidge.

Plate 9.7. Silver gulls on Carnac Island. Credit: I Abbott.

Plate 9.8. Quokka on Rottnest Island. December 2014. Credit: AA Burbidge.

Plate 9.9. View northward from summit of Penguin Island, showing blowouts. 1974. Credit: I Abbott.

Plate 9.10. Penguin Island. April 2009. Credit: AA Burbidge.

Plate 9.11. Picnic area, Penguin Island. April 2009. Credit: AA Burbidge.

Plate 9.12. Penguin Discovery Centre, showing little penguins. April 2009. Credit: AA Burbidge.

Plate 9.13. Boardwalk and Australian pelican colony, Penguin Island. December 2021.The boardwalks minimise vegetation destruction by visitors; however, the pelican breeding colony results in natural vegetation destruction. Credit: I Abbott.

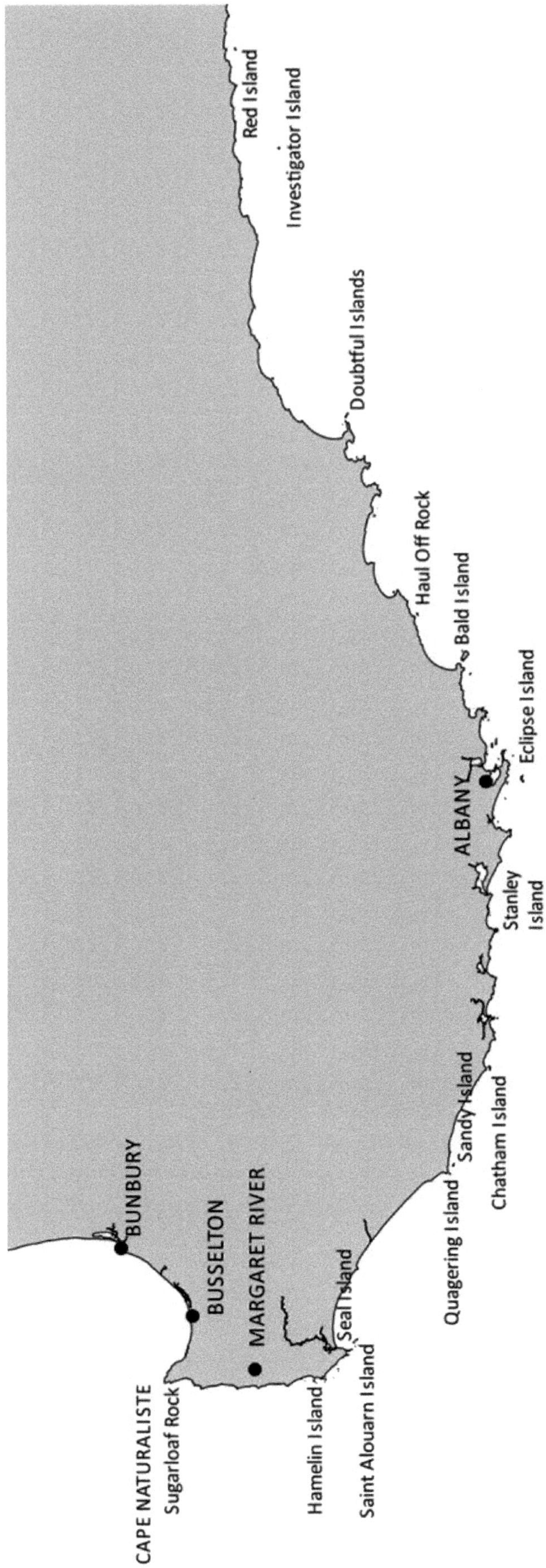

Map 10.1. Islands of the South-west corner Western Australia, from Cape Naturaliste to Red Island (East).

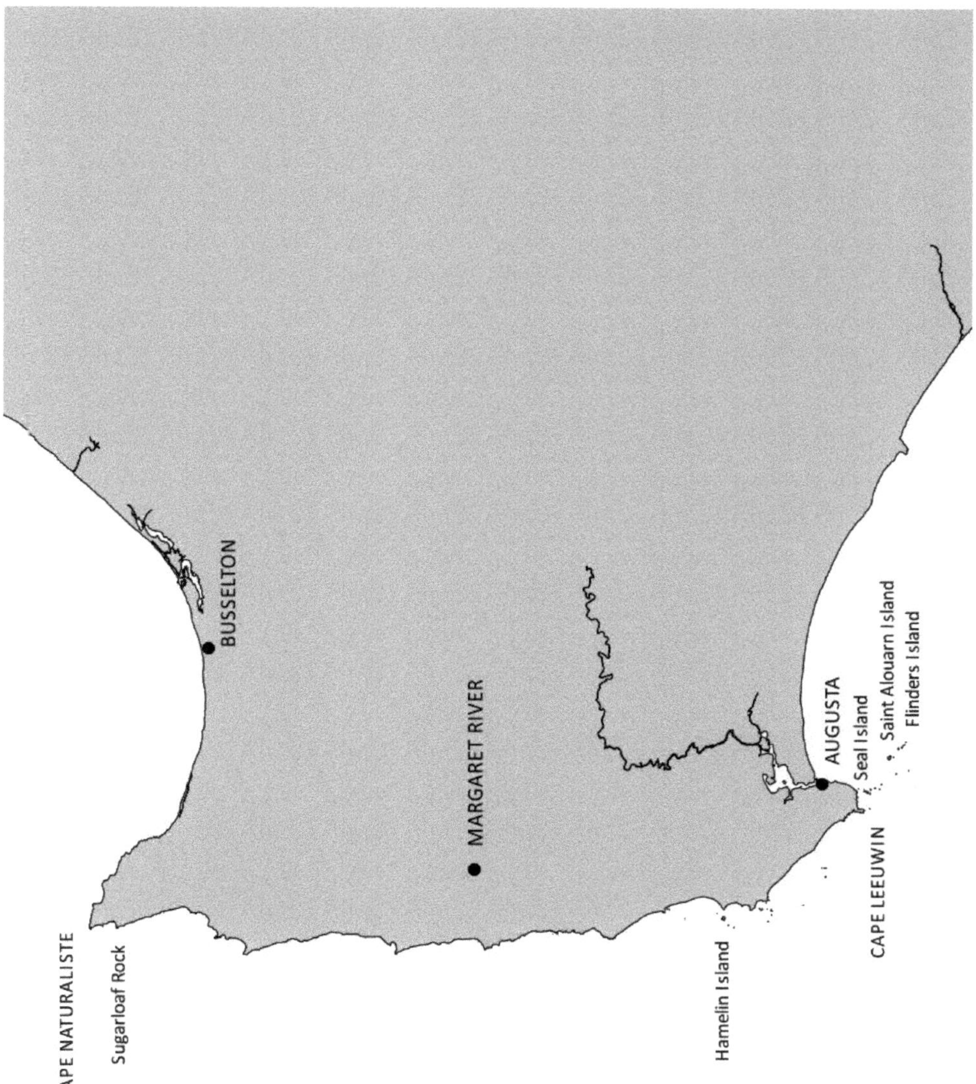

Map 10.2. Islands of the South-west corner of Western Australia, from Cape Naturaliste to Augusta.

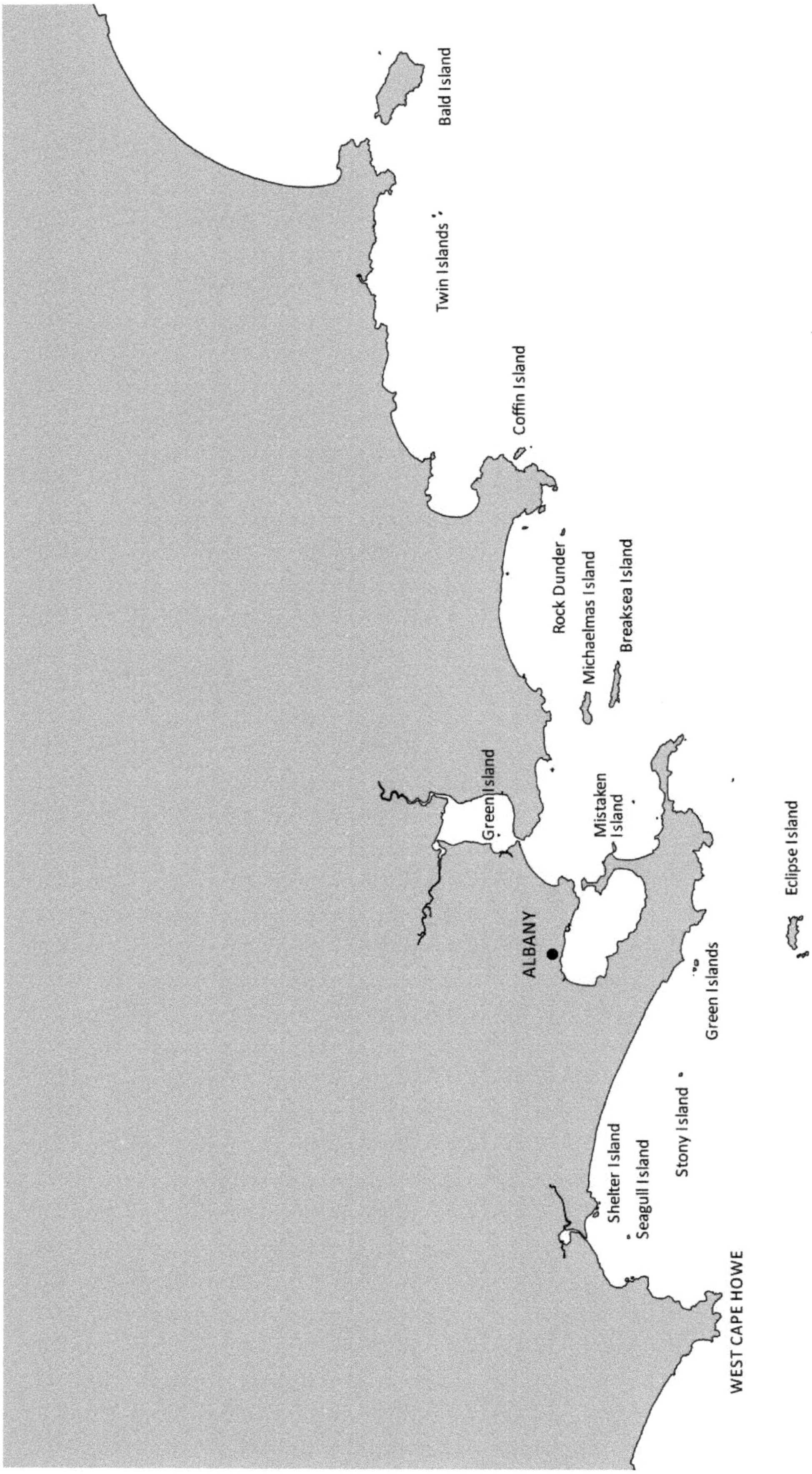

Map 10.3. Islands of the South-west corner of Western Australia, from West Cape Howe to Bald Island.

CHAPTER 10: SOUTH-WEST CORNER ISLANDS

'the surf broke sometimes with much violence on Break Sea Island, that the surf ranged to its elevated summit, during a continuance of the boisterous weather' (George Vancouver, 1791).

Breaksea Island *'is so exposed to the southern blasts & the heavy sea rolling against it, that it presents dreary barren rocks with very little soil or verdures'* (Archibald Menzies, 1791).

[Eclipse Island] *'well wooded to the top ... the higher parts ... are thickly wooded. The vegetation ... on its base is neither very plentiful or vivid'* (Robert Brown, 1801).

'The waves were so big out at sea, that little Breaksea Island sometimes disappeared under the torrents of foam they cast right up to its summit' (François Peron & Louis de Freycinet, 1803).

Mistaken Island *'is pleasant and more densely-wooded than the others'* (Nicolas Baudin, 1803).

[On Mistaken Island] *'a period of 3 Hours was afforded me to range around in pursuit of Flora, the subjects of which however I found very limited'* (Allan Cunningham, 1822).

'When steaming up the Sound and approaching the island [Breaksea,] *its granite sides are seen to slope in terrible manner to the water's edge, where a heavy swell sends spray far up the rocks. It is now calm. What must be the fury of the sea against these battlements in storm?'* (Archibald Campbell, 1889).

INTRODUCTION

For the purposes of this book, the coastline of the south-west corner of Western Australia (WA) commences at Becher Point, continues south to Capes Naturaliste and Leeuwin, then south-east to Point Nuyts, east to Bald Island, north-east to Hopetoun, and then east to (East) Red Island near Barker Inlet. This is a distance of c. 900 km.

There are 75 islands, of which 50 are vegetated. These are not evenly distributed, with many (32) forming an archipelago situated between West Cape Howe and Bald Island, a direct distance of c. 80 km.

Discovery and naming. Nearly all islands are visible from the mainland, and so are likely to have possessed Aboriginal names. None have been published, and may not have survived.

Unlike the west coast, no island bears a Dutch name even though Pieter Nuyts in *t'Gulden Zeepaert* traversed and crudely mapped the coast in 1626. Instead, the British first visited inshore and landed at (and named) King George Sound in 1791, during an expedition led by George Vancouver RN on HMS *Discovery* and William Broughton RN on HMS *Chatham*. Only five large islands were named. French navigators in 1792 and 1801 sailed along the coast but named only the Casuarina Isles, a group of two islands south of Walpole. Matthew Flinders RN on HMS *Investigator* visited in 1801-2, charting the coastline and naming another six islands. Western Australia's first Surveyor-General (JS Roe) named c. 12 islands in the 1830s. British Admiralty surveys in 1879 bestowed names on most of the rest.

Most names are descriptive, referring to colour (White, Green, Red, Black), shape (Sugarloaf, Peak, Flat, Saddle), geological appearance (Sandy, Stony, Smooth), location (Inner, West), utility (Shelter, Breaksea), and conspicuous fauna (Seal, Goose, Seagull, Gull). Some are named after ships (Chatham, Investigator), a notable event (Eclipse) or religious day (Michaelmas), and another two names express uncertainty about insularity (Mistaken, Doubtful). Many islands are named after people (Hamelin, Saint Alouarn, Flinders, Stanley, Richards, Coffin, Cheyne, Glasse); Migo is named for a distinguished Noongar man.

PHYSICAL CHARACTERISTICS

Hydrography. Most of the Southern Ocean near the WA coast is inadequately surveyed, as all large vessels have no need to come close inshore. It is therefore difficult to be precise about the time when some islands became separated from the mainland when sea level rose following the end of the most recent glacial maximum. Nearly all islands, however, lie within the 50 m isobath.

The oldest islands are those in the deepest waters of the continental shelf. The White-topped Rocks and Eclipse Island emerge from the 50-60 m isobath, followed by South West Island (located at 35.19° South, the most southern land in WA, situated south-west of Eclipse Island), Stony Island and Rock Dunder (40-50 m), Goose Island (30-40 m), Breaksea Island, Inner Island and Green Islands in Torbay (20-30 m), Glasse Island (15-20 m), and Flinders Island and Seal Island in King George Sound (10-15 m). The most recently-isolated islands are Saint Alouarn, Saddle and Michaelmas Islands and Gull Rock (5-10 m), Hamelin, Peak, the islets in Cosy Corner, and Seal Island near Cape Leeuwin (<5 m), and Mistaken Island in King George Sound and Green Island in Oyster Harbour (<2 m).

Geology, geomorphology, and soils. All islands lie on the continental shelf and thus their rocks and sandy soils closely resemble those on tors (also known as monadnocks) of the coastal mainland. The western coast islands (Hamelin and Peak Islands, together with those situated in Cosy Corner, south of Hamelin Island), are composed entirely of aeolianite and sit atop reefs. In contrast, all islands along the south coast consist of a base of ancient igneous and metamorphic rock (granite, gneiss, and adamellite) which is sometimes mantled with aeolianite. Exposed rock is conspicuous, which when combined with high cliffs on the southern aspect gives these islands a formidable appearance (Plates 10.1-10.3).

Soils derived from igneous and metamorphic rocks are acidic, whereas those derived from aeolianite are alkaline. Depth of soil varies with topography, being least in rocky parts and greatest in clefts in rocks and at the foot of rock faces. The only island which has had its soils comprehensively examined and described by an expert is Chatham Island.

These islands experience to varying degree the Leeuwin Current, a body of water that flows down the western coast of WA (globally unique) and east along the southern coast of WA. It consists of warm (>24° C) water flowing mainly in autumn and winter. It has comparatively low salinity and very low concentrations of nutrients.

Island area and elevation. Small pieces of land fully exposed to the ocean swell and <5 m (sometimes up to 10 m) in elevation are awash at all times. These have a black appearance and are conventionally classified as rocks and not islands. Examples include Cumberland, Vancouver, Northumberland, and Roe Rocks. As elevation increases, up to 33 m, islands usually lack soil as some waves wash over them during gales. Examples include Winjee Sam Rock (14 m, near Yallingup), Cow Island (formerly White-topped Rock, 33 m), and an unnamed rock west of Chatham Island (27 m). Because of swells formed by strong winds, combined with a considerable fetch provided by the Southern Ocean, waves break with great violence and run up the slope of the smooth, steep sides of the granitic islands for a distance of c. 10 m (Plate 10.4).

During gales, waves 10 m high break and send spume over all but the largest islands. Unlike the aeolianite islands of the west coast, there are no reefs and intertidal platforms to dissipate wave energy, and thus reduce the relentless impact of waves.

Islands sheltered from the ocean swell by a peninsula or other islands farther offshore, or by reefs, avoid being awash because the swell is refracted and loses force. Thus Flat Rock, which lies within King George Sound and is sheltered by Flinders Peninsula, supports some soil and vegetation even though it is only 3 m in height.

Most islands have a planar area of <5 ha. Fourteen islands have an area >20 ha; these are Bald (789 ha), Breaksea (107 ha), Eclipse (105 ha), Michaelmas (94 ha), Chatham (82 ha), Middle Doubtful (59 ha), West Doubtful (44 ha), 'East' Red (31 ha), Coffin (27 ha), Investigator (26 ha), Haul Off Rock (24 ha), Sandy (23 ha), Saddle (23 ha), and Glasse (21 ha). Another six islands have an area between 10 ha and 20 ha.

The largest islands are also the tallest, attaining maximum elevation of 311 m (Bald Island), 186 m (Chatham Island), 152 m (Michaelmas Island), 109 m (Eclipse Island), and 102 m (Breaksea Island). Few of the smaller islands exceed 50 m in height.

Climate. The warmer months experience little rainfall, whereas winter is the period of highest rainfall. This is the Mediterranean-type climate typical of south-west WA. Islands situated between Cape Naturaliste and Albany receive more rainfall annually (median 900-1200 mm) than those farther east (500-750 mm). Not much more specific information about the climatic characteristics of the islands can be provided, as only Breaksea and Eclipse Islands once had meteorological stations, with information being collected by lighthouse keepers. This ceased when the lighthouses were automated in 1926 and 1976 respectively.

Median rainfall was 834 mm for Breaksea Island (1890-1926) and 972 mm for Eclipse Island (1926-1973), more than that received on the adjacent mainland at Albany (923 mm, 1877-present). A more apposite comparison is one based on the same period. For the years 1891-1920, Albany experienced higher rainfall (median 966 mm) than Breaksea Island (807 mm). For the years 1931-1960, Albany had more rain (median 948 mm) than Eclipse Island (906 mm).

As expected, mean maximum temperature is lower on the islands, and mean minimum temperature is higher than at Albany. The differences are, however, minor: c. 1-1.5° C in maximum temperature in February, and <1° C in July. The difference in minimum temperature is insignificant in February but more pronounced (by c. 1-1.5° C) in July.

BIOLOGICAL CHARACTERISTICS

Biological discovery. A significant legacy of the influential English botanist and aristocrat Sir Joseph Banks (1743-1820) was the willingness of the Royal Navy to accommodate a naturalist (often the ship's medical officer) on board and facilitate his collecting of plant and animal specimens. King George Sound is the only safe natural harbour in the south-west corner and was thus the first landfall after the long voyage from Cape Town, possessing abundant supplies of wood and water.

Archibald Menzies botanised around King George Sound in 1791, followed by Robert Brown (1801-02) and Allan Cunningham (1818). French navigators also visited the Sound in 1803 and 1826. However, the large number of plant species endemic to the mainland around the Sound caused naturalists to focus their attention on the mainland and not on the islands. Only Seal Island in the Sound and Green Island in Oyster Harbour were frequently visited.

Once Albany was founded by the British in 1826 and then settled, local officials began to record some of the biological features of the islands. However, it was not until the brief visit in 1890 to Breaksea Island by Archibald Campbell that information about selected species began to be published. Comprehensive lists of bird species were made in 1909-10 (Mistaken Island), 1954 (Eclipse Island), 1959 (Bald Island and islands near Cape Leeuwin), 1975 (Breaksea Island, Michaelmas Island, Gull Rock, Seal Island, Green Island), 1976 (Sandy Island), and 1982 (Quagering [Flat] Island, Stanley Island). Compilation of complete lists of plant species began much later: 1959 (Bald Island and islands near Cape Leeuwin), 1975 (Chatham Island, and islands in and near King George Sound), and 2005 (Saddle Island). A partial list of plant species for Stanley Island (1982) is also available.

Despite these activities, these islands remain neglected compared with many WA islands. Emphasis has been on vascular flora, land birds, seabirds, and mammals. Except for Saddle Island, nothing is known about the invertebrates, lichens, mosses and fungi occurring on them.

Vegetation. The original vegetation of many islands has been described (and for the largest, mapped). Satellite imagery for inaccessible islands and others not yet visited by scientists has facilitated reliable recognition of their vegetation types.

Sailing in a small boat around small islands situated in the open ocean and rising steeply from a great depth is a fearsome experience, even on a calm day. Examples of such islands unable to retain soil and vegetation are Cow Island (0.9 ha, 33 m), Cliff Head (5.2 ha, 26 m), an unnamed islet east of Breaksea Island (1.1 ha, 20 m), Black Rock (1.7 ha, 12 m), and the southern of the Twin Islands (1.7 ha, 26 m).

Only the tallest small islands avoid being washed over and thus can retain a scanty cover of shallow soil, developed by erosion on the flat top of islands. Once soil is continually present, plant species can persist. On the smaller, taller, and most exposed islands, the only vegetation that can tolerate a saline environment is herbfield dominated by halophytes or succulents such as the prostrate species coastal pigface *Carpobrotus virescens* and round-leaved pigface *Disphyma crassifolium*. Examples of such islands with only this vegetation type are Sugarloaf Rock (0.9 ha, 42 m), Peak Island (0.2 ha, 13 m), Goose Island (2.7 ha, 46 m), South West Island (4.1 ha, 42 m), and Rock Dunder (6 ha, 60 m). This type of vegetation is also prevalent on the northern peninsula of Bald Island, southern peninsula of Breaksea Island, and north-eastern peninsula of Eclipse Island, all of which are exposed. Even on a calm day, the breaking swell sends a mist of saline droplets over the tallest islands, necessitating frequent cleaning of visitors' spectacles.

Large island area confers a buffer to destructive wave action and permits deeper soils to develop and persist. Tussock grassland (1 m tall) and low closed-shrubland (1-1.5 m tall) are then often present. The former, dominated by *Poa poiformis*, is extensive on Sandy Island

and on parts of Breaksea, Michaelmas, and Bald Islands. Shrubland, comprised mainly of *Rhagodia baccata*, is prevalent on Saint Alouarn, Chatham, and Breaksea Islands.

These islands, even large ones, usually have large areas of bare rock. Joint lines in the granite are filled with soil and support the third vegetation type: lithic complex. This comprises a variety of plant species, including the fern *Cheilanthes tenuifolia*, and *Andersonia sprengelioides*, *Thryptomene saxicola*, *Verticordia plumosa*, *Boronia alata*, and *Dodonaea ceratocarpa*.

The two remaining vegetation types are restricted to the very largest islands. Open heath on Bald Island consists of c. 15 plant species occurring in a mosaic pattern (Plate 10.5). Michaelmas Island is also rich in plant species, whereas on the more exposed Eclipse and Breaksea Islands *Thryptomene saxicola*, *Leucopogon revolutus*, and *Ficinia nodosa* are prominent.

Low closed-forest is formed by eight tree species: *Melaleuca lanceolata* (3-12 m); *Melaleuca microphylla* (6 m); *Taxandria marginata* (6 m), WA peppermint *Agonis flexuosa* (10 m), *Callitris preissii* (12 m); Bald Island marlock *Eucalyptus conferruminata* (8 m), yate *Eucalyptus cornuta* (6-12 m); and coastal ridge-fruited mallee *Eucalyptus angulosa* (12 m). On the more exposed large islands forests develop only on the sheltered side and then only where soil is deep as in declivities (Plates 10.6, 10.7).

Rock type has a major influence on the succulent and halophytic plant species present on the periphery of the larger islands. On the acidic soils derived from granite *Carpobrotus virescens*, *Disphyma crassifolium*, *Salicornia blackiana*, *Maireana oppositifolia*, *Apium prostratum*, *Sporobolus virginicus* and *Leucophyta brownii* are conspicuous, whereas on soils derived from aeolianite *Rhagodia baccata*, *Threlkeldia diffusa*, *Atriplex paludosa*, *Olearia axillaris*, and *Nitraria billardierei* are prominent.

Extent of exposure to waves and gales interacts with island area and maximum elevation to limit the extent of vegetation. For example, many small (<5 ha), low exposed islands have as little as 0.5-12% of their planar area occupied by vegetation. As area, elevation and degree of shelter increase, the proportion of the island vegetated exceeds c. 40% (for example, Saint Alouarn, Quagering, Sandy, Saddle, Stanley, Migo, Shelter, Eclipse, Seal in King George Sound, Mistaken, and Coffin Islands), and reaches c. 85% for some of the largest, for example Michaelmas and Bald Islands.

Size of island floras (native species). As expected small islands have fewer plant species than larger ones, but other factors are influential. These include elevation, topography, distance from larger islands and the mainland coast, time of isolation, degree of exposure to gales and swell, extent of bare rock, and rock type.

The five islands with area <10 ha and subjected to detailed study have only 11-14 native plant species. In contrast a small (0.2 ha) islet west of Mistaken Island and thus very sheltered and possibly at times joined to it has 20 species. Hamelin Island (11 ha), protected by reefs, had 43 species collected in 1959, 37 in 1977, and 43 in 1997, 2008 and 2009 (composite data, not separated by year).

Larger islands have more native plant species, but not always in the expected sequence: Mistaken Island (11 ha, 41 species); Saddle Island (23 ha, 24 species); Sandy Island (23 ha, 31 species); Coffin Island (27 ha, 24 species); Chatham Island (82 ha, 16 species); Michaelmas Island (94 ha, 68 species); Eclipse Island (105 ha, 40 species); Breaksea Island (107 ha, 41 species); and Bald Island (789 ha, 84 species).

Lichens, mosses, non-marine algae, and terrestrial invertebrates. Saddle Island was investigated for three hours in November 2005. Fourteen lichen species, two moss species, three non-marine alga species, and 14 terrestrial invertebrate species were recorded. Three species of trapdoor spiders have been collected on Bald Island.

Frogs. None recorded.

Reptiles. The only island in this region known to have snakes is Mistaken Island, with the tiger snake and dugite having been recorded. One gecko species, *Christinus marmoratus*, is known from Chatham, Eclipse, Breaksea, Bald, and Middle Doubtful Islands. King's skink occurs on Chatham, Eclipse, Michaelmas, Seal (in King George Sound), and Bald Islands. It is the largest species of skink in WA and preys on the eggs of seabirds (and domestic fowls when kept on Eclipse Island by lighthouse staff). The type specimen of this species was collected on Seal Island.

Four other skink species are present: *Ctenotus labillardieri* on Chatham, Eclipse, Michaelmas, Bald, and Middle Doubtful Islands; *Hemiergis peronii* on Chatham, Eclipse, Michaelmas, and Bald Islands; *Egernia nitida* on Eclipse and Bald Islands; and *Egernia napoleonis* on Bald Island.

Land birds. The number of land bird species able to persist on an island depends on the type of vegetation present and its areal extent. As mentioned previously, these attributes themselves are constrained by island area, elevation, and exposure to wave action. Importantly, without several visits it is difficult to distinguish resident (breeding) species from vagrants.

Small, low exposed, islands support only one vegetation type, herbfield. There, rock parrots forage for seeds of *Carpobrotus virescens*, *Calandrinia calyptrata* and *Leucophyta brownii*, as well as the exotic *Arctotheca populifolia* and *Sonchus oleraceus*. They also nest in cavities in rocks on some islands. These parrots form very mobile flocks and may occur at times in large numbers (up to 80). Welcome swallows feed above the islands and nest in sheltered cavities in rocks. Some larger, taller exposed islands also have tussock grassland and low closed-shrubland, attractive habitat for silvereyes, pipits, and brown quail. The largest and tallest exposed island, Bald Island, supports large areas of tall vegetation and thus has the most land bird species, including the brush bronzewing, white-breasted robin, western whistler, grey fantail, spotted scrubwren, inland thornbill, New Holland honeyeater, and red-eared firetail.

Small, low, sheltered islands (often close to the mainland) develop low closed-forest and support brush bronzewings and spotted scrubwrens. The largest (Michaelmas Island) is able to support populations of the white-breasted robin, western whistler, grey fantail, inland thornbill, Gilbert's honeyeater, New Holland honeyeater, inland thornbill, and brown honeyeater.

Two species of cuckoo have only been recorded from Michaelmas and Bald Islands, probably because their host species do not occur on other islands.

New Holland honeyeaters require year-round access to nectar, which is achieved only if a variety of species of Myrtaceae and Proteaceae are present. Eclipse Island has only two *Melaleuca* species and as expected lacks any resident honeyeater population. Michaelmas and Bald Islands have many bird-pollinated plant species, and it is only there that honeyeaters have been recorded in large numbers. The nomadic purple-crowned lorikeet has so far only been recorded on Bald Island, but its status there is unknown.

Seven species of terrestrial raptor have been recorded on the islands: nankeen kestrel, brown falcon, peregrine falcon, brown goshawk, collared sparrowhawk, swamp harrier, and wedge-tailed eagle. Only three of these species are known to nest on the islands: nankeen kestrel (Eclipse Island); brown goshawk (Bald Island); and wedge-tailed eagle (Michaelmas and Bald Islands).

Australian ravens occasionally visit some islands, probably to scavenge on the carcases of seabirds, as well as to prey opportunistically on their eggs.

The land bird species most recorded on islands are the rock parrot (16 islands), silvereye (15 islands), and welcome swallow (11 islands). Interesting records of vagrant land birds are the scarlet robin (Chatham Island), grey currawong and white-breasted robin (Mistaken Island), and red-eared firetail (Coffin Island).

Marine raptors. Eastern ospreys have bred on an unnamed islet close to the mouth of the Margaret River and on Hamelin Island, and have also been reported (not breeding) on Saint Alouarn and Eclipse Islands. White-bellied sea-eagles have been recorded, presumably only as occasional visitors, on Chatham, Eclipse, Coffin, and Bald Islands.

Shorebirds. Most islands in this region lack beaches. Pied oystercatchers have been recorded (breeding) only on Green Island in Oyster Harbour. The sooty oystercatcher forages on the rocky shoreline of many islands and nests among herbfield vegetation.

Waterbirds. All islands lack waterbodies and thus no ducks, grebes, herons, or darters breed or feed on them. The Australian pelican breeds on Green Island in Oyster Harbour. Cape Barren geese once probably occurred on many islands, at least as far west as Goose Island south of Nornalup Inlet. These birds graze on the foliage of succulent-leaved plants which are ubiquitous on islands. Small numbers were counted in January 1990 on West Island (2) and Hauloff Rock (1), in February 1992 on (East) Red Island (8, including 3 dead), and in April 1993 on Red Island (West, 4). The only rail so far recorded on any island is the spotless crake (Eclipse Island, vagrant).

Seabirds. Safe from terrestrial predation and Aboriginal hunting, all but one of 12 seabird species that occur here breed exclusively on islands. How, then, in 1830 did the Noongar Minang man Mokare know that Shelter Island 'abounded with mutton birds'? Perhaps he saw them flying to the island, which is very close to the mainland.

Information about which islands have seabird nesting-colonies is incomplete, as some islands have never been visited by naturalists. Five species construct burrows where soil is deep enough; otherwise they nest under dense vegetation. All either lay only one egg or rear only one chick. Little penguins breed on at least 12 islands. They come ashore each evening all year. On Breaksea Island they took 30 minutes to reach the top of the island, a distance of c. 200 m sloping steeply to c. 70 m height.

Flesh-footed shearwaters breed between September and May, and have been recorded on at least 18 islands. The largest colonies are on Saddle Island (c. 7000 pairs) and Sandy Island (c. 3000 pairs). This species once bred on the mainland coast near Shelter Island, Torbay. A bird captured on Breaksea Island was fitted with a small transmitter, and travelled c. 500 km from the island each day while foraging over the ocean. Once breeding has finished, birds travel west and north-westward into the Indian Ocean at least as far as 4000 km from Breaksea Island.

The little shearwater breeds on at least five islands from April to November, and the white-faced storm petrel breeds on at least three islands from August to April. The great-winged petrel (Plate 10.8) breeds from May to June and occurs on at least six islands.

The remaining seven species are surface nesters. The red-tailed tropicbird bred on Sugarloaf Rock from at least 1963 until 2014. The maximum number recorded was 34 pairs in 1969-70. One bird leg-banded there in January 2006 was recaptured and released on an island near Madagascar in April 2015, >9 years later. This is the longest movement recorded for this species.

Silver gulls breed on 14 islands, with large colonies (up to 80 pairs) on Hamelin Island, Gull Rock (Plate 10.9) and Seal Island in King George Sound, whereas the Pacific gull breeds in small numbers (a few pairs) on only six islands.

The four tern species mostly have a small breeding presence: Caspian tern (six islands, with a large colony on Seal Island near Cape Leeuwin), greater crested tern (seven islands, with the largest colonies on Seal Island near Cape Leeuwin and Gull Rock), fairy tern (Seal Island near Cape Leeuwin, Sandy Island, and Glasse Island), and bridled tern (stack near Hamelin Island, two islands near Cosy Corner (south of Hamelin Island), a large colony on Seal Island near Cape Leeuwin, and Investigator Island). Bridled terns were also seen on Saddle Island (10 birds) and nearby Goose Island (500 birds) in February 1990.

Box 10.1 John Warham MNZM (1919-2010)

Born in England, John became a professional photographer of nature and through this developed an interest in bird behaviour. In 1952 he migrated to WA, living from his photography and associated natural history articles. He published his first paper on WA birds in 1954, followed by c. 30 more until 1958. His research focus for the rest of his life began with his two visits in 1954 and another in 1956 to Eclipse Island near Albany. There he studied the nesting of shearwaters and petrels. These were pioneering studies for WA ornithology.

John made the transition from amateur to professional ornithologist in 1965 when he gained the degree of BSc with first class Honours from the University of Durham, followed three years later with an MSc degree. He was then appointed as lecturer in zoology at the University of Canterbury in New Zealand, and later visited many of its islands to study breeding penguins, albatrosses, and petrels. He obtained his PhD degree in 1973 and later the DSc degree.

His legacy is three important books on seabirds: a handbook of Australian seabirds, co-authored with Dominic and Vincent Serventy (1971), and two books on the ecology and behaviour of petrels (1990, 1996).

Bill Lane (1922-2000)
Bill was an amateur ornithologist who lived most of his life in Sydney, NSW. He bred finches when a boy, and joined the Royal Australasian Ornithologists Union in 1947. He began leg-banding birds in 1956 on Lion Island, and continued banding birds for most of his life. He was a prolific contributor to ornithological journals.

Bill trained many ornithologists to become banders, and helped found the Australian Bird Study Association. In 1973 he, Peter Fullagar, and Durno Murray formed a committee and initiated the Seabird Island series, published in the scientific journal *Corella*. By 1989 accounts for > 200 islands had been published. He visited 11 of the islands covered by this chapter, beginning in 1976 with Mistaken and Saint Alouarn Islands. Bill returned in 1981-2 and documented the occurrence of seabirds on Gull Rock, and Breaksea, Michaelmas, Seal (in King George Sound), Green (in Oyster Harbour), Coffin, Sandy, Quagering [formerly known as Flat], and Stanley Islands. In the same period he also visited 13 islands in the Archipelago of the Recherche, the subject of Chapter 11. All of his observations were promptly published, evidence of his serious approach.

Land mammals. Only three native species occur, distributed seemingly at random on five islands. The bush rat is present on Chatham and Michaelmas Islands, and the mardo occurs on Michaelmas and Middle Doubtful Islands. The only medium-sized species is the quokka, known only from Breaksea (where it no longer occurs) and Bald Islands. In 1976 scientists spent six days searching Bald Island for Gilbert's potoroo, locating none. They did, however, confirm that the quokka remained abundant.

Marine mammals. Two pinniped species occur. Australian sea lions have been recorded breeding on seven islands between Haul Off Rock and (West) Red Island, with <100 pups counted mostly on Red, West and Investigator Islands, and Haul Off Rock. The long-nosed fur seal has a wider distribution, breeding on islands from Flinders Island (near Cape Leeuwin) east to Investigator Island. The number of pups counted per island is <200, with most on Eclipse, Middle Doubtful and Investigator Islands, and Haul Off Rock.

Fur seals haul out only onto rocky shorelines, whereas sea lions prefer sandy beaches and will travel as far as 300 m inland from the water's edge. In doing so, they make well-defined paths. In 1821 a sea lion was seen on the summit (32 m above sea level) of Seal Island in King George Sound. Where *Nitraria billardierei* bushes are present, sea lions will tunnel under them and create hump-like thickets.

Notable biogeographic curiosities. Three fern species are known from the islands. Forked spleenwort *Asplenium aethiopicum* occurs only on Michaelmas and Bald Islands, whereas rock fern *Cheilanthes tenuifolia* is found on three large, exposed islands (Breaksea, Eclipse, Bald) and on two sheltered islands (Mistaken, Michaelmas).

The shore spleenwort *Asplenium decurrens* (previously known as *A. obtusatum* var. *northlandicum*) is a fern found on exposed cliffs. It is widely distributed in the Southern Hemisphere, including south-eastern Australia. In WA, however, it has a very limited occurrence. It was first collected in 1866 on Breaksea Island by the plant collector George Maxwell and again recorded in 1890. This was the only known locality of this species in WA until 1975, when two small clumps (<1 m^2 in area) were found on Chatham Island, but not close to the shoreline. A rock-climber, John Watson, then surveyed similar habitat along the adjacent mainland coast and found small populations at Peak Head and Waychinicup National Park (1990), and later rediscovered the population on the southern slope of Breaksea Island (1998). Another small population was discovered on the coast opposite Chatham Island by Isaac Hatch (2001). The habitat of this species is widespread on islands (as well as the mainland coast), and there is no obvious explanation why this fern has such a limited geographical distribution in WA.

Within this region, the cypress *Callitris preissii* occurs only on Bald Island. Its bark is thin and easily killed by all but the mildest fires. The tallest trees attain heights of 10-12 m and occur in scattered groves. Growth rate, estimated by counting tree rings, indicates that the oldest trees examined were 105-137 years old. No seedlings or other regeneration were observed. It seems that once in a century fire, ignited by lightning, is necessary for successful recruitment.

The quotation in the epigraph (above) by the botanist Allan Cunningham is the first remark about the poverty of plant species on the islands relative to the biodiverse adjacent mainland coast. Just as interesting are apparently chance occurrences on the islands. For example, restricted to Bald Island is *Banksia praemorsa*; other native plant species restricted to just one of these islands are *Exocarpos sparteus* (Eclipse), *Verticordia plumosa* (Eclipse), and *Paraserianthes lophantha* (Michaelmas).

There are numerous other inexplicable island distributions: orchids (four species) occur only on Mistaken, Michaelmas, Breaksea, and Eclipse Islands; *Hakea* (three species) only on Michaelmas and Bald Islands; *Clematis pubescens* only on Michaelmas, Breaksea, and Bald Islands; *Acacia* (four species) on Mistaken, Michaelmas, and Bald Islands; four pea species (Papilionaceae) on Mistaken, Breaksea, Michaelmas, and Coffin Islands; and *Agonis linearifolia* and *Taxandria marginata* are known only from Michaelmas and Bald Islands.

Another puzzling attribute is that neither great pied nor black-faced cormorants breed on these islands.

Speciation. There are no known species endemic to one or more islands (but absent from the mainland). In 1909 an ornithologist at the British Museum (Natural History) in London described a new bird species, *Zosterops shortridgii*, from 'Rabbit' [Mistaken] Island. Evidently, William Ogilvie-Grant did not realise the absurdity of this, as this island is distant only c. 150 m from the mainland. Also, silvereyes form flocks, are highly nomadic, and thus are not isolated on Mistaken Island.

Several seabird species have subspecies endemic to these islands (and in some cases to other WA islands) as breeding species, though they wander much more widely outside the breeding season. These are the great-winged petrel (subspecies *macroptera*), little shearwater (subspecies *tunneyi*), red-tailed tropicbird (subspecies *westralis*), Pacific gull (*georgii*), and bridled tern (*anaethetus*). The Cape Barren goose also has an endemic subspecies (*grisea*, which extends east to the Archipelago of the Recherche), as does the rock parrot (*petrophila*).

Species turnover. Detecting natural change in species composition on the same island is hampered by the scarcity of published lists. For annual plant species to be detected, islands require visits to be in late spring/early summer.

The only islands that has been censused for plant species at least twice are Hamelin Island, Saint Alouarn, and Seal (near Cape Leeuwin) Islands. Because visits were not always in the same season, it is prudent to consider only the perennial species. On Saint Alouarn Island, *Poa poiformis* has become extinct and *Leucophyta brownii* has immigrated. On Seal Island, *Threlkeldia diffusa* has become extinct, and *Olearia axillaris* and *Rhagodia baccata* have immigrated. The largest change has taken place on Hamelin Island. Between 1959 and 1977 seven perennial species were no longer present and eight species had colonised, and further changes were detected in the period from 1997 to 2009. These three islands are <10 ha which facilitates a thorough search.

Studies in Torndirrup National Park, part of Flinders Peninsula near Albany, demonstrate how many species have been lost from the larger islands: Five frog species; six skink species; one varanid; the carpet python; four other snake species; and five mammal species. Many species of land bird have also become naturally extinct following isolation of islands from the mainland.

There have been three natural extinctions and at least five natural immigrations of seabirds. The great-winged petrel population breeding on Mistaken Island became extinct between 1911 and 1921, and the flesh-footed shearwater became extinct there between 1906 and 1921. The red-tailed tropicbird colonised Sugarloaf Rock in the 1960s but became extinct as a breeding population by c. 2006. Bridled terns colonised islands near Cape Hamelin and Cape Leeuwin in the 1950s, and Investigator Island in the 2000s.

Whether any island populations of land bird species have become naturally extinct since Europeans began visiting these islands is often difficult to ascertain. Sandy, Chatham, Eclipse, Coffin, and Bald Islands have not experienced change in recent decades in their land bird faunas. Archibald Campbell's record of the little grassbird on Breaksea Island in 1890 may have referred to a residential population. This species was absent in 1975, and this may represent an extinction.

FACTORS LEADING TO CHANGE IN THE BIOTA

Natural factors. Large ocean swells sometimes double up, forming king waves. When these break against an obstruction (such as a cliff), a large volume of water is forced upwards to a great height. For example, a king wave photographed on the mainland coast opposite Eclipse Island reached land 40 m above sea level. Thus a small island could experience damage to vegetation, mortality of seabirds with eggs or young in the nest, and even depletion of soil.

The only known instance of a heatwave impacting on the biota occurred in February 1992 when three dead Cape Barren geese were found on (East) Red Island. In that month many dead geese were also found on islands farther east (Chapter 11).

The bridled tern is only a breeding visitor (September to March) to WA. In the 1840s it did not occur as a breeding species south of Houtman Abrolhos (Chapter 8). Fifty years later it had colonised the islands near Perth (Chapter 9). In 1955 it was recorded breeding on a stack near the northern end of Hamelin Island, and in 1959 was about to breed on Seal Island near Cape Leeuwin. In 1968 and 1969 it bred for the first (and only) time in South Australia (SA), on Baudin Rocks near Robe. Apparent vagrants were also recorded in SA in 1972, 1975, and 2001-02. These sightings are suggestive of undiscovered breeding on other islands along the southern coast of WA and SA. Perhaps the extension of its breeding range east from Cape Leeuwin reflects years with strong expression of the tropical Leeuwin Current and the associated elevation in ocean temperature.

Large breeding colonies of seabirds fertilise soil by depositing large quantities of nitrogen, phosphorus, and potassium over time. On Saint Alouarn Island, burrowing seabirds have caused the replacement of sclerophyllous shrubs by a few succulent and herbaceous species, thereby reducing biodiversity. On Bald Island some cypress trees have been destabilised by the burrowing of petrels. Breeding gulls and terns on Seal Island (near Cape Leeuwin) have caused changes in the abundance of several plant species within and close to their nests. This phenomenon has been described as floristic retrogression.

In 1791 Menzies noted that Michaelmas Island 'apparently bore the marks of the general conflagration of the country [on the mainland]'; presumably this fire had resulted from lightning strike as the Vancouver expedition was the first time that Europeans visited King George Sound.

Some 22 islands have become part of the mainland following deposition of sand by waves (tombolos) during the past seven millennia. Examples are to be found east of Rame Head, at Islet Point, near Beaufort inlet, and near Bremer Bay.

Human factors. Before European settlement and increased activity on islands, there were no exotic species present. Vine-cuttings and watercress were planted on Green Island (Oyster Harbour) in 1791, but were absent when the island was next inspected, in 1801. Collet Barker sowed seeds and planted potatoes on Michaelmas Island in 1830.

Flora. Exotic plant species are now a significant component of the flora of many islands, with >50 species recorded on islands between Cape Leeuwin and Bald Island. Three islands near Cape Leeuwin have been surveyed several times by botanists. In 1959, the proportion of exotic species was already 41% (Seal Island), 21% (Saint Alouarn), and 15% (Hamelin Island). By 2008-9, two exotic species had disappeared from Saint Alouarn Island, but iceplant *Mesembryanthemum crystallinum* had colonised. On Seal Island, iceplant had become dominant by 2008, and kikuyu grass *Cenchrus clandestinus*, sea spurge *Euphorbia paralias* and *Coprosma repens* had colonised. Silver gulls are probably the agent of introduction, as they scavenge in coastal settlements. Rock parrots feed on the seeds of iceplant, and they may be spreading this species from island to island.

Lighthouse families on Breaksea and Eclipse Islands established gardens, introducing pig's ears *Cotyledon orbiculata* to the former and arum lily *Zantedeschia aethiopica* to both islands. On Eclipse Island, arum lily now dominates much of the vegetation wherever the soil is deep enough (Plate 10.10). Silvereyes feed on its fruits and may have assisted in its spread there. North of the lighthouse buildings on Breaksea Island capeweed is conspicuously abundant; perhaps this area was used as a paddock to feed livestock.

Fauna. The laughing kookaburra, introduced to WA from 1899, has flown to Chatham, Eclipse, Breaksea, Michaelmas, and Bald Islands but has not established. Ten exotic mammal species are known (mostly in colonial times) to occur on the islands. Some were pet animals (domestic cat on Breaksea Island, early 1900s). There was one kangaroo on Green Island in Oyster Harbour in 1830. Dogs were present on Mistaken Island in 1887 and on Breaksea Island in the early 1900s. Livestock (goat) was reported on Mistaken Island in c. 1841, and sheep were once on Breaksea and Bald Islands. A donkey on Breaksea Island in the early 1900s was a working animal, used to haul cargo from the landing site to the lighthouse.

Rabbits have been reported on Mistaken (c. 1830), Breaksea (1881, one pair), Michaelmas (1889), Bald (<1896), Eclipse (1926), and Middle Doubtful Islands. They are no longer on Bald Island. It was common practice in colonial times to release rabbits on islands so that castaways would have ready access to food.

Two species of exotic rodent have been recorded on islands. The house mouse is present on Sugarloaf Rock and Green Island in Oyster Harbour, and the black rat occurred until recently on Shelter Island (in Torbay) and Mistaken Island. These islands are close to the mainland, within the swimming ability of these species. However, Green Island was used as a vegetable garden from 1827, and Mistaken Island was a quarantine station for a brief period in the 1880s. It is more likely that that these populations established from stowaways.

All of the expeditions that visited King George Sound before European settlement, in 1801, 1818, 1821, and 1822, killed seals on Seal Island. Little penguins were also killed in 1818 and 1822. A shooting party visiting Breaksea Island for one day in 1881 shot two quokkas, three brown quail, and 23 rabbits.

Following the publication in 1798 of Vancouver's 1791 voyage, sealers (from America) were at King George Sound in the period 1803-08, as were whalers (a British vessel being the first, 1800). In the 1820s sealers came from Sydney and ports in Tasmania. Seals were killed for food, skins, or both. Owing to the 'tragedy of the commons', it is probable that sealers operating in the 1820s landed on every island and killed all seals that were ashore. For example, no seals were seen in 1831 on Coffin Island on which Menzies had recorded sea lions in 1791; only their skeletons remained.

There has since been a slow recovery of sea lion and fur seal populations. The latter have regained their original breeding range, having established on rocks south of Flinders Island (south-east of Cape Leeuwin) in 1982, Chatham Island (absent in 1975, breeding by 1989), Saddle Island (present but not breeding, 1990), and Flinders Island (breeding by 1990). Australian sea lions have not yet recolonised Seal Island (King George Sound) or Green Island (Oyster Harbour).

The petrel or shearwater species (probably the flesh-footed shearwater) found nesting on Green Island (in Oyster Harbour) by Vancouver in 1791, Flinders in 1801, and Lockyer in 1827, became extinct once the island became a vegetable garden. Three dozen birds were collected there in 1827 to supplement the food of the military detachment based on the nearby mainland. Eight sealers, accompanied by Aboriginal women, lived on Breaksea Island for seven months in 1826 and subsisted on petrels and King's skinks. Some 500 great-winged petrels were killed on Coffin Island by three sealers during three days in June 1831.

Australian pelicans were recorded breeding on Green Island (Oyster Harbour) in 1826 by the d'Urville expedition. All the young birds were killed for food. This population only re-established in 1985.

Small populations (a few pairs) of Cape Barren geese were probably present on all islands, but were likely shot for food by sealers from the 1820s and by traders operating from Albany. They were last recorded on Seal Island in 1818, Green Island in Oyster Harbour in 1827, Goose Island in 1876, and Bald Island in 1889.

The population of brush bronzewings on Breaksea Island disappeared by the 1880s, possibly from being hunted for food by lighthouse keepers from 1858. This species disappeared from Mistaken Island sometime between the 1840s and 1975.

Habitat destruction. Few islands have been subjected to clearing of vegetation, and where it has occurred it has been of limited extent. Before automation, Eclipse and Breaksea Islands had lighthouse quarters (Plate 10.10, now partially restored on the latter, removed from the former). Tracks were cleared to facilitate the transportation of goods from the landing site to the lighthouse. These still exist. There are helipads on Breaksea and Eclipse Islands. Green Island in Oyster Harbour was completely cleared of its original vegetation in order to grow vegetables between 1828 and 1831.

Records of fire on the islands are few. Menzies in 1791 observed signs of recent fire on Mistaken Island. In 1803 a French expedition established an observatory on Mistaken Island and fire escaped during cooking and burnt much of the island. In 1826 sealers ignited a fire on Michaelmas Island, as did a soldier in 1830. There was 'a severe wildfire' on part of Eclipse Island in 1968. Bald Island must be one of the largest islands of WA with no fire for >100 years.

Because Aboriginal people lacked watercraft and did not swim (although there is a report of the Noongar man Migo swimming to the island later named for him), it is most unlikely that islands experienced any ignition by them. Aboriginal people wading in shallow water were observed by both Cunningham and King in 1818 and 1821. Thus it is plausible that Mistaken Island could have been waded to as the distance from the adjacent mainland is only c. 150 m and the water is shallow.

Shore-based whalers occupied one of the Doubtful Islands for ten years from 1837, and Migo Island in 1845. If trypots were in use, then vegetation would have provided fuel for their operations.

The most extreme change occurred on Fly Island in Princess Royal Harbour, which was artificially joined to the mainland to become Geak Point.

Ecological restoration. Although rabbits were poisoned three times on Mistaken Island in the period 1977-79, they were still present in 1978 and 1980. They had apparently swum the short distance from the mainland and recolonised the island. Finally, in March 1980, 180 kg of poisoned carrots were distributed there. Subsequent inspection in April and December found no live rabbits. It was reported that the condition of the vegetation had improved, with more moss, new vegetation growth on previously bare sand, and no new ringbarking of trees.

Education and heritage. From 1986 until 2000 Breaksea Island served as a locality for environmental studies by students from North Albany Senior High School. This research involved assessment of the impact of rabbit browsing on vegetation (using exclosures), and identifying plant, invertebrate, and vertebrate species.

The first, convict-built lighthouse on Breaksea Island was not fit for use and was replaced in 1902 by a low but elegant granite tower. The keepers' quarters, abandoned following automation in 1926, were later substantially damaged by storms. These buildings and the jetty were restored in 2006-10, and the island is now a tourist destination.

Island reservation and management. The first island in this region to be set aside as a nature reserve was Cheyne Island, in 1913. Then followed Doubtful Islands (1952), Green Island in Oyster Harbour (1957), and Sandy and Flat Islands (1958). Most of the rest were formally reserved in the 1960s and 1970s.

No specific management plan exists for these islands, and there appears to be very limited monitoring. Visits by scientists since 1985 are poorly documented and the results of any surveys (except of seals) remain unpublished. However, a report of recent visits to the three islands near Cape Leeuwin is available.

The population of black rats on Shelter Island was eradicated in 2003. Rabbits on other islands need to be eradicated.

Two islands have been used to improve the conservation status of two threatened species. Noisy scrub-birds were introduced to Bald Island in 1992-4 and have established, and Gilbert's potoroo was introduced to Bald Island in 2005 and Michaelmas Island in 2016. The latter introduction appears to have failed. We recommend that the quokka be re-established on Breaksea Island once rabbits are eliminated.

Sources of information

First impressions: Campbell 1901; Cornell 1974, 2003; Lamb 1984; Menzies 1791; Orchard & Orchard 2018; Vallance *et al*. 2001.
Introduction: Australia Pilot 1972, 1973; British Admiralty 1914; Department of Transport 1999, 2000; Locate 2021; Royal Australian Navy 1973, 1980, 1981.
Discovery and naming: Cornell 2003; Flinders 1814; Lamb 1984; Murray & Hercock 2008.
Hydrography: British Admiralty 1914; Department of Transport 1999, 2000; Royal Australian Navy 1973, 1980, 1981.
Geology, geomorphology, and soils: Abbott & Watson 1978; Cresswell 1990; Geological Survey of Western Australia 1975.
Island area and elevation: Locate 1921; Australia Pilot 1972, 1973.
Climate: BOM 2021.
Biological discovery: Abbott 1980; Abbott & Watson 1978; Abbott & Wills 2016; Abbott *et al*. 2006; Campbell 1901; Gillham 1963; Lane 1985 & pers. comm.; Menzies 1791; Orchard & Orchard 2018, 2020; Storr 1965; Vallance *et al*. 2001; Warham 1955.
Vegetation: Abbott 1980ab, 1981; Abbott & Watson 1978; Abbott *et al*. 2006; Gillham 1963; Lane 1985 ab; Locate 2021; Onton & Webb 2009.
Size of island floras (native species): Abbott 1980a, 1981; Abbott & Watson 1978; Gillham 1963;
Lichens, mosses, non-marine algae, and terrestrial invertebrates: Abbott *et al*. 2006.
Frogs & reptiles: Western Australian Museum collection.
Land birds: Abbott 1980; Abbott & Watson 1978; Fullagar & van Tets 1976; Gillham 1963; Johnstone & Storr 1998, 2004; Bill Lane pers. comm.; Smith & Kolichis 1980; Storr 1965; Warham 1955.
Marine raptors, shorebirds, waterbirds: Johnstone & Storr 1998; Shaugnessy & Haberley 1994; Singor 2017.
Seabirds: Abbott 1978b; Abbott & Watson 1978; Anon. 2016; Dunlop 2009; Fullagar & Murray 1973; Johnstone & Storr 1998; Peter Lambert pers. comm; Lavers 2015 ; Powell 2009; Shoobert 2005.
Land mammals: Burbidge & Abbott 2017; Kabay & Start undated.
Marine mammals: Abbott 1979b; Campbell *et al*. 2014; Goldsworthy *et al*. 2021; Shaugnessy *et al*. 1994.
Notable biogeographic curiosities: Abbott 1980ab; Johnstone & Storr 1998; McCaw 1997; Smith 1979.
Speciation: Abbott & Wills 2016; Ogilvie-Grant 1909.
Species turnover: Abbott 1980a; Abbott & Wills 2016; Campbell 1901; Gillham 1963; Onton & Webb 2009.
Factors leading to change in the biota. Natural factors: Bonnin 1982; Christie 2003; Dunlop 2009; Gillham 1963; McCaw 1997; Smith 1990; Underwood 2013.
Factors leading to change in the biota. Human factors: Abbott 1979, 1980, 2001, 2008; Abbott & Wills 2016; Anon. 1885; Burbidge & Abbott 2017; Cornell 1974; Gales & Lambert 1985; Gillham 1963; Murray & Hercock 2008; Onton & Webb 2009; Rosenman 1987, Shoobert 2005.
Habitat destruction: Abbott 2001; Abbott & Wills 2016; Cornell 1974; Flinders 1814; Fullagar & van Tets 1976; Gibbs 2010; King 1827; Menzies 1791; Orchard & Orchard 2018; Rosenman 1987.
Ecological restoration: Young 1981.
Education and heritage: Abbott & Wills 2016; North Albany Senior High School reports; Onton & Webb 2009; Searle 2013.
Island reservation and management: DBCA 2017; Tony Friend pers. comm.; Anne Greig pers. comm.

Plate 10.1. Chatham Island, viewed from the mainland coast opposite. March 2012. Credit: AA Burbidge.

Plate 10.2. Topography and geomorphology of Chatham Island, view near summit. May 1975. Credit: I Abbott.

Plate 10.3. Cliffs of Chatham Island. May 1975. Credit: I Abbott.

Plate 10.4. Bald Island from the west. 1970. Credit: AA Burbidge.

Plate 10.5. Southern cliff face of Breaksea Island, continually exposed to swell and gales from the Southern Ocean. The lighthouse is situated on the summit (102 m). 1975. Credit: I Abbott.

Plate 10.6. View towards the north-eastern peninsula of Eclipse Island from the lighthouse, showing woodland of moonah. Introduced arum lilies flowering in foreground. 1975. Credit: I Abbott.

Plate 10.7. Yate woodland (foreground) and coastal ridge-fruited mallee *Eucalyptus angulosa* woodland (background) on the plateau of Michaelmas island, viewed from the western dome (152 m high). September 1975. Credit: I Abbott.

Plate 10.8. Great-winged petrel, Bald Island. Credit: AA Burbidge.

Plate 10.9. Northern side of Gull Rock (10 m high), King George Sound, and its silver gull colony. The shallow depth of soil leads to water stress and all annual species of plants have died (brown vegetation). November 1975. Credit: I Abbott.

Plate 10.10. View from lighthouse on the summit of Eclipse Island, showing infrastructure associated with housing three light-keepers (entailing destruction of vegetation) and infestation of arum lilies (white flowers). 1975. Credit: I Abbott.

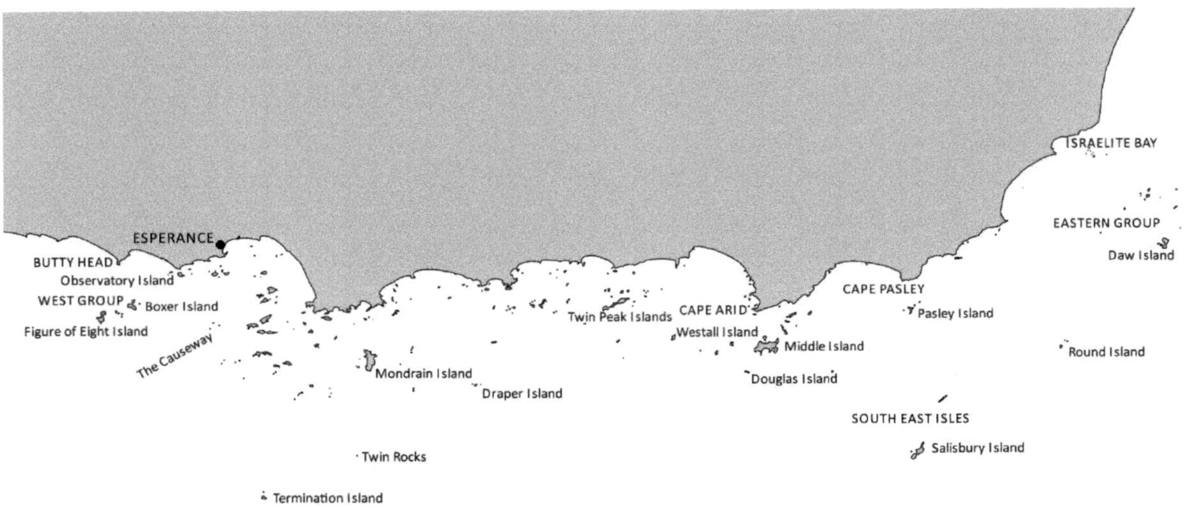

Map 11.1. The Archipelago of the Recherche.

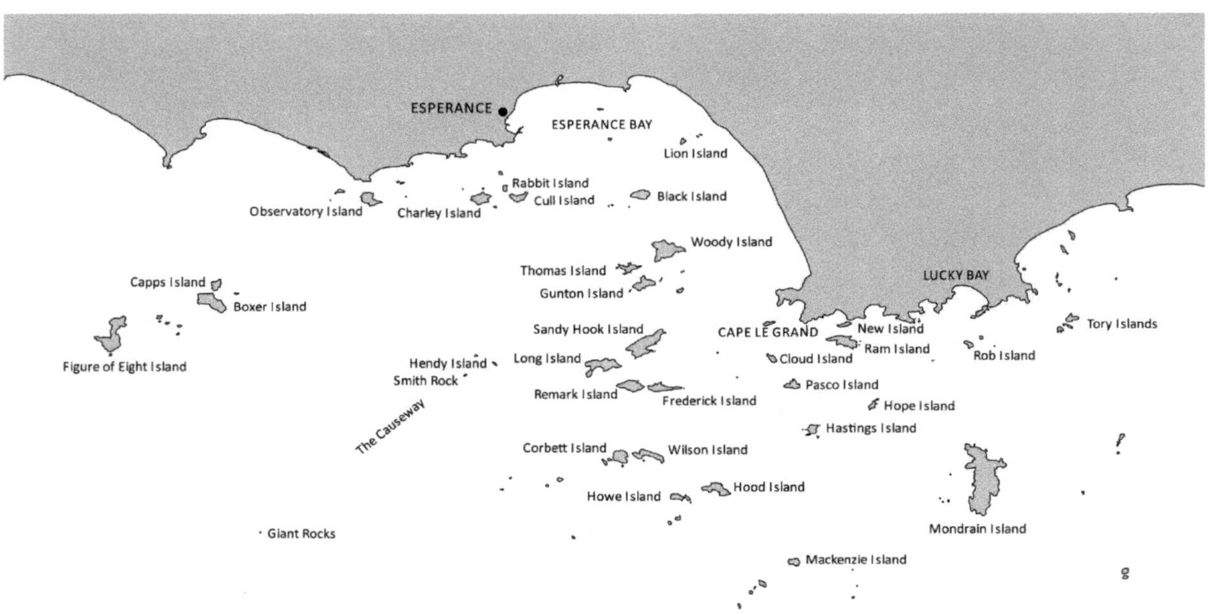

Map 11.2. Western islands of the Archipelago of the Recherche (West Group east to Tory Islands).

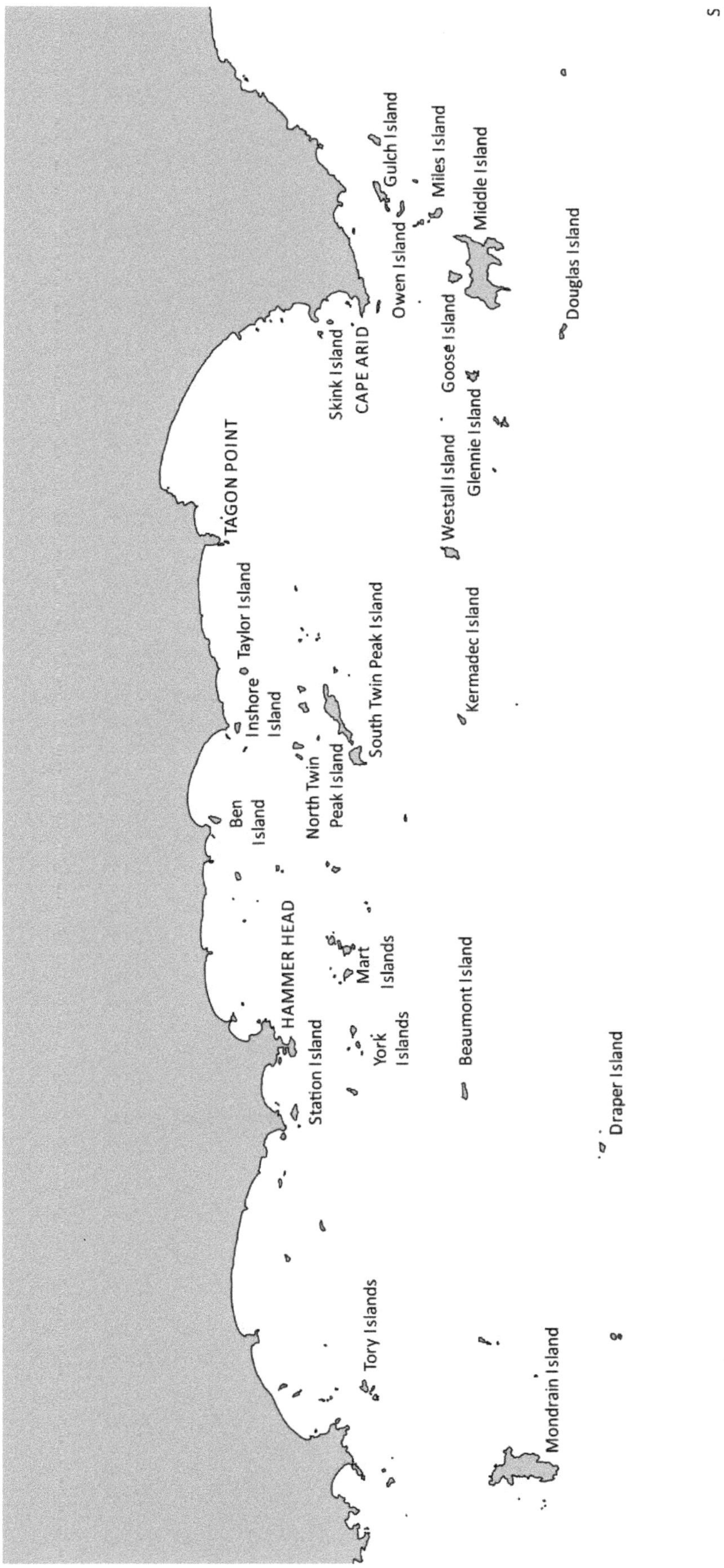

Map 11.3. Central islands of the Archipelago of the Recherche (Mondrain Island east to Middle Island).

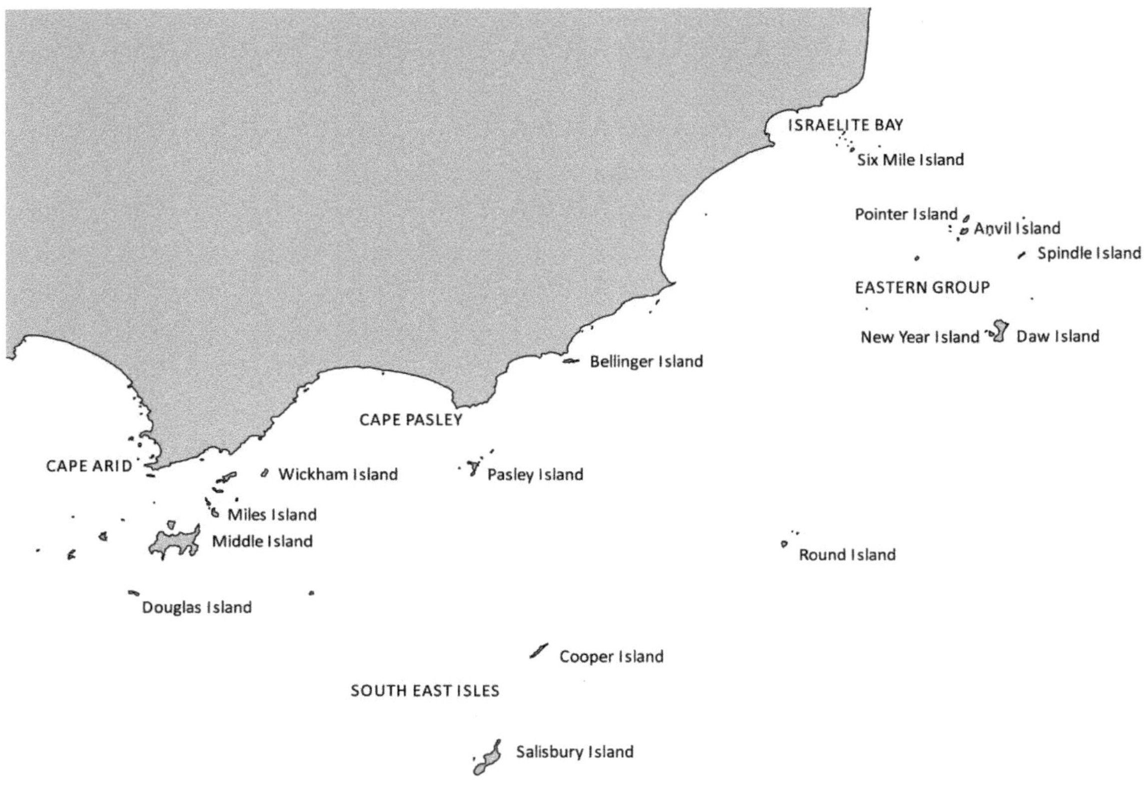

Map 11.4. Eastern islands of the Archipelago of the Recherche (Cape Arid east to Israelite Bay).

CHAPTER 11: ARCHIPELAGO OF THE RECHERCHE

'... this labyrinth of islands and rocks ... My list ... contained forty-five islands and clusters of rocks, independently of many patches of breakers where nothing above water appeared' (Matthew Flinders 1802).

'Unless proceeding to Esperance Bay ... the archipelago should be avoided at all times' (Australia Pilot 1973).

'The sea water sprays to the top of the highest islands, and is found at great height in crevices of rocks' (Bruny d'Entrecasteaux 1792 in Duyker & Duyker 2001).

'The botanical gentlemen employed the day in going around Middle Island, but they found very little to reward their labour' (Matthew Flinders 1802).

'In the evening I landed [on Middle Island, 1818] with the botanist [Allan Cunningham] and Mr. Roe, but we found little that was worthy of our attention' (Phillip King 1827).

INTRODUCTION

Geography. The Archipelago of the Recherche extends for 230 km from its most westerly island (Figure of Eight) to its most easterly island (Spindle). Its most southerly islands are Termination in the west (50 km offshore) and Salisbury (45 km offshore) in the east. Most islands occur within 30 km of the mainland coast (Map 11.1).

The archipelago comprises 345 islands and rocks at or above sea level, as well as numerous shoals and submerged rocks. Most islands are small, with 33% <1 ha in area and 30% with area between 1.0 and 4.9 ha. Only two islands exceed 500 ha in area, and another 11 are larger than 100 ha (Table 11.1).

Islands with large area tend to have higher maximum elevation than islands with smaller area. This is not surprising if islands are thought of as a series of cones of different basal diameter. Only two islands have maximum elevation exceeding 200 m: Mondrain (222 m) and Remark (220 m). Ten other large islands have maximum elevation >100 m: Middle Island (165 m); North Twin Peak (187 m); Sandy Hook (140 m); Ram (133 m); Daw (129 m); Salisbury (128 m); Woody (127 m); Figure of Eight (113 m); Charley (108 m); and Long (104 m). Only three islands with area <100 ha exceed 100 m in height: Corbett (127 m);

Termination (114 m); and Black (106 m). The height of many islands, however, remains to be measured. The lowest small island appears to be Smith Rock (1.3 ha), with elevation of 5 m.

Table 11.1. The number of islands and rocks, classified by their area, in the Archipelago of the Recherche.

Area (ha)	Number of islands	Names of islands with area >20 ha
0.1-0.9	115	
1.0-4.9	105	
5.0-9.9	29	
10.9-19.9	33	
20.0-29.9	14	Rob, New Year, New, Hope, York (E), Mart, Kermadec, Taylor, Cloud, Inshore, Pointer, Tunney, Owen, Round
30.0-39.9	12	Douglas, Hull, Mart (W), Mart (E), Tory (N), unnamed, Ben, Mackenzie, Beaumont, Station, Bellinger, Anvil
40.0-49.9	7	Hastings, Capps, Wickham, Glennie, Howe, Pasco, Miles
50.0-59.9	5	Thomas, Cull, Goose, Termination, Cooper
60.0-69.9	2	Pasley, Westall
70.0-79.9	2	Black, Gulch
80.0-89.9	6	Unnamed, Frederick, Corbett, Gunton, Charley, Observatory
90.0-99.9	2	Wilson, Hood
100.0-199.9	6	Remark, South Twin Peak, Ram, Long, Boxer, Woody
200.0-299.9	4	Daw, Sandy Hook, Figure of Eight, North Twin Peak
300.0-399.9	1	Salisbury
400.0-499.9		
500.0-599.9		
600.0-699.9		
700.0-799.9		
800.0-899.9	1	Mondrain
900.0-1000.0	1	Middle

When viewed from satellite imagery, many of the islands have a very irregular profile, comprising finger-like projections, headlands, peninsulas, and promontories (Plate 11.1). This reveals how the higher parts of the monadnocks (also known as tors) appeared before the sea level rose. Some were more or less conical; others consisted of plateaux with cols and spurs projecting in any direction, and a few were flat without any prominent features.

Discovery and naming. The archipelago was first encountered by Europeans in 1791. This was by George Vancouver RN at its south-western extremity. Here he named Termination Island because this is where he completed his examination of the coastline before heading for deeper waters on his way east to the Pacific Ocean. No landing was made.

Bruny d'Entrecasteaux was, in 1792, the next visitor in *La Recherche*. He was accompanied by Jean-Michel Huon de Kermadec in *L'Espérance*. The archipelago was named *Archipel de la Recherche*. During a gale these ships navigated a clear passage to a secure anchorage near Observatory Island. Only four islands (*Observatoire, Remarque, Mondrain, Milieu*) and *Groupe de l'Ouest* were named, as were *Rochers qui veillent* (The Watchmen) and *La Chausée* (The Causeway) – a chain of mostly underwater rocks.

In 1802 Matthew Flinders RN on HMS *Investigator* visited the archipelago. He preserved (but anglicised) the French names and named another seven islands (Draper, Station, Mart, Twin Peak, Goose, Douglas and Round Islands), the Twin Rocks, and the South-East Isles and the Eastern Group (but did not name the individual islands comprising these clusters). Members of the ship's crew landed on Middle Island and the adjacent Goose Island.

The last expedition to visit the archipelago before European settlement of Western Australia (WA) was that led in 1818 by Phillip King RN, commander of HMC *Mermaid*. A party landed on Middle Island for several hours, 16 years after Flinders' visit.

The British Admiralty conducted extensive hydrographic surveys in 1879-80 (William Archdeacon RN and Frederick Howard RN) and 1897-1901 (JW Combe RN, HMS *Waterwitch* and HM Surveying Ship *Penguin*). These resulted in 64 islands being named. Most names are descriptive (e.g. Figure of Eight, Black, Cliff, Finger, Little, Inshore, Lion, Dome); others are named after ships' officers (Glennie, Brewis, Miles, Nares, Wilson, Helby, Archdeacon, Corbett, Dailey), pioneer settlers (Taylor), admirals (Hood, Howe), and even a prime minister (Salisbury).

In the 1980s and 1990s scientists studying the biota on unnamed islands suggested appropriate names so that in their publications they could unambiguously link their studies to a designated island. These names honour ornithologists (Godman, Gould, Tunney, Hull, Ford), species (Cormorant Rock, Sea Lion Rock, Skink, Lichen), local people (Mackenzie and Daw, replacing the duplicated names 'Round' and 'Christmas'), and people associated with scientific studies (Lorraine, Libke). Several names are descriptive (Harlequin, Canard, Anvil, Pointer, Six Mile, Cranny, Spindle). Many small islands and rocks are still without a name.

PHYSICAL CHARACTERISTICS

Hydrography. Despite the large area covered by the archipelago, all islands lie within the 80 m isobath. Those separated first as sea levels rose following the waning of the last glacial maximum include Termination, Salisbury, and Cooper Islands as well as several small rocks (70 m isobath), estimated to have occurred c. 14 000 years ago. The next oldest are Mackenzie Island (50 m, c. 12 500 years ago), the West Group and Corbett, Wilson, Hood, and Howe Islands (40 m, <11 500 years ago), Thomas, Woody, Gunton, Long, Sandy Hook,

Remark, Frederick, Pasco, Mondrain, Tory, York, Mart, and Pasley Islands (30 m, c. 10 500 years ago), and most of the inshore islands, including North and South Twin Peak and Middle Islands (<20 m, c. 9 500 years ago). Many of these islands, however, remained joined to nearby islands until c. 7 000 years ago. These distinctions are best appreciated by examining the charts.

Particularly interesting is the large number of inselbergs that just failed to become islands as sea levels rose. There are 19 named reefs, shoals, patches, rocks or banks that lie 1.2 to 12.8 m below sea level. Those with depths of 7 m over them seldom break or break only during heavy swells and present a most dangerous hazard to ships.

Geology, geomorphology, and soils. The kinds of rocks that form the islands are the same as the inselbergs and bluffs, headlands, and capes present on parts of the adjacent mainland coast. Examples are Butty Head, Dempster Head, Cape Le Grand, Cape Arid, Cape Pasley, Point Malcolm, and Point Dempster. These rocks are migmatites composed of granite and gneiss in various proportions. They are part of the Albany-Fraser geological province (see also Chapter 10). Some islands have a cap of aeolianite rock, and this is particularly extensive on the south side of Middle Island, as well as on Boxer, Goose, and Salisbury Islands, and less so on Goose and Daw Islands (Plates 11.2-11.4). Some other islands nearby lack aeolianite (e.g. Figure of Eight, Sandy Hook, Mondrain, North Twin Peak, Westall).

The topography of islands is diverse. Some are bornhardts (large, dome-shaped, and steep-sided), well exemplified by Dome (90 m) and Remark (220 m) Islands. Larger islands have several peaks, with valleys and gullies running to lower levels (Plates 11.5, 11.6). Some small and flat islands and wave-washed rocks represent the rocky summits of monadnocks. Large boulders may also be present, as on Corbett, Wilson, Hastings, Termination, and Bellinger Islands.

Topographical diversity determines the spatial extent and depth of soil. On dome-shaped islands the extent of soil is often limited, owing to exposure to wave action, and to the slow process of ablation. Depth of soil is constrained by slope and degree of resistance caused by run-off of rain resulting from storms. These islands usually have extensive areas of exposed rock which lack soil because rain washes soil downslope, where it may accumulate, or into the sea.

Islands with peaks, valleys, gullies, dells, and plateaux present better opportunities for the retention of soil, the development of deeper soils, and proliferation of vegetation in the form of shrubs and trees. Large joints between blocks of granite also retain soil pockets, which enable vegetation to flourish and humus to accumulate. This results in a linear and trellis arrangement of vegetation on some islands (e.g. Pasco, Cloud, and Hastings Islands).

Exfoliated slabs of rock and scree may also help conserve soil from being washed into the sea.

The type of rock also determines soil pH, with those derived from granite and gneiss being acidic, and those from calcareous (aeolianite) rock being alkaline. This influences the floristic composition of vegetation. Knowledge of soil chemistry is limited to measurements of phosphate and calcium levels on Daw Island. Dense colonies of burrowing seabirds increase soil fertility via excretion of guano containing nitrate and phosphate compounds.

On some of the larger islands sand has been blown and deposited over rocks. Dune sand tends to support a different type of flora and vegetation, according to the depth of the sand.

Although beaches are extensive on the mainland coast, they are uncommon in the archipelago and occur only on sheltered coasts of islands. The largest beach is on the sheltered northern coast of Middle Island. Smaller beaches are present on Anvil, Daw, Bellinger, Wickham, Goose, Figure of Eight, and Sandy Hook Islands. Most islands are, however, too exposed to swells, currents, and longshore drift to accrete sufficient sand to form stable beaches.

More than 20 islands have been joined since isolation to the mainland as a result of wave refraction forming a sandy tombolo. The most spectacular are Hammer Head (65 ha, 76 m high) and Tagon Point (108 ha). Anvil Island consists of two islands so close together that they have become joined by a small sandy isthmus colonised by vegetation. At the western end of Wylie Bay several rocky islets are only c. 50 m from being joined to the mainland by a tombolo.

Climate. None of the islands have weather stations. Meteorological data from the mainland coast (Esperance) demonstrate that summer has mean maximum temperature of 20° C, with low rainfall, and winter experiences mean minimum temperatures of 10-13° C, and is wet. Median annual rainfall is 749 mm. It is likely that the western islands experience more rainfall than those farther east. In 1889 a flood was experienced on Middle Island, resulting in a well becoming briefly submerged under 1 m of water.

Unless sheltered by a nearby island, insular shore lines are exposed to strong winds at all times of the year, but from different directions. Between May and September, most winds come from the north-west in the morning, shifting to the west in the afternoon. From November to February, the predominant wind direction is from the south-east in the morning and afternoon. During storms, spray from breaking waves can ascend 30 m (recorded on Boxer Island). Spray from waves has been seen breaking over Hendy Island, which is 16 m high.

BIOLOGICAL CHARACTERISTICS

Biological discovery. Jacques-Julien Houtou de Labillardière and Claude Riche were naturalists on the d'Entrecasteaux expedition. Collections and observations of birds, seals, insects and plants were made in 1792 but it seems that few island specimens were collected. Only the plants *Eucalyptus cornuta, Leptospermum sericeum*, and *Chorizema ilicifolia* were collected on Observatory Island and later named. Although these naturalists were the first Europeans to discover the Cape Barren goose, it was not scientifically described until 1801 from specimens collected on islands in Bass Strait. Other animal species noted were the Australian sea lion, little penguin, and brush bronzewing.

Robert Brown was the naturalist on Flinders' expedition, which visited Middle and Goose Islands in 1802 and 1803 to take advantage of a secure anchorage. In 1802 'a few new plants' were found and a wallaby, two species of seal, a falcon, geese, and penguins were observed. Sailors landed on Mondrain Island in 1802 and collected a specimen of the rock-wallaby, which was later brilliantly painted by the ship's artist, Ferdinand Bauer. In 1818 Allan Cunningham (botanist on HMC *Mermaid*) observed and collected ten plant specimens on Middle Island. The presence of 'kangaroo-rats', geese, and nesting seabirds was noted.

There was then a long hiatus in biological activity in the archipelago. George Maxwell collected plant specimens on Middle and Gulch Islands in 1863. John Tunney in 1904 and 1906 collected bird specimens for the WA Museum on 11 islands. An expedition planned for 1915 by the Royal Australasian Ornithologists Union did not proceed probably as a consequence of priority to winning the Great War.

In 1921 a small party from the Australian Museum in Sydney collected animals from six islands and in 1922 published the first scientific paper on the avifauna. During the 1940s three brief visits were made to a few islands. The visit in 1950 by the Australian Geographic Society (see Box 11.1) was significant and comprehensive. There was then another hiatus in biological study until the 1970s. Many islands remain unexplored because of the time and cost of waiting on the mainland for suitable weather, the great expense involved in boat or helicopter travel from Esperance, difficulties in landing from a boat even when the swell is at a minimum, and the risk of being stranded should the weather change abruptly.

Box 11.1. The expedition of the Australian Geographical Society

The Society was formed in Melbourne in 1946, and seems to have done little since its formation, apart from becoming associated with the publication of *Walkabout*, a popular travel magazine. The value of an expedition to the Archipelago of the Recherche was suggested by Thomas Andrews, a member of the Society, who had lived on Middle Island as a teenager for eight months in 1889 with his elder brother (and for the first three months with his father).

The AGS expedition visited 20 islands in the period 7-30 November 1950. Most personnel were from Melbourne, including the leader (John Béchervaise MBE OAM, 1910-1998, a teacher at Geelong Grammar School), botanist (James Willis AM, 1910-1995, National Herbarium of Victoria), a photographer, and two assistants. Vincent Serventy AM, 1916-2007, was the zoologist, based in Perth. The other two members were the boat captain Don Mackenzie (1915-2012), who dexterously and safely managed all landings from a 12 foot rowboat, and a deckhand, both based in Esperance. Mackenzie later had an island named after him.

What was outstanding about the expedition was the timely publication of its results – nine papers between 1952 and 1954. Unusual for the time was the range of species collected by Willis (marine algae, mosses, macrofungi, lichens) and by Serventy (invertebrates, including molluscs, spiders, opilionids, plants, frogs, and reptiles). Observations were made by Serventy of birds and mammals. Publication also included a masterful review of historical aspects (by Béchervaise) and geomorphology (by Rhodes Fairbridge, 1914-2006, a geologist based in Perth, and Serventy).

The botanist's endeavours were highly praised by the expedition leader: 'No man of the whole company drove himself harder than Jim Willis...every day was alike...Jim was off with his bulky portfolio of drying papers at first light and at night...he would be busy in his tent by candle-light, recording the many plant species he had observed or collected'.

Vegetation. Vegetation maps have been published for Middle, Mondrain, and Woody Islands, and reliable descriptions are available for c. 30 other islands. The AGS expedition of 1950 identified 12 vegetation types, but these are more conveniently grouped into about five broad types characterised by a combination of structural and floristic features.

Small, low islands exposed to wave action (e.g. Six Mile, Spindle, Cave) have only rudimentary vegetation, comprising one type: herbfield. Typical plant species (all tolerant to salt spray, and thus widespread in the archipelago) include *Sporobolus virginicus*, the succulent *Disphyma clavellatum* and *Carpobrotus virescens*, *Lepidium foliosum*, *Apium annuum*, *Lobelia alata*, *Leucophyta brownii*, and *Malva preissiana*. Probably all of the islands with area <5 ha (most not yet examined botanically) have only this vegetation type.

Increasing area and height serves to buffer islands from the physically destructive process of continuous wave action and less frequent storm surges and king waves. These islands develop a second vegetation type: prostrate to low heath c. 20 cm tall, dominated by

Frankenia tetrapetala and *Atriplex cinerea*. Lower parts of New Year Island have this vegetation type. Higher up, around the summit, low open heath to 40 cm in height appears, dominated by *Olearia axillaris, Rhagodia baccata, Threlkeldia diffusa*, and *Myoporum insulare*. The low-spreading shrub *Nitraria billardierei* forms thickets (low closed heath) in sheltered sandy areas on five islands.

Islands with larger area (>20 ha) and more diversified topography develop tussock grassland dominated by *Poa poiformis/Stipa flavescens* (recorded on 17 and 15 islands respectively) and low open heath c. 50 cm tall. On Taylor Island, valleys contain low shrubland dominated by *Paraserianthes lophantha*. Exposure to wind and salt deposition has resulted in this species on Termination Island becoming prostrate. This species is widespread, being recorded on 13 islands.

Kwongan-type heath is restricted to the sandy plateaux of Mondrain and Sandy Hook Islands, and has many plant species characteristic of sandplain vegetation, e.g. *Xanthorrhoea platyphylla*. However, no *Banksia* species occurs there. (*Banksia speciosa* was observed on Woody Island in 1975 but not by other ecologists, and no specimen was collected.)

Trees occur on eleven islands (Plates 11.7, 11.8). Woodland up to 4-8 m tall and dominated by *Melaleuca lanceolata* (eight islands) and *M. globifera* (eight islands) occurs on the larger islands, but its extent and height are determined by the degree of shelter. In exposed situations these species occur only as tall shrubs no higher than 3 m. Eucalypt woodland is present on seven islands, comprised of up to six species: *Eucalyptus incrassata* (five islands), *E. cornuta* (eight islands), *E. utilis* (three islands) , *E. conferruminata* (six islands), *E. notactites* (two islands), and *E. insularis* (one island). On Middle and Woody Islands trees of *E. utilis* and *E. cornuta* attain heights of 15-20 m, but some are of mallee form. The other species seldom exceed 4-8 m, and in exposed situations may be as low as 3 m and of mallee form.

There are four other tree species. *Allocasuarina huegeliana* (three islands) grows to c. 5 m in height whereas *A. trichodon* (five islands) reaches only 2-4 m. *Callitris drummondii* (only on Middle Island) and *C. preissii* (on six islands) usually reach 3 m in height, sometimes 4-7 m on Woody Island.

Size of island floras (native species). The most recent census (1995) of plant species in the archipelago recorded 317 native plant species. The largest island (Middle) has the most plant species present, but is also the most studied island. Nonetheless, there is a strong correlation between island area and the size of an island's flora. Along a coast with waves of high energy, even relatively large and high islands (up to 1 ha) are devoid of plant species. The largest islands lacking any plant species are the most northern island in the eastern part of the Mart group (7.6 ha, height unknown), Tizard Island (6.2 ha, 9 m), George (6 ha, 7 m), Little Island (5.2 ha, 42 m), and an unnamed island south of Lichen Island (4.6 ha, 14 m).

Terrestrial invertebrates. Collections of invertebrates have so far been limited and of an incidental nature. Six butterfly species have been recorded on Middle Island, with four of these regarded as visitors. The two residential species are fringed heath-blue *Neolucia agricola* (common and widespread on the island) and golden-haired sedge-skipper *Hesperilla chrysotriche* (locally abundant). Caterpillars of the former feed on the foliage of a species of native pea, whereas those of the latter feed on the foliage of a sedge.

During the AGS expedition of 1950, 16 species of spiders and opiliones were collected; seven were described as species new to science. One land mollusc was also formally named as a new species. Incidental collections were also made of molluscs, centipedes, scorpions, and spiders on Salisbury Island in 1982. Honeybees have not yet been reported from the archipelago.

Frogs and reptiles. The herpetofauna of the archipelago has been well-studied, and staff at the Western Australian Museum have collated all records (up to 1995) for 47 islands, with one semi-aquatic frog species (*Litoria cyclorhyncha*) and 20 reptile species present (Plate 11.9). Tadpoles of the frog have been found in gnamma holes on the two largest islands (Middle and Mondrain).

The most common and widely-distributed reptiles are the geckoes (marbled gecko, 32 islands; barking gecko, 19 islands) and four species of skink (King's skink, 31 islands; *Egernia napoleonis*, 30 islands; *Ctenotus labillardieri*, 21 islands; and *Hemiergis peronii*, 19 islands). Most lizard species live on or under granitoid rocks. The heath goanna occurs only on the largest island, Middle Island.

The crowned snake is the most common and widely-distributed of the four recorded species of snake, occurring on nine islands. Next are the southern death adder (seven islands), dugite (two islands), and carpet python (two islands).

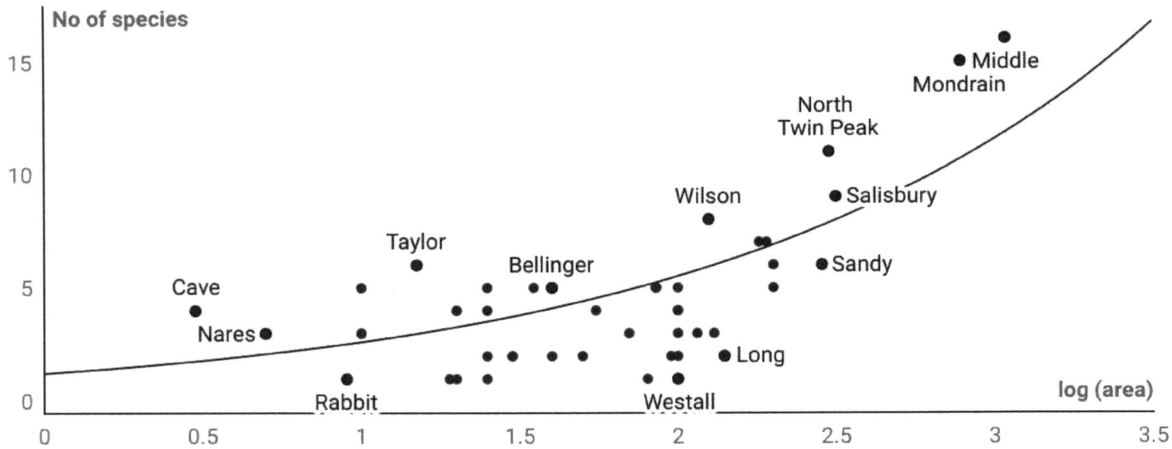

Figure 11.1. Relationship between island area and number of reptile species in the Archipelago of the Recherche.

The smallest island on which each species occurs varies. For example, it is 3 ha for the marbled gecko, 10 ha for King's skink and the crowned snake, 125 ha for the southern death adder, 200 ha for the dugite, 277 ha for the bobtail, and 780 ha for the carpet python. The most remote islands have two reptile species (Mackenzie, 36 ha; Douglas, 31 ha), five species (Anvil, 39 ha; Termination, 56 ha; Figure of Eight, 245 ha), seven species (Daw, 214 ha), and nine species (Salisbury, 350 ha). Three of the four largest islands (Middle, Salisbury, North Twin Peak, South Twin Peak) collectively support c. 90% of the archipelago's herpetofaunal diversity.

Land birds. This component of the avifauna consists of 23 species, many of which are distributed widely throughout the archipelago subject of course to constraints imposed by island area, habitat availability and quality, and distance from the mainland or large islands.

The eleven widely-distributed species are as follows: singing honeyeater (recorded on 48 islands); welcome swallow (43); rock parrot (33); Australian raven (26); silvereye (15); brown quail (14); brush bronzewing (13); Australian pipit (13); western whistler (10); New Holland honeyeater (8); and barn owl (8). Another 14 species occur only on a few islands, and with some it is difficult to determine if they are vagrant, visitor, breeding visitor, or resident: nankeen kestrel (5 islands); spotted scrubwren (4); fan-tailed cuckoo (3); black-faced cuckoo-shrike (3); red-eared firetail (3); wedge-tailed eagle (2); spotted pardalote (2); red wattlebird (2); tree martin (2); sacred kingfisher (2); peregrine falcon (2); brown goshawk (1); grey butcherbird (1); and rufous fieldwren (1).

The singing honeyeater is a species with a versatile diet, consisting of invertebrates, nectar, and soft fruits. In the archipelago such fruits are produced by several widely-distributed plant species, particularly *Rhagodia baccata, Enchylaena tomentosa, Threlkeldia diffusa,*

Tetragonia implexicoma, *Leucopogon* spp., and *Myoporum insulare.* Rock parrots feed on the flowers and flower buds of widespread species such as *Olearia axillaris*, *Leucophyta brownii*, and *Atriplex cinerea*, as well as fruits of species listed above.

Mist-netting on Woody and Mondrain Islands revealed that the silvereye and New Holland honeyeater are the most abundant bird species there, just as they are on the adjacent mainland. However, both species were less abundant on the islands.

Marine raptors. The white-bellied sea-eagle has been recorded on 16 islands and may nest on some. Eastern ospreys are absent from the islands and mainland.

Shorebirds and waterbirds. The most widespread breeding shorebird is the sooty oystercatcher, which feeds on the granite slopes close to water level. Surveys in 1991 and 1993 counted 229 and 359 respectively, with the highest counts on Wilson, Middle, and Owen Islands. The pied oystercatcher has only been recorded on Middle Island, which has the longest beach in the archipelago.

Pacific black ducks, grey teal, and Australian shelducks are vagrant. Thomas Andrews in 1889 observed a flood on Middle Island; this attracted ducks and other waterfowl (not identified). The white-faced heron has been observed on only two islands, and appears to be vagrant. Spotless crakes are known from eight islands, and the buff-banded rail has been recorded from two islands (Anvil, Daw) in the eastern part of the archipelago. Both species are probably vagrant. Nankeen night-herons have been found nesting on Lion Island.

The Cape Barren goose (Plate 11.10) feeds on plant matter on islands, and is not dependent on water for food. A survey in February 1991 of 88 islands recorded only 103 geese on 20 islands, but in April 1993 631 birds were observed on 79 islands, with the largest populations counted on Cull, Daw, Round, Wickham, Westall, Corbett, and New Year Islands. Geese visit the mainland in summer but return to the islands to breed between April and September.

Seabirds. Twelve species are known to breed in the archipelago. The six species that create burrows (or sometimes nest under dense vegetation) are the little penguin (recorded on 30 islands), little shearwater (16), flesh-footed shearwater (13, Plate 11.11), white-faced storm-petrel (12), great-winged petrel (8), and short-tailed shearwater (5). This last species was not known to breed in the archipelago until Bill Lane (Box 10.1) visited Figure of Eight Island in 1981; this was an extension in breeding range of 1100 km. These shearwaters had long been known to feed over the ocean around the islands but it was assumed that they bred on islands in south-eastern Australia.

The flesh-footed shearwater population that breeds on Woody Island has been studied intensively for three years. Adults return in September, dig new burrows or renovate existing ones, court and mate in October, and lay eggs in November and December. Incubation takes 54 days, with most eggs hatching in late January. The young fledge and depart in May. On rocky islands with little soil (e.g. Charley Island), burrows are restricted to gullies, whereas on larger islands (e.g. Mondrain Island) they are extensive.

The largest breeding colonies of the white-faced storm-petrel occur on Canning, Lorraine, and Mart Islands, and those of the great-winged petrel are to be found on Termination and Goose Islands.

Six seabird species make a nest on prostrate vegetation or rock. They are the Pacific gull (recorded on 18 islands), Caspian tern (15), silver gull (11), black-faced cormorant (4, Plate 11.12), crested tern (4), and fairy tern (2). Pacific gulls prise gastropods present in the intertidal zone off their substrate and drop the animals on 'anvil' sites to break open the shell and then extract the flesh. They prefer *Turbo torquata* and *Dicathais textilosa*.

Land mammals. Five native species have been recorded. The most widespread is the bush rat (12 islands). Another species of rodent, identified as the ash-grey mouse, is known only from Woody Island, where it was trapped on eight occasions on the margin of eucalypt thickets and open areas in 1975 by James Goodsell, Alan Tingay, and Susan Tingay. One of the specimens was drawn, but none was preserved and lodged in the WA Museum. Subsequent trapping has failed to relocate this species.

Three marsupial species occur: the black-flanked rock-wallaby (Mondrain, Salisbury, Wilson and Westall Islands, Plate 11.13); tammar wallaby (Middle and North Twin Peak Islands); and short-nosed bandicoot (Daw Island).

The possibility that the broad-faced and Gilbert's potoroo species might occur on one or more of the large islands in the archipelago was investigated in 1976. Five islands (Mondrain, North Twin Peak, Corbett, Wilson, and South Twin Peak Islands) were visited. Traps were set on the first two islands just listed, but no potoroos were found. These actions took place before Gilbert's potoroo was rediscovered, at Two Peoples Bay in 1994. Broad-faced potoroos are presumed to be extinct.

Marine mammals. Two seal species occur in the archipelago, the long-nosed fur seal (Plate 11.14) and the Australian sea lion (Plate 11.15). These islands are the Western Australian stronghold of both species. Fur seals confine themselves to the rocky shores of the islands, whereas sea lions will haul out into vegetated parts. For example, one sea lion was detected some 60 m above sea level on Kermadec Island (maximum elevation 79 m).

Australian sea lions are known to breed on 15 islands in the archipelago, based on surveys conducted in the period 1988-92. The number of pups recorded (261) ranged from one to 53 per island, with the largest number of pups recorded on Spindle (53), Six Mile (43), Kimberley (42), Glennie (24), Ford (24), and Round (20) Islands.

In the comprehensive survey of January 1990, 50 islands in the archipelago were visited, and sea lions were recorded on 31 of them. Large numbers (>20) were counted on nine islands: Ford (57), Round (42), Daw (41), Mackenzie (31), "High North" (34), Salisbury (28), Wickham and Beaumont (each with 26), and Kermadec (23). Two islands have been visited twice, and each showed contrasting change in the abundance of pups. Six Mile Island had 40 and 44 pups in 1991 and 2017, whereas on Kimberley Island the number decreased from 51 (1992) to 29 (2017). A count from a low-flying helicopter in April 1993 over all islands revealed a population of 1,207 sea lions.

The sea lion population in the archipelago differs genetically from populations elsewhere in WA and those farther east (in South Australia). This indicates that sea lions breed within the archipelago and do not visit islands outside.

In January 1989, 79 islands in the archipelago were inspected (mostly from a boat) for long-nosed fur seals. Haul-out sites and breeding colonies were detected on seven islands, with 181 animals counted. Breeding (presence of pups) was observed only on Hood Island and Seal Rock. This study updated the number of islands known to have breeding colonies to eleven.

A second, expanded survey, took place in January 1990 and resulted in a count of 2534 fur seals on 25 islands. Pups were seen on eight islands, with the largest numbers on Salisbury Island (451), Seal Rock (187), and Cooper Island (123). The survey of February 1991 focused on ten islands known to have breeding sites: 609 pups were counted, with the largest colonies on Seal Rock (172), and Libke (166), Beaumont (75), and Daw (70) Islands. Salisbury Island was not visited.

A count from a low-flying helicopter in April 1993 over all islands revealed a population of 1759 fur seals.

A survey of 11 islands in January 1999 yielded a count of 2416 pups. The islands with the most pups were Salisbury (974), Libke (340), Seal Rock (301), Cooper (291), and Cranny (140). Population size of pups on ten of these islands increased from 1990-91 to 1999. In the most recent survey (February 2011), 2753 pups were counted on the 11 known breeding colonies in the archipelago. Nearly half occurred on Salisbury Island.

Notable biogeographic curiosities. Although James Willis in 1950 identified 17 plant species limited to just one island, the current status of these species is uncertain because many islands remain to be adequately explored by botanists. There are also many discontinuities in distribution of plant, reptile, bird, and mammal species in the archipelago.

For example, the conspicuous plant species *Nitraria billardierei* is known from only five islands across the full geographical extent of the archipelago. The mistletoe *Amyema melaleucae* parasitises *Melaleuca lanceolata*, a species known from ten islands and common on several of them, yet it occurs only on the largest (Middle Island). The dugite is known only from two western islands, whereas the southern death adder and bobtail occur throughout but are unexpectedly missing from several large islands. No snakes have been recorded from Middle Island, much-visited since 1889. Tiger snakes occur on no island in the archipelago even though they are present on the adjoining mainland.

The singing honeyeater is absent from Salisbury, Daw, and Termination Islands, which are far from the mainland. Spotted scrubwrens are present on four islands (the smallest being 115 ha) but not on seven others larger than this.

Additional anomalies are presented by the terrestrial mammals. The bush rat is absent from the largest island (Middle), yet it occurs on 12 islands, the smallest of which is Bellinger Island (39 ha). Why has a short-nosed bandicoot persisted on only Daw Island (214 ha) and not on any of the six islands larger than 214 ha?

Even more perplexing is the contrasting distribution of the two wallaby species. Neither occurs together on one island. The smallest island with the tammar wallaby is North Twin Peak (277 ha), but this species is absent from the three islands larger than this island; however, all of these have the black-footed rock-wallaby on them. Is this attributable to interspecific competition? The smallest island with the rock-wallaby is Westall Island (69 ha), yet of the 17 islands larger, only four have populations of this species. One hypothesis to explain these differences is that the islands with rock-wallabies were isolated before those with tammar wallabies, and the islands may have differed in habitats or preferred food-plants at the time of isolation.

Speciation. No species is definitively known to be endemic to the archipelago. However, the spiders *Encoptarthria serventyi* and *Araneus recherchensis*, together with the millipede *Atelomastix dendritica*, may prove to be endemic once comprehensive collecting of terrestrial invertebrates is undertaken on more islands in the archipelago.

In contrast, five subspecies endemic to the archipelago have been described. These are *Eucalyptus insularis insularis* (known only from North Twin Peak Island and distinct from mainland populations), *Eucalyptus conferruminata recherche* (on six islands), the dugite

Pseudonaja affinis tanneri (on Figure of Eight and Boxer Islands, and distinct from the mainland), the Recherche Cape Barren goose (distinct from populations in South Australia and Tasmania), and the Recherche rock-wallaby (present on Mondrain, Wilson, and Westall Islands and distinct from populations on Salisbury Island and the mainland, which are the black-flanked rock-wallaby. Salisbury Island was isolated before the islands with the Recherche rock-wallaby on them, but why the latter occurs only on islands is a mystery.

Species turnover. It is at present difficult to determine the extent of this for the native flora. This is because most islands have been given insufficient attention by botanists, mainly as a result of access limiting the time spent ashore listing species and collecting specimens.

Three islands have been well explored botanically, but all are large and there is no certainty that some species have not been overlooked. On Middle Island, Robert Brown collected 47 species in 1802; James Willis recorded 132 species in 1950; and Angas Hopkins and Arthur Weston listed 208 species in 1973 and increased this to 235 species following five more visits up to 1978 (52 days of field work during six visits).

Mondrain Island has yielded 176 species from visits in 1950 and 1976. However, 40 of the species collected in 1950 were not found in 1976, and 18 species found in 1976 were unrecorded in 1950. Woody Island shows similar differences, with 85 species recorded in 1950 and 121 species listed in 1976. In 1976 27 species had not previously been recorded. Its total flora is 170 species.

The only definite example of species turnover is that of the red-eared firetail. Despite visits to Woody Island in 1950, 1975, and 1976, this finch was first recorded there in the 1980s. We accept this as a successful immigration.

FACTORS LEADING TO CHANGE IN THE BIOTA

Natural factors. Strong winds and associated deposition of salt spray, heatwaves, fire ignited by lightning strikes, browsing of vegetation by wallabies, and fertilisation of soil by breeding seabirds and sea lions are the principal natural factors influencing ecological changes on these islands. Aboriginal people (Njunga) on the mainland coast adjoining the archipelago possessed no watercraft and thus had no means of accessing the islands, many of which are visible from the mainland. They also lacked the ability to swim to islands close to shore.

Continual deposition of sodium chloride is a regular consequence of the swell breaking on island shore lines and is very important on small, low islands. The retention of sclerophyllous vegetation there is restricted to the most sheltered parts of small islands.

Less frequent, but more intense episodes of salt deposition connected with gales, influence vegetation structure on large, tall islands. This results in planation of vegetation, and restricts its height at lower levels and in more exposed situations. These impacts intensified once the islands became detached from the mainland thousands of years ago. Plant species unable to find refuge became locally extinct.

Heatwaves, that is high temperatures exceeding the average maximum temperature in summer, should result in the death of plants and animals. For example, of the 60 Cape Barren geese counted on 19 islands in 1991, 22 were dead. In February 1992, 19 birds (out of 200 counted on 56 islands) were found dead. Esperance recorded a maximum temperature of 44°C on 31 January and 46.7°C on 1 February of 1991, the highest maxima recorded to that year, although on 6 January 2010, the maximum temperature was 46.9°C. Prolonged high temperatures (usually associated with low rainfall) would be expected to reduce the population size of plant and animal species.

Lightning strikes often occur on the islands, with ten islands known to be affected in the 25 year period from 1972 to 1997. When the Flinders expedition visited Middle Island in January 1802, its botanist collected 11 plant species that were not found again until after a fire in 1972. This indicates that Middle Island experienced a natural fire in c. 1798-1800. Lightning that strikes rocky areas does not usually result in spread of fire, and so the vegetation of such areas is probably long unburnt. A lightning strike on Figure of Eight Island in the summer of 1919-20 resulted in a fire that burnt for several days.

In January 1944, seven fires (apparently ignited by lightning) with flames up to 20 m high in gullies resulted in c. 60% of Mondrain Island being burnt during several weeks. Many shearwaters were found incinerated. In March 2002 the peak of Mondrain Island was struck by lightning, and 90% of the island was burnt during several weeks, killing many shearwaters in their burrows and incinerating reptiles under granite slabs. Within two years regeneration exceeded 2 m. Vegetation and affected animal populations of most species probably recover within c. 5 years of fire.

Browsing of palatable plant species by rock-wallabies and tammar wallabies helps determine vegetation structure and its floristic composition. For example, Thomas Andrews in 1889 observed that the dense population of tammar wallabies on Middle Island had eaten all leaves to a height of c. 1 m. After a fire in 1973, tammar wallabies heavily browsed *Anthocercis genistoides*, *Kennedia nigricans*, *Commersonia corniculata,* and *Muehlenbeckia adpressa*. This level of herbivory also reduced seed production.

Burrowing seabirds elevate the fertility of deeper soils by excreting nitrogen-, phosphorus-, and potassium-rich salts. Many sclerophyllous plant species cannot tolerate these high nutrient levels, and are replaced by those species that can. This leads to distinctive assemblages of plant species where seabirds nest.

Human factors. Impacts on the fauna began in 1792 with landings by the French on Observatory Island. The tameness of fur seals and Cape Barren geese allowed sailors to walk up to them and strike animals with a club. Also eaten were little penguins. After months of living on a monotonous diet of preserved food, these animals were a welcome source of fresh food. Seals were also boiled down in 1803 for their oil when the Flinders expedition revisited Middle Island and adjacent islands.

Depredations intensified in the 1820s and 1830s when sealers arrived after slaughtering the seal populations on islands in Bass Strait (from 1798) and on Kangaroo Island (from 1803). They probably landed on all accessible islands in the archipelago and killed as many seals as they could, particularly the long-nosed fur seal, for their skins. The visit of HMS *Waterwitch* in the 1880s to survey the ocean 'almost exterminated' the geese, and the crew of HMS *Penguin* in 1900-01 ate eggs of the geese. During an open season of one month authorised by the government in 1919, sealers killed 900 seals for their skins.

In 1889 Thomas Andrews, aged 13 years, spent eight months on Middle Island with his brother (and for three months with his father). In 242 days they snared tammar wallabies and prepared 864 skins. They also had pigs, hens, and a dog. These snares attracted wedge-tailed eagles, 50-100 Australian ravens, and goannas. Raven and eagle nests were destroyed by cutting down trees with nests.

The next major disturbance was the deliberate or accidental use of fire, thereby setting the vegetation alight. Sailors on the Flinders expedition started fires on Mondrain and Goose Islands in January 1802. These fires temporarily removed the habitat of animals and also killed many. In 1971 a fisherman stranded on Figure of Eight Island lit a fire to attract the attention of rescuers. This fire burnt the entire island.

Destruction of habitat began as early as 1802 when crew of HMS *Investigator* collected firewood on Middle Island. On their return visit the next year trees were cut down to provide wood to enable seals to be boiled down for their oil. Once the adjacent coast was settled in the 1860s, pastoralists used some of the larger islands situated in or near Esperance Bay to keep sheep for several months each year, legally on Woody, Charley, and Cull Islands, and illegally on Figure of Eight, Thomas, Gunton, Rabbit, and Observatory Islands. Keeping sheep on islands continued until 1967 even though it was supposed to cease in 1958. Pastoralists used fire to provide nutritious green pick for the sheep.

The advantage to the pastoralists was the absence of dingoes from the islands and no need for fences to keep sheep from straying. However, a fence was constructed across Woody

Island to prevent sheep from browsing the foliage of poison plants (*Gastrolobium bilobum*). Middle Island was leased from 1937 to 1958 but it does not appear that sheep were ever kept there. Mondrain Island was also considered to have little pastoral value.

Many exotic plant species occur, with some 16% of the known flora of the archipelago being exotic species, either self-introduced from the mainland or introduced by human activity. Individual islands differ, with Woody (57 species), Middle (22), Mondrain (19), and Figure of Eight (19) having the most. Twenty one islands have 1-11 exotic plant species recorded. Termination, Spindle, and Six Mile Islands have none. Surprisingly, the largest island (Middle) has only c. 9% of its flora comprised of exotic plant species. In 1950 *Euphorbia paralias* was found only on part of Middle Island. By 1973 it had spread around the island and was common.

Six exotic mammal species have been introduced to several islands, but only three remain: the European rabbit (Goose Island); the goat (Cull Island); and the house mouse (Figure of Eight Island). The most unusual was the unsuccessful release of five deer in 1904-5 on Cull Island, during the heyday of the acclimatisation movement. Surprisingly, black rats are absent from the archipelago. Previous reports of them on Boxer, North Twin Peak, and Woody Islands are misidentifications of bush rats, a native species of rodent. There is a small number of western grey kangaroos on Woody Island, introduced sometime before 1975. It is fortunate that the only lighthouses in the archipelago (on Cull and Figure of Eight Islands) were built in 1965 and were automated; there was never a resident light-keeper on either island and thus no cats or dogs.

In 1841 and in dire straits, the explorer Edward Eyre saw a ship anchored in Rossiter Bay. He was welcomed aboard the French whaler *Mississippi* and recorded that pigs, sheep, and tortoises were being kept on an offshore island, not yet identified, and there were also gardens with potatoes and peas.

The only benefit of excessive grazing by goats on Cull and Rabbit Islands has been that it favours proliferation of the food plants preferred by Cape Barren geese. Large populations of geese were seen on these islands in 1969 and 1993.

The most recent calamity to affect some marine animal species was the bulk cargo carrier MV *Sanko Harvest*, bound for Esperance. It took a short cut on 14 February 1991 and struck a reef between Hastings and Hood Islands. It began to break up, leaking oil and phosphate fertiliser into the ocean, and sank 3-4 days later. The oil reached Hood Island and Seal Rock. Hundreds of oiled seal pups were cleaned by government officers, but at least 15 pups and 30 seabirds are known to have died. There do not seem to have been any long-term impacts.

Island-based tourism is at present confined to Woody Island, the only commercial lease, and Middle Island. Operations commenced on Woody Island with the construction of a wharf in 1973 and later a jetty and some buildings were constructed and camping sites for tents were developed. Authorities had approved this venture because of the report in 1950 of black rats on the island. Only recently was this shown to represent a misidentification of the native rat, and this discovery at the final hour terminated a plan to distribute poison baits on the island in order to eradicate the rats. Tourism on Middle Island involves a commercial operator licensed to conduct tours, with no approval to develop commercial facilities. There are flights over Middle Island to allow tourists to view the pink Lake Hillier.

No mining has taken place on any island in the archipelago. There was early interest in guano deposits, but these proved to be small. In 1980 two mining claims on Salisbury Island for guano, phosphate, and limestone were rejected after a rapid biological survey revealed its high conservation values.

In the early days whaling took place in the waters of the archipelago, initially by the French and Americans. In the 1860s whalers based in Albany cruised the archipelago. No records of their activities are available, and it seems likely that chimney bases, stone floors, and a stone-lined well on Middle Island represent the vestiges of a whaling station. Whale look-out shelters have been recorded on nearby Goose Island.

Salt was extracted at Lake Hillier, Middle Island, in the 1830s and in 1889. Remains of a tram line have been recorded there.

Island reservation, management, and monitoring.

In 1947 an expedition with nine members, led by the scientist and philanthropist Russell Grimwade, travelled from Melbourne to collect natural history information along the coast between Port Lincoln in South Australia and Perth in WA. When visiting Esperance its members were requested to lobby for the mainland east of Esperance and the offshore islands to be reserved as a sanctuary for wildlife. Upon reaching Perth, two members met the Minister for Lands, and the result was that land surrounding Cape Le Grand was set aside as a Class A reserve, and the islands of the archipelago were formally declared as Class C flora and fauna reserves in 1948. The islands were elevated to Class A status (the most protected) in 1969.

The first management intervention in the archipelago, other than the pollution event described above, was the removal of c. 70 African boxthorn plants from Woody Island in 1991. This species was eradicated in 1993 on Cull Island, an operation that took eight days, where it had proliferated because the resident goat population of c. 40 animals did not eat its foliage. The current status of boxthorn on Figure of Eight Island is unknown.

The next intervention was the translocation of 82 dibblers in 2015-17 to Gunton Island. Most of these animals were bred at Perth Zoo, and have successfully established and now breed there. This is now an important insurance population should the species become extinct on the mainland.

In 2017-18 ten Gilbert's potoroos were released on Middle Island. Monitoring in 2021 resulted in the trapping of three previously-known animals and three animals that had been born on the island. Images from camera traps included young-at-heel. These data indicate that potoroos have been able to survive for three years and conditions on the island are suitable for breeding. The persistence of this population thus appears promising.

There has been a recent 'trial' translocation of the noisy scrub-bird to Mondrain Island. No details are as yet publicly available, nor is the justification for this translocation so far beyond its recognised historical range.

A management plan for the archipelago was published in 2016. The area of the archipelago reserve is 7145 ha. Some 15 poorly-known plant species occur in the archipelago. Monitoring of vegetation and species is undertaken following significant fires on the islands, and it has been found that vegetation cover returns to pre-fire levels after eight years, and that regeneration is at least 2 m tall two years after fire.

Sources of information

Epigraph: Australia Pilot 1973; Duyker & Duyker 2001; Flinders 1814; King 1827.
Geography: Abbott & Wills 2016; Andrews 1959; Béchervaise 1951; BOM 2021; DAPW 2016; Google Earth 2021; Google Maps 2021; Locate 2021; Royal Australian Navy 1974b, 2009.
Discovery and naming: Béchervaise 1954; British Admiralty 1881, 1902; Duyker & Duyker 2001; Flinders 1814; King 1827; Lamb 1984; Marchant 1982; Murray & Hercock 2008.
Hydrography: i-Boating 2021; Lewis *et al.* 2013; Royal Australian Navy 1974, 2009.
Geology, geomorphology, and soils: Fairbridge & Serventy 1954; French undated; Lowry & Doepel 1974; Morgan & Peers 1973.
Biological discovery: Abbott & Wills 2016; Anon. 1913; Dell 2005; Flinders 1814; Glauert 1954; Hull 1922; King 1827; Labillardiere 1800; Mabberley 2017; Macpherson 1954; Main 1954; Orchard & Orchard 2020; Serventy 1952, 1953; Vallance *et al.* 2001; Whittell 1938; Willis 1953.
Box: Béchervaise 1951, Serventy 1951.
Vegetation: Abbott & Black 1978; Burbidge *et al.* 1982; French undated; Goodsell *et al.* 1976; Keighery 1995; Willis 1953.
Size of island floras (native species): Abbott & Black 1978; Burbidge *et al.* 1982; Goodsell *et al.* 1976; Google Earth 2021; Google Maps 2021; Keighery 1995; Locate 2021; Weston & Trudgen undated; Willis 1953.
Terrestrial invertebrates: Abbott & Black 1978; Burbidge *et al.* 1982; Macpherson 1954; Main 1954; Williams & Powell 2006.
Frogs and reptiles: Glauert 1954; Smith & Johnstone 1996.
Land birds & Marine raptors: Abbott & Black 1978; Burbidge *et al.* 1982; Goodsell *et al.* 1976; Hull 1922; Johnstone & Storr 1998, 2004; Kabay & Start undated; Serventy 1947; Serventy 1952; T Start pers. comm.; Tingay & Tingay undated.

Shorebirds & Waterbirds: Andrews 1959; Halse *et al.* 1995; Kabay & Start undated; Shaughnessy & Haberley 1994; Smith *et al.* 2005.

Seabirds: Lane 1983; Johnstone & Storr 1998; Powell *et al.* 2007; T Start pers. comm.

Land mammals: Burbidge & Abbott 2017; Serventy 1953.

Marine mammals: Burbidge *et al.* 1993; Campbell *et al.* 2008, 2014; Gales *et al.* 1994; Goldsworthy *et al.* 2021; Shaughnessy *et al.* 1994.

Notable biogeographic curiosities: Burbidge & Abbott 2017; Johnstone & Storr 1998, 2004; Smith & Johnstone 1996; Willis 1953.

Speciation: Edward & Harvey 2010; Johnstone & Storr 1998; M French pers. comm.; Main 1954; Smith & Johnstone 1996.

Species turnover: Abbott & Black 1978; Abbott & Wills 2016; E Rippey pers. comm.; Weston 1985; Willis 1953.

Natural factors leading to change in the biota: Abbott & Black 1978; Abbott & Wills 2016; Andrews 1959; Edwards 1981; Fowler 1945; Halse *et al.* 1995; Hopkins 1981; Shaughnessy & Haberley 1994; Weston 1985; Whitley 1944.

Human factors leading to change in the biota: Andrews 1959; Australian Transport Safety Bureau 1991; Bakker & Weston undated; Burbidge & Abbott 2017; Burbidge *et al.* 1982; Burbidge *et al.* 2012; Cumpston 1970; Edwards 1981; Eyre 1845; Flinders 1814; Ibbotson 2001; Keighery 1995; Labillardière 1800; Lohr & Keighery 2014; Pearson *et al.* 2004, 2005; E Rippey pers. comm.; Serventy 1952, 1953; Storrie & Pobar 1994; Willis 1953; Woodward 1908.

Island reservation, management, and monitoring: Anon. 1948ab; Comer *et al.* 2021; DPAW 2016; Friend 2017; Tony Friend pers. comm.; Friend *et al.* 2017; Anne Greig pers. comm.; Pearson *et al.* 2004; Weston 1985.

Plate 11.1. Aerial view of Thomas Island (92 m high). November 1973. Credit: AA Burbidge.

Plate 11.2. Northern peak (128 m high) of Salisbury Island, viewed from the southern peak (117 m high). March 1982. Credit: AA Burbidge.

Plate 11.3. Landscape and vegetation, Daw Island. March 1982. Credit: AA Burbidge.

Plate 11.4. New Year Island, viewed from Daw Island. March 1982. Credit: AA Burbidge.

Plate 11.5. View from the north-eastern peak of Mondrain Island. February 1976. Credit: I Abbott.

Plate 11.6. View of south-western valleys on Mondrain Island. February 1976. Credit: I Abbott.

Plate 11.7. Coastal moort woodland, former sheep paddock, and mist-net on Woody Island. February 1976. Credit: I Abbott.

Plate 11.8. Tourist path within coastal moort woodland, Woody Island. May 2011. Credit: AA Burbidge.

Plate 11.9. Common scaly-foot, a species of legless lizard. Mondrain Island. May 1976. Credit: AA Burbidge.

Plate 11.10. Flesh-footed shearwater chick, Daw Island. March 1982. Credit: AA Burbidge.

Plate 11.10. Cape Barren geese, Mondrain Island. May 1976. Credit: AA Burbidge.

Plate 11.12. Black-faced cormorant. May 2011. Credit: AA Burbidge.

Plate 11.13. Black-footed rock-wallaby, Salisbury Island. March 1982. Credit: AA Burbidge.

Plate 11.14. Colony of long-nosed fur seals, Salisbury Island. March 1982. Credit: AA Burbidge.

Plate 11.15. Migmatite rocks, Australian sea lion, and black-footed rock-wallaby, Salisbury Island. March 1982. Credit: AA Burbidge.

CHAPTER 12: SIMILARITIES AND DIFFERENCES BETWEEN REGIONS IN THE CHARACTERISTICS OF THEIR ISLANDS

'neither the nature of the soil, nor height of the land, nor the climate, nor the general character of the associated beings [species], *and therefore their action one on another, can differ much in the different islands'* [of the Galápagos Archipelago] (Charles Darwin).

'mere distance is one of the least important of the causes which have determined the likeness or unlikeness in the animals of different countries [islands such as Japan and Britain]' (Alfred Wallace).

'Islands offer...the closest approach in nature to the conditions of a [human]-controlled laboratory. They have one inherent advantage...about which the experimenters can only speculate, the presumably amassed influence of prolonged time' (Robert Murphy)

'An evolutionist's impression of Lilliputian or Brobdingnagian races [of animal species] *occurring on islands...may be determined by the particular island groups under study'* (George Watson).

INTRODUCTION

In Chapter 1 we outlined the themes which were to be considered region by region in Chapters 2 to 11. In this chapter we integrate all of this information and compare and contrast the physical, biological, and Aboriginal influences that have operated since the end of the last glacial maximum (LGM). What are the similarities and differences between the ten island regions of Western Australia (WA) in the patterns and idiosyncrasies that characterise their island biotas? How have the processes that have operated since the islands became disconnected from the mainland coast varied from region to region?

We examine regional insular similarities and differences in various historical, geographic, hydrographic and geomorphological features, discovery by humans, island area, maximum elevation, climate, rock type, degree of isolation from the mainland, and time since their separation from the mainland or their formation via accumulation of sand or coral rock on a reef. Differences in climatic attributes, landforms (topography), and soil retention or development are also assessed. How all of these factors interact and contribute to, and shape, the retention or development of a diversity of vegetation types, and consequently the extent of habitat suitable for animals, and the species composition of islands at the time of first visit by Europeans (from 1616) and subsequent colonisation (from 1826) in the various regions is a significant focus of this chapter.

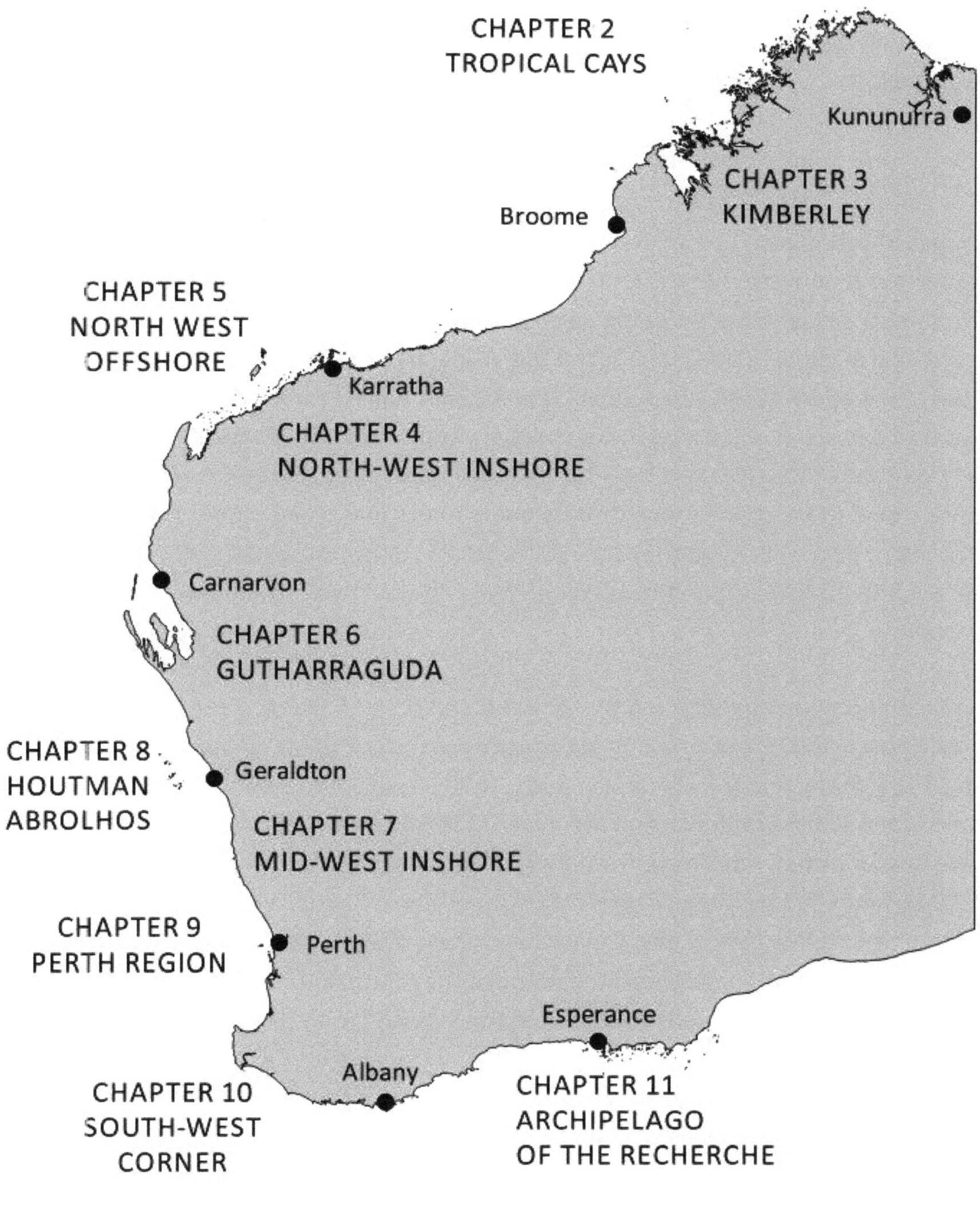

Map 12.1 Map of Western Australia, showing the link to each chapter of this book.

We also evaluate the impact of First Nations' use of WA's islands. In most regions Aboriginal people had no means of visiting islands, whereas in others they could travel long distances and access the resources available, even on more remote islands. To what extent did their use of fire, gathering of plant food, and hunting impact on island ecosystems?

We weigh up the collective influence of all factors and identify those most potent during the past 7000-15 000 years, always seeking to highlight insular disparities and commonalities between regions.

DISCOVERY AND NAMING

The names of WA islands are an important component of their cultural history. They constitute a rich heritage. Islands visited by Aboriginal people after separation had local names, sometimes with names being applied to different parts of the same island. It seems likely that conspicuous islands not visited but visible from the mainland coast would also have been named by Aboriginal people, as is the case with Rottnest Island. Unfortunately, few names have survived European colonisation, with some islands receiving their European names from the 1600s. This was well before there was any appreciation of, and familiarity with, Aboriginal languages. In recent years many Aboriginal island names have been revived and applied to islands in Kimberley region. Hopefully, many more of the original Australian names will be officially adopted throughout WA, and dual naming will become commonplace.

The earliest European names given to WA islands are Dutch: Dorre (1616), Rottnest (1697), and Dirk Hartog (1697). Most names, not surprisingly, are English, dating from Vancouver's visit in 1791 (Albany region), Flinders in 1801-1802 (Esperance region), King in 1819-1822 (Kimberley and Pilbara regions), and Stokes in 1838 (Kimberley region) and 1840 (Pilbara offshore and Houtman Abrolhos). Numerous islands have French names, following visits in 1792, 1801, and 1803 by the navigators d'Entrecasteaux, Baudin, Hamelin, and Freycinet. Apart from descriptive names, the English honoured royalty, aristocrats, bureaucrats, and patrons, whereas the French celebrated many of their illustrious scientists and literary figures, especially for the islands constituting the Institut group off the Kimberley coast.

Many small, insignificant islands have lacked names until recently and some still do. Studies on these by scientists and their need to refer unequivocally to them in publications have resulted in many prominent naturalists and scientists being commemorated. This applies particularly to between Dongara and Lancelin, in Houtman Abrolhos, and in the Archipelago of the Recherche. Australian plant names have been applied to previously-unnamed islands in the Montebellos, complementing English plant names given to islands by the British atomic weapons test group in the 1950s.

GEOGRAPHY

The ten regions recognised in this book show remarkable differences in the number of islands present (Table 12.1). Why is there such a lack of uniformity? At first glance this disparity in distribution and number appears haphazard, but several factors are relevant, including geography and geomorphology of the adjacent mainland coast. Most WA islands occur within nine named archipelagoes, more than any other state or territory of Australia.

Table 12.1. Number of islands in the coastal regions of Western Australia.

Chapter	Region	Number of islands	Rank
2	Tropical cays	c. 24	10
3	Kimberley	2633	1
4	North-west inshore	204	5
5	Pilbara offshore	c. 236	3
6	Shark Bay	42	8
7	Mid-west inshore	40	9
8	Houtman Abrolhos	210	4
9	Perth	121	6
10	South-west corner	75	7
11	Archipelago of the Recherche	c. 345	2

Let us travel along the WA coastline from its north-eastern extremity to its south-eastern terminus (Map 12.1). In the Kimberley region, there are few islands between the WA/Northern Territory border and Cape Londonderry. Then, along the north-west Kimberley coast, there is a large number in the Bonaparte and Buccaneer Archipelagoes. From Cape Leveque south to Broome few islands occur, while offshore from Eighty Mile Beach (from Broome to Cape Keraudren) there are none. There are very few from there to Port Hedland. Three large groups of islands are encountered to the south: Dampier Archipelago near Karratha, many sandy and mangrove islands from Cape Preston to Exmouth Gulf (some not named), and several clusters well offshore (Barrow and adjacent islands, the Lowendal Islands, and the Montebello Islands).

From North West Cape south to Point Quobba we cross the Tropic of Capricorn and very few islands occur. Then we encounter the archipelago in Shark Bay, with four large islands and 36 smaller islands. No islands occur adjacent to the Zuytdorp Cliffs between Shark Bay and Kalbarri. Offshore from Geraldton is another large archipelago comprising three island groups – Houtman Abrolhos. From Dongara to Yanchep there is a linear group of small islands, all inshore, while in the Perth region there are several clusters of islands centred on Rottnest Island and Shoalwater Bay.

Much of the coast between Becher Point and West Cape Howe has either no islands or very few. Between Torbay Head and Mermaid Point there is another, though moderately sized, unnamed archipelago. This includes four large islands and a few small ones. Then, from Mermaid Point to east of Shoal Cape, islands are few and small. From Shoal Cape to Butty Head there is a single island, Red Island (not to be confused with an island of the same name farther west). From Butty Head east to Israelite Bay another great archipelago occurs, the Archipelago of the Recherche. Finally, from Israelite Bay north-east to the WA/South Australian border, islands are absent.

Coasts that are flat with few hills tend to have few islands. After the LGM, rising sea levels flooded what became the inner continental shelf and the limited number of tall hills resulted in few islands being formed. In contrast, coasts with extensive topographic variety, including deep valleys, hills, mountains, peninsulas, promontories, and cliffs have a configuration suited to the fragmentation of higher portions of the land and thus the formation of many islands. In contrast, the steep limestone Baxter Cliffs where the Nullarbor Plain meets the Southern Ocean are subjected to intense erosion which is unproductive for the creation of islands. Coasts with limestone (aeolianite) ridges characteristically erode to form reefs and residual mushroom-shaped stacks.

AREA

About 70% of islands with an area of 100 ha or more occur along the Kimberley coastline (Table 12.2). There is a great regional disparity in the area of the largest island(s), with the northern regions having the largest islands. In contrast, there is less difference between regions in the area of their smallest islands. Any differences can be explained in terms of exposure to wave action, elevation, and rock type.

Table 12.2. The largest and smallest islands in each region.

Chapter	Region	Largest island(s) (ha)	Smallest vegetated island(s) (ha)
2	Tropical cays	Adele (218), West Lacepede (82), East Lacepede (63), North Turtle (50), West Ashmore (28)	Cartier (2), East Lacepede (<0.5), Splittgerber (0.1)
3	Kimberley	Wurrulgu (Augustus) (18 929); Wuuyuruu (Bigge) (17 128)	several <0.1 ha, some only a few square metres
4	North-west inshore	Murujuga (11 804), Dolphin (3281), Enderby (3241), West Intercourse (2367)	Great Sandy (18), Cohen (16), Maryanne (6)
5	Pilbara offshore	Barrow (23 569), Hermite (985), Trimouille (499), Northwest (123), Varanus (80)	unnamed W of Bluebell Island (0.02)
6	Shark Bay	Dirk Hartog (62 000)	Meade (0.19)

7	Mid-west inshore	Boullanger (34), Escape (27)	Edward (0.2), Osprey (0.1), Target Rock (0.08), Orton Rock (0.04), North Ronsard Rock (0.04)
8	Houtman Abrolhos	West Wallabi (587), East Wallabi (307), North (176), Pelsaert (166), Rat (61)	Many << 1 ha
9	Perth	Rottnest (1705), Garden (1054)	Gull Rock (0.04), East Shag (0.04), and many <0.01
10	South-west	Bald (809), Breaksea (104), Eclipse (99)	Flat Rock (0.3)
11	Archipelago of the Recherche	Middle (1036), Mondrain (810), Salisbury (368)	Harlequin (8.3), Slipper (7), Creak (7), SW of Lucky Bay (6), Roy (6), Button (5), N of Owen (5), Hector Rock (4), Limpet Rock (4), Cranny Island (4), islet north of North Twin Peaks Island (3), Ruby (3), John (3), unnamed island in Rossiter Bay (1.5)

MAXIMUM ELEVATION

There is considerable variation between regions in the height of the tallest islands, but less so with the lowest island(s) (Table 12.3). Most islands have low elevation. Differences can be explained in terms of wave action, geology and island area.

Table 12.3. The maximum and minimum elevation of islands by region.

Chapter	Region	Tallest island(s) (m)	Lowest vegetated island(s) (m)
2	Tropical cays	North Turtle (9), West Lacepede (7), Browse (6), Sand (6), Sandy (6), Bedout (4)	Cartier (1), West Ashmore (2.5), Bedwell (2), Middle Lacepede (2)
3	Kimberley	Adolphus (220), Bigge (143), Lacrosse (133)	West Montgomery (2), unnamed mangrove islands in Ord River (3)
4	North-west inshore	Depuch (157), Dolphin (137)	Low (1), Keast (2), Maryanne (3), Sandy (3)
5	Pilbara offshore	Barrow (65), Trimouille (43), Northwest (23), Varanus (23), South East (21)	Double (16), Flag (10)
6	Shark Bay	Dirk Hartog (185), Bernier (37)	Meade (2)
7	Mid-west inshore	Wedge (19), Lancelin (17), Favorite (16)	Edward (5), North Essex Rock (5), Osprey (4), Tern

			(4), unnamed south of Osprey (2)
8	Houtman Abrolhos	East Wallabi (15), West Wallabi (15), North (15)	Many 1-3
9	Perth	Garden (55), Rottnest (30), Penguin (13)	Many unnamed (1-2)
10	South-west	Bald (311), Chatham (186), Michaelmas (152), Sugarloaf Rock (42)	Green in Oyster Harbour (12), Gull Rock (10), Seal near Cape Leeuwin (9), Sandy (6), Whaleback (6), Flat Rock (3)
11	Archipelago of the Recherche	Mondrain (222), Remark (220), South Twin Peak (182), North Twin Peak (187)	Fur Rock (16), Wickham (12), Lorraine (10), Hector Rock (9), N of Owen (8), Ruby (8), Arid (5)

ISOLATION

Islands in the regions vary in their degree of isolation from the mainland (Table 12.4). Although it seems a simple exercise to measure distance from the mainland, some larger offshore islands probably act as a source area for the biota on nearby islands. Thus Barrow Island may be more appropriate as the source area for islands in the Montebello and Lowendal Groups, as these groups were joined only a few thousand years ago. Similarly, in the Archipelago of the Recherche, Middle and Mondrain Islands may act as source areas for nearby islands. Small islands between larger areas are termed 'stepping stones'.

Table 12.4. The most remote and least isolated islands in each region.

Chapter	Region	Most remote (km) from mainland	Least isolated (km) from mainland
2	Tropical cays	Cartier (290, but 188 to Roti, Indonesia), Ashmore Group (351 but 146 from Roti), Sandy - Scott Reef (310)	North Turtle (18), Lesueur (9), Sandy – Cape Talbot (1)
3	Kimberley	Champagny (30), Cassini (30), North Maret (30)	Reveley (1), Mary (1), Jar (1), Boongaree (1), Uwins (1), Storr (1), Koolan (1)
4	North-west inshore	Airlie (30), Bessieres (24), Thevenard (20), North Sandy (20)	Murujuga (joined at low water), South Mangrove (5), Depuch (3), Potter (1.4), Finucane (0.3)
5	Pilbara offshore	North West (90), Barrow (55), Varanus (60, but 11.5 from Barrow)	West Intercourse (1.3), Trimouille (1.2 to North West), Dixon (0.6), Middle (0.4 to Barrow), Dolphin (0.4), S of Varanus (0.1 to Varanus),
6	Shark Bay	Dorre (55), Bernier (40)	Dirk Hartog (1.5), islet near Eagle Point (0.23), Mead (0.18), unnamed in Brown Inlet (0.14)
7	Mid-west inshore	Beagle (9), South Fisherman (6), South Outer Rock (6)	Boullanger (1.2), Lancelin (0.6), South Ronsard Rock (0.6), Edward (0.1), Snag (0.1)
8	Houtman Abrolhos	West Wallabi (70), Eastern (70), Pelsaert (60), North (60)	Many <100 m to nearest larger island
9	Perth	Rottnest (18), Carnac (7, but 3.6 to Garden Island)	Garden (2), Penguin (0.7); many <100 m to nearest larger island
10	South-west	Investigator (20), Eclipse (6), Breaksea (4), (E) Red (3.9), Sandy (2.6), Michaelmas (2.2), Seal in King George Sound (2)	Bald (1.4), Chatham (1.2), Cheyne (1), Saddle (1), (W) Red (0.9), Seal near Cape Leeuwin (0.8), Quagering (0.8), Green (Cave Point) (0.7), Hamelin (0.7), Gull Rock (0.6), Flat Rock (0.3), Stanley (0.25), Coffin (0.2), West Doubtful (0.2), Mistaken (0.13), Shelter (0.13), Richards (0.1), Migo (0.06), Sugarloaf Rock (0.05), unnamed islet NW of Mistaken (0.03)
11	Archipelago of the Recherche	Termination (50), Salisbury (45), Daw (33)	Cliff (1.4), Observatory (0.9), NW Rossiter Bay (0.8), Ben (0.5), Barely (0.2), Lorraine (0.2), SW Rossiter Bay (0.1), SE Lucky Bay (0.03)

ROCKS AND LANDFORMS

Where the mainland coast is low and even if there are large rivers present, few islands resulted when sea level rose after the LGM because of infill by sediment. The existence of rugged cliffs and topographic diversity (high hills and mountains, particularly with elevated headlands, peninsulas, and promontories) predisposes the creation of islands (the summits of submerged mountains and hills) as the sea level rose.

The underlying geology – limestone (aeolianite), sandstone, and igneous or metamorphic rocks – also determines the abundance of islands and shapes the landform of these islands. Limestone rocks are situated on low ridges and on mostly submerged reefs. Through the processes of wind action, salt deposition and wave action, these platforms erode into small islets shaped like mushrooms. Many are found in Shark Bay, between Dongara and Lancelin, around Rottnest Island, and in Shoalwater Bay.

Strong winds can also have a creative action, as on Rottnest, Garden, Carnac, and Penguin Islands. There, winds have blown sand on top of the aeolianite rock to form sand dunes. In the tropical north, cyclones result in the flooding of rivers; these bring sand into the ocean and over time this is deposited on reefs, forming low islands on those reefs that are protected enough from the swell.

Sandstone islands occur abundantly in Kimberley region, and granite islands are prevalent between Cape Leeuwin and Israelite Bay on the south coast of WA. These rock types are responsible for the beautiful scenery of both the mainland coast and nearby islands. These agreeable features, together with beaches and wildflowers, are attractive to tourists.

CLIMATE

Mean annual rainfall in coastal WA is not uniform. It is greatest in the north-western part of Kimberley region and the small section of the south-western coast between Cape Leeuwin and Albany. These regions have a mean annual rainfall of more than 1000 mm. The driest islands are those between Dampier and Dirk Hartog Island, which receive on average 200-300 mm per annum. Between these extremes, the Buccaneer Archipelago (Kimberley region) receives between 800 and 900 mm, islands between Lancelin and Cape Naturaliste, and most of the Archipelago of the Recherche, have an average annual rainfall of more than 600 mm, with some islands between Cape Leeuwin and Albany receiving more than 1000 mm.

There is also a striking difference in seasonality of rainfall. The Kimberley region receives almost all of its rainfall during the wet season from December to March (when temperatures are high). This is opposite to the south-west (Dirk Hartog Island to Archipelago of the Recherche); there, most rain falls in late autumn, winter, and early spring (April-September). Between Carnarvon and Port Hedland there is a gradual transition from a winter- to a summer-rainfall pattern.

The combination of low rainfall and high maximum temperatures experienced by species in Kimberley region should necessitate adaptations of species different from those required by species living under the combination of high rainfall and low temperatures of the south-west.

EXPOSURE TO WAVE ACTION AND SALT SPRAY

Normal exposure. All islands are affected to some degree by deposition of salt generated by wind and by wave action. For example, islets situated in the sheltered bays and coves of Rottnest Island seldom experience wave heights more than 0.5 m, whereas waves in more exposed locations may reach 2.5 m. Low, small rocks have no chance of escaping wave action (Plate 12.1). This is also the fate of larger and higher islets (Plates 12.2, 12.3). However, small islands sheltered from severe wave action can often support a small area of vegetation if soil is present.

Most islands are exposed to the full force of the ocean swell. Waves breaking on small, low islands may have either removed any soil and vegetation or restricted them to the highest parts sheltered from the prevailing wind. High islands are less affected but usually show zonation of vegetation, with plant species least tolerant of salt occurring only at higher elevations. Halophytic species persist and dominate vegetation closest to the shoreline. At even higher elevations, physically destructive wave action is no longer relevant but the deposition of salt spray on the exposed side of an island will curtail the growth of trees and tall shrubs, and often cause planation of shrubs. Topographic variation in the form of gullies, valleys and ridges can further mitigate the deposition of salt spray and its impact on both vegetation structure and its floristic composition.

Extreme exposure and tidal surge. A major difference between tropical and temperate parts of WA is exposure to cyclones. These generate wind speeds exceeding 250 km/h, wave heights of 13 m or more, and tidal surge. Tropical cyclones seldom impact the south-west, where the main source of strong winds is winter low pressure systems from the west. These gales may generate wind speeds up to 120 km/h or more. In the period from 1829 to 1992 there were 28 severe gales with gusts exceeding 120 km/h.

An even more extreme source of wave action is the very rare occurrence of a tsunami. The north-western coastline of WA, situated south-east of the Sunda Arc, is fully exposed to forces generated by distant earthquakes and volcanic eruptions. No tsunami is known to have impacted significantly on WA islands in historic times, but this was not the case during recent millennia between Shark Bay and Kimberley region. The presence of gigantic boulders displaced from the coastline of Dirk Hartog, Dorre, Bernier, Koks, Barrow, Legendre, and Dolphin Islands has been interpreted as evidence of the occurrence of mega-tsunamis within the past 5000 years, possibly caused by offshore earthquakes, local faulting, or (less likely) underwater volcanism or asteroid impacts. Similar boulders on Kimberley islands, such as Lamarck Island, may have resulted from tsunamis generated by major earthquakes near Java. About 2900 years ago a tsunami completely washed over Legendre Island, which has maximum elevation of 17 m. Tsunamis would also have washed over small, low islands such as cays (Adele, Browse Islands) and sandy islands (e.g. Delambre, North Sandy, Great Sandy, Airlie, Thevenard, Bessieres, Serrurier and Muiron Islands). Vegetation and plant and animal species would have been removed and re-established later. It is unknown if Houtman Abrolhos was affected.

Tidal range differs greatly along the WA coast. In the Kimberley tides range up to 12 m, while in the south-west, ranges of around 1–1.5 m are usual. Large tides mean that some islands may be more accessible to mobile mainland species.

There is evidence that sea level rose c. 2 m above the current level c. 6000 years ago. This lasted for c. 1000 years and would have appreciably reduced the area of small, low islands, and presumably affected the extent of soil and vegetation, as well as reducing the size of or even eliminating populations of plant and animal species.

HYDROGRAPHY AND TIME SINCE ISLANDS WERE FORMED

The continental shelf gradually slopes from the mainland coast to the 100 m bathymetric line, after which it abruptly descends to great depths characteristic of the oceans off WA. The reason there are few islands close to the 100 m isobath is that mountain ranges and isolated hills would need to have been extraordinarily elevated above the then-coastal plain to achieve the later status of 'island'. This also explains why continental islands are more numerous closer to the mainland coast.

Some of the islands that lie far offshore were much larger when first isolated. Eclipse Island, for example, was isolated c. 12 500 years ago when sea level rose 50 m after the LGM and was a much larger island than it is now. Houtman Abrolhos was separated from the mainland c. 11 500 years ago, and the three groups comprising this archipelago had separated by 10 500 years ago.

From Cape Leeuwin north to about Cape Preston, most islands formed once the sea level rose to about only 5-10 m below current levels, i.e. c. 7000-7500 years ago. This includes some large islands, such as Garden, Rottnest, and Dirk Hartog Islands. These are therefore some of our 'youngest' islands. Note, however, that many inshore continental islands, and some in the Houtman Abrolhos, are cays, formed by accretion of sand and rubble on reefs. It is likely that some of these are as young as c. 5000 years.

The non-continental islands situated far offshore from the north-western coast of WA – cays such as Ashmore, Browse, Cartier, Cunningham, Adele, and Bedwell Islands – occur on top of coral reefs surrounded by depths of 100-300 m.

Why is it important to take an interest in bathymetry? Islands closest to the present mainland coast are the youngest, and thus have had the least time (c. 7000 years) for extinction of populations of plant and animal species to occur. All other factors being equal, the oldest islands have had more time for extinctions to take place than the youngest islands. Therefore young islands of comparable size to old islands are expected to support more species than older islands.

BIOLOGICAL DISCOVERY

First Nations people have a wide knowledge of the plants and animals of their country and have names for conspicuous or other important species, including those valuable as food. Animals are part of 'the dreaming': people's understanding of the world and its creation. Names and information about places and species were passed down orally from generation to generation.

Scientific knowledge about the terrestrial ecology and biodiversity of the islands of WA has been gained incrementally since William Dampier collected botanical specimens on Dirk Hartog and East Lewis Islands in 1699. Sailors marooned in the Houtman Abrolhos in 1627, followed by visits to many other islands by English and French navigators from 1791, also added new information about the insular occurrence of conspicuous species of plants, reptiles, mammals, and seabirds.

Although colonisation of WA by Europeans from 1826 enabled more opportunities for the acquisition of scientific information about natural history, this process continued for some time to be based on incidental observations. It still focused on conspicuous species and other easily observed species. The first attempts at recording all species of interest to an observer concerned birds: Archibald Campbell (Rat Island, Houtman Abrolhos, 1889); Robert Hall (several islands in Houtman Abrolhos, 1899), and Frederick Whitlock (Rottnest Island, 1903). Birds have continued to be the best-studied taxon.

Opportunistic interest in the herpetofauna of WA islands began in the 1840s on some islands in Houtman Abrolhos. In the early 1900s this expanded to Rottnest and Dirk Hartog Islands. The first attempts at comprehensive sampling began in 1913 (Houtman Abrolhos). Progress, however, was sporadic. More detailed attention to reptiles began later, in the

1950s in the Archipelago of the Recherche, in the 1960s with the islands between Dongara and Lancelin and Barrow Island, and in the 1970s for the Kimberley.

About the same time, careful study of island floras commenced, with Glen Storr prominent in compiling lists of plant species for many islands, including Rottnest and Penguin Islands.

The first biological survey of WA islands was of Houtman Abrolhos in 1913. Its focus was zoological. The second survey, of a selection of islands in the Archipelago of the Recherche in 1950, was more comprehensive and included plant species. Since the 1950s biological surveys of numerous islands have been undertaken, including Bernier and Dorre Islands (1959), Depuch Island (1964), Barrow Island (1960s), and many islands in Kimberley region (1970s). The biota of several of these islands has since been studied many times, and the information now available is quite detailed and geographical coverage is comprehensive. Many islands; however, remain unvisited by naturalists or biologists. Current knowledge of fungi and invertebrates on WA islands is very incomplete.

VEGETATION – ITS DIVERSITY AND EXTENT

The area of an island is the major determinant of the number of types of vegetation present and their extent. Large islands have more variety in their types of vegetation than small islands. The simplest categorisation of vegetation is by a combination of height and identity of the predominant plant species: rainforest; woodland; mangrove; shrubland; heath; and herbfield. The nuances involved with exposure of an island to wave action and strong winds were discussed earlier in this chapter. Most important is topographic variation dependent on the occurrence of several peaks. These peaks result in the formation of valleys and gullies sheltered from winds and waves, and these less exposed sites permit the development of taller vegetation dominated by trees. Patches of trees then aggregate to form eucalypt woodland and forest and rainforest. A mosaic of low and tall vegetation increases habitat complexity and diversity, enabling more animal species to persist.

Rock type determines the physical and chemical composition of soil, which then directly influences vegetation type. Islands with more than one type of soil support a more diverse flora because many plant species depend on a particular range of acidity or alkalinity (pH) and ionic composition of soil. These differences are clearly evident with the volcanic and sandstone soils of islands in Kimberley region, and with the granitoid and aeolianite soils of islands along the south coast of WA.

Vegetation on islands is usually similar to that on the adjacent mainland. This is unsurprising given their shared geology, soils, climate, and flora. Thus fig trees occur only on islands of Kimberley region, Barrow Island and the Dampier Archipelago. Mangroves occur (though not always continuously) on many islands between Exmouth and Cambridge Gulfs, with a remarkable outlier on at least eight islands in Houtman Abrolhos. The number of mangrove species is highest (15 species) in the north and least (one species) in the south. Vegetation dominated by *Callitris*, *Eucalyptus*, and *Melaleuca* trees occur (or formerly occurred) on the larger islands in the south-west between Perth and Cape Arid.

NATIVE PLANT AND ANIMAL SPECIES

Reduced number of species on islands. A key attribute of all islands is their depauperate biota. Why and when do native species become extinct on continental islands? Which species are more prone to extinction? How does the process of selective extinction occur?

The most obvious change is the immediate impact of isolation: movement of species from the mainland to the newly-formed island becomes more difficult, particularly for plant species and animal species unable to fly (all reptile species and all species of terrestrial mammals except bats). For many species, immigration of individual plants (as seeds) and animals of most species ceases.

In any ecological community most species are rare and are thus more vulnerable to local extinction than species that are abundant. Extended periods of drought may result in large demographic fluctuations, and may edge species closer to a tipping point. Once an island becomes isolated from the mainland, it obviously becomes a smaller piece of land than the adjacent mainland, and fewer individual organisms are able to share the reduced area. This can lead to rapid changes in the extent of habitat, its suitability, as well as increases or decreases in the abundance of some competitors (of either the same or another species) for resources such as food and shelter. Some predators may also respond to changes in the abundance of their prey. Such environmental and ecological changes may necessitate adjustment by species in their feeding ecology and tolerance of slight alteration of habitat or accommodation to a sub-optimal habitat. These processes can benefit some species (which may become more abundant); however, most of these changes are unfavourable and conspire to deplete the number of species in the newly-isolated insular ecological community.

Evidence for the likely rapidity of these changes comes from the fragmented landscape of the agricultural region of south-west WA known as the Wheatbelt. A 14 million hectare area of native vegetation present in 1829 had been reduced to 13% of its original area by 1987. The natural bushland left in the wheatbelt now comprises 639 remnants (including many nature reserves), with a median area of 114 ha. Detailed study of some of these remnants has revealed that 15 bird species disappeared from Kellerberrin district in a period of only 80 years after clearing 93% of the original bushland for farms. In smaller remnants (<100 ha), only 20 bird species dependent on native bushland have persisted for c. 50 years since their isolation. After an additional 15 years, four more species had become locally extinct.

On nature reserves, the number of passerine bird species dependent on remnant native vegetation in the wheatbelt now differs little from the number of passerine bird species present on offshore islands in the same size range. This demonstrates how rapidly the process of local extinction follows from isolation. In contrast, the number of lizard species on wheatbelt nature reserves is still higher than on islands of comparable area, indicating that lizard species have persisted longer than bird species on these remnant bushland reserves.

Some small patches of woodlands on farms (up to c. 20 ha in area) have been isolated for c. 60 years. These have no more than six lizard species per isolate, and this is similar to continental islands of comparable area off the western coast of WA. This is suggestive of rapid loss of species following disconnection from the mainland.

Some species show the reverse trend. Seals, seabirds, and sea turtles breed exclusively (or almost so) on islands. This is because terrestrial predators failed to persist on small islands.

Increased isolation from the nearest source area. Islands remote from the mainland (or a larger island) have fewer species than islands situated closer inshore. However this generalisation has exceptions. Some remote islands can retain many of their original species if their area is very large. Barrow Island is the outstanding example of this phenomenon in WA.

Hyperabundance. Lack of predators can result in prey populations becoming extremely abundant. Small islands in particular usually lack carnivorous marsupials, native rodents, and visits by Aboriginal people. They are able to support large breeding populations of seabirds and some have retained populations of herbivorous marsupials such as quokkas, rock-wallabies, tammar wallabies, and hare-wallabies, as well as omnivores such as bandicoots, and bettongs which are mostly fungivores.

Some predator species become more abundant on islands, capitalising on the ready availability and increased abundance of seabirds and wallabies. Examples include the dibbler, King's skink, and the carpet python.

Habitat changes. Some island populations live in habitats quite different from those on the mainland. A familiar example is the quokka on Rottnest Island, living in wattle scrub and low heath with very limited available water. On the mainland this species inhabits dense vegetation along streams. Spotless crakes, which usually inhabit freshwater wetlands, live in exposed samphire and mangroves on Pelsaert Island, tawny grassbirds live in beach spinifex on Adele Island, and the little grassbird inhabits dense *Nitraria* shrubland on Sandland Island.

Behavioural changes. Owing to the absence of predators, some species become very tame on islands, and can be easily approached. Examples include quokkas on Rottnest Island, Cape Barren geese on islands in the Archipelago of the Recherche, spotless crakes foraging around an observer's feet on Pelsaert Island (unlike on the mainland where this species feeds on mudflats close to vegetation and is difficult to observe), and buff-banded rails on Penguin Island foraging under picnic tables (Plate 12.4). Bridled terns breeding on Penguin Island in Shoalwater Bay south of Rockingham (much visited by tourists for at least 40 years) allow closer approach by an observer (1-2 m) before taking flight compared to Bridled Island (Lowendal Group) which is not visited by tourists (5-6 m).

Mammal species also illustrate this point. Hare-wallabies enter campsites on Bernier and Dorre Islands, golden bandicoots enter dining areas on Barrow Island, and spectacled hare-wallabies, brushtail possums, and boodies forage near the accommodation areas on Barrow Island. The hare-wallaby that originally occurred on Hermite Island could easily be run down. This tameness has meant that mammals translocated from islands to the Australian mainland are not wary of introduced predators such as dogs, foxes, and cats.

On Carnac Island, the (introduced?) tiger snake is docile compared to the adjacent mainland. At Herdsman Lake, tiger snakes continually aggravated exhibit the flat-neck posture, which is a typical warning signal. Furthermore, Carnac Island neonates that are reared in captivity, and thus are regularly confronted by human predators, become more defensive in comparison with neonates exposed to natural conditions on the island.

Bird song may also differ on islands, as demonstrated by populations of the singing honeyeater and red-capped robin on Rottnest Island. The length of breeding season may also differ. On Rottnest Island the quokka breeds for six months, half the length of time on the mainland.

Morphological changes. The following have been documented. An increase in body mass has occurred with the singing honeyeater on Rottnest Island and the perentie on Barrow Island relative to the adjacent mainland. Four species exhibit nanism, i.e. a decrease in body mass: the painted button-quail in Houtman Abrolhos; the euro on Barrow Island (about half as heavy as on the mainland, Plate 12.5); the bobtail on Rottnest Island; and the dugite in the Archipelago of the Recherche. The causes of such changes have yet to be ascertained.

Changes in body colour are also noteworthy. The white-winged fairy-wren on Dirk Hartog and Barrow Islands has black-and-white plumage instead of the blue-and-white plumage on the adjacent mainland. The singing honeyeater on Rottnest Island is darker in plumage than populations on the mainland coast opposite.

Genetic changes. All island populations examined so far (mostly mammals) show reduced genetic diversity in the form of low heterozygosity, with many alleles fixed. These reductions could result from genetic drift, a bottleneck in the past (possibly caused by a prolonged drought significantly decreasing population size), or both factors. Because of restricted extent of suitable habitat, only a small population of rock-wallabies exists on Barrow Island and genetic analysis has shown that it has unprecedented low levels of genetic variation. Similar low genetic diversity has been found in hare-wallabies on Bernier and Dorre Islands.

TERRESTRIAL PLANTS

Island floras are generally dissimilar to the flora of the nearest mainland coast (Table 12.5). Wattles are, however, well represented on islands, and often one or two species dominate part of the vegetation on large islands (e.g. Rottnest, Garden, Middle Islands) as well as on some small islands (e.g. Anchor and Airlie Islands). Eucalypt species are poorly represented even on large islands, and common coastal mainland species such as jarrah, marri and tuart are absent from nearby large south-west islands. Species belonging to large genera of Proteaceae (e.g. *Banksia*, *Dryandra*, *Grevillea*, *Hakea*) are missing from most islands. This is also the case for other speciose genera such as *Stylidium*, *Leucopogon*, *Melaleuca*, and *Eremophila*, as well as conspicuous species in the genera *Nuytsia*, *Macrozamia*, *Xanthorrhoea*, *Anigozanthos*, and *Kingia*.

Sensitivity to deposition of salt on foliage may be responsible for the absence of many plant species from islands or their restriction to just a few large islands (Table 12.5).

Table 12.5. Representation of various plant genera on all WA islands relative to the entire mainland coast of WA.

Taxon	No. species recorded on islands (% of species on coastal mainland)	No species recorded on adjacent coastal mainland
Acacia	66 (22%)	299
Eucalyptus	13 (11%)	118
Grevillea	10 (10%)	98
Drosera	10 (15%)	66
Corymbia	10 (56%)	18

TERRESTRIAL INVERTEBRATES

Despite comprising a large proportion of Earth's biodiversity, these animals have not been given the attention they deserve. Invertebrates were first collected from WA islands in the 1890s, but the first comprehensive studies were of butterflies in the 1970s on Koolan Island. This was followed by research on butterflies on some of the islands near Perth, termites on Barrow Island, land snails, spiders and earthworms on several islands in Kimberley region, and terrestrial isopods on Rottnest Island. Earthworms, pseudoscorpions, scorpions, spiders, opiliones, and ants have also attracted attention, but most collecting of these taxa has been opportunistic.

The terrestrial invertebrate fauna (including troglobitic animals) present on Barrow Island is the most completely known. This was achieved so as to provide a baseline prior to greatly-expanded industrial activity on the island. It serves to enhance monitoring essential to biosecurity arrangements recently put in place.

It is not yet feasible for a biogeographical study of invertebrates on WA islands to be undertaken.

TERRESTRIAL VERTEBRATES

Many species of passerine land bird, non-passerine land bird, skink, and gecko occur on WA islands (Table 12.6).

Table 12.6. Representation of various animal taxa on all WA islands relative to the entire mainland coast of WA.

Taxon	No. species recorded on islands (% of species on coastal mainland)	No species recorded on coastal mainland
Frogs	30 (42%)	72
Skinks	75 (51%)	147
Agamid lizards	16 (39%)	41
Varanid lizards	8 (50%)	16
Geckoes	36 (67%)	54
Pygopod lizards	13 (54%)	24
Snakes (non-marine)	32 (37%)	86
Land birds (non-passerine)	52 (53%)	99
Land birds (passerine)	95 (59%)	161
Seabirds	25 (95%)	1
Terrestrial mammals (excluding bats)	32 (35%)	91

BIOGEOGRAPHY

The biota of continental islands is almost always a simplified version of that present on the adjacent mainland coast. Thus islands of Kimberley region have many of the parrot, frog, lizard, bird and mammal species present on the mainland, but not all. The large size of some of these islands, combined with their well-developed vegetation (particularly rainforest, eucalypt woodland, and mangal) and proximity to the mainland, have acted together to form a fairly rich fauna. On small islands, depleted biotas result in unusual, anomalous, or unique combinations of species that differ from those on the mainland.

In contrast, cays and depositional islands have formed after the sea level stabilised at the conclusion of the LGM. These islands have gained their biota by individuals crossing the intervening ocean, either by flying (birds and some insects) or other means of aerial transport (spiders), transmarine transport on wood or vegetation washed out to sea (reptiles), and in the case of Kimberley region via Aboriginal watercraft (dingoes). In Houtman Abrolhos, ten small (<2 ha) islets formed by the accumulation of coral fragments have no more than two lizard species present.

Biogeographers seek to discover patterns of occurrence of species on islands, assess their randomness or orderliness, and explain why some species have persisted for millennia and others have not. Examples of idiosyncratic distributions include the red-capped robin on Rottnest Island. On the adjacent mainland coast this species is only an infrequent summer/autumn visitor from inland. Presumably just before the island disconnected from the mainland this species was a resident on the mainland. Other examples with different taxa include the different subspecies of rock-wallabies on islands that isolated at different times in the Archipelago of the Recherche.

Genera with many species (congeners) often produce interesting geographic patterns on islands. For example, goannas are well represented on WA islands, with three species present on many tropical islands: *Varanus acanthurus* (present on 26 islands); *V. glauerti* (26 islands); and *V. glebopalma* (15 islands). In south-west WA only two species are present and these occur only on a few islands: *V. gouldii* (two islands) and *V. rosenbergi* (one island). Focus on just the islands of Pilbara region reveals even more disparities. The Montebello Islands have *V. gouldii* and *V. acanthurus* but the Lowendal Group has only *V. acanthurus*. Barrow Island has *V. giganteus* and *V. acanthurus*; Potter Island has *V. brevicauda* but Thevenard and South Muiron Islands have only *V. acanthurus*. The islands of Dampier Archipelago have five goanna species and these have enigmatic distributions. In Kimberley region six species occur on islands, and these occur alone on some islands and in various combinations on larger islands.

Many similar examples of unexpected species combinations can be found with plant, frog, snake, land bird, wallaby, and bandicoot species. Whether these are the result of random extinctions or of island characteristics is yet to be established.

SPECIATION

Although many species and subspecies endemic to one or more islands have been described and named – birds (46), mammals (14), reptiles (12), and butterflies (3) – most are no longer considered valid by taxonomists. At present the number of endemic taxa considered to be valid is birds (eight subspecies), mammals (three subspecies), reptiles (two species and eight subspecies), and butterflies (one subspecies).

Molecular analyses of island and mainland specimens are becoming the standard means of appraising a taxon's endemic status, replacing the previous method based on differences in body size, plumage, and other morphological characteristics. Once these new methods are applied, it is likely that some currently-recognised endemic species and subspecies will be invalidated. However, previously unrecognised differences in phylogenetic lineage may result in the recognition of new endemic taxa.

SPECIES TURNOVER

In 1963 two American ecologists, Robert MacArthur and Edward Wilson, formally proposed a new theory explaining why islands have the number of species that they do. They postulated a dynamic equilibrium, achieved by the number of species becoming extinct on an island being balanced by the number of species migrating to the island. Their theory was greeted with enthusiasm by island biogeographers and stimulated new research. Some of this demonstrated that the theory was not as universal in scope as first thought.

Ascertaining the numbers of native species becoming extinct and colonising the same island is a fundamental way of testing the validity and scope of the theory. What do the data from WA islands tell us?

Plant species. A census of native plant species on 30 islets around Rottnest Island two years apart found no change in species on 15 islets (static equilibrium), immigration on three islets, and local extinctions on 12 islets. Of the larger islands in Perth region, three native species became extinct on Carnac Island sometime after 1839, and one species has disappeared from both Rottnest and Penguin Islands. There is thus little evidence of dynamic equilibrium.

Reptiles. Records are uncertain, with one confirmed local extinction of a lizard species on one island, and five suspected local extinctions of subspecies on two WA islands. However, these are believed to be caused by human activities and thus do not constitute a fair test of the theory.

Land bird species. There are 32 WA islands for which adequate records exist of at least two visits. Many reveal no extinctions or immigrations. This is the static form of equilibrium and not the dynamic form as postulated. For the 19 other islands, immigrations (32) exceed extinctions (13). This is evidence of a dynamic process but not of equilibrium.

Seabird species. More than 60 instances of turnover have been recorded on WA islands. Immigrations (establishment of breeding colonies) total 44, greatly outnumbering extinctions (22). Eight species have extended their occupation of WA islands since records by Europeans commenced. Extension of breeding range southward in some species is thought to be associated with increased sea temperature via the poleward-flowing Leeuwin Current.

Mammal species. Of the 54 insular populations and wallabies and bettongs on WA islands, only 5% have become extinct since European settlement of WA. None of these is attributable to natural causes. There has been no natural colonisation of islands by these mammals. This is evidence of static, not dynamic, equilibrium.

In summary, the evidence from these five groups is that native species exhibit minimal turnover, and application of the dynamic equilibrium theory of island biogeography does not provide a valid account of natural change on WA islands.

VEGETATION CHANGE

Changes with time in the structure of plant communities, including vegetation height and the identity of the dominant species, are difficult to ascertain without maps. Such vegetation maps require expert on-the-ground inspection of an island coupled with careful interpretation of aerial photographs taken several years apart but at the same time of year.

Reliable vegetation maps are available for the three largest islands in Perth region. During the period 1942-1976, Garden Island shows a retreat of Rottnest cypress *Callitris preissii* stands near its northern end, replacement of *Acacia cochlearis* thickets by *Acacia rostellifera* scrub near the southern end, invasion of grassland at both ends of the island by A. *rostellifera*, replacement of A. *rostellifera* scrub on the eastern side by Rottnest teatree *Melaleuca lanceolata*, and increase in the extent of cypress and teatree in the interior of the island.

Rottnest Island has experienced marked change in its vegetation. It seems that *Acacia rostellifera* scrub originally covered much of the island, with forest formed by three tree species restricted to the sheltered eastern sector of the island. A series of wildfires and increased density of the quokka have resulted in fragmentation of the forest and scrub vegetation types.

The vegetation of Carnac Island has been studied since 1951, and has also experienced dramatic changes in its vegetation types. These have been attributed to prolonged drought, salt toxicity amplified during gales, nesting seabirds, browsing by rabbits, and competition from exotic plant species.

Vegetation maps have also been prepared in recent decades for islands in Houtman Abrolhos, the largest islands near Albany, and the islets between Dongara and Lancelin. These islands need to be revisited and new vegetation maps prepared so that any change can be determined.

NATURAL FACTORS LEADING TO CHANGE IN THE BIOTA (NATIVE SPECIES)

Most WA islands have experienced no human interference since they were separated from the mainland following rising sea level after the LGM. WA islands have existed for 5-14 millennia, and have thus had a long time to be exposed to periods of unusually low rainfall (drought), high rainfall (flooding), and high winds (gales and/or tropical cyclones). These natural events are likely to have respectively caused short-term reductions in population size (with some terminating the occurrence of some species), erosion and perhaps loss of soil, and killing and/or removal of foliage.

Some islands have experienced infrequent but extraordinary incidents. As previously mentioned, tsunamis result in one or more gigantic waves washing over small, low islands. These will remove soil, animals, and vegetation and presumably cause extinctions of some species. Lightning strikes, when combined with prolonged drought, should result in intense wildfire which may be unstoppable and burn out an entire island. Again, some extinctions may result, but a positive consequence is that the process of secondary succession is initiated. This is likely to cause temporary change in the dominance of some plant species, and also produce an abundance of new, protein-rich, growth that benefits both leaf-eating insects and grazing and browsing mammals.

Given that most species in an ecosystem occur in low numbers, their populations are likely to be sparsely distributed, and thus vulnerable to local extinction or reduction in genetic fitness.

The formation of small islands invariably results in the local extinction of most predator species, especially mammals. Large reptile species, such as King's skink and snakes, are known to prey on eggs of seabirds. Their disappearance is thus favourable to nesting seabirds.

Another factor important on small islands is itinerant breeding of the great pied cormorant. This species nests colonially in large numbers (up to several hundred pairs). Their nutrient-rich faeces and excessive trampling of vegetation leads to temporary local destruction of woody perennials on which they nest and their replacement by soft annuals unsuited to provide platforms for nesting. These cormorants thus do not show the degree of site fidelity characteristic of most seabird species. Frequent relocation may also serve to reduce the impact of ectoparasites on young cormorants.

CHANGE IN THE BIOTA (NATIVE SPECIES) ASSOCIATED WITH ABORIGINAL ACCESS

The arrival of Europeans to the islands that are now part of WA in the 1600s is discussed in earlier chapters of this book. This extraordinary series of events, culminating in settlement in the 1820s, caused great destruction of the biota of some islands. This is discussed in Chapter 13.

In this Chapter, attention is instead focused on non-European humans. It is well known that non-Europeans who colonised islands had substantial impacts on islands in the Pacific and Indian Oceans. For example, on remote Easter Island and Madagascar they introduced rats, burned forests, felled trees, and hunted animal species.

Aboriginal people colonised Australia from Asia on some form of watercraft perhaps as long as 80 000 years ago. The extent of the coastline increased dramatically during the LGM, reaching its maximum area by c. 20 000 years ago. Then, as the sea level rose, much of this coast was reclaimed by the ocean. At some time, most coastal-dwelling Aboriginal groups discontinued use of watercraft. This resulted in the newly-created islands being inaccessible. It was only in Dampier Archipelago and Kimberley region that use of watercraft was retained.

The most striking impact of Aboriginal people on islands is on the diversity of wallaby (including bettong) species. Islands inaccessible to Aboriginals (through lack of watercraft) have more of these species than islands that were accessible. In addition, of the 21 islands accessible to Aboriginals and known to possess at least one species of wallaby, the median area is 1970 ha, and the smallest is 300 ha. This contrasts with the 21 islands with wallabies that were not accessible to Aboriginal people: the median area of these islands is 650 ha, and the smallest island area is only 70 ha. There is also a notable difference in the size of the smallest island with three wallaby species: 11 800 ha for accessible islands, and 4200 ha for inaccessible islands. Low rainfall Barrow Island (23 569 ha), inaccessible, has three wallaby species and a kangaroo; higher rainfall Wurrulgu (Augustus Island, 18 929 ha) and Wuuyuruu (Bigge Island, 17 128 ha), both accessible, each has only a single species of wallaby.

Another way of looking at this is to compare the proportion of islands with an area of 100 ha or more in terms of their accessibility to Aboriginal people. The proportion of islands with at least one wallaby species present is 50% for islands inaccessible to Aborigines, compared with only 12% for islands that were accessible. Predation by Aboriginals, and perhaps by their dingoes at times left behind, presumably eliminated populations of wallabies on the smaller islands. Aborigines would have also lit fires which may have reduced cover thus facilitating hunting of wallabies and bettongs. However, the impact was probably small because fires were managed in a mosaic pattern, in contrast to islands not visited by Aborigines. There fire ignited by lightning strike would have spread a great distance given the continuity of deep litter.

The second major impact of Aboriginal hunting on islands of Dampier Archipelago and Kimberley region was the seasonal harvesting of seabirds and sea turtles (and their eggs). The rudimentary watercraft (log or small raft) did not permit the transportation of large numbers of eggs back to the mainland. Consumption of killed seabirds, turtles, and their eggs would therefore have been constrained by the time Aborigines spent on these islands on each visit, which would have been largely restricted to periods after significant rainfall because of lack of permanent water supply in the dry season on most islands. (Some of the larger islands in Dampier, Buccaneer, and Bonaparte Archipelagoes did, however, have permanent supply of water). The largest populations of nesting seabirds tend to occur on cays beyond the regular reach of their watercraft, and this circumstance should have minimised human impact on this resource.

The third major imprint of Aboriginal access to islands is their use of fire. Setting fire to country is both a cultural activity and a means of readily collecting small animals (rodents, reptiles, small marsupials) either killed during the passage of the fire or sheltering in unburnt patches of vegetation or in burrows. Fire, mostly patchy, was also used to regenerate plant species used for food. Depending on the frequency, intensity, extent, and timing, fire causes plant species sensitive to fire to persist only in areas either less burnt or not at all. Rainforest vegetation on islands of Kimberley region is sensitive to intense fire, which may account for its patchy occurrence. However, people often protected rainforest patches from fire as they were an important source of food.

CONCLUSION

There are many resemblances among islands in most regions of WA. Wave action and wind speeds are physical phenomena with identical effects on islands except when extremes (tropical cyclones, tsunamis) occur. Island area and elevation influence biodiversity on islands in similar ways across all islands, except for extenuating circumstances (elevation too low and subdued to support any soil and vegetation, island too rocky to support much soil, different soil properties because of rock type, landform not conducive to the development of different plant communities).

Causes of major differences in island biotas are climate (the long-term pattern of occurrence of temperature and rainfall), the suite of species that were present on the mainland coast immediately before each island disconnected, the area of an island, the length of time that an island has been isolated, and whether Aboriginal people could access an island.

Sources of information

Chapters 2-11; Abbott & Wills 2016; Bonnet *et al.* 2005; Dunlop 1996; Kitchener & How 1982; Kitchener *et al.* 1980, 1982; Meathrel & Klomp 1990; Saunders 1989; Wooller & Dunlop 1981.

Plate 12.1. Northumberland Rock (elevation 4 m, area 0.35 ha) south of Bald Head, continuously exposed to swell. Credit: I Abbott.

Plate 12.2. Unnamed islet west of Chatham Island, devoid of soil and vegetation as a result of exposure to wave action at all times, despite its area (3.1 ha) and elevation (27 m). May 1975. Credit: I Abbott.

Plate 12.3. Unnamed granite islet east of Breaksea Island devoid of soil and vegetation as a result of exposure to extreme wave action during gales, despite its area (1.1 ha) and elevation (20 m). August 1975. Credit: I Abbott.

Plate 12.4. Buff-banded rail, a species exhibiting tameness at the picnic ground on Penguin Island. April 2009. Credit: AA Burbidge.

Plate 12.5. Barrow Island euro, a dwarf subspecies. April 1985. Credit: AA Burbidge.

Ian Abbott & Andrew A. Burbidge

CHAPTER 13: NATURE CONSERVATION ON WESTERN AUSTRALIAN ISLANDS

'Our home is girt by sea
Our land abounds in nature's gifts
Of beauty rich and rare'

From the Australian national anthem.

'Islands are natural workshops of evolution'

Richard Dawkins

WESTERN AUSTRALIAN ISLAND VALUES

Australia is Earth's unique island continent as well as a land of islands. More than 8300 islands occur around Australia's coastline and in its external territories. Western Australia's (WA) long coastline of 12 895 km plus another 7894 km around islands (Geoscience Australia data), has more islands than any other State or Territory. There are between 3550 and 3750 islands in the State[2], most (2633) in the Kimberley. Mainland WA comprises 32.9% of Australia, but has about 46% of Australian islands, with some significant differences from islands in the rest of Australia (Box 12.1). Most WA islands are small; however, more than 240 are larger than 100 ha, and four of these are larger than 10 000 ha. Despite the small contribution that islands make to the State's land area (0.12%), they are extraordinarily and disproportionally valuable for biodiversity conservation.

Most WA islands are continental in origin, having been isolated by rising sea levels after the last glacial maximum ended c. 18 000 to 20 000 years ago. Most were isolated between 10 000 and 6000 years ago. A few are cays, accumulations of sand or coral rock on reefs and these may be much younger than continental islands and, although often close to the mainland, may never have been connected to it. Some WA cays, e.g. Bedwell and Adele, are effectively oceanic islands (never connected to a continent and distant from one). Cartier Island, and the islands on Ashmore Reef are oceanic island cays, but are Commonwealth Territories, not legally part of WA. WA islands cover a greater latitudinal range, 21.4° (from 13.74° to 35.14°S), greater than those in other Australian states and include a wide range of landforms, rainfall, vegetation, and animals.

[2] Providing a precise number of islands in WA depends on how you define an island as there are numerous, often small, unvegetated rocks scattered along the coast. Geoscience Australia states that there are 3747 islands in WA. A careful count by one of us (IA) using 1:100 000 map sheets produced a figure of 3576, published in 2016. For this book, we did another count using currently available data and came up with 3590, excluding the three islands on Ashmore Reef and Cartier Island that are legally not part of WA.

The biodiversity conservation values of islands arise from their isolation from the mainland. Despite having lost many species since sea levels rose after the last glacial maximum, via the process of extinction of local populations due to their small size and changing climates, our islands harbour species now extinct or threatened on the mainland, protect terrestrial species that nest on the ground, provide breeding places for seabirds, seals and sea lions, indicate what elements of mainland ecosystems once looked like with limited disturbance, and have endemic species and subspecies. As well, some islands are valuable for recreation, and tourism, particularly ecotourism.

Global significance. Western Australia's islands and their plants and animals are of global significance. They cover a wide range of latitudes from the wet-dry tropics through the arid zone to the temperate, Mediterranean-climate of the south-west. They range in size from a few square metres to islands of over 1000 ha, with a few over 10 000 ha. While most are low, a few rise to many tens of metres above sea level. There are several archipelagoes, allowing comparisons between large and small islands with similar climates and landforms. WA's islands are vital for global biodiversity conservation, protect threatened species and seabirds, offer scientific research opportunities and increasingly attract visitors.

Many WA's islands have been largely unaffected by human activity since separation a few thousands of years ago. While First Nations people accessed Kimberley and Pilbara inshore islands, those south of the Pilbara never had human visitors prior to European settlement, and of those, most have never been degraded. This contrasts greatly with islands elsewhere in the world. In Europe, most islands have been dramatically changed through human occupation, farming, and introduction of animals. Polynesians have occupied most Pacific islands for centuries bringing Pacific rats and other changes with them. Almost all Asian islands are occupied. North and South American islands and African islands have been similarly impacted.

Reservation of Western Australia's islands for biodiversity conservation. The concept of setting aside land for the conservation of nature understandably was not a high priority in 1829. The urgent requirement then was to provide land for settlers so that they could secure their economic wellbeing, particularly in feeding their livestock and in planting crops to produce food. This would collectively allow the new colony to prosper and progress. The small number of colonists and limited clearing of native vegetation for the establishment of farms and villages had only localised impacts on native species for a few decades.

From c. 1800 whaling ships from Britain, France, and the USA had begun to kill whales in WA waters. Sealers arrived in the 1820s from Sydney and Hobart, and presumably landed on every island with seals. Both activities preceded the formal occupation by Europeans of WA, first at King George Sound in 1826 and then at Swan River three years later.

Whaling and sealing were unregulated, setting in train the 'tragedy of the commons', i.e. the first exploiters seized as much of a natural resource as they could in order to prevent subsequent exploiters benefiting from wise use of these resources.

On the mainland the colonists hunted animal species for food so as to bide them until their livestock increased in numbers. Once economic success was assured, settlers became less dependent on wildlife for food and then had the leisure time to hunt native animals ('game') for recreational shooting ('sport').

It was not until 1874 that WA legislated its first Game Act. This conferred (nominal) protection during the breeding season of hunted species. The only island species affected was the 'Rottnest snipe', the banded stilt found feeding in the saline lakes of Rottnest Island.

In 1883 the Destruction of Rabbits Act declared (*inter alia*) it illegal to keep rabbits on any island five miles or less offshore of mainland WA. This was to protect populations that had bred from rabbits placed on nine islands to provide succour for shipwrecked mariners. Two years later the Act was amended to render it illegal to keep rabbits on any WA island.

The natural advantages possessed by WA islands, including their (mostly) small size, isolation, infertile soils, paucity of edible vegetation for livestock, lack of permanent water, and absence of Aboriginal people except those with watercraft, served the conservation of insular biotas very well.

When industrial use of islands began in the 1870s, there was no strong legislative oversight. Turtles and their eggs were harvested with no monitoring by government. Mining of guano on islands in Shark Bay and the Houtman Abrolhos was also unmonitored (apart from the measurement of the amount extracted so as to provide a royalty payment to government). Plundering of seabirds and their eggs on remote cays of the Kimberley region by trepang fishers from Asia was unregulated. There were no safeguards in place to prevent pet cats being introduced or black rats from landing from ships collecting guano, or from vessels attending pearling camps on islands.

Effective legislation to protect island conservation values came well after inimical factors began to affect islands. The first island reserved for protection of native game under the *Game Act 1892* was in 1898 at Houtman Abrolhos, many years after guano extraction began. This forbade shooting and the taking of seabird eggs. This action was followed for Pelican Island (Shark Bay) in 1900 and Dorre Island in 1907 (rescinded in 1909, but reinstated after the lock hospitals were abandoned, Chapter 6).

The most significant early initiative was taken by the scientist Guy Shortridge, who visited WA in 1904-07 to collect mammal and bird specimens for the natural history museum in London. Before he departed he was interviewed in Perth by the leading newspaper. He advocated that islands should be protected because they retained populations of native mammals, many of which had become very rare on the adjacent mainland. The newspaper criticised the government for neglecting conservation values of islands.

Shortridge's suggestions were later championed by members of the local science society, resulting in the government reserving Barrow Island (1908), specifically for the protection of native mammal and bird species (Plate 13.1). Momentum then gradually increased with the reservation of more islands for nature conservation: Rottnest Island (1917); Houtman Abrolhos (1929), the islands constituting the Archipelago of the Recherche (1948), and Bernier and Dorre (1957). The 1965 report by the WA Sub-Committee of the Australian Academy of Science Committee on National Parks made wide-ranging recommendations for the reservation of areas within WA, including amendments to the Houtman Abrolhos reserve and the upgrading of the Archipelago of the Recherche reserve to Class A. In the 1970s and 1980s many islands where seabirds nest were reserved following initiatives of the then Department of Fisheries and Wildlife. It was not until 2004 that WA's largest island, Dirk Hartog, was declared a national park, having been a pastoral lease for more than 100 years.

South of the Kimberley almost all WA islands are now national parks, nature reserves or conservation parks, all managed by a single government agency – perhaps a situation unique in the world. The more than 2000 Kimberley islands are now mostly within Indigenous Protected Areas and managed by Traditional Owners – another near-unique occurrence. Although feral animals established on a few WA islands, most are now free of them due to management action.

Along with a stable government and a scientifically-based management agency, these characteristics also make WA's islands friendly places for scientists to work. Many are fascinating places to visit with nature-based tourism increasingly taking advantage of islands.

Islands conserving threatened species. In WA, some islands have proved to be essential for mammal conservation. Of the several thousand WA islands, 88 indigenous terrestrial mammal species have so far been recorded on 155. Four mammal species would be extinct were it not for populations persisting on WA islands (Table 13.1). Mammals that are listed as threatened species that occur on islands but also occur (now rarely) on the mainland are northern quoll, dibbler, golden bandicoot, quokka, rufous hare-wallaby (Plate 13.2), black-flanked rock-wallaby, nabarlek, and ghost bat. The threatened ngwayir (western ringtail possum) occurs on Molloy Island (not an offshore island) in Hardy Inlet, near Augusta. Butler's dunnart, which has not been detected on the WA mainland for decades, occurs on islands in the Northern Territory. Other mammal species that have declined on the mainland have populations on islands: scaly-tailed possum, spectacled hare-wallaby, monjon, nabarlek, northern leaf-nosed bat, bare-rumped sheath-tailed bat, djini (pale field-rat), and koorrawal (golden-backed tree-rat).

Elsewhere in Australia, apart from Tasmania, only a single Australian mammal has been protected from extinction by having a remnant population on a continental island: wopilkara survived on the Franklin Islands in South Australia. Tasmania harbours three mammal species that became extinct on the mainland: eastern quoll, eastern barred bandicoot, and eastern bettong.

While most seabird species nest exclusively on islands, few terrestrial birds occur and breed only on islands. The WA exceptions are the Barrow Island black-and-white fairy-wren (Chapter 4) and the Houtman Abrolhos painted button-quail (Chapter 8). Bird subspecies thought to be confined to Shark Bay Islands (Chapter 6) have been shown not to differ significantly in genetic composition from those on the mainland.

In Australia, reptiles have not declined to the same extent as mammals; however, the Lancelin Island skink is confined to Lancelin Island, the Abrolhos dwarf bearded dragon and the Abrolhos spiny-tailed skink occur only on the Houtman Abrolhos islands, the Rottnest Island bobtail is confined to the island of that name, and a subspecies of the dugite has been recorded from Boxer and Figure of Eight Islands in the Archipelago of the Recherche.

Species that breed on the ground. Some species that breed on the ground cannot withstand mammalian predators. Of particular importance are seabirds and sea turtles. When predators such as black rats and cats establish on islands the typical result is extirpation of most resident seabirds. Seals also breed almost only on islands, possibly because any mainland colonies that existed in the past were extirpated by humans.

Australian sea lions, endemic to Australia, breed from the Houtman Abrolhos south and east into South Australia, with 34 colonies having been recorded in WA, 32 of which are currently active. These are almost all on islands, although a few small colonies have been located at Twilight Cove in WA and in the Bunda Cliffs in South Australia, possibly because these were inaccessible to humans. Australian sea lions are declining in abundance with probably fewer than 10 000 mature individuals remaining.

The other seal that breeds in WA is the long-nosed fur seal, which breeds only on islands; in WA mostly along the south coast, but the species is recovering from past depredations and is extending its range westward and northward as far as Rottnest Island.

Almost all of the 28 species of seabirds that nest in WA breed only on islands, while some shorebirds breed mainly on islands. Very few seabirds breed on the mainland, where nests and chicks are often destroyed by foxes, cats and vehicles, and disturbed by people. The fairy tern, which has it major breeding colonies in WA on islands (Chapter 8), also nests on mainland beaches and spits. In eastern Australia, where breeding is mostly on the mainland, the species is declining and is threatened with local extinction. Shorebirds that breed mainly on islands include the pied and sooty oystercatchers. The eastern reef-egret breeds predominantly on islands.

White-bellied sea-eagles, eastern ospreys and brahminy kites nest on islands, as well as on the mainland. It is not known what proportion of these species nests on islands.

Sea turtles. The four species of sea (or marine) turtles (loggerhead, hawksbill, green, and flatback) that breed in WA nest on both mainland beaches and islands. However, mainland nesting beaches are under increasing threat from foxes, feral pigs and people. Thus, islands are becoming increasingly important in maintaining viable populations. In the Kimberley region, where beaches on the mainland are rare, almost all nesting is on islands, with a major flatback rookery at Cape Domett the exception.

Largely undisturbed ecosystems. One characteristic of WA's islands, often not appreciated, is that, south of the Pilbara (until very recently) humans did not access islands after they separated from the mainland, as Aboriginal people in these areas did not possess watercraft. Islands south of the Pilbara also are infrequently burned, as even when there are lightning strikes, the salty nature of the vegetation inhibits the spread of fires except under extreme fire weather. These combinations mean that many islands have ecosystems that differ from those on the adjacent mainland and can show how mainland, much more disturbed, ecosystems have changed. Islands are thus microcosms of biodiversity and offer excellent opportunities for ecological research.

Island endemics. As most WA islands are continental in origin, and the remainder are recently-developed cays, terrestrial species that occur on them are mostly genetically very similar to mainland populations, although some that have been isolated on islands for many generations have become inbred, losing much of their original genetic variability. Three species and a few subspecies of mammals, birds, reptiles and freshwater fish are endemic to islands (Table 13.1), although some have not yet been subjected to detailed genetic analysis.

Table 13.1 Endemic vertebrate species and subspecies on WA islands.

Species or subspecies	Location
Barrow Island euro	Barrow Island
Barrow Island spectacled hare-wallaby	Barrow Island
Shark Bay hare-wallaby	Bernier and Dorre Islands
Barrow Island golden bandicoot	Barrow Island
Recherche Cape Barren goose	Islands of the Archipelago of the Recherche, occasional on adjacent mainland
Houtman Abrolhos painted button-quail	Wallabi Islands, Houtman Abrolhos
Dirk Hartog western grasswren	Dirk Hartog Island
Barrow Island white-winged fairy-wren	Barrow Island
Barrow Island leopard skink	Barrow Island
Lancelin Island skink	Lancelin Island
Rottnest bobtail	Rottnest Island
Barrow Island blind snake *Anilios longissimus*	Barrow Island
Rottnest dugite	Rottnest Island
Recherche dugite	Boxer and Figure of Eight Islands, Archipelago of the Recherche
Barrow Island cave gudgeon	Barrow Island
subterranean eel (unnamed)	Barrow Island

Many invertebrate stygofauna and troglofauna species are known only from Barrow Island (Chapter 4) and some land snails in the family Camaenidae are restricted to one or two Kimberley islands (Chapter 3).

Four species that became extinct on mainland Australia survived on WA islands (Table 13.2). These populations, sometimes termed 'new endemics', have been the source for translocations (Plate 13.3).

Table 13.2. Species that became restricted to WA islands

Species	Island(s)
Shark Bay bandicoot	Bernier and Dorre Islands
maning (banded hare-wallaby)	Bernier and Dorre Islands
boodie	Bernier, Dorre, and Barrow Islands
djoongari (Shark Bay mouse)	Bernier Island

The nabarlek is now extremely rare on the mainland, but occurs on four Kimberley islands: Borda, Augustus, Hidden, and Long.

FACTORS LEADING TO CHANGE IN THE BIOTA

Non-indigenous species. Island animals and, to a lesser extent plants, are highly susceptible to invasive species. Worldwide, the loss of indigenous animal species from islands is primarily due to introduced predators. WA is no exception. Elsewhere, land clearing and hunting are major threats, but not in WA.

A 2017 review by the authors of this book revealed 266 records of 21 species of non-indigenous mammal species on 138 WA islands. Some of these introductions have led to the loss of indigenous island mammal populations (Table 13.3). Most extirpations are due to feral cats. It has been demonstrated that, in Australia, feral cats have been much more likely to cause mammal extinctions on islands with low annual rainfall, and almost all WA islands that have lost species to cats have a mean annual rainfall of c. 300 mm. Lack of information can make these correlations tricky, as Breaksea Island, with annual rainfall of 900 to 1000 mm may have lost quokkas, because skulls found there are in the WA Museum, and they were probably present in 1881. Lighthouse keepers on the island had pet dogs and cats, and may have shot quokkas for food, and sealers also lived on the island for a period. Rabbits occur there and their presence may have limited food availability for quokkas.

The influence of predation by feral cats is not limited to mammals. For example, hatchling green turtles are easy prey as they traverse bare sand between the hatching site and ocean (Plate 13.4).

A future threat is the possible arrival of cane toads on Kimberley islands. Cane toads have established on Adolphus Island in Cambridge Gulf, washed down the Ord River during floods, and may arrive on other islands, especially along the north-west Kimberley coast where floods are common. The threatened northern quoll occurs on at least 17 Kimberley islands and will likely disappear from some of them if cane toads arrive. Goannas die if they eat cane toads; however, research in Queensland shows that some goannas learn to avoid cane toads and numbers gradually return to pre-toad abundance. Large snakes also die after eating cane toads. There are many endemic land snails on Kimberley Islands and research has shown that cane toads eat more snails than do native frogs; however, many land snails live on vertical cliff faces which cane toads cannot access. Monitoring in the wild is necessary to demonstrate whether impact is significant and whether conservation action is required.

Just as important as predator impacts on prey are their impacts on ecosystem processes. If rats establish on an island, they will eliminate breeding populations of some seabird species. This in turn annuls the transfer of nutrients from the marine food of seabirds to the island as there are no longer any seabirds to manure the soil via their faeces. This process alters the floristic composition of the vegetation of the affected island.

Weeds can also have significant impact. African boxthorn and iceplant (Plate 13.5) are major weeds of the Houtman Abrolhos islands (Chapter 8) and other west coast islands (Chapter 7), affecting nesting seabirds. Buffel grass *Cenchrus ciliaris* and kapok bush *Aerva japonica* are major weeds of Pilbara islands and stinking passion flower *Passiflora foetida* is a major weed on some Kimberley islands.

Table 13.3. Known extirpations of indigenous species from Western Australian islands, with presumed cause.

Species	Island	Date of extinction	presumed reason for extirpation
chuditch	Dirk Hartog	probably late 19th Century	predation by feral cats
dibbler	Dirk Hartog	probably late 19th Century	predation by feral cats
golden bandicoot	Hermite	shortly before 1912	predation by feral cats, and possibly black rats
Shark Bay bandicoot	Dirk Hartog, Faure	probably late 19th Century	predation by feral cats
woylie	Dirk Hartog, Faure	probably late 19th Century	predation by feral cats
boodie	Boodie	1985–1988	predation by black rats; poisoning from rodenticide
	Dirk Hartog	probably late 19th Century	predation by feral cats
black-flanked rock-wallaby	Depuch	between 1962 and 1980	predation by red foxes
spectacled hare-wallaby	Trimouille	between 1840 and 1912	predation by feral cats
	Hermite	between 1912 and 1952	predation by feral cats

Species	Island	Date of extinction	presumed reason for extirpation
rakali (water-rat)	Hermite	between 1971 and 1994	predation by feral cats and/or black rats
	Alpha	between 1983 and 1994	predation by feral cats and/or black rats
wopilkara (greater stick-nest rat)	Dirk Hartog	probably late 19th Century	predation by feral cats
ngadji (pebble-mound mouse)	Murujuga	prior to 1994	predation by feral cats and/or red foxes
wildjin (desert mouse)	Bernier	between 1910 and 1961	cause unknown, possibly due to grazing and trampling by goats
	Dirk Hartog	probably late 19th Century	predation by feral cats
djoongari (Shark Bay mouse)	Dirk Hartog, Faure	probably late 19th Century	predation by feral cats
mootit (bush rat)	East Wallabi	between 1967 and 1977	cause unknown, persists on adjacent West Wallabi Island
dayang (heath mouse)	Dirk Hartog	probably late 19th Century	predation by feral cats
wedge-tailed shearwater	Rat Island	early 20th Century	predation by rats and cats, guano mining
brown noddy	Rat Island	early 20th Century	predation by rats and cats, guano mining
	Bedout Island	early 20th Century	predation by black rats
sooty tern	Bedout Island	early 20th Century	predation by black rats
	Rat Island	early 20th Century	predation by rats and cats, guano mining
Dirk Hartog western grasswren	Dirk Hartog	Between 1918 and 1920	predation by feral cats

There is evidence from Southern Ocean islands that house mice predate seabird eggs and chicks, but no relevant data are available from WA.

Human factors. Since Europeans arrived, some islands have been detrimentally affected. Barrow, Thevenard, Airlie and Varanus Islands in the Pilbara have (or had until recently) oil and gas infrastructure. Several islands have (or had) lighthouses and in early days had houses for lighthouse keepers and their families. In some cases this led to the introduction of pet animals, some of which became feral. Small islands in the Dampier Archipelago are now ports, linked to the mainland by causeways. Garden Island is a major defence base, also linked by a causeway to the mainland. Thevenard and Direction Islands, near Onslow, and Woody Island near Esperance have tourist accommodation, while Rottnest Island has changed considerably due to its closeness to the Perth metropolitan area (Chapter 9) and its use, firstly as a prison, and later as a popular tourist destination. As more people own boats, access to islands is increasing, and that increases the risk of accidental introductions of non-indigenous animals such as rats, mice, tramp ants, and weeds. Placing appropriate conditions on island developments by government is important.

It is well known that tourism has the potential to accelerate the introduction of non-native species to islands (Plate 13.6). One of the most extreme examples is the Galápagos Islands, Ecuador. Within nearly 200 years some 1600 species have been introduced by humans, resulting in the establishment of 1476 species. Most species of invertebrates and plants were introduced intentionally, and hundreds more came as either contaminants (as seeds, in soil) or stowaways. There is a strong correlation between the cumulative number of introduced species recorded and both the number of residents and the number of tourists.

Black rats arrived on a number of Pilbara islands, and the Lacepede Islands north of Broome, because pearling luggers beached there, sometimes inadvertently due to cyclones. Cats were introduced to Rat Island (Houtman Abrolhos) in the hope they would control the rats; what actually happened was that they mostly preyed on seabirds. How cats arrived in the Montebello Islands is a mystery, but it was possibly from a shipwreck. Rabbits and goats were introduced to some islands, as it was thought they would provide food for shipwrecked sailors. Goats remain on Cull Island near Esperance and rabbits still occur on Eclipse, Mistaken, Michaelmas, Breaksea and Goose Islands, islands in the delta of the Murray River in Peel Inlet, and possibly on Rabbit Island in the Archipelago of the Recherche. Where reinvasion is unlikely, these populations should be eradicated.

Despite pastoral leases being issued for several of the larger WA islands in the mid- to late-1800s, only two islands, Dirk Hartog and Faure in Shark Bay, have been significantly impacted. There, sheep and goats have degraded the vegetation, and cats and house mice have established (Chapter 6). Goats established on Bernier Island after a pastoral lease was abandoned, but fortunately six out of seven indigenous mammals recorded there survived.

Kimberley islands were frequently visited by traditional owners and some still are. It is probable that dingoes, taken to islands by people with some not being returned to the mainland (Chapter 3), led to the extirpation of larger mammals, as the only islands with large macropods (kangaroos and the agile wallaby) are within swimming distance of the mainland.

Fortunately, despite these impacts, most WA islands remain largely unaffected by people and feral animals.

The climate emergency. In 2007, the International Panel on Climate Change (IPCC) drew attention to the threats that climate change has had, and is expected to have, on islands, worldwide. The report stressed that islands have characteristics which make them especially vulnerable to the effects of climate change, particularly sea-level rise and extreme weather events. The 2021 IPCC Working Group I Report stated that 'It is unequivocal that human influence has warmed the atmosphere, ocean and land. Widespread and rapid changes in the atmosphere, ocean, cryosphere and biosphere have occurred.' Global mean sea level increased by c. 20 cm between 1901 and 2018 with the rate increasing from 1.3 mm/year between 1901 and 1971 to 3.7 mm/year between 2006 and 2018. The rate will continue to accelerate but vary from place to place due to ocean currents.

The longest sea level records in WA are from the Fremantle tide gauge, installed in 1897; however land subsidence due to groundwater abstraction means the data are unreliable. A 2021 study that combined several coastal tide gauge observations and satellite sea level measurements showed that during the past 50 years mean sea level around WA rose by c. 10 cm, with regional variability as well as year to year variability due to changing ocean current strength, such as that caused by the Leeuwin Current and ENSO (the El Niño Southern Oscillation).

Increasing global mean sea levels are already locked in due to past greenhouse gas releases and will continue to rise during this century and into the future, even with significant cuts in greenhouse gas emissions. The IPCC has predicted that, during the next 2000 years, global mean sea level will rise by c. 2 to 3 m if warming is limited to 1.5°C, 2 to 6 m if warming is limited to 2°C, and 19 to 22 m with 5°C of warming.

Biological evidence from the Southern Hemisphere is demonstrating phenological change. Many species are responding to warmer temperatures by advancing their timing of life-cycle events in spring. This includes the onset of breeding and timing of flowering.

In relation to WA's islands, sea level rise will exacerbate inundation, amplify storm surge, and accelerate coastal erosion. Some low WA islands will be particularly affected by rising sea levels and some are likely to disappear. Those affected include most islands in the Houtman Abrolhos, numerous low cays off the Pilbara coast, and cays off the Kimberley coast such as Mary, Adele, Browse, Cartier and Ashmore Reef Islands. The Houtman Abrolhos islands and some cays are currently protected from storm-driven waves by coral reefs. Normally, coral reefs can be expected to rise as sea levels rise; however, increasing ocean temperatures and acidity (due to dissolved CO_2) may lead to coral reef decline or death, exposing the islands to oceanic conditions.

About 90% of the excess heat arising from human activities has been absorbed by the upper levels of the ocean and, as a result, surface waters worldwide have warmed at an average rate of 0.15°C per decade over the past 40 years. This has led to marine heatwaves in several parts of the world, including WA. The terrestrial maximum temperatures extending from Shark Bay to Cape Leeuwin for all of March 2011 were 2°C higher than the long-term March average over the period 1971–2000, and sea surface temperatures were also abnormally high, driven partly by a very strong Leeuwin Current, causing a marine heatwave that led to 43% loss of kelp forests, significantly reduced breeding of little penguins at Penguin Island, and probably reduced breeding success in some other seabirds off the mid-west coast. Elsewhere in the world, marine heatwaves have caused the collapse of seabird populations.

By virtue of the high specific heat of water (in the ocean surrounding an island), change in maximum and minimum temperatures on islands should be buffered, meaning that the difference between mean maximum and minimum temperatures is less on an island than on the adjacent mainland coast. Increasing ocean temperatures may reduce this difference in the future. This is exemplified by Barrow Island, which has a mean maximum temperature of 30.2°C and the highest temperature recorded of 44.0°C (data from centre of island from 1969 to 1993), while Dampier has a mean maximum of 32.5°C with the highest temperature recorded of 48.2°C (data from 1994 to 2020). The weather station at the airport near the east coast of Barrow Island, operating from 2000 to 2020 had a mean maximum of 29.4°C and a highest temperature of 36.1°C; significantly lower than the middle of the island and significantly lower than Dampier, a difference probably more typical of most islands off WA, as they are smaller than Barrow.

Increasing ocean temperatures and acidity may affect food availability for seabirds. Some seabird species have already extended their breeding range southward on WA's west coast, but others may find it difficult to do so; for example, lesser noddies nest in mangrove trees and there are no significant stands of mangroves on islands south of Houtman Abrolhos.

Sea turtles may be dramatically affected as the beaches in which they nest disappear and as temperatures increase. Incubation temperature determines sex in sea turtles with lower nest temperatures producing males and higher temperatures producing females. While different species are affected slightly differently, the general picture is clear: increasing incubation temperatures in beach sand will lead to fewer males and more females. More than 99 per cent of juvenile and sub-adult green turtles studied at a site in the northern Great Barrier Reef are now female. Flatback turtles seem more adaptable and can change the season in which they nest, but there are limits.

Climate predictions suggest that tropical cyclones may become less frequent, but will intensify. The Pilbara coastal islands are often affected by tropical cyclones, but as sea levels rise, coastal erosion will increase. Severe cyclones rarely come as far south as WA's mid-west coast; however that may change as demonstrated by category 3 Cyclone Seroja that caused major damage to mid-west towns and coasts in April 2021.

Mean rainfall has already decreased significantly in the south-west of WA, mostly since the early 1970s, and is predicted to decrease further. This will undoubtedly affect island biodiversity and some species may be lost. Of particular concern are the Shark Bay islands that are of extremely high nature conservation value, which are expected to experience continuing winter rainfall decline. Whether increased summer rainfall compensates for this loss remains to be seen. Increasing summer maximum temperatures may negate any increase in summer rainfall as well as placing animals under increased stress.

Effects on island terrestrial biodiversity due to ongoing and increasing climate change are often difficult to predict, but it seems likely that some non-mobile species may, as temperature increases and rainfall decreases, disappear from islands.

Island reservation and management. A group of Northern Territory scientists, when describing changes in the mammals of the Sir Edward Pellew Islands in the Gulf of Carpentaria, labelled the group as both a 'refuge and death-trap'; refuge because they harboured mammals that were threatened on the mainland and death trap because once invasive species arrived, many of the native mammals disappeared. Some WA islands show a similar pattern: local extinctions after the arrival of invasive species (Table 12.1), demonstrating the need for surveillance and action.

Almost all islands south of the Kimberley are conservation reserves: national parks, conservation parks, or nature reserves. These are managed by the Parks and Wildlife Service, part of the Department of Biodiversity, Conservation and Attractions. Most Kimberley islands are now Indigenous Protected Areas and managed by Aboriginal Corporations, supported by Aboriginal rangers, with management being based on traditional knowledge and scientific advice (Chapter 3). This comprehensive protection and management of islands in WA is thought to be unique worldwide.

Ecological restoration. The management action that has had the most positive impact on WA islands has been invasive species eradications (Plates 13.7, 13.8). A widely-cited world-wide study in which one of us (AAB) participated, published in 2016, was titled 'Invasive mammal eradication on islands results in substantial conservation gains'. This paper has been widely cited. WA has been a leader in island eradications, commencing with the removal of rabbits from the 16 ha Carnac Island near Perth in 1969 and culminating in the recent eradication of sheep, goats and feral cats from the 62 000 ha Dirk Hartog Island.

In all, up to 2020, seven invasive species (sheep, goats, feral cats, red foxes, rabbits, black rats, and house mice) have been eradicated from at least 80 islands, with black rats having been eradicated from at least 50 islands. The total number of islands that had black rats in the Montebello Islands is unknown but exceeded 40, and all islands were treated (Table 13.4). In addition, other mammal species recorded from islands no longer occur there: monkey (species unknown), one-humped camel, horse, cattle, domestic dog, guinea pig, red deer, European hare, pig, western grey kangaroo (no longer on Boullanger Island, several individuals remain on Woody Island). Goats have either been removed from WA islands or have been eradicated, but remain only on Cull Island near Esperance.

Table 13.4 Successful eradications of invasive species on WA islands.

Species	**Islands** (listed north to south)
red fox	Legendre, Dolphin, Angel, Gidley
cat	Legendre, Hauy, Keast, Cohen, Gidley, unnamed north of Gidley, Angel, Hermite, Serrurier, Dirk Hartog, Rottnest
European rabbit	Wooded, Morley, Leo, Green Islets, Carnac
goat	Bernier, Dirk Hartog, Faure
sheep	Dirk Hartog, Faure
black rat	Lacepede Islands (4), Bedout, Montebellos Islands (>40), Barrow, Prince, Double, Boomerang, Middle (near Barrow), Pasco, Boodie, Rat, Penguin, Shelter
house mouse	Bridled, Varanus, 'Beacon' (Lowendal Islands), Barrow, Three Bays, Beacon (Houtman Abrolhos)

Translocations. New Zealand pioneered the eradication of invasive species from their islands and translocations of threatened species, mostly birds, to island havens. Western Australia, following New Zealand's lead, was the pioneer of similar actions in Australia. Here, eradications have allowed translocations of native animals to islands (Plates 13.9, 13.10), greatly improving their conservation status (Table 13.5). As well, eradications, have allowed species to recover on islands, or re-use islands where breeding once occurred. On Rat Island in the Houtman Abrolhos, several seabird species recommenced nesting after the eradication of feral cats and black rats (Chapter 8). Islands have also been used for introductions (sometime called marooning), of threatened species that did not previously occur there.

Table 13.5. Successful conservation translocations of native animals to WA islands, as at December 2021.

SPECIES	**ISLAND(S)**
dibbler	**Dirk Hartog,** *Escape, Gunton*
Shark Bay bandicoot	**Faure**, **Dirk Hartog**
golden bandicoot	**Hermite,** *Doole*
boodie	**Boodie**, *Alpha*, **Faure**
nilgyte (Gilbert's potoroo)	*Bald, Middle*
mala (rufous hare-wallaby)	*Trimouille, Dirk Hartog*
spectacled hare-wallaby	**Hermite**
Rothschild's rock-wallaby	*West Lewis**
maning (banded hare-wallaby)	**Faure**, **Dirk Hartog**
djoongari (Shark Bay mouse)	*North West*, **Faure**
wopilkara (greater stick-nest rat)	*Salutation*
noisy scrub-bird	*Bald*
black-and-white fairy-wren	Hermite[#]
spinifexbird	**Hermite**

Reintroductions are in bold, introductions are italicised

*rock-wallabies spread naturally to East Lewis Island, which is connected to West Lewis at low tide.

[#] uncertain whether an introduction or reintroduction, see Chapter 6.

All species in Table 13.5 are listed as threatened except Rothschild's rock-wallaby and spinifexbird.

In 2021, noisy scrub-birds were introduced to Mondrain Island in the Archipelago of the Recherche, but at the time of writing it is too early to say whether the introduction will be successful. Similarly, three translocations to Dirk Hartog Island, of Shark Bay mice, djoongari and wopilkara, commenced in 2021, but it is too early to state whether they will be successful.

Island biosecurity. The deleterious effects of non-indigenous species on island biotas are well established. It is of little value eradicating invasive species on islands if they re-establish, so biosecurity (quarantine) systems need to be in place before eradications occur. Well-developed biosecurity systems are in place for Barrow and Dirk Hartog Islands, but high-quality biosecurity systems are needed for several other high-biodiversity-value islands. The Barrow Island Quarantine Management System developed and implemented with independent advice by Chevron Australia for the Gorgon Gas project, is considered world class. From 2009 to 2015 more than 0.5 million passengers and 12.2 million tonnes of freight were transported to Barrow Island under the biosecurity system, requiring 1.5 million hours of inspections. Only a single non-indigenous species, a jumping spider, is known to have become established as a result of the Gorgon Gas Project. For most other islands, however, education of people visiting islands, particularly boat owners, is the best way of preventing unintentional 'hitch-hiking' of invasive species that are present on the adjacent mainland. Surveillance by island managers (and the general public) of islands, checking for any arrivals, is vital to allow response as soon as possible. Only a relatively small number of islands are regularly visited by people, so surveillance needs to be prioritised and focused on them.

THE FUTURE

Western Australia's islands and their biota are a state and national treasure, and are of global significance. There are many of them; they protect an amazing array of animal and plant species; very few have invasive species; many have had no human use since the sea level rose after the last glacial maximum; and almost all south of Kimberley are managed by a single government agency. Most Kimberley islands are native title, and most are Indigenous Protected Areas.

WA's islands often differ significantly from islands elsewhere in Australia (Box 13.1). These differences mean that many of their values are not replicated elsewhere.

Box 13.1. Major differences between WA's islands and those elsewhere in Australia.

- WA's largest island (Dirk Hartog) is 11th in the list of Australia's largest islands, which are distributed as follows: Tasmania (3); Northern Territory (3); Queensland (3); and South Australia (1).
- Of the 39 Australian islands with area >10 000 ha, only four are in WA (Dirk Hartog, Barrow, Augustus, and Bigge). Nearly 70% are from Queensland (14) and the Northern Territory (13). Tasmania has five such islands, followed by WA (4), Victoria (2), and South Australia (1). New South Wales lacks any large islands.
- All four of WA's islands with area >10 000 ha occur north of the Tropic of Capricorn.
- WA has nine archipelagoes, more than any other part of Australia.
- Only WA islands have spinifex (*Triodia*) hummock grassland as the dominant vegetation, with Barrow Island having the greatest extent of this vegetation type.
- Arid islands occur only in WA, resulting in some desert species being present, e.g. spinifexbird, perentie, and Main's frog *Cyclorana maini* occur on Barrow Island. Spinifexbird also occurs on Thevenard Island. WA has Australia's only island populations of spinifexbird.

- WA islands in the Kimberley region have rainforests, but (as on the mainland) only in small patches, unlike some inshore islands of Queensland (tropical rainforest) and Tasmania (temperate rainforest) with extensive tracts of these vegetation types.
- Dirk Hartog Island was the only Australian island with a population of a grass-wren species (an endemic subspecies, now extinct), but the species is in the process of being re-established (from the WA endemic subspecies present on the adjacent mainland) following the eradication of cats on the island).
- Dirk Hartog Island has one of only five insular populations of the southern emu-wren in Australia.
- Islands in the Kimberley of WA have a total of six goanna species with several islands having three species. Barrow Island has the only insular population of Earth's second-largest lizard (the perentie).
- Islands of Kimberley region have a total of 23 frog species, with Bigge Island having 10 species; this is only exceeded by some of the inshore islands of Queensland.
- Caemenid land snails on some Kimberley islands are a significant hotspot of diversity with many endemic species.
- WA has Australia's largest island populations of rock parrots.
- Nine species of native pigeons and doves, three cockatoo, and five parrot species breed on WA islands.
- Several WA islands have one or two of three species of fairy-wren present on the adjacent mainland, whereas islands along eastern Australia have only one of two species.
- Because WA lacks feral populations of introduced birds common in parts of eastern Australia (Eurasian skylark, common blackbird, song thrush, common mynah, common starling, house sparrow, European goldfinch), its islands have only a few exotic bird species (laughing kookaburra, spotted dove, laughing dove, rock dove), but only on a few islands in the south-west.

WA is in the unique position of having a globally-significant heritage of continental islands unaffected by ice and to a large extent by human contact. Many of these islands retain large breeding populations of seabirds, seals, and sea turtles. All have elements of an ancient fauna and flora that characterise the biota of Australia: marsupials, parrots, endemic genera of plants, etc. These islands, as well as being a critical part of the reserve system conserving WA's biodiversity, provide attractive opportunities for scientists to study the ecology, genetics, phylogeny, and biogeography of species. In particular, archipelagoes constitute natural laboratories where the processes of immigration and extinction have acted to unequal extents, resulting in unique combinations of species of plant, reptile, terrestrial bird, and mammal species. Some of these islands provide attractive opportunities for tourism.

WA, unlike most parts of the world with many islands, is a politically stable state, well-serviced by five universities, and with a sound record of government facilitating serious scientific study via the issue of permits to study nature in nature reserves and national parks and via cooperative programs. There is still much to learn about the biology of continental islands, and we hope that this book will entice scholars to consider conducting research on the islands of WA.

Much has been achieved to improve the biodiversity conservation status of WA's islands, particularly legally protecting them as conservation reserves, eradicating invasive species, and reintroducing or introducing threatened species. Translocations of locally extinct species have improved the status of several threatened species. Biological surveys of islands have greatly improved knowledge of what is where and what needs looking after.

Island managers have significant challenges due to increasing island visitation and the climate emergency. WA's islands are a publicly-owned, valuable asset and their effective management requires a well-funded, and scientifically-based management agency. In the Kimberley, Aboriginal Corporations need substantial financial support and scientific, as well as traditional owner advice. Without good management, many of the island values we have documented in this book may be lost (at worst) or compromised (at best).

Sources of information

Island data: Geoscience Australia.
General: Abbott & Wills 2016; Woinarski *et al.* 2014.
Terrestrial mammals: Burbidge & Abbott 2017.
Sea lions: Goldsworthy *et al.* 2021.
Eradications: Burbidge & Morris 2002, Miskelly & Powlesland 2013.
Translocations: Graham *et al.* 2018; Morris *et al.* 2015, Rayner *et al.* 2021.
Invasive species: Jones *et al.* 2016; Woinarski *et al.* 2011.
Eradications: Algar *et al.* 2019; Burbidge & Morris 2002.
Biosecurity: Nias *et al.* 2010, Scott *et al.* 2017.
Climate emergency: Chambers *et al.* 2013; Featherstone *et al.* 2015; Ruthrof *et al.* 2018, Oppenheimer *et al.* 2019; Lowe *et al.* 2021, IPCC 2021, Smith *et al.* 2021; NOAA 2021.

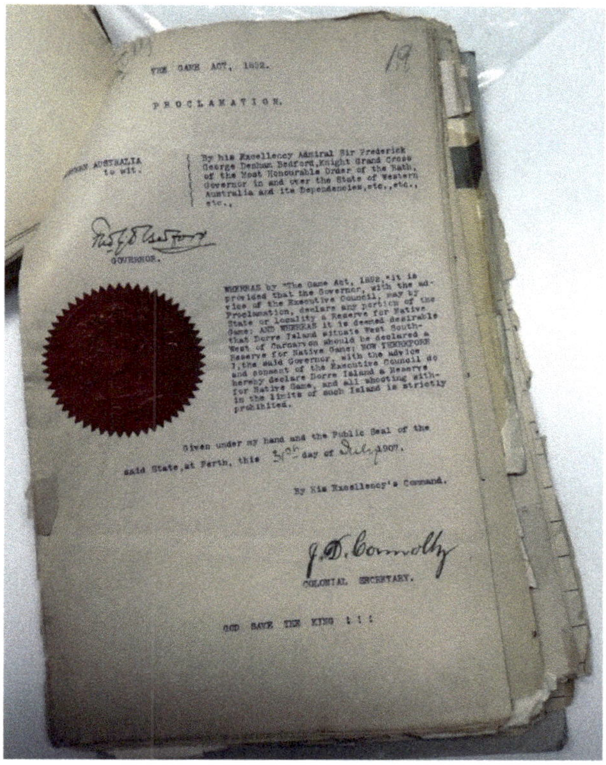

THE GAME ACT, 1892.

PROCLAMATION.

AUSTRALIA
to wit.

By his Excellency Admiral Sir Frederick George Denham Bedford, Knight Grand Cross of the Most Honourable Order of the Bath, Governor in and over the State of Western Australia and its Dependencies, etc., etc., etc.,

GOVERNOR.

WHEREAS by "The Game Act, 1892," it is provided that the Governor, with the advice of the Executive Council, may by Proclamation, declare any portion of the State or locality a Reserve for Native Game: AND WHEREAS it is deemed desirable that Dorre Island situate West South-West of Carnarvon should be declared a Reserve for Native Game; NOW THEREFORE I, the said Governor, with the advice and consent of the Executive Council do hereby declare Dorre Island a Reserve for Native Game, and all shooting within the limits of such Island is strictly prohibited.

Given under my hand and the Public Seal of the said State, at Perth, this 30th day of July 1907.

By His Excellency's Command.

COLONIAL SECRETARY.

GOD SAVE THE KING ! ! !

Plate 13.1. Proclamation by the Governor of Western Australia of Barrow Island as a reserve for game under the Game Act. File 014578F3102 volume 1, Department of Conservation and Land Management, State Records Office of Western Australia. The Proclamation was officially published in the WA Government Gazette in plain form on 18 February 1910 (see Chapter 5). Credit: I Abbott.

Plate 13.2. Shark Bay rufous hare-wallaby, a subspecies endemic to Bernier and Dorre Islands, Shark Bay. Credit: Bert & Babs Wells collection, courtesy of the Department of Biodiversity, Conservation and Attractions.

Plate 13.3. Shark Bay boodie, a subspecies endemic to Bernier and Dorre Islands, Shark Bay. Credit: Bert & Babs Wells collection, courtesy of the Department of Biodiversity, Conservation and Attractions.

Plate 13.4. Evidence of native animals killed and consumed by one feral cat. In this case, the prey is hatchling green turtles on a mainland beach at Jurabi on North West Cape. Credit: Lucy Clausen, Department of Biodiversity, Conservation and Attractions.

Plate 13.5. Iceplant, a pernicious introduced plant species smothering wedge-tailed shearwater burrows. This weed has so far been recorded on 148 Western Australian islands. Western end of Rottnest Island. December 2014. Credit: AA Burbidge.

Plate 13.6. Tourists landing at jetty on Woody Island. May 2011. Credit: AA Burbidge.

Plate 13.7. Feral animal eradication. Laying rodenticide in 1996, Montebello Islands; part of a successful archipelago-wide program to eliminate black rats and feral cats. Credit: AA Burbidge.

Plate 13.8. 13 km barrier fence erected in 2013 to divide Wirruwana (Dirk Hartog Island) roughly in half to facilitate the eradication of feral cats. November 2015. Credit: AA Burbidge.

Plate 13.9. Translocation of mala to Trimouille Island, Montebello Islands, in June 1998. Credit: AA Burbidge.

Plate 13.10. Recently released mala on Trimouille Island, Montebello Islands, in June 1998. Note radio-tracking antenna and ear tag. Credit: AA Burbidge.

APPENDIX 1. COMMON AND SCIENTIFIC NAMES OF MAMMALS, BIRDS AND WELL-KNOWN REPTILES MENTIONED IN THIS BOOK

agile wallaby *Notomacropus agilis*
ash-grey mouse *Pseudomys albocinereus*
Asian brown flycatcher *Muscicapa dauurica*
Arctic warbler *Phylloscopus borealis*
Australasian pipit *Anthus novaezeelandiae*
Australian magpie *Cracticus tibicen*
Australian pelican *Pelecanus conspicillatus*
Australian pied oystercatcher *Haematopus longirostris*
Australian raven *Corvus coronoides*
Australian sea lion *Neophoca cinerea*
Australian shelduck *Tadorna tadornoides*
Australian yellow white-eye *Zosterops luteus*
banded hare-wallaby *Lagostrophus fasciatus*
banded lapwing *Vanellus tricolor*
banded stilt *Cladorhynchus leucocephalus*
bare-rumped sheath-tailed bat *Saccolaimus saccolaimus*
barking gecko (southern) *Underwoodisaurus milii*
barn owl *Tyto alba*
Barrow Island white-winged (black-and-white) fairy-wren *Malurus leucopterus edouardi*
Barrow Island boodie *Bettongia lesueur* unnamed subspecies
Barrow Island euro *Osphranter robustus isabellinus*
Barrow Island golden bandicoot *Isoodon auratus barrowensis*
Barrow Island spectacled hare-wallaby *Lagorchestes conspicillatus conspicillatus*
bar-shouldered dove *Geopelia humeralis*
beach spinifex *Spinifex longifolius*
bar-tailed godwit *Limosa lapponica*
beach stone-curlew *Esacus magnirostris*
black rat *Rattus rattus*
black-eared cuckoo *Chalcites osculans*
black-faced cormorant *Phalacrocorax fuscescens*
black-faced cuckoo-shrike *Coracina novaehollandiae*
black-flanked rock-wallaby *Petrogale lateralis lateralis*
black-footed rock-wallaby *Petrogale lateralis subspecies??*
blue-winged kookaburra *Dacelo leachii*
boobook owl *Ninox boobook*
boodie *Bettongia lesueur*
brahminy kite *Haliastur indus*
bridled tern *Onychoprion anaethetus*
brown booby *Sula leucogaster*
brown falcon *Falco berigora*
brown goshawk *Accipiter fasciatus*
brown honeyeater *Lichmera indistincta*
brown noddy *Anous stolidus*
brown quail *Synoicus ypsilophorus*
brush bronzewing *Phaps elegans*

budgerigar *Melopsittacus undulatus*
buff-banded rail *Hypotaenidia philippensis*
bungarra (sand goanna) *Varanus gouldii*
bush rat *Rattus fuscipes*
bush stone-curlew *Burhinus grallarius*
California quail *Callipepla californica*
Cape Barren goose *Cereopsis novaehollandiae*
carpet python *Morelia spilota*
Caspian tern *Hydroprogne caspia*
cat *Felis catus*
coastal moort *Eucalyptus utilis*
coastal ridge-fruited mallee *Eucalyptus angulosa*
chuditch (western quoll) *Dasyurus geoffroii*
collared kingfisher *Todiramphus chloris*
collared sparrowhawk *Accipiter cirrocephalus*
common brushtail possum (south western subspecies) *Trichosurus vulpecula hypoleucos*
common greenshank *Tringa nebularia*
common pheasant *Phasianus colchicus*
common planigale *Planigale maculata*
common sandpiper *Actitis hypoleucos*
common scaly-foot *Pygopus lepidopodus*
common sheath-tailed bat *Taphozous georgianus*
crested bellbird *Oreoica gutteralis*
crowned snake *Elapognathus coronatus*
curlew sandpiper *Calidris ferruginea*
dayang (heath mouse) *Pseudomys shortridgei*
delicate mouse *Pseudomys delicatulus*
desert mouse *Pseudomys desertor*
dibbler *Parantechinus apicalis*
dingo *Canis familiaris*
Dirk Hartog Island white-winged fairy-wren *Malurus leucopterus leucopterus*
Dirk Hartog rufous fieldwren *Calamanthus campestris hartogi*
Dirk Hartog southern emu-wren *Stipiturus malachurus hartogi*
Dirk Hartog western grass-wren *Amytornis textilis carteri*
djini (pale field-rat) *Rattus tunneyi*
djoongari (Shark Bay mouse) *Pseudomys gouldii*
djoorri (common rock-rat) *Zyzomys argurus*
dog *Canis familiaris*
Dorre rufous fieldwren *Calamanthus campestris dorrie*
dugite *Pseudonaja affinis*
dusky gerygone *Gerygone tenebrosa*
eastern barred bandicoot *Perameles gunnii*
eastern bettong *Bettongia gaimardi*
eastern osprey *Pandion cristatus*
eastern quoll *Dasyurus viverrinus*
eastern reef egret *Egretta sacra*
euro *Osphranter robustus*

fairy martin *Petrochelidon ariel*
European rabbit *Oryctolagus cuniculus*
fairy tern *Sternula* (*Sterna*) *nereis nereis*
fan-tailed cuckoo *Cacomantis flabelliformis*
fat-tailed false antechinus *Pseudantechinus macdonnellensis*
Finlayson's cave-bat *Vespadelus finlaysoni*
flatback turtle *Natator depressus*
flesh-footed shearwater *Ardenna carneipes*
galah *Eolophus roseicapilla*
ghost bat *Macroderma gigas*
Gilbert's honeyeater *Melithreptus chloropsis*
Gilbert's potoroo *Potorous gilbertii*
goat *Capra hircus*
golden bandicoot *Isoodon auratus*
golden-backed tree-rat *Mesembriomys macrurus*
grassland melomys *Melomys burtoni*
great knot *Calidris tenuirostris*
greater frigatebird *Fregata minor*
great pied cormorant *Phalacrocorax varius*
greater crested tern *Thalasseus bergii*
great-winged petrel *Pterodroma macroptera*
green turtle *Chelonia mydas*
grey-bellied dunnart *Sminthopsis fuliginosus*
grey butcherbird *Cracticus torquatus*
grey currawong *Strepera versicolor*
grey fantail *Rhipidura albiscapa*
grey-headed honeyeater *Ptilotula keartlandi*
grey plover *Pluvialis squatarola*
grey teal *Anas gracilis*
grey-tailed tattler *Tringa brevipes*
hawksbill turtle *Eretmochelys imbricata*
heath goanna *Varanus rosenbergi*
heath mouse *Pseudomys shortridgei*
helmeted guineafowl *Numida meleagris*
horse *Equus caballus*
house mouse *Mus musculus*
Houtman Abrolhos lesser noddy *Anous tenuirostris melanops*
Houtman Abrolhos painted button-quail *Turnix varius scintillans*
iceplant *Mesembryanthemum crystallinum*
Indian peafowl *Pavo cristatus*
inland thornbill *Acanthiza apicalis*
island monarch *Monarcha cinerascens*
kerakenga (tropical or Lakeland Downs short-tailed mouse) *Leggadina lakedownensis*
Kimberley rock-rat *Zyzomys woodwardi*
King's skink *Egernia kingii*
koorrawal (golden-backed tree-rat) *Mesembriomys macrurus*
Lancelin Island skink *Ctenotus lancelini*

laughing dove *Spilopelia senegalensis*
laughing kookaburra *Dacelo novaeguineae*
lesser coucal *Centropus bengalensis*
lesser crested tern *Thalasseus benghalensis*
lesser frigatebird *Fregata ariel*
lesser long-eared bat *Nyctophilus geoffroyi*
lesser noddy *Anous tenuirostris*
letter-winged kite *Elanus scriptus*
little button-quail *Turnix velox*
little corella *Cacatua sanguineus*
little grassbird *Poodytes gramineus*
little long-tailed dunnart *Sminthopsis dolichura*
little penguin *Eudyptula minor*
little pied cormorant *Microcarbo melanoleucos*
little shearwater *Puffinus assimilis*
little tern *Sternula albifrons*
little woodswallow *Artamus minor*
loggerhead turtle *Caretta caretta*
long-nosed fur seal *Arctocephalus forsteri*
magpie-lark *Grallina cyanoleuca*
mangrove fantail *Rhipidura phasiana*
maning *Lagostrophus fasciatus*
marbled gecko *Christinus marmoratus*
mardo *Antechinus flavipes*
masked booby *Sula dactylatra*
masked owl, northern subspecies *Tyto novaehollandiae kimberli*
Mertens water monitor *Varanus mertensi*
monjon *Petrogale burbidgei*
moolpoo (Western chestnut mouse) *Pseudomys nanus*
mootit (bush rat) *Rattus fuscipes*
mulgara *Dasycercus cristicauda*
nabarlek *Petrogale concinna*
nankeen kestrel *Falco cenchroides*
nankeen night-heron *Nycticorax caledonicus*
New Holland honeyeater *Phylidonyris novaehollandiae*
ngadji (pebble-mound mouse) *Pseudomys chapmani*
noisy scrub-bird *Atrichornis clamosus*
northern leaf-nosed bat *Hipposideros stenotis*
northern mastiff-bat *Chaerephon jobensis*
northern quoll *Dasyurus hallucatus*
northern sooty oystercatcher *Haematopus fuliginosus ophthalmicus*
olive ridley turtle *Lepidochelys olivacea*
orange-footed scrubfowl *Megapodius reinwardt*
Pacific black duck *Anas superciliosa*
Pacific gull *Larus pacificus*
Pacific (or Polynesian) rat *Rattus exulans*
painted button-quail *Turnix varius*

pakooma or brush-tailed rabbit-rat *Conilurus penicillatus*
peregrine falcon *Falco peregrinus*
pied honeyeater *Certhionyx variegatus*
perentie *Varanus giganteus*
pied butcherbird *Cracticus nigrogularis*
pied imperial pigeon *Ducula bicolor*
pied oystercatcher *Haematopus longirostris*
pig *Sus scrofa*
purple-backed fairy-wren *Malurus assimilis*
purple-crowned lorikeet *Glossopsitta porphyrocephala*
quenda *Isoodon fusciventer*
quokka *Setonix brachyurus*
rabbit (European) *Oryctolagus cuniculus*
rainbow bee-eater *Merops ornatus*
rainbow pitta *Pitta iris*
rakali (water-rat) *Hydromys chrysogaster*
Recherche Cape Barren goose *Cereopsis noveahollandiae grisea*
Recherche rock-wallaby *Petrogale lateralis hackettii*
red wattlebird *Anthochaera carunculata*
red-backed button-quail *Turnix maculosus*
red-capped plover *Charadrius ruficapillus*
red-capped robin *Petroica goodenovii*
red-cheeked dunnart *Sminthopsis virginiae*
red-eared firetail *Stagonopleura oculata*
red-footed booby *Sula sula*
red-necked stint *Calidris ruficollis*
red-necked avocet *Recurvirostra novaehollandiae*
red-necked phalarope *Phalaropus lobatus*
red-tailed black-cockatoo *Calyptorhynchus banksii*
red-tailed tropicbird *Phaethon rubricauda*
rock dove *Columba livia*
rock parrot *Neophema petrophila*
roseate tern *Sterna dougallii*
rose-crowned fruit-dove *Ptilinopus regina*
Rothschild's rock-wallaby *Petrogale rothschildi*
ruddy turnstone *Arenaria interpres*
rufous fieldwren *Calamanthus campestris*
rufous hare-wallaby *Lagorchestes hirsutus*
rufous whistler *Pachycephala rufiventris*
sacred kingfisher *Todiramphus sanctus*
saltwater or estuarine crocodile *Crocodylus porosus*
sand goanna *Varanus gouldii*
sanderling *Calidris alba*
sandstone shrike-thrush *Colluricincla woodwardi*
sandy inland mouse *Pseudomys hermannsburgensis*
scaly-tailed possum *Wyulda squamicaudata*
scarlet robin *Petroica boodang*

Shark Bay bandicoot *Perameles bougainville*
Shark Bay hare-wallaby *Lagorchestes hirsutus bernieri*
Shark Bay mouse *Pseudomys gouldii*
sharp-tailed sandpiper *Calidris acuminata*
sheep *Ovis aries*
shining flycatcher *Myiagra alecto*
short-beaked echidna *Tachyglossus aculeatus*
short-eared rock-wallaby *Petrogale brachyotis*
short-tailed shearwater *Ardenna tenuirostris*
silver gull *Chroicocephalus novaehollandiae*
silvereye *Zosterops lateralis*
singing honeyeater *Lichenostomus virescens*
sooty oystercatcher *Haematopus fuliginosus*
sooty tern *Onychoprion fuscatus*
southern death adder *Acanthophis antarcticus*
spangled drongo *Dicruris bracteatus*
spectacled hare-wallaby *Lagorchestes conspicillatus*
spinifexbird *Poodytes carteri*
spinifex pigeon *Geophaps plumifera*
spiny-tailed goanna *Varanus acanthurus*
spiny-tailed skink *Egernia stokesii*
spotless crake *Zapornia tabuensis*
spotted dove *Spilopelia chinensis*
spotted pardalote *Pardalotus punctatus*
spotted scrubwren *Sericornis maculatus*
sulphur-crested cockatoo *Cacatua galerita*
swamp harrier *Circus approximans*
tammar wallaby *Notomacropus eugenii*
tawny grassbird *Megalurus timoriensis*
terek sandpiper *Xenus cinereus*
tiger snake *Notechis scutatus*
Torresian crow *Corvus orru*
tree martin *Petrochelidon nigricans*
variegated fairy-wren *Malurus lamberti*
wedge-tailed eagle *Aquila audax*
wedge-tailed shearwater *Ardenna pacifica*
welcome swallow *Hirundo neoxina*
western gerygone *Gerygone fusca*
western ringtail possum (ngawyir) *Pseudocheirus occidentalis*
western whistler *Pachycephala fuliginosa*
whimbrel *Numenius phaeopus*
white-bellied cuckoo-shrike *Coracina papuensis*
white-bellied sea-eagle *Haliaeetus leucogaster*
white-breasted robin *Eopsaltria georgiana*
white-breasted whistler *Pachycephala lanioides*
white-breasted woodswallow *Artamus leucorynchus*
white-browed scrubwren *Sericornis frontalis*

white-faced heron *Egretta novaehollandiae*
white-faced storm-petrel *Pelagodroma marina*
white-plumed honeyeater *Ptilotula penicillata*
white-quilled rock pigeon *Petrophassa albipennis*
white-striped mastiff-bat *Austronomus australis*
white-tailed tropicbird *Phaethon lepturus*
white-winged fairy-wren *Malurus leucopterus*
white-winged triller *Lalage tricolor*
wildjin (desert mouse) *Pseudomys desertor*
willie wagtail *Rhipidura leucophrys*
wopilkara *Leporillus conditor*
woylie *Bettongia penicillata*
yellow-throated miner *Manorina flavigula*
yellow oriole *Oriolus flavocinctus*
yellow-spotted goanna *Varanus panoptes*

APPENDIX 2. UNFAMILIAR WORDS EXPLAINED

adamellite. A coarse-grained porphyritic igneous rock (with at least one group of crystals obviously larger than another group), a variety of monzogranite and dominated by phenocrysts of orthoclase in a granular groundmass of perthite, plagioclase and quartz; usually paler than **granite**.

aeolianite. A rock formed by the lithification of sediments deposited by aeolian processes (i.e. the wind).

allele. (Short for allelomorph.) One of two or more forms of a given gene variant. For example, the ABO blood grouping is controlled by the ABO gene, which has six common alleles. Nearly every living human's phenotype (observable characteristics) for the ABO gene is some combination of just these six alleles.

alluvial. Material that has been transported by water or wind and deposited elsewhere, typically in a valley.

anchialine pool. A landlocked body of water with a subterranean connection to the ocean.

apex predator. A predator at the top of a food chain, without natural predators.

basalt. An igneous rock formed by the rapid cooling of certain types of lava.

biodiversity. The biological variety and variability of life on Earth; a measure of variation at the genetic, species, and ecosystem level.

biogeography. The study of the distribution of species and ecosystems in geographic space and through geological time.

biota. The plant and animal life of a region.

bornhardt. A dome-shaped rock, composed of granite, that rises abruptly from a plain.

bottleneck. A genetic or population bottleneck is due to a sharp reduction in population size leading to a reduction in genetic diversity.

carbonate ion. A salt (chemical compound) of carbonic acid (H_2CO_3).

cay. Islands formed by accumulations of sand or coral rock on reefs.

chert. A hard fine-grained sedimentary rock composed of very small crystals of quartz.

Class A. Under the WA Land Administration Act as Class A reserve can only be significantly amended or cancelled by an Act of Parliament.

commensal. Commensalism is a long-term biological interaction (symbiosis) in which members of one species gain benefits, while those of the other species neither benefit nor are harmed. Some species are commensal with humans.

conservation park. In Western Australia, an area reserved for conservation and recreation but that has a lower level of protection than a national park.

coprophilous. Living in or growing on dung.

drowned coastline. A coastline that has been inundated by the sea by a relative rise in sea levels.

ecosystem engineer. An animal that creates, significantly modifies, maintains or destroys a habitat.

ectoparasite. A parasite that lives outside the host's surface.

effective population size. The number of breeding individuals in a population.

El Niño. The warm phase of the El Niño–Southern Oscillation (ENSO). It is associated with a band of warm ocean water that develops in the central and east-central equatorial Pacific, which can affect the temperature and strength of the Leeuwin Current.

eluvial. Soils that are derived *in situ* (in the original place) by weathering of rock.

endemic. Found only in a circumscribed region, including an island, state, nation, or continent.

exotic. In relation to plants and animals: introduced, alien, not indigenous.

fauna. Animals.

flora. Plants.

fluctuating asymmetry. Refers to small, random deviations away from perfect bilateral symmetry reflecting genetic pressures that may be due to inbreeding. Examples in humans include unequal sizes of eyes, ears, etc.

gabbro. Coarse-grained, intrusive igneous rock formed from the slow cooling of magnesium-rich and iron-rich magma (molten rock deep beneath the Earth's surface).

genetic drift. The change in the frequency of an existing gene variant (allele) in a population due to random sampling of organisms.

genetic fitness. Individual reproductive success, equal to the average contribution to the gene pool of the next generation that is made by individuals.

geomorphology. The scientific study of the origin and evolution of topographic and bathymetric features created by physical, chemical or biological processes operating at or near the Earth's surface. Geomorphologists seek to understand why landscapes look the way they do.

gneiss. A type of metamorphic rock formed by high-temperature and high-pressure metamorphic processes acting on formations composed of igneous or sedimentary rocks.

goanna. Species of lizards in the genus *Varanus*, sometime called monitors

granite. A coarse-grained igneous rock composed mostly of quartz, alkali feldspar, and plagioclase. It forms from magma with a high content of silica and alkali metal oxides that slowly cools and solidifies underground.

granophyre. A subvolcanic rock that contains quartz and alkali feldspar in characteristic angular intergrowths.

guano. Material composed largely of seabird excrement.

heterozygosity. A measure of genetic variation; refers to having inherited different forms of a particular gene from each parent; a heterozygous genotype stands in contrast to a homozygous genotype, where an individual inherits identical forms of a particular gene from each parent.

HMSV. His (or Her) Majesty's Survey Vessel.

HMC. His (or Her) Majesty's Cutter.

Holocene. The current geological epoch. It began approximately 11,650 years before present, after the last glacial period, which concluded with the Holocene glacial retreat. The Holocene and the preceding Pleistocene together form the Quaternary period.

Igneous rock. Rock formed through the cooling and solidification of magma or lava. One of the three main types of rock; the others are sedimentary and metamorphic.

inselberg. An isolated steep hill that rises abruptly from a plain.

interglacial. A geological interval of warmer global average temperature lasting thousands of years that separates consecutive glacials or ice ages (the Earth is currently in an interglacial, known as the Holocene).

introduction (of a species). See **translocation**.

isobath. Depth contours.

karst. An area formed from the dissolution of soluble rocks such as limestone, dolomite, and gypsum; being characterised by underground drainage systems with sinkholes and caves.

last glacial maximum. The last time when glacial sheets were at their greatest extent covering much of North America, northern Europe and Asia, profoundly affected Earth's climate by causing drought, desertification, and a large drop in sea levels of c. 120 m lower than today's. At that time mainland Australia was joined to Tasmania, New Guinea and most continental islands. Maximum ice coverage was between 26 500 years and 19 000 – 20 000 years ago, when deglaciation commenced in the Northern Hemisphere, causing an abrupt rise in sea level. Decline of the West Antarctica ice sheet occurred between 14 000 and 15 000 years ago, consistent with evidence for another abrupt rise in the sea level c. 14 500 years ago.

Leeuwin Current. A warm ocean current that flows southwards near the western coast of Australia, then rounding Cape Leeuwin to enter the waters south of Australia where its influence extends as far as Tasmania; the Holloway Current, which flows south west along the Kimberley and Pilbara coasts, transports warmer, lower salinity water from northern Australia into the Leeuwin Current.

lianas. Long-stemmed woody plants that climb up trees.

LORAN. A long-range radio navigation system developed in the US during World War II.

mallee. Eucalypts that grow with multiple stems from underground lignotubers.

mangal. A mangrove swamp, a wet environment dominated by woody trees and shrubs adapted to saline conditions.

metapopulation. A group of spatially separated populations of the same species which interact at some level.

monadnock. An isolated hill, ridge, range, or mountain of erosion-resistant rock rising above a plain. The name is from an indigenous tribe in New Hampshire, USA. See also tor.

nanism. Dwarfism.

neonate. A newly born or newly hatched animal.

offset (environmental). A provision that is meant to compensate for the loss of native vegetation and/or species due to a development.

opilones. An order of arachnids colloquially known as harvestmen, harvesters, or daddy longlegs.

petroglyph. An image created by people by removing part of a rock surface by incising, picking, carving, or abrading, as a form of rock art.

phenology. The scientific study of periodic biological phenomena, such as flowering, breeding, and migration, in relation to climatic conditions.

Phytophthora. A genus of plant-damaging water moulds, whose member species are capable of causing enormous damage in natural ecosystems and economic losses on crops worldwide.

planar area. Two-dimensional area of an object, not taking topography into account.

planation. A planation surface is a large-scale surface that is almost flat with the possible exception of some residual hills.

Pleistocene. The geological epoch that lasted from c. 2 580 000 to 11 700 years ago, spanning Earth's most recent period of repeated glaciations.

poison pea. A member of the genus *Gastrolobium* of the pea family almost entirely endemic to south-west WA. Death results when an animal not native to this region eats the foliage, flowers or seeds. These contain the toxin monofluoroacetate.

potoroid. The Potoroidae is a marsupial family that includes potoroos and bettongs.

Quaternary. The current and most recent of the three periods of the Cenozoic Era. The Quaternary Period is divided into two epochs: the Pleistocene (2.588 million years ago to 11.7 thousand years ago) and the Holocene (11.7 thousand years ago to today, although a third epoch, the Anthropocene, has been proposed but is not yet officially recognised.

radionuclide. Also called a radioactive isotope, a substance that is radioactive.

raptor. Birds of prey, usually meant to include eagles, hawks, falcons, etc. and owls.

reintroduction (of a species). See **translocation**.

sclerophyllous. A type of vegetation evolved to reduce water loss that has hard leaves, short internodes (the distance between leaves along the stem) and leaf orientation parallel or oblique to direct sunlight.

short-range endemic species. Animals or plants that display restricted geographic distributions, nominally less than 10 000 km^2; many Australian short-range endemics have much smaller ranges than 10 000 km^2.

species turnover. The proportion of species either gained or lost relative to the total number of species observed across a time period.

specific heat. The quantity of heat required to raise the temperature of one gram of a substance by one Celsius degree.

statutory body. A statutory body (or authority) is a generic term for an authorisation by Parliament given to a person or group of people to exercise specific powers.

stepping stones. In island biogeography, plants or animals being able to establish on an island via other islands.

stygofauna. Animals that have adapted to and dwell only in water in cool, dark caves and in groundwater.

subfossil. The remnants, usually bones, of dead animals that are not fossilised. Subfossils of vertebrates are often found in caves or other shelters, where the remains have been preserved for hundreds or even thousands of years, often the result of predation by owls.

tor. A prominent hill consisting of unweathered bedrock and standing abruptly above its surroundings. The name is Anglo-Saxon. See also monadnock.

tramp ant. A diverse group of invasive ant species that have become established across the globe. They become dominant in a new habitat due to traits such as aggression toward other ant species, little aggression toward members of their own species, efficient recruitment and large colony size. Argentine (*Linepithema humile*), African big-headed (or coastal brown) (*Pheidole megacephala*) and Singapore (*Trichomyrmex destructor*) ants are tramp ants that are established in Western Australia.

translocation. The human-mediated movement of living organisms from one area, with release in another; includes **reintroduction** (translocation to a site where a species previously occurred) and **introduction** (translocation to a site where the species did not previously occur).

troglofauna. Terrestrial animals that dwell in and have adapted to cool, dark caves.

tsunami. A series of waves in a water body caused by the displacement of a large volume of water, generally in an ocean or a large lake. Earthquakes, volcanic eruptions and other underwater explosions (including detonations, landslides, glacier calvings, meteorite impacts and other disturbances) above or below water all have the potential to generate a tsunami.

vascular plant. Plants with vascular tissues (phloem and xylem) that conduct water and nutrients. Includes ferns, gymnosperms (conifers and cycads) and angiosperms (flowering plants).

APPENDIX 3. FURTHER READING

Abbott I 1977a Species richness, turnover and equilibrium in insular floras near Perth, Western Australia. *Australian Journal of Botany* **25**, 193-208.

Abbott I 1977b Observations on the distribution of bird species on small islands near Perth. *Western Australian Naturalist* **13**, 196-199.

Abbott I 1978a Factors determining the number of land bird species on islands around south-western Australia. *Oecologia* **33**, 221-233.

Abbott I 1978b Seabird Islands No. 55. Breaksea Island, King George Sound, Western Australia. *Corella* **2**, 24-25.

Abbott I 1979a The distribution and abundance of seabirds at sea and on islands near the mid- and north-western coasts of Australia. *Corella* **3**, 93-102.

Abbott I 1979b The past and present distribution and status of sea lions and fur seals in Western Australia. *Records of the Western Australian Museum* **7**, 375-390.

Abbott 1980c The transition from mainland to island, illustrated by the flora and landbird fauna of headlands, peninsulas and islands near Albany, Western Australia. *Journal of the Royal Society of Western Australia* **63**, 79-92.

Abbott I 1980a The floras of 37 south-western Australian islands. *Western Australian Herbarium Research Notes* **3**, 19-36.

Abbott I 1980b Aboriginal man as an exterminator of wallaby and kangaroo populations on islands round Australia. *Oecologia* **44**, 347-354.

Abbott I 1981 Vegetation maps of four large islands near Albany, Western Australia. *Western Australian Herbarium Research Notes* No. 5, 5-18.

Abbott I 1982 Birds recorded on 22 tropical island of Western Australia. *Corella* **6**, 119-122.

Abbott I 2001 Historic record of Australian pelican, *Pelecanus conspicillatus*, breeding in southwest Western Australia. *Western Australian Naturalist* **23**, 1-7.

Abbott I 2006 The islands of Western Australia. Changes over time in human use. *Early Days* **12**, 635-653.

Abbott I 2008 Historical perspectives of the ecology of some conspicuous vertebrate species in south-west Western Australia. *Conservation Science Western Australia* **6**, 1–214.

Abbott I 2011 Appendix 4. Fauna and ecology. In *Western Australian Exploration 1836-1845*. (eds M Hercock & S Milentis, with P Bianchi), pp. 412-418. Hesperian Press, Carlisle WA.

Abbott I & Black R 1978 An ecological reconnaissance of four islands in the Archipelago of the Recherche, Western Australia. *Journal of the Royal Society of Western Australia* **60**, 115-128. Addendum **63**, 31 (1980).

Abbott I & Black R 1980 Changes in species composition of floras on islets near Perth, Western Australia. *Journal of Biogeography* **7**, 399-410.

Abbott I, Cranfield R, Van Heurck & T Middleton 2006 The vegetation, flora and fauna of Saddle Island, near Walpole, Western Australia. *Western Australian Naturalist* **25**, 153-168.

Abbott I, Marchant N & Cranfield R 2000 Long-term change in the floristic composition and vegetation structure of Carnac Island, Western Australia. *Journal of Biogeography* **27**, 333-346.

Abbott I & Watson JR 1978 The soils, flora, vegetation and vertebrate fauna of Chatham Island, Western Australia. *Journal of the Royal Society of Western Australia* **60**, 65-70.

Abbott I & Wills A 2016 Review and synthesis of knowledge of insular ecology, with emphasis on the islands of Western Australia. *Conservation Science Western Australia* **11**.

Abjornsson A 1910 Carnac Island, 10th September, 1909. *Journal of the Natural History and Science Society of Western Australia* **3**(1), 38-39.

Akerman K 2015 A Review of the Indigenous watercraft of the Kimberley Region, Western Australia. *The Great Circle* **37**, 82-111.

Algar D, Angus GJ & Onus ML 2011 Eradication of feral cats on Rottnest Island, Western Australia. *Journal of the Royal Society of Western Australia* **94**, 439-443.

Algar D, Burbidge AA & Angus GJ 2002 Cat eradication on Hermite Island, Montebello Islands, Western Australia. In *Turning the Tide: The eradication of invasive species*. (eds C Veitch & Clout MN), pp. 14-18. IUCN SSC Invasive Species Specialist Group, Gland, Switzerland and Cambridge, UK.

Algar D, Johnston M, Tiller C, Onus M, Fletcher J, Desmond G, Hamilton N & Speldewinde P 2019 Feral cat eradication on Dirk Hartog Island, Western Australia. *Biological Invasions* **22**, 1037-1054.

Amos WH 1980 *Wildlife of the Islands*. Harry N. Abrams, Inc.

Andrews TC 1959 Correspondence relating to his life on Middle Island in 1889. Typescript held in library, Department of Biodiversity, Conservation and Attractions, Kensington WA.

Anon. 1878 Letter to editor. *The Western Australian Times*, 17 May, p.3.

Anon. 1882 Untitled. *The Inquirer & Commercial News*, 4 October, p. 5.

Anon. 1885 H.M.S. "Bacchante" at the Antipodes. By Prince Edward and Prince George of Wales. *The West Australian* 2.5.1885, 1.

Anon. 1914 A W.A. session. *Emu* **13**, 234.

Anon. 1948a We went west. *Wild Life* **10**, 15-17, 40.

Anon. 1948b On the south-west sand plains. *Wild Life* **10**, 175-178

Anon. 1952 A draper to fix a drain. *Wild Life* **16**, 9.

Anon. 1971 Goat eradication–Bernier Island. *SWANS State Wildlife Advisory News Service* **2**, 60-61.

Anon. 2016 Recovery round-up. Red-tailed tropicbird. *Corella* **40**, 23.

Anon. 2019 *Islands*. Australian Geographic, Sydney.

Appleyard RT & Manford T 1979 *The Beginning: European Discovery and Early Settlement of Swan River Western Australia.* University of Western Australia Press, Nedlands.

Australia Pilot 1972 *Volume 5. North, North-west, and West Coasts of Australia from the West Entrance of Endeavour Strait to Cape Leeuwin*. Hydrographic Department, Ministry of Defence, Taunton UK.

Australia Pilot 1973 *Volume 1. South coast of Australia from Cape Leeuwin to Green Point*. 6th edition. Ministry of Defence, Taunton UK.

Australian Transport Safety Bureau 1991 Grounding of bulk carrier Sanko Harvest. https://www.atsb.gov.au/pdf

Backhouse J 1993 Holocene vegetation and climate record from Barker Swamp, Rottnest Island, Western Australia. *Journal of the Royal Society of Western Australia*. **76**, 53-61.

Baglin D & Mullins B 1970 *Islands of Australia*. Horwitz Publications, North Sydney NSW.

Bakker HR & Weston AS undated Tammar grazing and the pyrosphere on Middle Island. Typescript held in library, Department of Biodiversity, Conservation and Attractions, Kensington WA.

Bancroft WJ, Garkaklis MJ & Roberts JD 2004 Continued expansion of the wedge-tailed shearwater, *Puffinus pacificus*, nesting colonies on Rottnest Island, Western Australia. *Emu* **104**, 79-82.

Barnes M, Desmond A, Carter R, Newell J, Davis R, Clarke RH, Burbidge AH & Garnett ST 2021 Houtman Abrolhos painted button-quail *Turnix varius scintillans*. In *The Action Plan for Australian Birds 2020*. pp. 311-313. CSIRO Publishing, Melbourne.

Basedow H 1918 Narrative of an expedition of exploration in north-western Australia. *Proceedings of the Royal Geographical Society of Australasia (South Australian Branch)* **18**, 105-295.

Bates D 1938 *The passing of the Aborigines*. John Murray, London.

Baudin N 1974 *The journal of Post Captain Nicolas Baudin*. Translated from the French by C Cornell. Libraries Board of South Australia, Adelaide.

Baynes A 1990 The mammals of Shark Bay, Western Australia. In *Research in Shark Bay. Report of the France–Australe Bicentenary Expedition Committee*. (eds P Berry, S Bradshaw & B Wilson), pp. 313-325. Western Australian Museum, Perth.

Baynes A 2008 The original non-volant mammal fauna of Faure Island, Shark Bay, Western Australia. *Records of the Western Australian Museum* Supplement No. 75, 25-31.

Beatty B 1965 *Next Door to Paradise. Australia's Countless Islands*. Cassell Australia, Melbourne.

Béchervaise J 1951 The Archipelago of the Recherche. *Walkabout* **17**(4), 10-17.

Béchervaise JM 1954 General history. Australian Geographical Society Report No. 1, Part 1a.

Bell S 2021 Red-tailed tropicbird breeding on Rottnest Island. *Western Australian Bird Notes* **178**, 4.

Bencini R, McCulloch C, Mills HR & Start AN 2001 Habitat and diet of the dibbler (*Parantechinus apicalis*) on two islands in Jurien Bay, Western Australia. *Wildlife Research* **28**, 465-468.

Bennison C, Friend A, Button T, Mills H, Lambert C & Bencini R 2016 Potential impacts of poison baiting for introduced house mice on native animals on islands in Jurien Bay, Western Australia. *Wildlife Research* **43**, 61-68.

Berrill NJ & Berrill M 1969 *The Life of Sea Islands*. McGraw-Hill, New York.

Berry PF 1993 Historical background, description of the physical environments of Ashmore Reef and Cartier Island and notes on exploited species. *Records of the Western Australian Museum* Supplement No. 44, 1-11.

Black AB, Burbidge AH & Garnett ST 2021 Dirk Hartog western grasswren *Amytornis textilis carteri*. In *The Action Plan for Australian Birds 2020*. pp. 562-563. CSIRO Publishing, Melbourne.

Blundell V, Doohan K, Vachon D, Allbrook M, Jebb MA & Bornman J 2017 *Barddabardda wodjenangorddee: We're telling all of you*. Dambimangari Aboriginal Corporation, Derby.

BOM 2021 – Bureau of Meteorology online. http://www.bom.gov.au/climate/data/

Bonnet X, Aubret F, Lourdais O, Ladyman M, Bradshaw D & Maumelat S 2005. Do 'quiet' places make animals placid? Island vs. mainland tiger snakes. *Ethology* **111**(6), 573-592.

Bonnin M 1982 The Baudin Rocks, 1968-1982. *South Australian Ornithologist* **29**, 23-24.

Bridge PJ (ed) 2020 *Due North. The Exploration of the North West of Western Australia*. Western Australian Explorers' Diaries Project. Hesperian Press, Carlisle WA.

British Admiralty 1880 North West Coast of Australia...with the off-lying islands and reefs 1: 1 916 000. Chart BA 475, updated to 1971.

British Admiralty 1881 Doubtful Island Bay to the Head of the Great Australian Bight 1:610 000. Chart 1059.

British Admiralty 1902 Cape Le Grand to Cape Pasley with part of the Archipelago of the Recherche 1:143 900. Chart 3189.

British Admiralty 1914 Cape Naturaliste to...Doubtful Island Bay 1: 603 000. Chart BA 1034.

Brockman J 1987 *He Rode Alone...Being the Adventures of Pioneer Julius Brockman from his Diaries*. Artlook books, Perth.

Brooker MG, Smith GT, Saunders DA, Ingram JA, Leone J & de Rebeira CPS 1995 A biological survey of Garden Island, Western Australia: 1. Birds and Reptiles. *Western Australian Naturalist* **20**, 169-184; Errata. **21**, 142-143 (1996).

Burbidge A, Blyth J, Fuller P, Kendrick P, Stanley F & Smith L 2000 The terrestrial vertebrate fauna of the Montebello Islands, Western Australia. *CALMScience* **3**, 95-107.

Burbidge A, Haberley B, Halse S, Lane J & Pearson G 1993 How many geese are enough? *Landscope* **9**(1), 28-33.

Burbidge AA 1971 The fauna and flora of the Monte Bello Islands. Department of Fisheries and Fauna Report No. 9. Department of Fisheries and Fauna, Perth.

Burbidge AA 2008 Towards monitoring rock-wallabies on Barrow Island, Western Australia. *CALMScience* **7**, 43–48.

Burbidge AA, Abbott I, Comer S, Adams E, Berry O & Penwarden KE 2012 Unforeseen consequences of a misidentified rodent: Case study from the Archipelago of the Recherche, Western Australia. *Australian Mammalogy* **34**, 55-58.

Burbidge AA & Abbott I 2017 Mammals on Western Australian islands: Occurrence and preliminary analysis. *Australian Journal of Zoology* **65**, 183-195.

Burbidge AA & Fuller PJ 1989 Numbers of breeding seabirds on Pelsaert Island, Houtman Abrolhos, Western Australia. *Corella* **13**, 57-61.

Burbidge AA & Fuller PJ 1996 The Western Australian Department of Conservation and Land Management seabird breeding islands database. In *The Status of Australia's Seabirds: Proceedings of the National Seabird Workshop, Canberra, 1993*. Pp. 73-137. Environment Australia Biodiversity Group.

Burbidge AA & Fuller PJ 2000 The breeding seabirds of Shark Bay, Western Australia. *CALMScience* **3**, 109-124.

Burbidge AA, Fuller PJ, Hopkins AJM, Kinnear JE & McKenzie NL 1982 The wildlife of Salisbury Island, Archipelago of the Recherche, Western Australia. Typescript held in library, Department of Biodiversity, Conservation and Attractions, Kensington WA.

Burbidge AA & George AS 1978 The flora and fauna of Dirk Hartog Island, Western Australia. *Journal of the Royal Society of Western Australia* **60**, 71-90.

Burbidge AA & McKenzie NL 1978 The islands of the north-west Kimberley, Western Australia. *Wildlife Research Bulletin Western Australia* No. 7.

Burbidge AA & Morris KD 2002 Introduced mammal eradications for nature conservation on Western Australian islands: A review. In *Turning the tide: The eradication of invasive species*. (eds CR Veitch & MN Clout), pp. 64-70. (IUCN SSC Invasive Species Specialist Group: Gland, Switzerland and Cambridge, UK.)

Burbidge AA, Johnstone RE & Fuller PJ 1996 The status of seabirds in Western Australia. In *The Status of Australia's Seabirds: Proceedings of the National Seabird Workshop, Canberra, 1993*. Pp. 57-71. Environment Australia Biodiversity Group.

Burbidge AA, McKenzie NL & Kenneally KF 1991 *Nature conservation reserves in the Kimberley*. Department of Conservation and Land Management, Perth.

Burbidge AA & Prince RIT 1972 The fauna, flora and planned usage of the Dampier Archipelago. Department of Fisheries and Fauna Western Australia Report No. 11.

Burbidge AH, Cowan S & Garnett ST 2021 Dorre Island rufous fieldwren *Calamanthus campestris dorre*. In *The Action Plan for Australian Birds 2020*. (eds ST Garnett & GB Baker), pp 656-667. CSIRO Publishing, Melbourne.

Burbidge AH, Dolman G, Ottewell K, Johnstone R & Burbidge M 2021a Genetic and morphological relationships of fieldwrens (*Calamanthus*): Implications for conservation status and management. *Emu - Austral Ornithology*, https://doi.org/10.1080/01584197.2021.1922293

Burbidge AH, Ford SS & Garnett ST 2021b Barrow Island white-winged fairy-wren *Malurus leucopteris edouardi*. In *The Action Plan for Australian Birds 2020*. (eds ST Garnett & GB Baker), pp 512-513. CSIRO Publishing, Melbourne.

Burbidge AH, Ford S, Blythman M & Garnett ST 2021c Dirk Hartog rufous fieldwren *Calamanthus campestris hartogi*. In *The Action Plan for Australian Birds 2020*. (eds ST Garnett & GB Baker), pp 667-669. CSIRO Publishing, Melbourne.

Burbidge AH, Ford S, Blythman M & Garnett ST 2021d Dirk Hartog southerm emu-wren *Stipiturus malachurus hartogi*. In *The Action Plan for Australian Birds 2020*. (eds ST Garnett & GB Baker), pp 518-520. CSIRO Publishing, Melbourne.

Bush TE & Lodge GA 1977 Birds of Bedout Island – A visit in May 1972. *Western Australian Naturalist* **13**, 189-190.

CALM 1990 Dampier Archipelago Nature Reserves Management Plan 1990-2000. Department of Conservation and Land Management, Kensington WA.

Cameron JMR (ed) 2006 *The Millendon Memoirs: George Fletcher Moore's Western Australian Diaries and Letters, 1830-1841*. Hesperian Press, Carlisle WA.

Campbell AJ 1890 A naturalist in Western Australia. Rottnest Island. *The Australasian* 3 May, p. 44.

Campbell AJ 1901 *Nests and Eggs of Australian Birds*... The author, Sheffield UK.

Campbell R 2005 Historical distribution and abundance of the Australian sea lion (*Neophoca cinerea*) on the west coast of Western Australia. Fisheries Research Report No. 148.

Campbell RA, Gales NJ, Lento GM & Baker CS 2008 Islands in the sea: Extreme female natal site fidelity in the Australian sea lion, *Neophoca cinerea*. *Biology Letters* **4**, 139-142.

Campbell R, Holley D, Collins P & Armstrong S 2014 Changes in the abundance and distribution of the New Zealand fur seal (*Arctocephalus forsteri*) in Western Australia: Are they approaching carrying capacity? *Australian journal of Zoology* **62**, 261-267.

Cannell BL, Chambers LE, Wooller RD & Bradley JS 2012 Poorer breeding by little penguins near Perth, Western Australia is correlated with above average sea surface temperatures and a stronger Leeuwin Current. Marine and Freshwater Research **63**, 914-925.

Carlquist S 1965 *Island Life: A natural history of the islands of the world*. American Museum of Natural History, New York.

Carter M, Clarke R, Pierce F, Dooley S, Swann G & Grant M 2010 Lesser coucal *Centropus bengalensis* on Ashmore Reef: First record for Australia. *Australian Field Ornithology* **27**, 109-118.

Carter M, Clarke RH & Swann G 2011 Island monarch *Monarcha cinerascens* on Ashmore Reef: First records for Australian Territory. *Australian Field Ornithology* **28**, 150-161.

Carter M & Swann G 2011 Untitled. *Western Australian Bird Notes* No. 137, 11.

Chambers LE & 15 others 2013 Phenological changes in the Southern Hemisphere. *PLOS ONE* **8**(10), e75514.

Churchill DM 1960 Late Quaternary changes in the vegetation on Rottnest Island. *Western Australian Naturalist* **7**, 160-166.

Clark D (translator & editor) 1994 *Baron Charles von Hügel New Holland Journal November 1833-October 1834.* Melbourne University Press, Carlton.

Clark WN 1842 Letter to editor. *The Perth Gazette and Western Australian Journal*, 20 August, p. 3.

Clarke RH, Carter M & Swann G 2009 Asian brown flycatchers *Muscicapa dauurica* on Ashmore Reef: First records for Australia. *Australian Field Ornithology* **26**, 123-131.

Clarke RH, Carter M, Swann G & Thomson J 2011 The status of breeding seabirds and herons at Ashmore Reef, off the Kimberley coast, Australia. *Journal of the Royal Society of Western Australia* **94**, 365-376.

Clarke RH, Swann G, Carter MJ, Mott RM & Herrod A 2017 The avifauna of Cartier Island Commonwealth Marine Reserve, north-western Australia. *Australian Field Ornithology* **34**, 18-25.

Cathcart B 1994 *Test of greatness: Britain's struggle for the atomic bomb*. John Murray, London.

Chester Q & McGregor A 1997 *Australia's Wild Islands*. Hodder Headline Australia, Rydalmere NSW.

Christensen J 2008 Shark Bay 1616-1991: The spread of science and the emergence of ecology in a World Heritage area. PhD Thesis, School of Humanities, The University of Western Australia.

Christie M 2003 Bridled terns in the South East of South Australia. *South Australian Ornithologist* **34**, 79-80.

Clarke RH, Swann G, Carter MJ, Mott RM & Herrod A 2017 The avifauna of Cartier Island Commonwealth Marine Reserve, north-western Australia. *Australian Field Ornithology* **34**, 18-25.

Coate K 1997 Seabird Islands No. 236. Adele Island, Western Australia. *Corella* **21**, 124-128.

Coate KH, Smith LA & Fontanini L 1994 The birds of Adele Island, Western Australia including notes on recently established breeding colonies of red-footed boobies (*Sula sula*) and great frigate birds (*Fregata minor*). *Western Australian Naturalist* **19**, 285-291.

Coate Y & Coate K 1986 *Lonely Graves of Western Australia*. Hesperian Press, Carlisle WA.

Coate Y & Coate K 2000 *More Lonely Graves of Western Australia & Burials at Sea*. Hesperian Press, Carlisle WA.

Coghlan JE 1885 Report upon the work carried out by the joint Admiralty and Colonial Marine Survey Department during the year 1884. Legislative Council Report No. 16. Government Printer, Perth.

Collie A 1830 On some particulars connected with the natural history of the kangaroo. *Zoological Journal* **5**, 238-241.

Collins LB, Zhu ZR, Wyrwoll K-H, Hatcher BG, Playford PE, Eisenhauer A, Chen JH, Wasserberg GJ & Bonani G 1993 Holocene growth history of a reef complex on a cool-water carbonate margin: Easter Group of the Houtman Abrolhos, Eastern Indian Ocean. *Marine Geology* **115**, 29-46.

Collis T 2013 The Lacepedes. The hidden treasure of the Kimberley. *Australian Birdlife* **2**(2), 22-25.

Comer S, Burbidge AH, Danks A & Garnett ST 2021 Noisy scrub-bird *Atrichornis clamosus*. In *The Action Plan for Australian Birds 2020*. (eds ST Garnett & GB Baker), pp. 480-483. CSIRO Publishing, Clayton South, Vic.

Connell GW 1983 Biogeography and community structure of insular herpetofaunas. BSc (Honours) thesis, University of Western Australia, Crawley WA.

Conservation Commission Western Australia & Department of Conservation and Land Management 2003 Carnac Island Nature Reserve Management Plan. *dpaw.wa.gov.au*

Cornell C 1974 *The Journal of Post Captain Nicolas Baudin*. Libraries Board of South Australia, Adelaide.

Cornell C 2003 *Voyage of Discovery to the Southern Lands... Second edition 1824. Book IV, comprising Chapters XXII to XXXIV*. Friends of the State Library of South Australia, Adelaide.

Cottesloe L (ed) 1979 *Diary & Letters of Admiral Sir CH Fremantle...relating to the Founding of the Colony of Western Australia 1829*. Fremantle Arts Centre Press, Fremantle.

Cowan M 2018 Reptiles of Dirk Hartog Island. Department of Biodiversity, Conservation and Attractions, Perth.

Crawford IM 1968 *The art of the Wanjina. Aboriginal cave paintings in Kimberley, Western Australia*. Oxford University Press, London.

Crawford IM 2001 *We won the victory: Aborigines and outsiders on the north-west coast of the Kimberley*. Fremantle Arts Centre Press, Fremantle.

Cresswell G 1990 *The Leeuwin Current. Corella* **14**, 113-118.

Cumpston JS 1970 Kangaroo Island 1800-1836. Roebuck Society Publication No. 1, Canberra.

Darwin C 1906. *The Voyage of the Beagle*. Dent, London. Originally published in 1839 as Journal and Remarks, 1832-1836, Volume 3 of the *Narrative of the Surveying Voyages of His Majesty's Ships Adventure and Beagle...* H Colburn, London.

Davis KL, Colefax AP, Tucker JP, Kelaher BP & Santos IR 2021a Global coral reef ecosystems exhibit declining calcification and increasing primary productivity. *Communications Earth & Environment* **2**.

Davis RA, Carter R & Burbidge AH 2021b Report on the first trial of remote camera monitoring for the Abrolhos Painted Button-Quail (*Turnix varius scintillans*). Unpublished report, Edith Cowan University, Joondalup.

DBCA 2017 Albany coast and parks reserves. Management plan 90. Department of Biodiversity, Conservation and Attractions, Perth.

DBCA 2020 Pilbara inshore islands nature reserves and proposed additions. Draft management plan. Department of Biodiversity, Conservation and Attractions. Kensington WA.

De Barros L & Coughran DK 2017 *Rottnest Island. Kingdom of the Quokka*. Sea Dog TV International, Marmion WA.

DEC & MPRA 2007 Rowley Shoals Marine Park Management Plan No. 56, 2007-2017. Department of Environment and Conservation/Marine Parks & Reserves Authority, Kensington WA.

Dell J 2005 The fauna encountered in south-west Western Australia: Its culinary value and scientific legacy. In Matthew Flinders and his Scientific Gentlemen: The Expedition of HMS *Investigator* to Australia, 1801-05. (eds J Wege, A George, J Gathe, K Lemson & K Napier), pp. 115-127. Western Australian Museum, Welshpool, WA.

Dell J & Cherriman S 1990 The birds of Faure Island, Shark Bay, Western Australia. *Records of the Western Australian Museum* Supplement No. 75, 55-70.

Department of Conservation and Land Management & National Parks and Nature Conservation Authority 1992 Shoalwater islands Management Plan 1992-2002

Department of Parks and Wildlife 2015 *Barrow group nature reserves. Management plan 82*. Department of Parks and Wildlife, Perth.

Department of Planning and Infrastructure 2002 Hydrographic chart 412, Rottnest Island.

Department of Transport WA 1995 Hydrographic chart 001, Ocean Reef to Cape Peron

Department of Transport Western Australia. 1999 Bremer Bay 1: 25 000. Chart WA 909.

Department of Transport Western Australia. 2000 Nornalup Inlet 1: 15 000. Chart WA 1046.

Donaldson M 2009 *Burrup Rock Art. Ancient Aboriginal Rock Art of Burrup Peninsula and Dampier Archipelago Western Australia*. Wildrocks Publications, Mount Lawley WA.

Doughty P, Palmer RA, Cowan MA & Pearson DJ 2012 Biogeographic patterns of frogs of the Kimberley islands, Western Australia. *Records of the Western Australian Museum, Supplement* **81**, 109-124.

DPAW 2016 Esperance and Recherche Parks and Reserves Management Plan 84. Department of Parks and Wildlife and Conservation Commission of Western Australia, Perth.

Drake-Brockman H 1963 *Voyage to disaster: The life of Francisco Pelsaert*. Angus & Robertson, Sydney.

DEC/CCWA 2004 Turquoise Coast and Island Nature Reserves. Management Plan No. 50. Department of Environment and Conservation/Conservation Commission of Western Australia.

Dickman CR, Daly SEJ & Connell GW 1991 Dietary relationships of the barn owl and Australian kestrel on islands off the coast of Western Australia. *Emu* **91**, 69-72

Dunlop JN 1979 The occurrence of breeding roseate terns, *Sterna dougalli*, at Lancelin Island, Western Australia. *Western Australian Naturalist* **14**, 118-119.

Dunlop JN 1996 Habituation to human disturbance by breeding bridled terns *Sterna anaethetus. Corella* **20**, 13-16.

Dunlop JN 2009 The population dynamics of tropical seabirds establishing frontier colonies on islands off south-western Australia. *Marine Ornithology* **37**, 99-105.

Dunlop JN & Greenwell CN 2020 Seasonal movements and metapopulation structure of the Australian fairy tern in Western Australia. *Pacific Conservation Biology* **27**, 47-60.

Dunlop JN, Rippey E, Bradshaw LE & Burbidge AA 2015. Recovery of seabird colonies on Rat Island (Houtman Abrolhos) following the eradication of introduced predators. *Journal of the Royal Society of Western Australia* **98**, 29-36.

Durlacher JS 2013 *Landlords of the Iron Shore*. Hesperian Press, Carlisle WA.

Dutton G (ed) 1986 *The Book of Australian Islands*. Macmillan, South Melbourne Vic.

Duyker E & Duyker M 2001 *Bruny d'Entrecasteaux Voyage to Australia & the Pacific 1791-1793*. Melbourne University Press, Carlton South Vic.

Edmiston RJ & White BJ 1974 The vegetation of Rottnest. Forests Department, Perth.
Edward KL & Harvey MS 2010 A review of the Australian millipede genus *Atelomastix* (Diplopoda: Spirostreptida: Iulomorphidae). *Zootaxa* **2371**, 1-63.
Edwards H 1966 *Islands of the angry ghosts*. Hodder & Stoughton, London. Reprinted several times by different publishers and available as an eBook.
Edwards H 1970 *The wreck on the Half-Moon Reef*. Rigby, Adelaide.
Edwards PI 1981 The journal of Peter Good, gardener on Matthew Flinders voyage to Terra Australis 1801-03. *Bulletin of the British Museum (Natural History), Historical Series* **9**, 1-213.
Eldridge MDB, King JM, Loupis AK, Spencer PBS, Taylor AC, Pope LC & Hall GP 1999 Unprecedented low levels of genetic variation and inbreeding depression in an island population of the black-footed rock-wallaby. *Conservation Biology* **13**, 531-541.
Eldridge MDB, Neaves LE & Spencer PBS 2018 Genetic analysis of three remnant populations of the rufous hare-wallaby (*Lagorchestes hirsutus*) in arid Australia. *Australian Mammalogy* **41**, 123-131.
Errington S 2012 Garden Island in the winter of 1829. *Early Days* **14**, 1-20.
Eyre EJ 1845 *Journals of Expeditions of Discovery*...T & W Boone, London.
Fairbridge RW 1948 Discoveries in the Timor Sea, north-west Australia. *Journal and Proceedings of the Royal Australian Historical Society* **34**, 193-218.
Fairbridge RW & Serventy VN 1954 Physiography. Australian Geographical Society Report No. 1, Part 1b.
Featherstone WE, Penna NT, Filmer MS & Williams SDP 2015 Nonlinear subsidence at Fremantle, a long-recording tide gauge in the Southern Hemisphere. *Journal of Geophysical Research: Oceans* **120**, 7004-7014.
Fischer SR 2012 *Islands from Atlantis to Zanzibar*. Reaktion Books, London.
Fisher CT 1992 The importance of early Victorian natural historians in the discovery and interpretation of the Australian fauna, with special reference to John Gilbert. PhD thesis, Liverpool Polytechnic.
Flinders M 1814 *A Voyage to Terra Australis*...G & W Nicol, London.
Foley JC 1957 Droughts in Australia. Review of records from earliest years of settlement to 1955. Bureau of Meteorology Bulletin No 43.
Ford J 1963 The reptilian fauna of the islands between Dongara and Lancelin, Western Australia. *Western Australian Naturalist* **8**, 135-142.
Ford J 1965a The reptilian fauna of the islands between Dongara and Lancelin, Western Australia: Additional notes. *Western Australian Naturalist* **9**, 174-175.
Ford J 1965b The avifauna of the islands between Dongara and Lancelin, Western Australia. *Emu* **64**, 129-144, 181-203.
Fowler S 1945 The fire on Mondrain Island, W.A. *Emu* **44**, 334-335.
Fraser C 1830 Remarks on the botany, &c. of the banks of Swan River, Isle of Buache, Baie Geographe, and Cape Naturaliste. *Botanical Miscellany* **1**, 221-236.
French M undated *Archipelago of the Recherche: The Islands of Esperance*. The author.
Friend T 2017 Island home: A new start for dibblers. *Landscope* **33**(1), 39-42.
Friend T, Button T & Bougher N 2018 Establishing a new insurance population of the world's rarest marsupial. Report on the first trial translocation of Gilbert's potoroo to Middle Island, July-August 2017. Unpublished report.
Fullagar P & Murray D The seabird island series. *Australian Bird Bander* **11**, 12-13.

Fullagar PJ & van Tets GF 1976 Bird notes from a winter visit to Eclipse Island, Western Australia. *Western Australian Naturalist* **13**, 136-144.

Fuller PJ & Burbidge AA 1981 The birds of Pelsaert Island, Western Australia. Report No. 44, Department of Fisheries and Wildlife, Western Australia.

Fuller PJ & Burbidge AA 1987 Discovery of the dibbler, Parantechinus apicalis, on islands of Jurien Bay. *Western Australian Naturalist* **16**, 177-181.

Fuller PJ & Burbidge AA 1998a Seabird Islands No. 239. Bedout Island, Pilbara Region, Western Australia. *Corella* **22**, 113-115.

Fuller PJ & Burbidge AA 1998b Seabird Islands No. 240. North Turtle Island, Pilbara Region, Western Australia. *Corella* **22**, 116-117.

Fuller PJ, Burbidge AA & Owens R 1994 Breeding seabirds of the Houtman Abrolhos, Western Australia: 1991-1993. *Corella* **18**, 97-113.

Gales NJ & Lambert PM 1985 Observations of the distribution of the New Zealand fur seal (*Arctocephalus forsteri*) in Western Australia. *Western Australian Naturalist* **16**, 70-71.

Gales NJ, Cheal AJ, Pobar GJ & Williamson P 1992 Breeding biology and movements of Australian sea-lions, *Neophoca cinerea*, off the west coast of Western Australia. *Wildlife Research* **19**, 405-416.

Gales NJ, Shaughnessy & Dennis T 1994 Distribution, abundance and breeding cycle of the Australian sea lion *Neophoca cinerea* (Mammalia: Pinnipedia). *Journal of Zoology* **234**, 353-370.

Garnett ST & Baker GB 2021 *The action plan for Australian birds*. CSIRO Publishing, Melbourne.

Geological Survey of Western Australia 1975 *The Geology of Western Australia*. Geological Survey of Western Australia, Memoir 2. Government Printer, Perth.

George AS 1971 The plants seen and collected in north-western Australia by William Dampier. *The Western Australian Naturalist* **11**, 173-177.

George AS 1990 *Willdampia*, a new generic name for Sturt Pea. *The Western Australian Naturalist* **11**, 173-178.

George AS 1999 *William Dampier in New Holland. Australia's first natural historian*. Bloomings Books, Hawthorn, Victoria.

Geoscience Australia. Border lengths. https://www.ga.gov.au/scientific-topics/national-location-information/dimensions/border-lengths#heading-1 Downloaded on 12 May 2021.

Geoscience Australia. Islands. https://www.ga.gov.au/scientific-topics/national-location-information/landforms/islands. Downloaded on 27 April 2021.

Geyle HM & 28 others 2021 Reptiles on the brink: Identifying the Australian terrestrial snake and lizard species most at risk of extinction. *Pacific Conservation Biology* **27**, 3-12.

Gibbs M 2010 *The Shore Whalers of Western Australia. Historical Archaeology of a Maritime Frontier. Studies in Australasian Historical Archaeology*, volume 2. Sydney University Press, Sydney.

Gibson LA & Kohler F 2012. Determinants of species richness and similarity of species composition of land snail communities on Kimberley islands. *Records of the Western Australian Museum, Supplement* **81**, 41-66.

Gibson LA & McKenzie NL 2012a Identification of biodiversity assets on selected Kimberley islands: Background and implementation. *Records of the Western Australian Museum, Supplement* **81**, 1-14.

Gibson LA & McKenzie NL 2012b Occurrence of non-volant mammals on islands along the Kimberley coast of Western Australia. *Records of the Western Australian Museum, Supplement* **81**, 15-39.

Gilbert AH 1829 Diary. Transcript copy. Acc. 6782A, Battye Library, Perth.

Gilbert J 1843 [Visit to Houtman Abrolhos]. *The Inquirer*, 19 April.

Gilbert J [no date]. Untitled. Australian Joint Copying Project, Microfilm reels M725-6. National Library of Australia, Canberra.

Gillespie RG & Clague DA 2009 *Encyclopedia of Islands*. University of California Press, Berkeley, California.

Gillham ME 1959 Diary. Volume X, pp. 85-95, 109-131 (26.11-2.12.1959). Glamorgan Archives, Cardiff, Wales.

Gillham ME 1961 Alteration of the breeding habit by sea-birds and seals in Western Australia. *Journal of Ecology* **49**, 289-300.

Gillham ME 1963 Association of nesting sea-birds and vegetation types on islands off Cape Leeuwin, south-western Australia. *Western Australian Naturalist* **9**, 29-46.

Glauert L 1954 Reptiles and frogs. Australian Geographical Society Report No. 1, Part 5.

Goldsworthy SD, Shaughnessy PD, Mackay AI, Bailleu F, Holman D, Lowther AD, Page B, Waples K, Raudino H, Bryars S & Anderson T 2021 Assessment of the status and trends in abundance of a coastal pinniped, the Australian sea lion *Neophoca cinerea*. *Endangered Species Research* **44**, 421-437.

Goodsell J, Tingay A & Tingay SR 1976 A resource survey of Woody Island, Archipelago of the Recherche. Report No. 21, Department of Fisheries and Wildlife, Perth.

Google Earth Pro 2021 https://google-earth-pro.en.lo4d.com/windows

Google Maps 2021 https://www.google.com.au/maps

Graham NAJ, Wilson SK, Carr P, Hoey AS, Jennings S & MacNeil MA 2018 Seabirds enhance coral reef productivity and functioning in the absence of invasive rats. *Nature* **559**, 250-253.

Green J 2018 The mystery of the missing VOC shipwreck in the Houtman Abrolhos Islands, Western Australia. Special Publication No. 20, Australian National Centre of Excellence for Maritime Archaeology.

Green N 2020 *Into the Arms of Death: The Impact of Introduced Disease on the Indigenous Population of Western Australia 1829-1950*. Hesperian Press, Carlisle WA.

Greenwell C 2021 Life history, ecology, and population dynamics of the Australian Fairy Tern and implications for their conservation. PhD thesis, Murdoch University.

Greenwell CN, Dunlop JN, Admiraal R & Loneragan NR 2020 The secret life of fairy terns: Breeding chronology and life history observations of *Sternula nereis nereis* in south-western Australia. *Pacific Conservation Biology* **27**, 143-154.

Halkyard B 2014 Exploiting Green and Hawksbill Turtles in Western Australia: The Commercial Marine Turtle Fishery. In *Historical Perspectives of Fisheries Exploitation in the Indo-Pacific.* (eds J. Christensen and M. Tull). Springer Science+Business Media, Dordrecht.

Hallermann J 2020 An annotated list of reptiles and amphibians from the 1905 Hamburg expedition to southwest Australia deposited in the Zoological Museum Hamburg. *Evolutionary Systematics* **4**, 61-70.

Halse SA, Burbidge AA, Lane JAK, Haberley B, Pearson GB & Clarke A 1995 Size of the Cape Barren goose population in Western Australia. *Emu* **95**, 77-83.

Hartley R & Pearson D 2009 Island home for rare skink. *Landscope* **24**(2), 47-49.

Harvey JM, Alford JL, Longman VM & Keighery GJ 2001. A flora and vegetation survey of the islands of the Houtman Abrolhos, Western Australia. *CALMScience* **3**, 521-623.

Henderson G & Henderson K-J 1988 *Unfinished Voyages – Western Australian Shipwrecks 1851-1880.* University of Western Australia Press, Nedlands WA.

Henson M, Kenneally KF, Griffin EA & Barrett RL 2014 Terrestrial flora. In *Ecological studies of the Bonaparte Archipelago and Browse Basin*. Pp. 19-102. INPEX Operations Australia Pty Ltd, Perth, Western Australia.

Hercock M 1998 The relationship between public policy and the environment of four Western Australian offshore islands: A legacy of perception and resource use. PhD thesis, University of Western Australia.

Hickman AH & Gibson DL 1982 Port Hedland-Bedout Island. 2nd edition. Geological survey of Western Australia. Australian Government Publishing Service, Canberra.

Hickman AH & Smithies RH 2001 Roebourne, Western Australia. 2nd edition. Geological Survey of Western Australia, Perth.

Hickman AH & Strong CA 2003 Dampier-Barrow Island, Western Australia. 2nd edition. Geological Survey of Western Australia, Perth.

Holsworth WN 1965 Nesting success of the osprey on Rottnest Island, Western Australia. *Western Australian Naturalist* **10**, 13-15.

Hopkins AJM 1981 Studies on Middle Island in the Recherche Archipelago. *S.W.A.N.S. State Wildlife Authority News Service* **11**(2), 6-10.

How RA, Cowan MA, Teale RJ & Schmitt LH 2020 Environmental correlates of reptile variation on the Houtman Abrolhos archipelago, eastern Indian Ocean. *Journal of Biogeography* **47**, 2017-2028.

How RA, Pearson DJ, Desmond A & Maryan B 2004 Reappraisal of the reptiles on the islands of the Houtman Abrolhos, Western Australia. *The Western Australian Naturalist* **24**, 172-178.

How RA, Schmitt L, Teale RJ & Cowan MA 2006 Appraising vertebrate diversity on Bonaparte Islands, Kimberley, Western Australia. *The Western Australian Naturalist* **25**, 92-110.

Hull AFB 1922 A visit to the Archipelago of the Recherche S.W. Australia. *Emu* **21**, 277-289.

Humphreys G 2014 Troglofauna. In *Ecological studies of the Bonaparte Archipelago and Browse Basin*. Pp. 187–201. INPEX Operations Australia Pty Ltd, Perth, Western Australia.

Humphreys G, Alexander J, Harvey MS & Humphreys WF 2013 The subterranean fauna of Barrow Island, north-western Australia: 10 years on. *Records of the Western Australian Museum Supplement* **83**, 145-158.

Hunt HA 1929 *Results of Rainfall Observations made in Western Australia... for all years of record up to 1927...and record of notable meteorological events...* Bureau of Meteorology, Melbourne.

Hunt S-J 1986 *Spinifex and Hessian*...University of Western Australia Press, Nedlands.

Ibbotson J 2001 Lighthouses of Australia: Images from the End of an Era. Australian Lighthouse Traders, Surrey Hills Vic.

i-Boating 2021 fishing-app.gpsnauticalcharts.com

IPCC 2021 Summary for Policymakers. *Climate Change 2021: The physical science basis. Contribution of Working Group I to the Sixth Assessment Report of the Intergovernmental Panel on Climate Change*. Cambridge University Press.

Johansen MP, Child DP, Cresswell T, Harrison JJ, Hotchkis MAC, Howel, NR, Johansen A, Sdraulig S, Thiruvoth S, Young E & Whiting SD 2019 Plutonium and other radionuclides persist across marine-to-terrestrial ecotopes in the Montebello Islands sixty years after nuclear tests. *Science of the Total Environment* **691**, 572-583.

Johnson B & Colombelli-Négrel D 2021 Breeding success in southern Australian little penguins is negatively correlated with high wind speeds and sea surface temperatures. *Ornithological Applications* **123**, 1-15.

Johnstone RE 1990 Mangroves and mangrove birds of Western Australia. *Records of the Western Australian Museum* Supplement No 32.

Johnstone RE & Storr GM 1998 *Handbook of Western Australian Birds. Vol. 1 Non-passerines (Emu to Dollarbird).* Western Australian Museum, Perth.

Johnstone RE & Storr G M 2004 *Handbook of Western Australian Birds. Volume II. Passerines (Blue-winged Pitta to Goldfinch)*. Western Australian Museum, Perth.

Jones HP, Holmes ND, Butchart SH, Tershy BR, Kappes PJ, Corkery I, Aguirre-Munoz A, Armstrong DP, Bonnaud E, Burbidge AA, Campbell K, Courchamp F, Cowan PE, Cuthbert RJ, Ebbert S, Genovesi P, Howald GR, Keitt BS, Kress SW, Miskelly CM, Oppel S, Poncet S, Rauzon MJ, Rocamora G, Russell JC, Samaniego-Herrera A, Seddon PJ, Spatz DR, Towns DR & Croll DA 2016 Invasive mammal eradication on islands results in substantial conservation gains. *Proceedings of the National Academy of Sciences of the United States of America* **113**, 4033-4038.

Joske P, Jeffery C & Hoffman L 1997 *Rottnest Island : A Documentary History*. Centre for Migration and Development Studies, University Of Western Australia.

Kabay ED & Start AN undated (1976?) Results of the search for the potoroo in south west and south coast of Western Australia 1975/76. Typescript held in library, Department of Biodiversity, Conservation and Attractions, Kensington WA.

Karadada J, Karadada L, Goonack W, Mangolamara G, Bunjuck W, Karadada L, Djanghara B, Mangolamara S, Oobagooma J, Charles A, Williams D, Karadada R, Saunders T & Wightman G 2011 *Uunguu Plants and Animals*. Wunambal Gaambera Aboriginal Corporation and Department of Natural Resources, Environment, The Arts and Sport, Kalumburu.

Keighery G 1995 Additions to the flora of the Recherche Archipelago. *Western Australian Naturalist* **20**, 133-138.

Keighery GJ, Alford JJ & Longman V 2002 A vegetation survey of the islands of the Turquoise Coast from Dongara to Lancelin, south-western Australia. *Conservation Science Western Australia* **4**, 13-62.

Keighery GJ, Gibson N, Kenneally KF & Mitchell AA 1995. Biological inventory of Koolan Island, Western Australia. 1. Flora and vegetation. *Records of the Western Australian Museum* **17**, 237-248.

Keighery G & Muir W 2008 Vegetation and vascular flora of Faure Island, Shark Bay, Western Australia. *Records of the Western Australian Museum* Supplement No. 75, 11-19.

Kendrick PG 2003 Mass deaths of sea turtles on the Montebello Islands, October 1953, following Operation Hurricane. *Western Australian Naturalist* **24**, 107-110.

Kendrick PG 2007 The non-volant vertebrate fauna of the Burrup Peninsula, Pilbara, Western Australia. *The Western Australian Naturalist* **25**, 197-241.

Kenneally KF 1993 Ashmore Reef and Cartier Island. *Flora of Australia* **50**, 43-47.

King PP 1827 *Narrative of a survey of the intratropical and western coasts of Australia, performed between the years 1818 and 1822*. John Murray, London. Available as an eBook on Gutenburg.net.au.

Kingsley MR, Lavers JL, Steeves TE & Burridge CP 2020 Genetic distinctiveness of masked booby (*Sula dactylatra*) on Bedout Island, Western Australia. *Emu* **120**, 150-155.

Kitchener DJ, Chapman A, Dell J & Muir BG 1980 Lizard assemblage and reserve size and structure in the Western Australian wheatbelt – Some implications for conservation. *Biological Conservation* **17**, 25-62.

Kitchener DJ, Dell J & Muir BG 1982 Birds in Western Australian wheatbelt reserves – Implications for conservation. *Biological Conservation* **22**, 127-163.

Kitchener DJ & How RA 1982 Lizard species in small mainland habitat isolates and islands off south-western Western Australia. *Australian Wildlife Research* **9**, 357-363.

Labbé AMT, Dunlop JN, Calver M, Shephard JM & Van Keulen M 2017 Restoration of native vegetation and re-introduction of *Malva preissiana* on Penguin Island – preliminary findings. *Journal of the Royal Society of Western Australia* **100**, 1-3.

Labillardiere M 1800 *Voyage in Search of La Pérouse*...J Stockdale, London.

Lamb WK (ed) 1984 *George Vancouver: A Voyage of Discovery to the north Pacific Ocean and round the World 1791-1795*. Hakluyt Society, London.

Lamont CR, Bamford MJ, Harvey MS & Fitzpatrick JJ 2014 Terrestrial fauna. In *Ecological studies of the Bonaparte Archipelago and Browse Basin*. Pp. 103-186. INPEX Operations Australia Pty Ltd, Perth, Western Australia.

Lane SG 1983 Short-tailed shearwater on Figure of Eight Island, Archipelago of the Recherche, Western Australia. *Emu* **83**, 37-38.

Lane SG 1984 Further notes on visits to islands of the south coast of Western Australia. *Corella* **8**, 64-66.

Lane SG 1985a Seabird Islands No. 148. Stanley Island, Western Australia. *Corella* **8**, 121-122.

Lane SG 1985b Seabird Islands No. 149. Flat [Quagering] Island, Western Australia. *Corella* **8**, 123-124.

Langford D & Burbidge AA 2001 Translocation of the mala (*Lagorchestes hirsutus*) from the Tanami Desert, Northern Territory to Trimouille Island, Western Australia. *Australian Mammalogy* **23**, 37-46.

Lavers JL 2015 Population status and threats to Flesh-footed shearwaters (*Puffinus carneipes*) in South and Western Australia. *ICES Journal of Marine Science* **72**, 316-327.

Lehmann C (ed) 1844-47 *Plantae Preissianae*... Meissner, Hamburg.

Lewis SE, Sloss CR, Murray-Wallace CV, Woodroffe CD & Smithers SG 2013 Post-glacial sea-level changes around the Australian margin: A review. *Quaternary Science Reviews* **74**, 115-138.

Lipfert O 1912 A visit to Bernier and Dorré Islands. *Records of the Western Australian Museum and Art Gallery* **1**(2), 98-101.

Locate 2021 https://maps.slip.wa.gov.au/landgate/locate/

Lohr C 2017 Pilbara Islands database. Department of Biodiversity, Conservation and Attractions. Kensington WA.

Lohr C, Passeretto K, Lohr M & Keighery G 2015 Prioritising weed management activities in a data deficient environment: The Pilbara islands, Western Australia. http://dx.doi.org/10.1016/j.heliyon.2015.e00044

Lohr MT & Keighery G 2014 The status and distribution of alien plants on the islands of the south coast of Western Australia. *Conservation Science Western Australia* **9**, 181-200.

Lohr MT, Lohr CA, Keighery G & Long V 2016. The status and distribution of non-native plants on the gazetted and territorial islands off the north coast of Western Australia. *Conservation Science Western Australia* **10**, 8 [online].

Lowe RJ, Cuttler MVW & Hansen JE 2021 Climatic drivers of extreme sea level events along the coastline of Western Australia. *Earth's Future* **9**.

Lowry DC 1974 Dongara-Hill River, Western Australia. Geological Survey of Western Australia, Perth. Australian Government Publishing Service, Canberra.

Lowry DC & Doepel JJG 1974 Malcolm-Cape Arid Western Australia. Sheet SI/51-7. Australian Government Publishing Service, Canberra.

Lyons MN, Keighery GJ, Gibson LA & Handasyde T 2014 Flora and vegetation communities of selected islands off the Kimberley coast of Western Australia. *Records of the Western Australian Museum*, Supplement **81**, 205-244.

Mabberley DJ 2017 *Painting by Numbers: The Life and Art of Ferdinand Bauer*. New South Publishing, Sydney.

Macpherson JH 1954 Molluscs (sea shells and snails). Australian Geographical Society Report No. 1, Part 7.

Main BY 1954 Spiders and Opiliones. Australian Geographical Society Report No. 1, Part 6.

Main AR, Shield JW & Waring H 1959 Recent studies on marsupial ecology. In *Biogeography and Ecology in Australia* (eds A Keast, RL Crocker, CS Christian), pp. 315-331. W Junk, Den Haag, The Netherlands.

Majer JD, Callan SK, Edwards K, Gunawardene NR & Taylor CK 2013 Baseline survey of the terrestrial invertebrate fauna of Barrow Island. *Records of the Western Australian Museum Supplement 83*.

Mangolamara S, Karadada L, Oobagooma J, Woolagoodja D & Karadada J (compiled by K Doohan) 2018 *We are coming to see you*. Wunambal Gaambera Aboriginal Corporation, Kalumburu.

Marchant LR 1988 *An Island unto Itself: William Dampier and New Holland*. Hesperian Press, Carlisle WA.

Marchant NG & Abbott I 1981 Historical and recent observations of the flora of Garden Island, Western Australia. *Western Australian Herbarium Research Notes* **5**, 49-62.

Maryan B, Stevenson C, Pearson D, How R & Schmitt L 2009. Additions to the reptiles known from islands in the Houtman Abrolhos, Western Australia. *The Western Australian Naturalist, 26*, 275-277.

Mather S 2020 Summer migrants – The importance of Rottnest Island for trans-equatorial bird species. *Stilt* **73**, 25-32.

Meathrel CE & Klomp NI 1990 Predation of little penguin eggs by King's skinks on Penguin Island, Western Australia. *Corella* **14**, 129-130.

McArthur WM 1957 Plant ecology of the coastal islands near Fremantle, WA. *Journal of the Royal Society of Western Australia* **40**, 46-63.

McCaw L 1997 *Callitris preissii* on Bald Island, Western Australia Preliminary observations on distribution, stand structure and tree age. Report lodged in library, Department of Biodiversity, Conservation and Attractions.

McKenzie NL & Bullen RD 2012 An acoustic survey of zoophagic bats on islands in the Kimberley, Western Australia, including data on the echolocation ecology, organisation and habitat relationships of regional communities. *Records of the Western Australian Museum, Supplement* **81**, 67-108.

McKenzie NL, Fontanini L, Lindus N & Williams MR 1995 Biological inventory of Koolan Island, Western Australia. 2. Zoological notes. *Records of the Western Australian Museum* **17**, 249-266.

McKenzie NL, Johnston RB & Kendrick PG (eds) 1991 *Kimberley rainforests of Australia.* Surrey Beatty and Sons, Canberra.

Menzies A 1791 Journal [Manuscript]. Microfilm reel M157, National Library of Australia, Canberra.

Michaelsen W & Hartmeyer R 1907-1930 *Die Fauna Südwest-Australiens.* G Fischer, Jena.

Milentis S (ed) 2022 Western Australian Exploration 1846-1859. Hesperian Press, Carlisle WA.

Miller S, Bencini R, Mills H & Moro D 2003 Food availability for the dibbler (*Parantechinus apicalis*) on Boullanger and Whitlock Islands, Western Australia. *Wildlife Research* **30**, 649-654.

Mills HR & Bencini R 2000 New evidence for facultative male die-off in island populations of dibblers, *Parantechinus apicalis. Australian Journal of Zoology* **48**, 501-510.

Milton DA 2005 Birds of Ashmore Reef National Nature Reserve: An assessment of its importance for seabirds and waders. *The Beagle* Supplement **1**, 133-141.

Montague PD 1913 The Monte Bello Islands. *The Geographical Journal* **42**, 34-44.

Montague PD 1914 A report on the fauna of the Monte Bello Islands. *Proceedings of the Zoological Society* **45**, 625-652.

Moor K & Ferguson I 1979 A day in the life of Beagle Islands. *Weekend News* [Perth], 26.5.1979, pp. 14-15.

Morgan KH & Peers R 1973 Esperance-Mondrain Island Western Australia. Sheet SI/51-6. Australian Government Publishing Service, Canberra.

Moro D 2002 Translocation of captive-bred dibblers *Parantechinus apicalis* (Marsupialia: Dasyuridae) to Escape Island, Western Australia. *Biological Conservation* **111**, 305-315.

Moro D & Lagdon R 2013 History and environment of Barrow Island. *Records of the Western Australian Museum* Supplement No. 83, 001-008.

Moro D, Palmer R, Greatwich B, Dickinson R & Anderson H 2019 Browse Island. A journey to Western Australia's most remote nature reserve. *Landscope* **34**(2), 22-27.

Morris KD 2002 The eradication of the black rat (*Rattus rattus*) on Barrow and adjacent islands off the north-west coast of Western Australia. In *Turning the tide: The Eradication of Invasive Species.* (eds CR Veitch & MN Clout), pp 219-225. IUCN SSC Invasive Species Specialist Group, Gland, Switzerland and Cambridge, UK.

Morris K, Page M, Kay R, Renwick J, Desmond A, Comer S, Burbidge AH, Kuchling G & Sims C 2015 Forty years of fauna translocations in Western Australia: Lessons learned. *Advances in Reintroduction Biology of Australian and New Zealand Fauna.* Pp 217-235.

Mott R, Hodgson JC, Herrod A & Clarke RH 2020 Nest-site fidelity in red-tailed tropicbirds informs costing a localised ant baiting strategy at Ashmore Reef. *Emu* **120**, 269-273.

Mulvaney K 2009 Dating the Dreaming: Extinct fauna in the petroglyphs of the Pilbara region, Western Australia. *Archaeology in Oceania* **44** Supplement, 40-48.

Mulvaney K 2015 *Murujuga Marni: Rock Art of the Macropod Hunters and Mollusc Harvesters.* UWA Publishing, Crawley WA.

Murphy RC 1938. The need of insular exploration as illustrated by birds. *Science* **88**, 533-539.

Murray I & Hercock M 2008 *Where on the coast is that?* Hesperian Press, Carlisle, WA.

Newspapers: *The Western Mail* 22.9.1900: 33; *Geraldton Express* 16.11.1900.

Nias RC, Burbidge AA, Ball D & Pressey RL 2010. Island arks: The need for an Australian national island biosecurity initiative. *Ecological Management & Restoration* **11**, 166-167.

Nicolle D, Brooker MIH & French ME 2014 A new subspecies of the threatened monocalypt *Eucalyptus insularis* (Myrtaceae) from Western Australia. *Nuytsia* **24**, 249-253.

NOAA 2021 tides and currents. https://tidesandcurrents.noaa.gov/sltrends/sltrends_global.shtml, downloaded October 2021

NPNCA 1999 Jurabi and Bundegi coastal parks, and Muiron islands. Management plan 1999-2000. National Parks and Nature Conservation Authority, Perth.

Ogilvie-Grant WR 1909 On a collection of birds from Western Australia. *Ibis* **3 (series 9)**, 650-689.

Onton K & Webb A 2009 South west islands survey report 2009. Department of Environment and Conservation, Bunbury WA.

Oppenheimer M, Glavovic BC, Hinkel J, van de Wal R, Magnan AK, Abd-Elgawad A, Cai R, Cifuentes-Jara M, DeConto R, Ghosh T, Hay J, Isla F, Marzeion B, Meyssignac B & Sebesvari Z 2019 Sea level rise and Implications for low-lying islands, coasts and communities. In *IPCC Special Report on the Ocean and Cryosphere in a Changing Climate.*

Orchard AE & Orchard TA 2018 *The Australian botanical journals of Allan Cunningham. The later King expeditions February 1819-September 1822.* Privately published, Weston Creek, ACT.

Orchard AE & Orchard TA 2020 Allan Cunningham's Plant Specimens, 1816-1822. The authors, Weston Creek ACT.

Palmer RA, Pearson DJ, Cowan MA & Doughty P 2013 Islands and scales: A biogeographic survey of reptiles on Kimberley islands, Western Australia. *Records of the Western Australian Museum, Supplement* **81**, 183-204.

Pearman GI 1971 An exploratory investigation of the growth rings of *Callitris preissii* trees from Garden Island and Naval Base. *Western Australian Naturalist* **12**, 12-17.

Pearson D & Bonnet X 2007 L'île des serpents. A twisted tale of 'tigers', Frenchmen and seagulls. *Landscope* **22 (3)**, 40-47.

Pearson D, Comer S & Cochrane A 2005 Return to Mondrain. *Landscope* **20**(4), 40-44.

Pearson D, Hopper S, Cochrane A, Comer S & Danks A 2004 Fire in the ark. *Landscope* **20**(1), 10-17.

Pearson D & Jones B 2000 Lancelin Island skink Recovery Plan. Department of Conservation and Land Management Wildlife Management Program No. 22.

Pearson D, Shine R & Williams A 2002 Geographic variation in sexual size dimorphism within a single snake species (*Morelia spilota*, Pythonidae). *Oecologia* **131**, 418-426.

Pearson DJ, Cowan MA & Caton W 2013 The avifauna of larger islands along the Kimberley coast, Western Australia. *Records of the Western Australian Museum, Supplement* **81**, 125-144.

Pearson M 1989 The Archipelago of the Recherche historical and archaeological reconnaissance. In Survey of the Eastern Group islands of the Recherche Archipelago. Report submitted to the WA heritage Commission. Typescript held in library, Department of Biodiversity, Conservation and Attractions, Kensington WA.

Pen LJ & Green JW 1983 Botanical exploration and vegetational changes on Rottnest Island. *Journal of the Royal Society of Western Australia*. **66**, 20-24.

Péron F & de Freycinet L 1824 *Voyage of Discovery to Southern Lands. Second edition. Books I to III.* Translation by C Cornell (2006). Friends of the State Library of South Australia, Adelaide.

Phillips V 2016 The demographics and ecology of the Rottnest Island quokka (*Setonix brachyurus*). PhD thesis, University of Western Australia.

Playford PE 2014 Recent mega-tsunamis in the Shark Bay, Pilbara, and Kimberley areas of Western Australia. *Journal of the Royal Society of Western Australia* **97**, 173–188.

Playford PE, Leech REJ & Kendrick GW 1977 Geology and hydrology of Rottnest Island. Geological Survey of Western Australia Report No. 6.

Playford PE, Cockbain AE, Berry PF, Roberts AP, Haines PW & Brooke BP 2013. The geology of Shark Bay. Geological Survey of Western Australia. Bulletin 146.

Powell CDL 2009 Foraging movements and the migration trajectory of flesh-footed shearwaters *Puffinus carneipes* from the south coast of Western Australia. *Marine Ornithology* **37**, 115-120.

Powell CDL, Wooller RD & Bradley JS 2007 Breeding biology of the flesh-footed shearwater *(Puffinus carneipes)* on Woody Island, Western Australia. *Emu* **107**, 275-283.

Pruett-Jones S & O'Donnell E 2004 Land birds on Barrow Island: Status, population estimates, and responses to an oil-field development. *Journal of the Royal Society of Western Australia* **87**, 101-108.

Rayner K, Lohr CA, Garretson S & Speldewinde P 2021. Two species, one island: Retrospective analysis of threatened fauna translocations with divergent outcomes. *PLoS One* **16**, e0253962.

Return to 1616. https://www.sharkbay.org/restoration/dirk-hartog-island-return-1616/

Richards JD & Wilson B (eds) 2008 A biological survey of Faure Island, Shark Bay world heritage property, Western Australia. *Records of the Western Australian Museum* Supplement No. 75.

Ride WDL, Mees GF, Douglas AM, Royce RD & Tyndale-Biscoe CH 1962 *The Results of an Expedition to Bernier and Dorre Island Shark Bay, Western Australia*. Fisheries Department, Perth.

Ride WDL & Neumann A (eds) 1964 Depuch Island. Western Australian Museum special publication No 2.

Rippey E, Hislop MC & Dodd J 2003 Reassessment of the vascular flora of Rottnest Island, Western Australia. *Journal of the Royal Society of Western Australia* **86**, 7-23.

Rippey ME & Hobbs RJ 2003 The effects of fire and quokkas (*Setonix brachyurus*) on the vegetation of Rottnest Island. *Journal of the Royal Society of Western Australia* **86**, 49-60.

Rippey E, Rippey JJ & Dunlop N 2002 Increasing numbers of pied cormorants breeding on the islands of Perth, WA and consequences for the vegetation. *Corella* **26**, 61-64.

Rippey E, Rippey J, Dunlop N, Durant C, Green B & Lord J 1998 The changing flora of the Shoalwater Bay islands. *Western Australian Naturalist* **22**, 81-103.

Rippey E, Rippey JJ, Green B & Dunlop N 2002 Comparison of the vegetation of the islands in Shoalwater Bay (Rockingham, Western Australia) with that of the coastal bushland. *Journal of the Royal Society of Western Australia* **85**, 169-179.

Rippey E & Rowland B 1995 *Plants of the Perth Coast and Islands*. University of Western Australia Press, Nedlands.

Robert WCH 1972 *The Explorations, 1696-1697, of Australia by Willem De Vlamingh*. Philo Press, Amsterdam.

Rosenman H 1987 *An Account in Two Volumes of Two Voyages to the South Seas by Jules S-C Dumont d'Urville…1826-1829… Volume 1. Astrolabe 1826-1829*. Melbourne University Press, Melbourne.

Rottnest Island Authority 2020 Rottnest Island Management Plan 2020-24. *ria.wa.gov.au*

Royal Australian Navy 1967a Hydrographic chart 114, Approaches to Rottnest Island. Hydrographic Service.

Royal Australian Navy 1967b Exmouth Gulf with Approaches. 1: 150 000. Chart Aus 744. Hydrographic Service.

Royal Australian Navy 1968 Monte Bello Islands to North West Cape. 1: 300 000. Chart Aus 328. Hydrographic Service.

Royal Australian Navy 1973 Port of Albany and approaches. 1: 50 000. Chart Aus 118. Hydrographic Service.

Royal Australian Navy 1974a Approaches to Port Hedland. 1: 50 000. Chart Aus 53. Hydrographic Service.

Royal Australian Navy 1974b Approaches to Esperance. 1: 75 000. Chart Aus 119. Hydrographic Service.

Royal Australian Navy 1979 Barrow Island to Onslow. 1: 150 000. Chart Aus 743. Hydrographic Service.

Royal Australian Navy 1980 Point d'Entrecasteaux to Point Hillier. 1: 150 000. Chart Aus 758. Hydrographic Service.

Royal Australian Navy 1981 Plans in Western Australia. West and south coasts. Chart Aus 116. Hydrographic Service.

Royal Australian Navy 1983 Port Hedland to Port Walcott. 1: 150 000. Chart Aus 740. Hydrographic Service.

Royal Australian Navy 2009 Cape Le Grand to Cape Pasley. 1:150 000. Chart Aus 763. Australian Hydrographic Office, Wollongong NSW.

Royal Australian Navy 2019 Dampier Archipelago. 1: 750 000. Chart Aus 57. Australian Hydrographic Office.

Ruthrof KX, Breshears DD, Fontaine JB, Froend RH, Matusick G, Kala J, Miller BP, Mitchell PJ, Wilson, SK & van Keulen M 2018 Subcontinental heat wave triggers terrestrial and marine, multi-taxa responses. *Scientific Reports* **8**, 1-9.

Sandland PT 1931 A trip to Seal Island, near Jurien Bay, W.A. *Emu* **30**, 296-299.

Sansom J, Blythman M. Dadour L & Rayner K 2019 Deployment of novel nest-shelters to increase nesting attempts in a small population of rock parrots *Neophema petrophila*. *Australian Field Ornithology* **36**, 74-78.

Saunders DA 1989 Changes in the avifauna of a region, district and remnant as a result of fragmentation of natural vegetation: The wheatbelt of Western Australia. A case study. *Biological Conservation* **50**, 99-135.

Saunders D & de Rebeira P 1993 *The Birds of Rottnest*. The authors, Guildford WA.

Scheffers AM, Scheffers SR, Kelletat DH, Squire P, Collins L, Feng Y, Zhao J.-X, Joannes-Boyau R, May SM, Schellmann G & Freeman H 2012 Coarse clast ridge sequences as suitable archives for past storm events? Case study on the Houtman Abrolhos, Western Australia. *Journal of Quaternary Science* **27**, 713-724.

Scott JK, McKirdy SJ, van der Merwe J, Green R, Burbidge AA, Pickles G, Hardie DC, Morris K, Kendrick PG, Thomas ML, Horton KL, O'Connor SM, Downs J, Stoklosa R, Lagdon R, Marks B, Nairn M & Mengersen K 2017 Zero-tolerance biosecurity protects high-conservation-value island nature reserve. *Scientific Reports* **7**, 772.

Searle DJ, Semeniuk V & Woods PJ 1988 Geomorphology, stratigraphy and Holocene history of the Rockingham-Becher Plain, South-western Australia. *Journal of the Royal Society of Western Australia* **70**, 89-109.

Searle G 2013 *First Order: Australia's Highway of Lighthouses*. Seaside Lights, Glenelg, South Australia.

Serventy DL 1947 Notes from the Recherche Archipelago, Western Australia. *Emu* **47**, 44-49.

Serventy DL 1952 The birds of the Sahul Shelf with remarks on the nesting seasons of Western Australian sea-birds. *Emu* **52**, 33-59.

Serventy DL & Marshall AJ 1964 A natural history reconnaissance of Barrow and Montebello Islands, 1958. Division of Wildlife Research Technical Paper No. 6. Commonwealth Scientific and Industrial Research Organisation, Melbourne.

Serventy V 1951 A zoologist in the Recherche Archipelago. *Walkabout* **17**(5), 29-32.

Serventy VN 1952 Birds. Australian Geographical Society Report No. 1, Part 2.

Serventy VN 1953 Mammals. Australian Geographical Society Report No. 1, Part 4.

Shark Bay Terrestrial Reserves and Proposed Reserve Additions. Management Plan No. 75. 2012. https://www.dpaw.wa.gov.au/parks/management-plans/approved-management-plans.

Shark Bay World Heritage Property. https://www.sharkbay.org/world-heritage/shark-bay-world-heritage-values/

Shaughnessy PD, Gales NJ, Dennis TE & Goldsworthy SD 1994 Distribution and abundance of New Zealand fur seals, *Arctocephalus forsteri*, in South Australia and Western Australia. *Wildlife Research* **21**, 667-695

Shaughnessy PD & Haberley B 1994 Surveys of Cape Barren geese *Cereopsis novaehollandiae* in Western Australia, 1987-1992. *Corella* **18**, 8-13.

Shoobert J (ed) 2005 *Western Australian Exploration Volume One December 1826-December 1835 The Letters, Reports & Journals of Discovery in Western Australia*. Hesperian Press, Carlisle WA.

Singor M 2017 Osprey along the Swan River and other sites. *Western Australian Bird Notes* **161**, 28-30.

Smith GG 1979 A new record of *Asplenium obtusatum* Forst. F. var. *obtusatum* in Western Australia. *Western Australian Naturalist* **14**, 123-125.

Smith GT & Kolichis N 1980 The flora and fauna of Coffin Island. *Western Australian Naturalist* **14**, 225-196.

Smith JA, Wright, L & Morris KD 2006 BiblioBarrowIsland – an annotated bibliography of the natural history of Barrow Island 1622 – 2004. *Conservation Science Western Australia* **5**, 296–364.

Smith KE, Burrows MT, Hobday AJ, Sen Gupta A, Moore PJ, Thomsen M, Wernberg T & Smale DA 2021 Socioeconomic impacts of marine heatwaves: Global issues and opportunities. *Science* **374**, eabj3593.

Smith LA & Johnstone RE 1996 Biogeography of the herpetofauna of the Archipelago of the Recherche, Western Australia. *Journal of the Royal Society of Western Australia* **79**, 165-173.

Smith LA, Johnstone RE & Dell J 2005 Vertebrate fauna of the Eastern Group, Archipelago of the Recherche, Western Australia. *Western Australian Naturalist* **24**, 232-246.

Smith VW 1990 The terrestrial vertebrate fauna of the Torndirrup National Park. *Western Australian Naturalist* **18**, 82-90.

Somaweera R, Yeoh PB, Jucker T, Clarke RH & Webber BL 2020 Historical context, current status, and management priorities for introduced Asian house geckos at Ashmore Reef, north-western Australia. *BioInvasions Records* **9**, 408-420.

Sparks J 1976 *Island Life*. Danbury Press, London.

Spencer, PBS, Sandover S, Nihill K, Wale CH, How RA & Schmitt LH 2017 Living in isolation: Ecological, demographic and genetic patterns in northern Australia's top marsupial predator on Koolan Island. *Australian Mammalogy* **39**, 17-27.

Start AN & McKenzie NL 1992 Conservation values of islands in Exmouth Gulf. A preliminary report. Department of Biodiversity, Conservation and Attractions Library, Kensington WA.

Start AN, Moro D, Adams M & Bencini RA 2006 Dunnarts from Boullanger Island: New evidence and reassessment of a taxonomic issue with resource implications. *Australian Mammalogy* **28**, 51-58.

Stevenson RE & Talbot FH (eds) 1994 *Islands*. RD Press, Surry Hills NSW.

Stirling J 1827 Letter to Governor Darling. *Historical Records of Australia*. **6 (series 3)**, 551-578.

Stokes JL 1846 *Discoveries in Australia*… T & W Boone, London.

Storr GM 1961 The flora of the Shoalwater Bay islands. *Western Australian Naturalist* **8**, 43-50.

Storr GM 1962 Annotated flora of Rottnest Island, Western Australia. *Western Australian Naturalist* **8**, 109-124.

Storr GM 1963 Some factors inducing change in the vegetation of Rottnest Island. *Western Australian Naturalist* **9**, 15-22.

Storr GM 1964-65 The avifauna of Rottnest Island, Western Australia. *Emu* **64**, 48-60; 105-113; 172-180.

Storr GM 1965 Notes on Bald Island and the adjacent mainland. *Western Australian Naturalist* **9**, 187-228.

Storr GM 1990 Birds of the Shark Bay area, Western Australia. In *Research in Shark Bay: Report of the France–Australe Bicentenary Expedition Committee.* (eds PF Berry, SD Bradshaw & BR Wilson), pp. 299-301. Western Australian Museum, Perth.

Storr GM & Harold G 1978. Herpetofauna of the Shark Bay region, Western Australia. *Records of the Western Australian Museum* **6**, 449-467.

Storr GM, Johnstone RE & Griffin P 1986 Birds of the Houtman Abrolhos, Western Australia. *Records of the Australian Museum, Supplement* **24**.

Storr GM, Green JW, & Churchill DM 1959 The vegetation of Rottnest Island. *Journal of the Royal Society of Western Australia* **42**, 70-71

Storrie A & Pobar G 1994 Harvest from the Sanko. *Landscope* **9**(4), 23-27.

Stuart EJ 1923 *A Land of Opportunities*...Bodley Head, London.

Surman C 2019 *Houtman Abrolhos. A natural history*. Halfmoon Biosciences - Persephone Publishing, Ocean Beach, Western Australia.

Surman CA & Nicholson LW 2009 A survey of the breeding seabirds and migratory shorebirds of the Houtman Abrolhos, Western Australia. *Corella* **33**, 81-98.

Surman CA, Burbidge AA & Fitzhardinge J 2016 Long-term population trends in the vulnerable Lesser Noddy *Anous tenuirostris melanops* at the Houtman Abrolhos, Western Australia. *Corella* **40**, 69-75.

Surman CA, Nicholson LW & Ayling S 2017 Foraging behaviour of the Lesser Noddy *Anous tenuirostris melanops* from the eastern Indian Ocean: Insights from micro-geologging. *Marine Ornithology* **45**, 123–128.

Swann G 1998 Unusual sightings. *Bird Observer* No. 784.

Taylor NEW 1987 *A Saga of the North Coast Yeera-Muk-A-Doo*. Hesperian Press, Carlisle WA.

Taylor S 2019 *Birding Australia's Islands*. John Beaufroy Publishing, Oxford UK.

Teichart C 1947 Contributions to the geology of Houtman's Abrolhos, Western Australia. *Proceedings of the Linnean Society of New South Wales* **71**, 145-196.

Tindale NB 1974 *Aboriginal Tribes of Australia: Their Terrain, Environmental controls, Distribution, Limits, and Proper Names.* University of California Press, Berkeley, California.

Tingay A & Tingay SR undated [1985?] The vertebrate fauna of the Archipelago of the Recherche. Typescript held in library, Department of Biodiversity, Conservation and Attractions, Kensington WA.

Toral-Granda MV, Causton CE, Jager H, Trueman M, Izurieta JC, Araujo E, Cruz M, Zander KK, Izurieta A & Garnett ST 2017. Alien species pathways to the Galapagos Islands, Ecuador. *PLoS One*, **12**(9), 1-21.

Tunney JT 1902 Field notes...made on Bedout Islands [sic], 30 miles N.W. of Condon, April, 1901. *Emu* **1**, 73.

Tyler P & Ross K 1977 Abraham Leeman – Castaway. *Westerly* No. 1 (March), 34-60.

Underwood R 2013 Lucky escapes in Torndirrup National Park. A personal account of the sometimes treacherous southern coastline. *Landscope* **29**(1), 53-56.

Vallance TG, Moore DT & Groves EW 2001 *Nature's Investigator: The Diary of Robert Brown in Australia, 1801-1805.* Australian Biological Resources Study, Canberra.

Van De Graaff WJE, Denman PD & Hooking RM 1982 Onslow, Western Australia. Geological Survey of Western Australia, Perth.

Veth P 1993 The Aboriginal occupation of the Montebello Islands. *Australian Aboriginal Studies* **1993**(2), 39-50.

Veth P & 23 others 2017. Early human occupation of a maritime desert, Barrow Island, North-West Australia. *Quaternary Science Reviews* **168**, 19-29.

Vigilante T, Toohey J, Gorring A, Blundel V, Saunders T, Mangolamara S, George K, Oobagooma J, Waina M, Morgan K & Doohan K 2013. Island country: Aboriginal connections, values and knowledge of the Western Australian Kimberley islands in the context of an island biological survey. *Records of the Western Australian Museum, Supplement* **81**, 145-182.

Walker J 1892 The bird-life of Adèle Island, north-west Australia. *Ibis* **4 (series 6)**, 254-261.

Wallace AR 1881 *Island Life*. Harper & Bros., New York.

Ward IA, Bateman MD, Larcombe P, Scott PM, Li T, Murai K, Khan NS, Veth P & Cullen P 2022 A pilot study into the geochronological and geomorphic context for the archaeology of Barrow Island, Western Australia. *Quaternary International*. https://doi.org/10.1016/j.quaint.2022.02.002

Warham J 1955 The birds of Eclipse Island. *Emu* **55**, 165-169.

Waring H 1956 Marsupial studies in Western Australia. *Australian Journal of Science* **18**, 66-73.

Watson GE 1964 Ecology and evolution of passerine birds in the islands of the Aegean Sea. PhD thesis. Yale University.

Weston AS 1985 Fire – and persistence of the flora on Middle Island, a southwestern Australian offshore island. In *Symposium on Fire Ecology and Management in Western Australian Ecosystems* (ed JR Ford), pp. 111-118. Western Australian Institute of Technology, Environmental Studies Group Report No. 14.

Weston AS & Trudgen undated The vascular flora [of Middle Island]. Typescript held in library, Department of Biodiversity, Conservation and Attractions, Kensington WA.

Wheeler T 2019 *Islands of Australia*. NLA Publishing, Canberra ACT.

Wickham JC 1842 Notes on Depuch Island. *Journal of the Royal Geographical Society* **12**, 79-83.

Whitecar WB 1864 Four years aboard the whaleship: Embracing cruises in the Pacific, Atlantic, Indian, and Antarctic Oceans, in the Years 1855-9. Forgotten Books, London.

Whitley GP 1944 Fire and petrels: The mystery of Mondrain Island. *Emu* **44**, 6-7.

Whitlock FL 1919 Notes on birds breeding in Dampier Archipelago, N.W. coast of Australia. *Emu* **18**, 240-253.

Whittell HM 1938 Notes on field-trips of J.T. Tunney. *Emu* **38**, 322-326.

Wilde SA & Low GH 1980 Pinjarra, Western Australia. Geological Survey of Western Australia, Perth.

Williams AAE 1997 The butterflies (Lepidoptera) of Garden and Rottnest Islands, Western Australia. *Australian Entomologist* **24**, 27-34.

Williams AAE & Powell R 2006 A visit to Middle Island. *Landscope* **22**(1), 44-50.

Williams IR 1968 Yarraloola, Western Australia. Geological Survey of Western Australia. Bureau of Mineral Resources, Geology and Geophysics, Canberra.

Willis JH 1953 Land flora. Australian Geographical Society Report No. 1, Part 3a.

Wilson TB 1835 *Narrative of a Voyage round the World*...Sherwood, Gilbert, & Piper, London.

Woinarski J, Ball D & Burbidge AA 2014 Islands. In *Ten Commandments Revisited* (eds D Lindenmayer, S Dovers & S Morton), pp. 117-128. CSIRO Publishing, Collingwood, Victoria.

Woinarski JCZ, Ward S, Mahney T, Bradley J, Brennan K, Ziembicki M & Fisher A 2011 The mammal fauna of the Sir Edward Pellew island group, Northern Territory, Australia: Refuge and death-trap. *Wildlife Research* **38**, 307-322.

Wolfe KM, Mills HR, Garkaklis MJ & Bencini R 2004 Post-mating survival in a small marsupials is associated with nutrient inputs from seabirds. *Ecology* **85**, 1740-1746.

Woodward HP 1908 Phosphate deposits of Christmas Island. *Journal of Agriculture Western Australia* **17**, 552-556, 589-592.

Woodward HP 1917 The phosphatic deposits of Western Australia. *Bulletin of the Geological Survey of Western Australia*. No. 74, 9-28.

Wooller RD & Dunlop JN 1981 Itinerant breeding by pied cormorants on Carnac Island, Western Australia. *Corella* **5**, 97.

Wunambal Gaambera Aboriginal Corporation 2010 *Wunambal Gaambera Healthy Country Plan – Looking after Wunambal Gaambera Country 2010 – 2020*. Wunambal Gaambera Aboriginal Corporation. Kalumburu Community, Wyndham WA 6740.

Wyrwoll K-H, Courtney J & Sandercock P 2000 The climatic environment of the Carnarvon Basin, Western Australia. *Records of the Western Australian Museum* Supplement 61, 13-27.

Wyrwoll K-H, Zhu Z, Kendrick G, Collins L & Eisenhauer A 1995 Chapter 43. Holocene sea-level events in Western Australia: Revisiting old questions. In *Journal of Coastal Research - Special Issue No. 17: Holocene cycles: Climate, sea levels, and sedimentation.* pp. 321-326.

Young C 1981 Rabbit eradication on islands off the W.A. coast. *SWANS* **11**(1), 13-16.

Zuiderbaan L 1977 Translation of Volkersen's journal, sailing with *Wackende Boey*, to the Southland 1658. Typescript. Maritime Archaeology Department, Western Australian Museum, Fremantle.

INDEX

A

Abbott, Ian, 18, 46, 144, 296
Aboriginal people, 6, 8, 23, 25, 40-41, 43, 45, 47-48, 52-53, 56-57, 71-72, 75, 77-79, 84, 93, 115, 117, 122, 147, 158, 182, 189, 191, 195, 211, 219, 226-227, 253, 268, 270, 282, 286, 290-292, 298, 301, 308, 312
Aboriginal watercraft, 1, 6, 23, 58, 71, 78, 147, 158, 182, 227, 253, 286, 290-291, 298, 301
abundance of populations, 8, 17, 19-23, 25-26, 51, 55, 72, 74, 77-78, 81, 94, 97-98, 100-101, 104, 106-107, 118-119, 123, 125-126, 128-130, 143-145, 147, 162-171, 185, 188, 190-192, 194-197, 196, 200, 218-221, 224-225, 247-251, 254-256, 281-282, 289-291, 300, 303
access by air or sea, 7, 11-12, 18, 22, 24-26, 45-46, 56, 79-80, 95, 107, 123, 127, 148-149, 167, 171, 186, 197, 199, 227-228, 244-245, 251, 255, 257, 304, 315
Abutilon Island, 101
Adele Island, 16-21, 23, 25-26, 28, 46-47, 54, 272-279, 282, 296, 306
Airlie Island, 68-69, 79, 80, 87, 275, 278, 284, 304
Alfred Wallace, 4, 268
alga, 217, 245
Algar, David, 76
Alpha Island, 96, 102, 104-106, 108, 114, 304, 309
Angel Island, 66, 70-71, 73, 78, 80, 88-90, 308
Anvil Island, 238, 240-241, 243, 248-249
Archipelago, 1, 4-6, 10, 41-42, 44, 46, 54, 67-76, 78-81, 83, 101, 139, 161, 211, 221, 223, 239-241, 243-253, 255-276, 278, 280, 283, 286, 290-291, 297-299, 300-301, 304-305, 309-311
Archipelago of the Recherche, 6, 10, 221, 223, 236-258, 270-277, 280, 283, 286, 299-301, 305, 309
Ashmore Islands, 16, 18, 20-21, 26, 31, 35
- East, 19, 24-26, 28
- Middle, 24
- West, 19, 26, 28, 272-273

Ashmore Reef, 17, 19-20, 24-25, 27, 168, 296, 306
atomic weapons, 94, 114, 270
Australian Geographical Society, 244
Australian Wildlife Conservancy, 128-130

B

Bald Island, 208, 210-211, 213-219, 221-228, 231-234, 273-275, 309
bandicoot
- golden, 53-55, 63, 82-83, 96-97, 100, 103, 107, 112, 283, 299, 301, 303, 309
- Shark Bay, 118-119, 123-124, 128-129, 133, 302-303, 309
- short-nosed, 250, 252

Banggoon (Hidden Island), 43, 46-47
Barrow Island, 3, 10, 53, 66, 73, 82-83, 92-102, 104, 107, 109-110, 113, 119, 126, 271-275, 278, 280, 282- 286, 290, 298-311, 313
Barrow Island euro, 110, 283, 295
Baston, George, 120
Batavia (city), 102, 160
Batavia (ship), 158, 163
Bathurst, HMSV, 42, 123, 162, 182, 184
Baudin Island, 114, 116, 131
Baudin Island spiny-tailed skink, 131, 137
Baudin, Nicolas, 24, 42-43, 71, 101-102, 122, 129, 182, 211
beach, 44, 51-52, 73-74, 80, 95, 98, 101,104-106, 116, 148, 163, 170, 181,

187, 192, 219, 221, 243, 249, 271, 276, 300-301, 305, 307

Beacon Island, 101-102, 154, 158, 170, 308
Beagle Islands, 138-140, 147
Beagle, HMS, 19, 24, 72, 94, 102, 140, 159, 160, 163-164, 169
Bedout Island, 17-19, 21, 23-26, 28, 31, 50, 66-67, 273, 304, 308
Bedwell Island, 16-17, 19, 22-23, 28, 273, 279, 296
behavioural change, 255, 283

Bernier Island, 10, 82-83, 107-108, 114-121, 123, 127, 273, 275, 278, 280, 283-284, 299, 301-305, 308
berry saltbush, 73, 141, 187, 190, 198, 216, 223, 246, 248
Bezout Island, 69, 71, 74, 79, 81
Bigge Island (Wuuyuruu), 10, 40, 43, 48, 58-59, 272-273, 291, 310-311
biogeography, 75-78, 85-86, 193, 221-222, 252, 268, 285-288, 311
biological characteristics, 18-22, 45-54, 73-78, 95-99, 101, 103-104, 118-120, 123-126, 129, 140-143, 162-169, 184-193, 214-222, 244-252

biological discovery, 18, 45-47, 94, 101, 118, 123, 162, 184-186, 214-215, 244-245, 279-280

biosecurity, 76, 285, 309
Bird Island, 191, 194, 197
Bluebell Island, 102-103, 106, 272
bobtail, 193, 248, 252, 283, 300-301
Bonaparte Archipelago, 38
booby
- brown, 20-21, 32
- masked, 21, 32
- red-footed, 20, 24, 35

Booby Island, 64,
boodie, 83, 96, 111, 119, 123-124, 302, 309, 314
Boodie Island, 83, 92-97, 99-100, 303, 308- 309
Boomerang Island, 99, 308
Boongaree (Sydney Aboriginal man), 72
Boongaree Island, see Wunandarra
Borda Island, 52-53, 60
Boullanger Island, 60, 138-143, 145-149, 152, 273, 275, 308
Boxer Island, 236, 240, 242-243, 253, 256, 300-301
boxthorn, African, 148-149, 170, 173, 257, 303
Breaksea Island, 210-219, 221-228, 232, 273, 275, 294, 302, 305
Bridled Island, 101-102, 166, 220, 223, 283, 308
bronzewing, brush, 164, 189, 194, 218, 227, 244, 248
Brown, Robert, 211, 214, 244
Browse Island, 10, 16-19, 22-26, 28, 36-37, 273, 278-279, 306
Buccaneer Archipelago, 10, 37, 41-42, 44, 46, 54, 271, 276
buffel grass, 56, 80, 100, 105
Buller Island, 144
Burbidge, Andrew, 49, 73, 94, 100-101, 103, 123, 162, 165, 170, 308
Butler, Harry, 94, 101, 103, 299
button-quail
- Houtman Abrolhos painted, 19, 75, 164, 171, 173, 194, 283, 300-301
- red-backed, 19
- little, 75
- painted, 50, 72, 75, 164, 171, 173, 194, 244, 283, 300-301

C

Cape Barren goose, 196, 219, 223-224, 227, 244, 249, 253-256, 265, 283, 301
Cape Inscription, 114, 116, 122-124, 134
Carlia Island, 52, 180-184, 186-192, 194-197

Carnac Island (Ngooloomayup), 181-184, 186-192, 194-197, 199, 204, 275-276, 283, 287, 289, 308
carpet python, 6, 163-164, 189, 223, 247-248, 282
Cartier Island, 16-19, 23, 25-26, 28, 272-273, 275, 279, 296, 306
cat, feral, 5, 8, 25, 54, 76, 81-82, 99, 103-104, 106-108, 115, 118-120, 124-130, 143, 149, 164, 169-170, 173, 196, 199, 225, 256, 283, 298, 300, 302-305, 308-309, 311, 314, 316
Cathedral Rocks, 192, 199
Cervantes Island, 139-142, 148
Chambers Island, 54, 312
Charley Island, 236, 239-240, 250, 255
Chatham Island, 15, 208, 213-215, 217-218, 222, 226, 230-231, 293
Chatham, HMS, 212
Cheyne Island, 212, 228, 275
Clarke, Rohan, 20
climate & weather, 1, 3, 8, 22, 37, 44-45, 55, 78, 116, 119, 129, 140, 146, 161-162, 172, 183-184, 214, 224, 243, 249, 253-254, 268, 276-277, 280-281, 284, 289, 292, 297, 302-303, 306
climate change, 80, 119, 169, 172, 198, 305, 307, 312
Cockatoo Island, 41, 45, 54
Coffin Island, 210, 212-213, 216- 219, 221-222, 224, 226, 275
Corbett Island, 236, 239-242, 249-250
cormorant
  black-faced, 266
  great pied, 126, 130, 135, 143, 147, 172, 183, 191, 193, 290
crake, spotless, 164, 219, 283
Cull Island, 236, 240, 249, 255-257, 305, 308
Cunningham Island, 23, 25, 28, 72, 123, 191, 227, 279
Cunningham, Allan, 72, 122-123, 184, 186, 211, 214, 222, 239, 244
cyclone, 21-23, 25, 67, 77-78, 81, 93, 116, 127, 162, 276-277, 289, 292, 305, 307

D

Darwin, Charles, 4, 72, 268
d'Entrecasteaux, Bruny, 239, 241
Dampier Archipelago, 6, 10, 66-69, 71-75, 78-81, 83, 271, 280, 286, 290-291, 304
Dampier Landing, 114, 122
Dampier Peninsula, 21, 41, 44, 51
Dampier, William, 17, 22, 42, 71, 117, 122- 123, 279
Daw Island, 236, 238-243, 248-252, 261, 264, 275
death adder
  Kimberley, 55, 65
  southern, 247-248, 252
Delambre Island, 66, 68, 70-71, 79-80, 278
Depuch Island, 66, 68, 70-73, 78, 82, 273, 275, 280, 303
dibbler, 124, 143, 146, 149, 153, 282, 299, 303, 309
dingo, 40, 48-49, 53, 71, 78-79, 82, 255, 286, 291, 305
Direction Island, 68-69, 79, 87, 304
Dirk Hartog, 122
Dirk Hartog homestead, 116
Dirk Hartog Island (Wirruwana), 3, 5, 10, 53, 83, 93, 114-119, 122-128, 130-131, 134-135, 159, 270, 272-273, 275-276, 278-279, 283, 299, 301, 303-305, 308-311, 316
Dirk Hartog Island National Park, 83, 126, 128, 131
Discovery, HMS, 212
Dixon Island, 66
djini (pale field-rat), 63, 75, 77, 129, 299
djoorri (common rock-rat), 77
Dolphin Island, 66-68, 70-71, 73, 75-78, 80, 272-273, 275, 278, 308
Doole Island, 69, 82, 83, 309
Dorre Island, 10, 108, 114-121, 123, 127, 270, 275, 278, 280, 283-284, 298-299, 301-302

Double Islands, 83, 92-95, 97, 99-100, 273, 308
Doubtful Islands, 208, 212-213, 217, 221, 226-228, 275
Drummond, James, 184
dugite, 6, 193, 217, 247-248, 252, 283, 300- 301
dunnart, grey-bellied, 143, 145, 149

E

eagle
 wedge-tailed, 218, 248, 255
 sea, see sea-eagle
Eagle Island, 136
East Wallabi Island, 154, 157, 162-164, 170-172, 273-274, 304
Easter Group, 154-155, 157-158, 160-161, 163, 165, 167, 169-171, 290
echidna, short-beaked, 47, 52-53, 71
Eclipse Island, 43, 210-222, 224-227, 233, 235, 273, 275, 278
ecological restoration, 56, 106-108, 128, 149, 173, 198-199, 228, 308
education, 146, 186, 228-229, 310
Egg Island, 114, 131, 136
egret, eastern reef, 19, 21, 104, 165-166
endemic species & subspecies, 5, 50-51, 55, 94, 96-99, 104, 119, 125, 129, 131, 145, 164, 185, 193, 214, 223, 252, 287, 297, 300-303, 311
Enderby Island, 66, 68, 70-73, 77, 79-80, 82, 88, 91, 272
ENSO (the El Niño Southern Oscillation), 168, 306
eradication, 21, 52, 99, 100, 102-103, 106-107, 120, 126-127, 170, 173, 199, 308-309, 311, 316
Escape Island, 139-142, 149, 273
Eucalyptus, 48, 54, 73, 95, 118, 163, 216, 244, 246, 252, 281, 284
European discovery, 17-18, 42, 94, 101-102, 117, 122, 128, 158-161, 182-183, 211-212, 240-241, 270
Exmouth Gulf, 66-69, 71, 73-74, 77, 80, 82- 83, 271
exposure to wave action, 5, 8, 17, 141, 145, 161-162, 169, 186, 188, 194, 213, 215-217, 224, 242-243, 245-246, 253-254, 272, 276-277, 280, 289, 254, 272, 277, 280, 289, 306
extinction, 4, 5, 8, 22, 48, 54, 56, 77- 78, 82, 124-125, 144, 161, 163-164, 167, 169-170, 193, 196, 215, 223-224, 243-245, 253, 276, 279, 281-282, 287-289, 297, 299-300, 302-303, 306, 308, 311

F

Factors leading to change in biota, 22-26, 29-30, 54-56, 78-80, 99-100, 102, 104-106, 120, 126, 129-130, 145-149, 169-172, 194-198, 224-227, 253-257, 288-291, 302-307
Faole Island, 172
Faure Island, 5, 114-116, 128-130, 303-305, 308-309
Favorite Island, 140, 142, 149, 153, 273
Figure of Eight Island, 236, 239, 240-243, 248-249, 253-257, 300-301
Finucane Island, 68, 70, 73, 76-77, 79-81, 275
fire, 8, 25-27, 56, 77, 80, 82, 118, 124, 149, 163, 188, 195, 197, 222, 225, 227, 253- 255, 258, 270, 291, 301
First Nations people, 6, 270, 279, 297
Fisherman Islands, 138, 140, 142, 144, 147, 275
Flat Island, 69, 212-213, 215, 221, 228, 273-275
Flinders, Matthew, 42, 212, 239, 241
Ford, Julian, 140, 142, 144-145, 147-148
Forestier, Group, 68, 71-72
fox, red, 48, 54, 67, 82, 99, 107, 125, 130, 148, 283, 300, 303-304, 308
Freycinet Island, 114
Freycinet, Louis de, 123, 182, 211
frigatebird
 great, 20-21, 34

lesser, 20-21, 33
frogs, 4, 8-9, 51, 53-55, 73, 98, 164, 184-185, 188, 217, 223, 245-247, 285-287, 303, 310-311
Fuller, Phillip, 162, 170
fungi, 185, 215, 245, 280
fur seal, long-nosed, 192, 221, 226, 250-251, 255, 267, 282, 300

G

Game Act 1892, 120, 172, 298
Garden Island (Meeandip), 6, 180-184, 186-199, 202, 273-276, 278, 284, 288, 304
Garlinju (Coronation Islands), 42-43
Garrlala (Mary Island), 43, 53, 56
genetic change, 119, 131, 284
geology, 6, 8, 17, 22, 39, 44, 48, 50, 52-53, 67-68, 73, 94, 110, 116-117, 129, 140-141, 157, 161, 164, 183-184, 186-187, 201, 211-213, 216, 242-243, 247, 249, 254, 268, 272-273, 276, 280, 292, 296
geomorphology, 40, 44, 48, 50, 145, 183, 186, 212-213, 216, 230-232, 242-243, 245-246, 268, 271-272, 276-277, 280, 292, 296-297
Gidley Island, 66, 69-71, 78, 308
Gilbert, John, 162-163, 173, 184, 186, 189-192, 194, 200, 218, 221, 228, 250, 258, 309
Gillham, Mary, 144, 146
Glauert Island, 43
goanna, 48, 55, 74, 76, 98, 101, 129, 247, 255, 286, 303, 311
goat, feral, 8, 49, 54, 56, 79, 81, 83, 115, 118, 120, 124-126, 128-130, 135, 170, 196, 225-257, 304-305, 308
Goodwyn Island, 69, 71
Goose Island (Casuarina Isles), 212, 215, 219-220, 227
Goose Island (Recherche), 237, 240-244, 250, 255-257, 305
Gorgon Gas Project, 100, 107, 126, 310
grassbird
little, 19, 143, 224, 282
tawny, 18-19, 282
Great Sandy Island, 44, 69, 79, 272, 278
Green Island (near Rottnest), 188, 203
Green Island (Oyster Harbour), 210, 212, 214- 215, 219, 225-228
Green Island (Torbay), 200, 212
Green Islands (mid-west coast), 138-140, 143, 212
guano, 22, 24-25, 36, 64, 79, 130, 146-147, 160, 162-163, 169-172, 195, 243, 257, 298, 304
Gulch Island, 240, 244
gull
Pacific, 130, 143, 166, 191, 194, 220, 223, 250
silver, 19, 50, 74, 101, 104, 130, 143, 148, 191, 198, 204, 234, 250
Gull Rock, 212, 215, 220-221, 234, 273-275
Gumalamala (Saint Andrew Island), 40, 43, 47, 55
Gun Island, 156, 159-160, 171
Gutharraguda (Shark Bay), 114-117, 128

H

Hamelin Island, 128-129, 208-209, 182, 212, 217, 219-220, 223-225, 270, 275
hare-wallaby
banded, 122, 123-124, 302, 309, 133
Barrow Island spectacled, 83, 96, 100, 102-103, 107, 111, 283, 299, 301, 303, 309
mala, 107, 119, 309, 317
rufous (Shark Bay), 107, 119, 124, 299, 309, 313
harrier, spotted, 143
Haüy Island, 71
Hayes, Weibbie, 158, 175
Hermite Island, 92, 96-97, 102-104, 106-108, 113, 272, 283, 303-304, 308-309
Heywood Islands, 17, 42
honeyeater

New Holland, 218
singing, 74, 98, 143, 193, 248, 252, 283
Hope Island, 67
house gecko, Asian, 26
Houtman Abrolhos, 3, 6, 10, 139, 144, 157-173, 192, 224, 270-271, 273-275, 278-281, 283, 286, 289, 298-301, 303, 305-307, 309
Houtman, Frederick de, 158
hydrography, 17, 139, 183, 212, 241, 278-279
hyperabundance, 282

I

iceplant, 148, 170, 225, 303, 315
Institut islands, 43
Intercourse Islands
East, 79-80
East Mid, 80
West, 66-67, 76, 78, 80, 272, 275
invertebrate, 4, 18, 23, 48, 51, 54, 56-67, 98-100, 145-146, 185-186, 188, 190, 196, 215, 217, 228, 245, 247-248, 252, 280, 285-286, 302-303, 305, 310-312
Investigator Island, 208, 212-213, 220-221, 223
Investigator, HMS, 212, 241, 255
island area, 8, 18, 28, 39, 48-49, 69-70, 74-77, 85-86, 102, 116-117, 139, 141, 143, 157, 181-183, 187-188, 213-217, 239-240, 245-246, 248, 272-273, 280, 290-292, 310

island elevation, 4, 8, 17, 28, 40, 67-68, 94, 117, 140-141, 181-183, 187-188, 213-214, 216-217, 239-240, 250, 268, 272-274, 277-278, 292

island isolation, 4-5, 8, 17, 24, 67-68, 75, 77, 80, 97, 117, 139, 216, 223, 243, 248, 252, 268, 270, 274-275, 281-282, 297-298

island management, 20, 26-27, 56-57, 75, 82-83, 100, 102, 108, 126-128, 130, 149-150, 173, 199-200, 228, 257-258, 299, 307-308, 310, 312
island naming, 4, 17, 40, 41-43, 72, 94, 101-103, 117, 122, 128, 140, 158, 160, 162, 182-183, 200, 211-212, 229, 240, 258, 270
island values, 3, 7-8, 19, 24, 74, 94, 99, 101, 115, 117, 120, 172, 213, 257, 297-299, 308, 310-312
Iwany (Sunday Island), 41-43, 45-47, 50, 54

J

Jar Island, 47, 275
Jirritmirii (Adolphus Island), 43, 47, 49, 55
Jones Island, 17, 24-26, 28, 50, 98, 108, 150, 312
Jurien Bay, 147-149

K

Kalumburu, 42, 44-45, 52
kapok, 40, 80, 100, 105, 303
Kimberley Craton, 44
King George Sound, 123, 212-217, 220-221, 225-226, 275, 297
King, Phillip Parker, 17, 22, 42, 49, 71, 94, 123, 162, 182, 239, 241
kite, brahminy, 74, 300
Koks Island, 114, 116-117, 120, 278
Koolan Island, 22, 41, 47, 54-56, 275, 285

L

Labillardière, Jacques-Julien Houtou de, 244
Lacepede Islands, 10, 16-19, 21-22, 24-26, 28-29, 44, 47, 51, 54, 56, 272-273, 305, 308
Lacrosse Island, 273
Lancelin Island, 138, 140, 142-143, 145, 147-149, 151-152, 165, 300-301
land birds, 8, 14, 18-20, 49-50, 73-76, 81, 97, 119, 125, 129, 142-144, 164, 189-190, 215, 217-218, 248-249, 285
Lane, Bill, 220, 229, 249

last glacial maximum, 39, 68, 117, 158, 241, 268, 272, 276, 278, 286, 289, 290, 296- 297, 310
Leeuwin Current, 145, 161, 168, 172, 195, 213, 224, 288, 306
Legendre Island, 66-67, 70-71, 77, 80, 278, 308
Lesueur Island, 17-19, 24-25, 28, 275
Lewis Islands, 66, 68, 70-71, 73, 77, 79-82, 258, 279, 309
lichen, 215, 217, 241, 245
lighthouse, 46, 79, 87, 134, 197, 214, 217, 225, 227-228, 232-233, 304
Lipfert, Otto, 118
Little Island, 180-181, 194, 247
Little Turtle Island, 17, 19, 24
lock hospital, 117-118, 120, 298
Lohr, Cheryl, 9, 57, 75, 84, 259
Long Island (Houtman Abrolhos), 154, 159
Long Island (Buccaneer Archipelago), 54
Low Rocks, 50
Lowendal Islands, 6, 92-93, 101-102, 271, 274, 283, 286, 308

M

mala, 107, 119, 309
Malgana people, 115, 122-123
Malus Islands, 66, 68, 70-72, 79
mammal
    marine, 123, 144, 176, 192-193, 221, 250-251, 267
  terrestrial, 18, 48-49, 52-55, 71, 74, 77, 81, 96, 98-100, 103, 110-112, 115, 118, 124, 126-127, 129, 133-145, 153, 163, 170, 176, 191-192, 204, 221, 245, 250-252, 266-267, 279, 281-283, 285, 287-289, 295, 298-299, 301-302, 305, 307-308, 313-314, 317
Mangrove Islands
  Middle, 69
  North, 69
mangrove, white, 129, 163
mangroves, 22, 40, 44, 53-54, 67, 69, 73-74, 96, 129, 163, 166, 168, 172, 271, 273, 275, 280-282, 286, 307
maning (banded hare-wallaby), 119, 122, 123-124, 127, 129, 302, 309, 133
Maret Islands, 44, 47, 51, 275
Marrgalarlg (Uwins Island), 43, 47, 49, 53, 55
McKenzie, Norman L, 46
Meade Island, 130
Meeandip (Garden Island), 182
Mermaid Reef, 17
Mermaid, HMC, 22, 42, 71, 94, 241, 244
Mewstone, 180-183
Michaelmas Island, 210, 212-218, 221-222, 225-228, 233, 274-275, 305
Middle Island (Houtman Abrolhos), 156, 160
Middle Island (Barrow Group), 92, 94, 97, 100
Middle Island (Recherche), 236-239, 241-244, 247, 252-258
Migo (Aboriginal man), 227
Migo Island, 212, 216, 227
Mistaken Island (Dampier Archipelago), 79-80
Mistaken Island (King George Sound), 210-212, 215, 217-218, 223, 225-228, 305
Miya (Byam Martin Island), 43, 47
Mondrain Island, 10, 236-237, 239-242, 244-247, 249-250, 253-256, 258, 262, 264-265, 273-274, 309
monitoring, 27, 107, 126-127, 149-150, 200, 228, 257-258, 285, 298, 303
monjon, 53, 55, 299
Montague, Paul, 102
Montalivet Islands, 44
Montebello Islands, 6, 10, 14, 66, 83, 92-94, 102-108, 113, 271, 274, 286, 305, 308, 316
Montgomery Islands, 273
Moolgoodna (Booby Island), 43, 50
Morley Island, 155, 162-163, 168, 170, 308

morphological change, 119, 131, 141, 283, 287, 311

Morris, Keith, 9, 83
moss, 215, 217, 228, 245
mouse
  dayang (heath mouse), 124, 304
  djoongari (Shark Bay mouse), 82-83, 107, 119, 128, 302, 304, 309
  house, 8, 22, 25-26, 81, 99, 100, 102, 130, 143, 149, 170, 189, 196, 226, 256, 304- 305, 308
  kerakenga (northern short-tailed mouse), 81-82
  mingkiri (sandy inland mouse), 77, 124, 128
  noodji (ash-grey mouse), 119, 124, 128, 250
  wildjin (desert mouse), 304
Muiron Islands, 66, 68, 70-71, 74, 80, 278, 286
Mulugudba (Kingfisher Island), 43, 47
Murray Island, 84, 108, 131, 150, 159, 173, 200-221, 229, 258, 305
Murujuga, 66, 68, 70-71, 73-76, 78-82, 272, 275, 304

N

nabarlek, 52-55, 299, 302

New Year Island, 238, 249, 261
Nganguru Archipelago, 42-43, 51
Ngooloomayup (Carnac Island), 182
Ngurraali (Middle Osborn Island), 43, 47, 49, 52
Niiwalarra (Sir Graham Moore Island), 40-44, 47, 52, 54, 56
Ninemba (Champagny Island), 42-43
nitre bush, 140
noddy
  black, 20
  brown, 20-21, 50, 143, 165, 178, 304
  Houtman Abrolhos lesser, 21, 26, 163, 166-169, 172, 174, 177, 307
Noomoonjoo (Caffarelli Island), 43, 46
North Island, 154, 161, 163-164, 167, 170-171, 173
North Turtle Island, 17-19, 23-25, 28, 66-67, 272-273, 275
North Twin Peak Island, 239-240, 242, 248, 250, 252, 256, 273-274
North West Island (Montebello Islands), 83, 92, 102, 107
northern quoll, 52-55, 62, 299, 303
Northumberland Rock, 293

O

Observatory Island, 236, 240-241, 244, 255, 275
Ooloogija (Lachlan Island), 43, 46-47
Oonggaliyan (Long Island), 42-43, 46-47
Osborn Islands
    Middle, 59

    South West, 61

osprey, eastern, 19, 74, 104, 130, 143, 149, 164-166, 190, 300
Oyster Harbour, 221, 226, 274
oystercatcher
  pied, 130, 249
  sooty, 26, 74, 104, 130, 143, 165, 167, 169, 170, 190, 219, 249, 300, 304

P

pale field-rat, see djini
parrot, rock, 142, 184, 196, 199, 217-218, 223, 248, 311
Pasco Island, 83, 92-94, 100, 160, 240, 242, 308
passionflower, stinking, 54
Pelican Island (Kimberley), 40, 50
Pelican Island (Shark Bay), 114, 128, 130-131, 298
pelican, Australian, 21, 50, 130-131, 191, 194, 207, 219, 227
Pelsaert Group, 154, 156, 158-163, 165, 170- 172

Pelsaert Island, 160-163, 165, 168-172, 174-175, 282-283
Pelsaert, Francisco, 158
Penguin Island, 3, 180-200, 205-207, 276, 280, 283, 287, 294, 306
penguin, little, 3, 183, 191, 194-196, 198, 206, 244, 249, 255, 306
perentie, 100, 113, 283, 310-311
Péron, François, 122
petrel, great-winged, 219, 223, 226, 234, 249-250
Picard Island, 68, 71
pigface, coastal, 215
plant species, 4-6, 8, 18, 23, 25, 45, 47-48, 54, 56, 71-73, 78, 80-81, 94-96, 118, 120, 123-124, 126, 129, 140-141, 144-146, 148-149, 162-163, 172, 184-188, 193-197, 211, 214-218, 222-225, 243, 245-249, 252-258, 277, 280-282, 284, 287, 291, 303, 310-311
Plym, HMS, 105
possum
  brushtail, 83, 96, 283
  scaly-tailed, 47, 53, 55, 299
  western ringtail, 299
Potter Island, 69, 76, 77, 275, 286
prahus, 24
Preiss, Ludwig, 184
Preston Island, 67-69, 71, 74, 80, 102, 160, 271, 278

Q

Quagering Island, 208, 215-216, 221, 275
Quoin Bluff South, 126, 131
quokka, 3, 6, 123, 182, 185, 191-192, 196-198, 221, 228, 282-283, 289, 299

R

rabbit, 45, 75, 81, 99, 118, 127, 130, 149, 170, 173, 196, 226, 228, 256, 289, 298, 305, 308
rail, buff-banded, 18, 164, 194, 199, 249, 283, 294
rakali, 52, 54, 96, 103, 304
rat
  black, 8, 25-26, 54, 56, 75, 83, 99-100, 103, 108, 119, 169-170, 173, 196, 200, 226, 228, 256-257, 298, 300, 303-304, 308-309, 316
  bush, 163-164, 169, 172, 221, 250, 252, 256, 304
  Pacific, 19, 25-26, 54, 297
Rat Island, 155, 162, 164, 167, 169-173, 279, 304-305, 309
Remark Island, 236, 239-240, 242, 274
reptiles, 4, 8-9, 18, 43, 46-47, 50-55, 57, 71, 74-75, 82, 98-99, 100-101, 104, 108, 113, 118, 120, 126, 129, 131, 137, 141-145, 150, 152, 164, 169-170, 173, 177, 184-185, 188-189, 193, 200, 217, 245, 247-248, 252, 254, 264, 279, 280-281, 283, 286-289, 291, 300-301, 310-311
Return to 1616, 123-124, 126, 131
Rippey, Elizabeth, 187-188
Rock Dunder, 210, 212, 215
rock lobster, western, 147, 171
rock-wallaby
  black-footed, 72, 82, 110, 252, 266-276
  Recherche, 253
    Rothschild's, 72, 75, 82, 91, 309
Rosemary Island, 66, 71, 80-81
Rottnest Island (Wadjemup), 3, 6, 10, 180-199, 201-204, 270-280, 282-288, 298-301, 304, 315
Rowley Shoals, 10, 17, 22, 27

S

Saddle Island, 212-213, 215-217, 219-220, 226, 275
Saint Alouarn Island, 208-209, 212, 216, 219, 221, 223-225
Saint-Aloüarn, François de, 122
Salisbury Island, 239-242, 247-248, 250-253, 257, 260, 266-267, 273, 275
Salutation Island, 114, 116, 130, 309
Sand Island (Jamieson Reef), 28

Sandy Hook Island, 236, 239, 240-243, 246
Sandy Island (Cape Talbot), 28
Sandy Island (Scott Reef), 50
scrub-bird, noisy, 258, 309
seabird, 4-5, 7-8, 17-30, 32-35, 46, 50, 71, 73-74, 78, 80-81, 97, 104, 115, 125, 130, 135, 141-143, 145-147, 149, 158-159, 161-163, 165-170, 172-174, 177-179, 183-188, 190-191, 194-195, 198-200, 204, 206-207, 215, 217-221, 223-224, 234, 243-244, 249-250, 253-256, 264, 266, 279, 282, 285, 288-291, 297-300, 303-307, 309, 311
sea level rise, 172, 306
sea lion, Australian, 3, 7, 123, 141, 144-145, 147, 149, 159, 163, 176, 183, 192, 194- 195, 221, 226, 244, 250-251, 253, 267, 297, 300, 312
sea-eagle, white-bellied, 19, 74, 104, 130, 143, 163-164, 166, 169, 190, 219, 249, 300
Serrurier Island, 68-69, 71, 81-82, 278
Shark Bay (Gutharraguda), 8, 83, 107, 115-116, 118, 119, 122-124, 127-131, 139, 271-273, 275-276, 278, 298, 300-307, 309
shearwater
  flesh-footed, 223, 226, 249-250, 264
  little, 143, 165-166, 191, 219, 223, 249
  short-tailed, 249
  wedge-tailed, 19, 74, 97, 101, 130, 143, 148, 164, 166, 169-170, 191, 193, 304
sheep, 3, 8, 79, 81, 115, 118, 120, 124-126, 128-130, 135, 170, 196, 225, 255-256, 305, 308
shelduck, Australian, 190, 249
Shelter Island, 210, 219, 226, 228
Shoalwater Bay, 180, 182-183, 185, 188, 191, 194, 197, 199, 271, 276, 283
Sholl Island, 70, 79, 92, 104
shorebird, 20, 73-74, 80, 97, 125, 129, 165, 190, 203, 219, 249, 300
Shortridge, Guy, 118, 298
silvereye, 143, 164, 189-190, 218, 225, 248- 249
Simpson Island, 69
Sir Frederick Island, 46, 54
Sir Graham Moore Island, 61
skink
  Abrolhos spiny-tailed, 164,169-170, 177, 300
  Baudin Island spiny-tailed, 131, 137
  Lancelin Island, 142, 145, 149, 300-301
  King's, 142, 188, 199, 217, 226, 247-248, 282, 289
Slope Island, 114, 131
snake, 6, 65, 74, 98, 126, 129, 141-142, 188-189, 217, 223, 247-248, 252, 283, 285, 287, 289, 301, 303
soil, 3, 5, 20, 23, 48, 53, 67, 94, 105, 116, 129, 140-141, 144-149, 163, 170-171, 183, 187, 191, 194-195, 197-198, 211-213, 215-216, 219, 224-225, 242-243, 250, 253-254, 268, 277-278, 280, 289, 292, 298, 303, 305
speciation, 142, 145, 193, 222, 252, 287
species immigration, 144, 193, 223, 253, 281, 287-288, 311
species turnover, 19, 144, 193-194, 223-224, 253, 287-288
Spindle Island, 238-239, 241, 245, 251, 256
spinifexbird, 75, 104, 309, 310
spleenwort, 221-222
Stanley Island, 208, 212, 215-216, 221, 275
Sterna Island, 50
stilt, banded, 190, 196, 203
Stokes, John Lort, 27, 72, 84, 94, 159-160, 162-164, 169, 270
storm-petrel, white-faced, 143, 166, 170, 191, 249-250

Storr, Glen M, 43, 185
Storr Island, see Yuwadon Island
Straggler Rocks, 181
stygofauna, 51, 99, 302
Sugarloaf Rock, 208-209, 212, 215, 220, 223, 226, 274-275
Sunday Island, 43, 54, 128
Sunday, East Island, 46
Surman, Christopher, 9, 162, 166, 168, 173- 174

T

tammar wallaby, 6, 159, 163-164, 171, 173, 176, 184, 189, 192, 250, 252, 254-255, 282
Tasman, Abel, 42
Termination Island, 236, 239-242, 246, 248, 250, 252, 256, 275
tern
  bridled, 22, 50, 101, 104, 126, 143, 148, 170, 191, 220, 223-224
  Caspian, 21, 50, 104, 126, 130, 143, 166, 191, 220, 250
  fairy, 19, 74-75, 94, 97-98, 104, 107, 125- 126, 130, 147, 167, 170, 184, 191, 220, 250, 300, 309, 311
  greater crested, 20, 22, 50, 104, 130, 143, 179, 191, 220
  lesser crested, 21, 101, 104, 130
  roseate, 21, 50, 104, 130, 143, 167, 179, 194
  sooty, 26, 165, 167, 169-170, 178, 304
Thevenard Island, 66, 68, 70-71, 73, 75, 79-82, 90-91, 275, 278, 286, 304, 310
Thomas Island, 260
thylacine, 89, 71
tiger snake, 6, 189, 252, 283
toad, cane, 8, 55, 302
tombolo, 140, 181, 225, 243
tourism, 3, 6, 8, 22, 26, 56, 80-81, 91, 100, 105, 128, 139, 172-173, 197-199, 206-207, 228, 257, 276, 283, 297, 299, 304-305, 311
translocation of species, 82-83, 107-108, 119, 127, 130, 258, 302, 309, 311-312, 317
tree-rat, golden-backed, 47, 52-54, 299
Trigg Island, 181
Trimouille Island, 92, 102-107, 272-273, 275, 303, 309, 317
troglofauna, 51, 99, 302
tropicbird
  red-tailed, 20, 22, 34, 166, 191, 220, 223
  white-tailed, 22
Troughton Island, 44-45, 50
tsunami, 8, 67, 78, 116, 132, 277, 289, 292
Tunney, John, 18, 94, 118, 244
Turtle Bay, 114, 116, 122, 126
turtle
  flatback, 51, 98
  green, 19, 22, 24, 51, 106, 112, 302, 307, 314
  hawksbill, 101, 104
  loggerhead, 51, 98, 126
Two Rocks, 138-139

U

Umpatayi (South West Osborn Island), 40, 43, 47, 52

V

Vancouver, George, 211-212, 240
Varanus Island, 101-102, 273, 304
vegetation, 4-5, 7-8, 18, 22-23, 28, 44-45, 47-48, 52, 54, 56, 73, 76-77, 79, 95-96, 118-119, 120, 122-124, 126, 129, 135, 140-142, 145-146, 149, 162-163, 171, 173, 184-188, 192, 194-195, 197-199, 211, 213, 215-219, 224-225, 227-228, 233-235, 242-245-246, 249-250, 253-255, 258, 263, 268, 277-278, 280-284, 286, 288-292, 296-298, 301, 303, 305, 310-311
Vlamingh, Willem de, 122-123, 162, 182, 191

W

Wadjemup, see Rottnest Island
Wallabi Group, 154, 157-159, 161, 164-165
Wallabi Islands, 154, 157, 159, 162-165, 171-172, 273-275, 304
Wanganiny (Bathurst Island), 43, 46
Warham, John, 45, 162, 220
Warnbro Sound, 182
waterbird, 190, 219, 249
Wedge Island, 138, 140, 147-148, 151
weeds, 26, 54, 56, 80, 83, 100, 105, 149, 198, 303, 304
welcome swallow, 74, 142, 218, 248
Westall Island, 236-237, 240, 242, 249-250, 252-253
white-winged fairy-wren, 75, 97, 125, 283, 301
Whitlock Island, 143, 145-146
Wickham, John Clements, 72, 84, 159-160, 240, 243, 249, 251, 274
Wirruwana, see Dirk Hartog Island
Wooded Island, 155, 163, 165-166, 168, 170, 172, 308
Woody Island, 236, 239, 240-241, 245-246, 249-250, 253, 255-257, 263, 304, 308, 315
wopilkara (greater stick-nest rat), 124, 128, 130, 299, 304, 309
World Heritage Property, 115, 131
woylie, 127, 129
Wreck Point, 139, 156, 160, 180-181
Wunandarra (Boongaree Island), 40, 43, 47- 48, 51, 55
Wurrulgu (Augustus Island), 40, 43-44, 47-49, 51, 53, 96, 189-190, 192, 272, 291, 302, 310

Y

Yuwadon (Storr Island), 43

Z

Zeewijk, 159-161, 163

www.ingramcontent.com/pod-product-compliance
Lightning Source LLC
LaVergne TN
LVHW071200210926
848482LV00014B/696